Semi-conductors	Dielectric Constants[c]	Band Gap[d] E_g (eV)	dE_g/dT (meV/K)	dE_g/dP (meV/GPa)
C	5.7	(5.48) 7.3	(−0.05)−0.6	(5)
Si	11.9	(1.11)3.48	(−0.28)	(−14)
Ge	16.2	(0.66)0.81	(−0.37)−0.4	(50)121
SiC (zincblende polytype)	9.72, 6.52	(2.4)7.0	(−0.3)	
AlN	8.5, 4.68	6.28		
AlP	9.8, 7.54	(2.45)3.62		
AlAs	10, 8.16	(2.15)3.14	(−0.4)−0.51	(−5.1)102
AlSb	11.22, 9.88	(1.63)2.219	(−0.35)	(−15)
GaN	9.5, 12.2; 5.35, 5.8	3.44	−0.48	40
GaP	11.1, 9.0	(2.27)2.78	(−0.52)−0.65	(−14)105
GaAs	13.1, 11.1	1.43	−0.395	115
GaSb	15.7, 14.4	0.70	−0.37	140
InN		2.09		
InP	12.4, 9.52	1.35	−0.29	108
InAs	14.6, 11.8	0.36	−0.35	98
InSb	17.9, 15.7	0.18	−0.29	157
ZnO	7.8, 8.75; 3.7, 3.75	3.445(4.2K)		25
ZnS	9.6, 5.13	3.68	−0.47	57
ZnSe	8.33, 5.9	2.7	−0.45	70
ZnTe	9.86, 7.28	2.26	−0.52	83
CdS	8.45, 9.12; 5.32, 5.32	2.485	−0.41	45
CdSe	9.3, 6.1	1.75	−0.36	50
CdTe	10.3, 6.9	1.43	−0.54	80
HgSe	25.6, 21	−0.061		
HgTe	21, 15.2 (77K)	−0.30(4.4K)		

[c] Low frequency dielectric constant for crystals with the diamond structure. For crystals with the zincblende structure the values represent, respectively, the low frequency and high frequency (ir) dielectric constants. For wurtzite structure crystals the first two values given in the table represent, respectively, the low frequency dielectric constant for electric field polarized along *a* and *c*. The next two values are the corresponding high frequency (ir) dielectric constants.

[d] Values given in parentheses are for indirect band gaps. The corresponding temperature and pressure coefficients in the adjacent columns are also put in parenthesis.

Fundamentals of Semiconductors

Advanced Texts in Physics

This program of advanced texts covers a broad spectrum of topics which are of current and emerging interest in physics. Each book provides a comprehensive and yet accessible introduction to a field at the forefront of modern research. As such, these texts are intended for senior undergraduate and graduate students at the MS and PhD level; however, research scientists seeking an introduction to particular areas of physics will also benefit from the titles in this collection.

Springer

Berlin
Heidelberg
New York
Barcelona
Hong Kong
London
Milan
Paris
Singapore
Tokyo

Peter Y. Yu Manuel Cardona

Fundamentals of Semiconductors

Physics and Materials Properties

Third, Revised and Enlarged Edition
With 250 Two-Color Figures,
52 Tables and 116 Problems

Springer

Professor Dr. Peter Y. Yu
University of California, Department of Physics
CA 94720-7300 Berkeley, USA
email: pyyu@lbl.gov

Professor Dr., Dres. h.c. Manuel Cardona
Max-Planck-Institut für Festkörperforschung, Heisenbergstrasse 1
70569 Stuttgart, Germany
email: cardona@cardix.mpi-stuttgart.mpg.de

ISBN 3-540-41323-5 3rd Edition
Springer-Verlag Berlin Heidelberg New York

ISBN 3-540-65352-X 2nd Edition
Springer-Verlag Berlin Heidelberg New York

Library of Congress Cataloging-in-Publication Data.
Yu, Peter Y., 1944 –. Fundamentals of semiconductors: physics and materials properties /Peter Y. Yu, Manuel Cardona. – 3rd, rev. and enlarged ed. p. cm. Includes bibliographical references and index. ISBN 3540413235 (alk. paper) 1. Semiconductors. 2. Semiconductors–Materials. I. Cardona, Manuel, 1934 –. QC611.Y88 2001
537.6'22–dc21 2001020462

Springer-Verlag Berlin Heidelberg New York
a member of BertelsmannSpringer Science + Business Media GmbH

http://www.springer.de

Cover picture: The crystal structure drawn on the book cover is a "wallpaper stereogram". Such stereograms are based on repeating, but offset, patterns that resolve themselves into different levels of depth when viewed properly. They were first described by the English physicist Brewster more than 100 years ago. See: *Superstereograms* (Cadence Books, San Francisco, CA, 1994)

Production editors: P. Treiber and C.-D. Bachem, Heidelberg
Typesetting: EDV-Beratung F. Herweg, Hirschberg
Computer-to-plate and printing: Mercedes-Druck, Berlin
Binding: Lüderitz & Bauer, Berlin

SPIN: 10778914 57/3141/ba - 5 4 3 2 1 0 – Printed on acid-free paper

Preface to the Third Edition

The support for our book has remained high and compliments from readers and colleagues have been most heart-warming. We would like to thank all of you, especially the many students who have continued to send us their comments and suggestions. We are also pleased to report that a Japanese translation appeared in 1999 (more details can be obtained from a link on our Web site: http://pauline.berkeley.edu/textbook). Chinese and Russian translations are in preparation.

Semiconductor physics and material science have continued to prosper and to break new ground. For example, in the years since the publication of the first edition of this book, the large band gap semiconductor GaN and related alloys, such as the GaInN and AlGaN systems, have all become important materials for light emitting diodes (LED) and laser diodes. The large scale production of bright and energy-efficient white-light LED may one day change the way we light our homes and workplaces. This development may even impact our environment by decreasing the amount of fossil fuel used to produce electricity. In response to this huge rise in interest in the nitrides we have added, in appropriate places throughout the book, new information on GaN and its alloys. New techniques, such as Raman scattering of x-rays, have given detailed information about the vibrational spectra of the nitrides, available only as thin films or as very small single crystals. An example of the progress in semiconductor physics is our understanding of the class of deep defect centers known as the DX centers. During the preparation of the first edition, the physics behind these centers was not universally accepted and not all its predicted properties had been verified experimentally. In the intervening years additional experiments have verified all the remaining theoretical predictions so that these deep centers are now regarded as some of the best understood defects. It is now time to introduce readers to the rich physics behind this important class of defects.

The progress in semiconductor physics has been so fast that one problem we face in this new edition is how to balance the new information with the old material. In order to include the new information we had either to expand the size of the book, while increasing its price, or to replace some of the existing material by new sections. We find either approach undesirable. Thus we have come up with the following solution, taking advantage of the Internet in this new information age. We assume that most of our readers, possibly all, are "internet-literate" so that they can download information from our Web site.

Throughout this new edition we have added the address of Web pages where additional information can be obtained, be this new problems or appendices on new topics. With this solution we have been able to add new information while keeping the size of the book more or less unchanged. We are sure the owners of the older editions will also welcome this solution since they can update their copies at almost no cost.

Errors seem to decay exponentially with time. We thought that in the second edition we had already fixed most of the errors in the original edition. Unfortunately, we have become keenly aware of the truth contained in this timeless saying: "to err is human". It is true that the number of errors discovered by ourselves or reported to us by readers has dropped off greatly since the publication of the second edition. However, many serious errors still remained, such as those in Table 2.25. In addition to correcting these errors in this new edition, we have also made small changes throughout the book to improve the clarity of our discussions on difficult issues.

Another improvement we have made in this new edition is to add many more material parameters and a Periodic Table revealing the most common elements used for the growth of semiconductors. We hope this book will be not only a handy source for information on topics in semiconductor physics but also a handbook for looking up material parameters for a wide range of semiconductors. We have made the book easier to use for many readers who are more familiar with the SI system of units. Whenever an equation is different when expressed in the cgs and SI units, we have indicated in red the difference. In most cases this involves the multiplication of the cgs unit equation by $(4\pi\varepsilon_0)^{-1}$ where ε_0 is the permittivity of free space, or the omission of a factor of $(1/c)$ where c is the speed of light.

Last but not least, we are delighted to report that the Nobel Prize in Physics for the year 2000 has been awarded to two semiconductor physicists, Zhores I. Alferov and Herbert Kroemer ("for developing semiconductor heterostructures used in high-speed- and opto-electronics") and a semiconductor device engineer, Jack S. Kilby ("for his part in the invention of the integrated circuit").

Stuttgart and Berkeley,
January 2001

Peter Y. Yu
Manuel Cardona

Preface to the Second Edition

We have so far received many comments and feedback on our book from all quarters including students, instructors and, of course, many friends. We are most grateful to them not only for their compliments but also for their valuable criticism. We also received many requests for an instructor manual and solutions to the problems at the end of each chapter. We realize that semiconductor physics has continued to evolve since the publication of this book and there is a need to continue to update its content. To keep our readers informed of the latest developments we have created a Web Page for this book. Its address (as of the writing of this preface) is: *http://pauline.berkeley.edu/textbook*. At this point this Web Page displays the following information:

1) Content, outline and an excerpt of the book.
2) Reviews of the book in various magazines and journals.
3) Errata to both first and second printing (most have been corrected in the second edition as of this date).
4) Solutions to selected problems.
5) Additional supplementary problems.

The solutions in item (4) are usually incomplete. They are supposed to serve as helpful hints and guides only. The idea is that there will be enough left for the students to do to complete the problem. We hope that these solutions will satisfy the need of both instructors and students. We shall continue to add new materials to the Web Page. For example, a list of more recent references is planned. The readers are urged to visit this Web Page regularly to find out the latest information. Of course, they will be welcomed to use this Web Page to contact us.

While the present printing of this book was being prepared, the 1998 International Conference on the Physics of Semiconductors (ICPS) was being held in Jerusalem (Israel). It was the 24th in a biannual series that started in 1950 in Reading (U.K.), shortly after the discovery of the transistor by Shockley, Bardeen and Brattain in 1948. The ICPS conferences are sponsored by the International Union of Pure and Applied Physics (IUPAP). The proceedings of the ICPS's are an excellent historical record of the progress in the field and the key discoveries that have propelled it. Many of those proceedings appear in our list of references and, for easy identification, we have highlighted in red the corresponding entries at the end of the book. A complete list of all conferences held before 1974, as well as references to their proceedings, can be found in the volume devoted to the 1974 conference which was held in Stuttgart [M. H. Pilkuhn, editor (Teubner, Stuttgart, 1974) p. 1351]. The next ICPS is scheduled to take place in Osaka, Japan from Sept. 18 to 22 in the year 2000.

The Jerusalem ICPS had an attendance of nearly 800 researchers from 42 different countries. The subjects covered there represent the center of the current interests in a rapidly moving field. Some of them are already introduced in this volume but several are still rapidly developing and do not yet lend themselves to discussion in a general textbook. We mention a few keywords:

Fractional quantum Hall effect and composite fermions.
Mesoscopic effects, including weak localization.
Microcavities, quantum dots, and quantum dot lasers.
III–V nitrides and laser applications.
Transport and optical processes with femtosecond resolution.
Fullerites, C_{60}-based nanotubes.
Device physics: CMOS devices and their future.

Students interested in any of these subjects that are not covered here, will have to wait for the proceedings of the 24th ICPS. Several of these topics are also likely to find a place in the next edition of this book.

In the present edition we have corrected all errors known to us at this time and added a few references to publications which will help to clarify the subjects under discussion.

Stuttgart and Berkeley,
November 1998

Peter Y. Yu
Manuel Cardona

Preface to the First Edition

I, who one day was sand but am today a crystal
by virtue of a great fire
and submitted myself to the demanding rigor
of the abrasive cut,
today I have the power
to conjure the hot flame.
Likewise the poet, anxiety and word:
sand, fire, crystal, strophe, rhythm.
– woe is the poem that does not light a flame

David Jou, 1983
(translated from the Catalan original)

The evolution of this volume can be traced to the year 1970 when one of us (MC) gave a course on the optical properties of solids at Brown University while the other (PYY) took it as a student. Subsequently the lecture notes were expanded into a one-semester course on semiconductor physics offered at the Physics Department of the University of California at Berkeley. The composition of the students in this course is typically about 50 % from the Physics Department, whereas the rest are mostly from two departments in the School of Engineering (Electrical Engineering and Computer Science; Materials Science and Mineral Engineering). Since the background of the students was rather diverse, the prerequisites for this graduate-level course were kept to a minimum, namely, undergraduate quantum mechanics, electricity and magnetism and solid-state physics. The Physics Department already offers a two-semester graduate-level course on condensed matter physics, therefore it was decided to de-emphasize theoretical techniques and to concentrate on phenomenology. Since many of the students in the class were either growing or using semiconductors in device research, particular emphasis was placed on the relation between physical principles and device applications. However, to avoid competing with several existing courses on solid state electronics, discussions of device design and performance were kept to a minimum. This course has been reasonably successful in "walking this tight-rope", as shown by the fact that it is offered at semi-regular intervals (about every two years) as a result of demands by the students.

One problem encountered in teaching this course was the lack of an adequate textbook. Although semiconductor physics is covered to some extent in all advanced textbooks on condensed matter physics, the treatment rarely provides the level of detail satisfactory to research students. Well-established books on semiconductor physics are often found to be too theoretical by experimentalists and engineers. As a result, an extensive list of reading materials initially replaced the textbook. Moreover, semiconductor physics being a mature field, most of the existing treatises concentrate on the large amount of well-established topics and thus do not cover many of the exciting new developments. Soon the students took action to duplicate the lecture notes, which developed into a "course reader" sold by the Physics Department at cost. This volume is approximately "version 4.0" (in software jargon) of these lecture notes.

The emphasis of this course at Berkeley has always been on simple physical arguments, sometimes at the expense of rigor and elegance in mathematics. Unfortunately, to keep the promise of using only undergraduate physics and mathematics course materials requires compromise in handling special graduate-level topics such as group theory, second quantization, Green's functions and Feynman diagrams, etc. In particular, the use of group theory notations, so pervasive in semiconductor physics literature, is almost unavoidable. The solution adopted during the course was to give the students a "five-minute crash course" on these topics when needed. This approach has been carried over to this book. We are fully aware of its shortcomings. This is not too serious a problem in a class since the instructor can adjust the depth of the supplementary materials to satisfy the need of the students. A book lacks such flexibility. The readers are, therefore, urged to skip these "crash courses", especially if they are already familiar with them, and consult the references for further details according to their background.

The choice of topics in this book is influenced by several other factors. Most of the heavier emphasis on optical properties reflects the expertise of the authors. Since there are already excellent books emphasizing transport properties, such as the one by K. H. Seeger, our book will hopefully help to fill a void. One feature that sets this book apart from others on the market is that the materials science aspects of semiconductors are given a more important role. The growth techniques and defect properties of semiconductors are represented early on in the book rather than mentioned in an appendix. This approach recognizes the significance of new growth techniques in the development of semiconductor physics. Most of the physics students who took the course at Berkeley had little or no training in materials science and hence a brief introduction was found desirable. There were some feelings among those physics students that this course was an easier way to learn about materials science! Although the course offered at Berkeley lasted only one semester, the syllabus has since been expanded in the process of our writing this book. As a result it is highly unlikely that the volume can now be covered in one semester. However, some more specialized topics can be omitted without loss of continuity, such as high field transport and hot electron effects, dynamic effective ionic charge, donor–acceptor pair transitions, resonant Raman and Brillouin scattering, and a few more.

Homework assignment for the course at Berkeley posed a "problem" (excuse our pun). No teaching assistant was allocated by the department to help with grading of the problem sets. Since the enrollment was typically over thirty students, this represented a considerable burden on the instructor. As a "solution" we provide the students with the answers to most of the questions. Furthermore, many of the questions "lead the student by the hand" through the calculation. Others have hints or references where further details can be found. In this way the students can grade their own solutions. Some of the material not covered in the main text is given in the form of "problems" to be worked out by the student.

In the process of writing this book, and also in teaching the course, we have received generous assistance from our friends and colleagues. We are especially indebted to: Elias Burstein; Marvin Cohen; Leo Esaki; Eugene Haller; Conyers Herring; Charles Kittel; Neville Smith; Jan Tauc; and Klaus von Klitzing for sharing their memories of some of the most important developments in the history of semiconductor physics. Their notes have enriched this book by telling us their "side of the story". Hopefully, future students will be inspired by their examples to expand further the frontiers of this rich and productive field. We are also grateful to Dung-Hai Lee for his enlightening explanation of the Quantum Hall Effect.

We have also been fortunate in receiving help from the over one hundred students who have taken the course at Berkeley. Their frank (and anonymous) comments on the questionnaires they filled out at the end of the course have made this book more "user-friendly". Their suggestions have also influenced the choice of topics. Many postdoctoral fellows and visitors, too numerous to name, have greatly improved the quality of this book by pointing out errors and other weaknesses. Their interest in this book has convinced us to continue in spite of many other demands on our time. The unusually high quality of the printing and the color graphics in this book should be credited to the following people: H. Lotsch, P. Treiber, and C.-D. Bachem of Springer-Verlag, Pauline Yu and Chia-Hua Yu of Berkeley, Sabine Birtel and Tobias Ruf of Stuttgart. Last but not the least, we appreciate the support of our families. Their understanding and encouragement have sustained us through many difficult and challenging moments. PYY acknowledges support from the John S. Guggenheim Memorial Foundation in the form of a fellowship.

Stuttgart and Berkeley,
October 1995

Peter Y. Yu
Manuel Cardona

A SEMI-CONDUCTOR

Contents

1. Introduction

CONTENTS

In textbooks on solid-state physics, a **semiconductor** is usually defined rather loosely as a material with electrical resistivity lying in the range of 10^{-2} – 10^9 Ω cm.[1] Alternatively, it can be defined as a material whose **energy gap** (to be defined more precisely in Chap. 2) for electronic excitations lies between zero and about 4 electron volts (eV). Materials with zero bandgap are metals or semimetals, while those with an energy gap larger than 3 eV are more frequently known as insulators. There are exceptions to these definitions. For example, terms such as semiconducting diamond (whose energy gap is about 6 eV) and semi-insulating GaAs (with a 1.5 eV energy gap) are frequently used. GaN, which is receiving a lot of attention as optoelectronic material in the blue region, has a gap of 3.5 eV.

The best-known semiconductor is undoubtedly **silicon** (Si). However, there are many semiconductors besides silicon. In fact, many minerals found in nature, such as zinc-blende (ZnS) cuprite (Cu_2O) and galena (PbS), to name just a few, are semiconductors. Including the semiconductors synthesized in laboratories, the family of semiconductors forms one of the most versatile class of materials known to man.

Semiconductors occur in many different chemical compositions with a large variety of crystal structures. They can be elemental semiconductors, such as Si, carbon in the form of C_{60} or nanotubes and selenium (Se) or binary compounds such as gallium arsenide (GaAs). Many organic compounds, e. g. polyacetylene $(CH)_n$, are semiconductors. Some semiconductors exhibit magnetic ($Cd_{1-x}Mn_xTe$) or ferroelectric (SbSI) behavior. Others become superconductors when doped with sufficient carriers (GeTe and $SrTiO_3$). Many of the recently discovered high-T_c superconductors have nonmetallic phases which are semiconductors. For example, La_2CuO_4 is a semiconductor (gap $\simeq 2$ eV) but becomes a superconductor when alloyed with Sr to form $(La_{1-x}Sr_x)_2CuO_4$.

---

[1] Ω cm is a "hybrid" SI and cgs resistivity unit commonly used in science and engineering. The SI unit for resistivity should be Ω m

1.1 A Survey of Semiconductors

The following is a brief survey of several types of the better-known semiconductors.

1.1.1 Elemental Semiconductors

The best-known semiconductor is of course the element Si. Together with germanium (Ge), it is the prototype of a large class of semiconductors with similar crystal structures. The crystal structure of Si and Ge is the same as that of diamond and α-tin (a zero-gap semiconductor also known as "gray" tin). In this structure each atom is surrounded by four nearest neighbor atoms (each atom is said to be **four-fold coordinated**), forming a tetrahedron. These **tetrahedrally bonded** semiconductors form the mainstay of the electronics industry and the cornerstone of modern technology. Most of this book will be devoted to the study of the properties of these tetrahedrally bonded semiconductors. Some elements from the groups V and VI of the periodical table, such as phosphorus (P), sulfur (S), selenium (Se) and tellurium (Te), are also semiconductors. The atoms in these crystals can be three-fold (P), two-fold (S, Se, Te) or four-fold coordinated. As a result, these elements can exist in several different crystal structures and they are also good glass-formers. For example, Se has been grown with monoclinic and trigonal crystal structures or as a glass (which can also be considered to be a polymer).

1.1.2 Binary Compounds

Compounds formed from elements of the groups III and V of the periodic table (such as GaAs) have properties very similar to their group IV counterparts. In going from the group IV elements to the III–V compounds, the bonding becomes partly ionic due to transfer of electronic charge from the group III atom to the group V atom. The ionicity causes significant changes in the semiconductor properties. It increases the Coulomb interaction between the ions and also the energy of the fundamental gap in the electronic band structure. The ionicity becomes even larger and more important in the II–VI compounds such as ZnS. As a result, most of the II–VI compound semiconductors have bandgaps larger than 1 eV. The exceptions are compounds containing the heavy element mercury (Hg). Mercury telluride (HgTe) is actually a zero-bandgap semiconductor (or a semimetal) similar to gray tin. While the large bandgap II–VI compound semiconductors have potential applications for displays and lasers, the smaller bandgap II–VI semiconductors are important materials for the fabrication of infrared detectors. The I–VII compounds (e. g., CuCl) tend to have even larger bandgaps ($>$3 eV) as a result of their higher ionicity. Many of them are regarded as insulators rather than semiconductors. Also, the increase in the cohesive energy of the crystal due to the Coulomb interaction between the ions favors the rock-salt structure containing six-fold coordinated atoms rather than tetrahedral bonds. Binary compounds formed from group IV and VI elements, such as lead sulfide (PbS), PbTe and tin sulfide (SnS), are also semiconductors. The large ionicity of these compounds also favors six-fold coordinated ions. They are similar to the mercury chalcogenides in that they have very small bandgaps in spite of their large ionicity. These small bandgap IV–VI semiconductors are also important as infrared detectors. GaN, a large bandgap III–V compound, and the mixed crystals $Ga_{1-x}In_xN$ are being used for blue light emitting diodes and lasers [1.1].

1.1.3 Oxides

Although most oxides are good insulators, some, such as CuO and Cu_2O, are well-known semiconductors. Since cuprous oxide (Cu_2O) occurs as a mineral (cuprite), it is a classic semiconductor whose properties have been studied extensively. In general, oxide semiconductors are not well understood with regard to their growth processes, so they have limited potential for applications at present. One exception is the II–VI compound zinc oxide (ZnO), which has found application as a transducer and as an ingredient of adhesive tapes and sticking plasters. However, this situation has changed with the discovery of superconductivity in many oxides of copper.

The first member of these so-called high-T_c superconductors, discovered by Müller and Bednorz[2], is based on the semiconductor lanthanum copper oxide (La_2CuO_4), which has a bandgap of about 2 eV. Carriers in the form of holes are introduced into La_2CuO_4 when trivalent lanthanum (La) is replaced by divalent barium (Ba) or strontium (Sr) or when an excess of oxygen is present. When sufficient carriers are present the semiconductor transforms into a superconducting metal. So far the highest superconducting transition temperature at ambient pressure ($T_c \simeq 135$ K) found in this family of materials belongs to $HgBaCa_2Cu_3O_{8+\delta}$. $HgBaCa_2Cu_3O_{8+\delta}$ reaches a $T_c \simeq 164$ K under high pressure [1.2]. At the time this third edition went into print this record had not yet been broken.

1.1.4 Layered Semiconductors

Semiconducting compounds such as lead iodide (PbI_2), molybdenum disulfide (MoS_2) and gallium selenide (GaSe) are characterized by their layered crystal structures. The bonding within the layers is typically covalent and much stronger than the van der Waals forces between the layers. These layered semiconductors have been of interest because the behavior of electrons in the layers is quasi-two-dimensional. Also, the interaction between layers can be mod-

[2] For this discovery, Bednorz and Müller received the Physics Nobel Prize in 1987.

ified by incorporating foreign atoms between the layers in a process known as intercalation.

1.1.5 Organic Semiconductors

Many organic compounds such as polyacetylene $[(CH_2)_n]$ and polydiacetylene are semiconductors. Although organic semiconductors are not yet used in any electronic devices, they hold great promise for future applications. The advantage of organic over inorganic semiconductors is that they can be easily tailored to the applications. For example, compounds containing conjugate bonds such as –C=C–C= have large optical nonlinearities and therefore may have important applications in opto-electronics. The bandgaps of these compounds can be changed more easily than those of inorganic semiconductors to suit the application by changing their chemical formulas. Recently new forms of carbon, such as C_{60} (fullerene), have been found to be semiconductors. One form of carbon consists of sheets of graphite rolled into a tube of some nanometers in diameter known as *nanotubes* [1.3,4]. These carbon nanotubes and their "cousin", BN nanotubes, hold great promise as nanoscale electronic circuit elements. They can be metals or semiconductors depending on their pitch.

1.1.6 Magnetic Semiconductors

Many compounds containing magnetic ions such as europium (Eu) and manganese (Mn), have interesting semiconducting and magnetic properties. Examples of these magnetic semiconductors include EuS and alloys such as $Cd_{1-x}Mn_xTe$. Depending on the amount of the magnetic ion in these alloys, the latter compounds exhibit different magnetic properties such as ferromagnetism and antiferromagnetism. The magnetic alloy semiconductors containing lower concentrations of magnetic ions are known as dilute magnetic semiconductors. These alloys have recently attracted much attention because of their potential applications. Their Faraday rotations can be up to six orders of magnitude larger than those of nonmagnetic semiconductors. As a result, these materials can be used as optical modulators, based on their large magneto-optical effects. The perovskites of the type $Mn_{0.7}Ca_{0.3}O_3$ undergo metal–semiconductor transitions which depend strongly on magnetic field, giving rise to the phenomenon of **collossal magneto-resistance** (CMR) [1.5].

1.1.7 Other Miscellaneous Semiconductors

There are many semiconductors that do not fall into the above categories. For example, SbSI is a semiconductor that exhibits ferroelectricity at low temperatures. Compounds with the general formula I–III–VI_2 and II–IV–V_2 (such as $AgGaS_2$, interesting for its nonlinear optical properties, $CuInSe_2$, useful for solar cells, and $ZnSiP_2$) crystallize in the chalcopyrite structure. The bonding in these compounds is also tetrahedral and they can be considered as analogs of the group III–V and II–VI semiconductors with the zinc-blende structure. Compounds formed from the group V and VI elements with formulas such as As_2Se_3 are semiconductors in both the crystalline and glassy states. Many of these semiconductors have interesting properties but they have not yet received much attention due to their limited applications. Their existence shows that the field of semiconductor physics still has plenty of room for growth and expansion.

1.2 Growth Techniques

One reason why semiconductors have become the choice material for the electronics industry is the existence of highly sophisticated growth techniques. Their industrial applications have, in turn, led to an increased sophistication of these techniques. For example, Ge single crystals are nowadays amongst the purest elemental materials available as a result of years of perfecting their growth techniques (see Appendix by E.E. Haller in p. 555). It is now possible to prepare almost isotopically pure Ge crystals (natural Ge contains five different isotopes). Nearly perfect single crystals of Si can be grown in the form of ingots over twelve inches (30 cm) in diameter. Isotopically pure ^{28}Si crystals have been shown to have considerably higher thermal conductivity than their natural Si counterparts [1.6]. **Dislocation** densities in these crystals can be as low as 1000 cm^{-3}, while impurity concentrations can be less than one part per trillion (10^{12}).

More recent developments in crystal growth techniques have made semiconductors even more versatile. Techniques such as Molecular Beam Epitaxy (MBE) and Metal-Organic Chemical Vapor Deposition (MOCVD) allow crystals to be deposited on a substrate one monolayer at a time with great precision. These techniques have made it possible to synthesize artificial crystal structures known as **superlattices** and **quantum wells** (Chap. 9). A recent advance in fabricating low-dimensional nanostructures takes advantage of either alignment of atoms with the substrate or strain between substrate and epilayer to induce the structure to *self-organize* into superlattices or quantum dots. Although a detailed discussion of all the growth techniques is beyond the scope of this book, a short survey of the most common techniques will provide background information necessary for every semiconductor physicist. The references given for this chapter provide further background material for the interested reader.

1.2.1 Czochralski Method

The Czochralski method is the most important method for growing **bulk** crystals of semiconductors, including Si. The method involves melting the raw ma-

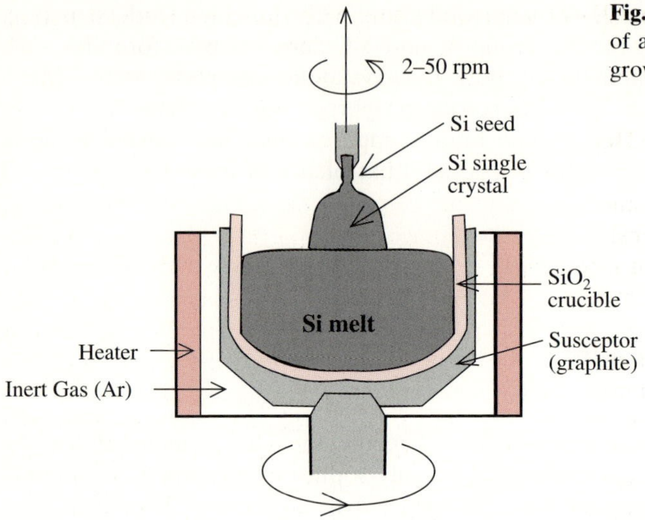

Fig. 1.1. Schematic diagram of a Czochralski furnace for growing Si single crystals

terial in a crucible. A seed crystal is placed in contact with the top, cooler region of the melt and rotated slowly while being gradually pulled from the melt. Additional material is solidified from the melt onto the seed. The most significant development in the **Czochralski technique** [1.7] (shown schematically in Fig. 1.1) is the discovery of the **Dash technique** [1.8,9] for growing dislocation-free single crystals of Si even when starting with a dislocated seed. Typical growth speed is a few millimeters per minute, and the rotation ensures that the resultant crystals are cylindrical. Silicon ingots grown by this method now have diameters greater than 30 cm.

The crucible material and gas surrounding the melt tend to contribute to the background impurities in the crystals. For example, the most common impurities in bulk Si are carbon (from the graphite crucible) and oxygen. Bulk GaAs and indium phosphide (InP) crystals are commonly grown by the Czochralski method but with the melt isolated from the air by a layer of molten boron oxide to prevent the volatile anion vapor from escaping. This method of growing crystals containing a volatile constituent is known as the **Liquid-Encapsulated Czochralski (LEC) Method**. As expected, LEC-grown GaAs often contains boron as a contaminant.

1.2.2 Bridgman Method

In the **Bridgman method** a seed crystal is usually kept in contact with a melt, as in the Czochralski method. However, a temperature gradient is created along the length of the crucible so that the temperature around the seed crystal is below the melting point. The crucible can be positioned either horizontally or vertically to control convection flow. As the seed crystal grows, the temperature profile is translated along the crucible by controlling the heaters along the furnace or by slowly moving the ampoule containing the seed crystal within the furnace.

1.2.3 Chemical Vapor Deposition

Both the Czochralski and Bridgman techniques are used to grow bulk single crystals. It is less expensive to grow a thin layer of perfect crystal than a large perfect bulk crystal. In most applications devices are fabricated out of a thin layer grown on top of a bulk crystal. The thickness of this layer is about 1 μm or less. Economically, it makes sense to use a different technique to grow a thin high quality layer on a lower quality bulk substrate. To ensure that this thin top layer has high crystalline quality, the crystal structure of the thin layer should be similar, if not identical, to the substrate and their lattice parameters as close to each other as possible to minimize **strain**. In such cases the atoms forming the thin layer will tend to build a single crystal with the same crystallographic orientation as the substrate. The resultant film is said to be deposited **epitaxially** on the substrate. The deposition of a film on a bulk single crystal of the same chemical composition (for example, a Si film deposited on a bulk Si crystal) is known as **homo-epitaxy**. When the film is deposited on a substrate of similar structure but different chemical composition (such as a GaAs film on a Si substrate), the growth process is known as **hetero-epitaxy**.

Epitaxial films can be grown from solid, liquid or gas phases. In general, it is easier to precisely control the growth rate in **gas phase epitaxy** by controlling the amount of gas flow. In **Chemical Vapor Deposition** (CVD) gases containing the required chemical elements are made to react in the vicinity of the substrate. The semiconductor produced as a result of the reaction is deposited as a thin film on a substrate inside the **reactor**. The temperature of the substrate is usually an important factor in determining the epitaxy and hence the quality of the resultant film. The most common reaction for producing a Si film in this way is given by

$$\underset{\text{(silane)}}{SiH_4} \xrightarrow{\text{heat}} \underset{\underset{\text{substrate}}{\downarrow}}{Si} + 2H_2\uparrow . \tag{1.1}$$

Highly pure Si can be produced in this way because the reaction by-product H_2 is a gas and can be easily removed. Another advantage of this technique is that dopants, such as P and As, can be introduced very precisely in the form of gases such as phosphine (PH_3) and arsine (AsH_3). III–V compound semiconductors can also be grown by CVD by using gaseous metal-organic compounds like trimethyl gallium [$Ga(CH_3)_3$] as sources. For example, GaAs films can be grown by the reaction

$$Ga(CH_3)_3 + AsH_3 \rightarrow 3CH_4\uparrow + GaAs. \tag{1.2}$$

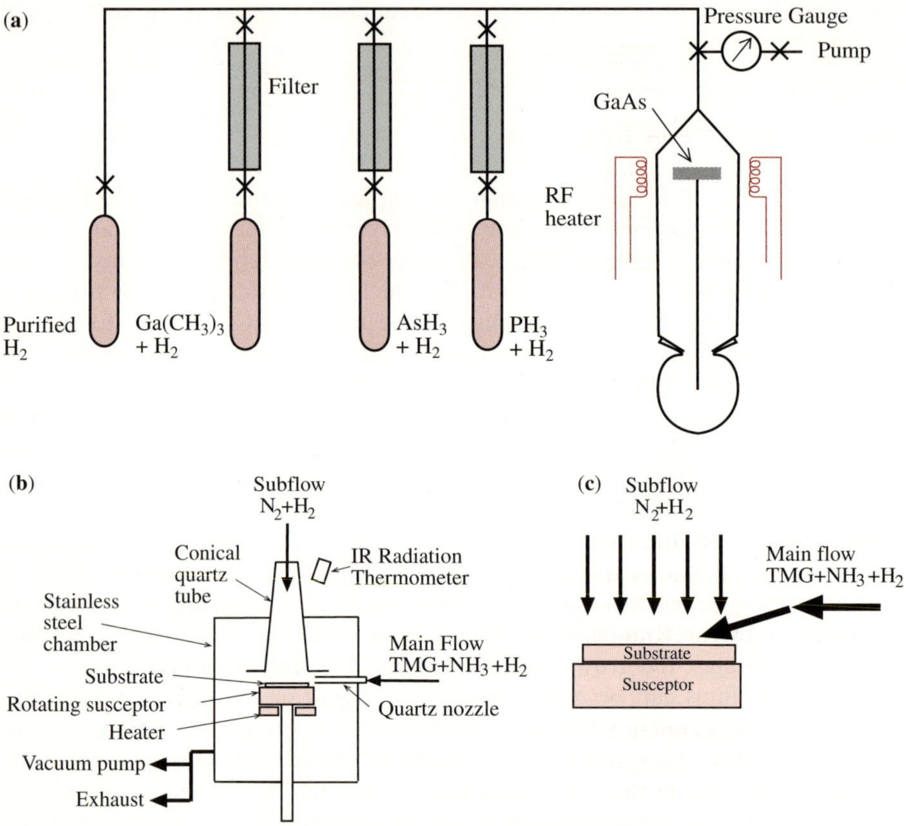

Fig. 1.2. (**a**) Schematic diagram of a MOCVD apparatus [1.10]. (**b**) Details of two-flow MOCVD machine introduced by Nakamura and co-workers for growing GaN. (**c**) Schematic diagram of the gas flows near the substrate surface [1.11]

This method of growing epitaxial films from metal-organic gases is known as **Metal-Organic Chemical Vapor Deposition** (MOCVD), and a suitable growth apparatus is shown schematically in Fig. 1.2a. A recent modification introduced for growing GaN is shown in Fig. 1.2b. Figure 1.2c shows the details of interaction between the two gas flows near the substrate [1.11].

1.2.4 Molecular Beam Epitaxy

In CVD the gases are let into the reactor at relatively high pressure (typically higher than 1 torr). As a result, the reactor may contain a high concentration of contaminants in the form of residual gases. This problem can be avoided by growing the sample under **UltraHigh Vacuum** (UHV) conditions. (Pressures below 10^{-7} torr are considered high vacuum, and a base pressure

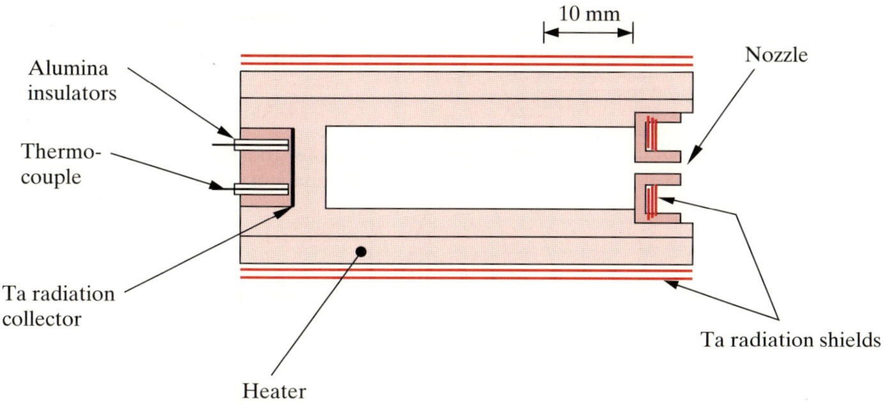

Fig. 1.3. Schematic diagram of an effusion (Knudsen) cell [1.10]

around 10^{-11} torr is UHV. See Sect. 8.1 for further discussion of UHV conditions and the definition of torr.) The reactants can be introduced in the form of **molecular beams**. A molecular beam is created by heating a source material until it vaporizes in a cell with a very small orifice. Such a cell is known as an **effusion** (or **Knudsen**) **cell** and is shown schematically in Fig. 1.3. As the vapor escapes from the cell through the small nozzle, its molecules (or atoms) form a well-collimated beam, since the UHV environment outside the cell allows the escaping molecules (or atoms) to travel ballistically for meters without collision. Typically several molecular beams containing the necessary elements for forming the semiconductor and for doping the sample are aimed at the substrate, where the film grows epitaxially. Hence this growth technique is known as **Molecular Beam Epitaxy** (MBE).

Figure 1.4 shows the construction of a typical MBE system. In principle, it is difficult to control the concentration of reactants arriving at the substrate, and hence the crystal stoichiometry, in MBE growth. The technique works because its UHV environment makes it possible to utilize electrons and ions as probes to monitor the surface and film quality during growth. The ion-based probe is usually **mass spectrometry**. Some of the electron based techniques are **Auger Electron Spectroscopy** (AES), **Low Energy-Electron Diffraction** (LEED), **Reflection High-Energy Electron Diffraction** (RHEED), and **X-ray** and **Ultraviolet Photoemission Spectroscopy** (XPS and UPS). These techniques will be discussed in more detail in Chap. 8. The one most commonly used in MBE systems is RHEED.

A typical RHEED system consists of an electron gun producing a high-energy (10–15 keV) beam aimed at a very large angle of incidence (**grazing incidence**) to the substrate surface (see Fig. 1.4). The reflected electron diffraction pattern is displayed on a phosphor screen (labeled RHEED screen in Fig. 1.4) on the opposite side. This diffraction pattern can be used to establish the surface geometry and morphology. In addition, the intensity of the zeroth-order diffraction beam (or specular beam) has been found to show damped

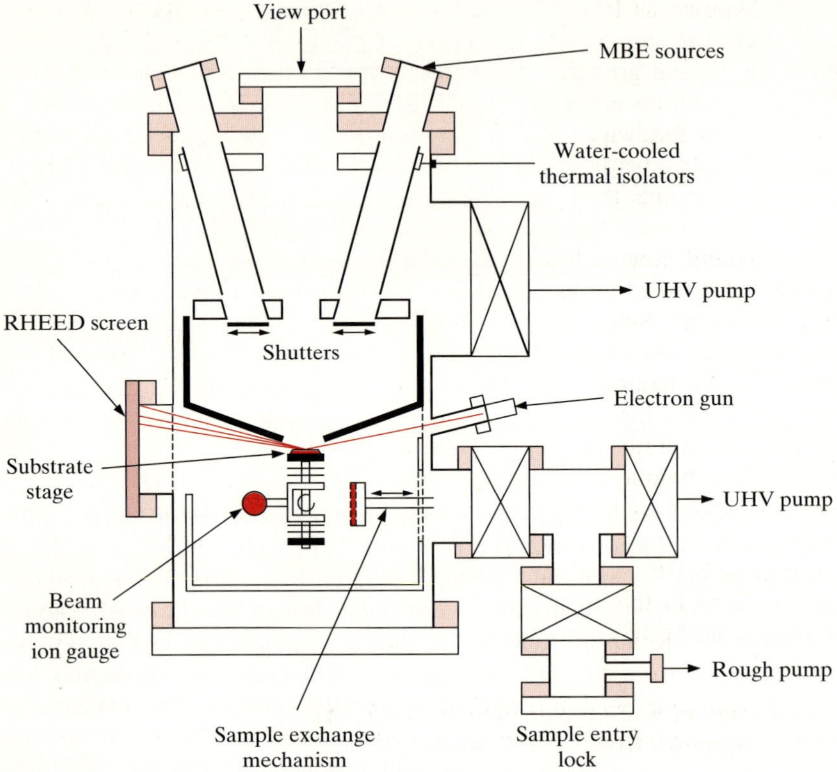

Fig. 1.4. Schematic diagram of a typical MBE system [1.10]

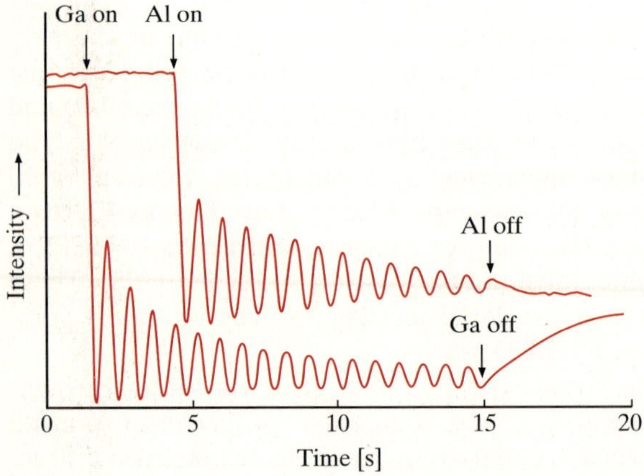

Fig. 1.5. Oscillations in the intensity of the specularly reflected electron beam in the RHEED pattern during the growth of a GaAs or AlAs film on a GaAs(001) substrate. One period of oscillation corresponds precisely to the growth of a single layer of GaAs or AlAs [1.10]

oscillations (known as RHEED oscillations) that allow the film growth rate to be monitored *in situ*. Figure 1.5 shows an example of RHEED oscillations measured during the growth of a GaAs/AlAs **quantum well**. Quantum wells are synthetic structures containing a very thin layer (thickness less than 10 nm) of semiconductor sandwiched between two thin layers of another semiconductor with a larger bandgap (see Chap. 9 for further discussion). Each oscillation in Fig. 1.5 corresponds to the growth of a single molecular layer of GaAs or AlAs.

To understand how such perfectly stoichiometric layers can be grown, we note that the Ga or Al atoms attach to a GaAs substrate much more readily than the As atoms. Since arsenic is quite volatile at elevated temperatures, any arsenic atoms not reacted with Ga or Al atoms on the substrate will not be deposited on a heated substrate. By controlling the molecular beams with shutters and monitoring the growth via RHEED oscillations, it is possible to grow a thin film literally one monolayer at a time.

The MBE technique is used for the growth of high-quality quantum wells. The only drawback of this technique in commercial applications is its slow throughput and high cost (a typical MBE system costs a least US $ 500 000). As a result, the MBE technique is utilized to study the conditions for growing high-quality films in the laboratory but large-scale commercial production of the films uses the MOCVD method.

1.2.5 Fabrication of Self-Organized Quantum Dots by the Stranski–Krastanow Growth Method

The epitaxial growth of a thin film A on a substrate B can occur in one of three main growth modes: (1) monolayer or two-dimensional growth; (2) three-dimensional growth or **Volmer–Weber** mode and (3) **Stranski–Krastanow mode** [1.12]. In mode (1) the atoms of A are attracted to the substrate more strongly than to each other. As a result the atoms first aggregate to form monolayer islands which then expand and coalesce to form the first monolayer. In mode (2) the atoms of A are attracted more strongly to each other than to the substrate. Thus they will first aggregate to form islands and as deposition continues these islands will grow and finally form a continuous film. In case of mode (3) the atoms of A will first grow two-dimensionally to form either a single monolayer or a small number of monolayers thin film. However, when growth proceeds further the additional atoms of A start to form three-dimensional islands on top of the thin film as in the Volmer–Weber mode. The continuous thin film is often referred to as the **wetting layer**.

One important factor which controls the growth of an epitaxial film is the **lattice mismatch** between the epitaxial layer A and the substrate B. Let us assume that the lattice mismatch between A and B is not too large, say only around 1% of their lattice constants. There are at least two possible ways for a thin film of A to grow on B. The first possibility is for atoms of A to line up on top of the corresponding atoms of B and to take on the lattice constant of B. In this case the film A is strained but **pseudomorphic** (a **pseudomorph** is an altered crystal form whose outward appearance is the same of another crystal species). In the second possibility the atoms of A retain their bulk lattice constant and therefore are out of registry with the substrate atoms. To minimize this mismatch between the two kinds of atoms, the thin film A will develop a kind of lattice defect known as a **dislocation** (see also Chap. 4) [1.13]. For example, if the lattice constant of A is smaller than B then the mismatch can be compensated by periodically inserting an extra plane of atoms of A into the film A to bring its atoms into alignment with the substrate atoms again. This kind of dislocation is known as a **misfit dislocation**. Since the lattice mismatch between A and B occurs in two directions lying within the surface, these dislocations form a two-dimensional network. The competition between these two growth modes for lattice-mismatched systems was studied by Frank and van der Merwe in 1949 [1.14]. The trade-off is between the strain energy in the strained pseudomorphic film and the energy required to form misfit dislocations in the unstrained film. The strain energy increases with the volume of the film while the dislocation energy depends only on the area of the film. As a result pseudomorphic growth dominates when the film thickness is small. However, as the film thickness increases it will become energetically more favorable for dislocations to form. One may expect this "cross-over" to occur at some **critical layer thickness**. The calculation of this critical thickness [1.13] is beyond the scope of this book.

The above consideration may suggest that the Stranski–Krastanow growth mode is undesirable for achieving an epitaxial film of uniform thickness. Recently it was found that this growth mode is a convenient and inexpensive way to produce nanometer structures known as **quantum dots** [1.15]. In this case the lattice constant of the epi-layer A has to be larger than that of the substrate B. The atoms of A can relax the tensile strain by "buckling" to form islands. The principle behind this island formation is similar to the buckling of a bi-metallic bar with increase in temperature, an effect used in making temperature sensors and thermostats. Since these quantum dots are formed spontaneously and can also be formed coherently, their formation is an example of a phenomenon in crystal growth known as **self-organization**. Figure 1.6 shows a plane-view transmission electro-microscope (TEM) image of a single sheet of InAs (film A with lattice constant 6.06 Å) quantum dots grown on GaAs substrate (B with lattice constant 5.64 Å).

Figure 1.7 shows the cross-sectional TEM image of a 25 layers thick stack of InGaAs quantum dots (the thicker part of the dark regions) grown on a GaAs substrate. Notice that the quantum dots are connected within the layers by thin dark regions representing the wetting layers. The various layers are separated by GaAs represented by the lighter regions. The quantum dots in Fig. 1.7 are aligned on top of each other to form arrays by the tensile strain which is transmitted through the thin GaAs layers. The coherence is gradually lost as the layers get farther from the substrate. In addition to quantum dot arrays, monolayer superlattices, such as GaP/InP, can also be grown by self-organization.

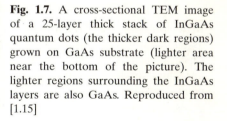

Fig. 1.6. A plane-view transmission electromicroscope (TEM) image of a single sheet of InAs quantum dots grown on a [100]-oriented GaAs substrate. Reproduced from [1.15]

Fig. 1.7. A cross-sectional TEM image of a 25-layer thick stack of InGaAs quantum dots (the thicker dark regions) grown on GaAs substrate (lighter area near the bottom of the picture). The lighter regions surrounding the InGaAs layers are also GaAs. Reproduced from [1.15]

1.2.6 Liquid Phase Epitaxy

Semiconductor films can also be grown epitaxially on a substrate from the liquid phase. This **Liquid Phase Epitaxy** (LPE) growth technique has been very successful in growing GaAs laser diodes. Usually a group III metal, such as Ga or In, is utilized as the solvent for As. When the solvent is cooled in contact with a GaAs substrate it becomes supersaturated with As and nucleation of

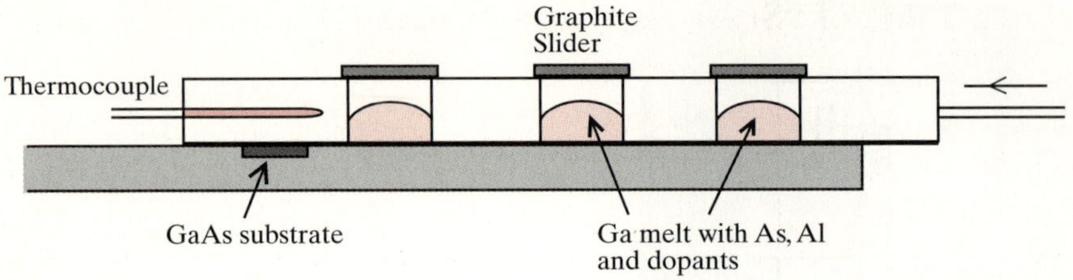

Fig. 1.8. Setup for LPE crystal growth

GaAs starts on the substrate. By using a slider containing several different solutes (as shown in Fig. 1.8), successive epitaxial layers (or epilayers in short) of different compositions and/or different dopants can be grown. The advantage of LPE is that the equipment required is inexpensive and easy to set up. However, it is difficult to achieve the level of control over the growth conditions possible with the MBE technique.

In summary, different techniques are employed to grow bulk single crystals and thin epilayers of semiconductors. The Czochralski or Bridgman techniques are used to grow bulk crystals. When feasible, the LPE method is prefered for growing thin films because of its low cost and fast growth rate. When epilayers of thickness less than 100 nm are required, it is necessary to utilize the MOCVD or MBE techniques.

In recent years optical mirror furnaces have become very popular for the growth of oxide semiconductors [1.16].

SUMMARY

In this chapter we have introduced the wide class of materials referred to as *semiconductors* and we have mentioned the large range of structural and physical properties they can have. Most of the semiconductors used in science and modern technology are single crystals, with a very high degree of perfection and purity. They are grown as bulk three-dimensional crystals or as thin, two-dimensional epitaxial layers on bulk crystals which serve as substrates. Among the techniques for growing bulk crystals that we have briefly discussed are the Czochralski and Bridgman methods. Epitaxial techniques for growing two-dimensional samples introduced in this chapter include chemical vapor deposition, molecular beam epitaxy, and liquid phase epitaxy. Self-organized two-dimensional lattices of quantum dots can also be grown with epitaxial techniques.

Periodic Table of "Semiconductor-Forming" Elements

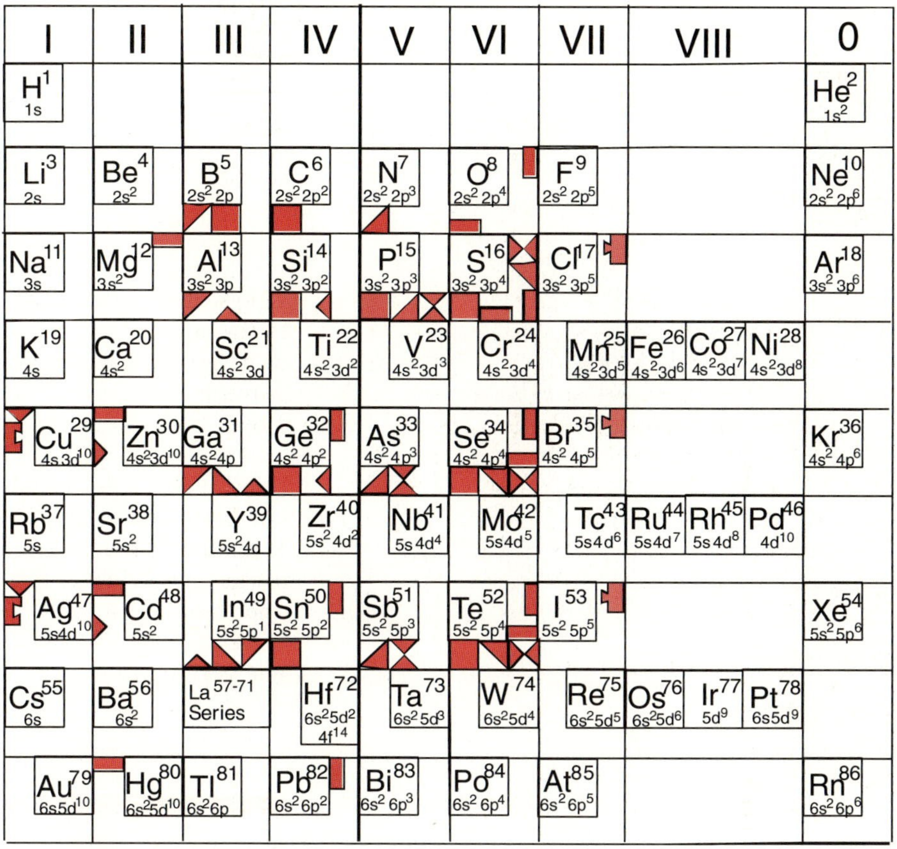

- Elements which crystallize as Semiconductors
- Elements forming Binary III-V Semiconductors
- Elements forming Binary III-VI Semiconductors
- Elements forming Binary II-VI Semiconductors
- Elements forming Binary I-VII Semiconductors
- Elements forming Binary IV-VI Semiconductors
- Elements forming I-III-VI_2 Chalcopyrite Semiconductors
- Elements forming II-VI-V_2 Chalcopyrite Semiconductors

2. Electronic Band Structures

CONTENTS

The property which distinguishes semiconductors from other materials concerns the behavior of their electrons, in particular the existence of gaps in their electronic excitation spectra. The microscopic behavior of electrons in a solid is most conveniently specified in terms of the electronic band structure. The purpose of this chapter is to study the band structure of the most common semiconductors, namely, Si, Ge, and related III–V compounds. We will begin with a quick introduction to the quantum mechanics of electrons in a crystalline solid.

The properties of electrons in a solid containing 10^{23} atoms/cm^3 are very complicated. To simplify the formidable task of solving the wave equations for the electrons, it is necessary to utilize the translational and rotational symmetries of the solid. **Group theory** is the tool that facilitates this task. However, not everyone working with semiconductors has a training in group theory, so in this chapter we will discuss some basic concepts and notations of group theory. Our approach is to introduce the ideas and results of group theory when applied to semiconductors without presenting the rigorous proofs. We will put particular emphasis on notations that are often found in books and research articles on semiconductors. In a sense, band structure diagrams are like maps and the group theory notations are like symbols on the map. Once the meaning of these symbols is understood, the band structure diagrams can be used to find the way in exploring the electronic properties of semiconductors.

We will also examine several popular methods of band structure computation for semiconductors. All band structure computation techniques involve approximations which tend to emphasize some aspects of the electronic properties in semiconductors while, at the same time, de-emphasizing other aspects. Therefore, our purpose in studying the different computational methods is to understand their advantages and limitations. In so doing we will gain insight into the many different facets of electronic properties in semiconductors.

We note also that within the past two decades, highly sophisticated techniques labeled "*ab initio*" have been developed successfully to calculate many properties of solids, including semiconductors. These techniques involve very few assumptions and often no adjustable parameters. They have been applied to calculate the total energy of crystals including all the interactions between the electrons and with the nuclei. By minimization of this energy as a function of atomic spacing, equilibrium lattice constants have been predicted. Other properties such as the elastic constants and vibrational frequencies can also be calculated. Extensions of these techniques to calculate excited-state properties have led to predictions of optical and photoemission spectra in good agreement with experimental results. It is beyond the scope of the present book to go into these powerful techniques. Interested readers can consult articles in [2.1].

2.1 Quantum Mechanics

The Hamiltonian describing a perfect crystal can be written as

$$\begin{aligned}\mathcal{H} = & \sum_i \frac{p_i^2}{2m_i} + \sum_j \frac{P_j^2}{2M_j} + \frac{1}{2}\sum_{j',j}{}' \frac{Z_j Z_{j'} e^2}{4\pi\varepsilon_0 |\boldsymbol{R}_j - \boldsymbol{R}_{j'}|} \\ & - \sum_{j,i} \frac{Z_j e^2}{4\pi\varepsilon_0 |\boldsymbol{r}_i - \boldsymbol{R}_j|} + \frac{1}{2}\sum_{i,i'}{}' \frac{e^2}{4\pi\varepsilon_0 |\boldsymbol{r}_i - \boldsymbol{r}_{i'}|}\end{aligned} \tag{2.1}$$

in the cgs system of units. (As mentioned in the preface to this edition, we have printed in red symbols which must be added to the cgs expression to convert them into Si units. ε_0 represents the permittivity of vacuum). In this expression $\boldsymbol{r}_i$ denotes the position of the ith electron, $\boldsymbol{R}_j$ is the position of the jth nucleus, Z_j is the atomic number of the nucleus, $\boldsymbol{p}_i$ and $\boldsymbol{P}_j$ are the momentum operators of the electrons and nuclei, respectively, and $-e$ is the electronic charge. $\sum'$ means that the summation is only over pairs of indices which are not identical.

Obviously, the many-particle Hamiltonian in (2.1) cannot be solved without a large number of simplifications. The first approximation is to separate electrons into two groups: **valence electrons** and **core electrons**. The core electrons are those in the filled orbitals, e. g. the $1s^2$, $2s^2$, and $2p^6$ electrons in the case of Si. These core electrons are mostly localized around the nuclei, so they can be "lumped" together with the nuclei to form the so-called **ion cores**. As a result of this approximation the indices j and j' in (2.1) will, from now on, denote the ion cores while the electron indices i and i' will label only the valence electrons. These are electrons in incompletely filled shells and in the case of Si include the $3s$ and $3p$ electrons.

The next approximation invoked is the **Born–Oppenheimer** or **adiabatic approximation**. The ions are much heavier than the electrons, so they move much more slowly. The frequencies of ionic vibrations in solids are typically less than 10^{13} s^{-1}. To estimate the electron response time, we note that the energy required to excite electrons in a semiconductor is given by its fundamental bandgap, which, in most semiconductors, is of the order of 1 eV. Therefore, the frequencies of electronic motion in semiconductors are of the order of 10^{15} s^{-1} (a table containing the conversion factor from eV to various other units can be found in the inside cover of this book). As a result, electrons can respond to ionic motion almost instantaneously or, in other words, to the electrons the ions are essentially stationary. On the other hand, ions cannot follow the motion of the electrons and they see only a time-averaged adiabatic electronic potential. With the Born-Oppenheimer approximation the Hamiltonian in (2.1) can be expressed as the sum of three terms:

$$\mathcal{H} = \mathcal{H}_{\text{ions}}(\boldsymbol{R}_j) + \mathcal{H}_{\text{e}}(\boldsymbol{r}_i, \boldsymbol{R}_{j0}) + \mathcal{H}_{\text{e-ion}}(\boldsymbol{r}_i, \delta\boldsymbol{R}_j), \tag{2.2}$$

where $\mathcal{H}_{\text{ion}}(\boldsymbol{R}_j)$ is the Hamiltonian describing the ionic motion under the influence of the ionic potentials plus the time-averaged adiabatic electronic potentials. $\mathcal{H}_{\text{e}}(\boldsymbol{r}_i, \boldsymbol{R}_{j0})$ is the Hamiltonian for the electrons with the ions frozen in their equilibrium positions $\boldsymbol{R}_{j0}$, and $\mathcal{H}_{\text{e-ion}}(\boldsymbol{r}_i, \delta\boldsymbol{R}_j)$ describes the change in the electronic energy as a result of the displacements $\delta\boldsymbol{R}_j$ of the ions from their equilibrium positions. $\mathcal{H}_{\text{e-ion}}$ is known as the **electron–phonon interaction** and is responsible for electrical resistance in reasonably pure semiconductors at room temperature. The vibrational properties of the ion cores and electron-phonon interactions will be discussed in the next chapter. In this chapter we will be mainly interested in the electronic Hamiltonian $\mathcal{H}_{\text{e}}$.

The electronic Hamiltonian $\mathcal{H}_{\text{e}}$ is given by

$$\mathcal{H}_{\text{e}} = \sum_i \frac{p_i^2}{2m_i} + \frac{1}{2}\sum_{i,i'}{}' \frac{e^2}{4\pi\varepsilon_0|\boldsymbol{r}_i - \boldsymbol{r}_{i'}|} - \sum_{i,j} \frac{Z_j e^2}{4\pi\varepsilon_0|\boldsymbol{r}_i - \boldsymbol{R}_{j0}|}. \tag{2.3}$$

Diagonalizing this Hamiltonian when there are $>10^{23}$ electrons/cm^3 in a semiconductor is a formidable job. We will make a very drastic approximation known as the **mean-field approximation**. Without going into the justifications, which are discussed in many standard textbooks on solid-state physics, we will assume that every electron experiences the same average potential $V(\boldsymbol{r})$. Thus the Schrödinger equations describing the motion of each electron will be identical and given by

$$\mathcal{H}_{1\text{e}}\Phi_n(\boldsymbol{r}) = \left(\frac{p^2}{2m} + V(\boldsymbol{r})\right)\Phi_n(\boldsymbol{r}) = E_n\Phi_n(\boldsymbol{r}), \tag{2.4}$$

where $\mathcal{H}_{1\text{e}}$, $\Phi_n(\boldsymbol{r})$ and E_n denote, respectively, the one-electron Hamiltonian, and the wavefunction and energy of an electron in an eigenstate labeled by n.

We should remember that each eigenstate can only accommodate up to two electrons of opposite spin (**Pauli's exclusion principle**).

The calculation of the electronic energies E_n involves two steps. The first step is the determination of the one-electron potential $V(\boldsymbol{r})$. Later in this chapter we will discuss the various ways to calculate or determine $V(\boldsymbol{r})$. In one method $V(\boldsymbol{r})$ can be calculated from *first principles* with the atomic numbers and positions as the only input parameters. In simpler, so-called semi-empirical approaches, the potential is expressed in terms of parameters which are determined by fitting experimental results. After the potential is known, it takes still a complicated calculation to solve (2.4). It is often convenient to utilize the symmetry of the crystal to simplify this calculation. Here by "symmetry" we mean geometrical transformations which leave the crystal unchanged.

2.2 Translational Symmetry and Brillouin Zones

The most important symmetry of a crystal is its invariance under specific translations. In addition to such **translational symmetry** most crystals possess some **rotational** and **reflection** symmetries. It turns out that most semiconductors have high degrees of rotational symmetry which are very useful in reducing the complexity of calculating their energy band structures. In this and the next sections we will study the use of symmetry to simplify the classification of electronic states. Readers familiar with the application of group theory to solids can omit these two sections.

When a particle moves in a periodic potential its wavefunctions can be expressed in a form known as **Bloch functions**. To understand what Bloch functions are, we will assume that (2.4) is one-dimensional and $V(x)$ is a periodic function with the translational period equal to R. We will define a **translation operator** T_R as an operator whose effect on any function $f(x)$ is given by

$$T_R f(x) = f(x + R). \tag{2.5}$$

Next we introduce a function $\Phi_k(x)$ defined by

$$\Phi_k(x) = \exp(\mathrm{i}kx)u_k(x), \tag{2.6}$$

where $u_\mathrm{k}(x)$ is a periodic function with the same periodicity as V, that is, $u_k(x + nR) = u_k(x)$ for all integers n. When $\Phi_k(x)$ so defined is multiplied by $\exp[-\mathrm{i}\omega t]$, it represents a plane wave whose amplitude is modulated by the periodic function $u_k(x)$. $\Phi_k(x)$ is known as a Bloch function. By definition, when x changes to $x + R$, $\Phi_k(x)$ must change in the following way

$$T_R\Phi_k(x) = \Phi_k(x + R) = \exp(\mathrm{i}kR)\Phi_k(x). \tag{2.7}$$

It follows from (2.7) that $\Phi_k(x)$ is an eigenfunction of T_R with the eigenvalue $\exp(\mathrm{i}kR)$. Since the Hamiltonian $\mathcal{H}_{1\mathrm{e}}$ is invariant under translation by R, $\mathcal{H}_{1\mathrm{e}}$ commutes with T_R. Thus it follows from quantum mechanics that the eigenfunctions of $\mathcal{H}_{1e}$ can be expressed also as eigenfunctions of T_R. We therefore conclude that an eigenfunction $\Phi(x)$ of $\mathcal{H}_{1e}$ can be expressed as a sum of Bloch functions:

$$\Phi(x) = \sum_k A_k \Phi_k(x) = \sum_k A_k \exp(\mathrm{i}kx) u_k(x), \tag{2.8}$$

where the A_k are constants. Thus the one-electron wavefunctions can be indexed by constants k, which are the **wave vectors** of the plane waves forming the "backbone" of the Bloch function. A plot of the electron energies in (2.4) versus k is known as the **electronic band structure** of the crystal.

The band structure plot in which k is allowed to vary over all possible values is known as the **extended zone scheme**. From (2.6) we see that the choice of k in indexing a wave function is not unique. Both k and $k + (2n\pi/R)$, where n is any integer, will satisfy (2.6). This is a consequence of the translation symmetry of the crystal. Thus another way of choosing k is to replace k by $k' = k - (2n\pi/R)$, where n is an integer chosen to limit k' to the interval $[-\pi/R, \pi/R]$. The region of k-space defined by $[-\pi/R, \pi/R]$ is known as the **first Brillouin zone**. A more general definition of Brillouin zones in three dimensions will be given later and can also be found in standard textbooks [2.2]. The band structure plot resulting from restricting the wave vector k to the first Brillouin zone is known as the **reduced zone scheme**. In this scheme the wave functions are indexed by an integer n (known as the **band index**) and a wave vector k restricted to the first Brillouin zone.

In Fig. 2.1 the band structure of a "nearly free" electron (i. e., $V \to 0$) moving in a one-dimensional lattice with lattice constant a is shown in both schemes for comparison. Band structures are plotted more compactly in the reduced zone scheme. In addition, when electrons make a transition from one state to another under the influence of a translationally invariant operator, k is conserved in the process within the reduced zone scheme (the proof of this

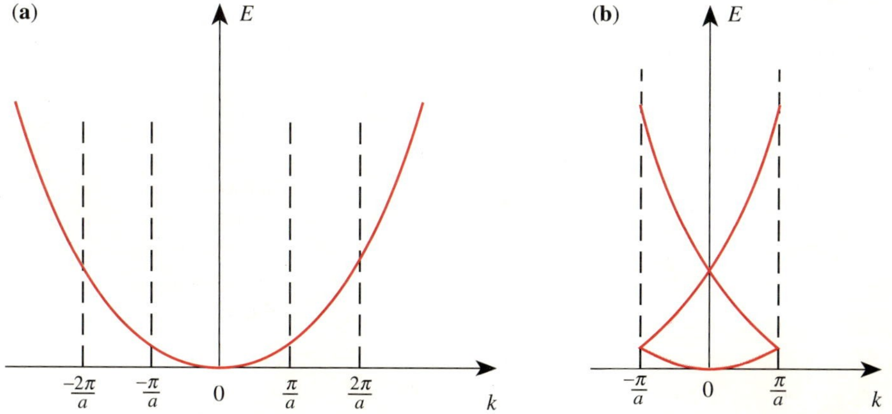

Fig. 2.1. The band structure of a free particle shown in (**a**) the extended zone scheme and (**b**) the reduced zone scheme

statement will be presented when matrix elements of operators in crystals are discussed, Sect. 2.3), whereas in the extended zone scheme k is conserved only to a multiple of (i. e. *modulo*) $2\pi/R$. Hence, the reduced zone scheme is almost invariably used in the literature.

The above results, obtained in one dimension, can be easily generalized to three dimensions. The translational symmetries of the crystal are now expressed in terms of a set of **primitive lattice vectors**: $\boldsymbol{a}_1$, $\boldsymbol{a}_2$, and $\boldsymbol{a}_3$. We can imagine that a crystal is formed by taking a minimal set of atoms (known as a **basis set**) and then translating this set by multiples of the primitive lattice vectors and their linear combinations. In this book we will be mostly concerned with the diamond and zinc-blende crystal structures, which are shown in Fig. 2.2a. In both crystal structures the basis set consists of two atoms. The ba-

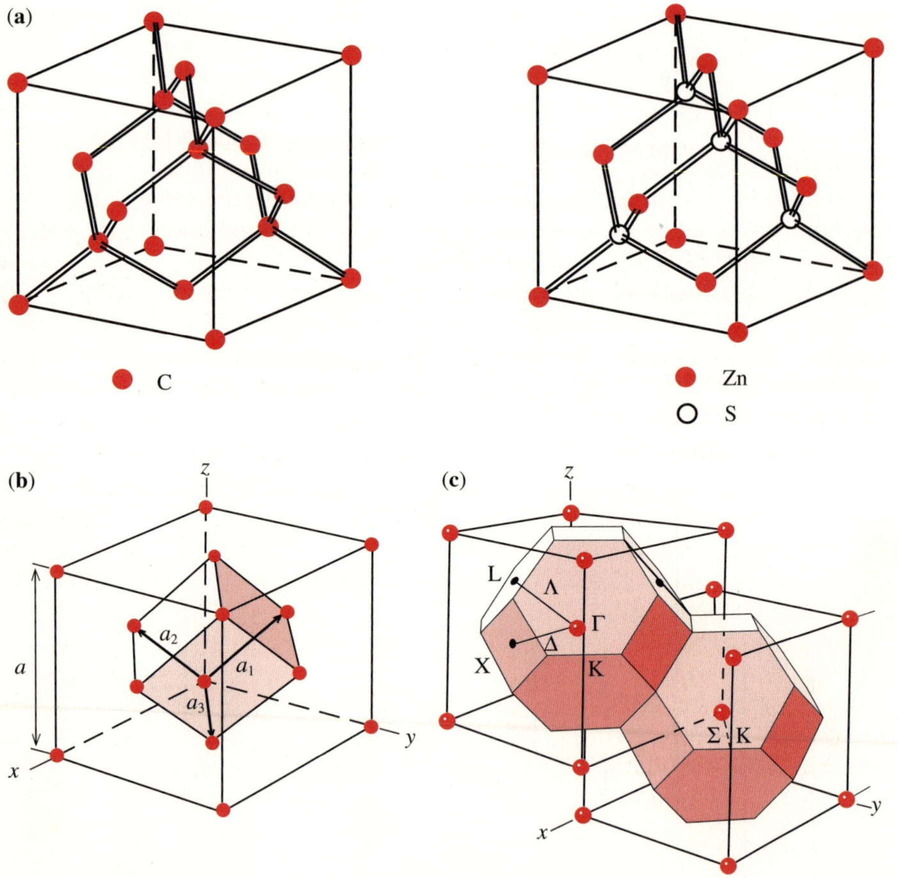

Fig. 2.2. (**a**) The crystal structure of diamond and zinc-blende (ZnS). (**b**) the fcc lattice showing a set of primitive lattice vectors. (**c**) The reciprocal lattice of the fcc lattice shown with the first Brillouin zone. Special high-symmetry points are denoted by Γ, X, and L, while high-symmetry lines joining some of these points are labeled as Λ and Δ

sis set in diamond consists of two carbon atoms while in zinc-blende the two atoms are zinc and sulfur. The lattice of points formed by translating a point by multiples of the primitive lattice vectors and their linear combinations is known as the **direct lattice**. Such lattices for the diamond and zinc-blende structures, which are basically the same, are said to be **face-centered cubic** (fcc) see Fig. 2.2b with a set of primitive lattice vectors. In general, the choice of primitive lattice vectors for a given direct lattice is not unique. The primitive lattice vectors shown in Fig. 2.2b are

$$\boldsymbol{a}_1 = (0, a/2, a/2),$$

$$\boldsymbol{a}_2 = (a/2, 0, a/2),$$

and

$$\boldsymbol{a}_3 = (a/2, a/2, 0),$$

where a is the length of the side of the smallest cube in the fcc lattice. This smallest cube in the direct lattice is also known as the unit cube or the **crystallographic unit cell**.

For a given direct lattice we can define a **reciprocal lattice** in terms of three **primitive reciprocal lattice vectors**: $\boldsymbol{b}_1$, $\boldsymbol{b}_2$, and $\boldsymbol{b}_3$, which are related to the direct lattice vectors $\boldsymbol{a}_1$, $\boldsymbol{a}_2$, and $\boldsymbol{a}_3$ by

$$\boldsymbol{b}_i = 2\pi \frac{(\boldsymbol{a}_j \times \boldsymbol{a}_k)}{(\boldsymbol{a}_1 \times \boldsymbol{a}_2) \cdot \boldsymbol{a}_3}, \tag{2.9}$$

where i, j, and k represent a cyclic permutation of the three indices 1, 2, and 3 and $(\boldsymbol{a}_1 \times \boldsymbol{a}_2) \cdot \boldsymbol{a}_3$ is the volume of the primitive cell. The set of points generated by translating a point by multiples of the reciprocal lattice vectors is known as the reciprocal lattice. The reason for defining a reciprocal lattice in this way is to represent the wave vector $\boldsymbol{k}$ as a point in **reciprocal lattice space**. The first Brillouin zone in three dimensions can be defined as the smallest polyhedron confined by planes perpendicularly bisecting the reciprocal lattice vectors. It is easy to see that the region $[-\pi/R, \pi/R]$ fits the definition of the first Brillouin zone in one dimension.

Since the reciprocal lattice vectors are obtained from the direct lattice vectors via (2.9), the symmetry of the Brillouin zone is determined by the symmetry of the crystal lattice. The reciprocal lattice corresponding to a fcc lattice is shown in Fig. 2.2c. These reciprocal lattice points are said to form a **body-centered cubic** (bcc) lattice. The primitive reciprocal lattice vectors $\boldsymbol{b}_1$, $\boldsymbol{b}_2$, and $\boldsymbol{b}_3$ as calculated from (2.9) are

$$\boldsymbol{b}_1 = (2\pi/a)\ (-1, 1, 1),$$

$$\boldsymbol{b}_2 = (2\pi/a)\ (1, -1, 1),$$

and

$$\boldsymbol{b}_3 = (2\pi/a)\ (1, 1, -1).$$

[Incidentally, note that all the reciprocal lattice vectors of the fcc lattice have the form $(2\pi/a)(i,j,k)$, where i, j, and k have to be either all odd or all even]. The first Brillouin zone of the fcc structure is also indicated in Fig. 2.2c. The symmetry of this Brillouin zone can be best visualized by constructing a model out of cardboard. A template for this purpose can be found in Fig. 2.27.

In Fig. 2.2c we have labeled some of the high-symmetry points of this Brillouin zone using letters such as X and Γ. We will conform to the convention of denoting high symmetry points and lines *inside* the Brillouin zone by **Greek** letters and points on the *surfaces* of the Brillouin zone by **Roman** letters. The **center** of the Brillouin zone is always denoted by Γ. The three high-symmetry directions [100], [110], and [111] in the Brillouin zone of the fcc lattice are denoted by:

$$[100]\ \text{direction}: \dot{\Gamma}\overline{\quad \Delta \quad}\dot{\text{X}}$$

$$[111]\ \text{direction}: \dot{\Gamma}\overline{\quad \Lambda \quad}\dot{\text{L}}$$

$$[110]\ \text{direction}: \dot{\Gamma}\overline{\quad \Sigma \quad}\dot{\text{K}}$$

The Brillouin zone of the fcc lattice is highly symmetrical. A careful examination of this Brillouin zone shows that it is unchanged by various rotations, such as a 90° rotation about axes parallel to the edges of the body-centered cube in Fig. 2.2c. In addition it is invariant under reflection through certain planes containing the center of the cube. These operations are known as **symmetry operations** of the Brillouin zone. The symmetry of the Brillouin zone results from the symmetry of the direct lattice and hence it is related to the symmetry of the crystal. This symmetry has at least two important consequences for the electron band structure. First, if two wave vectors $\boldsymbol{k}$ and $\boldsymbol{k}'$ in the Brillouin zone can be transformed into each other under a symmetry operation of the Brillouin zone, then the electronic energies at these wave vectors must be identical. Points and axes in reciprocal lattice space which transform into each other under symmetry operations are said to be **equivalent**. For example, in the Brillouin zone shown in Fig. 2.2c there are eight hexagonal faces containing the point labeled L in the center. These eight faces including the L points are equivalent and can be transformed into one another through rotations by 90°. Therefore it is necessary to calculate the energies of the electron at only one of the eight equivalent hexagonal faces containing the L point. The second and perhaps more important consequence of the crystal symmetry is that wave functions can be expressed in a form such that they have definite transformation properties under symmetry operations of the crystal. Such wave functions are said to be symmetrized. A well-known example of **symmetrized wave functions** is provided by the standard wave functions of electrons in atoms, which are usually symmetrized according to their transformation properties under rotations and are classified as *s*, *p*, *d*, *f*, etc. For example, an *s* wave function is unchanged by any rotation. The *p* wave functions are triply degenerate and transform under rotation like the three components of a vector. The *d* wave functions transform like the five components of a symmetric and traceless second-rank tensor. By classifying the wave functions in this way, some matrix elements of operators can be shown to vanish, i. e., **selection rules** can be deduced. Similarly, wave functions in crystals can be classified according to their transformation properties under symmetry operations of the crystal and selection rules can be deduced for operators acting on these wave functions. The mathematical tool for doing this is **group theory**. Many excellent textbooks have been written on group theory (see the reference list). It is desirable, but not necessary, to have a good knowledge of group theory in order to study semiconductor physics. Some elementary notions of group theory are sufficient to understand the material covered in this book. The next section contains an introduction to group theoretical concepts and notations. Students familiar with group theory can omit this section.

2.3 A Pedestrian's Guide to Group Theory

Since the purpose of this section is to introduce group theory terminology and notations, no effort will be made to prove many of the statements and theorems mentioned in it. At most we shall illustrate our statements with examples and refer the reader to books on group theory for rigorous proofs.

2.3.1 Definitions and Notations

The first step in studying the symmetry properties of any crystal is to determine its symmetry operations. For example: a square is unchanged under reflection about its two diagonals, or under rotation by 90° about an axis perpendicular to the square and passing through its center. One can generate other symmetry operations for a square which are combinations of these operations. One may say that it is possible to find an infinite number of symmetry operations for this square. However, many of these symmetry operations can be shown to consist of sequences of a few basic symmetry operations. The mathematical tool for systematically analyzing the symmetry operations of any object is group theory.

A **group** $\boldsymbol{G}$ is defined as a set of elements $\{a, b, c, \ldots\}$ for which an operation ab (which we will refer to as **multiplication**) between any two elements a and b of the group is defined. This operation must have these four properties:

- **Closure**: The result of the operation ab on any two elements a and b in $\boldsymbol{G}$ must also belong to $\boldsymbol{G}$.
- **Associativity**: for all elements a, b, and c in $\boldsymbol{G}$ $(ab)c = a(bc)$.
- **Identity**: $\boldsymbol{G}$ must contain an element e known as the **identity** or **unit element** such that $ae = a$ for all elements a in $\boldsymbol{G}$.
- **Inverse element**: for every element a in $\boldsymbol{G}$ there exists a corresponding element a^{-1} such that $a^{-1}a = e$. Element a^{-1} is known as the **inverse** of a.

Notice that the order in which one multiplies two elements a and b is important since ab is not necessarily equal to ba in general. If $ab = ba$ for all elements in $\boldsymbol{G}$, multiplication is **commutative** and $\boldsymbol{G}$ is said to be **Abelian**.

One can easily find many examples of groups. In particular, the set of symmetry operations of a crystal or a molecule can be shown to form a group. As an illustration, we will consider the molecule methane: CH_4. The structure of this molecule is shown in Fig. 2.3. It consists of a carbon atom surrounded by four hydrogen atoms forming the four corners of a regular tetrahedron.

To simplify the description of the symmetry operations of the methane molecule, we will introduce the **Schönflies notation**:

C_2: rotation by 180° (called a **two-fold rotation**);
C_3: rotation by 120° (called a **three-fold rotation**);
C_4: rotation by 90° (called a **four-fold rotation**);
C_6: rotation by 60° (called a **six-fold rotation**);
σ: **reflection** about a plane;
i: **inversion**;
S_n: rotation C_n followed by a reflection through a plane perpendicular to the rotation axis;
E: the **identity operation**.

For brevity, all the above operations are often denoted as rotations. To distinguish between a conventional rotation (such as C_3) from reflections (such as σ) or rotations followed by reflections (such as S_4) the latter two are referred to as **improper rotations**. Notice that the inversion is equal to S_2. This is not the only way to represent symmetry operations. An equally popular system is the **international notation**. The conversion between these two systems can be found in books on group theory [Ref. 2.3, p. 85].

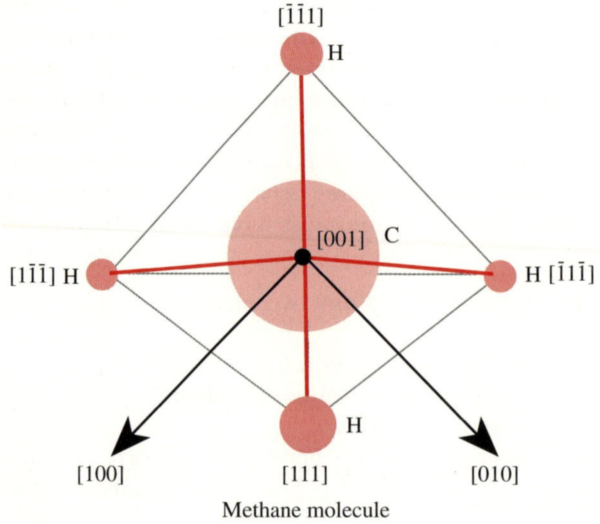

Fig. 2.3. A methane molecule (CH_4) displaying the bonds (*red lines*) and the coordinate axes (*black arrows*). The [001] axis is perpendicular to the paper

To specify a symmetry operation completely, it is also necessary to define the axis of rotation or the plane of reflection. In specifying planes of reflection we will use the notation (*kln*) to represent a plane that contains the origin and is perpendicular to the vector (k, l, n). (Readers familiar with crystallography will recognize that this notation is an "imitation" of the **Miller indices** for denoting lattice planes in cubic crystals). The corresponding simplified notation for the axis containing this vector is [*kln*]. In Fig. 2.3 we have first chosen the origin at the carbon atom for convenience. Using the coordinate system shown in Fig. 2.3, the four carbon–hydrogen bonds are oriented along the [111], $[1\bar{1}\bar{1}]$, $[\bar{1}1\bar{1}]$, and $[\bar{1}\bar{1}1]$ directions. We will now state without proof (the reader can check these results easily by constructing a balls-and-sticks model of the methane molecule) that the following operations are symmetry operations of the methane molecule:

E: the identity;
C_2: two-fold rotation about one of the three mutually perpendicular [100], [010] and [001] axes (three C_2 operations in total);
C_3: rotation by 120° in clockwise direction about one of the four C–H bonds (four operations in total);
C_3^{-1}: rotation by 120°, counterclockwise, about one of the four C–H bonds (four operations in total);
σ: reflection with respect to one of these six planes: (110), $(\bar{1}10)$, (101), $(\bar{1}01)$, (011), $(0\bar{1}1)$;
S_4: a four-fold clockwise rotation about one of the [100], [010], and [001] axes followed by a reflection on the plane perpendicular to the rotation axis (three operations in total);
S_4^{-1}: a four-fold counterclockwise rotation about one of the [100], [010], and [001] axes followed by a reflection on a plane perpendicular to the rotation axis (three operations in total).

It can be shown easily that the operations C_2 and σ are both the inverse of themselves. The inverse element of C_3 is C_3^{-1}, provided the axis of rotation is the same in the two operations. Similarly, the inverse element of S_4 is S_4^{-1}, provided the rotation axes remain the same. If we now define the multiplication of two symmetry elements a and b as a symmetry operation $c = ab$ consisting of first applying the operation b to the CH_4 molecule followed by the operation a, it can be shown easily that the 24 symmetry operations of CH_4 defined above form a group known as T_d. Such groups of symmetry operations of a molecule are known as **point groups**. As the name implies, point groups consist of symmetry operations in which at least one point remains fixed and unchanged in space. Point groups contain two kinds of symmetry operations: proper and improper rotations.

An infinite crystal is different from a molecule in that it has translational symmetry. Although in real life crystals never extend to infinity, the problems associated with the finite nature of a crystal can be circumvented by applying the so-called **periodic (or Born–von Kármán) boundary conditions** to the crystal. Equivalently, one can imagine that the entire space is filled with replicas of the finite crystal. It should be no surprise that the set of all symmetry operations of such an infinite crystal also forms a group. Such groups, which contain both translational and rotational symmetry operations, are known as **space groups**. There are 230 non-equivalent space groups in three dimensions.

Besides their translational invariance, crystals also possess rotational symmetries. Space groups can be divided into two types, depending on whether or not the rotational parts of their symmetry operations are also symmetry operations. Let us first consider the purely translational operations of an infinite crystal. It can be shown that these translational symmetry operations form a group (to be denoted by $\boldsymbol{T}$). $\boldsymbol{T}$ is known as **subgroup** of the space group $\boldsymbol{G}$ of the crystal. Let us now denote by R the set of all symmetry operations of $\boldsymbol{G}$ which involve either pure rotations (both proper and improper) only or rotations accompanied by a translation not belonging to $\boldsymbol{T}$. We will denote the elements of R as α, β, τ, etc. Such a subset of $\boldsymbol{G}$ is known as a **complex**. In general R is not a group. For example, if $\boldsymbol{G}$ contains a **screw axis** or **glide plane** (these will be defined later, see Fig. 2.4) then R will not form a group and the space group $\boldsymbol{G}$ is said to be **nonsymmorphic**. If no screw axis or glide planes are present, R is a group (and therefore a subgroup of $\boldsymbol{G}$): the space group $\boldsymbol{G}$ is then said to be **symmorphic**. The symmetry properties of symmorphic groups are simpler to analyze since both translational and rotational operations in such space groups form subgroups. In particular, it can be shown that the rotational symmetry operations of a symmorphic space group form point groups similar to those for molecules. However, there are restrictions on the rotational symmetry of a crystal as a result of its translational symmetry. For example, a crystal cannot be invariant under rotation by 72° (known as a five-fold rotation). However, a molecule can have this rotational symmetry. Point groups which are compatible with a lattice with translational symmetry are called **crystallographic point groups**. It can be shown that there are 32 distinct crystallographic point groups in three-dimensional space (see, e. g. [2.4]).

Of a total of 230 space groups there are only 73 symmorphic space groups. Thus the simpler, symmorphic space groups are more often the exception rather than the norm. We will now consider how to analyze the rotational symmetries of nonsymmorphic space groups. By definition, a nonsymmorphic space group must contain at least one symmetry operation that involves both translation and rotation such that the rotational operation is not a symmetry operation of $\boldsymbol{G}$ by itself. There are two possibilities for such an operation: the rotation can be either proper or improper. The axis for a proper rotation is called a **screw axis** while the plane that corresponds to a twofold improper rotation is known as a **glide plane**. In the case of a screw axis, the crystal is invariant under a rotation about this axis plus a translation along the axis. The crystal is invariant under reflection in a glide plane followed by a translation parallel to the glide plane.

Two simple examples of screw axes for a one-dimensional crystal are shown in Figs. 2.4a and b. From Fig. 2.4b it is clear that a simple three-fold rotation about the vertical axis is not a symmetry operation of this hypothet-

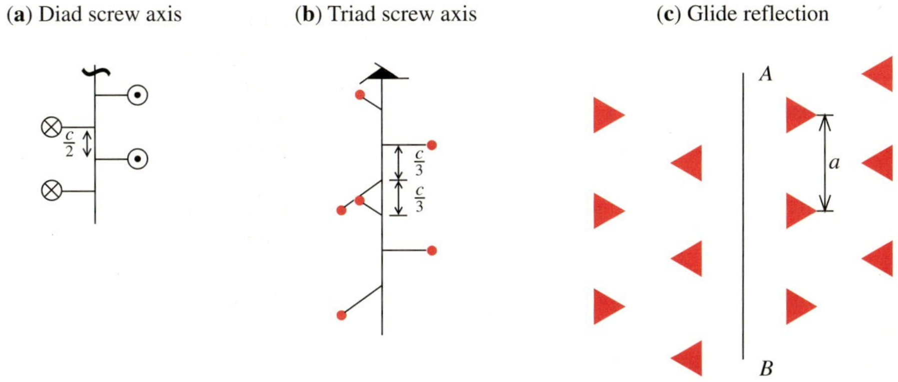

Fig. 2.4. Examples of (**a**) a diad screw axis, (**b**) a triad screw axis and (**c**) a glide plane. The crystals in (**a**) and (**c**) are assumed to be three dimensional, although only one layer of atoms is shown for the purpose of illustration. If they are two dimensional, the glide operation in (**c**) becomes equivalent to that of the diad screw in (**a**). ⊙ and ⊗ represent arrow pointing towards and away from the reader, respectively. Screw axis such as in (**b**) are found in the crystal structure of semiconductors like Se and Te [2.5]

ical crystal. However, if the crystal is translated by an amount ($c/3$) along the vertical axis after the three-fold rotation then the crystal is unchanged. the vertical axis is known in this case as a **triad screw axis**. An example of a diad screw axis is shown in Fig. 2.4a. A glide plane is shown in Fig. 2.4c. The plane labeled A–B in the figure is not a reflection plane. But if after a reflection in the A–B plane we translate the crystal by the amount ($a/2$) parallel to the A–B plane, the crystal will remain unchanged. This symmetry operation is known as a **glide** and the A–B plane is a glide plane. Now suppose R is the set of all pure rotational operations of $\boldsymbol{G}$ plus the glide reflection shown in Fig. 2.4c (which we will denote as m). R defined in this way is not a group since mm is a pure translation and therefore not an element of R.

To study the rotational symmetries of a space group independent of whether it is symmorphic or nonsymmorphic, we will introduce the concept of a **factor group**. Let $\boldsymbol{G}$ be the space group and $\boldsymbol{T}$ its subgroup consisting of all purely translational symmetry operations. Let $C = \{\alpha, \beta, \ldots\}$ be the complex of all the elements of $\boldsymbol{G}$ not in $\boldsymbol{T}$. Unlike the elements of the set R defined earlier, the translation operations in the elements of C can belong to $\boldsymbol{T}$. Next we form the sets $T\alpha$, $T\beta$, etc. The set $T\alpha$ consisting of operations formed by the product of a translation in $\boldsymbol{T}$ and an operation α not in $\boldsymbol{T}$ is known as a **right coset** of $\boldsymbol{T}$. As may be expected, the set αT is called a **left coset** of $\boldsymbol{T}$.

Let us first consider the case when $\boldsymbol{G}$ is a symmorphic group. For a symmorphic group we can decompose any symmetry operation α in C into the product of a translation α_t and a rotation α_r : $\alpha = \alpha_t\alpha_r$. Since multiplication is not necessarily commutative, we may worry about the order in which the two operations α_t and α_r occur. It can be shown that $\boldsymbol{T}$ has the property that the right coset Tx is equal to the left coset xT for every element x in $\boldsymbol{G}$. A subgroup with this property is known as an **invariant subgroup**. When we multiply α by another translation operation to form an element of the coset $T\alpha$ the resultant operation consists of a new translation but multiplied by the same rotation α_r. This suggests that we can establish a correspondence between the set of cosets $\{T\alpha, T\beta, \ldots\}$ and the set of rotational operations $R = \{\alpha_r, \beta_r, \ldots\}$. When $\boldsymbol{G}$ is symmorphic the set R is a subgroup of $\boldsymbol{G}$ so the set $\{T\alpha, T\beta, \ldots\}$ also forms a group. [In order that this set of cosets form a group, we have to define the product of two cosets $(T\alpha)(T\beta)$ as $T\alpha\beta$]. This group is known as the **factor group** of $\boldsymbol{G}$ with respect to $\boldsymbol{T}$ and is usually denoted by $\boldsymbol{G}/\boldsymbol{T}$. In establishing the factor group $\boldsymbol{G}/\boldsymbol{T}$ we have mapped all the elements of a coset $T\alpha$ into a single rotational operation α_r. Such a mapping of many elements in one set into a single element in another set is known as **homomorphism**. On the other hand, the mapping between the factor group $\boldsymbol{G}/\boldsymbol{T}$ and the subgroup $\boldsymbol{R}$ of $\boldsymbol{G}$ is one-to-one, and this kind of correspondence is known as **isomorphism**.

This isomorphism between the factor group $\boldsymbol{G}/\boldsymbol{T}$ and the point group $\boldsymbol{R}$ of a symmorphic space group can be extended to a nonsymmorphic space group. The main difference between the two cases is that while the rotational operations α_r, β_r, etc. are also elements in a symmorphic space group, this is not necessarily true for all rotations in a nonsymmorphic group. If α is a glide or screw then α_r is not an element in $\boldsymbol{G}$. We will still refer to the group $\boldsymbol{R}$ as the point group of a nonsymmorphic space group because $\boldsymbol{R}$ contains all the information about the rotational symmetries of the space group $\boldsymbol{G}$. However, special care must be exercised in studying the point groups of nonsymmorphic space groups since they contain elements which are not in the space group.

We will next study the symmetry operations of the zinc-blende and diamond crystal structures as examples of a symmorphic and a nonsymmorphic space group, respectively.

2.3.2 Symmetry Operations of the Diamond and Zinc-Blende Structures

Figure 2.2a shows the structures of the diamond and zinc-blende crystals. As pointed out in the previous section, both crystal structures consist of a fcc lattice. Associated with every lattice site there are two atoms which are displaced relative to each other by one quarter of the body diagonal along the [111] direction. The volume defined by the primitive lattice vectors and containing these two atoms forms a unit, known as the **primitive cell**, which is repeated at each lattice site. One simple way to construct these crystal structures is to start with two fcc sublattices, each containing only one atom located on every lattice site. Then one sublattice is displaced by one quarter of the body diagonal along the [111] direction with respect to the remaining sublattice. In the resulting crystal structure each atom is surrounded by four nearest neighbors forming a tetrahedron. The space group of the zinc-blende structure is symmorphic and is denoted by T_d^2 (or $F\bar{4}3m$ in international notation). Its translational symmetry operations are defined in terms of the three primitive lattice vectors shown in Fig. 2.2b. Its point group has 24 elements. These 24 elements are identical to the elements of the point group of a tetrahedron (or the methane molecule discussed in the last section and shown in Fig. 2.3) which is denoted by T_d.

The point group symmetry operations of the zinc-blende crystal are defined with respect to the three mutually perpendicular crystallographic axes with the origin placed at one of the two atoms in the primitive unit cell. With this choice of coordinates, the 24 operations are enumerated below (they are essentially identical to those of the methane molecule):

E:	identity
eight C_3 operations:	clockwise and counterclockwise rotations of 120° about the [111], [$\bar{1}$11], [1$\bar{1}$1], and [11$\bar{1}$] axes, respectively;
three C_2 operations:	rotations of 180° about the [100], [010], and [001] axes, respectively;
six S_4 operations:	clockwise and counterclockwise improper rotations of 90° about the [100], [010], and [001] axes, respectively;
six σ operations:	reflections with respect to the (110), (1$\bar{1}$0), (101), (10$\bar{1}$), (011), and (01$\bar{1}$) planes, respectively.

The diamond structure is the same as the zinc-blende structure except that the two atoms in the primitive unit cell are identical. If we choose the origin at the midpoint of these two identical atoms, we find that the crystal structure is invariant under inversion with respect to this origin. However, for the purpose of studying the point group operations, it is more convenient to choose the *origin at an atom*, as in the case of the zinc-blende structure. The crystal is no longer invariant under inversion with respect to this new choice of origin, but is unchanged under inversion plus a translation by the vector $(a/4)[1,1,1]$, where a is the length of the unit cube. This can be visualized by drawing the carbon atoms in the diamond structure along the [111] direction as shown in Fig. 2.5. **The space group of the diamond structure is nonsymmorphic**: it contains three glide planes. For example, the plane defined by $x = (a/8)$ is a glide plane since diamond is invariant under a translation by $(a/4)[0,1,1]$ followed by a reflection on this plane. In place of the three glide planes defined by $x = (a/8)$, $y = (a/8)$, and $z = (a/8)$, it is possible to use the "glide-like" operations:

$T(1/4,1/4,1/4)\sigma_x$:	reflection on the $x = 0$ plane followed by a translation of the crystal by the vector $a(1/4,1/4,1/4)$;
$T(1/4,1/4,1/4)\sigma_y$:	reflection on the $y = 0$ plane followed by a translation of the crystal by the vector $a(1/4,1/4,1/4)$; and

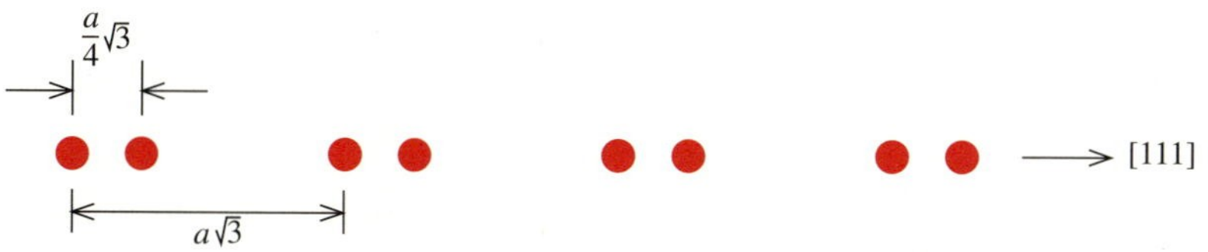

Fig. 2.5. Arrangement of atoms along the [111] direction of the diamond crystal. Notice that the crystal is invariant under inversion either with respect to the midpoint between the atoms or with respect to one of the atoms followed by an appropriate translation along the [111] axis

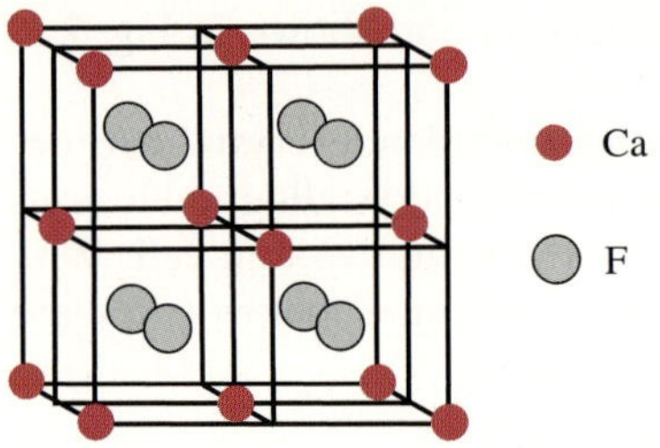

Fig. 2.6. Schematic crystal structure of CaF_2 (fluorite)

$T(1/4, 1/4, 1/4)\sigma_z$: reflection on the $z = 0$ plane followed by a translation of the crystal by the vector $a(1/4, 1/4, 1/4)$.

The factor group of the diamond lattice is isomorphic to the point group generated from the group T_d by adding the inversion operation. This point group has 48 elements and is denoted as O_h. While T_d is the point group of a tetrahedron, O_h is the point group of a cube. The space group of the diamond crystal is denoted by O_h^7 (or $Fd3m$ in international notation).

The CaF_2 (**fluorite**) structure shown in Fig. 2.6 is related to the diamond structure. This is the crystal structure of a family of semiconductors with the formula Mg_2X, where X = Ge, Si, and Sn. The lattice of CaF_2 is fcc as in diamond, but CaF_2 has three sublattices. The two fluorine sublattices are symmetrically displaced by one quarter of the body diagonal from the Ca sublattices, so there is inversion symmetry about each Ca atom. The space group of CaF_2 is symmorphic and its point group is also O_h, like diamond. This space group is denoted by O_h^5 (or $Fm3m$). It is clear that there is a one-to-one correspondence between the elements of the point group of CaF_2 and those of the factor group of diamond.

2.3.3 Representations and Character Tables

The effect of a symmetry operation, such as a rotation, on a coordinate system (x, y, z) can be represented by a transformation matrix. For example, under a four-fold rotation about the x axis the axes x, y, and z are transformed into x', y', and z' with $x' = x$, $y' = z$ and $z' = -y$. This transformation can be represented by the matrix $\boldsymbol{M}$:

$$\boldsymbol{M} = \begin{pmatrix} 1 & 0 & 0 \\ 0 & 0 & 1 \\ 0 & -1 & 0 \end{pmatrix}.$$

Similarly a three-fold rotation about the [111] axis will transform the axes x, y, and z into $x' = z$, $y' = x$, and $z' = y$. This transformation can be represented by

$$\begin{pmatrix} 0 & 0 & 1 \\ 1 & 0 & 0 \\ 0 & 1 & 0 \end{pmatrix}.$$

In the rest of this chapter we will use the abbreviated notation $(xyz) \rightarrow (xz\bar{y})$ to denote the transformation matrix for the four-fold rotation and $(xyz) \rightarrow (zxy)$ for the three-fold rotation. All the symmetry operations in a point group can be represented by transformation matrices similar to $\boldsymbol{M}$. It is easy to prove that the set of such transformation matrices corresponding to a group of symmetry operations is also a group. This group of matrices is said to form a **representation** of the group. There are actually an infinite number of such groups of matrices for a given group. The correspondence between a group and its representation is not, in general, an isomorphism but rather a homomorphism. A representation of a group $\boldsymbol{G}$ is defined as any group of matrices onto which $\boldsymbol{G}$ is homomorphic. Since representations of a group are not unique, we will be interested only in those of their properties that are common to all the representations of this group.

One way to generate a representation for a group is to choose some function $f(x, y, z)$ and then generate a set of functions $\{f_i\}$ by applying the symmetry operations O_i of the group to $f(x, y, z)$ so that $f_i = O_i[f]$.[1] By definition a group has to satisfy the closure requirement. This means that when the operation O is applied to f_i the resultant function $O[f_i]$ can be expressed as a linear combination of the functions f_i:

$$O[f_i] = \sum_j f_j a_{ji}. \tag{2.10a}$$

The coefficients a_{ji} form a square matrix, which will be referred to as a **transformation matrix**. The set of transformation matrices of the form $\{a_{ji}\}$ corresponding to all the operations in the group now forms a representation of the group. The functions $\{f_i\}$ used to generate this representation are said to form a set of **basis functions** for this representation. Clearly the choice of basis functions for generating a given representation is not unique.

If one uses the above method to generate a representation, then the dimension of the resulting transformation matrices will always be equal to the number of elements in the group (known as the **order** of a group). Some of the matrices will, however, be equal. Also many of the elements in these matrices will be zero. If the matrices in a given representation *for all the operations in a group* can be expressed in the following block form:

$$\begin{pmatrix} & & & 0 & 0 & \dots & 0 \\ & \alpha & & 0 & 0 & \dots & 0 \\ & & & 0 & 0 & \dots & 0 \\ 0 & 0 & 0 & \beta & 0 & \dots & 0 \\ 0 & 0 & 0 & & \gamma & \dots & 0 \\ 0 & 0 & 0 & & & \dots & 0 \\ 0 & 0 & 0 & & & \dots & 0 \\ 0 & 0 & 0 & 0 & 0 & \dots & \tau \end{pmatrix}, \tag{2.10b}$$

[1] In our applications $\{f_i\}$ will usually be a set of *degenerate* eigenfunctions corresponding to a given eigenvalue.

where α, β, ..., τ are square matrices, obviously the symmetry operations in this group can also be represented by the smaller matrices α and β, etc. While the matrices α_i and β_i for an operation i may not necessarily have the same dimension, the matrices α_i for all the operations i in the group must have the same dimension. A representation of the form of (2.10b) is said to be **reducible** otherwise the representation may be irreducible. As pointed out earlier, the choice of matrices to form a representation for a given group is not unique. Given one set of transformation matrices $\{A_i\}$, we can generate another set $\{A_i'\}$ by a **similarity transformation**: $A_i' = TA_iT^{-1}$, where T is an arbitrary nonsingular matrix with the same dimensionality as A_i. The transformed set of matrices $\{A_i'\}$ will also form a representation. The two sets of matrices $\{A_i\}$ and $\{A_i'\}$ are then said to be **equivalent**. Often the matrices of a representation may not appear to have the form given by (2.10b) and hence be regarded as irreducible. However, if by applying similarity transformation it is possible to express these matrices in the form of (2.10b) then this representation is also called reducible. Otherwise it is called **irreducible**.

Two axes of rotation or two reflection planes which transform into each other under a symmetry operation of a point group are said to be equivalent. It can be shown that the matrices of a representation which correspond to such equivalent rotations have identical **traces** (the trace of a matrix is the sum of the diagonal elements). Although the choice of irreducible representations for a group is not unique, the set of traces of these irreducible representations is unique since unitary transformations preserve the trace. This suggests that the set of all equivalent irreducible representations of a given group can be specified uniquely by their traces. For this reason the traces of the matrices in a representation are called its **characters**. The representations obviously contain more information than their characters; however, to utilize the symmetry of a given group it often suffices to determine the number of inequivalent irreducible representations and their characters.

The determination of the characters of an irreducible representation is simplified by these properties of a group:

- Elements in a group can be grouped into **classes**. A set of elements T in a group is said to form a class if for any element a in the group, $aT = Ta$. In a given representation all the elements in a class have the same character.
- The number of inequivalent irreducible representations of a group is equal to the number of classes.

These two properties suggest that if the elements of a group can be divided into j classes the characters of its j irreducible representations can be tabulated to form a table with j columns and j rows, which is known as a **character table**. Assume that a group has N elements and these elements are divided into j classes denoted by $C_1, C_2, \ldots, C_j$. The number of elements in each class will be denoted by $N_1, N_2, \ldots, N_j$. The identity operation E forms a class with only one element and, by convention, it is labeled C_1. This group also has j inequivalent irreducible representations (from now on the set of irreducible

Table 2.1. The character table of a group

	Classes			
Representations	$\{E\}$	$\{N_2C_2\}$	...	$\{N_jC_j\}$
R_1	$\chi_1(E)$	$\chi_1(2)$	...	$\chi_1(j)$
R_2	$\chi_2(E)$	$\chi_2(2)$	...	$\chi_2(j)$
.	.	.	...	.
.	.	.	...	.
R_j	$\chi_j(E)$	$\chi_j(2)$	...	$\chi_j(j)$

representations of a group will be understood to contain only inequivalent ones), which will be denoted by $R_1, R_2, \ldots, R_j$. The character of C_k in R_i will be denoted by $\chi_i(k)$. Since the identity operation E leaves any basis function invariant, its representations always consist of **unit matrices** (that is, diagonal matrices with unity as the diagonal elements). As a result, the character $\chi_i(E)$ is equal to the **dimension of the representation** R_i. Thus the character table of this group will have the form of Table 2.1.

In principle, the character table for the point group of a crystal can be calculated from the transformation matrices using a suitable set of basis functions. In practice, the character table can be obtained, in most cases, by inspection using the following two **orthogonality relations**:

$$\sum_k \chi_i(C_k)^* \chi_j(C_k) N_k = h\delta_{ij} \qquad (2.11)$$

$$\sum_i \chi_i(C_k)^* \chi_i(C_l) = (h/N_l)\delta_{kl}, \qquad (2.12)$$

where * denotes the complex conjugate of a character, h is the order of the group, N_k is the number of elements of class C_k, and δ_{ij} is the Kronecker delta.

As an illustration of the procedure used to obtain character tables we will consider two examples.

EXAMPLE 1 Character Table of the Point Group T_d

As we showed in Sect. 2.3.1, the point group T_d consists of 24 elements representing the proper and improper rotational symmetry operations of a tetrahedral methane molecule. In Sect. 2.3.2 we showed that this group is also the point group of the zinc-blende crystal. The 24 elements of this group can be divided into five classes

$$\{E\}, \quad \{8C_3\}, \quad \{3C_2\}, \quad \{6S_4\} \quad \text{and} \quad \{6\sigma\}$$

by noting that:

- rotations by the same angle with respect to **equivalent** axes belong to the same class and
- reflections on **equivalent** planes also belong to the same class.

Since the number of irreducible representations is equal to the number of classes, T_d has five irreducible representations, which are usually denoted by A_1, A_2, E, T_1 and T_2. Notice that the capital letter E has been used in the literature to denote a large number of entities varying from energy, electric field, the identity operation in group theory to an irreducible representation in the T_d group! To avoid confusion we will always specify what E stands for.

The next step is to construct the 5×5 character table using (2.11) and (2.12). First, we note again that the character of the class containing the identity operation $\{E\}$ is equal to the dimension of the representation. Substituting this result into (2.12) we find

$$\sum_i |\chi_i(E)|^2 = h. \tag{2.13}$$

Since the number of classes is usually small, this equation can often be solved by inspection. For T_d it is easily shown that the only possible combination of five squares which add up to 24 is: $2 \times 1^2 + 2^2 + 2 \times 3^2$. This result means that the group T_d has two irreducible representations of dimension one (denoted by A_1 and A_2), one irreducible representation of dimension two (denoted by E), and two irreducible representations of dimension three (denoted by T_1 and T_2). Next we note that a scalar will be invariant under all operations, so there is always a trivial **identity representation** whose characters are all unity. By convention this representation is labeled by the subscript 1, A_1 in the present case. So without much effort we have already determined one row and one column of the character table for T_d (Table 2.2).

The remaining characters can also be determined by inspection with the application of (2.12). For the classes other than $\{E\}$ the characters can be either positive or negative. The sign can be determined by inspection with some practice. For example, for the class $\{6\sigma\}$ the only combination of sums of squares satisfying (2.12) is $4 \times 1^2 + 0^2 = 24/6 = 4$. Applying (2.12) to the characters of $\{E\}$ and $\{6\sigma\}$ it can be easily seen that $A_2(6\sigma) = -1$, $E(6\sigma) = 0$

Table 2.2. Determining the character table for the T_d group by inspection

	$\{E\}$	$\{3C_2\}$	$\{6S_4\}$	$\{6\sigma\}$	$\{8C_3\}$
A_1	1	1	1	1	1
A_2	1	·	·	·	·
E	2	·	·	·	·
T_1	3	·	·	·	·
T_2	3	·	·	·	·

Table 2.3. Character table and basis functions of the T_d group

	$\{E\}$	$\{3C_2\}$	$\{6S_4\}$	$\{6\sigma\}$	$\{8C_3\}$	Basis functions
A_1	1	1	1	1	1	xyz
A_2	1	1	−1	−1	1	$x^4(y^2-z^2)+y^4(z^2-x^2)+z^4(x^2-y^2)$
E	2	2	0	0	−1	$\{(x^2-y^2), z^2-\frac{1}{2}(x^2+y^2)\}$
T_1	3	−1	1	−1	0	$\{x(y^2-z^2), y(z^2-x^2), z(x^2-y^2)\}$
T_2	3	−1	−1	1	0	$\{x, y, z\}$

while the two remaining characters for T_1 and T_2 contain 1 and −1. The final result for the character table of T_d is given in Table 2.3.

It is instructive to examine some possible basis functions for the irreducible representations of T_d. One choice of basis functions for the A_1 representation is a constant, as we have mentioned earlier. Another possibility would be the function xyz, which is also invariant under all symmetry operations of T_d. A_2 is very similar to A_1 except that under the operations S_4 and σ the character of A_2 is −1 rather than 1. This implies that the basis function for A_2 must change sign under interchange of any two coordinate axes, such as interchanging x and y. One choice of basis function for A_2 is $x^4(y^2-z^2)+y^4(z^2-x^2)+z^4(x^2-y^2)$. Similarly, the three-dimensional representations T_1 and T_2 differ only in the sign of their characters under interchange of any two coordinates. It can be shown that the three components x, y, and z of a vector transform as T_2. A corresponding set of basis functions for the T_1 representation would be $x(y^2-z^2)$, $y(z^2-x^2)$, and $z(x^2-y^2)$. The reader should verify these results by calculating the characters directly from the basis functions (Problem 2.2).

At the beginning of this chapter we pointed out the importance of notation in group theory. The notation we have used so far to label the irreducible representations of the T_d group: A_1, E, T_1, etc. is more commonly found in literature on molecular physics. We now introduce another notation used frequently in articles on semiconductor physics. The wave functions of a crystal with wave vector $\boldsymbol{k}$ at the center of the Brillouin zone (Γ point) always transform in the way specified by the irreducible representations of the point group of the crystal. Hence the Bloch functions at Γ of a zinc-blende crystal can be classified according to these irreducible representations. In semiconductor physics literature it is customary to use Γ plus a subscript i to label these irreducible representations of T_d. Unfortunately there are two different conventions in the choice of the subscript i for labeling the same irreducible representation. One of these conventions is due to Koster (more commonly used in recent research articles) while the other was proposed by Bouckaert, Smoluchowski and Wigner (BSW) and tends to be found in older articles. The correspondence between the different notations for the T_d point group is shown in Table 2.4.

Table 2.4. Commonly used notations for the irreducible representations of the T_d point group

Koster notation[a]	BSW notation	Molecular notation
Γ_1	Γ_1	A_1
Γ_2	Γ_2	A_2
Γ_3	Γ_{12}	E
Γ_4	Γ_{15}	T_2
Γ_5	Γ_{25}	T_1

[a] Note that Γ_4 and Γ_5 are sometimes reversed in the literature. We recommend the student to check it whenever he encounters this notation [2.4].

EXAMPLE 2 Character Table of O_h

We mentioned earlier in this section that the factor group of the diamond structure is O_h and that it is isomorphic to the point group derived from the T_d group by including the inversion operation i. It has therefore 48 elements: the 24 symmetry operations of T_d plus those of T_d followed by i. These include all 48 symmetry operations of a cube. From the properties of the group T_d, one can deduce that O_h has ten classes:

$\{E\}$: identity;
$\{3C_2\}$: C_2 rotation about each of the three equivalent [100] axes;
$\{6S_4\}$: two four-fold improper rotations about each of the three equivalent [100] axes;
$\{6\sigma_d\}$: reflection on each of the six equivalent (110) planes;
$\{8C_3\}$: two C_3 rotations about each of the four equivalent [111] axes;
$\{i\}$: inversion;
$\{3\sigma_h\}$: reflection on each of the three equivalent (100) planes;
$\{6C_4\}$: two C_4 rotations about each of the three equivalent [100] axes;
$\{6C_2'\}$: C_2 rotation about each of the six equivalent [110] axes;
$\{8S_6\}$: two three-fold improper rotations about each of the four equivalent [111] axes.

The first five classes are the same as those of T_d while the remaining five are obtained from the first five by multiplication with the inversion.

Correspondingly, there are ten irreducible representations. Five of them correspond to even transformations under those operations obtained from the T_d group operation followed by inversion, while the other five correspond to odd ones. Similarly, the basis functions of the irreducible representations of O_h are either even or odd under those operations. In the terminology of quantum mechanics, these basis functions are said to have even or odd **parity**. The characters for the O_h group are listed in Table 2.5, while a set of basis functions for its irreducible representations is given in Table 2.6. Table 2.5 has been purposedly presented in a way to show the similarity between the "unprimed" representations in the O_h group and those of the T_d group. For example, a scalar

Table 2.5. Character table of the O_h group presented in a way to highlight the similarity with Table 2.3 for the T_d group. BSW notation

	$\{E\}$	$\{C_2\}$	$\{S_4\}$	$\{\sigma_d\}$	$\{C_3\}$	$\{i\}$	$\{\sigma_h\}$	$\{C_4\}$	$\{C_2'\}$	$\{S_6\}$
Γ_1	1	1	1	1	1	1	1	1	1	1
Γ_2	1	1	−1	−1	1	1	1	−1	−1	1
Γ_{12}	2	2	0	0	−1	2	2	0	0	−1
Γ_{25}	3	−1	1	−1	0	−3	1	−1	1	0
Γ_{15}	3	−1	−1	1	0	−3	1	1	−1	0
$\Gamma_{1'}$	1	1	−1	−1	1	−1	−1	1	1	1
$\Gamma_{2'}$	1	1	1	1	1	−1	−1	−1	−1	−1
$\Gamma_{12'}$	2	2	0	0	−1	−2	−2	0	0	1
$\Gamma_{25'}$	3	−1	−1	1	0	3	−1	−1	1	0
$\Gamma_{15'}$	3	−1	1	−1	0	3	−1	1	−1	0

Table 2.6. Basis functions for the irreducible representations of the O_h group

Representation	Basis functions
Γ_1 :	1
Γ_2 :	$x^4(y^2 - z^2) + y^4(z^2 - x^2) + z^4(x^2 - y^2)$
Γ_{12} :	$\{[z^2 - (x^2 + y^2)/2],\ x^2 - y^2\}$
Γ_{25} :	$\{x(y^2 - z^2),\ y(z^2 - x^2),\ z(x^2 - y^2)\}$
Γ_{15} :	$\{x,\ y,\ z\}$
$\Gamma_{1'}$:	$xzy[x^4(y^2 - z^2) + y^4(z^2 - x^2) + z^4(x^2 - y^2)]$
$\Gamma_{2'}$:	xyz
$\Gamma_{12'}$:	$\{xyz[z^2 - (x^2 + y^2)/2],\ xyz(x^2 - y^2)\}$
$\Gamma_{25'}$:	$\{xy,\ yz,\ zx\}$
$\Gamma_{15'}$:	$\{yz(y^2 - z^2),\ zx(z^2 - x^2),\ xy(x^2 - y^2)\}$

still belongs to the Γ_1 representation while a vector belongs to the Γ_{15} representation in the O_h group. However, the relation between the "primed" and "unprimed" representations is not so clear. For example, a pseudo-scalar belongs to the $\Gamma_{2'}$ representation, while a pseudo-vector belongs to the $\Gamma_{15'}$ representation. Furthermore, some of the primed representations, e. g. $\Gamma_{15'}$ and $\Gamma_{25'}$, are even while others are odd under inversion.

When Table 2.5 is rearranged into Table 2.7, the correlations between the first five representations and the remaining five become clear. Note that sometimes a hybrid of the K and BSW notations is used: the primes are omitted and replaced by +, − superscripts to denote the parity. The student will find this notation in Chaps. 6 and 7.

Table 2.7. Character table of the O_h group rearranged to show the relationship between the even and odd parity representations. Both the Koster (K) and BSW notations are given [2.6]

K	BSW	$\{E\}$	$\{C_2\}$	$\{C_4\}$	$\{C_2'\}$	$\{C_3\}$	$\{i\}$	$\{\sigma_h\}$	$\{S_4\}$	$\{\sigma_d\}$	$\{S_6\}$
Γ_1^+	Γ_1	1	1	1	1	1	1	1	1	1	1
Γ_2^+	Γ_2	1	1	−1	−1	1	1	1	−1	−1	1
Γ_3^+	Γ_{12}	2	2	0	0	−1	2	2	0	0	−1
Γ_4^+	$\Gamma_{15'}$	3	−1	1	−1	0	3	−1	1	−1	0
Γ_5^+	$\Gamma_{25'}$	3	−1	−1	1	0	3	−1	−1	1	0
Γ_1^-	$\Gamma_{1'}$	1	1	1	1	1	−1	−1	−1	−1	−1
Γ_2^-	$\Gamma_{2'}$	1	1	−1	−1	1	−1	−1	1	1	−1
Γ_3^-	$\Gamma_{12'}$	2	2	0	0	−1	−2	−2	0	0	1
Γ_4^-	Γ_{15}	3	−1	1	−1	0	−3	1	−1	1	0
Γ_5^-	Γ_{25}	3	−1	−1	1	0	−3	1	1	−1	0

2.3.4 Some Applications of Character Tables

We will now describe some of the applications of character tables. Further applications will be found throughout this book.

a) Decomposition of Representation into Irreducible Components

A problem one often faces is this: when given a group $\boldsymbol{G}$ and a representation τ, how does one determine whether τ is reducible? If τ is reducible then how can it be decomposed into its irreducible components? These questions can be answered with the help of the character table of $\boldsymbol{G}$. Suppose $\chi_\tau(i)$ is the character of the given representation τ corresponding to the class $\{i\}$. If τ is an irreducible representation, the set of characters $\chi_\tau(i)$ must be equal to the characters of one of the irreducible representations of $\boldsymbol{G}$. If this is not the case then τ is reducible. Suppose τ is reducible into two irreducible representations α and β and $\chi_\alpha(i)$ and $\chi_\beta(i)$ are the characters of α and β, respectively. By definition $\chi_\alpha(i)$ and $\chi_\beta(i)$ must satisfy

$$\chi_\tau(i) = \chi_\alpha(i) + \chi_\beta(i) \tag{2.14}$$

for all classes $\{i\}$ in the group $\boldsymbol{G}$. The representation τ is said to be the **direct sum** of the two irreducible representations α and β. The direct sum will be represented by the symbol $\oplus$ as in $\tau = \alpha \oplus \beta$.

When the dimension of a reducible representation is not very large, it can often be reduced into a direct sum of irreducible representations by inspection. As an example, let us consider the group T_d with its character table given in Table 2.3 and a second-rank tensor $\{T_{ij}\}$ with components T_{xx}, T_{xy}, T_{xz}, T_{yx}, T_{yy}, T_{yz}, T_{zx}, T_{zy}, and T_{zz}. Using these components as basis functions we can generate a nine-dimensional representation of T_d, which we will denote as Γ. Obviously Γ must be reducible since no irreducible representation in T_d has dimensions larger than three. The way to decompose Γ into irreducible representations of T_d is to first determine the characters of Γ for all the classes in T_d. In principle this can be accomplished by applying the symmetry operations of T_d to the nine basis functions to produce the 9×9 matrices forming the representation Γ. A simpler and more direct approach is possible for this second-rank tensor. We note that a vector with three components x, y, and z forms a set of basis functions for the three-dimensional irreducible representation T_2 of T_d. Therefore, the 3×3 transformation matrices of a vector form a T_2 representation. By taking the **matrix product** of two such 3×3 transformation matrices we obtain a set of 9×9 matrices forming a representation for Γ. This suggests that the characters of Γ are equal to the squares of the characters for T_2:

	$\{E\}$	$\{3C_2\}$	$\{6S_4\}$	$\{6\sigma\}$	$\{8C_3\}$
χ_Γ :	9	1	1	1	0

When the matrices of a representation τ are equal to the matrix product of the matrices of two representations α and β, τ is said to be the **direct product** of α and β. Direct products are represented by the symbol $\otimes$ as in

$$\Gamma = T_2 \otimes T_2. \tag{2.15}$$

After determining the characters of Γ, the next step is to find the irreducible representations of T_d whose characters will add up to those of Γ. The systematic way of doing this is to apply the orthogonality relation (2.11). It is left as an exercise (Problem 2.3) to show that

$$T_2 \otimes T_2 = T_1 \oplus T_2 \oplus E \oplus A_1 \tag{2.16}$$

With practice this result can also be derived quickly by inspection. In the present example, one starts by writing down various combinations of representations with total dimensions equal to nine. Next one eliminates those combinations whose characters for the other classes do not add up to χ_Γ. Very soon it is found that the only direct sum with characters equal to that of Γ for all five classes of T_d is the one in (2.16). Once we realize that Γ can be decomposed into the direct sum of these four irreducible representations, we can use the basis functions for these representations given in Table 2.3 as a guide to deduce the correct linear combinations of the nine components of the second-rank tensor which transform according to these four irreducible representations:

$$\begin{aligned}
A_1 &: \; T_{xx} + T_{yy} + T_{zz} \\
E &: \; \{T_{xx} - T_{yy}, T_{zz} - (T_{xx} + T_{yy})/2\} \\
T_1 &: \; \{(T_{xy} - T_{yx})/2, \quad (T_{zx} - T_{xz})/2, \quad (T_{yz} - T_{zy})/2\} \\
T_2 &: \; \{(T_{xy} + T_{yx})/2, \quad (T_{xz} + T_{zx})/2, \quad (T_{yz} + T_{zy})/2\}.
\end{aligned}$$

b) Symmetrization of Long Wavelength Vibrations in Zinc-Blende and Diamond Crystals

The process we have described above is known as the **symmetrization** of the nine components of the second-rank tensor. This method can also be applied to symmetrize wave functions. When a Hamiltonian is invariant under the symmetry operations of a group, its wave functions can be symmetrized so as to belong to irreducible representations of this group. Just as atomic wave functions are labeled *s*, *p*, and *d* according to their symmetry under rotation, it is convenient to label the electronic and vibrational wave functions of a crystal at the point $\boldsymbol{k}$ in reciprocal space by the irreducible representations of the group of symmetry operations appropriate for $\boldsymbol{k}$. We will now explain this statement with an example drawn from the vibrational modes of zinc-blende and diamond crystals.

Although we will not discuss lattice vibrations in semiconductors until the next chapter, it is easier to demonstrate their symmetry properties than those of electrons. First, we can argue that vibrations of atoms in a crystal can be described by waves based on its translational symmetry, just as its electrons can be described by Bloch functions. For example, sound is a form of such vibration. Thus atomic motions in a crystal can be characterized by their **displacement vectors** (in real space) plus their wave vectors $\boldsymbol{k}$ (in reciprocal lattice space). The symmetry of a vibration is therefore determined by the effects of symmetry operations of the crystal on both vectors. Due to the discrete location of atoms in a crystal, a wave with wave vector equal to $\boldsymbol{k}$ or $\boldsymbol{k}$ plus a reciprocal lattice vector are indistinguishable (this point will be discussed further in Chap. 3). Thus an operation which transforms $\boldsymbol{k}$ into another wave vector $\boldsymbol{k}'$ differing from $\boldsymbol{k}$ by a reciprocal lattice vector also belongs to the group of symmetry operations of $\boldsymbol{k}$. This group is known as the **group of the wave vector** $\boldsymbol{k}$. In particular, the group of the Γ point or zone center is always the same as the point group of the crystal.

A long wavelength (that is, $\boldsymbol{k}$ near the Brillouin zone center) vibration in a crystal involves nearly uniform displacements of identical atoms in different unit cells. For a zinc-blende crystal with two atoms per primitive unit cell, a zone-center vibrational mode can be specified by two vectors representing the displacements of these two atoms. We have already pointed out that the three components of a vector transform under the symmetry operations of T_d according to the T_2, also called Γ_4 representation (see Tables 2.3 and 2.4). To discuss properties in the zinc-blende crystal we will switch to the Koster notation. For brevity we will refer to the vector as "**belonging**" to the Γ_4 representation. Two vectors, one associated with each atom in the primitive cell, give rise to a six-dimensional representation. Since irreducible representations in T_d have at most three dimensions, this representation is reducible. To reduce it one can calculate its characters by applying the symmetry operations of T_d to the two vectors. An alternative method is to consider the two atoms as the basis of a two-dimensional representation R. The characters of R are obtained by counting the number of atoms which are unchanged by the symmetry operations of T_d (since each atom satisfying this condition contributes one unity diagonal element to the representation matrix). The two atoms in the zinc-blende lattice are not interchanged by the operations of T_d, therefore all the characters of R are simply two. Thus R is reducible to two Γ_1 representations. The representations of the two displacement vectors in the unit cell of the zinc-blende crystal are equal to the direct product of R and Γ_4, which is equal to $2\Gamma_4$. These two Γ_4 representations correspond to the **acoustic** and **optical phonon** modes (see Chap. 3 for further details). In the acoustic mode the two atoms in the primitive cell move in phase while in the optical mode they move 180° out of phase.

As pointed out in Sect. 2.3.2, the factor group of the diamond crystal is isomorphic to the point group O_h. We should remember that the origin has been chosen to be one of the carbon atoms. The space group operation which corresponds to inversion in O_h is inversion about the origin plus a translation by $(a/4)(1,1,1)$ (for brevity this operation will be denoted here by i'). From Table 2.6 one finds that a vector belongs to the Γ_{15} representation of the O_h point group. As in the case of the zinc-blende crystal, we can obtain the characters of the six-dimensional representation by determining the characters of R and then calculating the direct product of R and Γ_{15}. The characters of R now depend on whether the symmetry operations include i'. For all symmetry operations which already exist for the zinc-blende structure and therefore do not involve i', the characters are equal to two, as in the zinc-blende crystal. For all other operations, the two atoms inside the primitive unit cell are interchanged by i', so their characters are zero. By inspection of Table 2.5 one concludes that R reduces to $\Gamma_1 \oplus \Gamma_{2'}$. Thus the displacement vectors of the two atoms in the primitive unit cell of diamond transform as Γ_{15} and $\Gamma_{25'}$. The displacement vectors of the acoustic phonon change sign (the parity is said to be odd) under i' and therefore have symmetry Γ_{15}. On the other hand the optical phonon parity is even and has symmetry $\Gamma_{25'}$. The effects of i' on the long-wavelength acoustic and optical phonons propagating along the body diagonal of the diamond crystal are shown in Fig. 2.7.

c) Symmetrization of Nearly Free Electron Wave Functions in Zinc-Blende Crystals

As an example of application of character tables in symmetrizing electronic wave functions, we will consider a **nearly free electron** in a zinc-blende crystal. By nearly free we mean that the electron is moving inside a crystal with a vanishingly small periodic potential of T_d symmetry, so that its energy E and wave function Φ are essentially those of a free particle:

$$\Phi(x,y,z) = \exp\left[\mathrm{i}(k_x x + k_y y + k_z z)\right] \tag{2.17}$$

and

$$E = \hbar^2 k^2/2m. \tag{2.18}$$

However, because of the periodic lattice, its wave vector $\boldsymbol{k}$ can be restricted to the first Brillouin zone in the reduced zone scheme.

We will assume that the crystal has the zinc-blende structure and that $\boldsymbol{k} = (2\pi/a)(1,1,1)$, where a is the length of an edge of the unit cube in the zinc-blende lattice. By applying the C_3 symmetry operations of zinc-blende we can show that all the eight points $(2\pi/a)(\pm1,\pm1,\pm1)$ in the Brillouin zone are equivalent. Furthermore, from the definition of the primitive reciprocal lattice vectors given in Sect. 2.2 all eight points differ from the zone center by

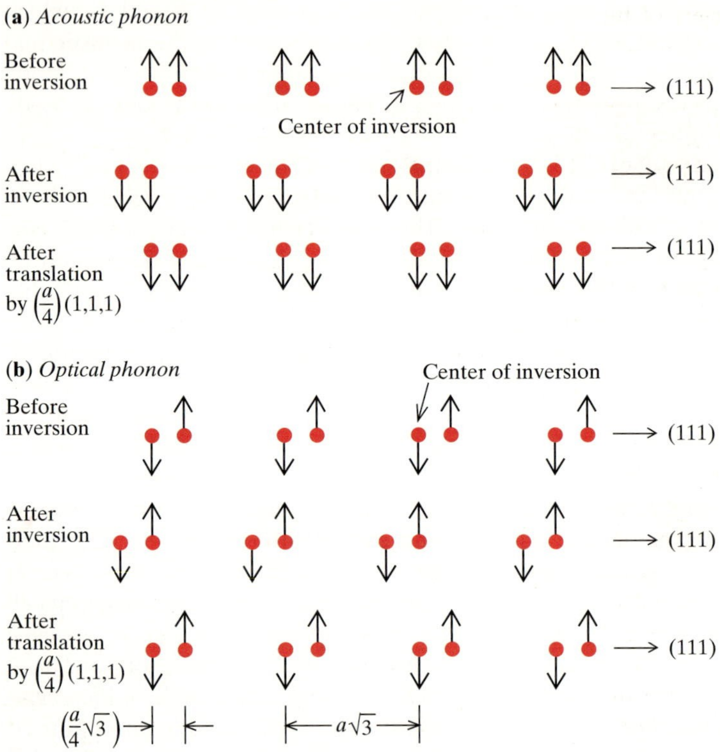

Fig. 2.7. Schematic diagrams of the transformation of (**a**) the acoustic phonon and (**b**) the zone-center optical phonon in diamond under inversion plus translation by $(a/4)(1,1,1)$ where a is the size of the unit cube of diamond

a primitive reciprocal lattice vector. Hence all eight points will map onto the zone center in the reduced zone scheme. The group of the wave vector $\boldsymbol{k} = (2\pi/a)(1,1,1)$ is therefore T_d. To simplify the notation we will represent the electronic wave functions $\exp[\mathrm{i}(k_x x + k_y y + k_z z)]$ as $\{k_x k_y k_z\}$. The eight wave functions $\{111\}, \{\bar{1}11\}, \{1\bar{1}1\}, \{11\bar{1}\}, \{\bar{1}\bar{1}1\}, \{\bar{1}1\bar{1}\}, \{1\bar{1}\bar{1}\}$ and $\{\bar{1}\bar{1}\bar{1}\}$ are degenerate but the degeneracy will be lifted by perturbations such as a nonzero crystal potential. Our goal now is to form symmetrized linear combinations of these eight wave functions with the aid of the character table of T_d in Table 2.3.

We note first that these eight wave functions form the basis functions of an eight-dimensional representation. Obviously this representation is reducible. Unlike the cases given in (a) and (b), there are no shortcuts in determining the characters of this eight-dimensional representation. Since characters are the sums of the diagonal elements, they can be deduced by determining the number of wave functions unchanged by the symmetry operations. The characters calculated in this way are given in Table 2.8.

Table 2.8. Characters of the representations formed by the nearly free electron wave functions with wave vectors equal to $(2\pi/a)(1,1,1)$ and $(2\pi/a)(2,0,0)$ in a zinc-blende crystal

Class	Transformation	Characters [111]	Characters [200]
E	xyz	8	6
$3C_2$	$x\bar{y}\bar{z}$	0	2
$6S_4$	$\bar{x}z\bar{y}$	0	0
6σ	yxz	4	2
$8C_3$	yzx	2	0

Table 2.9. Symmetrized nearly free electron wave functions in a zinc-blende crystal with wave vectors equal to $(2\pi/a)(\pm1,\pm1,\pm1)$

Representation	Wave function
Γ_1	$(1/\sqrt{8})(\{111\} + \{1\bar{1}\bar{1}\} + \{\bar{1}1\bar{1}\} + \{\bar{1}\bar{1}1\} + \{\bar{1}\bar{1}\bar{1}\} + \{\bar{1}11\} + \{1\bar{1}1\} + \{11\bar{1}\}) = (\sqrt{8})\cos(2\pi x/a)\cos(2\pi y/a)\cos(2\pi z/a)$
Γ_1	$(\sqrt{8})\sin(2\pi x/a)\sin(2\pi y/a)\sin(2\pi z/a)$
Γ_4	$(\sqrt{8})\{\sin(2\pi x/a)\sin(2\pi y/a)\cos(2\pi z/a)$; $\sin(2\pi x/a)\cos(2\pi y/a)\sin(2\pi z/a)$; $\cos(2\pi x/a)\sin(2\pi y/a)\sin(2\pi z/a)\}$
Γ_4	$(\sqrt{8})\{\sin(2\pi x/a)\cos(2\pi y/a)\cos(2\pi z/a)$; $\cos(2\pi x/a)\sin(2\pi y/a)\cos(2\pi z/a)$; $\cos(2\pi x/a)\cos(2\pi y/a)\sin(2\pi z/a)\}$.

By using the orthogonality relations or the method of "inspection", we found from Table 2.3 (using Table 2.4 to convert to the Koster notation) that the only combination of irreducible representations giving rise to the set of characters in Table 2.8 is the direct sum $2\Gamma_1 \oplus 2\Gamma_4$. Thus the eight $\{(\pm)1(\pm)1(\pm)1\}$ free electron wave functions can be expressed as two wave functions belonging to the one-dimensional Γ_1 representation and two wave functions belonging to the three-dimensional Γ_4 representation. The proper linear combinations of wave functions which transform according to these irreducible representations can be obtained systematically by using **projection operators** (see any one of the references on group theory for further details). In many simple cases this can be done by inspection also. The proper linear combinations of the [111] wave functions can be shown to be (see Problem 2.4) those in Table 2.9.

Similarly one can show that the six degenerate $\{(\pm)200\}$, $\{0(\pm)20\}$ and $\{00(\pm)2\}$ wave functions form a six-dimensional representation whose characters are given in Table 2.8. Using these characters one can decompose this six-

Table 2.10. Symmetrized nearly free electron wave functions in a zinc-blende crystal with wave vectors equal to $(2\pi/a)(\pm 2, 0, 0)$, $(2\pi/a)(0, \pm 2, 0)$, and $(2\pi/a)(0, 0, \pm 2)$

Representation	Wave function
Γ_1	$\cos(4\pi x/a) + \cos(4\pi y/a) + \cos(4\pi z/a)$
Γ_3	$\cos(4\pi y/a) - \cos(4\pi z/a)$; $\cos(4\pi x/a) - (1/2)[\cos(4\pi y/a) + \cos(4\pi z/a)]$
Γ_4	$\sin(4\pi x/a)$; $\sin(4\pi y/a)$; $\sin(4\pi z/a)$

dimensional representation into the direct sum $\Gamma_1 \oplus \Gamma_3 \oplus \Gamma_4$. The symmetrized wave functions are given in Table 2.10, while the proof is left as an exercise (Problem 2.4).

d) Selection Rules

In atomic physics one learns that optical transitions obey **selection rules** such as: in an electric-dipole transition the orbital angular momentum can change only by ± 1. These selection rules result from restrictions imposed on matrix elements of the **electric-dipole operator** [see (6.29,30)] by the rotational symmetry of the atomic potential. One may expect similar selection rules to result from the symmetry of potentials in crystals. To see how such selection rules can be derived, we will consider the following example.

Let $\boldsymbol{p}$ be the electron momentum operator and Ψ_1 be a wave function in a zinc-blende-type crystal with the point group T_d. Since $\boldsymbol{p}$ is a vector its three components p_x, p_y, and p_z belong to the irreducible representation T_2 (Γ_{15} or Γ_4 according to Table 2.4). Let us assume that Ψ_1 is a triply degenerate wave function belonging to T_2 also. Operating with $\boldsymbol{p}$ on Ψ_1 results in a set of nine wave functions, which we will label Ψ_3. These nine wave functions generate a nine-dimensional reducible representation which can be reduced to the direct sum $T_1 \oplus T_2 \oplus E \oplus A_1$ as shown in (2.16). Next we form the matrix element $M = \langle \Psi_2 | \boldsymbol{p} | \Psi_1 \rangle = \langle \Psi_2 | \Psi_3 \rangle$ between Ψ_3 and another wave function Ψ_2. Suppose Ψ_2 belongs to an irreducible representation B which is *not* one of the irreducible representations in the direct sum $T_1 \oplus T_2 \oplus E \oplus A_1$ of the wave function Ψ_3. From the orthogonality of the basis functions for different irreducible representations one concludes that the matrix element M is zero. In general, it can be proved that the matrix element between an operator p and two wave functions Ψ_1 and Ψ_2 can differ from zero only when the direct product of the representations of p and Ψ_1 contains an irreducible representation of Ψ_2. This important group theoretical result is known as the **matrix-element theorem**.

When applied to atoms the matrix-element theorem leads to the familiar selection rules for electric-dipole transitions. For instance, if Ψ_1 and Ψ_2 are atomic wave functions they will have definite parities under inversion, and the parity of their direct product is simply the product of their parities. The electric-dipole operator has odd parity so its matrix element is zero between two states of the same parity according to the matrix-element theorem. If Ψ_1 and Ψ_2 both have *s* symmetry then their direct product also has *s* symmetry. Since the dipole operator has *p*-symmetry its matrix element between two *s* states is zero. On the other hand if one of these two wave functions has *p* symmetry its direct product will contain a component with *p* symmetry, and the electric-dipole transition will be nonzero. Thus, application of the matrix-element theorem leads to selection rules for optical transitions in systems with spherical symmetry.

Using the matrix-element theorem we can also obtain very general selection rules for optical transitions in zinc-blende-type and diamond-type crystals. In Chap. 6 we will show that electric-dipole transitions in a crystal are determined by the matrix element of the electron momentum operator $\boldsymbol{p}$. In a zinc-blende-type crystal $\boldsymbol{p}$ belongs to the Γ_4 irreducible representations. To derive the selection rules for optical transitions involving zone-center wave functions we need to know the direct product between Γ_4 and all the irreducible representations of T_d. The results are summarized in Table 2.11.

From this table we can easily determine whether or not electric-dipole transitions between any two bands at the zone center of the zinc-blende crystal are allowed. For example, dipole transitions from a Γ_4 valence band to conduction bands with Γ_1, Γ_3, Γ_4, and Γ_5 symmetries are all allowed. Using Table 2.11 one can derive selection rules for optical excitation of phonons by photons in the infrared (to be discussed further in Chap. 6). The ground state of the crystal with no phonons should have Γ_1 symmetry. In zinc-blende crystals only Γ_4 optical phonons can be directly excited by an infrared photon via an electric-dipole transition. Such phonons are said to be **infrared-active**. On the other hand the $\Gamma_{25'}$ optical phonon of the diamond structure is not infrared-active because of the parity selection rule (Ge, Si, and diamond are highly transparent in the infrared!). The ionic momentum operator has symmetry Γ_{15} for the O_h group and odd parity under the operation i' of the diamond crystal. Hence electric-dipole transitions can only connect states with opposite parity.

Selection rules for higher order optical processes, such as Raman scattering can also be obtained from Table 2.11. As will be shown in Chap. 7, Raman scattering involves the excitation of a phonon via two optical transitions. If both optical transitions are of the electric-dipole type in a zinc-blende crystal, the excited phonon must belong to one of the irreducible representations of the direct product $\Gamma_4 \otimes \Gamma_4 = \Gamma_4 \oplus \Gamma_5 \oplus \Gamma_3 \oplus \Gamma_1$. Phonons which can be excited optically in Raman scattering are said to be **Raman-active**. Thus the Γ_4 optical phonon in the zinc-blende crystal is Raman-active in addition to being

Table 2.11. Direct products of the Γ_4 representation with all the representations of T_d

Direct product	Direct sum
$\Gamma_4 \otimes \Gamma_1$	Γ_4
$\Gamma_4 \otimes \Gamma_2$	Γ_5
$\Gamma_4 \otimes \Gamma_3$	$\Gamma_4 \oplus \Gamma_5$
$\Gamma_4 \otimes \Gamma_4$	$\Gamma_4 \oplus \Gamma_5 \oplus \Gamma_3 \oplus \Gamma_1$
$\Gamma_4 \otimes \Gamma_5$	$\Gamma_4 \oplus \Gamma_5 \oplus \Gamma_3 \oplus \Gamma_2$

infrared-active. Similarly, the symmetries of Raman-active phonons in crystals with the O_h point group can be shown to be $\Gamma_{25'}$, Γ_{12}, and Γ_1 (see Chap. 7 for further details). Hence the $\Gamma_{25'}$ optical phonon of the diamond structure, while not infrared-active, is Raman-active. In crystals with inversion symmetry (said to be **centrosymmetric**), an infrared-active phonon must be odd while a Raman-active phonon must be even under inversion, therefore a phonon cannot be both infrared-active and Raman-active in such crystals.

2.4 Empty Lattice or Nearly Free Electron Energy Bands

We now apply the group theoretical notations to the electron energy band structure of the diamond- and zinc-blende-type semiconductors. Since the electrons move in the presence of a crystal potential, their wave functions can be symmetrized to reflect the crystal symmetry, i. e., written in a form such that they belong to irreducible representations of the space group of the crystal. However, in order to highlight the symmetry properties of the electron wave function, we will assume that the crystal potential is vanishingly small. In this **empty lattice** or **nearly free electron model**, the energy and wave functions of the electron are those of a free particle as given by (2.18) and (2.17), respectively. The electron energy band is simply a parabola when plotted in the extended zone scheme. This parabola looks much more complicated when replotted in the reduced zone scheme. It looks especially intimidating when the wave functions are labeled according to the irreducible representations of the point group of the crystal. Such complications have resulted from using the crystal symmetry which was supposed to simplify the problem! The simplification, however, occurs when we consider the band structure of electrons in a non-empty lattice in the remaining sections of this chapter. In this section we will use group theory to analyze the symmetry properties of nearly free electron band structures in both zinc-blende- and diamond-type crystals.

2.4.1 Nearly Free Electron Band Structure in a Zinc-Blende Crystal

Figure 2.8 shows the energy band of a nearly free electron plotted in the reduced zone scheme for wave vectors along the [111] and [100] directions only. To analyze this band diagram we will consider the symmetry and wave functions at a few special high-symmetry points in reciprocal space.

$\boldsymbol{k} = (0, 0, 0)$

As pointed out in Sect. 2.3.4, the group of the $\boldsymbol{k}$ vector at the Γ point is always isomorphic to the point group of the lattice. Since the wave function is a constant for $k = 0$, it has the symmetry Γ_1.

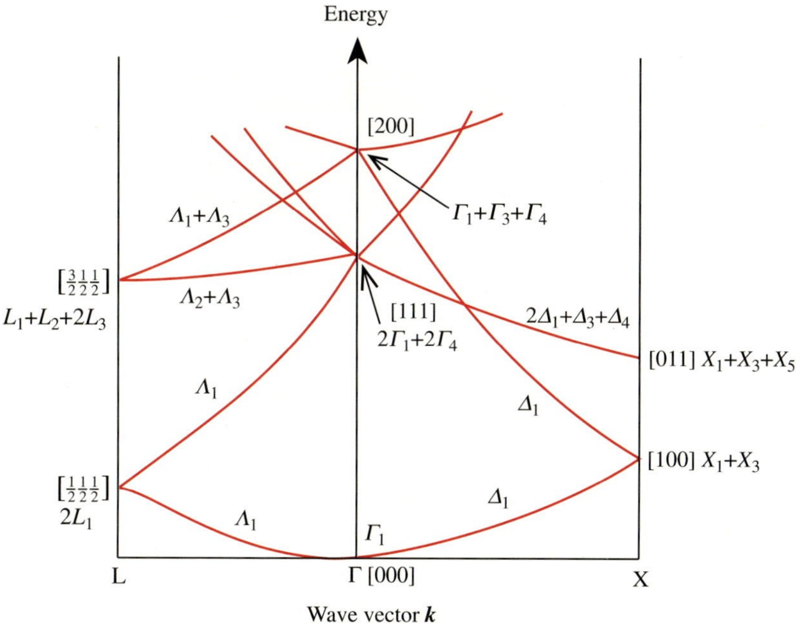

Fig. 2.8. Band structure of nearly free electrons in a zinc-blende-type crystal in the reduced zone scheme. The numbers in *square brackets* denote corresponding reciprocal lattice vectors in the extended zone scheme in units of $(2\pi/a)$, a being the size of the unit cube. *Note*: To conform to notations used in the literature, we will use + instead of $\oplus$ to represent the direct sum of two representations in all figures

$\boldsymbol{k} = (b, b, b)$, $b \neq (\pi/a)$

In the eight equivalent [111] directions the bands are labeled Λ, according to the Brillouin zone notations in Fig. 2.2c. The wave functions for $\boldsymbol{k} \neq 0$ are classified according to the group of the wave vector $\boldsymbol{k}$. The group of a wave vector along the [111] direction inside the Brillouin zone is C_{3v} and contains six elements divided into three classes:

$\{E\}$: identity;
$\{C_3, C_3^{-1}\}$: two three-fold rotations about the [111] direction;
$\{m_1, m_2, m_3\}$: three reflections in the three equivalent (110) planes containing the [111] axis.

The characters and basis functions for the irreducible representations of Λ are summarized in Table 2.12.

The free-electron wave function given by $\exp[(i\pi\zeta/a)(x + y + z)]$, where $0 < \zeta < 1$, is invariant under all the symmetry operations of the group of Λ so it belongs to the Λ_1 representation. We can also obtain this symmetry of the electron wave function by using the so-called **compatibility relations**. The symmetry of the endpoints of an axis in the Brillouin zone is higher than or equal to that of a point on the axis. Therefore the group of a point on an axis

Table 2.12. Characters and basis functions of the irreducible representations of the group of Λ (C_{3v}) in a zinc-blende-type crystal

	$\{E\}$	$\{2C_3\}$	$\{3m\}$	Basis functions
Λ_1	1	1	1	1 or $x+y+z$
Λ_2	1	1	-1	$xy(x-y)+yz(y-z)+zx(z-x)$
Λ_3	2	-1	0	$\{(x-y); \sqrt{\frac{2}{3}}(z-\frac{1}{2}[x+y])\}$

is either equal to or constitutes a subgroup of the group of the endpoints. In the latter case, a representation belonging to the group of the endpoints of an axis can be reduced to irreducible representations of the group of the axis. The procedure for this reduction is the same as that described in Sect. 2.3.4. The difference is that only symmetry operations common to both groups need be considered now. When a representation of the group of an axis is contained in a representation of one of the group's endpoints, the two representations are said to be **compatible**. For points lying on the [111] axis of a zinc-blende-type crystal, the group of Λ is a subgroup of Γ but is identical to the group of L. From the character tables for Γ and Λ it is clear that Γ_1 is compatible with Λ_1 only. Thus when the band starts out at the zone center with symmetry Γ_1, the symmetry of the band along the [111] direction must be Λ_1. This case illustrates a rather trivial application of the compatibility relations. Compatibility relations provide very useful consistency checks on band-structure calculations. Further applications of the compatibility relations can be found in Problem 2.6 at the end of this chapter.

$\boldsymbol{k} = (\pi/a)(1,1,1)$

In the zinc-blende structure the symmetry operations in the group of the L point are identical to those of the Λ axis. So the Λ_1 representation is compatible with the L_1 representation only. For free electrons the wave function is doubly degenerate at $(\pi/a)(1,1,1)$ since $(\pi/a)(1,1,1)$ and $(-\pi/a)(1,1,1)$ differ by $(2\pi/a)(1,1,1)$, a reciprocal lattice vector of the zinc-blende structure. Using the compatibility relations one can show that the next higher energy band along the Λ axis also has Λ_1 symmetry.

$\boldsymbol{k} = (2\pi/a)(1,1,1)$

The point $\boldsymbol{k} = (2\pi/a)(1,1,1)$ is equivalent to Γ since it differs from Γ by a reciprocal lattice vector. As shown in Sect. 2.3.4, the eight degenerate wave functions of the form $\exp[(\mathrm{i}2\pi/a)(\pm x \pm y \pm z)]$ can be symmetrized into two wave functions with Γ_1 symmetry and two sets of three wave functions with Γ_4 symmetry. The symmetries of the higher energy bands in the [111] direction are given in Fig. 2.8. They can be deduced using Table 2.12 and checked by the compatibility relations. The reader is urged to verify this as an exercise.

Table 2.13. Symmetry operations and classes of the group of Δ (C_{2v}) in the zinc-blende structure

Class	Symmetry operations
$\{E\}$	xyz
$\{C_4^2\}$	$x\bar{y}\bar{z}$
$\{m_d\}$	xyz
$\{m_d'\}$	$x\bar{z}\bar{y}$

Table 2.14. Characters of the irreducible representations of the group of Δ (C_{2v})

	$\{E\}$	$\{C_4^2\}$	$\{m_d\}$	$\{m_d'\}$
Δ_1	1	1	1	1
Δ_2	1	1	−1	−1
Δ_3	1	−1	1	−1
Δ_4	1	−1	−1	1

$\boldsymbol{k} = (c, 0, 0)$, $c \neq (2\pi/a)$

Wave vectors in the [100] and equivalent directions are denoted by Δ. The group of Δ (C_{2v}) contains four elements divided into the four classes listed in Table 2.13. The irreducible representations and characters of the group of Δ are summarized in Table 2.14. The symmetry of the wave function in the [100] direction is Δ_1 since this is the only representation compatible with Γ_1. The Δ axis ends at the X point on the surface of the Brillouin zone.

$\boldsymbol{k} = (2\pi/a)(1, 0, 0)$

The group of X contains twice as many symmetry operations as the group of Δ since the wave vectors $(2\pi/a)(1,0,0)$ and $(2\pi/a)(-1,0,0)$ differ by the reciprocal lattice vector $(2\pi/a)(2,0,0)$. The eight elements of the group of X (D_{2d}) are divided into five classes:

$\{E\}$:	identity;
$\{C_4^2(x)\}$:	two-fold rotation about the x axis;
$\{2C_4^2(y,z)\}$:	two-fold rotations about the y and z axes;
$\{2S_4\}$:	two four-fold improper rotations about the x axis;
$\{2m_d\}$:	two mirror reflections on the [011] and [0$\bar{1}\bar{1}$] planes.

The irreducible representations of the group of X and their characters are given in Table 2.15. The wave functions at the X point with $\boldsymbol{k} = (2\pi/a)(\pm 1, 0, 0)$ are doubly degenerate in the nearly-free electron model. From the compatibility relations it can be found that these wave functions belong to either the X_1 or X_3 representations.

Table 2.15. Characters of the irreducible representations of the group of X (D_{2d}) in the zinc-blende structure

	$\{E\}$	$\{C_4^2(x)\}$	$\{2C_4^2(y,z)\}$	$\{2S_4\}$	$\{2m_d\}$
X_1	1	1	1	1	1
X_2	1	1	1	−1	−1
X_3	1	1	−1	−1	1
X_4	1	1	−1	1	−1
X_5	2	−2	0	0	0

$\boldsymbol{k} = (2\pi/a)(0,0,2)$

The points $\boldsymbol{k} = (2\pi/a)(\pm 2,0,0)$, $(2\pi/a)(0,\pm 2,0)$, and $(2\pi/a)(0,0,\pm 2)$ differ from the zone center by reciprocal lattice vectors. As already shown in example (c) in Sect. 2.3.4, the six degenerate wave functions

$$\exp[\pm i4\pi x/a]; \; \exp[\pm i4\pi y/a]; \quad \text{and} \quad \exp[\pm i4\pi z/a]$$

can be symmetrized to transform like the Γ_1, Γ_3, and Γ_4 irreducible representations.

2.4.2 Nearly Free Electron Energy Bands in Diamond Crystals

Obviously, the band structure of a free electron is the same whether it is in a zinc-blende or a diamond crystal. Therefore, in order to obtain the symmetrized wave functions specific to the diamond structure, we have to assume first that the diamond crystal potential is nonzero and symmetrize the electron wave functions accordingly. Afterwards the crystal potential is made to approach zero. The band structure of nearly free electrons in a diamond-type crystal obtained in this way is shown in Fig. 2.9. It serves as an important guide to the band structure of Si (shown in Fig. 2.10 for comparison) calculated by more sophisticated techniques to be discussed later in this chapter.

The symmetries of the bands in diamond are very similar to those of zinc-blende because both crystals have a fcc lattice and tetrahedral symmetry. However, there are also important differences resulting from the existence of glide planes in the diamond structure as discussed in Sect. 2.3.2. We pointed out in that section that, if we choose the origin at one of the carbon atoms in diamond, the crystal is invariant under all the symmetry operations of the point group T_d plus three "glide-like" operations: $T(1/4,1/4,1/4)\sigma_x$, $T(1/4,1/4,1/4)\sigma_y$, and $T(1/4,1/4,1/4)\sigma_z$ (for brevity, we will now denote these three operations as $T\sigma_x$, $T\sigma_y$, and $T\sigma_z$, respectively). However, the factor group of the space group of diamond is isomorphic to the point group O_h. In symmetrizing the electronic wave functions in the diamond structure, one has to consider the effect of $T\sigma_x$ on the Bloch functions. In this subsection we shall pay special attention to the electron wave functions at the points Γ, L, and X of the Brillouin zone of the diamond crystal.

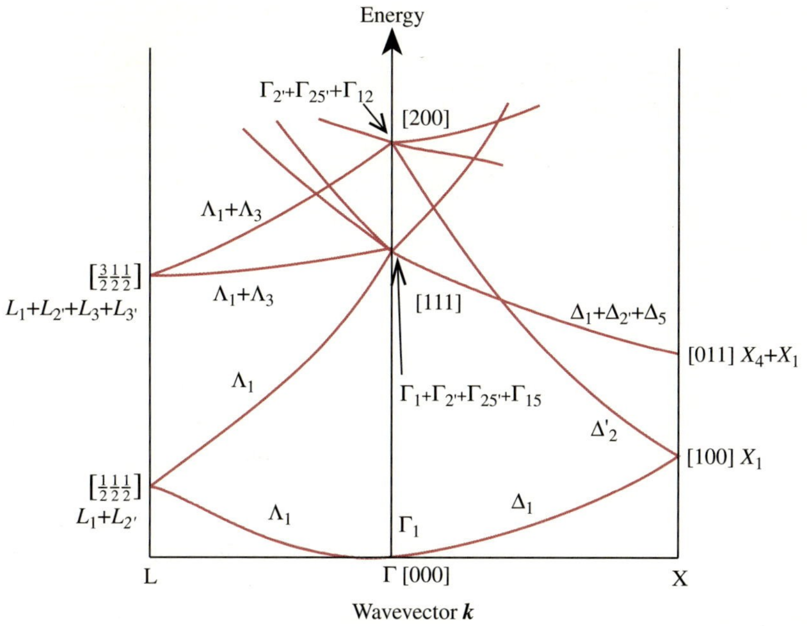

Fig. 2.9. Band structure of nearly free electrons for a diamond-type crystal in the reduced zone scheme

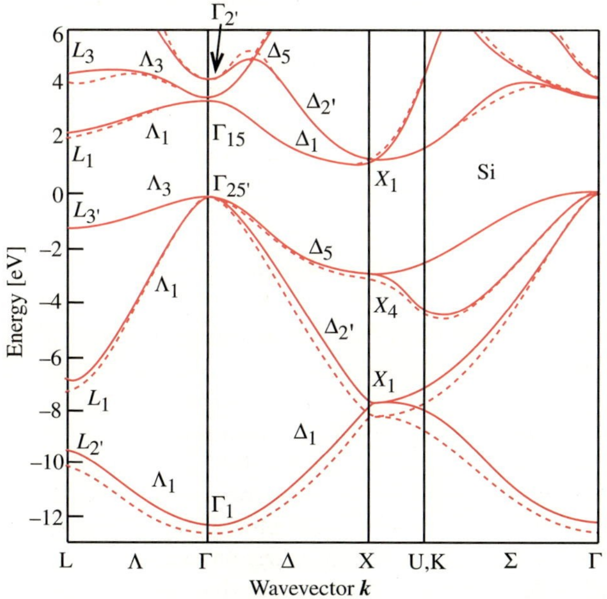

Fig. 2.10. Electronic band structure of Si calculated by the pseudopotential technique. The *solid* and the *dotted lines* represent calculations with a **nonlocal** and a **local pseudopotential**, respectively. [Ref. 2.8, p. 81]

$\boldsymbol{k} = (0, 0, 0)$

From (2.8) the Bloch functions at the zone center can be written as $u(\boldsymbol{r})$, where u has the periodicity of the lattice. We define C as a set formed from the group T_d plus all the operations obtained by multiplying each element of T_d by $T\sigma_x$. C defined this way is not a group because operations involving the glide, such as $(T\sigma_x)^2 = T(0, 1/2, 1/2)$ are not a member of C [for brevity, we will denote the operation $T(0, 1/2, 1/2)$ by Q]. Let us now generate a set $\{Cu\}$ consisting of 48 functions by applying the operations of C to $u(\boldsymbol{r})$. For any two symmetry operations, a and b of C we define the operation multiplication between the corresponding two elements au and bu in $\{Cu\}$ as $(au)(bu) = (ab)u$. The set of operations in $\{Cu\}$, defined by their effect on the function u, can be easily shown to form a group. In particular, $Qu(\boldsymbol{r}) = u(\boldsymbol{r})$ because $u(\boldsymbol{r})$ has the translational symmetry of the crystal and hence Qu is now an element of $\{Cu\}$. In this group, it is convenient to introduce the element $i'u$ where $i' = T(1/4, 1/4, 1/4)i$ was introduced in Sect. 2.3.4 (i is the inversion operation with respect to the origin). As pointed out in Sect. 2.3.4, the diamond crystal is not invariant under inversion with respect to one of the carbon atoms; it is however invariant under the combined operation of inversion followed by the translation $T(1/4, 1/4, 1/4)$. One can show that the 48 operations in $\{Cu\}$ are isomorphic to the O_h group. The character table of the group of wave functions of Γ is given in Table 2.16. It can be compared with the character table for the O_h group (Table 2.5). Note that the classes are listed in different orders in Tables 2.5 and 2.16. In Table 2.16 the five classes of symmetry operations in the point group T_d are listed first. The remaining five classes are obtained by multiplying the T_d operations by i'.

The effects of i' on the symmetry of wave functions at different high-symmetry points of the diamond crystal are not the same. For example, points along Λ are not invariant under i', so their symmetries are the same as in the zinc-blende crystal. On the other hand, the L point is invariant under i', therefore the wave functions at L have definite parity under i'.

Table 2.16. Characters of the irreducible representations of the group of Γ in the diamond structure. The notation is that of Koster (BSW notation in parentheses)

	$\{E\}$	$\{C_2\}$	$\{S_4\}$	$\{\sigma_d\}$	$\{C_3\}$	$\{i'\}$	$\{i'C_2\}$	$\{i'S_4\}$	$\{i'\sigma_d\}$	$\{i'C_3\}$
$\Gamma_1^+ (\Gamma_1)$	1	1	1	1	1	1	1	1	1	1
$\Gamma_2^+ (\Gamma_2)$	1	1	−1	−1	1	1	1	−1	−1	1
$\Gamma_3^+ (\Gamma_{12})$	2	2	0	0	−1	2	2	0	0	−1
$\Gamma_4^+ (\Gamma_{15'})$	3	−1	1	−1	0	3	−1	1	−1	0
$\Gamma_5^+ (\Gamma_{25'})$	3	−1	−1	1	0	3	−1	−1	1	0
$\Gamma_1^- (\Gamma_{1'})$	1	1	−1	−1	1	−1	−1	1	1	−1
$\Gamma_2^- (\Gamma_{2'})$	1	1	1	1	1	−1	−1	−1	−1	−1
$\Gamma_3^- (\Gamma_{12'})$	2	2	0	0	−1	−2	−2	0	0	1
$\Gamma_4^- (\Gamma_{15})$	3	−1	−1	1	0	−3	1	1	−1	0
$\Gamma_5^- (\Gamma_{25})$	3	−1	1	−1	0	−3	1	−1	1	0

$\boldsymbol{k} = (\pi/a)(1, 1, 1)$

The group of the L point in the diamond structure is isomorphic to the group of L in the fcc Bravais lattice (that is, a crystal formed by putting only one atom at each lattice point of a fcc lattice). The characters and basis functions for the irreducible representations in the group of L (D_{3d}) are shown in Table 2.17.

$\boldsymbol{k} = (2\pi/a)(1, 1, 1)$

It has been pointed out already in the case of the zinc-blende crystal that $\boldsymbol{k} = (2\pi/a)(1, 1, 1)$ is equivalent to Γ. This is, of course, also true for the diamond crystal. From the eight symmetrized wave functions for the zinc-blende crystal given in Table 2.9 it can be shown readily that, for the diamond crystal, the eight equivalent (111) wave functions are symmetrized to transform according to the irreducible representations in Table 2.18.

The symmetry of the wave functions in the diamond structure along the [001] directions are quite different from those of zinc-blende. We will first consider the X point since it presents an especially interesting case.

Table 2.17. Characters and basis functions for the irreducible representations of the group of L in the diamond structure

	$\{E\}$	$\{2C_3\}$	$\{3C_2\}$	$\{i'\}$	$\{2i'C_3\}$	$\{3i'C_2\}$	Basis functions
L_1	1	1	1	1	1	1	1
L_2	1	1	−1	1	1	−1	$xy(x^2-y^2)+yz(y^2-z^2)+zx(z^2-x^2)$
L_3	2	−1	0	2	−1	0	$\{z^2-1/2(x^2+y^2); (x^2-y^2)\}$
$L_{1'}$	1	1	1	−1	−1	−1	$(x-y)(y-z)(z-x)$
$L_{2'}$	1	1	−1	−1	−1	1	$x+y+z$
$L_{3'}$	2	−1	0	−2	1	0	$\{(x-z); (y-1/2[x+z])\}$

Table 2.18. Symmetrized nearly free electron wave functions in the diamond crystal with wave vectors equal to $(2\pi/a)(\pm1, \pm1, \pm1)$. The origin of coordinates has been taken to coincide with an atomic site.

Representation	Wave function
$\Gamma_1^+ (\Gamma_1)$	$\cos(2\pi x/a)\cos(2\pi y/a)\cos(2\pi z/a)$ $+\sin(2\pi x/a)\sin(2\pi y/a)\sin(2\pi z/a)$
$\Gamma_2^- (\Gamma_{2'})$	$\cos(2\pi x/a)\cos(2\pi y/a)\cos(2\pi z/a)$ $-\sin(2\pi x/a)\sin(2\pi y/a)\sin(2\pi z/a)$
$\Gamma_5^+ (\Gamma_{25'})$	$\sin(2\pi x/a)\cos(2\pi y/a)\cos(2\pi z/a)$ $+\cos(2x/a)\sin(2\pi y/a)\sin(2\pi z/a)$; plus two cyclic permutations
$\Gamma_4^- (\Gamma_{15})$	$\sin(2\pi x/a)\cos(2\pi y/a)\cos(2\pi z/a)$ $-\cos(2x/a)\sin(2\pi y/a)\sin(2\pi z/a)$; plus two cyclic permutations

$\boldsymbol{k} = (2\pi/a)(0, 0, 1)$

A very special property of the wave functions at the X point of the diamond structure is that all *relevant* irreducible representations of the group of the X point are doubly degenerate, but they do not have definite parity under i'. To understand this peculiar property, let us first enumerate all the symmetry operations of the group of the X point. We will start with the eight symmetry operations of the group of the point $(2\pi/a)(0, 0, 1)$ in the Brillouin zone of the zinc-blende structure:

$$\{E,\ C_4^2(z),\ 2C_4^2(x, y),\ 2S_4,\ 2m_d\}.$$

Next we will consider the combined effect of these operations and the operation $T\sigma_z$ on a wave function at the X point:

$$\phi = \exp(\mathrm{i}2\pi z/a)u(\boldsymbol{r}). \tag{2.19}$$

At first we may expect that we can construct a group for the X point by taking the above eight elements and adding to them their products with the operation $T\sigma_z$. This should result in a set of sixteen elements. It turns out that these sixteen elements do not form a group because translation and rotation do not necessarily commute. For example, consider the combined effect of $C_4^2(x)T\sigma_z$ on a vector (x, y, z):

$$(x, y, z) \xrightarrow{\sigma_z} (x, y, \bar{z}) \xrightarrow{T(1/4,1/4,1/4)} (x + 1/4a, y + 1/4a, -z + 1/4a)$$
$$\xrightarrow{C_4^2(x)} (x + 1/4a, -y - 1/4a, z - 1/4a).$$

If we interchange the order of $C_4^2(x)$ and $T\sigma_z$ we find that

$$[T\sigma_z C_4^2(x)](x, y, z) = (x + 1/4a, -y + 1/4a, z + 1/4a), \tag{2.20a}$$

so the operation C_4^2 does not commute with $T\sigma_z$. In particular,

$$[T\sigma_z C_4^2(x)]\phi = T(0, 1/2, 1/2)[C_4^2]T\sigma_z\phi = Q[C_4^2]T\sigma_z\phi. \tag{2.20b}$$

In order that the set $\{E\phi, C_4^2(z)\phi, \ldots, 2m_d\phi, T\sigma_z\phi, \ldots, T\sigma_z 2m_d\phi\}$ forms a group, the operation Q has to be included also. Taking the 16 operations mentioned above and their products with Q, a group with 32 elements is obtained. This group can be divided into 14 classes:

$C_1 = \{E\}$
$C_2 = \{C_4^2(x),\ C_4^2(y),\ QC_4^2(x),\ QC_4^2(y)\}$
$C_3 = \{C_4^2(z)\}$
$C_4 = \{QT\sigma_z\sigma_x,\ T\sigma_z\sigma_y\}$
$C_5 = \{T\sigma_z S_4,\ T\sigma_z S_4^{-1},\ QT\sigma_z S_4,\ QT\sigma_z S_4^{-1}\}$
$C_6 = \{T\sigma_z,\ QT\sigma_z\}$
$C_7 = \{T\sigma_z C_4^2(x),\ T\sigma_z C_4^2(y),\ QT\sigma_z C_4^2(x),\ QT\sigma_z C_4^2(y)\}$
$C_8 = \{T\sigma_z C_4^2(z),\ QT\sigma_z C_4^2(z)\}$
$C_9 = \{\sigma_x,\ \sigma_y\}$
$C_{10} = \{S_4,\ S_4^{-1},\ QS_4,\ QS_4^{-1}\}$
$C_{11} = \{Q\sigma_x,\ Q\sigma_y\}$

$C_{12} = \{QT\sigma_z\sigma_y,\ T\sigma_z\sigma_x\}$
$C_{13} = \{QC_4^2(z)\}$
$C_{14} = \{Q\}$

The characters of the corresponding 14 irreducible representations are given in Table 2.19. However, not all of these representations are acceptable for wave functions at the X point of the Brillouin zone in the diamond crystal. Since $(a/2)(0,1,1)$ is a lattice vector of the fcc lattice the operation Q will leave the periodic part of the X-point wave function invariant. The sinusoidal envelope $\exp(\mathrm{i}2\pi z/a)$ of the Bloch function changes sign under the translation Q, so overall the X-point wave functions must be odd under Q. Of the 14 irreducible representations only four are odd under the translation Q (or C_{14}). These are labeled X_1, X_2, X_3, and X_4 in Table 2.19. The interesting point is that these four representations are all doubly degenerate. This degeneracy results from the glide reflection and the fact that the two atoms in the unit cell of the diamond structure are identical. The degeneracy in the X_1 and X_2 states is lifted in the zinc-blende structure, where the two atoms in the primitive cell are different (see Problem 2.8). Some examples of symmetrized wave functions at the X point are

$\boldsymbol{k} = (2\pi/a)(0,0,1)$:

$X_1 : \{\cos(2\pi z/a);\ \sin(2\pi z/a)\}$

$\boldsymbol{k} = (2\pi/a)(\pm 1, \pm 1, 0)$:

$X_1 : \{\cos(2\pi x/a)\ \cos(2\pi y/a);\ \sin(2\pi x/a)\sin(2\pi y/a)\}$
$X_4 : \{\sin(2\pi x/a)\ \cos(2\pi y/a);\ \cos(2\pi x/a)\sin(2\pi y/a)\}$

$\boldsymbol{k} = (\xi\pi/a)(0,0,1)$ where $0 < \xi < 2$.

Table 2.19. Irreducible representations and characters of the group of symmetry operations on the wave functions at the X point $(2\pi/a)(0,0,1)$ of the Brillouin zone of the diamond structure [Ref. 2.7, p. 162]

	C_1	$4C_2$	C_3	$2C_4$	$4C_5$	$2C_6$	$4C_7$	$2C_8$	$2C_9$	$4C_{10}$	$2C_{11}$	$2C_{12}$	C_{13}	C_{14}
M_1	1	1	1	1	1	1	1	1	1	1	1	1	1	1
M_2	1	1	1	−1	−1	1	1	1	−1	−1	−1	−1	1	1
M_3	1	−1	1	−1	1	1	−1	1	−1	1	−1	−1	1	1
M_4	1	−1	1	1	−1	1	−1	1	1	−1	1	1	1	1
M_5	2	0	−2	0	0	2	0	−2	0	0	0	0	−2	2
M_1'	1	1	1	1	1	−1	−1	−1	−1	−1	−1	1	1	1
M_2'	1	1	1	−1	−1	−1	−1	−1	1	1	1	−1	1	1
M_3'	1	−1	1	−1	1	−1	1	−1	1	−1	1	−1	1	1
M_4'	1	−1	1	1	−1	−1	1	−1	−1	1	−1	1	1	1
M_5'	2	0	−2	0	0	−2	0	2	0	0	0	0	−2	2
X_1	2	0	2	0	0	0	0	0	2	0	−2	0	−2	−2
X_2	2	0	2	0	0	0	0	0	−2	0	2	0	−2	−2
X_3	2	0	−2	2	0	0	0	0	0	0	0	−2	2	−2
X_4	2	0	−2	−2	0	0	0	0	0	0	0	2[a]	2	−2

[a] An error in [2.7] has been corrected.

We will denote the Bloch function along the Δ direction as $\psi = \exp(\mathrm{i}\xi\pi z/a)u(\boldsymbol{r})$ as in (2.19). It is invariant under the following space group operations of the diamond structure:

$$\{E,\ \sigma_x,\ C_x^2(z),\ \sigma_y,\ T\sigma_z C_4^2(x),\ T\sigma_z S_4,\ T\sigma_z C_4^2(y),\ T\sigma_z S_4^{-1}\},$$

which can be divided into 5 classes:

$$\begin{aligned}
C_1 &= \{E\} \\
C_2 &= \{C_4^2(z)\} \\
C_3 &= \{T\sigma_z S_4,\ T\sigma_z S_4^{-1}\} \\
C_4 &= \{\sigma_x,\ \sigma_y\} \\
C_5 &= \{T\sigma_z C_4^2(x),\ T\sigma_z C_4^2(y)\}
\end{aligned}$$

The representations generated by these operations acting on ψ are isomorphic with the group of Δ for a cubic lattice. The corresponding characters are shown in Table 2.20.

Table 2.20. Irreducible representations and characters of the group of symmetry operations on the wave functions at the Δ point of the Brillouin zone of the diamond structure [Ref. 2.7, p. 158]

	C_1	C_2	$2C_3$	$2C_4$	$2C_5$
Δ_1	1	1	1	1	1
Δ_2	1	1	−1	1	−1
$\Delta_{2'}$	1	1	−1	−1	1
$\Delta_{1'}$	1	1	1	−1	−1
Δ_5	2	−2	0	0	0

2.5 Band Structure Calculations by Pseudopotential Methods

In Fig. 2.10 we have shown the electronic band structure of Si. It has been calculated with a sophisticated method known as the **pseudopotential technique**, which will be discussed in this section. Comparing these results with the nearly free electron band structure in Fig. 2.9 we notice that there are many similarities between the two. The nearly free electron band structure is basically a parabola redrawn in the reduced zone scheme. In the other case the band structure is computed by large-scale numerical calculations using supercomputers. The question is now: why do the two band structures, obtained by completely different methods, look so similar qualitatively? The answer to this question lies in the concept of pseudopotentials.

The electronic configuration of a Si atom is $1s^2 2s^2 2p^6 3s^2 3p^2$. When Si atoms form a crystal we can divide their electrons into core electrons and valence electrons as pointed out in Sect. 2.1. In crystalline Si the $1s$, $2s$, and $2p$

orbitals are completely occupied and form the core shells. The outer 3*s* and 3*p* shells are only partially filled. Electrons in these shells are called valence electrons because they are involved in bonding with neighboring Si atoms. The crystal structure of Si at ambient pressure is similar to that of diamond. The tetrahedral arrangement of bonds between a Si atom and its four nearest neighbors can be understood if one of the electrons in the 3*s* shell is "promoted" to the 3*p* shell so that the four valence electrons form **hybridized** $\boldsymbol{sp^3}$ **orbitals**. This sp^3 hybridization is well known from the bonding of carbon atoms and is responsible for the tetrahedral structure found in many organic molecules. But carbon atoms are more versatile than silicon atoms in that they can form double and triple bonds also. As a result, carbon atoms are crucial to all known forms of life while silicon atoms are important only to the highest form of life, namely human beings. It is these valence electrons in the outermost shells of a Si atom that are nearly free. These electrons are not affected by the full nuclear charge as a result of screening of the nucleus by the filled core shells. In the core region the valence electron wave functions must be orthogonal to those of the core. Thus the true wave functions may have strong spatial oscillations near the core, which make it difficult to solve the wave equation. One way to overcome this difficulty is to divide the wave functions into a smooth part (the **pseudo-wave function**) and an oscillatory part. The kinetic energy from the latter provides an "effective repulsion" for the valence electrons near the core (alternatively one can regard the valence electrons as being expelled from the core due to Pauli's exclusion principle). Thus we can approximate the strong true potential by a weaker "effective potential" or **pseudopotential** for the valence electrons. Since the "smooth" parts of the valence electron wave functions have little weight in the core region, they are not very sensitive to the shape of the pseudopotential there. Figure 2.11 shows qualitatively how the pseudopotential in Si varies with distance *r* from the nucleus. At large values of *r* the pseudopotential approaches the unscreened Coulomb potential of the Si^{4+} ion. This concept of replacing the true potential with a pseudopotential can be justified mathematically. It can

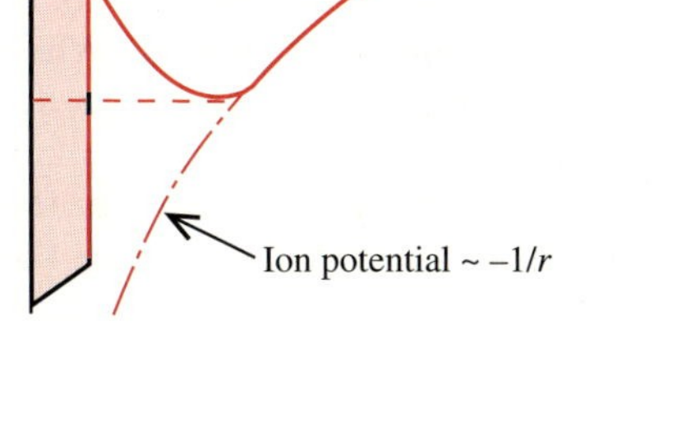

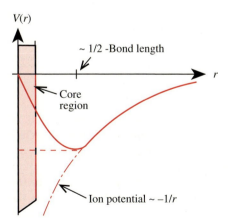

Fig. 2.11. Schematic plot of the atomic pseudopotential of Si in real space [Ref. 2.8, p. 17]. The *solid curve* in which $V(r) \rightarrow 0$ in the core region is said to be a "soft core" pseudopotential. The *broken curve* in which $V(r) \rightarrow$ constant is a "hard core" pseudopotential

be shown to reproduce correctly both the conduction and valence band states while eliminating the cumbersome, and in many cases irrelevant, core states [Ref. 2.8, p. 16].

Using the pseudopotential concept, the one-electron Schrödinger equation (2.4) can be replaced by the **pseudo-wave-equation**

$$\left[\frac{p^2}{2m} + V(\boldsymbol{r}_i)\right] \psi_{\boldsymbol{k}}(\boldsymbol{r}_i) = E_{\boldsymbol{k}}\psi_{\boldsymbol{k}}(\boldsymbol{r}_i), \tag{2.21}$$

where ψ is the pseudo-wave-function. This function is a good approximation to the true wave function outside the core region and therefore can be used to calculate the physical properties of the semiconductors which are dependent on the valence and conduction electrons only. Since pseudopotentials are weak perturbations on the free-electron band structure, a good starting point for diagonalizing (2.21) is to expand $\psi_{\boldsymbol{k}}$ as a sum of plane waves:

$$\psi_{\boldsymbol{k}} = \sum_{\boldsymbol{g}} a_{\boldsymbol{g}} |\boldsymbol{k} + \boldsymbol{g}\rangle, \tag{2.22}$$

where the vectors $\boldsymbol{g}$ are the reciprocal lattice vectors and $|\boldsymbol{k}\rangle$ represents a plane wave with wave vector $\boldsymbol{k}$. The coefficients $a_{\boldsymbol{g}}$ and the eigenvalues $E_{\boldsymbol{k}}$ can be determined by solving the secular equation

$$\det |[(\hbar^2k^2/2m) - E_{\boldsymbol{k}}]\delta_{\boldsymbol{k},\boldsymbol{k}+\boldsymbol{g}} + \langle \boldsymbol{k} | V(\boldsymbol{r}) | \boldsymbol{k}+\boldsymbol{g}\rangle | = 0, \tag{2.23}$$

The matrix elements of the pseudopotential $V(\boldsymbol{r})$ are given by

$$\langle \boldsymbol{k} | V(\boldsymbol{r}) | \boldsymbol{k}+\boldsymbol{g}\rangle = \left[\frac{1}{N}\sum_{\boldsymbol{R}} \exp(-\mathrm{i}\boldsymbol{g}\cdot\boldsymbol{R})\right] \frac{1}{\Omega}\int_{\Omega} V(\boldsymbol{r}) \exp[-\mathrm{i}\boldsymbol{g}\cdot\boldsymbol{r}]d\boldsymbol{r}, \tag{2.24}$$

where $\boldsymbol{R}$ is a direct lattice vector and Ω the volume of a primitive cell. As a result of summation over all the lattice vectors inside the bracket, the pseudopotential matrix element is zero unless $\boldsymbol{g}$ is a reciprocal lattice vector. In other words, the matrix elements of the pseudopotential are determined by Fourier components of the pseudopotential ($V_{\boldsymbol{g}}$) defined by

$$V_{\boldsymbol{g}} = \frac{1}{\Omega}\int_{\Omega} V(\boldsymbol{r}) \exp[-\mathrm{i}\boldsymbol{g}\cdot\boldsymbol{r}]d\boldsymbol{r}, \tag{2.25}$$

where $\boldsymbol{g}$ is a reciprocal lattice vector.

If there is only one atom per primitive cell these Fourier components of the pseudopotential are known as the **pseudopotential form factors**. When there are several different atoms in the primitive cell, it is convenient to define for each kind of atom a pseudopotential form factor and a structure factor which depends only on the positions of one particular kind of atom in the primitive cell. For example, let there be two kinds of atoms α and β in the crystal and let their positions inside the primitive cell be denoted by $\boldsymbol{r}_{\alpha i}$ and

$\boldsymbol{r}_{\beta i}$. The **structure factor** $S_{g\alpha}$ of atom α is defined as

$$S_{g\alpha} = \frac{1}{N_\alpha} \sum_i \exp(-\mathrm{i}\boldsymbol{g} \cdot \boldsymbol{r}_{\alpha i}), \tag{2.26}$$

where N_α is the number of α atoms in the primitive cell. The structure factor of atom β is defined similarly. The pseudopotential form factor $V_{g\alpha}$ for atom α can be defined as in (2.25) except that V is now the potential of one α atom and the integration is performed over Ω_α, which is the volume corresponding to one α atom. The pseudopotential $V(\boldsymbol{r})$ can be expressed in terms of the structure and form factors by

$$V(\boldsymbol{r}) = \sum_g (V_{g\alpha} S_{g\alpha} + V_{g\beta} S_{g\beta}) \exp(\mathrm{i}\boldsymbol{g} \cdot \boldsymbol{r}). \tag{2.27}$$

From (2.24) we conclude that the pseudopotential mixes the free-electron states whose $\boldsymbol{k}$'s differ by a reciprocal lattice vector. If these states are degenerate, the degeneracy may be split by the pseudopotential provided the corresponding form factor is nonzero. For example, consider the free-electron states with $\boldsymbol{k} = (2\pi/a)(\pm1, \pm1, \pm1)$ at the Γ point in the diamond structure. The $\boldsymbol{k}$'s of these eight-fold degenerate states differ by reciprocal lattice vectors $(2\pi/a)(2,0,0)$, $(2\pi/a)(2,2,0)$, and $(2\pi/a)(2,2,2)$. These eight states are degenerate when the electron is free. With the introduction of the pseudopotential, they become coupled and their degeneracy is partly lifted, producing energy gaps (compare Figs. 2.9 and 2.10). When an energy gap opens up at the Fermi level (highest occupied energy level) a semiconductor is obtained. This opening of energy gaps in the nearly-free-electron band structure by the pseudopotential form factors can be explained by Bragg reflection of the free-electron plane waves by the crystal potential with the formation of standing waves. When the pseudopotential form factors are small, their effect on the band structure is weak so the actual band structure is not too different from the free-electron band structure. This is the reason why the nearly free electron bands drawn in the reduced zone scheme are a good starting point for understanding the band structure of most semiconductors.

2.5.1 Pseudopotential Form Factors in Zinc-Blende- and Diamond-Type Semiconductors

The main reason why pseudopotentials are so useful is because only a small number of these form factors are sufficient for calculating a band structure. In semiconductors with the diamond structure, such as Si and Ge, just three pseudopotential form factors are sufficient. In semiconductors with the zinc-blende structure the number of required pseudopotential form factors doubles to six. To show this we first note that there are two atoms a and b in the unit cell. We will denote the **atomic pseudopotentials** of these two atoms by $V_\mathrm{a}(r - r_\mathrm{a})$ and $V_\mathrm{b}(r - r_\mathrm{b})$, where r_a and r_b are the positions of the two atoms in the unit cell. Substituting these potentials into (2.25) we obtain the Fourier components of the crystal pseudopotential

$$V_g = \frac{1}{\Omega}\int [V_\mathrm{a}(\boldsymbol{r}-\boldsymbol{r}_\mathrm{a}) + V_\mathrm{b}(\boldsymbol{r}-\boldsymbol{r}_\mathrm{b})] \ \exp[-\mathrm{i}\boldsymbol{g}\cdot\boldsymbol{r}]d\boldsymbol{r} \tag{2.28}$$

$$= \frac{1}{\Omega}\int [V_\mathrm{a}(\boldsymbol{r})\exp(-\mathrm{i}\boldsymbol{g}\cdot\boldsymbol{r}_\mathrm{a}) + V_\mathrm{b}(\boldsymbol{r})\exp(-\mathrm{i}\boldsymbol{g}\cdot\boldsymbol{r}_\mathrm{b})] \times \exp[-\mathrm{i}\boldsymbol{g}\cdot\boldsymbol{r}]d\boldsymbol{r}. \tag{2.29}$$

Without loss of generality we can take the midpoint between the two atoms in the unit cell as the origin, so that $\boldsymbol{r}_\mathrm{a} = (a/8)(1,1,1) = \boldsymbol{s}$ and $\boldsymbol{r}_\mathrm{b} = (-a/8)(1,1,1) = -\boldsymbol{s}$. We can now write

$$V_\mathrm{a}(\boldsymbol{r})\exp(-\mathrm{i}\boldsymbol{g}\cdot\boldsymbol{r}_\mathrm{a}) + V_\mathrm{b}(\boldsymbol{r})\exp(-\mathrm{i}\boldsymbol{g}\cdot\boldsymbol{r}_\mathrm{b}) = (V_\mathrm{a}+V_\mathrm{b})\cos(\boldsymbol{g}\cdot\boldsymbol{s}) - \mathrm{i}(V_\mathrm{a}-V_\mathrm{b})\sin(\boldsymbol{g}\cdot\boldsymbol{s}). \tag{2.30}$$

Next we define the symmetric and antisymmetric components of the pseudopotential form factor by

$$V_g^\mathrm{s} = \frac{1}{\Omega}\int (V_\mathrm{a}+V_\mathrm{b})\exp(-\mathrm{i}\boldsymbol{g}\cdot\boldsymbol{r})d\boldsymbol{r} \tag{2.31}$$

and

$$V_g^\mathrm{a} = \frac{1}{\Omega}\int (V_\mathrm{a}-V_\mathrm{b})\exp(-\mathrm{i}\boldsymbol{g}\cdot\boldsymbol{r})d\boldsymbol{r}. \tag{2.32}$$

Substituting the results in (2.30–32) back into (2.29) we arrive at

$$V_g = V_g^\mathrm{s}\cos(\boldsymbol{g}\cdot\boldsymbol{s}) - \mathrm{i}V_g^\mathrm{a}\sin(\boldsymbol{g}\cdot\boldsymbol{s}). \tag{2.33}$$

By symmetrizing the pseudopotential form factors in this way, it is clear that the antisymmetric form factors V_g^a vanish in the diamond structure. The factor $\cos(\boldsymbol{g}\cdot\boldsymbol{s})$ is just the structure factor of diamond defined in (2.26). In the III–V semiconductor, where the difference in the potentials of the anion and cation is small, V_g^a is expected to be smaller than V_g^s and furthermore V_g^s should be almost the same as in their neighboring group–IV semiconductors. For example, consider the pseudopotential form factors in Ge and the III–V semiconductor GaAs formed from its neighbors in the periodic table. In the diamond and zinc-blende structures, the reciprocal lattice vectors in order of increasing magnitude are (in units of $2\pi/a$):

$$\begin{aligned}
\boldsymbol{g}_0 &= (0,0,0);\\
\boldsymbol{g}_3 &= (1,1,1),\ (1,-1,1),\ \ldots,\ (-1,-1,-1);\\
\boldsymbol{g}_4 &= (2,0,0),\ (-2,0,0),\ \ldots,\ (\ \ 0,\ \ 0,-2);\\
\boldsymbol{g}_8 &= (2,2,0),\ (2,-2,0),\ \ldots,\ (\ \ 0,-2,-2);\\
\boldsymbol{g}_{11} &= (3,1,1),\ (-3,1,1),\ \ldots,\ (-3,-1,-1).
\end{aligned}$$

We can neglect pseudopotential form factors with $g^2 > 11(2\pi/a)^2$ because typically V_g decreases as g^{-2} for large $\boldsymbol{g}$. Figure 2.12 shows a schematic plot of a pseudopotential as a function of the magnitude of $\boldsymbol{g}$ ($\boldsymbol{g}$ is assumed to be spherically symmetrical as in the case of a free atom).

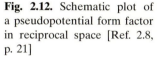

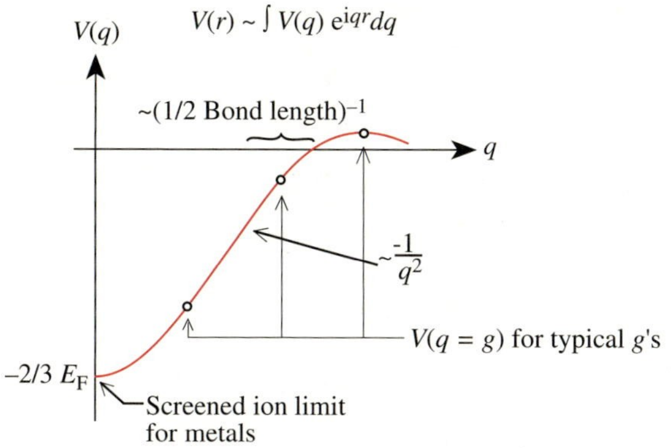

Fig. 2.12. Schematic plot of a pseudopotential form factor in reciprocal space [Ref. 2.8, p. 21]

The pseudopotential form factor V_0 corresponding to $\boldsymbol{g}_0$ is a constant potential, which merely shifts the entire energy scale: it can therefore be set equal to zero or any other convenient value (see Fig. 2.12). The pseudopotential form factors for all the equivalent reciprocal lattice vectors with the form $(\pm1, \pm1, \pm1)$ and magnitude $3(2\pi/a)$ are equal by symmetry and will be denoted by V_3. The structure factor corresponding to $\boldsymbol{g}_4$ is zero because $\cos(\boldsymbol{g} \cdot \boldsymbol{s}) = 0$ for $\boldsymbol{g} = (2\pi/a)(2, 0, 0)$. Thus we conclude that there are only three important pseudopotential form factors for Ge: V_3^{s}, V_8^{s} and V_{11}^{s}. In GaAs, V_8^{a} vanishes because $\sin(\boldsymbol{g} \cdot \boldsymbol{s}) = 0$, so only six pseudopotential form factors are required: V_3^{s}, V_8^{s}, V_{11}^{s}, V_3^{a}, V_4^{a}, and V_{11}^{a}. The pseudopotential form factors of Ge, GaAs, and a few other semiconductors are listed in Table 2.21. One should keep in mind that the sign of the antisymmetric form factors depends on whether the anion or cation is designated as atom a. The sign of the antisymmetric form factors in Table 2.21 are all positive because the cation has been chosen to be atom a and the anion (which has a more negative atomic pseudopotential) to be atom b. Note also that the magnitude of the form factor V_3^{s} is the largest and furthermore it is negative in sign, as shown schematically in Fig. 2.12. In the III–V and II–VI compounds, V_3^{s} is comparable to the corresponding V_3^{s} in the group-IV semiconductors

Table 2.21. Pseudopotential form factors of several group-IV, III–V and II–VI semiconductors (in units of Rydbergs = 13.6 eV) [2.9,10]. Note that the sign of V_i^{a} depends on the positions chosen for the anion and the cation (see text).

	V_3^{s}	V_8^{s}	V_{11}^{s}	V_3^{a}	V_4^{a}	V_{11}^{a}
Si	−0.211	0.04	0.08	0	0	0
Ge	−0.269	0.038	0.035	0	0	0
GaAs	−0.252	0	0.08	0.068	0.066	0.012
GaP	−0.249	0.017	0.083	0.081	0.055	0.003
InAs	−0.27	0.02	0.041	0.078	0.038	0.036
InSb	−0.25	0.01	0.044	0.049	0.038	0.01
ZnSe	−0.23	0.01	0.06	0.18	0.12	0.03
CdTe	−0.245	−0.015	0.073	0.089	0.084	0.006

and always larger than V_3^a in magnitude. Nevertheless, as the ionicity increases in going from the III–V semiconductors to the II–VI semiconductors, the antisymmetric pseudopotential form factors become larger. Some band structures of diamond- and zinc-blende-type semiconductors calculated by the pseudopotential method are shown in Figs. 2.13–15. These band structure calculations include the effect of **spin–orbit coupling**, which will be discussed in Sect. 2.6. As a result of this coupling, the irreducible representations of the electron wave functions must include the effects of symmetry operations on the spin wave function. (For example, a rotation by 2π will change the sign of the wave function of a spin-1/2 particle). The notations used in Figs. 2.13–15, including this feature, are known as the **double group notations** and will be discussed in Sect. 2.6.

The effect of ionicity on the band structures of the compound semiconductors can be seen by comparing the band structure of Ge with those of GaAs and ZnSe as shown in Figs. 2.13–15. Some of the differences in the three band structures result from spin–orbit coupling. Otherwise most of these differences can be explained by the increase in the antisymmetric components of the pseudopotential form factors as the ionicity increases along the sequence Ge, GaAs, ZnSe. One consequence of this increase in ionicity is that the energy gap between the top of the valence band and the bottom of the conduc-

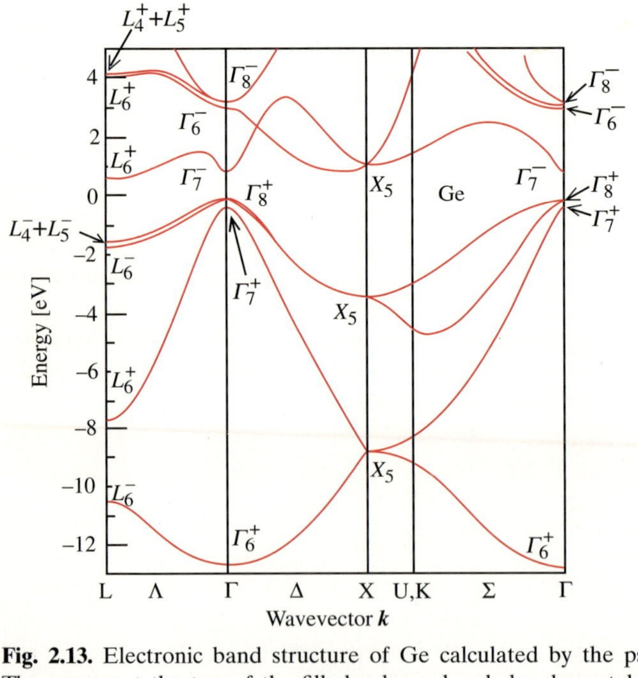

Fig. 2.13. Electronic band structure of Ge calculated by the pseudopotential technique. The energy at the top of the filled valence bands has been taken to be zero. Note that, unlike in Fig. 2.10, the double group symmetry notation is used [Ref. 2.8, p. 92]

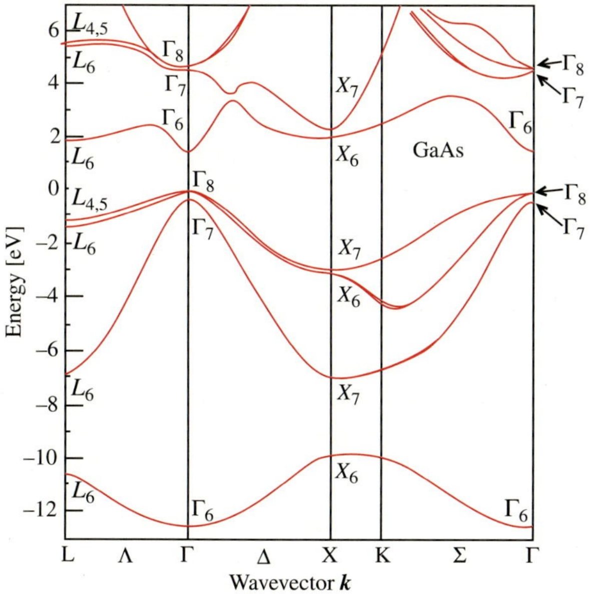

Fig. 2.14. Electronic band structure of GaAs calculated by the pseudopotential technique. The energy scale and notation (double group) are similar to those for Fig. 2.13 [Ref. 2.8, p. 103]

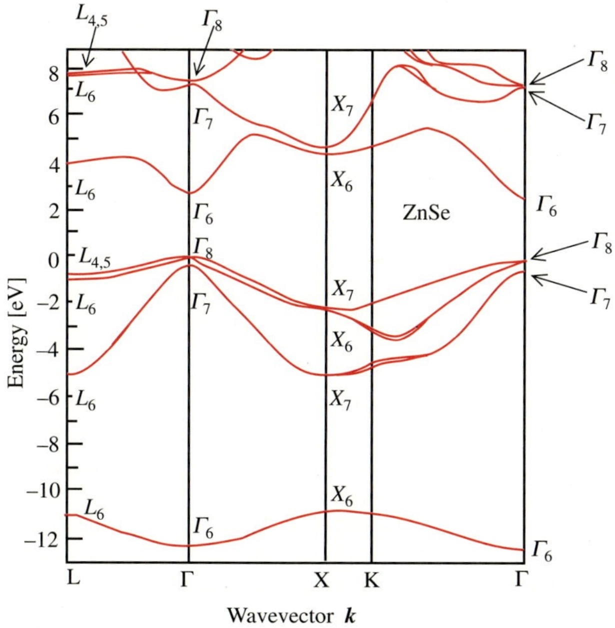

Fig. 2.15. Electronic band structure of ZnSe calculated by the pseudopotential technique. The energy scale and notation (double group) are similar to those for Fig. 2.14 [Ref. 2.8, p. 113]

tion band at Γ increases monotonically in going from Ge to ZnSe. Another consequence is that some of the doubly degenerate states in Ge at the X point of the Brillouin zone are split in the III–V and II–VI compounds, as pointed out in Sect. 2.4.2. For example, the lowest energy X_1 conduction band state in Ge is split into two spin doublets of X_6 and X_7 symmetry (X_1 and X_3 without spin–orbit coupling) in GaAs and ZnSe. The explicit dependence of this splitting on the antisymmetric pseudopotential form factors is calculated in Problem 2.8.

2.5.2 Empirical and Self-Consistent Pseudopotential Methods

There are two approaches to calculating pseudopotential form factors. Since the number of relevant pseudopotential form factors is small, they can be determined by fitting a small number of experimental data, such as the position of peaks in optical reflectivity spectra (Chap. 6) or features in the photoelectron spectra (Chap. 8). This approach is known as the **Empirical Pseudopotential Method** (EPM). The flow diagram for calculating the band structure with the EPM is as follows:

V_g

↓

$$V(\boldsymbol{r}) = \sum_g V_g \exp(-\mathrm{i}\boldsymbol{g}\cdot\boldsymbol{r})$$

↓

$H = (p^2/2m) + V(\boldsymbol{r})$

Solve $H\psi_{\boldsymbol{k}}(\boldsymbol{r}) = E_{\boldsymbol{k}}\psi_{\boldsymbol{k}}(\boldsymbol{r})$ to obtain $\psi_{\boldsymbol{k}}(\boldsymbol{r})$ and $E_{\boldsymbol{k}}$

↓

Calculate reflectivity, density of states, etc., and compare with experiments

↓

Alter V_g if agreement between theory and experiment is not satisfactory.

The disadvantage of the EPM is that it requires experimental inputs. However, this is not a major disadvantage since atomic pseudopotential form factors are often "transferable" in the sense that once they are determined in one compound they can be used (sometimes after suitable interpolation) in other compounds containing the same atom. For example, the atomic pseudopotential form factors for Ga determined empirically from GaAs can be used to calculate the band structure of other Ga compounds such as GaSb and GaP. With the availability of high-speed computers, however, it is possible to determine the pseudopotential form factors from first principles without any experimental input. These first-principles pseudopotential methods are known as **self-consistent** or ***ab initio*** **pseudopotential methods**. These methods use atomic pseudopotentials and a model for the crystal structure (from which an ionic potential V_{ion} can be constructed) as the starting point of the calculation. After the wave functions have been obtained the contribution of the valence electrons to the potential is calculated. It is then used to evaluate the total one-electron potential, which is compared with the starting potential. Self-consistency is achieved when the calculated one-electron potential agrees with the starting potential. The flow diagram for such a calculation is shown below. The **exchange and correlation** term V_{xc}, which takes into account the many-body effects, is usually calculated with approximations such as the **Local Density Approximation** (LDA)[2]. In this approximation, V_{xc} is assumed to be a function of the local charge density only. The LDA gives good results for the ground state properties such as the cohesive energies and charge density of the valence electrons. However, it gives poor results for the excitation energies. For instance, it typically underestimates the fundamental energy gap by about 1 eV. Thus it predicts semiconductors like Ge to be semimetals. The band structures shown in Figs. 2.10 and 2.13–15 have been calculated with the EPM since this method gives better overall agreement with experiments. This shortcoming of the LDA can be overcome by many-body techniques such as the **quasiparticle approach** [2.13].

Choose $V(r)$

$\downarrow$

Solve $(H + V)\psi = E\psi$

$\downarrow$

Calculate charge density $\varrho = \psi^*\psi$

$\downarrow$

Solve $\nabla^2 V_{\text{Hartree}} = 4\pi\varrho \left(\frac{1}{4\pi\varepsilon_0}\right)$

$\downarrow$

Calculate $V_{\text{xc}} = f[\varrho(r)]$

$\downarrow$

$V_{\text{sc}} = V_{\text{Hartree}} + V_{\text{xc}}$

$\downarrow$

Model structure $V_{\text{ion}} \rightarrow V = V_{\text{sc}} + V_{\text{ion}}$

In recent years the *ab initio* pseudopotential method has been refined so as to be able to handle *semicore* electrons such as the 3d electrons of the copper halides. The pseudopotentials used are very smooth near the core (*ultrasoft pseudopotentials*) and reduce the number of plane waves required for the expansion of the wavefunctions to converge. The method is particularly useful for CuCl and diamond [2.14].

---

[2] For this development of the local density functional method of calculating electronic structures W. Kohn was awarded the Nobel prize for Chemistry in 1998 [2.11, 12].

2.6 The $k \cdot p$ Method of Band-Structure Calculations

The pseudopotential method is not the only method of band structure calculation which requires a small number of input parameters obtainable from experimental results. In the empirical pseudopotential method the inputs are usually energy gaps. In optical experiments one typically determines both energy gaps and oscillator strengths of the transitions. Thus it can be an advantage if the optical matrix elements can also be used as inputs in the band structure calculation. In the $\boldsymbol{k} \cdot \boldsymbol{p}$ method the band structure over the entire Brillouin zone can be extrapolated from the zone center energy gaps and optical matrix elements. The $\boldsymbol{k} \cdot \boldsymbol{p}$ method is, therefore, particularly convenient for interpreting optical spectra. In addition, using this method one can obtain analytic expressions for band dispersion and effective masses around high-symmetry points.

The $\boldsymbol{k} \cdot \boldsymbol{p}$ method can be derived from the one-electron Schrödinger equation given in (2.4). Using the Bloch theorem the solutions of (2.4) are expressed, in the reduced zone scheme, as

$$\Phi_{n\boldsymbol{k}} = \exp(\mathrm{i}\boldsymbol{k} \cdot \boldsymbol{r}) u_{n\boldsymbol{k}}(\boldsymbol{r}), \tag{2.34}$$

where n is the band index, $\boldsymbol{k}$ lies within the first Brillouin zone, and $u_{n\boldsymbol{k}}$ has the periodicity of the lattice. When $\Phi_{n\boldsymbol{k}}$ is substituted into (2.4) we obtain an equation in $u_{n\boldsymbol{k}}$ of the form[3]

$$\left(\frac{p^2}{2m} + \frac{\hbar \boldsymbol{k} \cdot \boldsymbol{p}}{m} + \frac{\hbar^2 k^2}{2m} + V \right) u_{n\boldsymbol{k}} = E_{n\boldsymbol{k}} u_{n\boldsymbol{k}}. \tag{2.35}$$

At $\boldsymbol{k}_0 = (0,0,0)$, (2.35) reduces to

$$\left(\frac{p^2}{2m} + V \right) u_{n\boldsymbol{0}} = E_{n\boldsymbol{0}} u_{n\boldsymbol{0}} \quad (n = 1,2,3,\ldots). \tag{2.36}$$

Similar equations can also be obtained for $\boldsymbol{k}$ equal to any point $\boldsymbol{k}_0$. Equation (2.36) is much easier to solve than (2.4) since the functions $u_{n\boldsymbol{0}}$ are periodic. The solutions of (2.36) form a complete and orthonormal set of basis functions. Once $E_{n\boldsymbol{0}}$ and $u_{n\boldsymbol{0}}$ are known, we can treat the terms $\hbar \boldsymbol{k} \cdot \boldsymbol{p}/m$ and $\hbar^2 k^2/(2m)$ as perturbations in (2.35) using either **degenerate** or **nondegenerate perturbation theory**. This method for calculating the band dispersion is known as the **$\boldsymbol{k} \cdot \boldsymbol{p}$ method**. Since the perturbation terms are proportional to k, the method works best for small values of k [2.15]. In general, the method can be applied to calculate the band dispersion near any point $\boldsymbol{k}_0$ by expanding (2.35) around $\boldsymbol{k}_0$ provided the wave functions (or the matrix elements of $\boldsymbol{p}$ between these wave functions) and the energies at $\boldsymbol{k}_0$ are known. Furthermore, by using a sufficiently large number of $u_{n\boldsymbol{0}}$ to approximate a complete set of basis

[3] Equation (2.35) is rigorously valid only if V is a *local* potential, i.e., it depends only on one spatial coordinate r. This is not strictly true in the case of pseudopotential [2.8]

functions, (2.35) can be diagonalized with the help of computers to calculate the band structure over the entire Brillouin zone [2.16]. Only a limited number of energy gaps and matrix elements of $\boldsymbol{p}$ determined experimentally are used as input in the calculation.

As examples of application of the $\boldsymbol{k} \cdot \boldsymbol{p}$ method we will derive the band dispersion and effective mass for a nondegenerate band and for a three-fold degenerate (or nearly degenerate) p-like band. The nondegenerate band case is applicable to the conduction band minimum in direct-bandgap semiconductors with the zinc-blende and wurtzite structures (examples of the latter semiconductor are CdS and CdSe). The nearly degenerate band is a model for the top valence bands in many semiconductors with the diamond, zinc-blende, or wurtzite structures.

2.6.1 Effective Mass of a Nondegenerate Band Using the $k \cdot p$ Method

Let us assume that the band structure has an extremum at the energy $E_{n\mathbf{0}}$ and the band is nondegenerate at this energy. Using standard nondegenerate perturbation theory, the eigenfunctions $u_{n\boldsymbol{k}}$ and eigenvalues $E_{n\boldsymbol{k}}$ at a neighboring point $\boldsymbol{k}$ can be expanded to second order in k in terms of the unperturbed wave functions $u_{n\mathbf{0}}$ and energies $E_{n\mathbf{0}}$ by treating the terms involving k in (2.35) as perturbations.

$$u_{n\boldsymbol{k}} = u_{n\mathbf{0}} + \frac{\hbar}{m} \sum_{n' \neq n} \frac{\langle u_{n\mathbf{0}} \,|\, \boldsymbol{k} \cdot \boldsymbol{p} \,|\, u_{n'\mathbf{0}} \rangle}{E_{n\mathbf{0}} - E_{n'\mathbf{0}}} u_{n'\mathbf{0}} \tag{2.37}$$

and

$$E_{n\boldsymbol{k}} = E_{n\mathbf{0}} + \frac{\hbar^2 k^2}{2m} + \frac{\hbar^2}{m^2} \sum_{n' \neq n} \frac{|\langle u_{n\mathbf{0}} \,|\, \boldsymbol{k} \cdot \boldsymbol{p} \,|\, u_{n'\mathbf{0}} \rangle|^2}{E_{n\mathbf{0}} - E_{n'\mathbf{0}}}. \tag{2.38}$$

The linear terms in k vanish because $E_{n\mathbf{0}}$ has been assumed to be an extremum. It is conventional to express the energy $E_{n\boldsymbol{k}}$, for small values of k, as

$$E_{n\boldsymbol{k}} = E_{n\mathbf{0}} + \frac{\hbar^2 k^2}{2m^*}, \tag{2.39}$$

where m^* is defined as the **effective mass** of the band. Comparing (2.38) and (2.39) we obtain an expression for this effective mass:

$$\frac{1}{m^*} = \frac{1}{m} + \frac{2}{m^2 k^2} \sum_{n' \neq n} \frac{|\langle u_{n\mathbf{0}} \,|\, \boldsymbol{k} \cdot \boldsymbol{p} \,|\, u_{n'\mathbf{0}} \rangle|^2}{E_{n\mathbf{0}} - E_{n'\mathbf{0}}}. \tag{2.40}$$

Formula (2.40) can be used to calculate the effective mass of a nondengenerate band. Also it shows that an electron in a solid has a mass different from that of a free electron because of coupling between electronic states in different bands via the $\boldsymbol{k} \cdot \boldsymbol{p}$ term. The effect of neighboring bands on the effective mass of a band depends on two factors.

- A wave function $u_{n'\mathbf{0}}$ can couple to u_{n0} only if the matrix element $\langle u_{n'\mathbf{0}} | \boldsymbol{p} | u_{n\mathbf{0}} \rangle$ is nonzero. In Sect 2.3.4 we pointed out that, using the matrix element theorem and group theory, it is possible to enumerate all the symmetries $u_{n'\mathbf{0}}$ can have. For example, $\boldsymbol{p}$ has Γ_4 symmetry in the zinc-blende structure. If the conduction band has Γ_1 symmetry, as in GaAs, its effective mass will be determined only by coupling with bands having Γ_4 symmetry. On the other hand a valence band with Γ_4 symmetry can be coupled via $\boldsymbol{p}$ to bands with Γ_1, Γ_3, Γ_4, and Γ_5 symmetries.
- The energy separation $E_{n'\mathbf{0}} - E_{n\mathbf{0}}$ between the two bands n and n' determines the relative importance of the contribution of n' to the effective mass of n. Furthermore, bands with energies less than $E_{n\mathbf{0}}$ will contribute a positive term to $1/m^*$, making m^* smaller than the free electron mass. Conversely, bands with energies higher than $E_{n\mathbf{0}}$ tend to increase m^* or even cause m^* to become *negative* as in the case of the top valence bands in the diamond- and zinc-blende-type semiconductors.

These two simple results can be used to understand the trend in the conduction band effective mass m_c^* in many of the group-III–V and II–VI semiconductors with direct bandgaps. In these semiconductors the lowest conduction band at the zone center has Γ_1 symmetry. From the above considerations, its effective mass will be determined mainly by its coupling, via the $\boldsymbol{k} \cdot \boldsymbol{p}$ term, to the nearest bands with Γ_4 symmetry. They include both valence and conduction bands. As we will show in the next section, the conduction bands in the group-IV, III–V, and II–VI semiconductors have antibonding character, while the valence bands have bonding character. What this means is that in the diamond-type structure the $\Gamma_{2'}$ (or Γ_2^-) conduction band and its nearest Γ_{15} (or Γ_4^-) conduction band both have odd parity and the momentum matrix element between them vanishes because of the parity selection rule. In III–V semiconductors, the antisymmetric pseudopotential breaks the inversion symmetry. As a result, the momentum matrix element between the Γ_1 conduction band and its nearest Γ_4 conduction band in III–V semiconductors is nonzero, but still much smaller than its momentum matrix element with the top Γ_4 valence bands [2.17]. The separation between the Γ_1 conduction band and the Γ_4 valence band is just the direct band gap E_0, so m_c^* can be approximated by

$$\frac{1}{m_c^*} = \frac{1}{m} + \frac{2|\langle \Gamma_{1c} | \boldsymbol{k} \cdot \boldsymbol{p} | \Gamma_{4v} \rangle|^2}{m^2 E_0 k^2}. \tag{2.41}$$

It is customary to represent the three Γ_4 wave functions as $|X\rangle$, $|Y\rangle$, and $|Z\rangle$. From the T_d symmetry it can be shown that the only nonzero elements of $\langle \Gamma_{1c} | \boldsymbol{k} \cdot \boldsymbol{p} | \Gamma_{4v} \rangle$ are

$$\langle X | p_x | \Gamma_1 \rangle = \langle Y | p_y | \Gamma_1 \rangle = \langle Z | p_z | \Gamma_1 \rangle = \mathrm{i}P. \tag{2.42}$$

Without loss of generality we can assume that the wave functions $|X\rangle$, $|Y\rangle$, $|Z\rangle$, and $|\Gamma_1\rangle$ are all real. Since the operator $\boldsymbol{p}$ is equal to $-\mathrm{i}\hbar\boldsymbol{\nabla}$ the matrix element in (2.42) is purely imaginary and P is real. With these results (2.41) simplifies to

$$\frac{m}{m_{\rm c}^*} \approx 1 + \frac{2P^2}{mE_0}. \tag{2.43}$$

It turns out that the matrix element P^2 is more or less constant for most group-IV, III–V and II–VI semiconductors, with $2P^2/m \approx 20$ eV. The reason is that the values of P^2 for these semiconductors are very close to those calculated for nearly free electron wave functions: $P = 2\pi\hbar/a_0$ (see Problem 2.9). Since E_0 is typically less than 2 eV, $2P^2/(mE_0) \gg 1$ and (2.43) further simplifies to

$$\frac{m}{m_{\rm c}^*} \approx \frac{2P^2}{mE_0}. \tag{2.44}$$

In Table 2.22 we compare the values of $m_{\rm c}^*$ calculated from (2.44) with those determined experimentally for several group-IV, III–V, and II–VI semiconductors. The values of E_0 are from experiment.

Equation (2.44) can be extended to estimate the increase in m_c^* away from the band minimum (non-parabolicity) which can be qualitatively described by an increase in E_0. See problem 6.15.

Table 2.22. Experimental values of the Γ_1 conduction band effective masses in diamond- and zinc-blende-type semicondutors compared with the values calculated from (2.44) using the values of E_0 obtained from experiment [2.18]

	Ge	GaN	GaAs	GaSb	InP	InAs	ZnS	ZnSe	ZnTe	CdTe
E_0 [eV]	0.89	3.44	1.55	0.81	1.34	0.45	3.80	2.82	2.39	1.59
$m_{\rm c}^*/m$ (exp)	0.041	0.17	0.067	0.047	0.073	0.026	0.20	0.134	0.124	0.093
$m_{\rm c}^*/m$ ((2.44))	0.04	0.17	0.078	0.04	0.067	0.023	0.16	0.14	0.12	0.08

2.6.2 Band Dispersion near a Degenerate Extremum: Top Valence Bands in Diamond- and Zinc-Blende-Type Semiconductors

To apply the $\boldsymbol{k} \cdot \boldsymbol{p}$ method to calculate the band dispersion near a degenerate band extremum we consider the highest energy $\Gamma_{25'}$ (Γ_4) valence bands at the zone center of semiconductors with the diamond (zinc-blende) structure. As pointed out in the previous section, these valence band wave functions are p-like, and they will be represented by the eigenstates $|X\rangle$, $|Y\rangle$, and $|Z\rangle$. The electron **spin** is 1/2, so the spin states will be denoted by α and β to correspond to spin-up and spin-down states, respectively. In atomic physics it is well-known that the electron spin can be coupled to the **orbital angular momentum** via the **spin–orbit interaction**. The spin–orbit coupling is a relativistic effect (inversely proportional to c^2) which scales with the atomic number of the atom. Thus for semiconductors containing heavier elements, such as Ge, Ga, As, and Sb, one expects the spin–orbit coupling to be significant and has to include it in the unperturbed Hamiltonian, in particular for states near $k = 0$. The Hamiltonian for the spin–orbit interaction is given by

$$H_{so} = \frac{\hbar}{4c^2m^2}(\nabla V \times \boldsymbol{p}) \cdot \boldsymbol{\sigma}, \tag{2.45a}$$

where the components of $\boldsymbol{\sigma}$ are the **Pauli spin matrices**:

$$\sigma_x = \begin{pmatrix} 0 & 1 \\ 1 & 0 \end{pmatrix}; \quad \sigma_y = \begin{pmatrix} 0 & -\mathrm{i} \\ \mathrm{i} & 0 \end{pmatrix}; \quad \sigma_z = \begin{pmatrix} 1 & 0 \\ 0 & -1 \end{pmatrix}. \tag{2.45b}$$

(In crystals with the diamond structure the "vector" $\nabla V \times \boldsymbol{p}$ is an example of a pseudovector with symmetry $\Gamma_{15'}$). The Hamiltonian H_{so} operates on the spin wave functions so the symmetry of H_{so} should depend also on the symmetry properties of the spin matrices. As is known from quantum mechanics, spin behaves differently than classical properties of particles such as the orbital angular momentum. For example, a spatial wave function is invariant under a rotation of 2π about any axis. However, under the same rotation the spin wave functions of a spin-1/2 particle will change sign. Let us denote a rotation of 2π about a unit vector $\hat{\boldsymbol{n}}$ as $\hat{E}$ (Problem 2.10). For a spinless particle $\hat{E}$ is equal to the identity operation. For a spin-1/2 particle $\hat{E}$ is an additional symmetry operation in the point group of its spin-dependent wave function. Thus, if $\boldsymbol{G}$ is the point group of a crystal neglecting spin, then the corresponding point group including spin effects will contain $\boldsymbol{G}$ plus $\hat{E}\boldsymbol{G}$ and is therefore twice as large as $\boldsymbol{G}$. Groups containing symmetry operations of spin wave functions are known as **double groups**. It is beyond the scope of this book to treat double groups in detail. Interested readers should refer to references listed for this chapter at the end of the book [Refs. 2.4, p. 103; 2.5; 2.7, p. 258].

Although many band diagrams in this book use the double group notation (for example, Figs. 2.13–15), in most cases it is sufficient to know only the irreducible representations for the double group at the zone center (Γ point) of zinc-blende-type crystals. Since the single group of Γ in zinc-blende-type crystals contains 24 elements, one expects the double group to contain 48 elements. However, the number of classes in a double group is not necessarily twice that of the corresponding "single group". The reason is that a class C in the single group may or may not belong to the same class as $\hat{E}C$ in the double group. For example, two sets of operations C_i and $\hat{E}C_i$ belong to the same class if the point group contains a two-fold rotation about an axis perpendicular to $\hat{\boldsymbol{n}}_i$ (the rotation axis of C_i). In the case of the group of Γ of a zinc-blende-type crystal, elements in $\{3C_2\}$ and $\{3\hat{E}C_2\}$ belong to the same class in the double group. This is also true for the elements in $\{6\sigma\}$ and $\{6\hat{E}\sigma\}$. As a result, the 48 elements in the double group of Γ in zinc-blende-type crystals are divided into eight classes. These eight classes and the eight irreducible representations of the double group of Γ are listed in Table 2.23.

Instead of using Table 2.23 to symmetrize the p-like valence band wave functions in zinc-blende-type crystals including spin–orbit coupling, we will utilize their similarity to the atomic p wave functions. We recall that, in atomic physics, the orbital electronic wave functions are classified as s, p, d, etc., ac-

Table 2.23. Character table of the double group of the point Γ in zinc-blende-type semiconductors

	$\{E\}$	$\{3C_2/3\hat{E}C_2\}$	$\{6S_4\}$	$\{6\sigma/6\hat{E}\sigma\}$	$\{8C_3\}$	$\{\hat{E}\}$	$\{6\hat{E}S_4\}$	$\{8\hat{E}C_3\}$
Γ_1	1	1	1	1	1	1	1	1
Γ_2	1	1	−1	−1	1	1	−1	1
Γ_3	2	2	0	0	−1	2	0	−1
Γ_4	3	−1	−1	1	0	3	−1	0
Γ_5	3	−1	1	−1	0	3	1	0
Γ_6	2	0	$\sqrt{2}$	0	1	−2	$-\sqrt{2}$	−1
Γ_7	2	0	$-\sqrt{2}$	0	1	−2	$\sqrt{2}$	−1
Γ_8	4	0	0	0	−1	−4	0	1

cording to the orbital angular momentum l. The p states correspond to $l = 1$ and are triply degenerate. The three degenerate states can be chosen to be eigenstates of l_z, the z component of $\boldsymbol{l}$. The eigenvalues of l_z are known as the **magnetic quantum numbers** (usually denoted as m_l). For the p states $m_l = 1, 0, -1$. The wave functions of the orbital angular momentum operator are known as **spherical harmonics**. The spherical harmonics corresponding to the $l = 1$ states can be represented as (except for a trivial factor of $(x^2 + y^2 + z^2)^{-1/2}$):

$$|lm_l\rangle = \begin{cases} |1\,1\rangle & = -(x + \mathrm{i}y)/\sqrt{2}, \\ |1\,0\rangle & = z, \\ |1\,{-1}\rangle & = (x - \mathrm{i}y)/\sqrt{2}. \end{cases} \tag{2.46}$$

The spin-orbit interaction in atomic physics is usually expressed in terms of $\boldsymbol{l}$ and the spin $\boldsymbol{s}$ as

$$H_{\mathrm{so}} = \lambda \boldsymbol{l} \cdot \boldsymbol{s} \tag{2.47}$$

The constant λ is referred to as the spin–orbit coupling. The eigenfunctions of (2.47) are eigenstates of the **total angular momentum** $\boldsymbol{j} = \boldsymbol{l} + \boldsymbol{s}$ and its z component j_z. For $l = 1$ and $s = 1/2$ the eigenvalues of $\boldsymbol{j}$ can take on two possible values: $j = l + s = 3/2$ and $j = l - s = 1/2$. The eigenvalues of j_z (denoted by m_z) can take on the $2j + 1$ values $j, j-1, \ldots, -j+1, -j$. The eigenfunctions of j and j_z can be expressed as linear combinations of the eigenfunctions of the orbital angular momentum and spin (α = spin-up, β = spin-down):

$$|jm_j\rangle = \begin{cases} |3/2,\ 3/2\rangle = |1,\ 1\rangle\alpha \\ |3/2,\ 1/2\rangle = (1/\sqrt{3})(|1,\ 1\rangle\beta + \sqrt{2}\,|1,\ 0\rangle\alpha) \\ |3/2,\ -1/2\rangle = (1/\sqrt{3})(|1,\ -1\rangle\alpha + \sqrt{2}\,|1,\ 0\rangle\beta) \\ |3/2,\ -3/2\rangle = |1,\ -1\rangle\beta \end{cases} \tag{2.48}$$

$$\begin{cases} |1/2,\ 1/2\rangle = (1/\sqrt{3})(|1,\ 0\rangle\alpha - \sqrt{2}\,|1,\ 1\rangle\beta) \\ |1/2,\ -1/2\rangle = (1/\sqrt{3})(|1,\ 0\rangle\beta - \sqrt{2}\,|1,\ -1\rangle\alpha) \end{cases} \tag{2.49}$$

The spin–orbit interaction in (2.47) splits the $j = 3/2$ states in (2.48) from the $j = 1/2$ states in (2.49). This splitting Δ_0 is known as the **spin–orbit splitting** of the valence band at Γ, and in the case of the $j = 3/2$ and $j = 1/2$ states $\Delta_0 = 3\lambda/2$.

Using the atomic physics results as a guideline we can similarly symmetrize the six electronic states $|X\rangle\alpha$, $|X\rangle\beta$, $|Y\rangle\alpha$, $|Y\rangle\beta$, $|Z\rangle\alpha$, and $|Z\rangle\beta$ in the diamond- and zinc-blende-type semiconductors. First, we make use of the similarity between the p-like Γ_4 states and the atomic p states to define three "$(l = 1)$-like" states in the zinc-blende-type crystals:

$$\begin{aligned} |1, \ \ 1\rangle &= -(|X\rangle + \mathrm{i}|Y\rangle)/\sqrt{2}, \\ |1, \ \ 0\rangle &= |Z\rangle, \\ |1, -1\rangle &= (|X\rangle - \mathrm{i}|Y\rangle)/\sqrt{2}. \end{aligned} \qquad (2.50)$$

Next we define $(j = 3/2)$-like and $(j = 1/2)$-like states in the diamond- and zinc-blende-type crystals by substituting the expressions in (2.50) into (2.48) and (2.49). From now on we will refer to these $(j = 3/2)$-like and $(j = 1/2)$-like states as the $j = 3/2$ states and $j = 1/2$ states in the case of semiconductors.

From the characters of the double group of Γ in Table 2.23, one easily concludes that the four-fold degenerate $j = 3/2$ states belong to the Γ_8 representation, since this is the only four-dimensional representation. The two-fold degenerate $j = 1/2$ states must belong to either the Γ_6 or Γ_7 representations. A way to decide between these two representations is to calculate the character of the representation matrix generated by $j = 1/2$ states under an S_4 operation. Using the result of Problem 2.10 it can be shown that the $j = 1/2$ states belong to the Γ_7 representation. As in the atomic case, the Γ_8 and Γ_7 states are split by the spin–orbit Hamiltonian in (2.45a). Typically, the magnitude of the spin–orbit splitting Δ_0 in a semiconductor is comparable to the Δ_0 of its constituent atoms. For example, semiconductors containing heavier atoms, such as InSb and GaSb, have $\Delta_0 \approx 1$ eV, which is as large as or larger than the bandgap. When the anion and cation in the compound semiconductor have different Δ_0 the anion contribution tends to be weighted more, reflecting its larger influence on the p-like valence bands. In semiconductors containing lighter atoms, such as Si and AlP, $\Delta_0(\approx 0.05$ eV$)$ is negligible for many purposes. The values of Δ_0 in some diamond- and zinc-blende-type semiconductors are given in Table 2.24. The values of Δ_0 in Table 2.24 are all positive, and as a result the $j = 3/2$ (Γ_8) valence band has higher energy than the $j = 1/2$ (Γ_7) valence band states (Figs. 2.13–15). In some zinc-blende-type crystals, such as CuCl, where there is a large contribution to the valence bands from the core d-electrons, Δ_0 can be negative, leading to a reversal in the ordering of the Γ_8 and Γ_7 valence bands.

In Sect. 2.3.4d it was shown that the operator $\boldsymbol{p}$ couples a state with Γ_4 symmetry to states with Γ_1, Γ_3, Γ_4, and Γ_5 symmetries. By examining the band structure of several semiconductors calculated by the pseudopotential method (Figs. 2.10, 2.13–15) we find that the bands which have the above symmetries and are close to the Γ_4 valence bands are typically the lowest

Table 2.24. Valence band parameters A and B in units of $(\hbar^2/2m)$ and $|C|^2$ in units of $(\hbar^2/2m)^2$. The spin–orbit splitting of the valence bands Δ_0 is given in units of eV. The averaged experimental [exp] and theoretical [th, obtained from A, B, C^2 with (2.67, 69)] values of the effective masses of the heavy hole (hh), light hole (lh) and spin–orbit split-off hole (so) valence bands are in units of the free electron mass. [2.16, 18]

	A	B	$\|C\|^2$	Δ_0 [eV]	m_{hh}/m_0 exp	m_{hh}/m_0 th	m_{lh}/m_0 exp	m_{lh}/m_0 th	m_{so}/m_0 exp	m_{so}/m_0 th
C	−2.5	0.2	4.6	0.013[a]		0.66[b]		0.29[b]		0.39[b]
Si	−4.28	−0.68	24	0.044	0.54	0.45	0.15	0.14	0.23	0.24
Ge	−13.38	−8.5	173	0.295	0.34	0.43	0.043	0.041	0.095	0.1
SiC[c]	−2.8	−0.31	4.8	0.014		1.4		0.36		0.55
GaN[d]	−5.05	−1.2	34	0.017		0.5		0.13		0.2
GaP	−4.05	−0.98	16	0.08	0.57	0.5	0.18	0.17		0.25
GaAs	−6.9	−4.4	43	0.341	0.53	0.78	0.08	0.08	0.15	0.17
GaSb	−13.3	−8.8	230	0.75	0.8	0.9	0.05	0.04		0.15
InP	−5.15	−1.9	21	0.11	0.58	0.53	0.12	0.12	0.12	0.2
InAs	−20.4	−16.6	167	0.38	0.4	0.4	0.026	0.026	0.14	0.10
InSb	−36.41	−32.5	43	0.81	0.42	0.48	0.016	0.013		0.12
ZnS	−2.54	−1.5		0.07						
ZnSe	−2.75	−1.0	7.5	0.43		1.09		0.145		
ZnTe	−3.8	−1.44	14.0	0.93						
CdTe	−4.14	−2.18	30.3	0.92						

[a] See: J. Serrano, M. Cardona, and T. Ruf, Solid State Commun. **113**, 411 (2000)
[b] See: M. Willatzen, M. Cardona, N.E. Christensen, *Linear Muffin-tin-orbital and* $\boldsymbol{k}\cdot\boldsymbol{p}$ *calculation of band structure of semiconducting diamond.* Phys. Rev. B**50** 18054 (1994)
[c] See: M. Willatzen, M. Cardona, N.E. Christensen: *Relativistic electronic structure of 3C-SiC.* Phys. Rev. B**51**, 13150 (1995).
[d] See [1.1].

conduction bands with symmetries Γ_1 and Γ_4. For the conduction band Γ_{1c} we have already shown that the only significant momentum matrix elements are $\langle X|p_x|\Gamma_1\rangle = \langle Y|p_y|\Gamma_1\rangle = \langle Z|p_z|\Gamma_1\rangle = \mathrm{i}P$, see (2.42). One can also use symmetry arguments to show that the nonzero matrix elements of $\boldsymbol{p}$ between the Γ_4 valence bands and the Γ_4 conduction band states are

$$\begin{aligned}\langle X|p_y|\Gamma_{4c}(z)\rangle &= \langle Y|p_z|\Gamma_{4c}(x)\rangle = \langle Z|p_x|\Gamma_{4c}(y)\rangle = \mathrm{i}Q,\\ \langle X|p_z|\Gamma_{4c}(y)\rangle &= \langle Y|p_x|\Gamma_{4c}(z)\rangle = \langle Z|p_y|\Gamma_{4c}(x)\rangle = \mathrm{i}Q\end{aligned} \tag{2.51}$$

(details of the proof are left for Problem 2.11).

The Γ_{4v} valence bands together with the Γ_{1c} and Γ_{4c} conduction bands now form a set of 14 unperturbed wave functions which are coupled together by the $\boldsymbol{k}\cdot\boldsymbol{p}$ term of (2.35). The resultant 14×14 determinant can be diagonalized either with the help of computers or by using approximations. Löwdin's perturbation method is most commonly used to obtain analytic expressions for the dispersion of the valence bands. In this method the 14×14 matrix is divided into two parts: the wave functions of interest and their mutual interactions are treated exactly while the interaction between this group of wave functions and the remaining wave functions is treated by perturbation theory. For example, in the present case the six Γ_{4v} valence bands (including spin degeneracy) are of interest and their mutual coupling via the $\boldsymbol{k}\cdot\boldsymbol{p}$ and spin–orbit interactions will be treated exactly. The coupling between these valence band states and the conduction bands will be treated as a perturbation by defining an effective matrix element between any two valence band wave function as

$$H'_{ij} = H_{ij} + \sum_{\substack{k \neq \text{ the } \Gamma_4 \\ \text{valence bands}}} \frac{H_{ik}H_{kj}}{E_i - E_k}. \tag{2.52}$$

Within this approximation the 14×14 matrix reduces to a 6×6 matrix of the form $\{H'_{ij}\}$, where i and j run from 1 to 6. To simplify the notation we will number the six Γ_{4v} valence band wave functions as

$$\begin{aligned}
\Phi_1 &= |3/2,\ 3/2\rangle \\
\Phi_2 &= |3/2,\ 1/2\rangle \\
\Phi_3 &= |3/2,\ -1/2\rangle \\
\Phi_4 &= |3/2,\ -3/2\rangle \\
\Phi_5 &= |1/2,\ 1/2\rangle \\
\Phi_6 &= |1/2,\ -1/2\rangle
\end{aligned}$$

and the doubly degenerate Γ_{1c} and six-fold degenerate Γ_{4c} conduction band wave functions as Φ_7 to Φ_{14}.

The calculation of all the matrix elements H'_{ij} is left for Problem 2.14a. Here we will calculate only the matrix element H'_{11} as an example. According to (2.52) the effective matrix element H'_{11} is given by

$$\begin{aligned}
H'_{11} = &\left\langle \Phi_1 \left| \frac{\hbar^2 k^2}{2m} + \frac{\hbar \boldsymbol{k}\cdot\boldsymbol{p}}{m} \right| \Phi_1 \right\rangle \\
&+ \sum_j \left| \left\langle \Phi_1 \left| \frac{\hbar^2 k^2}{2m} + \frac{\hbar \boldsymbol{k}\cdot\boldsymbol{p}}{m} \right| \Phi_j \right\rangle \right|^2 \frac{1}{(E_1 - E_j)}.
\end{aligned} \tag{2.53}$$

To simplify the notation again we introduce the following symbols: E_0, energy separation between Γ_{1c} and the $j = 3/2$ valence bands; and E'_0, energy separation between Γ_{4c} and the $j = 3/2$ valence bands. Using these symbols we can express H'_{11} as

$$\begin{aligned}
H'_{11} = &\frac{\hbar^2 k^2}{2m} + \left\langle \Phi_1 \left| \frac{\hbar \boldsymbol{k}\cdot\boldsymbol{p}}{m} \right| \Phi_1 \right\rangle - \left(\left| \left\langle \Phi_1 \left| \frac{\hbar \boldsymbol{k}\cdot\boldsymbol{p}}{m} \right| \Gamma_{1c} \right\rangle \right|^2 \frac{1}{E_0} \right) \\
&- \left(\left| \left\langle \Phi_1 \left| \frac{\hbar \boldsymbol{k}\cdot\boldsymbol{p}}{m} \right| \Gamma_{4c} \right\rangle \right|^2 \frac{1}{E'_0} \right).
\end{aligned} \tag{2.54}$$

In principle, the term $\hbar\boldsymbol{k}\cdot\boldsymbol{p}/m$ can give rise to a term linear in k in the band dispersion. In the diamond-type semiconductors this term vanishes exactly because of the parity selection rule. In zinc-blende-type crystals the linear $\boldsymbol{k} \cdot \boldsymbol{p}$ term can be shown to be zero within the basis used. While the k linear term is strictly zero in diamond-type crystals because of the parity selection rule, this is *not* true in crystals without a center of inversion symmetry. In zinc-blende- and wurtzite-type crystals, it has been demonstrated [2.19, 20] that both the conduction and valence bands can possess small k-linear terms. However, these k-linear terms do not come from the $\boldsymbol{k} \cdot \boldsymbol{p}$ term alone, instead they involve also spin-dependent terms which have been neglected here. Since the k linear terms are relatively unimportant for the valence bands of most semiconductors they will not be considered further here.

To simplify the notation we define

$$\begin{aligned} L &= \frac{-\hbar^2 P^2}{m^2 E_0}; \\ M &= \frac{-\hbar^2 Q^2}{m^2 E_0'}; \\ N &= L + M; \\ L' &= \frac{-\hbar^2 P^2}{m^2 (E_0 + \Delta_0)}; \\ M' &= \frac{-\hbar^2 Q^2}{m^2 (E_0' + \Delta_0)}. \end{aligned}$$

With these definitions, the term

$$-\left(\left| \left\langle \Phi_1 \left| \frac{\hbar \boldsymbol{k} \cdot \boldsymbol{p}}{m} \right| \Gamma_{1c} \right\rangle \right|^2 \frac{1}{E_0} \right)$$

in (2.54) can easily be shown to be equal to

$$-\left(\left| \left\langle \Phi_1 \left| \frac{\hbar \boldsymbol{k} \cdot \boldsymbol{p}}{m} \right| \Gamma_{1c} \right\rangle \right|^2 \frac{1}{E_0} \right) = \frac{1}{2} L (k_x^2 + k_y^2) \tag{2.55}$$

while

$$-\left(\left| \left\langle \Phi_1 \left| \frac{\hbar \boldsymbol{k} \cdot \boldsymbol{p}}{m} \right| \Gamma_{4c} \right\rangle \right|^2 \frac{1}{E_0'} \right)$$

is given by

$$-\left(\left| \left\langle \Phi_1 \left| \frac{\hbar \boldsymbol{k} \cdot \boldsymbol{p}}{m} \right| \Gamma_{4c} \right\rangle \right|^2 \frac{1}{E_0'} \right) = \frac{1}{2} M (k_x^2 + k_y^2 + 2k_z^2). \tag{2.56}$$

The result is

$$H_{11}' = \frac{\hbar^2 k^2}{2m} + \frac{1}{2} N (k_x^2 + k_y^2) + M k_z^2. \tag{2.57}$$

Similarly we can show that the remaining matrix elements are

$$
\begin{aligned}
H'_{12} &= -\frac{N}{\sqrt{3}}(k_x k_z - \mathrm{i}k_y k_z)\\
H'_{13} &= -\frac{1}{2\sqrt{3}}[(L-M)(k_x^2-k_y^2) - 2\mathrm{i}Nk_x k_y]\\
H'_{14} &= 0\\
H'_{15} &= \frac{1}{\sqrt{2}}H'_{12}\\
H'_{16} &= -\sqrt{2}H'_{13}
\end{aligned}
$$

$$
\begin{aligned}
H'_{22} &= \frac{\hbar^2k^2}{2m} + \frac{1}{3}(M+2L)k^2 - \frac{1}{2}(L-M)(k_x^2+k_y^2)\\
H'_{23} &= 0\\
H'_{24} &= H'_{13}\\
H'_{25} &= \frac{1}{\sqrt{2}}(H'_{22} - H'_{11})\\
H'_{26} &= \sqrt{\frac{3}{2}}H'_{12}
\end{aligned}
$$

$$
\begin{aligned}
H'_{33} &= H'_{22}\\
H'_{34} &= -H'_{12}\\
H'_{35} &= -(H'_{26})^*\\
H'_{36} &= H'_{25}
\end{aligned}
$$

$$
\begin{aligned}
H'_{44} &= H'_{11}\\
H'_{45} &= -\sqrt{2}(H'_{13})^*\\
H'_{46} &= -(H'_{15})^*
\end{aligned}
$$

$$
\begin{aligned}
H'_{55} &= \frac{\hbar^2k^2}{2m} + \frac{1}{3}(2M'+L')k^2 - \Delta_0\\
H'_{56} &= 0
\end{aligned}
$$

$$
H'_{66} = H'_{55}.
$$

The matrix $\{H'_{ij}\}$ is Hermitian, i. e., $H'_{ij} = [H'_{ji}]^*$. This 6×6 matrix can be diagonalized numerically without further simplification. Readers with access to a personal computer and a matrix diagonalization program are encouraged to calculate the valence band structure of GaAs by diagonalizing this 6×6 matrix $\{H'_{ij}\}$ (Problem 2.14b).

The matrix $\{H'_{ij}\}$ can be diagonalized analytically with some approximations. We will now restrict k to values small enough that the matrix elements which couple the $J = 3/2$ and $J = 1/2$ bands, such as H'_{15}, H'_{16}, and H'_{25}, are negligible compared with the spin–orbit coupling. With this assumption, and limiting the expansion of the eigenvalue to terms of the order of k^2 only, the 6×6 matrix reduces to a 4×4 and a 2×2 matrix. The 2×2 matrix gives the energy of the doubly degenerate $j = 1/2$ Γ_7 band as

$$\begin{aligned} E_{\mathrm{so}} &= H'_{55} = \frac{\hbar^2k^2}{2m} + \frac{1}{3}(2M' + L')k^2 - \Delta_0 \\ &= -\Delta_0 + \frac{\hbar^2k^2}{2m}\left[1 - \frac{2}{3}\left(\frac{P^2}{m(E_0+\Delta_0)} + \frac{2Q^2}{m(E'_0+\Delta_0)}\right)\right]. \end{aligned} \tag{2.58}$$

Thus, within the above approximation, the constant energy surface for the $j = 1/2$ split-off valence band is spherical and the band dispersion parabolic, with an effective mass m_{so} given by

$$\frac{m}{m_{\mathrm{so}}} = 1 - \frac{2}{3}\left(\frac{P^2}{m(E_0+\Delta_0)} + \frac{2Q^2}{m(E'_0+\Delta_0)}\right). \tag{2.59}$$

The dispersion of the $j = 3/2$ bands is obtained by diagonalizing the 4×4 matrix

$$\begin{vmatrix} H'_{11} & H'_{12} & H'_{13} & 0 \\ (H'_{12})^* & H'_{22} & 0 & H'_{13} \\ (H'_{13})^* & 0 & H'_{22} & -H'_{12} \\ 0 & (H'_{13})^* & -(H'_{12})^* & H'_{11} \end{vmatrix}.$$

The secular equation for this matrix reduces to two identical equations of the form

$$(H'_{11} - E)(H'_{22} - E) = |H'_{12}|^2 + |H'_{13}|^2 \tag{2.60}$$

and their solutions are

$$E_{\pm} = \tfrac{1}{2}(H'_{11} + H'_{22}) \pm \tfrac{1}{2}[(H'_{11} + H'_{12})^2 - 4(H'_{11}H'_{22} - |H'_{12}|^2 - |H'_{13}|^2)]^{\frac{1}{2}}. \tag{2.61}$$

Substituting the matrix elements H'_{ij} as defined earlier into (2.61) $E_{\pm}$ can be expressed as

$$E_{\pm} = Ak^2 \pm [B^2k^4 + C^2(k_x^2k_y^2 + k_y^2k_z^2 + k_z^2k_x^2)]^{\frac{1}{2}}, \tag{2.62}$$

an equation first derived by *Dresselhaus* et al. [2.21]. The constants A, B, and C in (2.62) are related to the electron momentum matrix elements and energy gaps by

$$\frac{2m}{\hbar^2}A = 1 - \frac{2}{3}\left[\left(\frac{P^2}{mE_0}\right) + \left(\frac{2Q^2}{mE'_0}\right)\right] \tag{2.63}$$

$$\frac{2m}{\hbar^2}B = \frac{2}{3}\left[\left(\frac{-P^2}{mE_0}\right) + \left(\frac{Q^2}{mE'_0}\right)\right] \tag{2.64}$$

$$\left(\frac{2m}{\hbar^2}C\right)^2 = \frac{16P^2Q^2}{3mE_0mE'_0}. \tag{2.65}$$

Equations (2.63–65) show that it is more convenient to define the constants A, B, and C in units of $\hbar^2/2m$. Note that in the literature [2.17] the definitions of A, B, and C may contain a small additional term R, which is the matrix element of the electron momentum operator between the Γ_{4v} valence band and a higher energy Γ_{3c} conduction band. Inclusion of R is particularly important for large bandgap materials such as diamond [2.22]

The dispersion of the $\Gamma_8(J = 3/2)$ bands near the zone center is given by (2.62); this equation has been derived after much simplification and is valid only for energies small compared to the spin–orbit splitting. We note that both A and B are negative since the dominant term in both (2.63) and (2.64) is $2P^2/(3mE_0)$, which is $\gg 1$. As a result, the effective masses of these bands are negative. In many cases we have to consider the properties of a semiconductor in which a few electrons are missing from an otherwise filled valence band. Instead of working with electrons with negative masses, it is more convenient to introduce the idea of a **hole**. A filled valence band with one electron missing can be regarded as a band (known as a **hole band**) containing one hole. If the energy of the missing electron in the valence band is E (assuming that $E = 0$ is the top of the valence band) then the energy of the corresponding hole is $-E$ and is positive. With this definition the effective mass of a hole in the valence band is opposite to that of the corresponding missing electron and is positive also. Since the valence band represented by E_+ has a smaller dispersion and hence larger mass, it is generally referred to as the **heavy hole** band, while the band represented by E_- is known as the **light hole** band. From now on the energies of these two hole bands will be written as E_{hh} and E_{lh} with the corresponding hole energies defined as

$$E_{hh} = -Ak^2 - [B^2k^4 + C^2(k_x^2k_y^2 + k_y^2k_z^2 + k_z^2k_x^2)]^{\frac{1}{2}}, \tag{2.66a}$$

$$E_{lh} = -Ak^2 + [B^2k^4 + C^2(k_x^2k_y^2 + k_y^2k_z^2 + k_z^2k_x^2)]^{\frac{1}{2}}. \tag{2.66b}$$

Constant energy surfaces represented by (2.66a) and (2.66b) are shown in Fig. 2.16. The shapes of these constant energy surfaces are referred to as "warped" spheres. The warping occurs along the [100] and [111] directions because of the cubic symmetry of the zinc-blende crystal. In fact one can argue that these warped spheres are the only possible shapes for constant energy surfaces described by a second-order equation in cubic crystals. Assuming that odd-order terms in k are

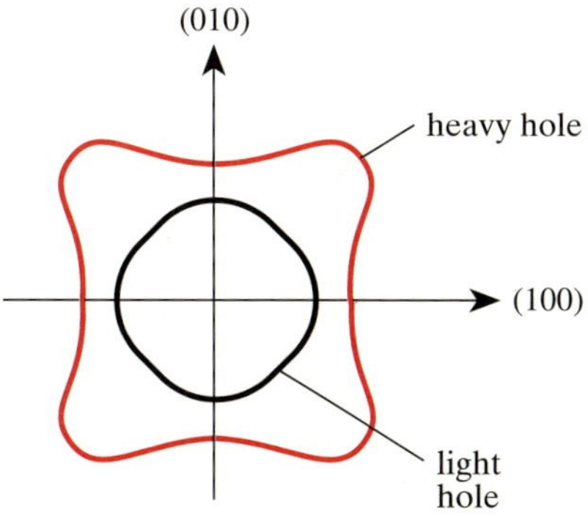

Fig. 2.16. Constant energy surfaces of the $J = 3/2(\Gamma_8)$ bands in diamond- and zinc-blende-type semiconductors

either zero or negligible, the lowest order terms even in k consistent with the cubic symmetry are k^2 and $[\alpha k^4 + \beta(k_x^2 k_y^2 + k_y^2 k_z^2 + k_z^2 k_x^2)]^{\frac{1}{2}}$. If we neglect higher order terms, the most general expression for the k dependence of the energy of the $\Gamma_8(j = 3/2)$ component of a Γ_4 state in a cubic crystal is of the form of (2.62), where A, B, and C are linearly independent parameters related to the electron momentum matrix elements. One may notice from the definitions of the coefficients A, B, and C in (2.63–65) that C can be expressed in terms of A and B. This is a result of neglecting in our model the coupling between the $\Gamma_{4\mathrm{v}}$ bands and higher conduction bands (such as $\Gamma_{3\mathrm{c}}$), for the inclusion of the lowest $\Gamma_{3\mathrm{c}}$ state see Problem 2.15d.

The hole band dispersions along the [100] and [111] directions are parabolic, but the hole effective masses are different along the two directions:

$$k\|(100) \qquad \frac{1}{m_{\mathrm{hh}}} = \frac{2}{\hbar^2}(-A + B), \tag{2.67a}$$

$$\frac{1}{m_{\mathrm{lh}}} = \frac{2}{\hbar^2}(-A - B), \tag{2.67b}$$

$$k\|(111) \qquad \frac{1}{m_{\mathrm{hh}}} = \frac{2}{\hbar^2}\left[-A + B\left(1 + \frac{|C|^2}{3B^2}\right)^{\frac{1}{2}}\right], \tag{2.68a}$$

$$\frac{1}{m_{\mathrm{lh}}} = \frac{2}{\hbar^2}\left[-A - B\left(1 + \frac{|C|^2}{3B^2}\right)^{\frac{1}{2}}\right]. \tag{2.68b}$$

From the above expressions we see that the warping of the valence bands is caused by the term $|C|^2$, which is proportional to Q^2. If the term B^2 is much larger than $|C|^2/3$ warping can be neglected and we can obtain the approximate result that $m_{\mathrm{lh}} \approx 3m_{\mathrm{c}}^*/2$ and $m_{\mathrm{so}} \approx 3m_{\mathrm{c}}^*$. Note that Q^2 is crucial to m_{hh}. If we put $Q^2 = 0$ we obtain the incorrect result $m_{\mathrm{hh}} = -m_0$ (even the sign is wrong!). Often, for simplicity, it is expedient to assume that the valence band masses are isotropic. In such cases average heavy and light hole masses m_{hh}^* and m_{lh}^* can be obtained by averaging (2.67) and (2.68) over all possible directions of $\boldsymbol{k}$ (Problem 4.4):

$$\frac{1}{m^*_{\mathrm{hh}}} = \frac{1}{\hbar^2}\left[-2A + 2B\left(1 + \frac{2|C|^2}{15B^2}\right)\right], \tag{2.69a}$$

$$\frac{1}{m^*_{\mathrm{lh}}} = \frac{1}{\hbar^2}\left[-2A - 2B\left(1 + \frac{2|C|^2}{15B^2}\right)\right]. \tag{2.69b}$$

In Table 2.24 we have listed what we judge to be reliable values of the constants A, B, and $|C|^2$ for several semiconductors obtained from data in [2.18]. In this table the three valence band effective masses calculated from (2.67–69) using these values of A, B, and $|C|^2$ and experimental energy gaps are compared with the experimentally determined effective masses.

We note that the constant energy surfaces for the valence bands as described by (2.62) have inversion symmetry: $E(\boldsymbol{k}) = E(-\boldsymbol{k})$, even though the crystal may not have such symmetry. This is a consequence of the electron Hamiltonian we have used being invariant under time reversal (**time-reversal symmetry**). A Bloch wave traveling with wave vector $\boldsymbol{k}$ is transformed into a Bloch wave with wave vector $-\boldsymbol{k}$ under time reversal. If the Hamiltonian is invariant under time reversal, these two Bloch waves will have the same energy.

Finally we point out that there is an alternate equivalent approach often used in the literature to represent the valence band dispersion in diamond- and zinc-blende-type semiconductors. Using group theory it is possible to derive an effective $\boldsymbol{k} \cdot \boldsymbol{p}$ Hamiltonian which is appropriate for the Γ_4 valence bands. An example of such a Hamiltonian was proposed by *Luttinger* [2.23]:

$$H_{\mathrm{L}} = \frac{\hbar^2}{2m}\left[\left(\gamma_1 + \frac{5}{2}\gamma_2\right)\nabla^2 - 2\gamma_3(\nabla \cdot \boldsymbol{J})^2 + 2(\gamma_3 - \gamma_2)(\nabla_x^2 J_x^2 + \text{c. p.})\right], \tag{2.70}$$

where the parameters γ_1, γ_2, and γ_3 are known as the **Kohn–Luttinger parameters**; $\boldsymbol{J} = (J_x, J_y, J_z)$ is an operator whose effects on the Γ_8 valence bands are identical to those of the angular momentum operator on the $J = 3/2$ atomic states, and c. p. stands for cyclic permutations. This approach facilitates the diagonalization of H_{L} together with additional perturbations applied to the crystal. In Chap. 4 we will see an application of this Hamiltonian to calculate the energies of acceptor states. The first two terms in (2.70) have spherical symmetry while the last represents the effect of the lower, cubic symmetry. It is thus clear that the warping of the valence band is directly proportional to the difference between γ_2 and γ_3. The Kohn–Luttinger parameters can be shown to be related to the coefficients A, B, and C in (2.62) by

$$(\hbar^2/2m)\gamma_1 = -A \tag{2.71a}$$

$$(\hbar^2/2m)\gamma_2 = -B/2 \tag{2.71b}$$

$$(\hbar^2/2m)\gamma_3 = \left[(B^2/4) + (C^2/12)\right]^{1/2} \tag{2.71c}$$

The proof of these results is left as an exercise (Problem 2.15).

2.7 Tight-Binding or LCAO Approach to the Band Structure of Semiconductors

The pseudopotential approach to calculating the band structure of semiconductors discussed in Sect. 2.5 starts with the assumption that electrons are nearly free and their wave functions can be approximated by plane waves. In this section we will approach the problem from the other extreme. We will assume that the electrons are tightly bound to their nuclei as in the atoms. Next we will bring the atoms together. When their separations become comparable to the lattice constants in solids, their wave functions will overlap. We will approximate the electronic wave functions in the solid by linear combinations of the atomic wave functions. This approach is known as the **tight-binding** approximation or **Linear Combination of Atomic Orbitals** (LCAO) approach. One may ask: how can two completely opposite approaches such as the pseudopotential method and the tight-binding method both be good starting points for understanding the electronic properties of the same solid? The answer is that in a covalently bonded semiconductor there are really two kinds of electronic states. Electrons in the conduction bands are delocalized and so can be approximated well by nearly free electrons. The valence electrons are concentrated mainly in the bonds and so they retain more of their atomic character. The valence electron wave functions should be very similar to **bonding orbitals** found in molecules. In addition to being a good approximation for calculating the valence band structure, the LCAO method has the advantage that the band structure can be defined in terms of a small number of **overlap parameters**. Unlike the pseudopotentials, these overlap parameters have a simple physical interpretation as representing interactions between electrons on adjacent atoms.

2.7.1 Molecular Orbitals and Overlap Parameters

To illustrate the tight-binding approach for calculating band structures, we will restrict ourselves again to the case of tetrahedrally bonded semiconductors. The valence electrons in the atoms of these semiconductors are in s and p orbitals. These orbitals in two identical and isolated atoms are shown schematically in Figs. 2.17a, 2.18a, and 2.19a. The p_z orbitals are not shown since

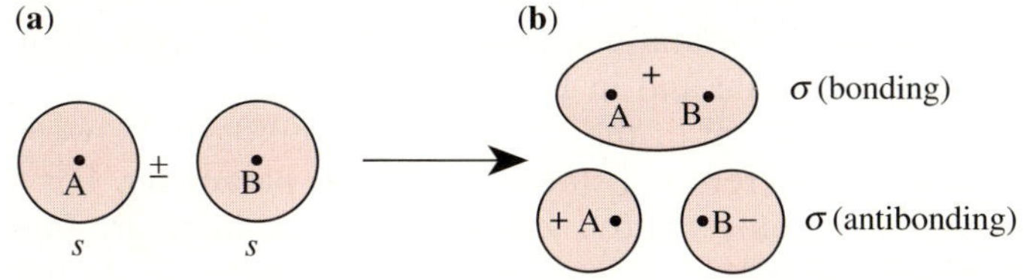

Fig. 2.17a,b. Overlap of two s orbitals to form bonding and antibonding σ orbitals

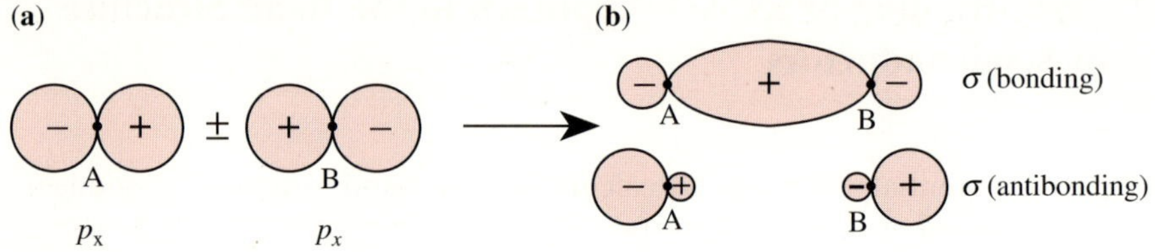

Fig. 2.18a,b. Overlap of two p_x orbitals along the x axis to form bonding and antibonding σ orbitals

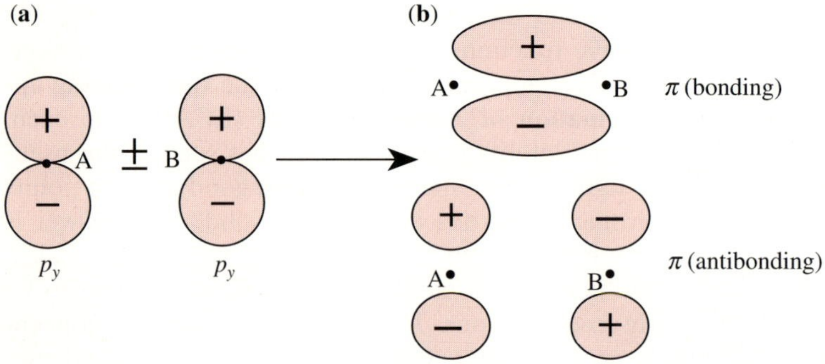

Fig. 2.19a,b. Overlap of two p_y orbitals to form bonding and antibonding π orbitals

their properties are similar to those of the p_y orbitals. Figures 2.17b, 2.18b, and 2.19b show schematically what happens to the atomic orbitals when the two atoms are brought together along the x direction until the atomic orbitals overlap to form a diatomic molecule. The interaction between the two atomic orbitals produces two new orbitals. One of the resultant orbitals is symmetric with respect to the interchange of the two atoms and is known as the **bonding orbital** while the other orbital, which is antisymmetric, is known as the **antibonding orbital**. In the case of p orbitals there are two ways for them to overlap. When they overlap along the direction of the p orbitals, as shown in Fig. 2.18b, they are said to form σ **bonds**. When they overlap in a direction perpendicular to the p orbitals they are said to form π **bonds**, as shown in Fig. 2.19b.

The interaction between the atomic orbitals changes their energies. Typically the antibonding orbital energy is *raised* by an amount determined by the interaction Hamiltonian H. The energy of the bonding orbital is *decreased* by the same amount. The changes in orbital energies are shown schematically in Fig. 2.20a for a homopolar molecule and in Fig. 2.20b for a heteropolar one. In both cases V is the matrix element of the interaction Hamiltonian between the atomic orbitals and is usually referred to as the **overlap parameter**. For a homopolar molecule containing only s and p valence electrons, there are four nonzero overlap parameters. To derive this result we will denote the atomic orbital on one of the atoms as $|\alpha\rangle$ and that on the second atom as $|\beta\rangle$. These

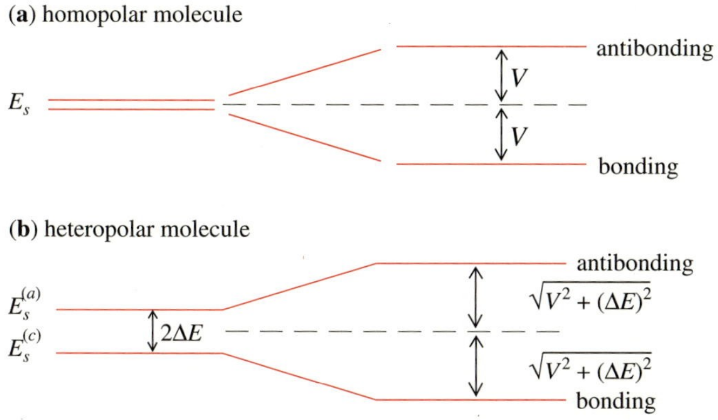

Fig. 2.20. Effect of orbital overlap on the energy levels in (**a**) a diatomic homopolar molecule and (**b**) a diatomic heteropolar molecule. V represents the matrix element of the interaction Hamiltonian

orbitals can be expressed as products of a radial wave function and a spherical harmonic $Y_{lm}(\theta, \phi)$ with the atom chosen as the origin. We will denote the vector going from the first atom (designated as A in Fig. 2.21) to the second atom (B) as $\boldsymbol{d}$. For both orbitals $|\alpha\rangle$ and $|\beta\rangle$ we will choose the coordinate axes such that the z axes are parallel to $\boldsymbol{d}$ and the azimuthal angles ϕ are the same (see Fig. 2.21). In these coordinate systems the spherical harmonic wave functions of the two atoms A and B are $Y_{lm}(\theta, \phi)$ and $Y_{l'm'}(\theta', \phi)$, respectively. The Hamiltonian H has cylindrical symmetry with respect to $\boldsymbol{d}$ and therefore cannot depend on ϕ. Thus the matrix element $\langle\alpha\,|\,H\,|\,\beta\rangle$ is proportional to the integral of the azimuthal wave functions $\exp[\mathrm{i}(m' - m)\phi]$. This integral vanishes except when $m = m'$. As a result of this selection rule we conclude that

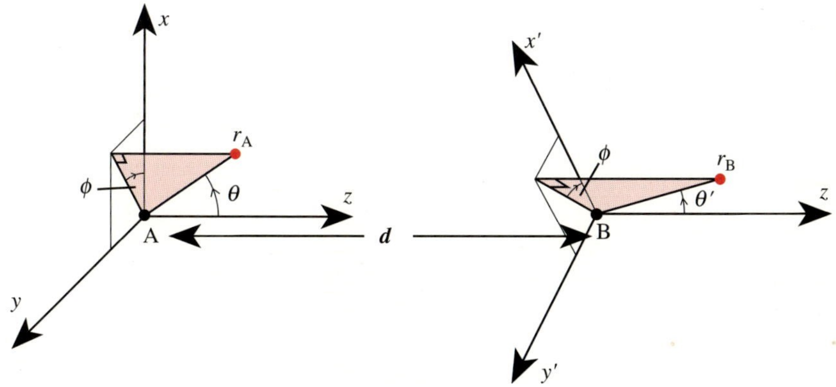

Fig. 2.21. Choice of the polar coordinate systems for the two atoms A and B in a diatomic molecule in order that the z axis be parallel to the vector joining the two atoms A and B and the azimuthal angle ϕ be identical for both atoms

there are four nonzero and linearly independent overlap parameters between the s and p electrons:

$$\langle s | H | s \rangle = V_{ss\sigma}; \ \langle s | H | p_z \rangle = V_{sp\sigma}; \ \langle p_z | H | p_z \rangle = V_{pp\sigma};$$
$$\text{and } \langle p_x | H | p_x \rangle = V_{pp\pi}.$$

We notice that $\langle p_x | H | p_y \rangle = 0$ and $\langle p_y | H | p_y \rangle = \langle p_x | H | p_x \rangle$ as a result of symmetry. The overlap parameters are usually labeled σ, π and δ for ($l = 2$ wave functions), depending on whether m is 0, 1, or 2 (in analogy with the s, p, and d atomic wave functions).

The concept of bonding and antibonding orbitals introduced for molecules can be easily extended to crystals if one assumes that the orbitals of each atom in the crystal overlap with those of its nearest neighbors only. This is a reasonable approximation for most solids. The results of orbital overlap in a solid is that the bonding and antibonding orbitals are broadened into bands. Those occupied by electrons form valence bands while the empty ones form conduction bands. Figure 2.22 shows schematically how the s and p orbitals evolve into bands in a tetrahedral semiconductor. In this case the bonding orbitals are filled with electrons and become the valence bands while the antibonding orbitals become the conduction bands. As may be expected, the crystal structure affects the overlap between atomic orbitals. For example, in a tetrahedrally coordinated solid each atom is surrounded by four nearest neighbors. The vectors $\boldsymbol{d}$ linking the central atom to each of its nearest neighbors are different, so it is not convenient to choose the z axis parallel to $\boldsymbol{d}$. Instead it is more convenient to choose the crystallographic axes as the coordinate axes. The spherical harmonics $Y_{lm}(\theta,\phi)$ of the atomic orbitals are then defined with respect to this fixed coordinate system. In calculating the overlap parameter for any pair of neighboring atoms, one expands the spherical harmonics defined with respect to $\boldsymbol{d}$ in terms of $Y_{lm}(\theta,\phi)$. An example of this expansion is shown schematically in Fig. 2.23.

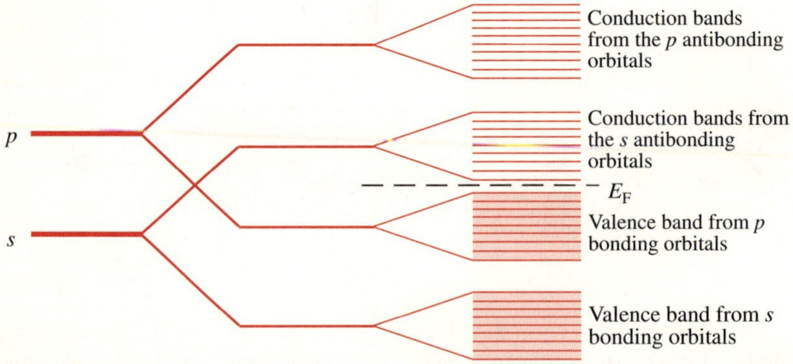

Fig. 2.22. Evolution of the atomic s and p orbitals into valence and conduction bands in a semiconductor. E_F is the Fermi energy

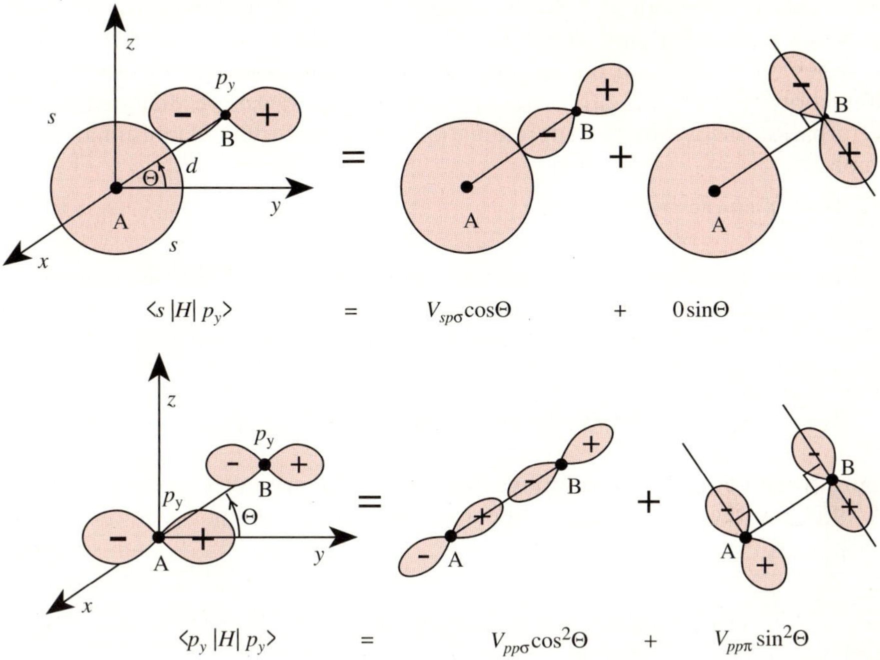

Fig. 2.23. Projection of the overlap parameter between an s and a p_y orbital, and between p_y orbitals, along the vector $\boldsymbol{d}$ joining the two atoms and perpendicular to $\boldsymbol{d}$

2.7.2 Band Structure of Group-IV Elements by the Tight-Binding Method

After this introduction to the interaction between atomic orbitals we are ready to perform a quantitative calculation of the electronic band structure using the method of Linear Combination of Atomic Orbitals (LCAO). While the method has been utilized by many authors [Ref. 2.25, p. 75], the approach we will describe follows that of *Chadi* and *Cohen* [2.25].

The position of an atom in the primitive cell denoted by j will be decomposed into $\boldsymbol{r}_{jl} = \boldsymbol{R}_j + \boldsymbol{r}_l$, where $\boldsymbol{R}_j$ denotes the position of the jth primitive cell of the Bravais lattice and $\boldsymbol{r}_l$ is the position of the atom l within the primitive cell. For the diamond and zinc-blende crystals $l = 1$ and 2 only. Let $h_l(\boldsymbol{r})$ denote the Hamiltonian for the isolated atom l with its nucleus chosen as the origin. The Hamiltonian for the atom located at $\boldsymbol{r}_{jl}$ will be denoted by $h_l(\boldsymbol{r} - \boldsymbol{r}_{jl})$. The wave equation for h_l is given by

$$h_l \phi_{ml}(\boldsymbol{r} - \boldsymbol{r}_{jl}) = E_{ml} \phi_{ml}(\boldsymbol{r} - \boldsymbol{r}_{jl}), \tag{2.72}$$

where E_{ml} and ϕ_{ml} are the eigenvalues and eigenfunctions of the state indexed by m. The atomic orbitals $\phi_{ml}(\boldsymbol{r} - \boldsymbol{r}_{jl})$ are known as **Löwdin orbitals**. They are different from the usual atomic wave functions in that they have been constructed in such a way that wave functions centered at different atoms are orthogonal to each other. Next we assume that the Hamiltonian for the crystal $\mathcal{H}$ is equal to the sum of the atomic Hamiltonians and a term $\mathcal{H}_{\mathrm{int}}$ which describes the interaction between the different atoms. We further assume the interaction between the atoms to be weak so that $\mathcal{H}$ can be diagonalized by perturbation theory. In this approximation the unperturbed Hamiltonian $\mathcal{H}_0$ is simply

$$\mathcal{H}_0 = \sum_{j,l} h_l(\boldsymbol{r} - \boldsymbol{r}_{jl}) \tag{2.73}$$

and we can construct the unperturbed wave functions as linear combinations of the atomic wave functions. Because of the translational symmetry of the crystal, these unperturbed wave functions can be expressed in the form of Bloch functions:

$$\Phi_{ml\boldsymbol{k}} = \frac{1}{\sqrt{N}} \sum_j \exp\left(\mathrm{i}\boldsymbol{r}_{jl} \cdot \boldsymbol{k}\right) \phi_{ml}(\boldsymbol{r} - \boldsymbol{r}_{jl}), \tag{2.74}$$

where N is the number of primitive unit cells in the crystal. The eigenfunctions $\Psi_{\boldsymbol{k}}$ of $\mathcal{H}$ can then be written as linear combinations of $\Phi_{ml\boldsymbol{k}}$:

$$\Psi_{\boldsymbol{k}} = \sum_{m,l} C_{ml} \Phi_{ml\boldsymbol{k}}. \tag{2.75}$$

To calculate the eigenfunctions and eigenvalues of $\mathcal{H}$, we operate on $\Psi_{\boldsymbol{k}}$ with the Hamiltonian $\mathcal{H} = \mathcal{H}_0 + \mathcal{H}_{\mathrm{int}}$. From the orthogonality of the Bloch functions we obtain a set of linear equations in C_{ml}:

$$\sum_{m,l} \left(H_{ml,m'l'} - E_{\boldsymbol{k}} \delta_{mm'} \delta_{ll'} \right) C_{m'l'}(\boldsymbol{k}) = 0, \tag{2.76}$$

where $H_{ml,m'l'}$ stands for the matrix element $\langle \Phi_{ml\boldsymbol{k}} | \mathcal{H} | \Phi_{m'l'\boldsymbol{k}} \rangle$ and $E_{\boldsymbol{k}}$ are the eigenvalues of H. To simplify the solution of (2.76) we introduce the following approximations.

- We include only the s^2 and p^6 electrons in the outermost partially filled atomic shells. We neglect spin–orbit coupling (although it can be included easily). The two atomic orbitals of s symmetry for the two atoms in the unit cell will be denoted by $S1$ and $S2$, respectively. Correspondingly, the atomic orbitals with p symmetry will be denoted by: $X1$, $X2$, $Y1$, $Y2$, $Z1$ and $Z2$, respectively. In the following equations the index m will represent the s, p_x, p_y, and p_z orbitals.
- When we substitute the wave functions $\Phi_{ml\boldsymbol{k}}$ defined in (2.74) into (2.76) we obtain

$$H_{ml,m'l'}(\boldsymbol{k}) = \sum_j^N \sum_{j'}^N \frac{\exp\left[\mathrm{i}(\boldsymbol{r}_{jl} - \boldsymbol{r}_{j'l'}) \cdot \boldsymbol{k}\right]}{N} \times \langle \phi_{ml} | (\boldsymbol{r} - \boldsymbol{r}_{jl}) | H | \phi_{m'l'}(\boldsymbol{r} - \boldsymbol{r}_{j'l'}) \rangle \tag{2.77}$$

$$= \sum_j^N \exp\left[\mathrm{i}(\boldsymbol{R}_j + \boldsymbol{r}_l - \boldsymbol{r}_{l'}) \cdot \boldsymbol{k}\right] \times \langle \phi_{ml}(\boldsymbol{r} - \boldsymbol{r}_{jl}) | H | \phi_{m'l'}(\boldsymbol{r} - \boldsymbol{r}_{jl'}) \rangle. \tag{2.78}$$

Instead of summing j over all the unit cells in the crystal, we will sum over the nearest neighbors only. In the diamond and zinc-blende crystals this means j will be summed over the atom itself plus four nearest neighbors. These atoms will be denoted as $j = 1, 2, 3, 4, 5$. If needed, one can easily include second neighbor or even further interactions.

Within the above approximation the collection of matrix elements of the form in (2.72) constitutes an 8×8 matrix (note that the dimensions of the matrix depend only on the number of basis functions, not the number of neighbors included). Applying symmetry arguments allows the number of nonzero and linearly independent matrix elements of $\mathcal{H}_{\text{int}}$ to be greatly reduced. As an example, we will consider the matrix element $H_{s1,s2}$. From (2.78) this matrix element is given by

$$H_{s1,s2} = [\exp(\mathrm{i}\boldsymbol{k}\cdot\boldsymbol{d}_1) + \exp(\mathrm{i}\boldsymbol{k}\cdot\boldsymbol{d}_2) + \exp(\mathrm{i}\boldsymbol{k}\cdot\boldsymbol{d}_3) + \exp(\mathrm{i}\boldsymbol{k}\cdot\boldsymbol{d}_4)] \times \langle S1 \,|\, \mathcal{H}_{\text{int}} \,|\, S2\rangle, \tag{2.79}$$

where we have assumed that atom 1 is located at the origin and $\boldsymbol{d}_\alpha$ ($\alpha = 1$ to 4) are the positions of its four nearest neighbors, with

$$\boldsymbol{d}_1 = (1, 1, 1)(a/4);$$
$$\boldsymbol{d}_2 = (1, -1, -1)(a/4);$$
$$\boldsymbol{d}_3 = (-1, 1, -1)(a/4);$$

and

$$\boldsymbol{d}_4 = (-1, -1, 1)(a/4).$$

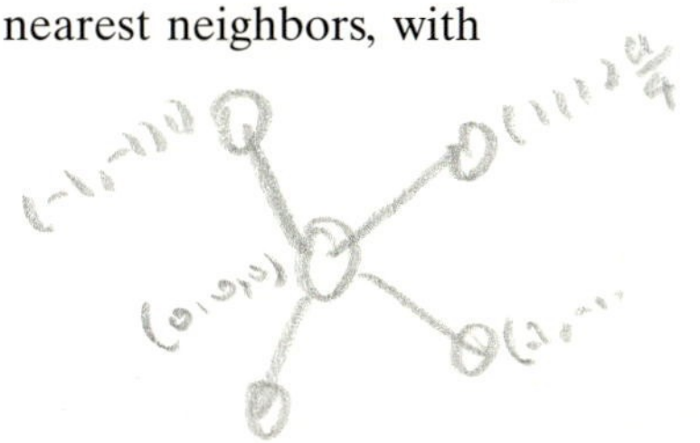

The matrix element $\langle S1 \,|\, \mathcal{H}_{\text{int}} \,|\, S_2\rangle$ is basically the same overlap parameter $V_{ss\sigma}$ as we have defined for molecules. The other matrix elements H_{s1,p_x2}, and H_{p_x1,p_x2}, etc., can also be expressed in terms of the overlap parameters $V_{sp\sigma}$, $V_{pp\sigma}$, and $V_{pp\pi}$. For example H_{s1,p_x2} can be shown to contain four terms involving the four phase factors $\exp(\mathrm{i}\boldsymbol{k}\cdot\boldsymbol{d}_\alpha)$ and the matrix element $\langle S1 \,|\, \mathcal{H}_{\text{int}} \,|\, X2\rangle$. However, for each nearest neighbor $\langle S1 \,|\, \mathcal{H}_{\text{int}} \,|\, X2\rangle$ has to be decomposed into σ and π components as shown in Fig. 2.23. This decomposition introduces a factor of $\cos\Theta = \pm(1/\sqrt{3})$. The $-$ or $+$ sign depends on whether the s orbital lies in the direction of the positive or negative lobe of the p_x orbital. As a result, it is convenient to introduce a new set of four overlap parameters appropriate for the diamond lattice:

$$V_{ss} = 4V_{ss\sigma}, \tag{2.80a}$$

$$V_{sp} = 4V_{sp\sigma}/\sqrt{3}, \tag{2.80b}$$

$$V_{xx} = (4V_{pp\sigma}/3) + (8V_{pp\pi}/3), \tag{2.80c}$$

$$V_{xy} = (4V_{pp\sigma}/3) - (4V_{pp\pi}/3), \tag{2.80d}$$

With this notation the matrix element $\langle S1(\boldsymbol{r}) \,|\, \mathcal{H}_{\text{int}} \,|\, X2(\boldsymbol{r} - \boldsymbol{d}_1)\rangle$ is given by $(V_{sp\sigma})/\sqrt{3} = V_{sp}/4$. The remaining three matrix elements are related to $\langle S1(\boldsymbol{r}) \,|\, \mathcal{H}_{\text{int}} \,|\, X2(\boldsymbol{r} - \boldsymbol{d}_1)\rangle$ by symmetry. For example, a two-fold rotation about the y axis will transform (x,y,z) into $(-x,y,-z)$, so $\boldsymbol{d}_1$ is transformed into $\boldsymbol{d}_3$. The s-symmetry wave function $|S1\rangle$ is unchanged while the p-symmetry wave function $|X2\rangle$ is transformed into $-|X2\rangle$ under this rotation. As a result, $\langle S1(\boldsymbol{r})|\mathcal{H}_{\text{int}}|X2(\boldsymbol{r}-\boldsymbol{d}_3)\rangle = -\langle S1(\boldsymbol{r})|\mathcal{H}_{\text{int}}|X2(\boldsymbol{r}-\boldsymbol{d}_1)\rangle$. By applying similar symmetry operations we can show that

$$\sum_\alpha \exp[\mathrm{i}(\boldsymbol{d}_\alpha\cdot\boldsymbol{k})]\langle S_1(\boldsymbol{r})|\mathcal{H}_{\text{int}}|X2(\boldsymbol{r}-\boldsymbol{d}_\alpha)\rangle = \tfrac{1}{4}V_{sp}\{\exp[\mathrm{i}(\boldsymbol{d}_1\cdot\boldsymbol{k})] + \exp[\mathrm{i}(\boldsymbol{d}_2\cdot\boldsymbol{k})] - \exp[\mathrm{i}\boldsymbol{d}_3\cdot\boldsymbol{k})] - \exp[\mathrm{i}(\boldsymbol{d}_4\cdot\boldsymbol{k})]\} \quad (2.81)$$

In the zinc-blende structure, because the atoms 1 and 2 are different, $\langle S1|\mathcal{H}_{\text{int}}|X2\rangle$ is, in principle, different from $\langle S2|\mathcal{H}_{\text{int}}|X1\rangle$. They are, however, often assumed to be equal [Ref. 2.24, p. 77]. The case of the zinc-blende crystal is left as an exercise in Problem 2.15. Here we will restrict ourselves to the case of the diamond structure.

The 8×8 matrix for the eight s and p bands can be expressed as in Table 2.25. E_s and E_p represent the energies $\langle S1|\mathcal{H}_0|S1\rangle$ and $\langle X1|H_0|X1\rangle$, respectively. The four parameters g_1 to g_4 arise from summing over the factor $\exp[\mathrm{i}(\boldsymbol{k}\cdot\boldsymbol{d}_\alpha)]$ as in (2.81). They are defined by

$$\begin{aligned}
g_1 &= (1/4)\{\exp[\mathrm{i}(\boldsymbol{d}_1\cdot\boldsymbol{k})] + \exp[\mathrm{i}(\boldsymbol{d}_2\cdot\boldsymbol{k})] + \exp[\mathrm{i}(\boldsymbol{d}_3\cdot\boldsymbol{k})] + \exp[\mathrm{i}(\boldsymbol{d}_4\cdot\boldsymbol{k})]\},\\
g_2 &= (1/4)\{\exp[\mathrm{i}(\boldsymbol{d}_1\cdot\boldsymbol{k})] + \exp[\mathrm{i}(\boldsymbol{d}_2\cdot\boldsymbol{k})] - \exp[\mathrm{i}(\boldsymbol{d}_3\cdot\boldsymbol{k})] - \exp[\mathrm{i}(\boldsymbol{d}_4\cdot\boldsymbol{k})]\},\\
g_3 &= (1/4)\{\exp[\mathrm{i}(\boldsymbol{d}_1\cdot\boldsymbol{k})] - \exp[\mathrm{i}(\boldsymbol{d}_2\cdot\boldsymbol{k})] + \exp[\mathrm{i}(\boldsymbol{d}_3\cdot\boldsymbol{k})] - \exp[\mathrm{i}(\boldsymbol{d}_4\cdot\boldsymbol{k})]\},\\
g_4 &= (1/4)\{\exp[\mathrm{i}(\boldsymbol{d}_1\cdot\boldsymbol{k})] - \exp[\mathrm{i}(\boldsymbol{d}_2\cdot\boldsymbol{k})] - \exp[\mathrm{i}(\boldsymbol{d}_3\cdot\boldsymbol{k})] + \exp[\mathrm{i}(\boldsymbol{d}_4\cdot\boldsymbol{k})]\}.
\end{aligned}$$

If $\boldsymbol{k} = (2\pi/a)(k_1,k_2,k_3)$ the g_j's can also be expressed as

$$\begin{aligned} g_1 = {} & \cos(k_1\pi/2)\cos(k_2\pi/2)\cos(k_3\pi/2)\\ & -\mathrm{i}\sin(k_1\pi/2)\sin(k_2\pi/2)\sin(k_3\pi/2), \end{aligned} \quad (2.82\text{a})$$

$$\begin{aligned} g_2 = {} & -\cos(k_1\pi/2)\sin(k_2\pi/2)\sin(k_3\pi/2)\\ & +\mathrm{i}\sin(k_1\pi/2)\cos(k_2\pi/2)\cos(k_3\pi/2), \end{aligned} \quad (2.82\text{b})$$

Table 2.25. Matrix for the eight s and p bands in the diamond structure within the tight binding approximation

	S1	S2	X1	Y1	Z1	X2	Y2	Z2
S1	E_s-E_k	$V_{ss}g_1$	0	0	0	$V_{sp}g_2$	$V_{sp}g_3$	$V_{sp}g_4$
S2	$V_{ss}g_1^*$	E_s-E_k	$-V_{sp}g_2^*$	$-V_{sp}g_3^*$	$-V_{sp}g_4^*$	0	0	0
X1	0	$-V_{sp}g_2$	E_p-E_k	0	0	$V_{xx}g_1$	$V_{xy}g_4$	$V_{xy}g_3$
Y1	0	$-V_{sp}g_3$	0	E_p-E_k	0	$V_{xy}g_4$	$V_{xx}g_1$	$V_{xy}g_2$
Z1	0	$-V_{sp}g_4$	0	0	E_p-E_k	$V_{xy}g_3$	$V_{xy}g_2$	$V_{xx}g_1$
X2	$V_{sp}g_2^*$	0	$V_{xx}g_1^*$	$V_{xy}g_4^*$	$V_{xy}g_3^*$	E_p-E_k	0	0
Y2	$V_{sp}g_3^*$	0	$V_{xy}g_4^*$	$V_{xx}g_1^*$	$V_{xy}g_2^*$	0	E_p-E_k	0
Z2	$V_{sp}g_4^*$	0	$V_{xy}g_3^*$	$V_{xy}g_2^*$	$V_{xx}g_1^*$	0	0	E_p-E_k

$$g_3 = -\sin(k_1\pi/2)\cos(k_2\pi/2)\sin(k_3\pi/2) + \mathrm{i}\cos(k_1\pi/2)\sin(k_2\pi/2)\cos(k_3\pi/2), \tag{2.82c}$$

$$g_4 = -\sin(k_1\pi/2)\sin(k_2\pi/2)\cos(k_3\pi/2) + \mathrm{i}\cos(k_1\pi/2)\cos(k_2\pi/2)\sin(k_3\pi/2), \tag{2.82d}$$

The valence and lowest conduction band energies of the diamond-type crystals can be obtained by diagonalizing the 8×8 matrix of Table 2.25, provided the four parameters V_{ss}, V_{sp}, V_{xx}, and V_{xy} are known. These four parameters can be determined by comparing the calculated band structure with a first principles or empirical band structure calculation. For example *Chadi and Cohen* [2.25] obtained the tight-binding parameters for C, Si, and Ge by comparison with empirical pseudopotential calculations. Their results are shown in Table 2.26. Note that the signs of V_{ss} etc. are, in part, arbitrary and are determined by the choice of the relative phases of the two overlaping atomic orbitals. The signs in Table 2.26 correspond to the choices shown in Figs. 2.17a and 2.23. The magnitudes of the interaction parameters decrease in the sequence C to Ge. We will show later that this trend can be understood from the increase in the lattice constant along this sequence. When the second-nearest neighbor interactions are included, only V_{xx} decreases somewhat. Since V_{xx} is the smallest interaction, the overall band structure is not significantly affected.

Table 2.26. Tight-binding interaction parameters (in eV) for C, Si, and Ge obtained by *Chadi* and *Cohen* [2.25] when only nearest-neighbor interactions are included

	$E_p - E_s$	V_{ss}	V_{sp}	V_{xx}	V_{xy}
C	7.40	−15.2	10.25	3.0	8.3
Si	7.20	−8.13	5.88	3.17	7.51
Ge	8.41	−6.78	5.31	2.62	6.82

To gain some insight into the band structure obtained with the tight-binding approach, we will calculate the band energies at the $k = 0$ point. From (2.82a–d) we find $g_2 = g_3 = g_4 = 0$ and $g_1 = 1$ at $k = 0$. Thus the 8×8 matrix simplifies into a 2×2 matrix for the s electrons and three identical 2×2 matrices for the p levels:

$$\begin{vmatrix} E_s - E(0) & V_{ss} \\ V_{ss} & E_s - E(0) \end{vmatrix} \tag{2.83a}$$

and

$$\begin{vmatrix} E_p - E(0) & V_{xx} \\ V_{xx} & E_p - E(0) \end{vmatrix}. \tag{2.83b}$$

These two matrices can be easily diagonalized to yield four energies:

$$E_{s\pm}(0) = E_s \pm |V_{ss}| \tag{2.84a}$$

and

$$E_{p\pm}(0) = E_p \pm |V_{xx}| \quad (2.84b)$$

As a result of the overlap of the atomic orbitals the two s and p levels of the two atoms inside the primitive cell are split by an amount equal to $2|V_{ss}|$ and $2|V_{xx}|$, respectively. The level E_{s+} is raised in energy and its wave function is antisymmetric with respect to the interchange of the two atoms. This state corresponds to the antibonding s state in a diatomic molecule. The level E_{s-} corresponds to the bonding s state. From Table 2.5 we expect the antisymmetric antibonding state to have $\Gamma_{2'}$ symmetry and the symmetric bonding state to have Γ_1 symmetry. Using a similar analogy, the triply degenerate antisymmetric Γ_{15} conduction band states correspond to the antibonding p orbitals while the symmetric $\Gamma_{25'}$ valence band states are identified with the bonding p orbitals.

In Fig. 2.24 the valence band structure of Si calculated by the tight-binding method is compared with that obtained by the empirical pseudopotential method. Figure 2.24 also compares the valence band density of states obtained by the two methods (We will define density of states of a band in Sect. 4.3.1 and also in Chap. 8, where this concept will be utilized). In this tight-binding calculation one second-nearest-neighbor interaction has been included in addition to the nearest-neighbor interactions. The agreement between the two methods is quite good for the valence bands. Figure 2.25 shows a comparison between the band structure of Ge calculated by the tight-binding method, the empirical pseudopotential method, and the nearly free electron model. While the valence bands are well reproduced by the tight-binding method with the simple sp^3 base used here, this is not true for the conduction bands since the

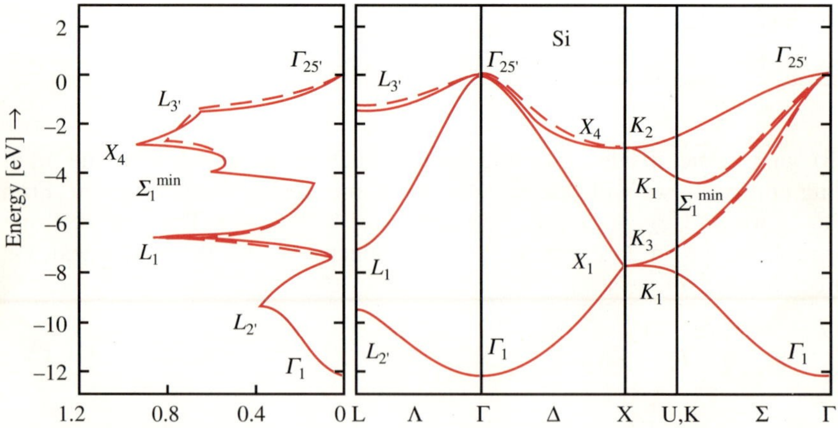

Fig. 2.24. The valence band structure and density of states (see Sect. 4.3.1 for definition) of Si calculated by the tight-binding method (*broken curves*) and by the empirical pseudopotential method (*solid lines*) [2.25]

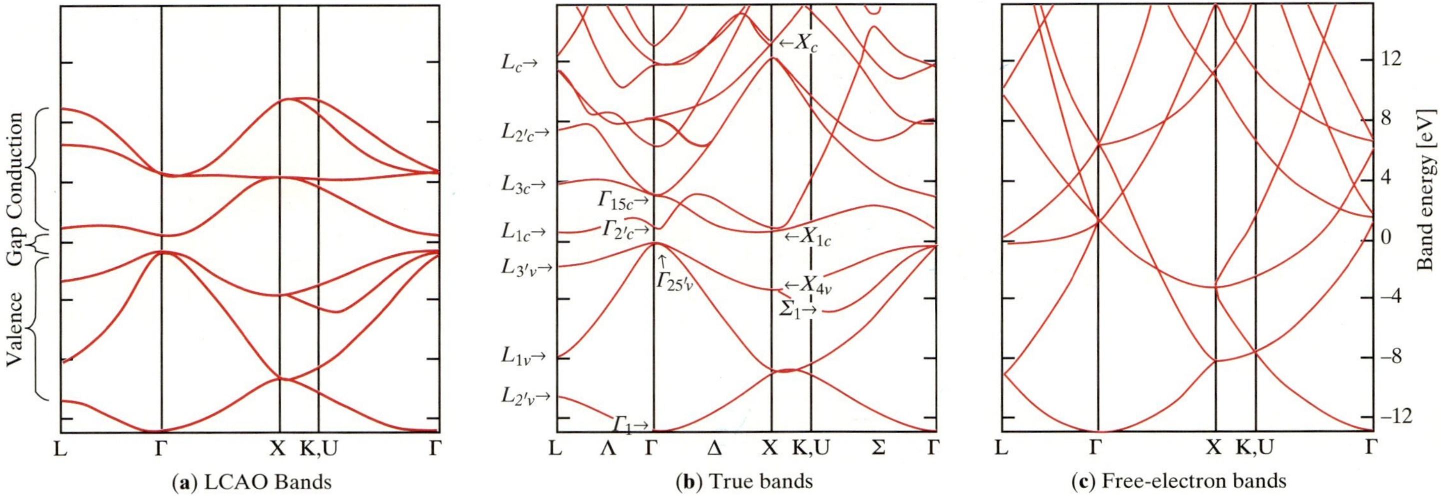

Fig. 2.25. A comparison between the band structure of Ge calculated by (**a**) the tight-binding method, (**b**) the empirical pseudopotential method, and (**c**) the nearly free electron model [Ref. 2.24, p. 79]

conduction band electrons are more delocalized. The accuracy of the conduction bands in the tight-binding calculations can be improved by introducting additional overlap parameters. However, there is another shortcoming in the tight-binding model presented here. There are only four conduction bands in this model because we have included only four *s* and *p* orbitals. To correct this problem additional orbitals and overlap parameters are required; unfortunately they destroy the simplicity of this model.

2.7.3 Overlap Parameters and Nearest-Neighbor Distances

So far we have shown that the advantage of the tight-binding approach is that the valence band structures of semiconductors can be calculated in terms of a small number of atomic energies and overlap parameters. Now we will demonstrate that these overlap parameters in different semiconductors can be expressed as a simple function of the nearest-neighbor distance multiplied by a geometric factor. These results combined make the tight-binding method very powerful for predicting the properties of many compounds (not just semiconductors) with only a small number of parameters [Ref. 2.24, p. 49].

One may expect some relationship between the overlap parameters and the interatomic distance based on the following simple argument. Figures 2.20 and 2.22 show that the atomic energy levels broaden into bands due to overlap of the atomic orbitals. The width of the band is essentially $2V$, where V is the relevant overlap parameter. At the same time the electron wave functions become delocalized over a distance given by the nearest-neighbor separation (i. e., the bond length) d as a result of this overlap. Using the uncertainty principle the momentum of the delocalized electron is estimated to be $(\hbar\pi/d)$, so the electron kinetic energy is given by $\hbar^2\pi^2/(2md^2)$. This result suggests that the overlap parameters depend on d as d^{-2}. This simple heuristic argument can be made more rigorous by comparing the band structures calculated by the tight-binding method and by the nearly free electron model. As an example, we will consider the lowest energy valence band in a crystal with the simple cubic structure. This band can be identified with the bonding *s* orbitals and its dispersion along the [100] direction can be shown to be given by $E_s - 4V_{ss\sigma} - 2V_{ss\sigma}\cos kx$ (Problem 2.16). Thus the width of this band is equal to $4V_{ss\sigma}$. On the other hand the nearly free electron model gives the band width as $\hbar^2\pi^2/(2md^2)$. Equating the band widths obtained by these two different methods we get

$$4V_{ss\sigma} = \frac{\hbar^2\pi^2}{2md^2}. \tag{2.85}$$

In general, all four overlap parameters for the *s* and *p* orbitals can be expressed in the form

$$V_{ll'm} = \tau_{ll'm}\frac{\hbar^2}{md^2} \tag{2.86}$$

where $\tau_{ll'm}$ is a factor which depends on the crystal symmetry. From (2.85) we see that $\tau_{ss\sigma} = \pi^2/8$ in crystals with the simple cubic structure. Table 2.27 lists the values of $\tau_{ll'm}$ for the simple cubic and diamond lattices.

For the diamond and zinc-blende crystals, *Harrison* [2.26] has treated the factors $\tau_{ll'm}$ as adjustable parameters in fitting the energy bands of Si and Ge. He found excellent agreement between the calculated values and the adjusted values for three of the parameters. The only exception is $\tau_{pp\pi}$, where the fitted value of -0.81 is somewhat lower than the calculated one.

Table 2.27. The geometric factor $\tau_{ll'm}$ relating the overlap parameters for the s and p bands to the free electron band width $\hbar^2/(md^2)$ as shown in (2.86). The last column represents the adjusted values obtained by fitting the energy bands of Si and Ge [Ref. 2.24, p. 49]

	Simple cubic	Diamond and zinc–blende	Adjusted values
$\tau_{ss\sigma}$	$-\pi^2/8 = -1.23$	$-9\pi^2/64 = -1.39$	-1.40
$\tau_{sp\sigma}$	$(\pi/2)[(\pi^2/4) - 1]^{1/2} = 1.90$	$(9\pi^2/32)[1 - (16/3\pi^2)]^{1/2} = 1.88$	1.84
$\tau_{pp\sigma}$	$3\pi^2/8 = 3.70$	$21\pi^2/64 = 3.24$	3.24
$\tau_{pp\pi}$	$-\pi^2/8 = -1.23$	$-3\pi^2/32 = -0.93$	-0.81

Table 2.27 together with (2.86) and the lattice constants are all that is needed to calculate the overlap parameters for computing the valence bands and the lowest conduction bands in many zinc-blende- and diamond-type semiconductors. Even without any detailed calculations, we can understand qualitatively the symmetries of the conduction and valence bands at the Brillouin zone center of the three group-IV elements Si, Ge, and gray tin (or α-Sn). The lattice constant increases from Si to α-Sn. This results in a decrease in the overlap parameters $|V_{ss}|$ and $|V_{xx}|$ (the variation from C to Ge is shown in Table 2.26). The decrease is larger for $|V_{ss}|$ than for $|V_{xx}|$. As a result, the ordering of the s and p orbitals changes from Si to α-Sn in the manner shown in Fig. 2.26. The Fermi level is located by filling the bands with the eight valence electrons available. In this way it is easily seen that the lowest conduction band at zone center in Si is p-like while the corresponding band in Ge is s-like. In this scheme α-Sn turns out to be a semi-metal because of the lower energies of the bands derived from the s orbitals. It was first shown by *Herman* [2.27] that relativistic effects are responsible for this in gray tin (and also in HgTe and HgSe. Note, however, that the s-p reversal for HgSe has recently been the object of controversy; see [2.28]).

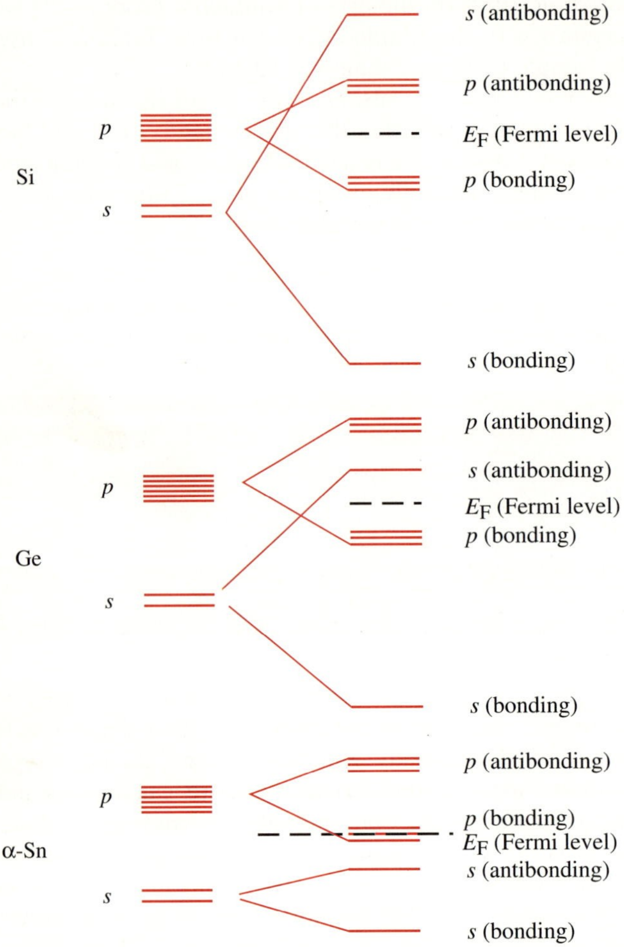

Fig. 2.26. Evolution of s and p atomic orbitals into the conduction and valence bands at zone center within the tight-binding approximation for Si, Ge, and α-Sn. The band ordering for diamond is similar to that of Si.

PROBLEMS

2.1 *Template of an fcc Brillouin Zone*

Construct a model of the Brillouin zone of the fcc lattice by pasting a copy of the template shown in Fig. 2.27 on cardboard and cutting along the solid lines. Score along the broken lines. Tape the edges together.

2.2 *Group Theory Exercises*

a) Verify the character table of the T_d point group as given in Table 2.3.

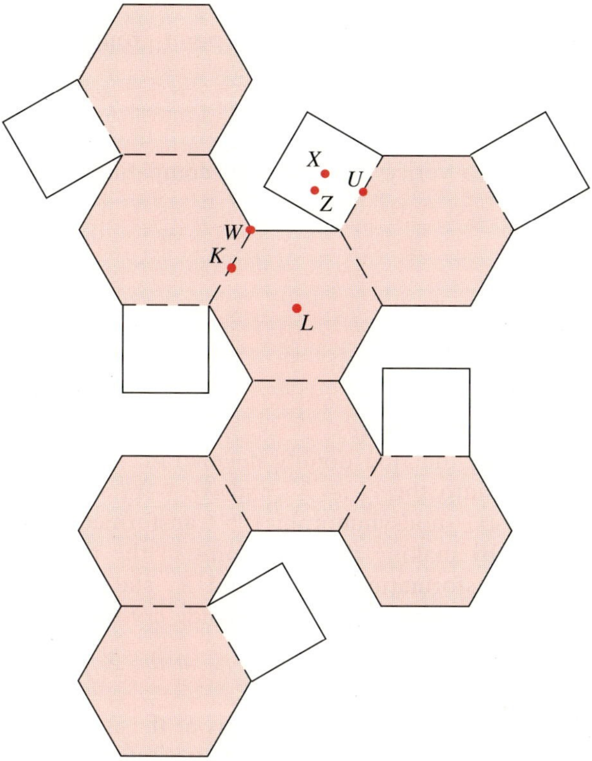

Fig. 2.27. Template for constructing a model of the Brillouin zone of the fcc lattice. Paste this sheet on thin cardboard and cut along *solid lines*. Score along *broken lines* and tape the joints

b) By applying the symmetry operations of T_d to the basis functions in Table 2.3 show that the functions transform according to their respective irreducible representations.

2.3 *Group Theory Exercises*
a) By using the character table of T_d show that $T_2 \otimes T_2 = T_1 \oplus T_2 \oplus E \oplus A_1$.

b) Verify that the symmetrized linear combinations of the matrix elements of a second-rank tensor given in Sect. 2.3.4 transform according to the irreducible representations T_1, T_2, E, and A_1.

2.4 *Symmetrized Wave Functions: Transformation Properties*
Verify that the symmetrized wave functions in Tables 2.9 and 2.10 transform according to their respective irreducible representations.

2.5 *Characters of C_{3v} and C_{2v} Point Groups*

Deduce by inspection the characters for the C_{3v} and C_{2v} point groups in Tables 2.12 and 2.14, respectively.

2.6 *Compatibility Relations*

Use Tables 2.3, 2.4, 2.12 and 2.14 to verify the following compatibility relations:

Γ_1	Δ_1	Λ_1
Γ_2	Δ_2	Λ_2
Γ_3	$\Delta_1 \oplus \Delta_2$	Λ_3
Γ_4	$\Delta_1 \oplus \Delta_3 \oplus \Delta_4$	$\Lambda_1 \oplus \Lambda_3$
Γ_5	$\Delta_2 \oplus \Delta_3 \oplus \Delta_4$	$\Lambda_2 \oplus \Lambda_3$.

2.7 *Representations of Nonsymmorphic Groups*

a) Using Tables 2.15 and 2.19 show that the doubly degenerate X_1 and X_2 states in the diamond crystal split into the $X_1 \oplus X_3$ and $X_2 \oplus X_4$ states, respectively, when the diamond crystal (nonsymmorphic) is transformed into a zinc-blende crystal (symmorphic) by making the two atoms in the primitive cell different. Under the same transformation the X_3 and X_4 states remain doubly degenerate and become the X_5 state in the zinc-blende crystal.

b) *Some Insight into the Doubly Degenerate Wavefunctions at the X point of the Brillouin Zone in the Diamond Structure.*
Within the free electron approximation, the wave functions at the X point of the Brillouin Zone can be written as: $\exp[\mathrm{i}\boldsymbol{k} \cdot \boldsymbol{r}]$ where $\boldsymbol{k} = (2\pi/a)(\pm1, 0, 0)$, $(2\pi/a)(0, \pm1, 0)$, or $(2\pi/a)(0, 0, \pm1)$. Let us consider the wave functions $\psi_1 = \sin[(2\pi/a)x]$ and $\psi_2 = \cos[(2\pi/a)x]$.

Assume that the crystal structure of diamond simply has inversion symmetry I: $(x, y, z) \rightarrow (-x, -y, -z)$; then by applying this symmetry operation to ψ_1 we obtain $-\psi_1$. Since the crystal is invariant under I we expect ψ_1 and $I\psi_1$ to have the same energy. We find that this is trivially satisfied since ψ_1 and $I\psi_1$ are linearly dependent. Thus we cannot conclude that the states ψ_1 and ψ_2 should be degenerate.

Now we take into account that the inversion operation in the diamond lattice is not simply I but rather I': $(x, y, z) \rightarrow (-x + (a/4), -y + (a/4), -z + (a/4))$. Applying I' to ψ_1 we find that: $I'\psi_1 = \sin[(2\pi/a)(-x + (a/4)] = \sin[(2\pi/a)(-x) + (\pi/2)] = -\cos[(2\pi/a)(-x)] = -\cos[(2\pi/a)x] = -\psi_2$. Since ψ_1 and ψ_2 are not linearly independent we have to conclude that ψ_1 and ψ_2 are degenerate from the fact that the crystal is invariant under I'. Similarly one can show that all the other plane wave states at the X point are doubly degenerate because of this symmetry operation I'.

2.8 *Pseudopotential Band Structure Calculation by Hand*

The purpose of this exercise is to show how pseudopotentials lift degeneracies in the nearly-free-electron band structure and open up energy gaps. Since the pseudopotentials are weak enough to be treated by perturbation theory, rather accurate band energies can be evaluated with a pocket calculator without resorting to a large computer.

We will consider only the six lowest energy wave functions at the X point of a zinc-blende-type semiconductor. In the nearly-free-electron model, the electron wave functions are given by $\exp(\mathrm{i}\boldsymbol{k}\cdot\boldsymbol{r})$, where $\boldsymbol{k} = (2\pi/a)(\pm 1, 0, 0)$ and $(2\pi/a)(0, \pm 1, \pm 1)$. For brevity these six wave functions will be denoted by $|100\rangle$, $|\bar{1}00\rangle$, $|011\rangle$, $|0\bar{1}1\rangle$, $|01\bar{1}\rangle$, and $|0\bar{1}\bar{1}\rangle$.

a) Show that these six wave functions can be symmetrized according to the following irreducible representations:

$$\psi_1 = (1/\sqrt{2})[|011\rangle + |0\bar{1}\bar{1}\rangle] \text{ and } \psi_2 = (1/\sqrt{2})[|011\rangle - |0\bar{1}\bar{1}\rangle] \leftrightarrow X_5$$
$$\psi_3 = (1/2)\{[|011\rangle - |0\bar{1}\bar{1}\rangle] + \mathrm{i}[|01\bar{1}\rangle + |0\bar{1}1\rangle]\} \leftrightarrow X_3$$
$$\psi_4 = (1/2)\{[|011\rangle - |0\bar{1}\bar{1}\rangle] - \mathrm{i}[|01\bar{1}\rangle + |0\bar{1}1\rangle]\} \leftrightarrow X_1$$
$$\psi_5 = (1/2)\{[|100\rangle + |\bar{1}00\rangle] + \mathrm{i}[|100\rangle - |\bar{1}00\rangle]\} \leftrightarrow X_1$$
$$\psi_6 = (1/2)\{[|100\rangle + |\bar{1}00\rangle] - \mathrm{i}[|100\rangle - |\bar{1}00\rangle]\} \leftrightarrow X_3$$

It should be noted that the pseudopotential form factors in Table 2.21 have been defined with the origin chosen to be the midpoint between the two atoms in the primitive cell. In order to conform with this coordinate system, the symmetry operations for the group of X have to be defined differently from those in Sect. 2.3.2. Taking the axes and planes of the point group to intersect at the midpoint some of the symmetry operations must involve a translation.

b) Calculate the matrix elements of the pseudopotential between these wave functions. This task can be greatly simplified by using the matrix element theorem. Since the pseudopotential V has the full symmetry of the crystal, it has Γ_1 symmetry. The only states that are coupled by V are then the X_3 states ψ_3 and ψ_6 and the X_1 states ψ_4 and ψ_5. Show that the resulting 6×6 matrix $\{V_{ij}\}$ is

$$\left|\begin{array}{cccccc} -v_8^s & 0 & 0 & 0 & 0 & 0 \\ 0 & -v_8^s & 0 & 0 & 0 & 0 \\ 0 & 0 & v_8^s + 2v_4^a & 0 & 0 & \mathrm{i}\sqrt{2}(-v_3^a - v_3^s) \\ 0 & 0 & 0 & v_8^s - 2v_4^a & \mathrm{i}\sqrt{2}(-v_3^a + v_3^s) & 0 \\ 0 & 0 & 0 & -\mathrm{i}\sqrt{2}(-v_3^a + v_3^s) & -v_4^a & 0 \\ 0 & 0 & -\mathrm{i}\sqrt{2}(-v_3^a - v_3^s) & 0 & 0 & +v_4^a \end{array}\right|$$

c) Diagonalize the secular determinant

$$\left|\left(\frac{\hbar^2 k^2}{2m} - E\right)\delta_{ij} + V_{ij}\right| = 0$$

by reducing it to three 2×2 determinants. Show that the resultant energy levels are

$$E(X_5) = \frac{4\pi^2\hbar^2}{ma^2} - v_8^s,$$

$$E(X_1) = \frac{1}{2}\left(\frac{6\pi^2\hbar^2}{ma^2} + v_8^s - 3v_4^a\right) \pm \frac{1}{2}\left[\left(\frac{2\pi^2\hbar^2}{ma^2} + v_8^s - v_4^a\right)^2 + 8(v_3^a - v_3^s)^2\right]^{1/2},$$

$$E(X_3) = \frac{1}{2}\left(\frac{6\pi^2\hbar^2}{ma^2} + v_8^s + 3v_4^a\right) \pm \frac{1}{2}\left[\left(\frac{2\pi^2\hbar^2}{ma^2} + v_8^s + v_4^a\right)^2 + 8(v_3^a + v_3^s)^2\right]^{1/2}.$$

d) Calculate the energies of the X_1, X_3, and X_5 levels in GaAs by substituting into the expression in (c) the pseudopotential form factors for GaAs. Take the lattice parameter a to be 5.642 Å. In Fig. 2.28 these results are compared with the nearly free electron energies and with the energies obtained by the EPM.

e) If you want to improve on the present calculation, what are the plane wave states and pseudopotential form factors you should include?

Note: Often in the literature, the origin of the coordinates adopted by the authors is not specified.[4] The symmetry of the band structure at the X point of the zinc-blende-type crystal depends on the choice of origin and this has

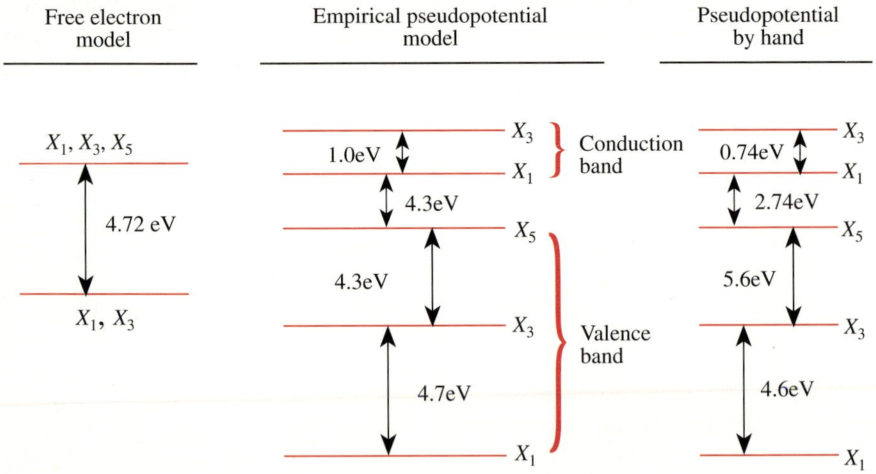

Fig. 2.28. The lowest energy bands at the X point of GaAs computed by the nearly free electron model, the EPM, and the perturbation approach of Problem 2.8. The X_1–X_3 notation corresponds to $v_j^a > 0$, i. e., to placing the cation at $(a/4)(111)$ and the anion at the origin.

[4] We assume, implicitly, that the origin is also the common point of the point group axes which specify the symmetry.

caused considerable confusion, see [2.29]. For example, if the origin is chosen at the anion the conduction band with the X_1 symmetry is mainly composed of the anion s wave function and cation p wave function. On the other hand, the X_3 conduction band state is made up of the cation s wave function and the anion p wave function. In all zinc-blende-type semiconductors with the exception of GaSb [2.30, 31] the X_1 state has lower energy than the X_3 state. If the origin is chosen at the cation, the signs of v_j^a and, correspondingly, the roles of X_1 and X_3 are reversed.

2.9 *Wave Functions of the L-Point of Zinc-Blende*
Using the symmetrized $\boldsymbol{k} = (2\pi/a)(1,1,1)$ wave functions in the nearly free electron model for zinc-blende-type crystals:

Γ_1: $\sqrt{8}\cos(2\pi x/a)\cos(2\pi y/a)\cos(2\pi z/a)$;
$\Gamma_4(x)$: $\sqrt{8}\sin(2\pi x/a)\cos(2\pi y/a)\cos(2\pi z/a)$,
and similar wave functions for $\Gamma_4(y)$ and $\Gamma_4(z)$ in Table 2.9,

show that the matrix elements of the momentum operator $\boldsymbol{p}$ between the Γ_1 and Γ_4 functions are given by

$$|\langle\Gamma_1|p_x|\Gamma_4(x)\rangle|^2 = |\langle\Gamma_1|p_y|\Gamma_4(y)\rangle|^2 = |\langle\Gamma_1|p_z|\Gamma_4(z)\rangle|^2 = (2\hbar\pi/a)^2$$

while all the other matrix elements of p_i such as $|\langle\Gamma_1|p_x|\Gamma_4(y)\rangle|^2$ are equal to 0.

2.10 *Double Group Representations*
In many quantum mechanics textbooks one can find the following result. The effect of a rotation by an infinitesimal amount $\delta\theta$ with respect to an axis defined by the unit vector $\hat{\boldsymbol{n}}$ on an orbital wave function $f(\boldsymbol{r})$ can be obtained by applying the operator $\exp[-\mathrm{i}\delta\Theta\boldsymbol{n}\cdot\boldsymbol{l}/\hbar]$ to $f(\boldsymbol{r})$. For a spin $s = 1/2$ particle the corresponding operator on the spin wave functions due to rotation by an angle Θ is given by $\exp[-\mathrm{i}\Theta\hat{\boldsymbol{n}}\cdot\boldsymbol{\sigma}/2]$. Using this operator, show that:

a) The effect of a 2π rotation on the wave functions α and β of a spin 1/2 particle is to change the sign of α and β, and hence the corresponding trace of $\hat{E}$ is -2;

b) the traces corresponding to the symmetry operations in Table 2.23 within the basis α and β are

$\{E\}$	$\{3C_2\}$ $\{3\hat{E}C_2\}$	$\{6S_4\}$	$\{6\sigma\}$ $\{6\hat{E}\sigma\}$	$\{8C_3\}$	$\{\hat{E}\}$	$\{6\hat{E}S_4\}$	$\{8\hat{E}C_3\}$
2	0	$\sqrt{2}$	0	1	-2	$-\sqrt{2}$	-1

c) *The Double Group at the X Point of the Zinc-Blende Structure*
As an additional exercise on the calculation of double group character table, we shall consider the X point of the zinc-blende structure.

The first step is to decide what are the classes in the double group. In this case we need only to compare the single group and double group classes at the zone center since the classes of X form a subset of these classes. It should not be difficult to see that there are now 7 classes:

$$\{E\},\ \{C_4^2(x), \hat{E}C_4^2(x)\},\ \{2C_4^2(y,z), 2\hat{E}C_4^2(y,z)\},\ \{2S_4\},\ \{2m_d\},\ \{\hat{E}\} \text{ and } \{\hat{E}S_4\}.$$

Using the results of Problem 2.10 one can show that the characters of these operations on the two spin wavefunctions are:

E	$C_4^2(x), \hat{E}C_4^2(x)$	$2C_4^2(y,z), 2\hat{E}C_4^2(y,z)$	$2S_4$	$2m_d$	$\hat{E}$	$\hat{E}S_4$
2	0	0	$\sqrt{2}$	0	-2	$-\sqrt{2}$

Using this result we can show that the character table for the double group of the X point in the zinc-blende crystal is:

	E	$C_4^2(x), \hat{E}C_4^2(x)$	$2C_4^2(y,z), 2\hat{E}C_4^2(y,z)$	$2S_4$	$2m_d$	$\hat{E}$	$\hat{E}S_4$
X_1	1	1	1	1	1	1	1
X_2	1	1	1	-1	-1	1	-1
X_3	1	1	-1	-1	1	1	-1
X_4	1	1	-1	1	-1	1	1
X_5	2	-2	0	0	0	2	0
X_6	2	0	0	$\sqrt{2}$	0	-2	$-\sqrt{2}$
X_7	2	0	0	$-\sqrt{2}$	0	-2	$\sqrt{2}$

Using these characters the reader should show that the X_1, X_3, and X_5 representations in the zinc-blende structure (see Problem 2.10) go over to the $X_6 \otimes X_1 = X_6$, $X_6 \otimes X_3 = X_7$ and $X_6 \otimes X_5 = X_6 \oplus X_7$ representations in the double group (see, for example, the band structure of GaAs in Fig. 2.14).

2.11 *The Structure Factor of Bond Charges in Si*

The intensity of x-ray scattering peaks from a crystal depends on the structure factor S of the crystal.

The structure factor of the Si crystal (face centered cubic or fcc lattice), in particular, is discussed in many standard textbooks on solid state physics, such as Kittel's "Introduction to Solid State Physics" (Chap. 2 in 6th Edition). The basis of the fcc structure is usually taken to be the cubic unit cell with four atoms per unit cube. These four atoms can be chosen to have the locations at $(0,0,0)$; $(0,1/2,1/2)$; $(1/2,0,1/2)$ and $(1/2,1/2,0)$ [in units of the size of the cube: a]. The structure factor $S_{\mathrm{fcc}}(hkl)$ for a wave vector (h,k,l) in reciprocal space then vanishes if the integers h, k and l contain a mixture of even and odd numbers. In the case of the Si crystal there are now 8 atoms per unit cube since there are two interpenetrating fcc sublattices displaced from each other by the distance $(1/4,1/4,1/4)$. As a result, the structure factor of the Si crystal

$S_{\text{Si}}(hkl)$ is given by:

$$S_{\text{Si}}(hkl) = S_{\text{fcc}}(hkl)[1 + \exp{(\text{i}\pi/2)(k + k + l)}].$$

This implies that $S_{\text{Si}}(hkl)$ will be zero if the sum $(h + k + l)$ is equal to 2 times an odd integer. When combining the above two conditions one obtains the result that $S_{\text{Si}}(hkl)$ will be non-zero only if (1) (k, k, l) contains only even numbers and (2) the sum $(h + k + l)$ is equal to 4 times an integer. See, for example, Kittel's "Introduction to Solid State Physics" (Chap. 2 in 6th Edition), Problem 5 at the end of Chap. 2. Based on this result one expects that the diffraction spot corresponding to (2, 2, 2) in the x-ray diffraction pattern of Si will have zero intensity since $h + k + l = 6$.

It has been known since 1959 that the so-called forbidden (2, 2, 2) diffraction spot in diamond has non-zero intensity (see Ref. [3.23] or Kittel's "Introduction to Solid State Physics", p. 73 in 3rd Edition). It is now well established that the presence of this forbidden (2, 2, 2) diffraction spot can be explained by the existence of bond charges located approximately mid-way between the atoms in diamond or silicon. What is the structure factor of the bond charges in the Si crystal if one assumes that they are located exactly mid-way between two Si atoms?

2.12 *Matrix Elements of* $\boldsymbol{p}$

a) Show that all matrix elements of $\boldsymbol{p}$ between the Γ_4 valence bands and the Γ_4 conduction bands of zinc-blende-type semiconductors of the form $\langle X | p_x | \Gamma_{4c}(z) \rangle$, $\langle Z | p_y | \Gamma_{4c}(z) \rangle$, or $\langle X | p_y | \Gamma_{4c}(y) \rangle$, where at least two of the labels x, y, or z are identical, vanish as a result of the requirement that the crystal is invariant under rotation by 180° with respect to one of the three equivalent [100] axes.

b) As a result of (a), the only nonzero matrix elements of $\boldsymbol{p}$ are of the form $\langle X | p_y | \Gamma_{4c}(z) \rangle$. Using the three-fold rotational symmetries of the zinc-blende crystal, show that

$$\langle X | p_y | \Gamma_{4c}(z) \rangle = \langle Y | p_z | \Gamma_{4c}(x) \rangle = \langle Z | p_x | \Gamma_{4c}(y) \rangle$$

and

$$\langle X | p_z | \Gamma_{4c}(y) \rangle = \langle Y | p_x | \Gamma_{4c}(z) \rangle = \langle Z | p_y | \Gamma_{4c}(x) \rangle.$$

c) Finally, use the reflection symmetry with respect to the (110) planes to show that

$$\langle X | p_y | \Gamma_{4c}(z) \rangle = \langle Y | p_x | \Gamma_{4c}(z) \rangle.$$

2.13 *Linear Terms in* $\boldsymbol{k}$

Show that the k linear term due to the $\boldsymbol{k} \cdot \boldsymbol{p}$ interaction is zero in the zinc-blende crystal at the Γ-point.

2.14 ***k · p** Method*
a) Use (2.52) to calculate the elements of the 6×6 matrix $\{H'_{ij}\}$.

b) Use a computer and a matrix diagonalization program to calculate the valence band structure of GaAs from these parameters for GaAs:
$P^2/(m_0) = 13$ eV; $Q^2/(m_0) = 6$ eV; $E_0 = 1.519$ eV; $E'_0 = 4.488$ eV; $\Delta = 0.34$ eV and $\Delta'_0 = 0.171$ eV.

2.15 *Valence Bands; **k · p** Hamiltonian*
a) Calculate the 4×4 matrix obtained by taking matrix elements of the Luttinger Hamiltonian in (2.70) between the $J_z = \pm 3/2$ and $\pm 1/2$ states of the $J = 3/2$ manifold.

b) Diagonalize this 4×4 matrix to obtain two sets of doubly degenerate levels with energies

$$E_\pm = \frac{\hbar^2}{2m}\{\gamma_1 k^2 \pm [4\gamma_2^2 k^4 + 12(\gamma_3^2 - \gamma_2^2)(k_x^2 k_y^2 + k_y^2 k_z^2 + k_z^2 k_x^2)]^{1/2}\}.$$

c) By comparing the results in (b) with (2.66) derive (2.71).

d) Calculate the contributions of the lowest Γ_3^- conduction band term to γ_1, γ_2, and γ_3. Show that it is not negligible for silicon and diamond [2.22].

2.16 *Energy Bands of a Semiconductor in the Tight-Binding Model*
a) Derive the 8×8 matrix for the s and p band energies in a zinc-blende-type semiconductor using the tight-binding model.

b) Show that at $k = 0$ the energies of the s and p bands are given by

$$E_{s\pm}(0) = \tfrac{1}{2}(E_{s1} + E_{s2}) \pm \tfrac{1}{2}[(E_{s1} - E_{s2})^2 + 4\,|\,V_{ss}\,|^2]^{1/2}$$

and

$$E_{p\pm}(0) = \tfrac{1}{2}(E_{p1} + E_{p2}) \pm \tfrac{1}{2}[(E_{p1} - E_{p2})^2 + 4\,|\,V_{xx}\,|^2]^{1/2}$$

instead of (2.84a) and (2.84b). E_{s1} and E_{s2} are the atomic s level energies $\langle S1\,|\,\mathcal{H}_0\,|\,S1\rangle$ and $\langle S2\,|\,\mathcal{H}_0\,|\,S2\rangle$, respectively, while E_{p1} and E_{p2} are the corresponding energies for the atomic p levels.

2.17 *Tight Binding Overlap Integrals*
Evaluate the geometric factors $\tau_{ll'm}$ in Table 2.27.

2.18 *Tight Binding Hamiltonian*
Given two p orbitals, one located at the origin and the other at the point $d(\cos\Theta_x, \cos\Theta_y, \cos\Theta_z)$, where d is the distance between the two p orbitals and $\cos\Theta_x$, $\cos\Theta_y$, and $\cos\Theta_z$ are the directional cosines of the second p orbital, show that the overlap parameters V_{xx} and V_{xy} are given by

$$V_{xx} = V_{pp\sigma}\cos^2\Theta_x + V_{pp\pi}\sin^2\Theta_x,$$

$$V_{xy} = [V_{pp\sigma} - V_{pp\pi}]\cos\Theta_x\cos\Theta_y.$$

2.19 *Conduction and Light Hole Bands in Small Band Gap Semiconductors* Write down the 2×2 Hamiltonian matrix which describes the conduction and the light hole band of a narrow gap semiconductor such as InSb. Diagonalize it and discuss the similarity of the resulting expression with the relativistic energy of free electrons and positrons [4.28]. Use that expression to estimate non-parabolicity effects on the conduction band mass.

SUMMARY

A semiconductor sample contains a very large number of atoms. Hence a quantitative quantum mechanical calculation of its physical properties constitutes a rather formidable task. This task can be enormously simplified by bringing into play the symmetry properties of the crystal lattice, i. e., by using group theory. We have shown how wave functions of electrons and vibrational modes (phonons) can be classified according to their behavior under symmetry operations. These classifications involve *irreducible representations* of the group of symmetry operations. The translational symmetry of crystals led us to *Bloch's theorem* and the introduction of *Bloch functions* for the electrons. We have learnt that their eigenfunctions can be indexed by wave vectors (Bloch vectors) which can be confined to a portion of the reciprocal space called the first *Brillouin zone*. Similarly, their energy eigenvalues can be represented as functions of wave vectors inside the first Brillouin zone, the so-called electron *energy bands*. We have reviewed the following main methods for calculating energy bands of semiconductors: the *empirical pseudopotential* method, the *tight-binding* or *linear combination of atomic orbitals* (LCAO) method and the $\boldsymbol{k} \cdot \boldsymbol{p}$ method. We have performed simplified versions of these calculations in order to illustrate the main features of the energy bands in diamond- and zinc-blende-type semiconductors.

3. Vibrational Properties of Semiconductors, and Electron-Phonon Interactions

CONTENTS

We will start the discussion of the vibrational properties of semiconductors by reviewing the theory of the dynamics of a crystalline lattice. The Hamiltonian describing a perfect crystal has already been given in (2.1). We note that the electrons have been separated into two groups. The core electrons are assumed to move rigidly with the nucleus to form what has been referred to as the ion. The valence electrons interact with these ions via the pseudopotentials. The part of the Hamiltonian in (2.1) which involves the nuclear motions is given by

$$H_{\mathrm{ion}}(\boldsymbol{R}_1,\ldots,\boldsymbol{R}_n)=\sum_j \frac{P_j^2}{2M_j}+\sum_{j,j'}{}' \frac{1}{2}\frac{Z_j Z_{j'} e^2}{4\pi\varepsilon_0|\boldsymbol{R}_j-\boldsymbol{R}_{j'}|}-\sum_{i,j}\frac{Z_j e^2}{4\pi\varepsilon_0|\boldsymbol{r}_i-\boldsymbol{R}_j|}, \quad (3.1)$$

where $\boldsymbol{R}_j$, $\boldsymbol{P}_j$, Z_j and M_j are, respectively, the nuclear positions, momentum, charge, and mass, $\boldsymbol{r}_i$ is the position of the electron and $\sum'$ means summation over pairs of indices j and j' where j is not equal to j'.

The appearance of the electronic coordinates in this Hamiltonian makes it difficult to solve for the nuclear motion since electronic motion is coupled to ionic motion. As pointed out already in Sect. 2.1, one way to separate the two is by introducing the Born–Oppenheimer or adiabatic approximation. In this approximation the electrons are assumed to follow the ionic motion adiabatically. As a result we can solve the electronic part of the Hamiltonian H in (2.1) to obtain the energies of the electrons as functions of the ion positions. On the other hand, the ions cannot follow the electronic motion and therefore they see only a time-averaged *adiabatic* electronic potential. Hence the Hamiltonian for the ions can be written as

$$H_{\mathrm{ion}}=\sum_j \frac{P_j^2}{2M_j}+E_{\mathrm{e}}(\boldsymbol{R}_1,\ldots,\boldsymbol{R}_n), \quad (3.2)$$

where $E_e(\boldsymbol{R}_1, \ldots, \boldsymbol{R}_n)$, which represents the **total energy** of the valence electrons with ions held stationary at positions $\boldsymbol{R}_1, \ldots, \boldsymbol{R}_n$, is treated as the interaction between ions via the electrons. With the help of supercomputers, it is now possible to calculate $E_e(\boldsymbol{R}_1, \ldots, \boldsymbol{R}_n)$ and then solve *ab initio* for the motion of the ions.

Since much of the work on lattice dynamics was carried out before supercomputers were available, these early investigations had to rely on a phenomenological approach. In this approach typically an equation of motion for the ions is obtained by expanding H_{ion} as a function of their displacements $\delta\boldsymbol{R}_j$ from their equilibrium positions $\boldsymbol{R}_{j0}$:

$$H_{\text{ion}} = H_0(\boldsymbol{R}_{10}, \ldots, \boldsymbol{R}_{n0}) + H'(\delta\boldsymbol{R}_{10}, \ldots, \delta\boldsymbol{R}_{n0}). \tag{3.3}$$

In (3.3), $H_0(\boldsymbol{R}_{10}, \ldots, \boldsymbol{R}_{n0})$ is the Hamiltonian of the crystal with all the nuclei at their equilibrium positions and $H'(\delta\boldsymbol{R}_{10}, \ldots, \delta\boldsymbol{R}_{n0})$ is the change in H_{ion} due to displacements of the nuclei by small amounts $\delta\boldsymbol{R}_{10}, \ldots, \delta\boldsymbol{R}_{n0}$ from the equilibrium positions. To diagonalize H', one expands H_{ion} around $\boldsymbol{R}_{10}, \ldots, \boldsymbol{R}_{n0}$. Since the $\boldsymbol{R}_{j0}$ are the equilibrium positions of the ions, the first-order terms in $\delta\boldsymbol{R}_j$ vanish. In addition, when all the $\delta\boldsymbol{R}_j$ are identical, the crystal is uniformly displaced and not distorted. Thus the lowest order terms in the expansion of H_{ion} relevant to the vibration of the crystal are second order in $\delta(\boldsymbol{R}_j - \boldsymbol{R}_k)$. If we keep only the terms quadratic in H', the motions of the nuclei are described by a collection of simple harmonic ocillators, and hence this approach is known as the **harmonic approximation**. We will restrict our treatment of the lattice dynamics of semiconductors to the harmonic approximation only. One important limitation of this approach is that we cannot explain some phenomena, for example, thermal expansion. In particular, it is now well established that the coefficient of linear expansion in many diamond- and zinc-blende-type semiconductors changes sign twice when their temperature is increased from liquid helium temperature to room temperature [3.1].

To simplify the notations we will denote the displacement from equilibrium of the ion k in the unit cell l by $\boldsymbol{u}_{kl}$. Within the harmonic approximation H' can be expressed as

$$H'(\boldsymbol{u}_{kl}) = \frac{1}{2}M_k\left(\frac{d\boldsymbol{u}_{kl}}{dt}\right)^2 + \frac{1}{2}\sum_{k'l'} \boldsymbol{u}_{kl} \cdot \Phi(kl, k'l') \cdot \boldsymbol{u}_{k'l'}. \tag{3.4}$$

In this equation $H'(\boldsymbol{u}_{kl})$ represents the change in the ion Hamiltonian induced by a displacement of the ion (kl) while all the other atoms are kept in their equilibrium position. The matrix $\Phi(kl, k'l')$ contains the **force constants** describing the interaction between the ions denoted by (kl) amd $(k'l')$. For example, the force on ion (kl) due to the displacement $\boldsymbol{u}_{k'l'}$ of the $(k'l')$ ion is given by $-\Phi(kl, k'l') \cdot \boldsymbol{u}_{k'l'}$. The force constants contain two parts. The first part is the direct ion–ion interaction due to their Coulomb repulsion, the second is an indirect interaction mediated by the valence electrons. The motion of one ion causes the electron distribution to change. This rearrangement of the electrons produces a force on the neighboring ions. In Sect. 3.2 we will consider the various models for calculating the force constants. At this point we will summarize the procedures for obtaining the phonon dispersion curves assuming the force constants are known.

The determination of the lattice dynamics described by the Hamiltonian in (3.4) can be carried out in two steps. First we will treat the Hamiltonian classically and solve the equation of motion. In this classical approximation (3.4) describes the energy of a collection of particles executing small-amplitude oscillations. As is well known in classical physics, these oscillations can be expressed in terms of **normal modes** which are independent of each other [3.2]. In the second step we quantize the energies of these normal modes. Each quantum of lattice vibration is called a **phonon**. The procedures for determining the quantized energy levels of one-dimensional simple harmonic oscillators can be found in most quantum mechanics textbooks, so we will not repeat them here. Instead we will concentrate on the determination of the normal modes of vibrations described by (3.4).

Since $\Phi(kl, k'l')$ possesses translational symmetry, we expect that the atomic displacements which diagonalize (3.4) can be expressed in terms of plane waves similar to the Bloch functions for electrons in a crystal [as defined by (2.6) in Sect. 2.2]. If $\boldsymbol{u}_{kl}$ is the displacement of the kth ion in the lth unit cell specified by the lattice vector $\boldsymbol{R}_l$, it can be related to the displacement $\boldsymbol{u}_{k0}$ of a corresponding ion in the unit cell located at the origin by a Bloch wave of the form

$$\boldsymbol{u}_{kl}(\boldsymbol{q}, \omega) = \boldsymbol{u}_{k0} \exp \mathrm{i}(\boldsymbol{q} \cdot \boldsymbol{R}_l - \omega t), \tag{3.5}$$

where $\boldsymbol{q}$ and ω are, respectively, the wave vector and frequency of the wave. There is, however, one important difference between phonon and electron Bloch waves. While an electron can be anywhere in the crystal, ion positions, in the classical approximation, are discrete. Since the $\boldsymbol{R}_l$ in (3.5) are lattice vectors, two waves whose wave vectors differ by a reciprocal lattice vector are equivalent. In terms of Brillouin zones introduced in Sect. 2.2, this result can be stated as: the phonon frequencies in the first Brillouin zone are identical to those in the other Brillouin zones. Thus the frequency versus wave vector plots for lattice vibrations in crystals (or **phonon dispersion curves**) are always presented in the reduced zone scheme. Another important consequence of (3.5) is that the degrees of freedom or the number of independent waves is equal to three times the number of atoms in the crystal. By substituting (3.5) for $\boldsymbol{u}_{kl}$ into (3.4) and using the resultant expression in the classical Hamilton equation [3.2] we obtain the equation for $\boldsymbol{u}_{k0}$:

$$M_k \omega^2 \boldsymbol{u}_{k0} = \sum_{k', m} \Phi(km, k'0) \exp(-\mathrm{i}\boldsymbol{q} \cdot \boldsymbol{R}_m) \boldsymbol{u}_{k'0}. \tag{3.6}$$

Introducing a *mass-modified Fourier transform* of Φ as

$$D_{kk'}(\boldsymbol{q}) = \sum_{m} \Phi(mk, 0k') (M_k M_{k'})^{-1/2} \exp\left[-\mathrm{i}\boldsymbol{q} \cdot \boldsymbol{R}_m\right], \tag{3.7}$$

(3.6) can be written as

$$\sum_{k'} \left[D_{kk'}(\boldsymbol{q}) - \omega^2 \delta_{kk'}\right] \boldsymbol{u}_{k'0} = 0. \tag{3.8}$$

$D_{kk'}(\boldsymbol{q})$ is known as the **dynamical matrix**. The vibrational frequencies ω are found by solving the secular equation

$$|D_{kk'}(\boldsymbol{q}) - \omega^2 \delta_{kk'}| = 0. \tag{3.9}$$

The vibrational amplitudes $\boldsymbol{u}_{k0}$ are obtained by substituting the solutions of (3.9) into (3.8). Readers should refer to textbooks on linear algebra or classical mechanics for further details of procedures for diagonalizing the matrix in (3.8). The main differences between the above classical treatment and a quantum mechanical calculation are: (1) in the quantum case the energy levels of a vibrational mode of frequency ω are quantized as $[n + (1/2)]\hbar\omega$ and (2) the creation and annihilation operators for a quantum of vibration (or phonon) are expressed in terms of the vibration amplitudes $\boldsymbol{u}_{k0}$.

The organization of the rest of this chapter is as follows: in the next section we discuss qualitative features of phonon dispersion curves in diamond- and zinc-blende-type semiconductors, in Sect. 3.2 we study models for computing these dispersion curves from (3.9), while the last section concentrates on interactions between electrons and phonons.

Within the harmonic approximation (3.4) the phonons have an infinite lifetime. Inclusion of higher order, anharmonic terms results in a finite lifetime which can be calculated by *ab initio* techniques [3.3].

3.1 Phonon Dispersion Curves of Semiconductors

Phonon dispersion curves in crystals along high-symmetry directions of the Brillouin zone can be measured quite precisely by inelastic neutron scattering[1] and more recently by high resolution inelastic x-ray scattering. Figures 3.1–3.3 show the dispersion curves of Si, GaAs, and GaN. These results can be regarded as representatives of semiconductors with the diamond-, zinc-blende-, and wurtzite-type lattices, respectively. In the diamond- and zinc-blende-type lattices there are two atoms per primitive unit cell, and hence there are six phonon branches. These are divided into three **acoustic phonon** branches (the three lower energy curves) and three **optical phonon** curves. Along high-symmetry directions (such as the [100] and [111] directions in Si and GaAs) the phonons can be classified as **transverse** or **longitudinal** according to whether their displacements are perpendicular or parallel to the direction of the wave vector $\boldsymbol{q}$.

In a solid the long wavelength transverse acoustic (abbreviated as TA) phonons are shear sound waves while the longitudinal acoustic (LA) phonons correspond to compressional sound waves. The velocities of these sound waves are determined by the shear and bulk elastic moduli, respectively. Since it is

[1] B.N. Brockhouse and C.G. Shull were awarded the Nobel Prize in 1994 for their development of neutron scattering spectroscopy, *Brockhouse* and *Iyengar* first measured the phonon dispersion in Ge by inelastic neutron scattering [Phys. Rev. **111**, 747 (1958)].

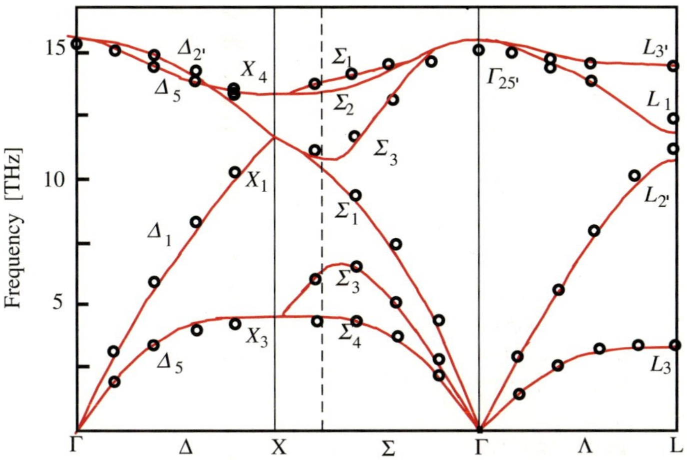

Fig. 3.1. Phonon dispersion curves in Si along high-symmetry axes. The *circles* are data points from [3.4]. The continuous curves are calculated with the adiabatic bond charge model of *Weber* [3.5]

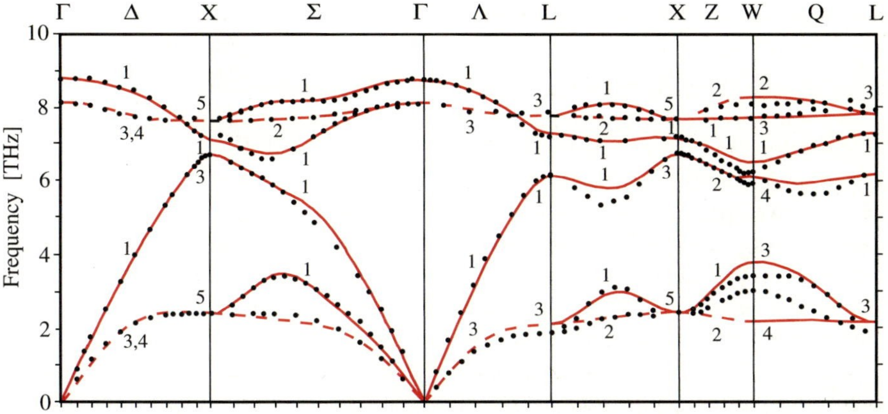

Fig. 3.2. Phonon dispersion curves in GaAs along high-symmetry axes [3.6]. The experimental data points were measured at 12 K. The *continuous lines* were calculated with an 11-parameter rigid-ion model. The numbers next to the phonon branches label the corresponding irreducible representations

usually easier to shear than to compress a crystal, the TA phonons travel with lower velocities than the LA phonons. Two special features of the TA phonons in the diamond- and zinc-blende-type semiconductors are: (1) their dispersion curves are relatively flat near the zone edge; and (2) their energies are much lower than the LA phonon energy near the zone edge. We will later show that these features are related to the covalent nature of bonds in these crystals.

In Si the transverse optical (TO) phonons and the longitudinal optical (LO) phonons are degenerate at the zone center. In GaAs and other zinc-blende-type semiconductors, the LO phonon has higher energy than the TO

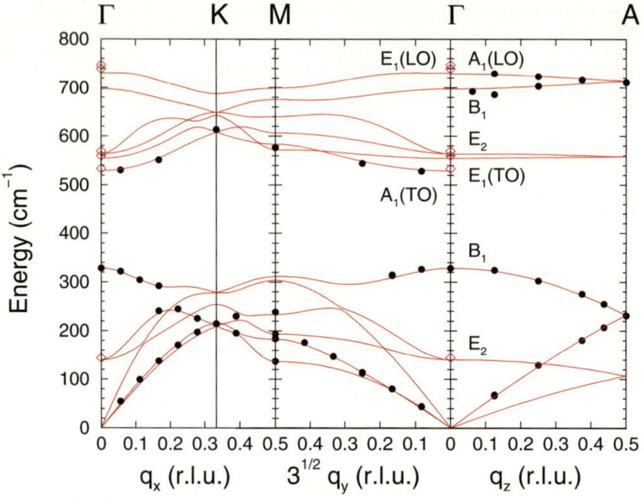

Fig. 3.3. Phonon dispersion along high symmetry directions in the wurtzite structure semiconductor GaN [3.7]. The experimental points have been obtained by Raman scattering (open circles) and by high resolution inelastic x-ray scattering (closed circles). The continuous curves are obtained by an *ab initio* calculation

phonons near the zone center. *Exactly at the zone center*, the TO and LO phonons in the zinc-blende crystals must also be degenerate because of the cubic symmetry of the zinc-blende structure. This degeneracy and dispersion of the zone-center optical phonons in zinc-blende crystals will be taken up again in Sect. 6.4 when we study the interaction between the TO phonons and infrared radiation. At wave vectors *near* but not exactly at the zone center, the LO phonon frequency in GaAs and other zinc-blende crystals is higher than that of the TO phonons. The reason lies in the partially ionic nature of the bonding in zinc-blende crystals. For example, in GaAs the As atoms contribute more electrons to the bond than the Ga atoms. As a result, the electrons in the covalent bond spend, on average, somewhat more time near the As atoms than near the Ga atoms, so the As atoms are slightly negatively charged while the Ga atoms are slightly positively charged. Let us assume that a long-wavelength TO phonon propagating along the [111] direction is excited. The positive and negative ions lie on separate planes perpendicular to the [111] axis. In a TO mode the planes of positive and negative ions essentially slide pass each other. The situation is similar to sliding the two plates of a parallel-plate capacitor relative to each other while keeping their separation constant. The energy of the capacitor is not changed by such motion. On the other hand, the energy of the charged capacitor is increased when the two plates are pulled apart because there is an additional restoring force due to the Coulomb attraction between the positively and negatively charged plates. Similarly, an additional Coulombic restoring force is present in long-wavelength LO phonon modes but not in the TO phonon modes.

The analogy between the optical phonons and the displacements of the capacitor plates is shown in Fig. 3.4. This additional restoring force (F) aris-

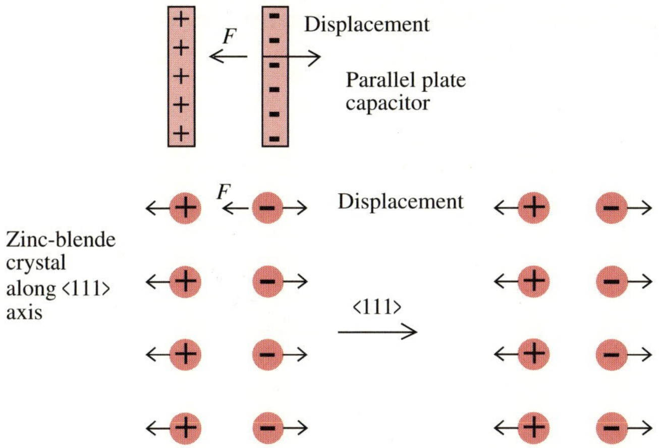

Fig. 3.4. Comparison between the displacement of atoms in an ionic crystal during a long-wavelength longitudinal optical vibration and in an infinite parallel plate capacitor. *F* represents the restoring force which results from displacements of the charges shown

ing from the displacement of the ions increases the frequencies of the long-wavelength LO phonons above those of the corresponding TO phonons. In Sect. 6.4.4 we shall show explicitly that there is a longitudinal electric field which depends on the atomic displacements in a LO phonon but not in a TO phonon. This longitudinal electric field results in an additional interaction between LO phonons and electrons (Sect. 3.3.5).

In Si the two atoms in the unit cell are identical so the bonding is purely covalent and the two atoms do not carry charge. As a result there is no additional restoring force associated with LO phonons and the zone-center optical phonons are degenerate.[2]

Vibrational modes in a crystal can be symmetrized according to the space group symmetry of the crystal just like the electronic states. A phonon mode is defined by the displacements of the atoms inside the unit cell. Thus the symmetry of the phonon must belong to the direct product of the representation of a vector and the respresentation generated by a permutation of the positions of equivalent atoms in the unit cell. The symmetry of the long-wavelength phonons in Si and GaAs has already been considered in Sect. 2.3.2a. An example of how to determine the symmetries of long-wavelength phonons in another cubic crystal, Cu_2O with six atoms per primitive unit cell, can be found in Problem 3.1. The corresponding phonons in a non-cubic crystal structure, such as the wurtzite structure, can be found in Problem 3.7b.

[2] Note, however, that if more than two atoms of the same kind are present within each primitive cell, infrared active modes and LO–TO splittings are possible. See the case of selenium and tellurium [3.8]

3.2 Models for Calculating Phonon Dispersion Curves of Semiconductors

To calculate the phonon dispersion curves from (3.9) it is necessary to know the force constants. In most calculations these force constants are obtained by first modeling the interactions between ions in terms of a number of parameters and then fitting some experimental quantities, such as sound velocity, zone-center phonon frequencies, bulk modulus, etc., by adjusting these parameters. Even after the force constants are known, numerical computations are necessary in solving (3.9) to obtain the phonon frequencies. Hence we can discuss only qualitatively the features of various models proposed for semiconductors.

3.2.1 Force Constant Models

The *Born–von Kármán* model [Ref. 3.9, p. 55] represents the first attempt to calculate the phonon dispersion in semiconductors such as diamond and Si. The atoms are assumed to be hard spheres connected by springs. The spring constants $\Phi(kl, k'l')$ are determined by fitting experimental results. *Born* [3.10] tried to fit the experimental results for C and Si with only two spring constants: α and β which determine the restoring force on each atom due to its own displacement and the displacements of its nearest neighbors. This simple model was applied by *Hsieh* [3.11] to calculate the phonon dispersion in Si. The calculated curve failed to fit the experimental dispersion curve at short wavelengths. The Si lattice turned out to be unstable under shear stress in this simple model. The flattening of the TA phonon dispersion near the zone edge cannot be explained without introducing long-range interatomic interactions. *Herman* [3.12] showed later that by extending interactions to the fifth-nearest neighbors and by using as many as 15 force constants, a good fit to the phonon dispersion curves of Ge could be achieved. This model is not easy to understand from a physical point of view since some of the distant-neighbor force constants were found to be larger than their nearer-neighbor ones.

3.2.2 Shell Model

One can argue that a model in which the atoms are regarded as point masses connected by springs will be a poor approximation for semiconductors. After all, the valence electrons in covalent semiconductors such as Ge and Si are not rigidly attached to the ions. In a model proposed by *Cochran* [3.13] each atom is assumed to consist of a rigid ion core surrounded by a shell of valence electrons (shown schematically in Fig. 3.5) that can move relative to the cores. This is the basis of the **shell model**.

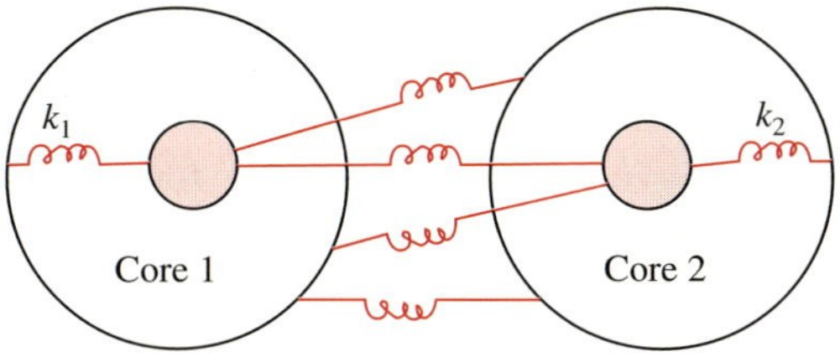

Fig. 3.5. Typical interactions between two deformable atoms in the shell model

In the shell model the interactions between the two atoms inside the unit cell of Si are represented schematically by springs, as shown in Fig. 3.5. One important feature introduced by the shell model is that long-range Coulomb interaction between atoms can be included. This is achieved by assigning charges to the shells so that dipole moments are produced when the shells are displaced relative to the ions. By using the interaction between the induced dipoles to simulate the long-range interaction, the short-range interaction can be limited to the nearest neighbors. With the shell model *Cochran* [3.13] was able to fit the phonon dispersion in Ge with five adjustable parameters. *Dolling and Cowley* [3.14] were able to fit the phonon dispersion curves of Si using an 11-parameter shell model. In this model the short-range interactions have been extended to the next-nearest neighbors. Similar 11-parameter shell models have been successfully used to fit the phonon dispersion curves even in III–V compounds. With 14-parameters the agreement between theory and experiment is quite good. The main criticism of the shell model is that the valence electron distributions in the diamond- and zinc-blende-type semiconductors are quite different from spherical shells. As a result, the parameters determined from the shell model have no obvious physical meaning and have limited applications beyond fitting the phonon dispersion curves. *Phillips* [3.15] has pointed out that the most serious problems of the shell model appear when applying it to covalent solids. The shell model artificially divides the valence charges between the two atoms involved in the covalent bond. In reality the valence electrons are "time-shared" between the two atoms in that they all spend part of their time on each atom.

3.2.3 Bond Models

It is well known that valence electrons in diamond- and zinc-blende-type semiconductors form highly directed bonds. These valence electrons are important for explaining cohesion in these semiconductors so they must also play an important role in determining the vibrational frequencies. The vibrational properties of molecules formed from covalent bonds have been extensively studied by chemists. These vibrational modes are usually analyzed in terms of **valence force fields** for stretching the bonds and for changing the angles between bonds (bond bending). The force constants can be determined in a straightforward manner from these valence force fields since the displacements of the ions are related to the bond coordinates. One advantage of this approach is that the force constants for bond stretching and bond bending are often characteristics of the bonds and can be transferred from one molecule to another containing the same bonds.

To see what kind of parameters are involved in this **valence force field method** (VFFM) of calculating lattice dynamics let us consider a crystal with two atoms, A and B, per unit cell. The potential energy of the valence bonds about the equilibrium positions can be expanded phenomenologically in terms of the valence bond coordinates as

$$\begin{aligned} V = \frac{1}{2} \Bigg[& \sum_{i,j} \sigma(\delta \boldsymbol{r}_{ij})^2 + \sum_{i,k} \mu(\delta \boldsymbol{r}_{ik})^2 + \sum_{\mathrm{BAB}} k_\Theta r_0^2 (\delta\Theta_{ijk})^2 \\ & + \sum_{\mathrm{ABA}} k'_\Theta r_0^2 (\delta\Theta_{jkl})^2 + \sum_{\mathrm{BAB}} k_{r\Theta} r_0 (\delta\Theta_{ijk})(\delta r_{ij}) \\ & + \sum_{\mathrm{ABA}} k'_{r\Theta} r_0 (\delta\Theta_{jkl})(\delta r_{jk}) + \dots \Bigg] . \end{aligned} \tag{3.10}$$

The first two terms in (3.10) correspond to bond stretching forces: j and k denote, respectively, the nearest and next-nearest neighbors of an atom i. The remaining terms give rise to bond bending forces. The bond motions for these terms are shown in Fig. 3.6. This method of calculating phonon dispersion works best when only a small number of valence force fields are sufficient to explain the phonon dispersion. In ionic crystals it is necessary to introduce additional long-range forces due to Coulomb interactions in order to reproduce the LO–TO phonon splittings near the zone center.

Musgrave and Pople [3.16] first applied this method to study the lattice dynamics of diamond. They included only two kinds of valence force fields in the potential energy: bond stretching and bond bending about a common apex atom. The model requires five parameters, which translates into six force constants involving the nearest and next-nearest neighbors. The model does not reproduce well the elastic constants nor the zone-center optical phonon fre-

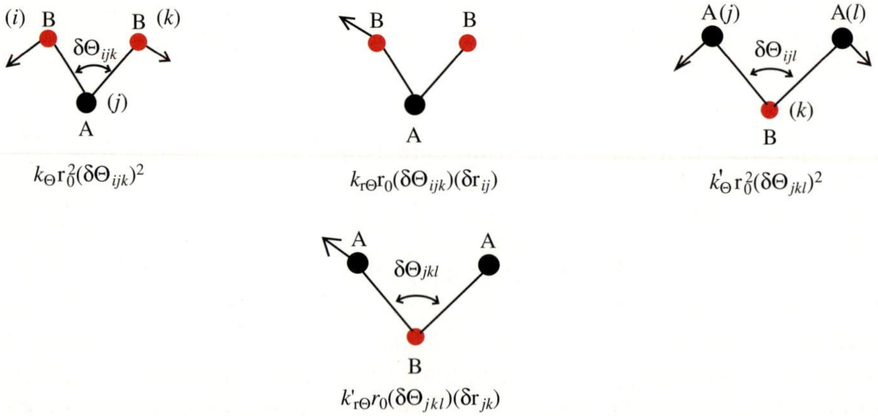

Fig. 3.6. Bond-bending configurations for a crystal with two atoms, A and B, per unit cell

quencies. Subsequent work showed that better results could be achieved by introducing additional parameters involving the change of two bond angles with a common bond. In ionic crystals the number of adjustable parameters is increased to eight to include the Coulomb interactions. Thus the number of adjustable parameters in the VFFM necessary for a good fit to the experimental results is comparable to that of the shell model, and there is no major advantage of one method over the other. The phonon dispersion curve in a wurtzite-type semiconductor, CdS, has been calculated by *Nusimovici and Birman* [3.17] using the VFFM with eight adjustable parameters. Since the experimental phonon dispersion curves of CdS were not known at that time,[3] these parameters were adjusted to fit the experimental zone center optical phonon energies. A simplified version of the VFFM has been introduced by *Keating* [3.22]. There are only two parameters, α and β, in this model for covalent semiconductors, with an additional charge parameter for ionic compounds. The Keating parameter α is equivalent to the bond-stretching term σ in (3.10) while β is equivalent to the bond-bending term k_Θ. Because of its simplicity and the clear physical meaning of its parameters, the Keating model has been widely used to study the elastic and static properties of covalent semiconductors [3.23a, 24]. The phonon dispersion curves calculated from the Keating model show reasonably good agreement with experimental results except for the problematic TA branch. If the parameters α and β are determined from the elastic constants, the zone-edge (X-point) TA phonon energies tend to be too high and cannot reproduce the flat dispersion in the experimental curves (see Fig. 3.1).

3.2.4 Bond Charge Models

To understand the motivation for the bond charge model, let us return to the Hamiltonian for the ions given in (3.2). The force constants can be obtained, in principle, by differentiating the total energy E_e with respect to the ion coordinates (this is known as the **frozen phonon approximation**) [3.21]. However, this approach is very computation intensive and requires the use of supercomputers. On the other hand, with some insight and approximations, phonon dispersion curves of diamond- and zinc-blende-type semiconductors can be calculated without supercomputers.

The most difficult part of this calculation is how to handle the Coulomb interaction between ions and electrons. This interaction causes the effective charge of an ion seen by other ions to be reduced, an effect known as **screening**. A simple way to introduce screening effects is to calculate the **dielectric function** ε (for a definition see Sect. 6.1) and then divide the ionic potential by ε. There are several approaches to approximating the screening of the ions by the valence electrons. One obvious simplification is to assume that the valence electrons are free so that the dielectric function of the semiconductor can be

[3] Neutron scattering cannot be performed on compounds containing ^{113}Cd because of its very large thermal neutron absorption cross section. Phonon dispersion curves in isotopically enriched in ^{114}CdS [3.18], ^{114}CdTe [3.19], and in CdSe [3.20] have however now been reported.

replaced by that of a metal. Using this approach, *Martin* [3.23b] found that the Coulomb repulsion between the ions is very strongly screened: the silicon lattice becomes unstable against a long-wavelength shear distortion. The TA modes have *imaginary* energies as a result. One way to avoid this problem is to localize some of the valence electrons so that not all of them can contribute to the screening of the ions. To handle these localized valence electrons Martin used the idea of **bond charges** introduced by *Phillips* [3.15]. X-ray scattering measurements had already suggested that there was a pile-up of charges along bonds in Si and diamond crystals which could not be explained by a spherical charge distribution such as in a shell model. This pile-up of charge in the covalent bond is known as the bond charge and is well known in the formation of covalent molecules. Experimentally, *Göttlicher* and *Wolfel* [3.25] observed a diffraction peak in the X-ray spectra of diamond corresponding to the (2,2,2) reciprocal lattice vector. This diffraction peak is forbidden by Bragg's diffraction law in a crystal with the diamond structure (see Problem 2.11). Those authors noted that this forbidden diffraction peak could be explained by assuming that approximately 0.4 of an electronic charge was located at the middle of each bond. More recently, the distribution of the bond charge in Si has been mapped out by *Yang* and *Coppens* [3.26] also using X-ray diffraction. Their experimental results are shown in Fig. 3.7a and are in excellent agreement with the theoretical charge distribution in Fig. 3.7b calculated by *Chelikowsky* and *Cohen* [3.27].

Martin [3.23b] introduced the bond charges into the lattice dynamics calculation of semiconductors in a very simple phenomenological manner. He assumed that bond charges of magnitude $Z_b e$ are located *exactly* midway between two adjacent atoms. As a first approximation he postulated that $Z_b e$ was given by

$$Z_b e = -\frac{2e}{\varepsilon}, \tag{3.11}$$

where $2e$ represents the two electrons involved in the covalent bond and the dielectric constant ε results from the screening of the bond charge by the re-

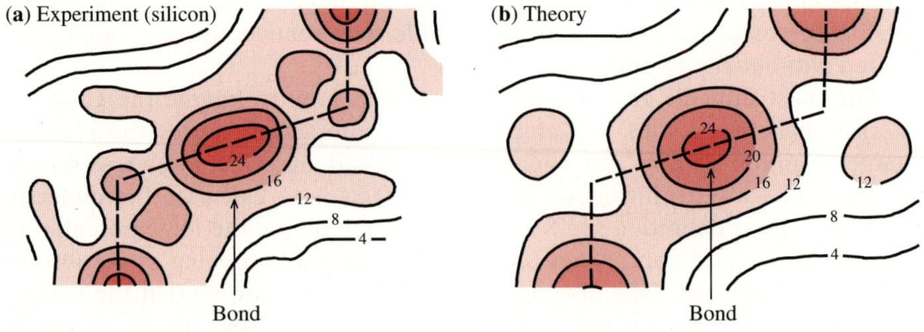

Fig. 3.7. Valence charge distribution in Si showing the constant charge density contours (**a**) determined experimentally from X-ray diffraction [3.26] and (**b**) calculated with the empirical pseudopotential method [3.27]. The numbers in the figure are in units of electrons per unit cell volume

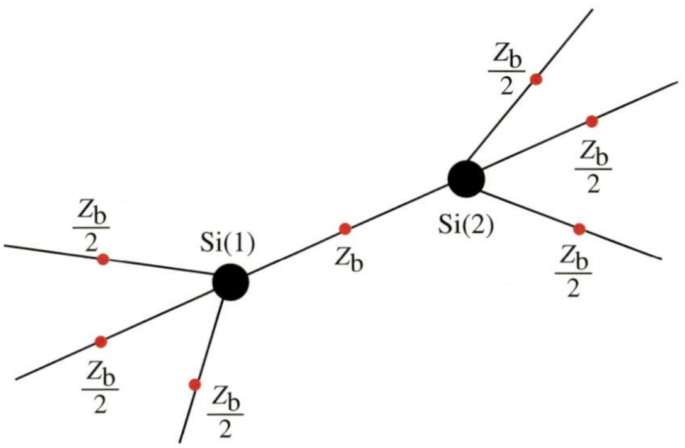

Fig. 3.8. Schematic diagram of the bond charge model proposed by *Martin* [3.23b] for Si. Z_b represents the bond charge

maining valence electrons. For simplicity ε was assumed to be given by the dielectric constant for small wave vectors and at low frequencies (see Sect. 6.1 for further discussion of the wavevector and frequency dependence of ε). For diamond, at least, this approximation worked rather well. The dielectric constant of diamond is equal to 5.7, so $Z_b e$ is equal to $0.35e$ in reasonable agreement with the value of $0.4e$ needed to fit the phonon dispersion curves [3.25]. This simple model for the Si crystal is shown schematically in Fig. 3.8. Each Si ion has a charge of $+4e$. The four valence electrons from each Si atom are divided into localized bond charges and nearly free electrons. Each Si atom contributes four $Z_b/2$ electron to the four bonds it forms with its four nearest neighbors. These bond charges are localized, and therefore do not contribute to screening of the Si ions. The remaining $(4 - 2Z_b)$ valence electrons from each Si atom are assumed to be free and can screen the ions. The forces which determine the phonon frequencies are:

- Coulomb repulsion between the bond charges;
- Coulomb attraction between the bond charges and the ions;
- Coulomb repulsion between the ions;
- a non-Coulombic force between the ions to be approximated by a spring.

Using this approach Martin was able to calculate the phonon dispersions and elastic constants of Si with no other adjustable parameters.

A further refinement of the bond charge models of Phillips and Martin was the **adiabatic bond charge model** (ABCM) proposed by *Weber* [3.5]. This model combines the features found in the bond charge model, in the shell model and in the Keating model. The ABCM uses the bond charge model of Martin as its starting point. A conceptual improvement in the ABCM is to treat the bond charges not as rigidly located in the middle of the bonds, as Martin did, but instead to allow the bond charges to follow the motion of the ions adiabatically as in the shell model. As a result of their Coulomb attraction to the ions, the bond charges are unstable against any small perturbation which moves them closer to one ion than to another. To stabilize the bond charges, Weber introduced two additional types of forces: (1) short range re-

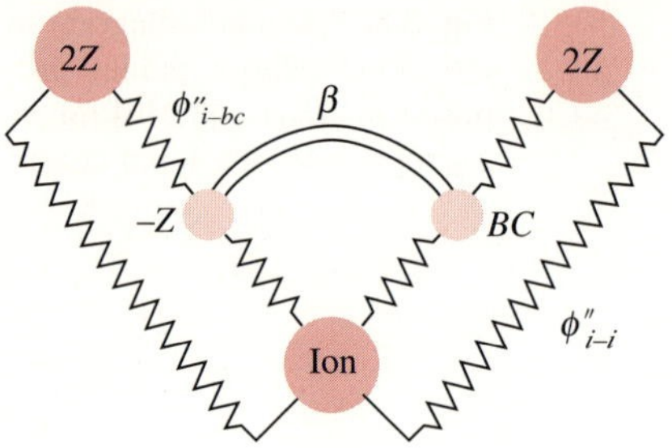

Fig. 3.9. Schematic diagram of the adiabatic bond charge model of *Weber* [3.5]. See text for explanations of the symbols

pulsive forces between the bond charges and the ions and (2) bond-bending forces as in the Keating model. The various interactions between nearest-neighbor ions and bond charges in the ABCM are shown schematically in Fig. 3.9. There are four adjustable parameters in the model:

- $\phi''_{\text{i-i}}$, the potential of central forces between ions;
- $\phi''_{\text{i-bc}}$, the potential of central forces between ions and bond charges;
- Z^2/ε, representing the Coulomb interaction between the bond charges;
- β, the bond-bending parameter in the Keating model.

Some of these parameters can be determined from the elastic constants or long-wavelength phonon dispersion curves while others have to be deduced from the zone-edge phonon energies. The phonon dispersion curves of the group-IV elements calculated with the ABCM are in good agreement with experiment. As an example, Figs. 3.1, 3.10 and 3.11 show, respectively, comparisons between the experimental results and the calculated phonon dispersion curves in Si, α-Sn (or gray tin), and diamond. The phonon dispersion curves of diamond pose a special problem for the ABCM since, unlike the other group-IV elements, the zone-edge TA phonon energies in diamond are quite high.

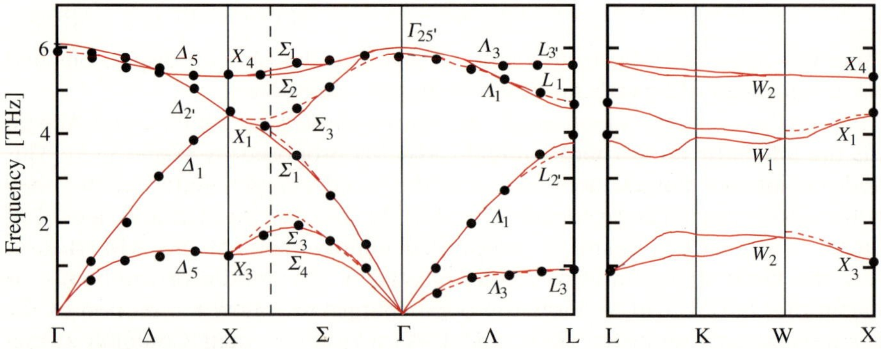

Fig. 3.10. Phonon dispersion curves of α-Sn. The solid lines were calculated with the ABCM of Weber; the broken lines are calculated by a valence force field model while the solid circles are experimental points. (From [3.5])

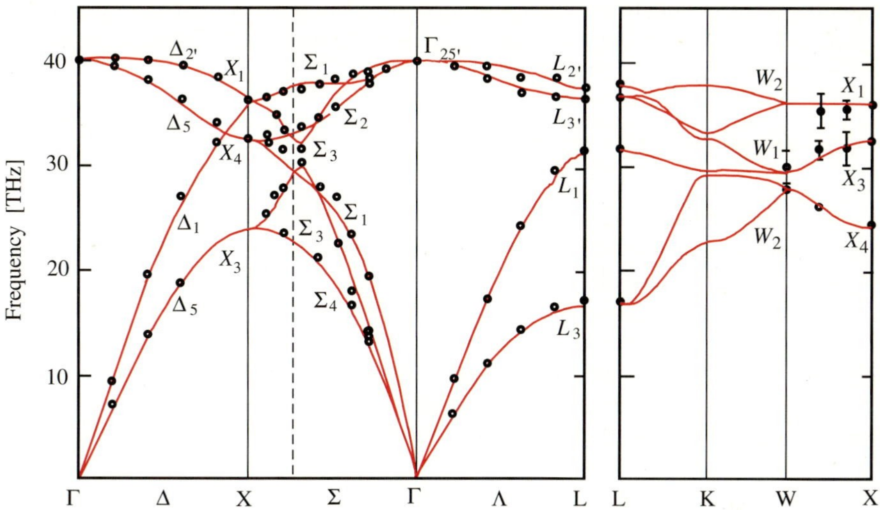

Fig. 3.11. Phonon dispersion curves of diamond. The *solid lines* were calculated with the ABCM of Weber while the *circles* represent experimental points. (From [3.5])

In addition, there are features in the optical phonon branches which cannot be reproduced by a four-parameter ABCM. To obtain a satisfactory fit to the experimental data in diamond, Weber introduced, in an *ad hoc* manner, an additional adjustable parameter in the bond-bending term. A comparison between the experimental phonon dispersion curves in diamond and the results calculated from this five-parameter ACBM is shown in Fig. 3.11. A minor but interesting feature which is not obvious in this figure is that the maximum energy of the optical phonon branch occurs along the [100] direction instead of at Γ as in Si and Ge. A more recent comparison between experimental lattice dynamical results and a first-principles calculation can be found in [3.28].

3.3 Electron–Phonon Interactions

In Sect. 2.1 we pointed out that, within the Born–Oppenheimer approximation, we can decompose the Hamiltonian of a crystal into three terms: $H_{\mathrm{ion}}(\boldsymbol{R}_j)$, $H_{\mathrm{e}}(\boldsymbol{r}_i, \boldsymbol{R}_{j0})$, and $H_{\mathrm{e-ion}}(\boldsymbol{r}_i, \delta\boldsymbol{R}_j)$. The first two terms deal separately with the motions of the ions and the electrons. In Chap. 2 and in Sect. 3.2 we discussed how to solve those two Hamiltonians to obtain, respectively, the electronic band structure and the phonon dispersion curves. We will now consider the third term, which describes the interaction between the electron and the ionic motion, i.e., the **electron–phonon interaction**. Within the spirit of the Born–Oppenheimer approximation we will assume that the electrons can respond instantaneously to the ionic motion so that the electron–phonon interaction Hamiltonian can be expressed as a Taylor series expansion of the electronic Hamiltonian $H_\mathrm{e}(\boldsymbol{r}_i, \boldsymbol{R}_j)$:

$$H_{\mathrm{e-ion}}(\boldsymbol{r}_i, \delta\boldsymbol{R}_j) = \sum_j \left(\frac{\partial H_\mathrm{e}}{\partial \boldsymbol{R}_j}\right)\bigg|_{\boldsymbol{R}_{j0}} \cdot \delta\boldsymbol{R}_j + \dots \,. \tag{3.12}$$

Usually the electronic Hamiltonian $H_\mathrm{e}(\boldsymbol{r}_i, \boldsymbol{R}_j)$ is not known, and therefore approximations are needed to calculate the electron–phonon interaction. For simplicity, we shall consider in detail mostly the long-wavelength phonons in diamond- and zinc-blende-type crystals with two atoms per unit cell. As discussed in Sect. 3.1 there are four kinds of long-wavelength (i.e., $\boldsymbol{k}$ near Γ) phonons: TA, LA, TO, and LO phonons. Their interactions with electrons will be treated separately. Interaction between electrons and large wave-vector phonons will be discussed at the end of this section.

3.3.1 Strain Tensor and Deformation Potentials

Let us assume that the electronic energies of a non-degenerate band $E_{n\boldsymbol{k}}$ (where n is the band index and $\boldsymbol{k}$ the wave vector) are known so that the expectation value of $(\partial H_\mathrm{e}/\partial \boldsymbol{R}_j)$ can be approximated by

$$\left(\frac{\partial H_\mathrm{e}}{\partial \boldsymbol{R}_j}\right)\bigg|_{\boldsymbol{R}_{j0}} \cdot \delta\boldsymbol{R}_j \approx \left(\frac{\partial E_{n\boldsymbol{k}}}{\partial \boldsymbol{R}_j}\right)\bigg|_{\boldsymbol{R}_{j0}} \cdot \delta\boldsymbol{R}_j. \tag{3.13}$$

The constant $(\partial E_{n\boldsymbol{k}}/\partial \boldsymbol{R}_j)$ represents simply the shift of the electronic band energy caused by a static displacement of the atoms. In the case of the long-wavelength acoustic phonons, the atomic displacements can correspond to a deformation of the crystal (**deformation potential theorem**). Such deformations will change the electronic energies at different points in the Brillouin zone; the parameters which describe these changes in the electronic energies induced by static distortions of the lattice are known as **deformation potentials**. Thus the coefficient $\partial E_{n\boldsymbol{k}}/\partial \boldsymbol{R}_j$ is related to the deformation potentials of the crystal. We shall now express the electron–phonon interactions in a semiconductor explicitly in terms of deformation potentials.

Within the limit of zero wave vector or infinite wavelength, an acoustic phonon becomes a uniform translation of the crystal. Obviously such translations will not alter the electronic band structure, hence, if all $\delta\boldsymbol{R}_j$'s are identical, the change in $E_{n\boldsymbol{k}}$ is zero. Thus we have to assume that an acoustic phonon has a nonzero but small wave vector in order to couple to electrons and consider the gradient of the atomic displacements:

$$d_{ij} = \frac{\partial(\delta\boldsymbol{R}_i)}{\partial \boldsymbol{R}_j}. \tag{3.14}$$

d_{ij} is a second-rank tensor which can be decomposed into the sum of a symmetric tensor e_{ij} and an antisymmetric one f_{ij}:

$$e_{ij} = \frac{1}{2}\left(\frac{\partial \delta \boldsymbol{R}_i}{\partial \boldsymbol{R}_j} + \frac{\partial \delta \boldsymbol{R}_j}{\partial \boldsymbol{R}_i}\right) \tag{3.15a}$$

and

$$f_{ij} = \frac{1}{2}\left(\frac{\partial \delta \boldsymbol{R}_i}{\partial \boldsymbol{R}_j} - \frac{\partial \delta \boldsymbol{R}_j}{\partial \boldsymbol{R}_i}\right). \tag{3.15b}$$

The antisymmetric tensor f_{ij} describes a rotation of the crystal and does not change the electron energies. However, the symmetric tensor e_{ij} describes a *strain* induced in the crystal by the atomic displacements and is known as the **strain tensor**. Such strains in a crystal can shift the electronic energies. This suggests that we can derive the electron–phonon interactions for long-wavelength acoustic phonons by expanding $H_e(\boldsymbol{r}_i, \boldsymbol{R}_j)$ in terms of the strain tensor e_{ij}.

As a symmetric second-rank tensor, e_{ij} contains six independent elements. One can interpret these elements as corresponding to different ways in which the crystal can be deformed. One question which arises is: how many linearly independent deformation potentials are needed to describe all the possible changes in the energy of a given electronic state induced by strain? This question can be answered in general by applying group theory. The answer depends on the space group of the crystal, and the wave vector and symmetry (in terms of the irreducible representations of the group of the wave vector) of the particular electronic state being considered. As an illustration of the principles involved, we will discuss the specific cases of electrons and acoustic phonons in diamond- and zinc-blende-type crystals.

Since electron and phonon properties are invariant under symmetry operations of a crystal, we expect the electron–phonon interaction to remain unchanged also. Thus the first step is to symmetrize the strain tensor and the electronic state in terms of the irreducible representations of the crystal. The symmetrization of electron states has already been discussed in Chap. 2. To symmetrize the strain tensor, we note that a second-rank tensor can be constructed from the tensor product of two vectors. A vector has Γ_4 symmetry in zinc-blende crystals, so a second-rank tensor can be decomposed into these irreducible representations (see also Problem 2.3):

$$\Gamma_4 \otimes \Gamma_4 = \Gamma_1 \oplus \Gamma_3 \oplus \Gamma_4 \oplus \Gamma_5. \tag{3.16}$$

To decompose the symmetric tensor e_{ij} according to the irreducible representations in (3.16), we can either use projection operators, as described in any of the references on group theory given for Chap. 2, or use the method of inspection. From the form of the basis functions for the T_d group given in Table 2.3 (see Sect. 2.3.3) we can guess that the three diagonal elements of a second-rank tensor transform as $\Gamma_1 \oplus \Gamma_3$ while the six off-diagonal elements transform as $\Gamma_4 \oplus \Gamma_5$. If we form symmetric and antisymmetric combinations of the off-diagonal elements, the symmetric combination belongs to a representation which has a character equal to 1 under the symmetry operation involving the reflection σ [under this reflection about the (110) plane $xy \to yx$, $xz \to yz$, and $yz \to xz$], while the antisymmetric combination leads to the character of -1 under the same operation. From Table 2.3 we see that the off-diagonal elements of a symmetric tensor (like the strain tensor) should belong to the Γ_4 representation while those of an antisymmetric tensor (like the one describing a rotation, a pseudovector) belong to the Γ_5 representation. Thus the elements of the (symmetric) strain tensor can be combined into these irreducible representations:

$\Gamma_1 : e_{11} + e_{22} + e_{33}$;
$\Gamma_3 : e_{11} - e_{22},\ \ e_{33} - (e_{11} + e_{22})/2$; and
$\Gamma_4 : e_{12},\ \ e_{23},\ \ e_{31}$.

Given any strain tensor e_{ij} in matrix form we can always decompose it into the sum of three matrices:

$$[e_{ij}(\Gamma_1)] = \frac{1}{3}\begin{bmatrix} e_{11}+e_{22}+e_{33} & 0 & 0 \\ 0 & e_{11}+e_{22}+e_{33} & 0 \\ 0 & 0 & e_{11}+e_{22}+e_{33} \end{bmatrix},$$

$$[e_{ij}(\Gamma_3)] = \frac{1}{3}\begin{bmatrix} 2e_{11}-(e_{22}+e_{33}) & 0 & 0 \\ 0 & 2e_{22}-(e_{33}+e_{11}) & 0 \\ 0 & 0 & 2e_{33}-(e_{11}+e_{22}) \end{bmatrix},$$

$$[e_{ij}(\Gamma_4)] = \begin{bmatrix} 0 & e_{12} & e_{13} \\ e_{12} & 0 & e_{23} \\ e_{13} & e_{23} & 0 \end{bmatrix}.$$

Note that the matrix with Γ_1 symmetry has a nonzero trace while the other two matrices are traceless. Using the definition of the strain tensor components in (3.15a), it can be shown that the trace $(e_{11} + e_{22} + e_{33})$ of the strain tensor is equal to the fractional volume change $(\delta V/V)$ or **volume dilation** associated with a strain pattern. On the other hand, a traceless strain matrix describes a **shear** of the medium. In a zinc-blende crystal, $e_{ij}(\Gamma_3)$ corresponds to the shear component of the strain produced by a uniaxial stress applied along a [100] direction of the crystal while $e_{ij}(\Gamma_4)$ corresponds to a [111] uniaxial stress. The proof of these results is left for Problem 3.2 at the end of this chapter.

Let us consider a long wavelength acoustic vibration described by a plane wave with frequency ω and wave vector $\boldsymbol{q}$:

$$\delta\boldsymbol{R} = \delta\boldsymbol{R}_0 \ \sin(\boldsymbol{q}\cdot\boldsymbol{r} - \omega t). \tag{3.17}$$

The strain tensor e_{ij} associated with this phonon, according to the definition in (3.15a), is given by

$$e_{ij} = \tfrac{1}{2}[q_i\delta R_{0j} + q_j\delta R_{0i}] \ \cos(\boldsymbol{q}\cdot\boldsymbol{r} - \omega t). \tag{3.18}$$

In a longitudinal mode, the displacement $\delta\boldsymbol{R}$ is parallel to the direction of propagation $\boldsymbol{q}$. Thus the nonzero strain tensor components for LA phonons are simply (in the limit of both q and ω approaching zero)

$$e_{ii} = q_i \delta R_{0i}. \tag{3.19}$$

The strain tensor associated with a long-wavelength LA is therefore a diagonal tensor. An examination of the trace of this tensor shows that the LA phonon always produces an oscillatory dilation ($\delta V/V$) with amplitude equal to $\boldsymbol{q} \cdot \delta \boldsymbol{R}_0$ *plus a shear*. This is consistent with our expectation that an acoustic wave causes periodic expansion and compression of a medium. A small uniform expansion of the crystal by δV will shift the energy of an electronic band extremum $E_{n\boldsymbol{k}}$ by an amount

$$\delta E_{n\boldsymbol{k}} = a_{n\boldsymbol{k}}(\delta V/V), \tag{3.20}$$

where $a_{n\boldsymbol{k}}$ is known as the **volume deformation potential** of the energy level $E_{n\boldsymbol{k}}$. In principle, this deformation potential can be determined by measuring the shift in $E_{n\boldsymbol{k}}$ induced by hydrostatic pressure. In practice, there are very few experimental techniques capable of measuring the volume deformation potential *directly*. Often optical measurements on samples under hydrostatic pressure are applied to determine the volume deformation potentials. In these optical experiments, usually only *energy differences* between two band extrema are measured. As a result, only *relative* volume deformation potentials between two band extrema are derived from such optical experiments while *absolute* ones are required in (3.20).

For non-degenerate bands, we can neglect the effect of the shear strain associated with LA and write down the electron–LA phonon interaction Hamiltonian $H_{\text{e-LA}}$ for small phonon wave vectors $\boldsymbol{q}$ as

$$H_{\text{e-LA}} = a_{n\boldsymbol{k}}(\boldsymbol{q} \cdot \delta \boldsymbol{R}), \tag{3.21}$$

where $\delta \boldsymbol{R}$ can be expressed in terms of **phonon creation and annihilation operators** c_q^+ and c_q by using the standard result obtained from quantum mechanics [Ref. 3.29, p. 107]

$$\begin{aligned} \delta \boldsymbol{R} = \sum_{\boldsymbol{q}} \left(\frac{\hbar}{2NV\varrho\omega} \right)^{1/2} \boldsymbol{e}_q \{ c_q^+ \; \exp[\mathrm{i}(\boldsymbol{q} \cdot \boldsymbol{r}_j - \omega t)] \\ + c_q \; \exp[-\mathrm{i}(\boldsymbol{q} \cdot \boldsymbol{r}_j - \omega t)] \}, \end{aligned} \tag{3.22}$$

where N is the number of unit cells in the crystal, V and ϱ are, respectively, the volume of the primitive cell and the density, and $\boldsymbol{e}_q$ is the phonon polarization unit vector. The Hamiltonian $H_{\text{e-LA}}$ in (3.21) is valid for a nondegenerate band extremum such as the Γ_1 conduction band minimum in GaAs and other zinc-blende semiconductors. The values of the volume deformation potentials $a(\Gamma_{1\text{c}})$ for the conduction band and the relative volume deformation potentials $a(\Gamma_{1\text{c}}) - a(\Gamma_{15\text{v}})$ in these semiconductors are listed in Table 3.1.

As pointed out earlier, only the relative volume deformation potentials $a(\Gamma_{1\text{c}}) - a(\Gamma_{15\text{v}})$ between the conduction and valence bands are measured in an optical experiment under hydrostatic stress. In cases where the absolute deformation potentials for the conduction band are known [3.31] one finds that the Γ_1 conduction band deformation potential is typically about ten times

Table 3.1. Deformation potentials for the conduction and valence band extrema in diamond and zinc-blende semiconductors (in eV). a denotes the volume deformation potential for the lowest energy Γ_{1c} conduction band minimum or the highest energy Γ_{15v} valence band maximum (zinc-blende notation). b and d are the shear deformation potentials for the Γ_{15v} valence band maximum. Ξ_d and Ξ_u denote deformation potentials at zone boundaries. Most of the data are taken from [3.30]

	Ξ_d	Ξ_u	$a(\Gamma_{1c})$	$a(\Gamma_{1c}) - a(\Gamma_{15v})$	b	d
Si	$\approx 5^{a}$	8.77^{a}		−10	−2.2	−5.1
Ge	-12.3^{b}	16.3^{b}		−12	−2.3	−5.0
GaP		13		−9.3	−1.8	−4.5
GaAs	6.5^{a}	14.5^{b}	−8.6	−9	−2.0	−5.4
GaSb				−8.3	−1.8	−4.6
InP			−7	−6.4	−2.0	−5.0
InAs				−6.0	−1.8	−3.6
InSb				−7.7	−2.0	−5.0
ZnS				−4.0	−0.62	−3.7
ZnSe				−5.4	−1.2	−4.3
ZnTe				−5.8	−1.8	−4.6
CdTe				−3.4	−1.2	−5.4

[a][100] valleys;

[b][111] valleys, D.N. Mirlin, V.F. Sapega, I.Ya. Karlik, R. Katilius: Hot luminescence investigation of L-valley spliting in GaAs. Solid State Commun. **61**, 799–805 (1987)

larger than that of the Γ_{15} (or $\Gamma_{25'}$) valence bands. The relative volume deformation potentials of the conduction and valence bands at the zone center in diamond- and zinc-blende-type semiconductors can be calculated quite easily within the tight-binding approximation. This is left as an exercise (Problem 3.9). However, the calculation of the absolute deformation potentials is not trivial [3.32].

One thing to notice about $H_{\mathrm{e-LA}}$ is the explicit and implicit dependence of its matrix element on q. From (3.22) the LA phonon displacement $\delta\boldsymbol{R}$ is proportional to $\omega^{-\frac{1}{2}}$. Since ω is linear with q for acoustic phonons (in the limit of long wavelength) and $H_{\mathrm{e-LA}}$ depends on $\boldsymbol{q}\cdot\delta\boldsymbol{R}$ in (3.21), $H_{\mathrm{e-LA}}$ varies explicitly with q as $q^{\frac{1}{2}}$. On the other hand, in (3.22) $\delta\boldsymbol{R}$ is expressed in terms of the phonon creation and annihilation operators, whose matrix elements depend on the **phonon occupation number**. The probability that a phonon state with energy $\hbar\omega$ is excited at a temperature T is known as its occupation number. Phonons are bosons and their phonon occupation number $N_{\mathrm{ph}}(\hbar\omega)$ is given by the **Bose–Einstein distribution function**

$$N_{\mathrm{ph}}(\hbar\omega) = \{\exp[\hbar\omega/(k_{\mathrm{B}}T)] - 1\}^{-1},$$

where k_{B} is the Boltzmann constant. The magnitudes squared of the matrix elements of the phonon creation and annihilation operators are $N_{\mathrm{ph}}(\hbar\omega) + 1$ and $N_{\mathrm{ph}}(\hbar\omega)$, respectively. For $k_{\mathrm{B}}T \gg \hbar\omega$, $N_{\mathrm{ph}}(\hbar\omega) \approx k_{\mathrm{B}}T/(\hbar\omega) \gg 1$. As a result, the phonon occupation term in the matrix element (squared) of $H_{\mathrm{e-LA}}$ depends on q as q^{-1}. Hence, the explicit and implicit dependences of the matrix element (squared) of $H_{\mathrm{e-LA}}$ on q cancel each other. These results will be used in Chap. 5 in calculating the mobility of electrons as a function of temperature.

3.3.2 Electron–Acoustic-Phonon Interaction at Degenerate Bands

In the previous section we concentrated on LA phonons because they always produce a change in the volume of the crystal which affects all energy bands. One can easily prove this based on the matrix-element theorem (see Sect. 2.3.4). A volume dilation does not change the symmetry of the crystal so the Hamiltonian in (3.21) must belong to the identity representation. When acting on an electronic band at the zone center (i. e., Γ), H_{e-LA} must contain a term belonging to the Γ_1 representation. If the electron symmetry is Γ_i, then $\Gamma_i \otimes \Gamma_1 = \Gamma_i$ and therefore the matrix element of $\langle \Gamma_i | H_{e-LA} | \Gamma_i \rangle$ is nonzero. Compared to their dilation component the shear components of LA phonons are usually less important. On the other hand transverse acoustic (TA) phonons contain only shear waves. To first order, a shear strain does not affect the energy of a nondegenerate band in a cubic crystal. Shear strains, however, lower the symmetry of a cubic crystal. As a result, the most important effect of a shear strain on a cubic crystal is to lift some of the degeneracy of energy bands at high symmetry points of the Brillouin zone. Again the matrix-element theorem can be applied to predict whether a particular shear strain of symmetry, say Γ_s, will split the degeneracy of a state of symmetry Γ_i. In this section we will consider the effect of strain and acoustic phonons on two important cases of degenerate bands in diamond- and zinc-blende-type semiconductors. These are the degenerate heavy and light hole bands and the degenerate conduction band minima in Si and Ge.

a) Degenerate Heavy and Light Hole Bands at Γ

In the previous section we showed that any strain tensor in a zinc-blende crystal can be decomposed into three separate *irreducible* tensors transforming as Γ_1, Γ_3, and Γ_4. This result suggests that we require three deformation potentials to describe the effect of a general strain on a general band extremum at Γ. In the case of the degenerate Γ_{15} valence bands, it is convenient to regard these six (including spin) bands as transforming like the eigenstates of a $J = 3/2$ and a $J = 1/2$ angular momentum operator (see Sect. 2.6.2). The hole–strain interaction Hamiltonian can be derived by symmetrizing the angular momentum operator J through multiplication by the appropriate components of the strain tensor (**method of invariants**). The procedure for doing this can be derived from group theory and has been described in detail by *Kane* [3.33] and by *Pikus* and *Bir* [3.34, 35]. The **Pikus and Bir effective strain Hamiltonian** for the $J = 3/2$ valence bands in the zinc-blende semiconductors is given by (see also the discussion in Problem 3.8)

$$\begin{aligned} H_{\mathrm{PB}} =& a(e_{xx} + e_{yy} + e_{zz}) + b\left[(J^2{}_x - J^2/3)e_{xx} + \text{c.p.}\right] \\ &+ \frac{2d}{\sqrt{3}}\left(\frac{1}{2}(J_xJ_y + J_yJ_x)e_{xy} + \text{c.p.}\right), \end{aligned} \tag{3.23}$$

where a, b, and d are the three deformation potentials corresponding to strain tensors with symmetries Γ_1, Γ_3, and Γ_4, respectively, and c.p. stands for cyclic permutation. It is no coincidence that H_{PB} is very similar to the Kohn–Luttinger Hamiltonian in (2.70). After all, both of them have been deduced based on the symmetry of the zinc-blende crystal, and the strain tensor e_{ij} has the same symmetry as $k_i k_j$ or $\nabla_i \nabla_j$.

The deformation potentials b and d determine the splitting of the four-fold degenerate $J = 3/2$ valence bands at Γ under [100] and [111] uniaxial stress, respectively. Although (3.23) has been derived for a static uniform strain, it is reasonable to assume that this Hamiltonian and the deformation potentials are valid also for long-wavelength acoustic phonons. By applying static uniaxial stress to the diamond- and zinc-blende-type semiconductors listed in Table 3.1, their shear deformation potentials can be determined. In particular, it can be shown (see Problem 3.8) that the splittings δE of the $J = 3/2$ valence bands under a uniaxial compressive stress of magnitude X along the [100] or [111] directions are given by

$$\delta E = \begin{cases} 2b(S_{11} - S_{12})X, & [100] \text{ stress}, \qquad (3.24a) \\ (1/\sqrt{3})dS_{44}X, & [111] \text{ stress}, \qquad (3.24b) \end{cases}$$

where S_{11}, S_{12} and S_{44} are components of the fourth-rank **compliance tensor** S_{ijkl}, which relates the strain tensor e_{ij} to the **stress tensor** X_{kl}. The definitions of the stress tensor and of the compliance tensor can be found in Problem 3.2. The shear deformation potentials (in the absence of spin–orbit interaction) can be calculated within the tight-binding approximation as shown in Problems 3.10–13.

b) Degenerate Conduction Band Minima along Δ in Si and at L in Ge

The second example of the interaction between phonons and degenerate electron bands is that of the degenerate conduction band minima in Si and Ge. In general, we expect that a shear strain will deform a crystal and lift some of the degeneracies in the conduction band as in the case of the degenerate valence bands discussed in the previous section. For example, Fig. 2.10 shows that the lowest conduction band minimum in Si occurs at a point $[k_{x0}, 0, 0]$ along the [100] direction of the Brillouin zone. The symmetry of that band in the group of the wave vector $\boldsymbol{k}$ along the [100] direction is Δ_1. While this state is nondegenerate for a given $\boldsymbol{k}$, there are five other $\boldsymbol{k}$ vectors (in the $[\bar{1}00]$, [010], $[0\bar{1}0]$, [001] and $[00\bar{1}]$ directions) which are equivalent to the [100] wave vector by symmetry, so that the Δ_1 conduction band minimum is six-fold degenerate as a result of degeneracy in reciprocal space. A uniaxial stress applied along the [100] direction will make the [100] and $[\bar{1}00]$ directions different from the remaining four equivalent directions. Thus, from symmetry arguments one expects that a [100]-oriented uniaxial stress will split the six equivalent conduction band minima in Si into a doublet and a quadruplet. On the other hand a uniaxial stress applied along the [111] direction will affect all six minima in the same way and leave the degeneracy unchanged.

To derive the strain Hamiltonian we have to consider the symmetries of the strain tensor, and the wave function of the band and of the equivalent wave vectors $\boldsymbol{k}$. For example, in the case of Si, the wave function of the conduction band has symmetry Δ_1 while the six equivalent k vectors form a six-fold reducible representation. If the wave function of the band for each given $\boldsymbol{k}$ is nondegenerate, the form of the strain Hamiltonian is much simpler. Since this is the case for the conduction band minima in both Si and Ge, this is the only case we will consider here. Treatment of the more difficult case of degenerate bands can be found in, for example, the article by *Kane* [3.33]. The strain Hamiltonian for the nondegenerate band was first derived by *Herring* and *Vogt* [3.37] by generalizing the definition of the volume deformation potential in (3.20) into

$$\delta E_{nk} = \sum_{j=1}^{6} \Xi_j e_j, \tag{3.25}$$

where the Ξ_j are the deformation potentials and the e_j are the components of the strain tensor. To simplify the notation this second-rank strain tensor has been contracted into a six-component array (see Problem 3.3 for a discussion of this contracted notation). In a general crystal without considering symmetry, six deformation potentials are required to describe the strain-induced energy shift of a nondegenerate electronic state at a point $\boldsymbol{k}$. This number is greatly reduced by symmetry considerations. For example, the strain tensor of a diamond crystal can be decomposed into three tensors belonging to the irreducible representations $\Gamma_1(\Gamma_1^+)$, $\Gamma_{12}(\Gamma_3^+)$ and $\Gamma_{25'}(\Gamma_5^+)$ (Tables 2.5 and 2.7; note that because of the inversion symmetry of the diamond lattice, all its second-rank tensors must be even under the inversion operation). As a result, we need no more than three deformation potentials to describe the strain-induced energy shift of a given nondegenerate electronic state in the diamond crystal. The number is further reduced to two for $\boldsymbol{k}$ pointing along high symmetry directions such as [100] and [111]. As pointed out in Sect. 3.3.1 and in Problem 3.4, the traceless strain tensors with Γ_3 and Γ_4 symmetries in the zinc-blende crystal correspond to the shear components of the strains produced by uniaxial stress applied along the [100] and [111] directions, respectively. Since all the equivalent [100] valleys in Si appear symmetrical with respect to a [111] uniaxial stress, their degeneracy cannot be split by such a stress. On the other hand, a [100] stress will split the [100] valley from the [010] and [001] valleys. As a result, only two deformation potentials are required to describe the strain effect on the [100] conduction valleys in Si. *Herring* and *Vogt* [3.37] have expressed this result as a strain Hamiltonian of the form

$$H_{\mathrm{HV}} = \Xi_d(\mathrm{Tr}\{\boldsymbol{e}\}) + \Xi_u(\hat{\boldsymbol{k}} \cdot \boldsymbol{e} \cdot \hat{\boldsymbol{k}}), \tag{3.26}$$

where $\mathrm{Tr}\{\boldsymbol{e}\}$ is the trace of the strain tensor $\boldsymbol{e}$, and $\hat{\boldsymbol{k}}$ is a unit vector along the direction of one of the equivalent [100] conduction band minima in reciprocal space. In (3.26) Ξ_u is a shear deformation potential associated with a uniaxial strain along the [100] direction and $\Xi_d + \Xi_u$ is the volume deformation potential (the reader is urged to verify this). Sometimes the deformation potentials Ξ_d and Ξ_u are also denoted by E_1 and E_2 following the notation introduced by *Brooks* [3.38].

It is straightforward to show that the above arguments for Si can also be applied to derive a strain Hamiltonian for the L_6 conduction band minima in Ge which occur along the four equivalent [111], $[\bar{1}11]$, $[1\bar{1}1]$, and $[11\bar{1}]$ directions. In this case a [001] uniaxial stress will not split the equivalent [111] valleys just as a [111] uniaxial strain will not split the [100] valleys. Instead a [111] stress will split the [111] valley (which will form a singlet) from the other three valleys, which remain degenerate (triplet). The result can be expressed in terms of a strain Hamiltonian similar to H_{HV} in (3.26). However, the shear deformation potential Ξ_u is now the shear deformation potential for a [111] uniaxial strain rather than a [100] strain (readers should check this also) and $\hat{\boldsymbol{k}}$ is a unit vector along one of the equivalent [111] directions.

Table 3.2 summarizes the relationship between the deformation potentials Ξ_j and the deformation potentials Ξ_d and Ξ_u in cubic semiconductors for $\boldsymbol{k}$ along high-symmetry directions. For equivalent valleys along the [110] directions, we note that three deformation potentials (Ξ_d, Ξ_u, and Ξ_p) are required. The proof of this result, which does not follow from (3.26), is left as an exercise in Problem 3.14. The values of the deformation potentials for the conduction band valleys in Si and Ge are given in Table 3.1. The electron–acoustic-phonon interaction Hamiltonians for Si and Ge are obtained by substituting (3.18), the strain tensor associated with an acoustic phonon, into (3.26).

Table 3.2. Relation between the deformation potentials Ξ_j and the deformation potentials Ξ_d, Ξ_u and Ξ_p in cubic semiconductors at high symmetry points [3.37]

Direction of k:	100	111	110
Ξ_1	$\Xi_d + \Xi_u$	$\Xi_d + (1/3)\Xi_u$	$\Xi_d + \Xi_u - (1/2)\Xi_p$
Ξ_2	Ξ_d	$\Xi_d + (1/3)\Xi_u$	$\Xi_d + \Xi_u - (1/2)\Xi_p$
Ξ_3	Ξ_d	$\Xi_d + (1/3)\Xi_u$	$\Xi_d - \Xi_u + \Xi_p$
Ξ_4	0	$(1/3)\Xi_u$	0
Ξ_5	0	$(1/3)\Xi_u$	0
Ξ_6	0	$(1/3)\Xi_u$	$(1/2)\Xi_p$

3.3.3 Piezoelectric Electron–Acoustic-Phonon Interaction

In noncentrosymmetric crystals, a stress can induce a macroscopic electric polarization field $\boldsymbol{E}$. This phenomenon is known as the **piezoelectric effect** [Ref. 3.39, p. 110]. This phenomenon can also be described as a strain inducing an electric field. The induced field will be proportional to the strain provided it is small. Since the strain tensor $\boldsymbol{e}$ has rank two and the induced electric field is a vector (or tensor of rank one) this constant of proportionality can be expressed as a third-rank **electromechanical tensor** $\boldsymbol{e}_{\mathrm{m}}$. In a medium with dielectric constant ε_∞ the strain-induced field can be expressed as

$$\boldsymbol{E} = (-4\pi)\frac{(\boldsymbol{e}_{\mathrm{m}} \cdot \boldsymbol{e})}{\varepsilon_\infty}\left(\frac{1}{4\pi\varepsilon_0}\right). \tag{3.27}$$

This result for a static strain can be extended to the case of an oscillating strain field associated with long-wavelength acoustic phonons. If $\boldsymbol{q}$ and $\delta\boldsymbol{R}$ [defined in (3.17)] are, respectively, the wave vector and atomic displacements associated with the acoustic phonon, the strain tensor corresponding to the acoustic phonon is $\mathrm{i}\boldsymbol{q}\delta\boldsymbol{R}$ from (3.15a). Substituting this result into (3.27) we obtain the sinusoidal macroscopic piezoelectric field induced by an acoustic phonon:

$$\boldsymbol{E}_{\mathrm{pe}} = (4\pi)\mathrm{i}\boldsymbol{e}_{\mathrm{m}} \cdot \frac{\boldsymbol{q}\delta\boldsymbol{R}}{4\pi\varepsilon_0\varepsilon_\infty}. \tag{3.28}$$

The longitudinal component of this electric field can be expressed in terms of a scalar piezoelectric potential ϕ_{pe}

$$\phi_{\mathrm{pe}} = -\boldsymbol{q} \cdot \boldsymbol{E}_{\mathrm{pe}}/(\mathrm{i}q^2). \tag{3.29}$$

In the presence of this potential, the energy of an electron will be changed by $-|e|\phi_{\mathrm{pe}}$, therefore the **piezoelectric electron–phonon Hamiltonian** can be written as

$$H_{\mathrm{pe}} = -|e|\phi_{\mathrm{pe}} = (4\pi)\frac{|e|}{4\pi\varepsilon_0 q^2\varepsilon_\infty}\,\boldsymbol{q} \cdot \boldsymbol{e}_{\mathrm{m}} \cdot (\boldsymbol{q}\delta\boldsymbol{R}). \tag{3.30}$$

The form of the electromechanical tensors in crystals with the zinc-blende and wurtzite structure can be determined by symmetry (see Problem 3.15). Using (3.30), explicit expressions for H_{pe} can be deduced for any acoustic phonons. As an example, the calculation of H_{pe} for acoustic phonons in wurtzite crystals is left as an exercise (see Problem 3.16 and paper by *Mahan* and *Hopfield* [3.40]). If we compare the deformation-potential electron–acoustic-phonon interaction in (3.21) with the piezoelectric electron–acoustic-phonon interaction in (3.30), we notice that (3.30) contains an additional $(1/q)$ dependence. This extra term arises from the Coulomb interaction in the piezoelectric electron–acoustic-phonon interaction. As a result, H_{pe} becomes stronger for small q or long-wavelength acoustic phonons. This is why H_{pe} is said to be a *long-range interaction* while the deformation potential interaction is a *short-range interaction*: the Fourier transform of a function of long range in q is of short range in r and vice versa.

The values of electromechanical tensor components in tetrahedrally coordinated semiconductors can be estimated within empirical models such as the tight-binding model. We will not discuss these calculations here and interested readers should consult [3.41, 42]. Instead, we will list the values of electromechanical tensor components for some typical semiconductors in Table 3.3. Note that the more-ionic wurtzite semiconductors such as ZnO, CdS, and CdSe have electromechanical tensor components that are larger and also different in sign from those of the zinc-blende semiconductors.

3.3.4 Electron–Optical-Phonon Deformation Potential Interactions

In crystals with two or more atoms per unit cell, a long-wavelength optical phonon involves relative displacements of atoms within the primitive unit cell.

Table 3.3. Values of the nonzero and linearly independent components of the electromechanical tensor in some zinc-blende and wurtzite-type (labeled as W) semiconductors [4.43, 44]. Note that the electromechanical tensor is defined in terms of the polarization. The cgs unit for polarization is statcoulomb/cm^2. The corresponding SI unit is Coulomb/m^2. The unit of the electromechanical tensor components in this table is 10^4 statcoulomb/cm^2. One statcoulomb/cm^2 is equal to $1/(3 \times 10^5)$ Coulomb/m^2

Semiconductor	$(e_m)_{14}$	$(e_m)_{33}$	$(e_m)_{31}$	$(e_m)_{15}$
AlAs	6.7			
AlSb	2.04			
GaP	−3.0			
GaAs	4.8			
GaSb	3.78			
InP	1.2			
InAs	1.38			
InSb	2.13			
ZnS	5.1			
ZnSe	1.35			
ZnTe	0.81			
CdTe	1.02			
ZnO(W)		33	−4.8	−9.3
CdS(W)		14.7	−7.5	−6.3
CdSe(W)		10.4	−4.8	−4.14

Unlike acoustic phonons, a long-wavelength optical phonon does not involve a macroscopic strain of the crystal, since there is no macroscopic distortion of the lattice. Instead, optical phonons can be regarded as "microscopic distortions" within the primitive unit cell. Optical phonons can change the energy of an electronic band in two ways, similar to the acoustic phonons. In nonpolar crystals optical phonons alter the electronic energies by changing the bond lengths and/or the bond angles. This electron–optical-phonon interaction is the analog of the deformation potential interaction of acoustic phonons and is known accordingly as the **deformation-potential electron–optical-phonon interaction**. In polar crystals a long-wavelength longitudinal optical (LO) phonon involves uniform displacements of the charged atoms within the primitive cell. Such relative displacement of oppositely charged atoms generates a macroscopic electric field. This electric field can then interact with electrons in a way similar to the piezoelectric electric field of acoustic phonons. This electron–longitudinal-optical-phonon interaction is known as the **Fröhlich interaction**. In this section we will first consider the deformation-potential interaction.

Let us define the distance between the two atoms inside the primitive unit cell of a diamond- or zinc-blende-type semiconductor as a_0 and the relative displacement between these two atoms associated with a zone-center optical phonon as $\boldsymbol{u}$. As in the case of the acoustic phonons, we will define a phenomenological *optical phonon deformation potential* to describe the electron–optical-phonon interaction:

$$H_{\mathrm{e-OP}} = D_{n,\boldsymbol{k}}(u/a_0), \tag{3.31}$$

where $D_{n,\boldsymbol{k}}$ is now the **optical phonon deformation potential** for the energy band indexed by n and $\boldsymbol{k}$. Since this deformation potential interaction does not depend on the phonon wave vector it is also a short-range interaction. In zinc-blende-type (or diamond-type) semiconductors the optical phonon displacement $\boldsymbol{u}$ has symmetry Γ_4 ($\Gamma_{25'}$ for diamond structure), so the matrix element of (3.31) between two nondegenerate s-like Γ_1 (or $\Gamma_{2'}$) conduction band states is zero. Thus there is *no* deformation-potential interaction between the lowest conduction band electrons and optical phonons in direct bandgap semiconductors such as GaAs and InP. The case of indirect bandgap semiconductors like Si or Ge is more complicated and will be discussed below together with the interaction between optical phonons and the p-like valence bands in diamond- and zinc-blende-type semiconductors.

In some crystals, macroscopic strains are symmetry compatible with displacements of atoms involved in zone-center optical phonons. In such crystals an optical phonon may be described in terms of an **internal strain** [3.45, 46]. For example, the optical phonons in diamond- and zinc-blende-type semiconductors can be represented by a relative displacement of the two atoms within the primitive cell along the [111] body diagonal. (Although a macroscopic [111] shear strain will also produce such relative displacement of the atoms within the primitive cell, this displacement is not uniquely defined by the strain tensor, see Problem 3.11.) In such cases the quantity $2u/a_0$ is known as the *internal strain* (see Problem 3.13) and it is possible to deduce the electron–optical phonon interaction by considering the effect of a [111] uniaxial stress. For example, if the spin–orbit couplings in the valence bands of diamond- and zinc-blende-type semiconductors are neglected, the optical phonon deformation potentials for the p-like valence band (usually denoted by d_0) can be related to that of the splitting of the valence bands under a [111] uniaxial stress. The calculation of d_0 within the tight-binding approximation is left as an exercise (Problem 3.12), see [3.30] for more details. For a more general discussion of the calculation of the optical-phonon deformation potentials in tetrahedrally bonded semiconductors, readers are referred to [3.47].

There are relatively few experimental techniques capable of determining the optical-phonon deformation potentials in semiconductors. The value of d_0 in Ge and GaAs has been measured to be 36 and 41 eV, respectively, by Raman scattering [3.48]. Other methods available for estimating d_0 are based on the temperature dependence of the hole mobilities in p-type samples and of the linewidth of the direct optical transition from the split-off valence band to the conduction band. These measurements involve phenomena to be discussed in Chaps. 5–7. Typically one finds that d_0 is of the order of 40 eV in most tetrahedrally coordinated semiconductors (see experimental values compiled in [3.47]).

3.3.5 Fröhlich Interaction

In a polar or partly ionic crystal with two atoms per unit cell, the long-wavelength longitudinal optical (LO) phonon can induce an oscillating macroscopic polarization, leading to an electric field $\boldsymbol{E}_{\mathrm{LO}}$, see, e.g. [Ref. 3.9, p. 86])

$$\boldsymbol{E}_{\mathrm{LO}} = -F\boldsymbol{u}_{\mathrm{LO}}, \tag{3.32}$$

where

$$F = -\left[4\pi N\mu\omega_{\mathrm{LO}}^2(\varepsilon_\infty^{-1} - \varepsilon_0^{-1})\right]^{1/2}(4\pi\varepsilon_0)^{-1/2}. \tag{3.33}$$

In (3.32, 33) the phonon amplitude $\boldsymbol{u}_{\mathrm{LO}}$ is defined as the displacement of the positive ion relative to the negative ion, N is the number of unit cells per unit volume of the crystal, μ is the reduced mass of the primitive cell defined by

$$\mu^{-1} = M_1^{-1} + M_2^{-1}, \tag{3.34}$$

M_1 and M_2 are the masses of the two atoms inside the primitive cell, ω_{LO} is the LO phonon frequency, and ε_∞ and ε_0 are, respectively, the high- and low-frequency dielectric constants. We will delay the derivation of (3.33) until Chap. 6, when we study the effect of long-wavelength transverse optical (TO) phonons on the infrared optical properties of semiconductors. This effect is due to the fact that the TO phonons associated with the LO phonons in polar crystals produce transverse electric dipole moments which couple to photons.

The longitudinal field in (3.32) can be expressed in terms of a scalar potential ϕ_{LO} in the same way as for the piezoelectric acoustic phonons in (3.29):

$$\phi_{\mathrm{LO}} = (F/\mathrm{i}q)u_{\mathrm{LO}}. \tag{3.35}$$

The interaction between an electron of charge $-|e|$ and this macroscopic Coulomb potential is known as the **Fröhlich interaction**. The Hamiltonian for this interaction is given by the simple expression

$$H_{\mathrm{Fr}} = (-e)\phi_{\mathrm{LO}} = (\mathrm{i}eF/q)u_{\mathrm{LO}}. \tag{3.36}$$

When combined with an expression for the displacement u_{LO} analogous to (3.22),

$$u_{\mathrm{LO}} = (\hbar/2N\mu\omega_{\mathrm{LO}})^{1/2}\{c_q^+ \exp[\mathrm{i}(\boldsymbol{q}\cdot\boldsymbol{r} - \omega_{\mathrm{LO}}t)] + \mathrm{c.c.}\}, \tag{3.37}$$

the Fröhlich Hamiltonian can be written as

$$H_{\mathrm{Fr}} = \sum_{\boldsymbol{q}}(\mathrm{i}C_{\mathrm{F}}/q)\{c_q^+ \exp[\mathrm{i}(\boldsymbol{q}\cdot\boldsymbol{r} - \omega_{\mathrm{LO}}t)] - \mathrm{c.c.}\} \tag{3.38}$$

where the coefficient C_{F} is given by

$$C_{\mathrm{F}} = e\left[\frac{2\pi\hbar\omega_{\mathrm{LO}}}{NV}(\varepsilon_\infty^{-1} - \varepsilon_0^{-1})\right]^{1/2}(4\pi\varepsilon_0)^{-1/2}. \tag{3.39}$$

Notice the change in the sign of the complex conjugate terms inside the two brackets in (3.37) and (3.38). This is necessary to ensure that the Fröhlich Hamiltonian is Hermitian.

While the deformation potentials for optical phonons and acoustic phonons are difficult to calculate, the Fröhlich interaction can be calculated in terms of macroscopic parameters such as ε_∞ and ε_0. Note that the Fröhlich interaction depends on the phonon wave vector as q^{-1}. Hence it diverges, in principle, as q decreases to zero. This is not possible in intraband electron–LO-phonon scattering because the LO phonon frequency is nonzero even at $q = 0$. Energy and momentum conservation prevents electrons from undergo-

Table 3.4. Summary of electron–phonon interactions in Si and GaAs. DP and PZ stand for deformation potential and piezoelectric interactions, respectively. Symbols in parentheses represent the commonly used notations for these interactions

Phonon	Si		GaAs	
	Conduction	Valence	Conduction	Valence
TA	DP (Ξ_u)	DP (b, d)	PZ	DP (b, d), PZ
LA	DP (Ξ_d, Ξ_u)	DP (a_v, b, d)	DP (a_c), PZ	DP (a_v, b, d), PZ
TO		DP (d_0)		DP (d_0)
LO		DP (d_0)	Fröhlich	DP (d_0), Fröhlich

ing intraband scattering via $q \equiv 0$ optical phonons. Nevertheless, depending on the electron band dispersion, q can be quite small, and this scattering mechanism can dominate at temperatures where a significant number of LO phonons are excited (i. e., $k_B T \geq \hbar\omega_{LO}$).

From the above discussion we see that there are many different ways for electrons to interact with long-wavelength phonons. Table 3.4 summarizes these different kinds of interactions in a representative polar (GaAs) and nonpolar (Si) semiconductor for electrons at either the lowest conduction band or the top valence bands. Note that while a TA phonon involves only shear strain and no volume dilation, an LA phonon can produce both.

3.3.6 Interaction Between Electrons and Large-Wavevector Phonons: Intervalley Electron–Phonon Interaction

So far we have considered the interaction between electrons and zone-center phonons. Interactions between electrons and zone-edge or near zone-edge phonons has been found to play an important role in optical absorption at indirect energy gaps and in phenomena involving hot electrons (Chap. 5). For example, a zone-edge phonon can scatter an electron from a band minimum at the zone center to a band minimum at the zone edge. This kind of electron–phonon scattering is known as **intervalley scattering**. In Chap. 5 we will show that intervalley scattering is responsible for the **Gunn effect** [5.33]. For indirect bandgap semiconductors, such as Si, where the conduction band minimum occurs at a point either inside the Brillouin zone or at the zone edge, there are several equivalent conduction band valleys. In these cases electrons can be scattered from one valley to another via a large wave-vector phonon.

There are several qualitative differences between the interaction of electrons with zone-center phonons and with zone-edge phonons. Zone-edge phonons cannot generate long-range electric fields [remember the q^{-1} dependence in (3.35)] so they have no analog of the Fröhlich or piezoelectric interaction. Intervalley electron–phonon interactions are always short-range and usually approximately independent of phonon wave vector. Whereas the energy difference between zone-center acoustic and optical phonon modes can be quite large, this difference may be insignificant for zone-edge phonons. We can express the intervalley electron-phonon Hamiltonian as

$$H_{\mathrm{iv}} = \boldsymbol{e}_{b\boldsymbol{q}} \cdot \frac{\partial H_{\mathrm{e}}}{\partial \boldsymbol{R}} u, \tag{3.40}$$

where $\boldsymbol{e}$, u, $\boldsymbol{q}$, and b respresent, respectively, the phonon polarization vector, amplitude, wave vector, and branch number. In practice, this Hamiltonian is often expressed in terms of its matrix element between two electronic states:

$$D_{ij}u = \langle n_i, \boldsymbol{k}_i \mid H_{\mathrm{iv}} \mid n_j, \boldsymbol{k}_j \rangle, \tag{3.41}$$

with D_{ij} known as the **intervalley deformation potential**. i and j denote, respectively, the initial and final valleys in the scattering, n and $\boldsymbol{k}$ are the electron band index and wave vector, respectively. The electron wave vectors $\boldsymbol{k}_i$ and $\boldsymbol{k}_j$ are related to $\boldsymbol{q}$ by wave vector conservation: $\boldsymbol{k}_j - \boldsymbol{k}_i = \boldsymbol{q}$. Notice that in (3.41) the intervalley deformation potential has the dimension of energy per unit length instead of energy. Table 3.5 lists intervalley deformation potentials (in units of eV/Å) of some zinc-blende semiconductors calculated by *Zollner* et al. [3.49].

Often more than one zone-edge phonon can participate in intervalley scattering. Possible selection rules can be deduced from group theory. For example, in scattering an electron from the Γ_{1c} valley to the L_{1c} valley in GaAs, the LA and LO phonons at the L point of the Brillouin zone are allowed because they both have L_1 symmetry. On the other hand, the transverse phonons at the L point have L_3 symmetry and are not allowed. Intervalley scattering plays an important role in determining the mobility of electrons in indirect bandgap semiconductors such as Si, in the relaxation of hot electrons in a direct bandgap semiconductor like GaAs, and in the optical absorption at indirect bandgaps. However, most of these phenomena involve scattering of electrons by several different phonon modes so it is difficult to deduce the values of the symmetry-allowed intervalley deformation potentials for the individual zone-edge phonons. The determination of intervalley deformation potentials from experiments will be made clear in Chap. 5 and 6, where these phenomena are discussed.

Table 3.5. Calculated intervalley deformation potentials for the conduction bands in a few representative zinc-blende semiconductors (in eV/Å). (From [3.49])

	Γ–X$_1$		Γ–X$_3$		Γ–L	
Semiconductor	LA	LO	LA	LO	LA	LO
GaP	1.5	0	0	1.2	1.2	1.0
GaAs	0	4.1	4.7	0	4.1	0.6
GaSb	0	4.5	2.5	0	2.8	2.7
InP	2.3	0	0	3.7	1.6	3.0
InAs	3.2	0	0	2.8	2.5	1.4
InSb	0	4.9	3.3	0	4.3	1.1

PROBLEMS

3.1 *Symmetry of Zone-Center Phonons in Cu_2O*

Figure 3.12 shows the crystal structure of cuprite (Cu_2O), which was one of the earliest minerals known to be a semiconductor. There are two molecules, i. e., six atoms, in the unit cell. The space group is O_h^4. This is a nonsymmorphic group which is homomorphic to the point group O_h. The crystal has inversion symmetry with respect to the copper atoms. When considering the point group symmetry it is more convenient to take an oxygen atom as the origin. The point group symmetry is clearly tetrahedral in this case. With the oxygen atom as the origin, the inversion operation has to be followed by a translation of the crystal by $(a/2)\,[1,1,1]$, where a is the size of the unit cube. Thus the factor group of Cu_2O is isomorphic to the factor group of Si except for the fact that the inversion operation in Si has to be followed by $(a/4)\,[1,1,1]$. There are ten classes in the group. Five of them do not involve the inversion operation while the other five do. Since there are six atoms per unit cell we expect fifteen optical phonon branches and three acoustic branches.

a) The permutation of the six atoms in the unit cell of Cu_2O by its symmetry operations defines a representation G. Determine the characters of G. This can be done by counting the number of atoms which are left unchanged by the symmetry operations of the group since each unchanged atom will contribute *one* to the trace of the transformation matrix.

b) Using the character table for the irreducible representations of the group of Γ in the diamond structure (Table 2.16) show that G can be decomposed into the direct sum of these irreducible representations: $G = 2\Gamma_1^+ \oplus \Gamma_1^- \oplus \Gamma_5^+$.

c) Show, by combining G with the Γ_4^- representation (that of a vector), that the 18 zone-center phonon modes in Cu_2O have the symmetries $\Gamma_1^- \oplus \Gamma_3^- \oplus 3\Gamma_4^- \oplus \Gamma_5^- \oplus \Gamma_5^+$. Of the three sets of triply degenerate Γ_4^- phonons, one set corresponds to the acoustic phonons while the other two are the infrared-active optical modes. The Γ_5^+ mode is Raman-active (see Chap. 7 for a discussion of Raman scattering). Optical phonons which are neither infrared-active nor

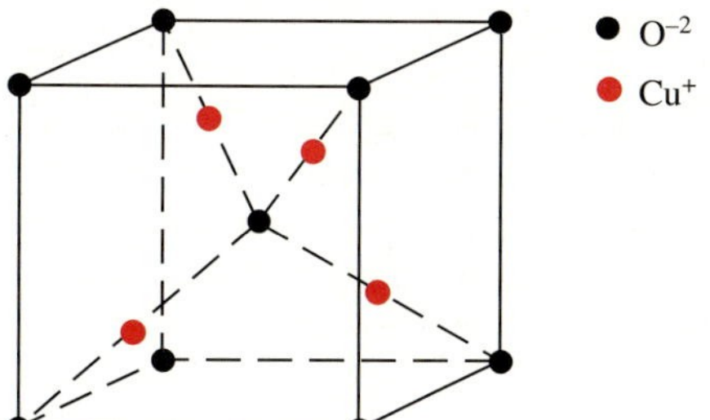

Fig. 3.12. Crystal structure of Cu_2O (cuprite). (From [3.50])

Raman-active are said to be silent. The Γ_1^- and Γ_3^- modes in Cu_2O are examples of silent modes.

3.2 *Stress Tensor*
In (3.15a) the strain tensor e_{ij} was defined. Strain can be induced in a solid by the application of stress. **Stress** can be defined as the force per unit area applied to a face of an elementary cube within a solid. Since both the direction of the force and orientation of the face to which the force is applied are important, it takes nine components to specify the stress completely: $F(X)_x$, $F(Y)_x$, $F(Z)_x$, ..., $F(Z)_z$. Here $F(X)_y$ denotes a force F per unit area applied along the x axis to a face whose normal is along the y axis. It is understood that these forces are applied in pairs so that a force $F(X)_y$ will be balanced by another force $-F(X)_y$, producing no acceleration of a solid in equilibrium. In addition, the off-diagonal components are equal: $F(X)_y = F(Y)_x$, $F(X)_z = F(Z)_x$, and $F(Y)_z = F(Z)_y$, so that there is also no net torque. As a result of these restrictions the stress components form a symmetric second-rank tensor $\boldsymbol{X}$, known as the **stress tensor**, defined by $X_{yz} = F(Y)_z$ and so on.

a) Show that a hydrostatic pressure P is specified by the diagonal stress tensor

$$\boldsymbol{X} = \begin{pmatrix} -P & 0 & 0 \\ 0 & -P & 0 \\ 0 & 0 & -P \end{pmatrix}.$$

Note that our sign convention for stress is that tensile stresses are positive while compressive stresses are negative.

b) Show that a tensile uniaxial stress X applied along the [100] direction is represented by the tensor

$$\boldsymbol{X} = \begin{pmatrix} X & 0 & 0 \\ 0 & 0 & 0 \\ 0 & 0 & 0 \end{pmatrix}.$$

c) Show that a tensile uniaxial stress X applied along the [111] direction is represented by the tensor

$$\boldsymbol{X} = (X/3) \begin{pmatrix} 1 & 1 & 1 \\ 1 & 1 & 1 \\ 1 & 1 & 1 \end{pmatrix}.$$

3.3 *Elastic Compliance Tensor in Diamond- and Zinc-Blende-Type Crystals*
In linear elasticity theory the strain induced in a medium is proportional to the applied stress. The constants of proportionality can be expressed as a fourth-rank tensor known as the **compliance tensor** (S_{ijkl}), which is defined by

$$e_{ij} = \sum S_{ijkl} X_{kl}.$$

Using the symmetry operations of a zinc-blende crystal, show that the nonzero and linearly independent components of the compliance tensor of a zinc-blende crystal are

$$S_{xxxx} = S_{yyyy} = S_{zzzz},$$
$$S_{xxyy} = S_{xxzz} = S_{yyzz} = S_{yyxx} = S_{zzxx} = S_{zzyy},$$
$$S_{xyxy} = S_{yxyx} = S_{xzxz} = S_{zxzx} = S_{yzyz} = S_{zyzy}.$$

Since both the strain and stress tensors are symmetric tensors, each can be described by six matrix elements rather than nine. One often finds in the literature a contracted notation in which the strain and stress tensors are represented by a six-component array [Ref. 3.39, p. 134]. For example the strain tensor is written as

$$\boldsymbol{e} = (e_1, e_2, e_3, e_4, e_5, e_6).$$

The numbering of 1 to 6 in this six-dimensional array is related to the 3×3 matrix by the convention

$$e = \begin{pmatrix} e_1 & e_6/2 & e_5/2 \\ e_6/2 & e_2 & e_4/2 \\ e_5/2 & e_4/2 & e_3 \end{pmatrix}.$$

One advantage of this contracted notation is that fourth-rank tensors, such as the compliance tensor, can be expressed as a 6×6 matrix. For example, S_{ijkl} in a zinc-blende-type crystal has the compact form

$$\begin{pmatrix} S_{11} & S_{12} & S_{12} & 0 & 0 & 0 \\ S_{12} & S_{11} & S_{12} & 0 & 0 & 0 \\ S_{12} & S_{12} & S_{11} & 0 & 0 & 0 \\ 0 & 0 & 0 & S_{44} & 0 & 0 \\ 0 & 0 & 0 & 0 & S_{44} & 0 \\ 0 & 0 & 0 & 0 & 0 & S_{44} \end{pmatrix}.$$

with $S_{mn} = 2S_{ijkl}$ when either m or $n = 4, 5$ or 6.
$S_{mn} = 4S_{ijkl}$ when both m and $n = 4, 5$ or 6.

3.4 *Strain Tensors for Hydrostatic and Uniaxial Stresses*

a) A hydrostatic pressure P is applied to a zinc-blende crystal of volume V. Show that the fractional change in volume is given by

$$\frac{\delta V}{V} = -3P(S_{11} + 2S_{12}).$$

The negative sign implies that the volume decreases for $P > 0$. The **bulk modulus** of a medium is defined as $B = -dP/d(\ln V)$, so for a zinc-blende crystal

$$B = [3(S_{11} + 2S_{12})]^{-1}.$$

b) A tensile uniaxial stress X is applied along the [100] axis of a zinc-blende-type crystal. Show that the resultant strain tensor is equal to

$$\begin{pmatrix} S_{11}X & 0 & 0 \\ 0 & S_{12}X & 0 \\ 0 & 0 & S_{12}X \end{pmatrix}$$

and that this matrix can be expressed as the sum of a diagonal matrix and a traceless shear strain tensor

$$[(S_{11}+2S_{12})X/3]\begin{pmatrix} 1 & 0 & 0 \\ 0 & 1 & 0 \\ 0 & 0 & 1 \end{pmatrix} + [(S_{11}-S_{12})X/3]\begin{pmatrix} 2 & 0 & 0 \\ 0 & -1 & 0 \\ 0 & 0 & -1 \end{pmatrix}.$$

Notice that the fractional change in volume of the crystal is equal to $(S_{11} + 2S_{12})X$ under this uniaxial stress.

c) Show that the strain tensor corresponding to a tensile uniaxial stress applied along the [111] direction of a zinc-blende-type crystal is given by

$$\frac{X}{3}\begin{pmatrix} S_{11}+2S_{12} & S_{44}/2 & S_{44}/2 \\ S_{44}/2 & S_{11}+2S_{12} & S_{44}/2 \\ S_{44}/2 & S_{44}/2 & S_{11}+2S_{12} \end{pmatrix}$$

Decompose this tensor into the sum of a multiple of a unit tensor and a traceless tensor as in part (b). Show that the fractional volume change is also equal to $(S_{11} + 2S_{12})X$ under a [111] stress.

d) Determine the strain tensor for a uniaxial stress applied along the [110] direction of a zinc-blende-type crystal.

3.5 *Elastic Stiffness Tensor*

In Problem 3.3 the second-rank strain tensor was expressed in terms of the stress tensor via a fourth-rank compliance tensor. Inversely, the stress tensor can be expressed in terms of the strain tensor by another fourth-rank tensor C_{ijkl} known as the **stiffness tensor**:

$$X_{ij} = \sum_{k,l} C_{ijkl} e_{kl}.$$

Show that in zinc-blende-type crystals the stiffness tensor, when written in the compact form introduced in Problem 3.3, is given by [Ref. 3.39, p. 140]

$$\begin{pmatrix} C_{11} & C_{12} & C_{12} & 0 & 0 & 0 \\ C_{12} & C_{11} & C_{12} & 0 & 0 & 0 \\ C_{12} & C_{12} & C_{11} & 0 & 0 & 0 \\ 0 & 0 & 0 & C_{44} & 0 & 0 \\ 0 & 0 & 0 & 0 & C_{44} & 0 \\ 0 & 0 & 0 & 0 & 0 & C_{44} \end{pmatrix},$$

where $C_{44} = (1/S_{44})$, $C_{11} - C_{12} = (S_{11} - S_{12})^{-1}$, and $C_{11} + 2C_{12} = (S_{11} + 2S_{12})^{-1}$.

Table 3.6 lists the stiffness constants of some semiconductors.

Table 3.6a. Stiffness constants C_{11}, C_{12}, and C_{44} in some diamond- and zinc-blende-type semiconductor (in 10^{12} dyne/cm^2). [Note that stiffness constants and pressure have the same units: dyne/cm^2 (cgs units) or pascal (= 10 dyne/cm^2)] Reproduced from *Madelung* et al. [3.43]. The values in italics are theoretical values. See A.F. Wright, Elastic Properties of zinc-blende and wurtzite AlN, GaN, and InN. J. Appl. Phys. **82**, 2833 (1997)

Semiconductor	C_{11}	C_{12}	C_{44}
C	10.76	1.25	5.76
Si	1.66	0.639	0.796
Ge	1.285	0.483	0.680
AlN	*3.04*	*1.6*	*1.93*
GaN	*2.93*	*1.59*	*1.55*
GaP	1.412	0.625	0.705
GaAs	1.181	0.532	0.594
GaSb	0.885	0.404	0.433
InN	*1.87*	*1.25*	*0.86*
InP	1.022	0.576	0.46
InAs	0.833	0.453	0.396
InSb	0.672	0.367	0.302
ZnS	1.046	0.653	0.461
ZnSe	0.81	0.488	0.441
ZnTe	0.713	0.407	0.312
CdTe	0.535	0.368	0.199

Table 3.6b. Stiffness constants C_{11}, C_{12}, C_{13}, C_{33}, C_{44}, and C_{66} in some wurtzite-type semiconductors (units in 10^{12} dynes/cm^2. To convert the unit to the SI unit of Pa, note that 10^{12} dynes/cm^2 is equal to 10^{11} Pa or 100 GPa)

Semiconductor	C_{11}	C_{12}	C_{13}	C_{33}	C_{44}	C_{66}
AlN	4.11	1.49	0.99	3.89	1.25	
GaN	3.9	1.45	1.06	3.98	1.05	
InN	1.90	1.04	1.21	1.82	0.1	
CdS	0.858	0.533	0.462	0.937	0.149	0.163
CdSe	0.74	0.452	0.393	0.836	0.132	0.145

3.6 *Elastic Waves in Zinc-Blende-Type Crystals*
Let $\boldsymbol{u}$ be the displacement induced by an elastic wave traveling in a continuum. The equation governing $\boldsymbol{u}$ as a function of time t is obtained from Newton's equation of motion:

$$\varrho \frac{\partial^2 \boldsymbol{u}}{\partial t^2} = \nabla \cdot \boldsymbol{X},$$

where ϱ is the density of the crystal and $\boldsymbol{X}$ the stress tensor.

a) Use the results of Problem 3.5 to express the stress tensor in terms of the strain tensor. Next, by using the definition of the strain tensor $\boldsymbol{e} = \nabla \boldsymbol{u}$, show that the wave equations governing the propagation of elastic waves in a zinc-blende-type crystal are

$$\varrho\frac{\partial^2 u_x}{\partial t^2} = C_{11}\frac{\partial^2 u_x}{\partial x^2} + C_{44}\left(\frac{\partial^2 u_x}{\partial y^2} + \frac{\partial^2 u_x}{\partial z^2}\right) + (C_{12}+C_{44})\left(\frac{\partial^2 u_y}{\partial x \partial y} + \frac{\partial^2 u_z}{\partial x \partial z}\right),$$

$$\varrho\frac{\partial^2 u_y}{\partial t^2} = C_{11}\frac{\partial^2 u_y}{\partial y^2} + C_{44}\left(\frac{\partial^2 u_y}{\partial x^2} + \frac{\partial^2 u_y}{\partial z^2}\right) + (C_{12}+C_{44})\left(\frac{\partial^2 u_x}{\partial x \partial y} + \frac{\partial^2 u_z}{\partial y \partial z}\right),$$

$$\varrho\frac{\partial^2 u_z}{\partial t^2} = C_{11}\frac{\partial^2 u_z}{\partial z^2} + C_{44}\left(\frac{\partial^2 u_z}{\partial x^2} + \frac{\partial^2 u_z}{\partial y^2}\right) + (C_{12}+C_{44})\left(\frac{\partial^2 u_y}{\partial z \partial y} + \frac{\partial^2 u_x}{\partial x \partial z}\right),$$

b) Assume that a longitudinal (sound) wave propagating along the [100] axis is represented by a solution to the above wave equations of the form $u_x = u_0 \exp[\mathrm{i}(kx - \omega t)]$ and $u_y = u_z = 0$, where k and ω are, respectively, the wave vector and frequency of the acoustic wave. Show that the resultant sound wave has a velocity $v_l = \omega/k = (C_{11}/\varrho)^{1/2}$. Repeat this calculation for a transverse shear wave propagating along the [100] direction by assuming a solution of the form $u_y = u_0 \exp[\mathrm{i}(kx - \omega t)]$, $u_x = u_z = 0$; and show that the transverse wave has a velocity $v_\mathrm{t} = (C_{44}/\varrho)^{1/2}$.

c) Repeat the calculations in (b) for waves propagating along the [111] direction and show that $v_\mathrm{l} = [(C_{11} + 2C_{12} + 4C_{44})/3\varrho]^{1/2}$ and $v_\mathrm{t} = [(C_{11} - C_{12} + C_{44})/3\varrho]^{1/2}$.

d) For waves propagating along the [110] direction, the transverse waves polarized along the [001] and $[1\bar{1}0]$ directions have different velocities. Show that the elastic waves have the following velocities along the [110] direction:

$v_\mathrm{l} = [(C_{11} + C_{12} + 2C_{44})/2\varrho]^{1/2}$;
$v_\mathrm{t} = (C_{44}/\varrho)^{1/2}$ when polarized along [001] and
$v'_\mathrm{t} = [(C_{11} - C_{12})/2\varrho]^{1/2}$ when polarized along $[1\bar{1}0]$.

The two quantized transverse waves are referred to as the **fast** (usually v_t) and **slow** (usually v'_t) TA **phonons**.

e) Discuss the eigenvectors (and the corresponding symmetries) of the TA and LA phonons propagating along [100], [111], and [110] directions of k-space.

3.7 *Elastic Waves and Optical Phonons in Wurtzite-Type Crystals*
The *wurtzite* crystal structure shown in Fig. 3.13 is a variation of the zinc-blende structure. The bonding between nearest neighbors is also tetrahedral. The relationship between these two structures can be visualized most easily by viewing the zinc-blende structure along the [111] direction. In this direction the zinc-blende structure can be considered as consisting of layers of atoms arranged in regular hexagons stacked together. This hexagonal symmetry of the atoms within the layers is, of course, the origin of the C_3 rotational symmetry of the zinc-blende crystal. Within each layer the atoms are identical and

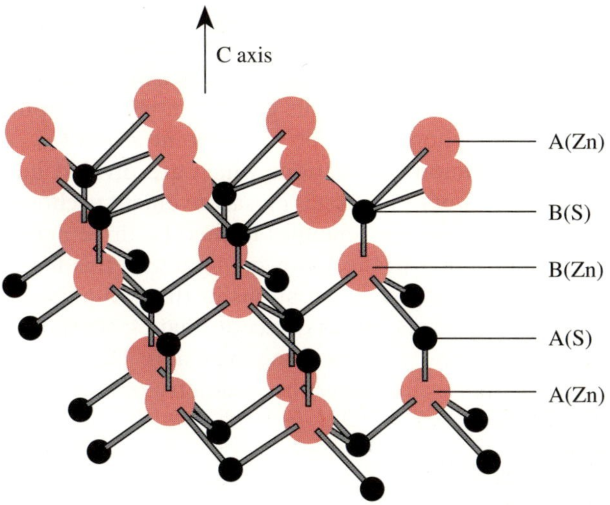

Fig. 3.13. Crystal structure of ZnS in the wurtzite form

the layers alternate between Zn and S. In the zinc-blende crystal the stacking order of the Zn and S atomic layers is the sequence

A(S)A(Zn)B(S)B(Zn)C(S)C(Zn)A(S)A(Zn)B(S)B(Zn) ...

as shown in Fig. 3.14a. If the stacking order of the layers is changed to

A(S)A(Zn)B(S)B(Zn)A(S)A(Zn)B(S)B(Zn)A(S)A(Zn) ...

as shown in Fig. 3.14b the wurtzite structure is obtained. The arrangement of the tetrahedra is said to be "staggered" in the zinc-blende structure and "eclipsed" in the wurtzite structure. The reason for these descriptions becomes clear in Fig. 3.14. The Bravais lattice of the wurtzite structure is hexagonal, with the axis perpendicular to the hexagons usually labeled as the *c axis*.

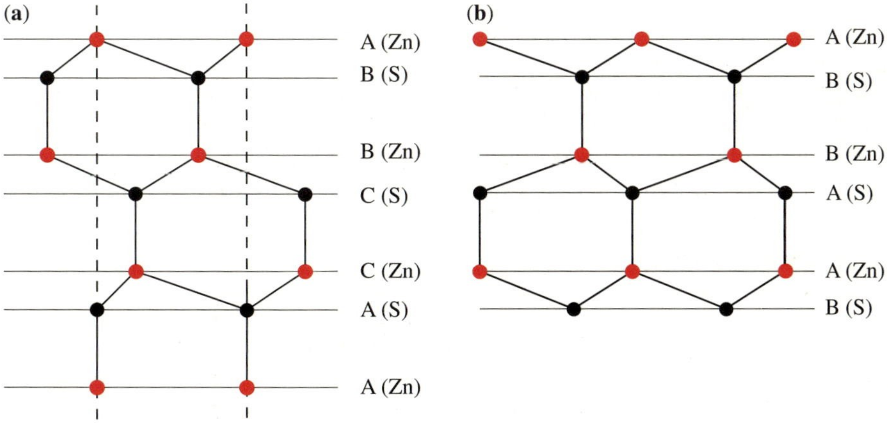

Fig. 3.14. Arrangement of atoms (**a**) along the [111] direction of the zinc-blende structure, showing the "staggered" stacking sequence, and (**b**) along the *c* axis of the wurtzite structure with the "eclipsed" geometry

The cohesive energy of the wurtzite structure is very close to that of the zinc-blende structure. As a result some group-II–VI semiconductors, such as ZnS and CdS, and also GaN, can crystallize in both zinc-blende and wurtzite structures. The space group of the wurtzite structure is C_{6v}^4. This group is nonsymmorphic with a screw axis along the c axis. It is homomorphic to the point group C_{6v}.

As in the latter there are 12 symmetry operations which are divided into 6 classes. To enumerate these symmetry operations it is most convenient to choose the origin in the center of one of the isosceles triangles formed by three neighboring atoms, such as the point labeled O in the following figure. In this figure the filled circles and the open circles represent the atoms in the two adjacent planes of the hcp structure. We shall also choose as our coordinate system the x and y axis as shown in the figure while the z-axis is along the c-axis. The reasons for our choice will become clear when we consider the symmetry operations.

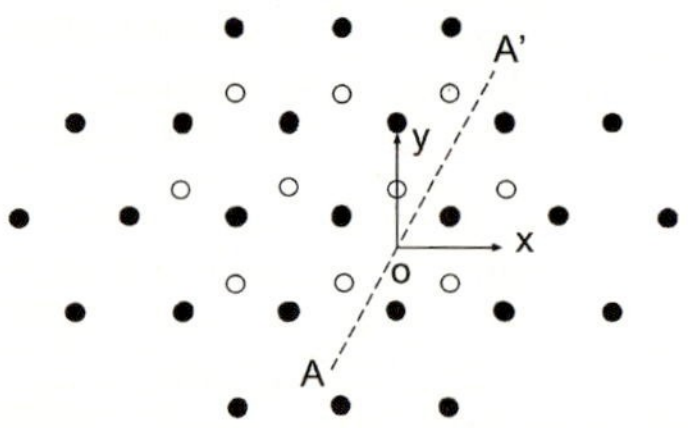

a) Show that the space group of wurtzite contains the following symmetry elements divided into the following classes:

$\{E\}$: identity;

$\{C_2\}$: a two-fold rotation about the c-axis followed by a translation by the vector $(0, 0, c/2)$. In other words this is a screw axis operation;

$\{C_3, C_3^{-1}\}$: three-fold rotations about the c-axis;

$\{C_6, C_6^{-1}\}$: a six-fold rotation about the c-axis followed by a translation by the vector $(0, 0, c/2)$. In other words this is also a screw axis operation. Note that this operation is a symmetry operation mainly because of our choice of the origin. This operation is not a symmetry operation if we simply choose the origin at one of the atoms;

$\{3\sigma_d\}$: reflection in the yz-plane plus two other reflection planes obtained from the yz-plane by three-fold rotations about the c-axis;

$\{3\sigma_d'\}$: reflection in the plane denoted by AA' in the figure followed by a translation by the vector $(0, 0, c/2)$ plus two other glide planes obtained from AA' by three-fold rotations about the c-axis.

b) Using the symmetry properties of the wurtzite crystal show that its stiffness tensor is given by (the c axis has been chosen to be the z axis)

$$\begin{pmatrix} C_{11} & C_{12} & C_{13} & 0 & 0 & 0 \\ C_{12} & C_{11} & C_{13} & 0 & 0 & 0 \\ C_{13} & C_{13} & C_{33} & 0 & 0 & 0 \\ 0 & 0 & 0 & C_{44} & 0 & 0 \\ 0 & 0 & 0 & 0 & C_{44} & 0 \\ 0 & 0 & 0 & 0 & 0 & (C_{11} - C_{12})/2 \end{pmatrix} .$$

c) Show that the wave equations for acoustic phonons in wurtzite crystals are given by

$$\varrho\frac{\partial^2 u_x}{\partial t^2} = C_{11}\frac{\partial^2 u_x}{\partial x^2} + \frac{(C_{11}-C_{12})}{2}\frac{\partial^2 u_x}{\partial y^2} + C_{44}\frac{\partial^2 u_x}{\partial z^2} + \frac{(C_{11}+C_{12})}{2}\frac{\partial^2 u_y}{\partial x\partial y} + (C_{13}+C_{44})\frac{\partial^2 u_z}{\partial x\partial z},$$

$$\varrho\frac{\partial^2 u_y}{\partial t^2} = C_{11}\frac{\partial^2 u_y}{\partial y^2} + \frac{(C_{11}-C_{12})}{2}\frac{\partial^2 u_y}{\partial x^2} + C_{44}\frac{\partial^2 u_y}{\partial z^2} + \frac{(C_{11}+C_{12})}{2}\frac{\partial^2 u_x}{\partial x\partial y} + (C_{13}+C_{44})\frac{\partial^2 u_z}{\partial y\partial z},$$

$$\varrho\frac{\partial^2 u_z}{\partial t^2} = C_{33}\frac{\partial^2 u_z}{\partial z^2} + C_{44}\left(\frac{\partial^2 u_z}{\partial x^2} + \frac{\partial^2 u_z}{\partial y^2}\right) + (C_{13}+C_{44})\left(\frac{\partial^2 u_y}{\partial z\partial y} + \frac{\partial^2 u_x}{\partial x\partial z}\right).$$

d) Assume a solution for an acoustic wave in the wurtzite crystal of the form

$$\boldsymbol{u}(t) = (u_1, u_2, u_3)\ \exp[\mathrm{i}(\boldsymbol{q}\cdot\boldsymbol{r} - \omega t)].$$

Because of the hexagonal symmetry about the c axis, the y axis can be chosen, without loss of generality, such that the wave vector $\boldsymbol{q}$ lies in the yz plane. Show that the three components of the amplitudes satisfy the equations

$$\begin{aligned}\varrho\omega^2 u_1 &= \tfrac{1}{2}(C_{11}-C_{12})(q\sin\Theta)^2 u_1 + C_{44}(q\cos\Theta)^2 u_1,\\ \varrho\omega^2 u_2 &= C_{11}(q\sin\Theta)^2 u_2 + C_{44}(q\cos\Theta)^2 u_2 + (C_{13}+C_{44})q^2\sin\Theta\cos\Theta\ u_3,\\ \varrho\omega^2 u_3 &= C_{33}(q\cos\Theta)^2 u_3 + C_{44}(q\sin\Theta)^2 u_3 + (C_{13}+C_{44})q^2\sin\Theta\cos\Theta\ u_2,\end{aligned}$$

where Θ is the angle between $\boldsymbol{q}$ and the c axis. Note that u_1 is decoupled from u_2 and u_3, so it is always possible to obtain a transverse acoustic wave perpendicular to the plane containing $\boldsymbol{q}$ and the c axis. This TA wave is usually referred to as TA_2.

e) Show that for $\Theta = 0°$ the longitudinal and transverse acoustic waves have velocities $v_\mathrm{l} = (C_{33}/\varrho)^{1/2}$ and $v_\mathrm{t} = (C_{44}/\varrho)^{1/2}$, respectively.

f) For $\Theta = 90°$ there are two transverse acoustic waves: one polarized along the c axis (TA_1) and one polarized along the x axis (TA_2). Show that the longitudinal wave has velocity $v_\mathrm{l} = (C_{11}/\varrho)^{1/2}$ while the velocities for the two transverse waves are $(C_{44}/\varrho)^{1/2}$ for TA_2 and $[(C_{11}-C_{12})/2\varrho]^{1/2}$ for TA_1.

g) Assume a Bloch function has the form: $\Psi_\mathrm{k}(\boldsymbol{r}) = u_\mathrm{k}(\boldsymbol{r})\exp(\mathrm{i}\boldsymbol{k}\cdot\boldsymbol{r})$ where $u_k(\boldsymbol{r})$ has the full periodicity of the lattice. By applying the symmetry operations of wurtzite to this wave function show that the corresponding character table is given by [see I. Rashba, Soviet Physics-Solid State, 1, 386 (1959)]:

	$\{E\}$	$\{C_2\}$	$\{2C_3\}$	$\{2C_6\}$	$\{3\sigma'_d\}$	$\{3\sigma_d\}$
Γ_1	1	η	1	η	η	1
Γ_2	1	η	1	$-\eta$	$-\eta$	-1
Γ_3	1	$-\eta$	1	$-\eta$	$-\eta$	1
Γ_4	1	$-\eta$	1	$-\eta$	η	-1
Γ_5	2	-2η	-1	η	0	0
Γ_6	2	2η	-1	$-\eta$	0	0

where $\eta = \exp[ik_z c/2]$ is the phase factor arising from the translation by the vector $(0, 0, c/2)$ in the screw and glide operations. In the particular case of $k = 0$ then $\eta = 1$ and we find that the character table of the wurtzite structure is identical to that of the point group C_{6v}. In the following table we show the correspondence between the two groups for this case:

Point Group	Space Group	$\{E\}$	$\{C_2\}$	$\{2C_3\}$	$\{2C_6\}$	$\{3\sigma'_d\}$	$\{3\sigma_d\}$
A_1	Γ_1	1	1	1	1	1	1
A_2	Γ_2	1	1	1	1	-1	-1
B_1	Γ_3	1	-1	1	-1	-1	1
B_2	Γ_4	1	-1	1	-1	1	-1
E_1	Γ_5	2	-2	-1	1	0	0
E_2	Γ_6	2	2	-1	-1	0	0

h) Show that the long wavelength acoustic phonons in the wurtzite structure have the symmetries: Γ_1 (or A_1) plus Γ_5 (or E_1). The direction of the displacement of the atoms for the Γ_1 representation is parallel to the c-axis while the displacements are perpendicular to the c-axis for the two-fold degenerate representation Γ_5.

i) Show that the nine zone-center optical phonons in the wurtzite structure belong to the irreducible representations: $\Gamma_1 \oplus 2\Gamma_3 \oplus \Gamma_5 \oplus 2\Gamma_6$ or A_1, $2B_1$, E_1 and $2E_2$.

3.8 *Pikus–Bir Strain Hamiltonians*

There are two forms of the *Pikus–Bir* strain Hamiltonian [3.34, 35] for the valence bands in the zinc-blende-type semiconductors (similar results can be obtained for diamond-type crystals) depending on whether spin–orbit interaction is included or not. If spin–orbit interaction is neglected, the p-like Γ_4 valence band wave functions labeled $|X\rangle$, $|Y\rangle$, and $|Z\rangle$ (Sect. 2.6.1) at the zone center are degenerate. From symmetry we can show that a uniaxial stress along the [100] direction will split the band labeled $|X\rangle$ off from the other two bands. Thus a [100] stress will split the three bands into a singlet and a doublet. If spin–orbit interaction is included and assumed to be much larger than any strain-induced effect, the valence bands at the zone center will be split into a four-fold degenerate Γ_8 (or $J = 3/2$) state and a doubly degenerate Γ_7 ($J = 1/2$) state (Sect. 2.6.2). For a sufficiently small stress we can neglect the effect of strain on the spin–orbit interaction. Since strain is not expected to affect the spin degeneracy of the electrons (*Kramers degeneracy*) we expect that a uniaxial stress will split the $J = 3/2$ states into two doublets. The strain Hamiltonian $\mathrm{H_{PB}}$ in (3.23) is valid in this limit of large spin-orbit interaction and care should be exercised in using it to analyze experimental results. For theoretical considerations, one may want to simplify the problem by "turning off" the spin–orbit interaction. In this case the total angular momentum operator $\boldsymbol{J}$ in (3.23) can be replaced by the orbital angular momentum operator $\boldsymbol{L}$. However, the deformation potentials b and d should also be changed to b^* and d^* since the expression for the strain-induced splitting will be different in the two cases. In Problems 3.11 and 3.12 we will perform theoretical calculations of b^* and d^*.

a) Use the results of Problem 3.4 to show that the Pikus–Bir strain Hamiltonian (3.23) reduces to

$$H_{\mathrm{PB}}(X) = a(S_{11} + 2S_{12})X + b(S_{11} - S_{12})X[J_x^2 - (J^2/3)]$$

for a stress X applied along the [100] direction.

b) Show similarly that the strain Hamiltonian is given by

$$H_{\mathrm{PB}}(X) = a(S_{11} + 2S_{12})X + (dS_{44}X/3\sqrt{3})[(J_xJ_y + J_yJ_x) + \text{c.p.}]$$

for a stress X applied along the [111] axis.

c) Use the result of parts (a) and (b) to show that the four-fold degenerate $J = 3/2$ valence bands will be split into two doubly degenerate bands with the splitting δE given by

$$\delta E = \begin{cases} 2b(S_{11} - S_{12})X, & [100]\ \text{stress}, \\ (d/\sqrt{3})S_{44}X, & [111]\ \text{stress}. \end{cases}$$

d) Let us rewrite the Pikus–Bir strain Hamiltonian in the absence of spin–orbit interaction as

$$\begin{aligned} H_{\mathrm{PB}} &= a(e_{xx} + e_{yy} + e_{zz}) + 3b^*[(L_x^2 - L^2/3)e_{xx} + \text{c.p.}] \\ &+ \frac{6d^*}{\sqrt{3}}\left[\frac{1}{2}(L_xL_y + L_yL_x)e_{xy} + \text{c.p.}\right], \end{aligned} \tag{3.41}$$

where b^* and d^* are equal to b and d in the absence of spin–orbit interaction. The prefactors of the b^* and d^* terms in (3.41) have been chosen to agree with those in (3.77) of [3.36]. Show that the stress-induced splittings are given by

$$\delta E = \begin{cases} 3b^*(S_{11} - S_{12})X, & [100]\ \text{stress}, \\ (\sqrt{3}/2)d^*S_{44}X, & [111]\ \text{stress}. \end{cases}$$

Show that if the spin–orbit splitting implicit in the Hamiltonian of (3.23) does not depend an strain, $b = b^*$ and $d = d^*$. Explain the origin of the difference of those prefactors in (3.23) and (3.41). Now include spin–orbit coupling and assume that the $J = 3/2$ and $J = 1/2$ states are separated by Δ. Use the Pikus–Bir Hamiltonian involving $\boldsymbol{L}$ to calculate the stress-induced splitting of these levels to second order in X for stress parallel to both [100] and [111].

3.9 *Calculation of Volume Deformation Potentials Using the Tight-Binding Model*

The calculation of the absolute volume deformation potential of a band edge in a finite solid is a difficult task because in a finite solid the band edges are dependent on surface properties (for example, see discussion in [3.32]). This difficulty does not appear when calculating the relative volume deformation potentials between two bands, such as between an s-like antibonding conduction band and a p-like bonding valence band. In this case the deformation potential can be calculated easily with the tight-binding model.

Assume that the zone-center $\Gamma_{2'}$ conduction and $\Gamma_{25'}$ valence band edges in a diamond-type semiconductor are given by (2.84a) and (2.84b), respectively. Furthermore, assume that the overlap parameters V_{ss} and V_{xx} both depend on the nearest neighbor distance d as d^{-2}. Show that the relative deformation potential $a_c - a_v$ is given by $(-2/3)(|V_{ss}| + |V_{xx}|)$. Use the empirical values of V_{ss} and V_{xx} for Si and Ge given in Table 2.26 to calculate the theoretical values of $a_c - a_v$ for Si and Ge. Compare these values with the experimental values for Si and Ge given in Table 3.1 under $a(\Gamma_{1c}) - a(\Gamma_{15v})$

3.10 *Calculation of the Shear Deformation Potential b^* for the Γ_{15} Valence Bands Using the Tight-Binding Model*

The shear deformation potentials b^* and d^* for the Γ_{15} (or $\Gamma_{25'}$) valence bands in zinc-blende (or diamond) type semiconductors can also be calculated using the tight-binding model. This problem involves the calculation of the shear deformation potential b^*, leaving the calculation of d^* to the next problem.

Assume a stress X is applied along the [100] crystallographic axis. According to Problem 3.4b, the strain induced in the crystal can be decomposed into a multiple of a unit matrix plus a traceless matrix of the form

$$\alpha \begin{pmatrix} 2 & 0 & 0 \\ 0 & -1 & 0 \\ 0 & 0 & -1 \end{pmatrix},$$

where $\alpha = (X/3)(S_{11} - S_{12})$. The effect of this shear strain is to lower the cubic symmetry of the crystal to tetragonal symmetry by making the x direction inequivalent to the y and z directions. As a result, we expect that the originally triply degenerate Γ_{15} wave functions $|X\rangle$, $|Y\rangle$, and $|Z\rangle$ will split into a doublet and a singlet. To calculate this splitting it is necessary to evaluate the overlap parameters V_{xx} and $V_{yy}(= V_{zz})$ in the presence of the shear strain.

a) Suppose that the locations of the four nearest neighbors of an atom located at the origin in the unstrained lattice are

$$\begin{aligned}\boldsymbol{d}_1 &= (1,1,1)a/4;\\ \boldsymbol{d}_2 &= (1,-1,-1)a/4;\\ \boldsymbol{d}_3 &= (-1,1,-1)a/4;\end{aligned}$$

and

$$\boldsymbol{d}_4 = (-1,-1,1)a/4.$$

Show that in the strained lattice these four nearest-neighbor locations are changed to

$$\begin{aligned}\boldsymbol{d}_1' &= (1+2\alpha, 1-\alpha, 1-\alpha)a/4;\\ \boldsymbol{d}_2' &= (1+2\alpha, -1+\alpha, -1+\alpha)a/4;\\ \boldsymbol{d}_3' &= (-1-2\alpha, 1-\alpha, -1+\alpha)a/4;\end{aligned}$$

and

$$\boldsymbol{d}_4' = (-1-2\alpha, -1+\alpha, 1-\alpha)a/4.$$

Show that, to first order in α, the bond lengths are unchanged by the strain but the bond angles are changed.

b) Use the results of Problem 2.16 to show that the overlap parameters in the strained lattice are

$$V_{xx}' = \frac{4}{3}V_{pp\sigma}(1+4\alpha) + \frac{8}{3}V_{pp\pi}(1-2\alpha)$$

and

$$V_{yy}' = \frac{4}{3}V_{pp\sigma}(1-2\alpha) + \frac{8}{3}V_{pp\pi}(1+\alpha).$$

c) Calculate the energy of the bonding orbitals with p-symmetry at zone-center in the strained lattice. Show that the splitting between the doublet with p_y and p_z symmetries and the singlet with p_x symmetry is given by $-8\alpha(V_{pp\sigma} - V_{pp\pi})$.

d) By combining the result of part (c) with those of Problem 3.9d show that

$$b = -\frac{8}{9}(V_{pp\sigma} - V_{pp\pi}) = -\frac{2}{3}V_{xy}.$$

Substitute the values of V_{xy} for Si and Ge given in Table 2.26 into the above expression to evaluate the theoretical values of b^* for Si and Ge. You will find that the calculated values are larger than the experimental values by about a factor of two but the signs agree. The agreement between theory and experiment can be improved by including an additional higher energy s orbital in the tight-binding model (the resultant model is referred to as the sp^3s^* *model*). See [3.30] for more details.

3.11 *Calculation of the Shear Deformation Potential d^* for the Γ_{15} Valence Bands Using the Tight-Binding Model*

The calculation of the shear deformation potential d^* is complicated by the following considerations.

Let us assume that a uniaxial stress is applied along the [111] axis of a diamond or zinc-blende crystal. As shown in Problem 3.4c, the resultant strain tensor can be decomposed into a multiple of a unit tensor and the traceless shear strain tensor (of Γ_{15} symmetry)

$$\alpha \begin{pmatrix} 0 & 1 & 1 \\ 1 & 0 & 1 \\ 1 & 1 & 0 \end{pmatrix}, \qquad \text{where } \alpha = S_{44}X/6.$$

This strain tensor describes a trigonal distortion of the cubic lattice along the [111] direction. The symmetry of this distortion is the same as that of a displacement of the two sublattices in the diamond or zinc-blende lattice along the body diagonal. As a result, it is possible to "mix" these two types of distortions. The displacement of the two neighboring atoms in the diamond or zinc-blende crystals along the body diagonal corresponds to the zone-center (Γ_{15}) optical phonon. A macroscopic Γ_{15} shear strain along the [111] direction can, therefore, change the nearest-neighbor distance along the [111] direction in addition to producing a trigonal distortion in the cubic lattice. To specify the change in the nearest-neighbor distance it is necessary to define an **internal strain parameters** ζ in terms of the relative displacement $\boldsymbol{u}$ between the two sublattices. The internal strain parameter will be defined in Problem 3.13. In this problem we will assume that the [111] stress does not produce an internal strain (i. e., $\zeta = 0$). The resultant deformation potential will be denoted by d' so as to distinguish it from the deformation potential d for the case when the internal strain is not zero.

From symmetry considerations, it is clear that the effect of a [111] stress on the p orbitals will be different from that of a [100] stress. While a [100] stress makes the p_x orbital different from the p_y and p_z orbitals, a [111] stress should affect the three orbitals in exactly the same way. Instead, a [111] stress will mix the three orbitals and produce a new set of three orbitals which should be symmetrized according to the trigonal symmetry (C_{3v} point group). Such a set of three orthogonal and symmetrized orbitals is given by

$$|X1'\rangle = [|X1\rangle - |Y1\rangle]/\sqrt{2},$$
$$|Y1'\rangle = [|X1\rangle + |Y1\rangle - 2|Z1\rangle]/\sqrt{6},$$
$$|Z1'\rangle = [|X1\rangle + |Y1\rangle + |Z1\rangle]/\sqrt{3}.$$

Similarly symmetrized p orbitals are defined for the second atom within the primitive cell. Under [111] stress the orbitals $|Z1'\rangle$ and $|Z2'\rangle$ which are parallel to the stress direction (and of Λ_1 symmetry) will split from the other two sets of two orbitals (of Λ_3 symmetry) which are perpendicular to the stress direction.

a) Show that under a [111] shear strain the locations of the four nearest neighbors in the diamond and zinc-blende lattice are changed to

$$\boldsymbol{d}_1' = \{(a/4) + (a\alpha/2)\}(1,1,1);$$
$$\boldsymbol{d}_2' = (a/4)(1-2\alpha,-1,-1);$$
$$\boldsymbol{d}_3' = (a/4)(-1,1-2\alpha,-1);$$
$$\boldsymbol{d}_4' = (a/4)(-1,-1,1-2\alpha).$$

b) Based on the result of (a), show that the effects of the [111] stress are: (1) to change the lengths of the nearest-neighbor distance by the amount (to first order in α)

$$(\delta d_1)/d_1 = 2\alpha;$$
$$(\delta d_2)/d_2 = (\delta d_3)/d_3 = (\delta d_4)/d_4 = -2\alpha/3$$

and (2) to change the orientations of the three bonds $\boldsymbol{d}_2$, $\boldsymbol{d}_3$, and $\boldsymbol{d}_4$. The directional cosines of the bond $\boldsymbol{d}_2'$ in the strained lattice are given by

$$\boldsymbol{d}_2' = \{3[1-(4\alpha/3)]\}^{-1/2}(1-2\alpha,-1,-1)$$

with similar results for $\boldsymbol{d}_3'$ and $\boldsymbol{d}_4'$.

c) Assuming that the overlap parameters $V_{pp\sigma}$ and $V_{pp\pi}$ both depend on bond distance d as d^{-2}, show that, to first order in α, the [111] shear strain does not change the matrix elements $\langle X1|\mathcal{H}_{\rm int}|X2\rangle$, $\langle Y1|\mathcal{H}_{\rm int}|Y2\rangle$, etc. On the other hand (use the results of Problem 2.16), the matrix elements $\langle X1|\mathcal{H}_{\rm int}|Y2\rangle$, $\langle X1|\mathcal{H}_{\rm int}|Z2\rangle$, etc., in the strained lattice are all changed to

$$V_{xy}' = (-8\alpha/9)(V_{pp\sigma} - V_{pp\pi}).$$

Thus a [111] shear strain couples the wave function $|X1\rangle$ with $|Y2\rangle$ and $|Z2\rangle$, $|Y1\rangle$ with $|X2\rangle$ and $|Z2\rangle$, and so on. The new 6×6 matrix for the six p orbitals of the two atoms within the primitive cell is now given by

	X1	Y1	Z1	X2	Y2	Z2
X1	$E_{p1}-E_k$	0	0	V_{xx}	V_{xy}'	V_{xy}'
Y1	0	$E_{p1}-E_k$	0	V_{xy}'	V_{xx}	V_{xy}'
Z1	0	0	$E_{p1}-E_k$	V_{xy}'	V_{xy}'	V_{xx}
X2	V_{xx}	V_{xy}'	V_{xy}'	$E_{p2}-E_k$	0	0
Y2	V_{xy}'	V_{xx}	V_{xy}'	0	$E_{p2}-E_k$	0
Z2	V_{xy}'	V_{xy}'	V_{xx}	0	0	$E_{p2}-E_k$

where E_{p1} and E_{p2} are the energies of the p orbitals of the two atoms in the primitive cell.

d) Assume that the two atoms in the primitive cell are identical (thus the results we will obtain from now on are only valid for the diamond structure) and diagonalize the 6×6 matrix in part (c) to obtain the eigenfunctions and energies in the [111]-strained lattice. The more general results for the case when the two atoms are different are given in [3.30].

Hint: Since the Hamiltonian is symmetric with respect to interchange of the atoms in the primitive cell, define linear combinations of the wave functions

$$|X_+\rangle = [|X1\rangle + |X2\rangle]/\sqrt{2}; \quad |X_-\rangle = [|X1\rangle - |X2\rangle]/\sqrt{2},$$
$$|Y_+\rangle = [|Y1\rangle + |Y2\rangle]/\sqrt{2}; \quad |Y_-\rangle = [|Y1\rangle - |Y2\rangle]/\sqrt{2},$$
$$|Z_+\rangle = [|Z1\rangle + |Z2\rangle]/\sqrt{2}; \quad |Z_-\rangle = [|Z1\rangle - |Z2\rangle]/\sqrt{2}.$$

Show that the 6×6 matrix decomposes into two 3×3 matrices:

X_+	Y_+	Z_+	X_-	Y_-	Z_-
$E_p-E_k+V_{xx}$	V'_{xy}	V'_{xy}	0	0	0
V'_{xy}	$E_p-E_k+V_{xx}$	V'_{xy}	0	0	0
V'_{xy}	V'_{xy}	$E_p-E_k+V_{xx}$	0	0	0
0	0	0	$E_p-E_k-V_{xx}$	$-V'_{xy}$	$-V'_{xy}$
0	0	0	$-V'_{xy}$	$E_p-E_k-V_{xx}$	$-V'_{xy}$
0	0	0	$-V'_{xy}$	$-V'_{xy}$	$E_p-E_k-V_{xx}$

Note that the wave functions with the + (−) subscript are antisymmetric (symmetric) under interchange of the two atoms inside the primitive cell so they correspond to the antibonding (bonding) orbitals. The two resultant 3×3 matrices can be diagonalized by a transformation to two new sets of symmetrized wave functions, one for the bonding orbitals and one for the antibonding orbitals. The set of eigenfunctions for the antibonding orbitals is given by

$$|X'_+\rangle = [|X_+\rangle - |Y_+\rangle]/\sqrt{2},$$
$$|Y'_+\rangle = [|X_+\rangle + |Y_+\rangle - 2|Z_+\rangle]/\sqrt{6},$$
$$|Z'_+\rangle = [|X_+\rangle + |Y_+\rangle + |Z_+\rangle]/\sqrt{3}.$$

The corresponding energies are given by

$$\langle Z'_+|\mathcal{H}_{\text{int}}|Z'_+\rangle = E_p + V_{xx} + 2V'_{xy}$$

and

$$\langle X'_+|\mathcal{H}_{\text{int}}|X'_+\rangle = \langle Y'_+|\mathcal{H}_{\text{int}}|Y'_+\rangle = E_p + V_{xx} - V'_{xy}.$$

The energies for the bonding orbitals are given by

$$\langle Z'_-|\mathcal{H}_{\text{int}}|Z'_-\rangle = E_p - V_{xx} - 2V'_{xy}$$

and

$$\langle X'_-|\mathcal{H}_{\text{int}}|X'_-\rangle = \langle Y'_-|\mathcal{H}_{\text{int}}|Y'_-\rangle = E_p - V_{xx} + V'_{xy}.$$

These results show that a [111] uniaxial stress splits both the antibonding (Γ_{15} conduction band) and bonding ($\Gamma_{25'}$ valence band) orbitals into a doublet and a singlet with an energy splitting equal to $3V'_{xy}$.

e) Using the results in part (d) and in Problem 3.8d, show that the deformation potential d' is given by

$$d' = (8/3^{5/2})(V_{pp\sigma} - V_{pp\pi}) = (2/3^{3/2})V_{xy}$$

in the tight-binding model. The values of d' obtained this way for Si and Ge are, respectively, 2.9 eV and 2.6 eV. These theoretical values of d' are not only quite different but also have signs *opposite* to those of the experimental values of d since we have neglected the internal strains, which are usually not equal to zero in stressed crystals.

3.12 *Calculation of the Optical Deformation Potential d_0 for the Γ_{15} Valence Bands Using the Tight-Binding Model*

As pointed out in Problem 3.11, a relative displacement $\boldsymbol{u}$ between the two sublattice in the diamond and zinc-blende crystals corresponds to the zone-center optical phonon with Γ_{15} symmetry. The deformation potential describing the electron–optical-phonon interaction is denoted by d_0 and is defined by

$$\delta E_{111} = \frac{d_0 u}{a},$$

where δE_{111} is the shift in energy of the singlet component ($[|X\rangle + |Y\rangle + |Z\rangle]/\sqrt{3}$) of the p orbital induced by the sublattice displacement u. d_0 can also be calculated within the tight-binding model in a way very similar to that adopted in Problem 3.11.

a) Assuming that the relative displacement of the two sublattices is given by $\boldsymbol{u} = (a\alpha/4)(1,1,1)$, show that the locations of the four nearest neighbors in the diamond or zinc-blende lattice are changed to

$$\boldsymbol{d}'_1 = (a/4)(1+\alpha)(1,1,1);$$
$$\boldsymbol{d}'_2 = (a/4)(1+\alpha, -1+\alpha, -1+\alpha);$$
$$\boldsymbol{d}'_3 = (a/4)(-1+\alpha, 1+\alpha, -1+\alpha);$$

and

$$\boldsymbol{d}'_4 = (a/4)(-1+\alpha, -1+\alpha, 1+\alpha).$$

b) Based on the result of (a), show that the effects of the relative displacement $\boldsymbol{u}$ are: (1) to change the lengths of the nearest-neighbor distances by the amount (to lowest order in α)

$$(\delta d_1)/d_1 = \alpha;$$
$$(\delta d_2)/d_2 = (\delta d_3)/d_3 = (\delta d_4)/d_4 = -\alpha/3$$

and (2) to change the orientations of the three bonds $\boldsymbol{d}_2$, $\boldsymbol{d}_3$, and $\boldsymbol{d}_4$. The directional cosines of the bond $\boldsymbol{d}_2'$ are given by

$$\boldsymbol{d}_2' = \{3[1-(\alpha/3)]\}^{-1/2}(1+\alpha, -1+\alpha, -1+\alpha)$$

with similar results for $\boldsymbol{d}_3'$ and $\boldsymbol{d}_4'$.

c) As in the case of Problem 3.11c, show that, to first order in α, the sublattice displacement does not change the matrix elements $\langle X1|\mathcal{H}_{\text{int}}|X2\rangle$, $\langle Y1|\mathcal{H}_{\text{int}}|Y2\rangle$, etc. On the other hand the matrix elements $\langle X1|\mathcal{H}_{\text{int}}|Y2\rangle$, $\langle X1|\mathcal{H}_{\text{int}}|Z2\rangle$, etc., are all equal to $V_{xy}' = (-16\alpha/9)(V_{pp\sigma} - V_{pp\pi})$. Set up the 6×6 matrix for calculating the energies of the p orbitals in the displaced sublattices as in Problem 3.11.

d) Diagonalize the 6×6 matrix in part (c) and show that the splitting between the doublet and singlet in the displaced sublattices is given by

$$\delta E = \frac{16\alpha}{3}(V_{pp\sigma} - V_{pp\pi}) = 4\alpha V_{xy}.$$

The optical phonon deformation potential is then given by

$$d_0 = \frac{32}{3\sqrt{3}}V_{xy} = 16d'.$$

Note that d_0 is much larger than d'.

3.13 *Internal Strain and the Deformation Potential d*

In Problems 3.11 and 3.12 the tight-binding model is used to calculate the deformation potentials for the diamond- and zinc-blende-type crystals in two extreme cases. In Problem 3.11 the crystal is assumed to be a continuum and the displacements of the atoms are described by the macroscopic strain tensor. In Problem 3.12 the atomic structure of the crystal is taken into account. The fact that the crystal is constructed from two sets of atoms located at two sublattices suggests that one can displace the two sublattices relative to each other in a way such that the symmetry of the primitive cell of the crystal is unchanged. For the diamond and zinc-blende structures this is the case when the two atoms within the primitive cell are displaced relative to each other along the body diagonal of the fcc lattice. Such microscopic relative displacements of the sublattices in a crystal cannot be described by the macroscopic strain tensor and are known as an **internal strain**.

By definition internal strain can only occur in crystals with more than one atom per primitive cell. The number of possible internal strain patterns for a given crystal structure can be determined by group theory and crystal symmetry [3.46]. In the diamond and zinc-blende structure the only possible internal strain is the relative displacement of the two atoms in the primitive cell along the [111] axis (or equivalent ones, e.g., $[1\bar{1}\bar{1}]$) described above. As noted in Problem 3.12, this displacement pattern is identical to that of the zone-center optical phonon in these structures. This is no coincidence. *Anastassakis* and *Cardona* [3.46] have shown that internal strains can be regarded as "frozen" Raman-active optical phonons.

It was pointed out by *Kleinman* [3.45] that under a [111] strain the bond length between the two atoms inside the primitive cell of Si is undetermined due to the possibility of an internal strain. Assume the [111] strain is given by the strain tensor

$$\alpha \begin{pmatrix} 0 & 1 & 1 \\ 1 & 0 & 1 \\ 1 & 1 & 0 \end{pmatrix}.$$

The internal strain is specified by choosing one atom as the origin and describing the displacement $\boldsymbol{u}$ of its nearest neighbor atom at $(a/4)(1,1,1)$ as

$$\boldsymbol{u} = (-a\alpha/2)(\zeta, \zeta, \zeta),$$

where ζ is the *internal strain parameter* mentioned in Problem 3.11. For $\zeta = 0$, there is no internal strain and the change in the bond length is specified completely by the macroscopic strain tensor as in Problem 3.11. For $\zeta = 1$ the internal strain exactly cancels the change in the bond length induced by the [111] strain so that the bond length in the strained crystal is unchanged.

a) Show that for a [111] shear strain with an internal strain parameter ζ the shear deformation potential d is given by

$$d = d' - (1/4)\zeta d_0.$$

b) The value of ζ has recently been deduced from experimental results for Si (0.58), Ge (0.56) [3.51] and a few zinc-blende semiconductors such as GaAs (0.65) [3.52]. Since $d_0 \simeq 16d'$ and the value of ζ is fairly close to 1, the deformation potential d is determined mainly by d_0. Calculate the value of d for Si using the value of d_0 obtained in Problem 3.12 and compare it with the experimental value of $d(-5.1\text{ eV})$ in Table 3.1.

3.14 *Stress Induced Splitting of* [110] *Valleys*
Consider a nondegenerate electronic state with wave vector $\boldsymbol{k}\|[110]$ in a cubic crystal. There are six equivalent valleys along the [110], $[\bar{1}10]$, [101], $[\bar{1}01]$, [011], and $[0\bar{1}1]$ directions (it is not necessary to consider the six opposite directions). Show from symmetry arguments that:

a) A [100] uniaxial strain will split these six valleys into a doublet {[011] and $[0\bar{1}1]$ valleys} and a quadruplet {the remaining four valleys} (this splitting is described by the shear deformation potential Ξ_u in Table 3.2);

b) A [111] uniaxial strain will split the six valleys into two triplets: {[110], [011], and [101] valleys} and {$[\bar{1}01]$, $[\bar{1}10]$, and $[0\bar{1}1]$} (this splitting is described by the shear deformation potential Ξ_p in Table 3.2).

c) Explain, using group theory, why the deformation potentials Ξ_u given in Table 3.2 are, in general, different for [100] and for [111] strain. Explain why two different *shear* deformation potentials Ξ_u and Ξ_p are needed for $\boldsymbol{k}$ along [110].

3.15 *Electromechanical Tensors in Zinc-Blende- and Wurtzite-Type Semiconductors*

a) Use the symmetry properties of the zinc-blende crystal described in Chap. 2 to show that the nonzero and linearly independent components of its electromechanical tensor $\boldsymbol{e}_{\mathrm{m}}$ are

$$(e_{\mathrm{m}})_{14} = (e_{\mathrm{m}})_{25} = (e_{\mathrm{m}})_{36}.$$

b) Use the symmetry properties of the wurtzite crystal described in Problem 3.7 to show that its electromechanical tensor has the form

$$\begin{pmatrix} 0 & 0 & 0 & 0 & (e_{\mathrm{m}})_{15} & 0 \\ 0 & 0 & 0 & (e_{\mathrm{m}})_{15} & 0 & 0 \\ (e_{\mathrm{m}})_{31} & (e_{\mathrm{m}})_{31} & (e_{\mathrm{m}})_{33} & 0 & 0 & 0 \end{pmatrix}.$$

3.16 *Piezoelectric Electron–Phonon Interactions in Wurtzite-Type Semiconductors*

a) Combine the results in Problems 3.7 and 3.15 to calculate the electric field induced by a phonon displacement $\boldsymbol{u}$, $\boldsymbol{E} = \boldsymbol{e}_m \cdot (\nabla \boldsymbol{u})/\varepsilon_\infty$, for all three acoustic modes. In particular show that for the TA_1 and LA phonons $\boldsymbol{E}$ is given by

$$E_x = 0,$$
$$E_y = \mathrm{i}q\left[(e_{\mathrm{m}})_{15} \sin\Theta\, u_z + (e_{\mathrm{m}})_{15} \cos\Theta\, u_\perp\right]/\varepsilon_\infty,$$

and

$$E_z = \mathrm{i}q\left[(e_{\mathrm{m}})_{31} \sin\Theta\, u_\perp + (e_{\mathrm{m}})_{33} \cos\Theta\, u_z\right]/\varepsilon_\infty,$$

where Θ is the angle between the phonon wave vector $\boldsymbol{q}$ and the c axis of the wurtzite crystal, u_z is the projection of $\boldsymbol{u}$ on the c axis, and $u_\perp$ is the component of $\boldsymbol{u}$ perpendicular to the c axis.

b) Substituting the results of part (a) into (3.29) and (3.30), show that the piezoelectric electron–phonon interaction in wurtzite crystals is given by [3.40]

$$\begin{aligned} H_{\mathrm{pe}} = {} & (4\pi|e|/\varepsilon_\infty)[(e_{\mathrm{m}})_{15} \sin\Theta(u_z \sin\Theta + u_\perp \cos\Theta) \\ & + (e_{\mathrm{m}})_{31} \sin\Theta \cos\Theta\, u_\perp + (e_{\mathrm{m}})_{33} \cos^2\Theta\, u_z]. \end{aligned}$$

3.17 *Phonons in Selenium and Tellurium*

In the harmonic approximation, the phonon frequencies are independent of strain. In nature, however, external strains result in small changes in the frequencies of phonons (anharmonic effects). The simplest of these effects are represented by the so-called mode Grüneisen parameters $\gamma_\omega = -d\ln\omega/d\ln V$, where V is the crystal volume. Each phonon frequency has a specific mode Grüneisen parameter γ_ω.

a) The γ_ω of optical phonons of tetrahedral semiconductors are $\sim$+1. Give a qualitative explanation.

b) The thermal expansion is a manifestation of anharmonicity. Derive the following expression for the *linear* thermal expansion coefficient $\alpha = (1/3)(d \ln V/dT)$:

$$\alpha = \langle\gamma\rangle C/3B, \tag{3.42}$$

where C is the specific heat, B the bulk modulus (p. 131) and $\langle\gamma\rangle$ an average Grüneisen parameter of all phonon frequencies.

c) The γ_ω of TA modes near the edge of the BZ is often negative [3.53]. Discuss the effect of a negative γ_ω on α [3.54].

3.18 *Dependence of Phonon Frequencies on Volume: Grüneisen Parameters*

Selenium and tellurium have three equal atoms per unit cell, hence 6 optical and 3 acoustic phonon branches. Discuss the eigenvectors of the optical phonons at the center of the BZ, their symmetries and degeneracies [2.5].

SUMMARY

Although the atoms in semiconductors are not stationary, their motion is so slow compared to that of electrons that they were regarded as static in Chap. 2. In this chapter we have analyzed the motion of atoms in semiconductors in terms of *simple harmonic oscillations*. Instead of calculating from first principles the *force constants* for these quantized oscillators or *phonons*, we have studied models based on which these force constants can be deduced from experimental results. The usefulness of these models is judged by the minimum number of parameters they require to describe experimental *phonon dispersion curves*. The more successful models typically treat the interaction between the electrons and ions in a realistic manner. The *shell model* assumes that the valence electrons are localized in deformable shells surrounding the ions. *Bond models* regard the solid as a very large molecule in which atoms are connected by bonds. Interactions between atoms are expressed in terms of *bond stretching* and *bond bending* force constants. In covalent semiconductors charges are known to pile up in regions between adjacent atoms, giving rise to *bond charges*. So far, models based on bond charges have been most successful in fitting experimental results.

In this chapter we have also studied the different ways electrons can be affected by phonons, i. e., *electron–phonon interactions*. These interactions have a significant effect on the optical and transport properties of electrons in semiconductors. We showed how long-wavelength acoustic phonons can change the energy of electrons via their strain field. These interactions can be described in terms of *deformation potentials*. Optical phonons can be regarded as giving rise to "internal strain" and their interactions with electrons can likewise be described by *optical-phonon deformation potentials*. In polar semiconductors both long-wavelength acoustic and optical phonons can generate electric fields through the charges associated with the moving ions. These fields can interact very strongly with electrons, giving rise to *piezoelectric electron–phonon interactions* for acoustic phonons and the *Fröhlich interaction* for optical phonons. Electrons located at band extrema near or at zone boundaries can be scattered from one valley to another equivalent valley via *intervalley electron–phonon interactions*.

4. Electronic Properties of Defects

CONTENTS

One reason why semiconductors are so useful for device applications is that their electrical properties can be modified significantly by the incorporation of small amounts of impurities or other kinds of defects. However, while one type of defect can make a semiconductor useful for fabricating a device, another type can have undesirable effects which render the device useless. The quantity of defects necessary to change the properties of a semiconductor is often considerably less than one defect atom per million host atoms.

As a result, our ability to control the defects in a semiconducting material often determines whether it can be used in device applications. To control the amount and nature of defects in a material typically involves developing a process for growing a relatively defect-free sample. Then the desired amounts of defects are introduced either during the growth process or after growth. There is extensive literature devoted to the study of defects in semiconductors. It would take more than a whole book to review all the properties of defects, so in this chapter we will limit ourselves to the study of the electronic properties of defects.

We shall begin by classifying the different kinds of defects found in semiconductors. Next we will separate defects into two broad categories. Impurities whose electronic energies can be calculated by means of the "effective mass approximation" are referred to as **shallow impurities**, while defects whose energies cannot be calculated with that approximation are known as *deep centers*. One method capable of calculating the energy levels of deep centers is the *Green's function method*. As simple illustrations of this method we will study deep centers in tetrahedrally bonded semiconductors.

4.1 Classification of Defects

In general, defects are classified into **point defects** and **line defects**. As the name implies, point defects usually involve isolated atoms in localized regions of a host crystal. Line defects, on the other hand, involve rows of atoms, and typical examples of line defects are **dislocations**. In addition to point and line defects, there are defects which are composed of a small number of point defects. These are referred to as **complexes**. Line defects are always detrimental to devices. Hence semiconductor wafers used in fabricating devices have to be as free of such defects as possible. The surface which terminates a three-dimensional crystal can also be considered as a two-dimensional "defect". However, the electronic states introduced by such surfaces are usually called "surface states" rather than defect states. In this chapter we shall not consider surface states or dislocations. Instead we will concentrate on the properties of point defects and complexes only, since they tend to determine the properties of semiconductor devices. For a brief discussion of surface states and their energies see Sect. 8.3.

Point defects are often further classified into the following kinds with special nomenclature and notations:

Vacancy: the vacancy created by a missing atom A is denoted by V_A.
Interstitial: an atom A occupying an interstitial site is denoted by I_A.
Substitutional: an atom C replacing a host atom A is denoted by C_A.
Antisite: a special kind of substitutional defect in which a host atom B occupies the site of another host atom A.
Frenkel defect pair: a complex V_A–I_A formed by an atom A displaced from a lattice site to a nearby interstitial site.

Vacancies and antisite defects are **intrinsic** or **native** defects since they do not involve foreign atoms. Their concentrations cannot be determined by chemical analysis or mass spectrometry. Defects involving foreign atoms (i. e., impurities) are referred to as **extrinsic** defects.

Many important defects are electrically active. Defects which can contribute free electrons to the host crystal are known as **donors**, while defects which can contribute holes (i. e., remove free electrons) are known as **acceptors**. Examples of donors in Si are substitutional group-V atoms such as P, As, and Sb or interstitial monovalent atoms such as Li and Na. The group-V atoms have one more valence electron than the Si atoms they replace. Furthermore, this extra electron is loosely bound to the group-V atom in Si so that it can be easily excited into the conduction band of the host Si crystal. Substitutional group-VI atoms such as S, Se, and Te in Si can contribute up to two conduction electrons so they are known as **double donors**. Examples of acceptors in Si are substitutional group-III atoms such as B, Al, Ga, and In. Substitutional group-II atoms (such as Be and Zn) in Si are **double acceptors**. When a substitutional impurity atom has the same valence as the host atom, it is referred to as an **isoelectronic** or **isovalent** center. Examples of such centers are C_{Si} in Si and N_P in GaP. As we will show later (Sect. 4.3.3), isovalent centers can behave as donors or acceptors or remain electrically inactive.

A term which is commonly found in the literature but rarely defined is **deep center**. In the past, some authors have used the term "deep centers" to mean defects whose electronic levels are located near the middle of the bandgap. Now we know that there are many defects with properties similar to those of deep levels but whose energies are not near the center of the gap. As a result, the term has been broadened to apply to any center which cannot be classified as shallow! Thus, to understand deep centers we have to understand first what shallow centers are. In the next section we shall study the properties of *shallow or hydrogenic impurities* in semiconductors with the diamond and zinc-blende structures. After that we shall study the properties of some deep centers.

4.2 Shallow or Hydrogenic Impurities

Let us consider a substitutional donor atom such as P_{Si} in Si. Compared to the Si nucleus, the P nucleus has one extra positive charge, which is balanced by the extra valence electron in the P atom as compared to the Si atom. The attractive potential between this extra valence electron and the P nucleus is not equal to that of an isolated P nucleus since the Coulomb potential of the P nucleus in Si will be screened, not only by the core electrons of the P atom, but also by the remaining four valence electrons of the P atom, and all the valence electrons of neighboring Si atoms. This screening effect allows us to approximate the attractive Coulomb potential seen by the extra valence electron in P by the Coulomb potential of a proton screened by the valence electrons of the Si host. Thus a P impurity in Si behaves effectively like a *hydrogen atom embedded in Si*, except that the mass of the P nucleus is so much heavier than the mass of the proton that we can assume it to be infinite. In addition, the Coulomb attraction between the electron and the positive charge in this "hydrogen-like" impurity is much weaker than the Coulomb attraction in the hydrogen atom since it is strongly screened by the large number of valence electrons in Si. As a result, the extra valence electron in the P atom is only loosely bound to the P atom when the atom is embedded in Si. This loosely bound electron can be ionized easily by thermal or electrical excitations. For this reason P in Si is known as a *donor* and the extra valence electron which it can "donate" to the Si conduction band is referred to as the *donor electron*. In the case of an acceptor impurity atom (such as a B atom replacing a Si atom) in Si, there is a deficiency of a valence electron when the acceptor bonds with its four nearest-neighbor Si atoms. Instead of regarding an acceptor atom as short of a valence electron, we think of it as possessing an *extra hole* which is loosely bound to a negatively charged B nucleus with an infinite mass. While a donor atom can be compared with a hydrogen atom, an acceptor atom is analogous to a positron bound to a negatively charged muon.

To calculate exactly the screened Coulomb potential between the donor electron and the donor ion is very difficult, since it depends on the many-body interactions between the electrons in the impurity atom with the valence electrons of the host. One simple approach to circumventing this problem is to assume that the positive charge on the donor ion is screened by the dielectric constant of the host crystal. With this approximation the Coulomb potential of the donor ion can be expressed as

$$V_S = +\frac{|e|}{4\pi\varepsilon_0\varepsilon_0 r}, \tag{4.1}$$

where ε_0 is the dielectric constant of the host crystal. If we assume further that the donor electron is not too localized near the donor ion then we can use the static (i. e., zero-frequency) dielectric constant as ε_0 in (4.1). More detailed studies of the dielectric constants of semiconductors will be presented in Chap. 6.

Since the donor electron is moving inside a semiconductor, its motion is affected by the crystal potential in addition to the impurity potential (4.1). The Schrödinger equation of the donor electron is given by

$$(H_0 + U)\Psi(\boldsymbol{r}) = E\Psi(\boldsymbol{r}), \tag{4.2}$$

where H_0 is the one-electron Hamiltonian of the perfect crystal, U is the potential energy of the electron in the screened Coulomb potential V_S

$$U = -|e|V_S \tag{4.3}$$

and $\Psi(\boldsymbol{r})$ is the donor electron wave function. In principle, one way to solve (4.2) is to expand $\Psi(\boldsymbol{r})$ in terms of the Bloch functions $\psi_{n\boldsymbol{k}}(\boldsymbol{r})$ of the perfect crystal, since they form a complete orthonormal set. This approach requires extensive numerical calculations. Since the defect breaks down the translational symmetry of the crystal we cannot take advantage of the Bloch theorem to simplify the problem. The most common approach to solving (4.2) is to utilize the *effective mass approximation*. This approximation makes use of the known electronic band structure parameters, such as effective masses, of the perfect crystal and is useful not only for calculating defect energy levels but also for studying properties of electrons under any weak external perturbation. We will make a digression to discuss this approximation in the next section.

4.2.1 Effective Mass Approximation

There are two approaches to deriving the effective mass approximation. One approach involves introducing the concept of **Wannier functions**. The other utilizes Bloch functions only. The Wannier functions are Fourier transforms of the Bloch functions, so the two approaches will eventually produce the same results. The Wannier function $a_n(\boldsymbol{r}; \boldsymbol{R}_i)$ is related to the Bloch function $\psi_{n\boldsymbol{k}}(\boldsymbol{r})$ by

$$a_n(\boldsymbol{r}; \boldsymbol{R}_i) = N^{-1/2} \sum_{\boldsymbol{k}} \exp(-\mathrm{i}\boldsymbol{k} \cdot \boldsymbol{R}_i)\psi_{n\boldsymbol{k}}(\boldsymbol{r}), \tag{4.4a}$$

$$\psi_{n\boldsymbol{k}}(\boldsymbol{r}) = N^{-1/2} \sum_{\boldsymbol{R}_i} \exp(\mathrm{i}\boldsymbol{k} \cdot \boldsymbol{R}_i)a_n(\boldsymbol{r}; \boldsymbol{R}_i), \tag{4.4b}$$

where $\boldsymbol{R}_i$ is a lattice vector, n the band index, $\boldsymbol{k}$ the wave vector in the reduced zone scheme, and N the number of unit cells in the crystal. While Bloch functions are indexed by the wave vectors in reciprocal lattice space, *Wannier functions are indexed by lattice vectors in real space*. While Bloch functions are more convenient for representing extended states in crystals, Wannier functions are more appropriate for localized states. We note that Wannier functions are similar to the Löwdin orbitals defined in (2.72) for the tight-binding model (Sect. 2.7.2). For very localized electrons, one can think of Wannier functions as atomic orbitals. In this chapter we will approach the problem of defects by using Wannier functions. An excellent reference for this approach is the classic text by Wannier himself [4.1]. An equally excellent reference for the alternative approach based on Bloch functions is the review article by *Kohn* [4.2].

Some of the properties of Wannier functions are summarized below

- $a_n(\boldsymbol{r}; \boldsymbol{R}_i)$ is a function of $\boldsymbol{r} - \boldsymbol{R}_i$ only. This can be easily shown by noting that $a_n(\boldsymbol{r}; \boldsymbol{R}_i) = a_n(\boldsymbol{r} + \boldsymbol{s}; \boldsymbol{R}_i + \boldsymbol{s})$ for any vector $\boldsymbol{s}$. From now on we will represent $a_n(\boldsymbol{r}; \boldsymbol{R}_i)$ as $a_n(\boldsymbol{r} - \boldsymbol{R}_i)$.
- Wannier functions $a_n(\boldsymbol{r} - \boldsymbol{R}_i)$ where $\boldsymbol{R}_i$ varies over all the lattice vectors inside the crystal, form a complete and orthonormal set just like the Bloch functions.
- Wannier functions are eigenfunctions of a **"lattice vector operator"** $\boldsymbol{R}_{\mathrm{op}}$ defined by

$$\boldsymbol{R}_{\mathrm{op}} a_n(\boldsymbol{r} - \boldsymbol{R}_i) = \boldsymbol{R}_i a_n(\boldsymbol{r} - \boldsymbol{R}_i). \tag{4.5}$$

The effect of $\boldsymbol{R}_{\mathrm{op}}$ on a wave function

$$\Psi(\boldsymbol{r}) = \sum_{n,\boldsymbol{k}} A_n(\boldsymbol{k})\psi_{n\boldsymbol{k}}(\boldsymbol{r})$$

can be represented approximately as

$$\boldsymbol{R}_{\mathrm{op}}\Psi(\boldsymbol{r}) \approx \sum_{n,\boldsymbol{k}} \left(\mathrm{i}\frac{\partial}{\partial \boldsymbol{k}} A_n(\boldsymbol{k})\right) \psi_{n\boldsymbol{k}}(\boldsymbol{r}). \tag{4.6}$$

The proof of (4.6) involves expanding $\Psi(\boldsymbol{r})$ first in terms of Bloch functions and then in terms of Wannier functions:

$$\begin{aligned}\Psi(\boldsymbol{r}) &= \sum_{n,\boldsymbol{k}} A_n(\boldsymbol{k})\psi_{n\boldsymbol{k}}(\boldsymbol{r}) \\ &= \sum_{n,\boldsymbol{k}} A_n(\boldsymbol{k}) \sum_{\boldsymbol{R}_i} (N^{-1/2}) \exp(\mathrm{i}\boldsymbol{k} \cdot \boldsymbol{R}_i)\, a_n(\boldsymbol{r} - \boldsymbol{R}_i). \end{aligned} \tag{4.7}$$

Operating on both sides of (4.7) with $\boldsymbol{R}_{\mathrm{op}}$ we obtain

$$\boldsymbol{R}_{\mathrm{op}}\Psi(\boldsymbol{r}) = \sum_{n,\boldsymbol{k}} A_n(\boldsymbol{k}) \sum_{\boldsymbol{R}_i} (N^{-1/2}) \exp(\mathrm{i}\boldsymbol{k}\cdot\boldsymbol{R}_i)\boldsymbol{R}_i a_n(\boldsymbol{r}-\boldsymbol{R}_i). \tag{4.8}$$

Formula (4.8) can be rewritten as

$$\boldsymbol{R}_{\mathrm{op}}\Psi(\boldsymbol{r}) = \sum_{n,\boldsymbol{k}} A_n(\boldsymbol{k}) \sum_{\boldsymbol{R}_i} (N^{-1/2}) \left(-\mathrm{i}\frac{\partial}{\partial \boldsymbol{k}}\right) \exp(\mathrm{i}\boldsymbol{k}\cdot\boldsymbol{R}_i)\, a_n(\boldsymbol{r}-\boldsymbol{R}_i) \tag{4.9}$$

$$= \sum_{n,\boldsymbol{k}} \left(-\mathrm{i}\frac{\partial}{\partial \boldsymbol{k}}\right) A_n(\boldsymbol{k})\psi_{n\boldsymbol{k}}(\boldsymbol{r}) - \sum_{n,\boldsymbol{k}} \left[\left(-\mathrm{i}\frac{\partial}{\partial \boldsymbol{k}}\right) A_n(\boldsymbol{k})\right] \psi_{n\boldsymbol{k}}(\boldsymbol{r}). \tag{4.10}$$

The summation over $\boldsymbol{k}$ on the right hand side of (4.10) can be approximated by an integral over $\boldsymbol{k}$ in the limit that the crystal volume becomes infinite. After integration, the first term on the right hand side of (4.10) involves

$$A_n(\boldsymbol{k}_2)\,\psi_{n\boldsymbol{k}_2}(\boldsymbol{r}) - A_n(\boldsymbol{k}_1)\,\psi_{n\boldsymbol{k}_1}(\boldsymbol{r}),$$

where $\boldsymbol{k}_1$ and $\boldsymbol{k}_2$ are two equivalent points on opposite surfaces of the Brillouin zone. Because of the periodicity of the Bloch functions $A_n(\boldsymbol{k})\psi_{n\boldsymbol{k}}(\boldsymbol{r})$, this term vanishes and we obtain (4.6) [4.1]. The $\approx$ sign in (4.6) is a reminder that this equation is an approximation because the values of $\boldsymbol{k}$ are discrete. The interpretation of (4.6) is as follows: the effect of operating with $\boldsymbol{R}_{\mathrm{op}}$ on any function $\Psi(\boldsymbol{r})$ is equivalent to applying the operator $(\mathrm{i}\partial/\partial\boldsymbol{k})$ to the coefficients $A_n(\boldsymbol{k})$ of the expansion of $\Psi(\boldsymbol{r})$ in terms of Bloch functions. One can show that $\hbar\boldsymbol{k}$ and $\boldsymbol{R}_{\mathrm{op}}$ are conjugate operators just like momentum $\boldsymbol{p}$ and $\boldsymbol{r}$ in the sense that

$$\boldsymbol{R} \leftrightarrow (\mathrm{i}\partial/\partial\boldsymbol{k}) \quad \text{and} \quad \boldsymbol{k} \leftrightarrow (-\mathrm{i}\partial/\partial\boldsymbol{R}). \tag{4.11}$$

From now on we will drop the subscript op in $\boldsymbol{R}_{\mathrm{op}}$ and, for simplicity, $\boldsymbol{R}$ will be understood to represent both the lattice vector $\boldsymbol{R}$ and its operator $\boldsymbol{R}_{\mathrm{op}}$. One should keep in mind that $\boldsymbol{k}$ and $\boldsymbol{R}$ are discrete variables while $\boldsymbol{p}$ and $\boldsymbol{r}$ are truly continuous variables. The correspondences in (4.11) represent approximations which are valid only under the various assumptions discussed above.

We will now utilize the result in (4.6) to simplify the Schrödinger equation (4.2). Let us expand the wave function $\Psi(\boldsymbol{r})$ in (4.2) as a linear combination of Wannier functions:

$$\Psi(\boldsymbol{r}) = N^{-1/2} \sum_{n,i} C_n(\boldsymbol{R}_i)\, a_n(\boldsymbol{r}-\boldsymbol{R}_i), \tag{4.12}$$

where n is again the band index and the $C_n(\boldsymbol{R}_i)$ are coefficients analogous to $A_n(\boldsymbol{k})$ in (4.7). Thus the $C_n(\boldsymbol{R}_i)$ can be regarded as the amplitudes of Wannier functions. We will show later that they are also solutions to a wave equation, so they are known as **envelope wave functions**. Let the eigenvalues of the unperturbed Hamiltonian $H_0(\boldsymbol{r})$ in (4.2) be represented as $W_n(\boldsymbol{k})$. Then (4.11) allows us to convert $W_n(\boldsymbol{k})$ into an operator on Wannier functions:

$$\langle n,\boldsymbol{R}_i \,|\, H_0(\boldsymbol{r}) \,|\, n',\boldsymbol{R}_j\rangle \leftrightarrow \delta_{nn'}\delta_{ij} W_n(-\mathrm{i}\partial/\partial\boldsymbol{R}). \tag{4.13}$$

where $|\,n',\boldsymbol{R}_j\rangle$ denotes the Wannier function $a_{n'}(\boldsymbol{r}-\boldsymbol{R}_j)$ and the index for $\boldsymbol{R}$ has been dropped, as in (4.11), to simplify the notation.

To obtain the operator corresponding to $U(\boldsymbol{r})$ in (4.2), we will assume that U is a slowly varying function of $\boldsymbol{r}$, so that the change in U within one lattice constant (a_0) is small compared with U, i. e.

$$a_0 \,|\, \nabla U(\boldsymbol{r}) \,| \ll U(\boldsymbol{r}). \tag{4.14}$$

The matrix elements of U between two Wannier functions are given by

$$\langle n, \boldsymbol{R}_i \,|\, U(\boldsymbol{r}) \,|\, n', \boldsymbol{R}_j \rangle = \int a_n^*(\boldsymbol{r} - \boldsymbol{R}_i)\, U(\boldsymbol{r})\, a_{n'}(\boldsymbol{r} - \boldsymbol{R}_j)\, d\boldsymbol{r}. \tag{4.15}$$

Suppose $\boldsymbol{r}$ lies within the primitive cell indexed by the lattice vector $\boldsymbol{R}$. Because U is a slowly varying function of $\boldsymbol{r}$ we can expand U about $U(\boldsymbol{R})$ to just the first order in $\nabla_{\boldsymbol{R}} U$:

$$U(\boldsymbol{r}) = U(\boldsymbol{R}) + (\boldsymbol{r} - \boldsymbol{R}) \nabla_{\boldsymbol{R}} U(\boldsymbol{R}). \tag{4.16}$$

Substituting (4.16) into (4.15) we obtain

$$\begin{aligned} \langle n, \boldsymbol{R}_i \,|\, U(\boldsymbol{r}) \,|\, n', \boldsymbol{R}_j \rangle &= U(\boldsymbol{R}) \delta_{nn'} \delta_{ij} \\ &\quad + \nabla U(\boldsymbol{R}) \int a_n^*(\boldsymbol{r} - \boldsymbol{R}_i)(\boldsymbol{r} - \boldsymbol{R}) a_{n'}(\boldsymbol{r} - \boldsymbol{R}_j) d\boldsymbol{r}. \end{aligned} \tag{4.17}$$

Using the inequality (4.14), the second term in (4.17) can be neglected compared with the first term, so that (4.17) reduces to

$$\langle n, \boldsymbol{R}_i \,|\, U(\boldsymbol{r}) \,|\, n', \boldsymbol{R}_j \rangle \approx \delta_{nn'} \delta_{ij} U(\boldsymbol{R}). \tag{4.18}$$

Combining (4.13) and (4.18) we obtain

$$\langle n, \boldsymbol{R}_i \,|\, H_0 + U(\boldsymbol{r}) \,|\, n', \boldsymbol{R}_j \rangle \leftrightarrow \delta_{nn'} \delta_{ij} [W_n(-\mathrm{i}\partial/\partial \boldsymbol{R}) + U(\boldsymbol{R})]. \tag{4.19}$$

We should note that while the lattice vectors $\boldsymbol{R}_i$ on the left hand side are discrete, at the right hand side of (4.19) they are treated as continuous variables. To be exact (4.19) should be replaced by a set of difference equations [4.3]. Replacing (4.19) into (4.2) and using (4.12) we arrive at a very useful equation:

$$[W_n(-\mathrm{i}\partial/\partial \boldsymbol{R}) + U(\boldsymbol{R})] C_n(\boldsymbol{R}) \approx E C_n(\boldsymbol{R}). \tag{4.20}$$

As an illustration of how to apply (4.20) to solve for the energies of a donor electron, we will assume that the lowest conduction band for the semiconductor is *isotropic, nondegenerate, and parabolic*, with the band minimum located at the zone center. Its energy is thus given by

$$W_n(\boldsymbol{k}) = E_c(\boldsymbol{0}) + \frac{\hbar^2 \boldsymbol{k}^2}{2m^*}, \tag{4.21}$$

where m^* is the effective mass of this conduction band and $E_c(\boldsymbol{0})$ the band edge. Since we consider only one conduction band the band index n will be dropped. We note that both Si and Ge do not satisfy these assumptions since their conduction band minima are degenerate and do not occur at the zone center. However, these assumptions are valid for many semiconductors with the zinc-blende structure, such as GaAs and InP. Substituting the expression for $W_n(\boldsymbol{k})$ in (4.21) into (4.20) we obtain

$$\left[-\left(\frac{\hbar^2}{2m^*}\right)\frac{\partial^2}{\partial \boldsymbol{R}^2}+U(\boldsymbol{R})\right]C(\boldsymbol{R}) \approx [E-E_c(\mathbf{0})]C(\boldsymbol{R}), \tag{4.22}$$

which is equivalent to the Schrödinger equation for a particle with effective mass m^* moving in a potential U. In other words, the net effect of the crystal potential on the donor electron inside the crystal is to change the electron mass from the value in free space to the effective mass m^* and also to contribute the factor ε_0 in (4.1). As a result this approach is known as the **effective mass approximation**. We should remember that (4.22) enables only the envelope function $C(\boldsymbol{R})$ to be calculated. This envelope function has to be multiplied by the Wannier function $a(\boldsymbol{r}-\boldsymbol{R})$ in order to arrive at the final electron wave function $\Psi(\boldsymbol{r})$.

The effective mass approximation will be used throughout this book in calculating the transport and optical properties of electrons in semiconductors (these topics will be covered in Chaps. 5 and 6, respectively). In order for this approximation to be valid the perturbing potential must be weak (so that no electrons are excited from one band to another) and slowly varying in space (so that $\boldsymbol{R}$ can be regarded as continuous). In the next section we will apply it to calculate the energies and wave functions of shallow donors.

4.2.2 Hydrogenic or Shallow Donors

By treating $\boldsymbol{R}$ as a continuous vector in real space rather than as a discrete lattice vector (4.22) can be solved as a differential equation. Equation (4.2) then essentially becomes the Schrödinger equation for a particle moving in a Coulomb potential and the motion of the donor electron becomes equivalent to that of the electron in the hydrogen atom. Donors whose electrons can be described by the solutions of (4.22) are said to be *hydrogenic* or *shallow*. The solutions to the Schrödinger equation for the electron in the hydrogen atom are well known. They can be found in many textbooks on quantum mechanics, hence we will not repeat them here. Instead we will state the results.

- There are both discrete and continuous eigenvalues. The continuum states of the donor electron are now the delocalized conducting states. Note that in the vicinity of the bandgap these conducting states are not the same as the conduction band states in the absence of the defect potential. Only in the limit that U approaches zero does the donor electron energy E in (4.22) approach the conduction band energy E_c and the wave functions of the donor electron become equal to those of the nearest conduction band electrons. The bound states of the donor electron are classified according to their *principal quantum number* N, *angular momentum* L, and *spin*. In atomic physics these bound states are denoted as 1s, 2s, 2p, etc. Similar notations are used to denote the bound states of shallow impurities. The energies of these bound states are given by the **Rydberg series**:

$$E - E_c(\mathbf{0}) = -R/N^2 \quad (N = 1, 2, 3, \ldots). \tag{4.23}$$

R is the **Rydberg constant** for the donor electron and is related to the Rydberg constant for the hydrogen atom $[e^4 m_0/(2\hbar^2)]$ by

$$R = \left(\frac{m^*}{m_0}\right)\left(\frac{1}{\varepsilon_0^2}\right)\left(\frac{e^4 m_0}{2\hbar^2}\right)\frac{1}{(4\pi\varepsilon_0)^2}, \tag{4.24}$$

m_0 being the free electron mass. A schematic diagram of some of the bound states of a donor atom near a simple parabolic conduction band is shown in Fig. 4.1.

- The extent of the bound-state electron wave functions in real space is measured in terms of a *donor Bohr radius* a^*. It is related to the Bohr radius in the hydrogen atom $[\hbar^2/(m_0 e^2)]$ by

$$a^* = \left(\frac{\varepsilon_0 m_0}{m^*}\right)\left(\frac{\hbar^2}{m_0 e^2}\right)(4\pi\varepsilon_0). \tag{4.25}$$

In particular, the wave function of the 1s state is given by

$$C_{1s}(\boldsymbol{R}) = \left(\frac{1}{\pi}\right)^{1/2}\left(\frac{1}{a^*}\right)^{3/2}\exp\left(\frac{-R}{a^*}\right). \tag{4.26}$$

In order that $\boldsymbol{R}$ can be considered continuous rather than discrete, we require $a^* \gg a_0$. This condition also ensures that it is meaningful to approximate the entire conduction band structure by an effective mass m^*. The reason is that the extent in $\boldsymbol{R}$ of an envelope function $C(\boldsymbol{R})$ corresponding to the electron wave function $\Psi(\boldsymbol{r})$ scales as a^*. On the other hand, the extent in $\boldsymbol{k}$-space of Bloch functions (which are indexed by $\boldsymbol{k}$) to be summed over in the reciprocal space to construct $\Psi(\boldsymbol{r})$ can be small. This is because of an "uncertainty principle" for two variables that are related by Fourier

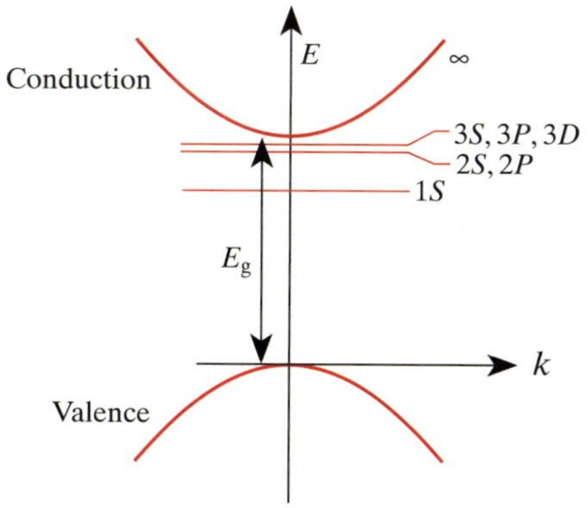

Fig. 4.1. Schematic diagram of the $n=1$, 2, and 3 bound states of a shallow donor electron near a nondegenerate and parabolic conduction band (corresponding to $n = \infty$). E_g is the bandgap

transformations. For example, if a function $f(t)$ of the time t has an extent Δt, its Fourier transform $g(\omega)$ is a function of angular frequency ω and has an extent $\Delta\omega$. Then there is a relation between Δt and $\Delta\omega$ given by $\Delta t \Delta\omega \approx 1$. Similarly, we expect that $\Delta \boldsymbol{k} \cdot \Delta \boldsymbol{R} \approx |\Delta \boldsymbol{k}| a^* \approx 1$ or $|\Delta \boldsymbol{k}| \leq (1/a^*)$. Hence only conduction band states over a small region of reciprocal space around the band minimum contribute to the defect wave function if $a^* \gg a_0$. This justifies the expansion of the conduction band energy in (4.21) to only the quadratic term, which involves the effective mass. We note that while the binding energy of the donor electron decreases as $(1/N^2)$, the extent of its wave function increases as N^2. It is therefore possible for the higher excited states of a donor electron to be well described by the hydrogenic model even if this is not true for its 1s ground state.

- The relative errors in eigenvalues introduced by using the effective mass approximation are of the order of $[a_0/(2\pi a^*)]^2$, where a_0, the lattice constant of the semiconductor, is usually a few angstroms. To obtain an order of magnitude estimate for a^* and R, let us assume some typical values for ε_0 and m^* in semiconductors, such as $\varepsilon_0 \approx 10$ and $m^* \approx 0.1 m_0$. Substituting these values into (4.24) and (4.25) gives an a^* of about 50 Å and a binding energy of about 14 meV for a donor electron. Since a^*, as shown by this estimate, is generally much larger than a_0, donor electrons in most semiconductors with conduction band minimum at Γ, can be described rather well by the effective mass approximation.

As pointed out earlier, the function $C(\boldsymbol{R})$ obtained by solving (4.22) is only the envelope function and not the complete wave function. It has to be multiplied by the Wannier functions to obtain the complete wave function. We shall now consider the donor electron wave function $\Psi(\boldsymbol{r})$ in the limit of a very large Bohr radius a^*. We first express the wave function in terms of Bloch functions. Starting with

$$\Psi(\boldsymbol{r}) = \sum_i C(\boldsymbol{R}_i) a(\boldsymbol{r} - \boldsymbol{R}_i) \tag{4.27}$$

we substitute (4.4a) for $a(\boldsymbol{r} - \boldsymbol{R}_i)$ and obtain

$$\Psi(\boldsymbol{r}) = \sum_i C(\boldsymbol{R}_i) N^{-1/2} \sum_{\boldsymbol{k}} \exp(-\mathrm{i}\boldsymbol{k} \cdot \boldsymbol{R}_i) \psi_{\boldsymbol{k}}(\boldsymbol{r}). \tag{4.28}$$

The Bloch function $\psi_{\boldsymbol{k}}(\boldsymbol{r})$ can be expressed in terms of the periodic functions $u_{\boldsymbol{k}}(\boldsymbol{r})$ using (2.6):

$$\psi_{\boldsymbol{k}}(\boldsymbol{r}) = \exp(\mathrm{i}\boldsymbol{k} \cdot \boldsymbol{r}) u_{\boldsymbol{k}}(\boldsymbol{r}). \tag{4.29}$$

Substituting (4.29) into (4.28) we arrive at

$$\Psi(\boldsymbol{r}) = \sum_i C(\boldsymbol{R}_i) N^{-1/2} \sum_{\boldsymbol{k}} \exp[-\mathrm{i}\boldsymbol{k} \cdot (\boldsymbol{R}_i - \boldsymbol{r})] u_{\boldsymbol{k}}(\boldsymbol{r}). \tag{4.30}$$

As pointed out earlier, when a^* is large, the summation over $\boldsymbol{k}$ in (4.30) can be restricted to a small region near the band minimum at the zone center. This allows us to assume the periodic function $u_{\boldsymbol{k}}(\boldsymbol{r})$ to be independent of $\boldsymbol{k}$ and equal to the function at $\boldsymbol{k} = 0$. Equation (4.30) can therefore be approximated by

$$\begin{aligned}\Psi(\boldsymbol{r}) &\approx u_0(\boldsymbol{r})\sum_i C(\boldsymbol{R}_i)N^{-1/2}\sum_{\boldsymbol{k}}\exp[-\mathrm{i}\boldsymbol{k}\cdot(\boldsymbol{R}_i-\boldsymbol{r})]\\ &= u_0(\boldsymbol{r})\sum_i C(\boldsymbol{R}_i)\delta(\boldsymbol{R}_i-\boldsymbol{r}) \qquad (4.31)\\ &= u_0(\boldsymbol{r})C(\boldsymbol{r}). \qquad (4.32)\end{aligned}$$

It should be remembered that (4.32) is only an approximate result. However, it provides a very simple picture of the defect wave function. The wave function of a conduction electron is constructed by multiplying u_0, a periodic function, with a plane wave. To construct the donor electron wave function, u_0 is multiplied by the envelope function $C(\boldsymbol{r})$ which is localized around the defect. Figure 4.2 shows a schematic wave function for a donor electron in the $1s$ state.

In Table 4.1 we list the binding energies for the $1s$ level of donors in several zinc-blende-type semiconductors calculated with (4.23) and (4.24) using the experimentally measured values of m^* and ε_0. These theoretical binding

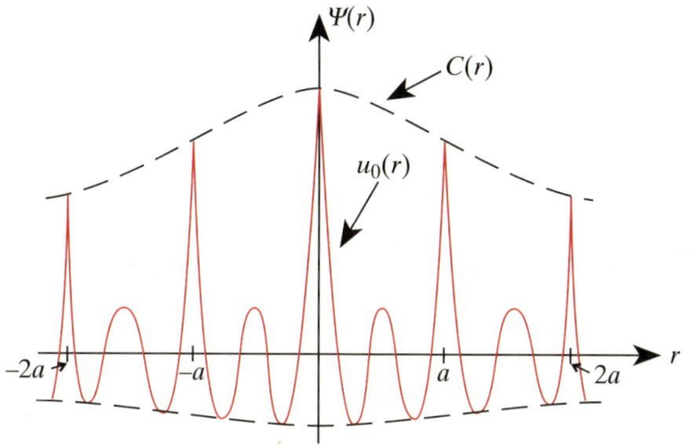

Fig. 4.2. Schematic diagram of a shallow donor electron wave function in real space. $u_0(r)$ is the Bloch function part and $C(r)$ is the envelope function; a is the distance between the lattice sites

Table 4.1. Experimental binding energies of the $1s$ state of shallow donors in some zinc-blende-type semiconductors (from [Ref. 4.4, p. 224]) compared with the predictions of (4.24)

Semiconductor	Binding energy from (4.24) [meV]	Experimental binding energy of common donors [meV]
GaAs	5.72	Si_{Ga}(5.84); Ge_{Ga}(5.88) S_{As}(5.87); Se_{As}(5.79)
InP	7.14	7.14
InSb	0.6	Te_{Sb}(0.6)
CdTe	11.6	In_{Cd}(14); Al_{Cd}(14)
ZnSe	25.7	Al_{Zn}(26.3); Ga_{Zn}(27.9) F_{Se}(29.3); Cl_{Se}(26.9)

energies are compared with the experimental values for some of the common donors in each semiconductor. This table shows that the effective mass approximation can predict the shallow donor binding energies rather well. However, there are also a large number of donors whose binding energies do not agree with the prediction of the effective mass theory. For example, Cu in GaAs is a donor with a binding energy of 70 meV. Centers whose binding energy cannot be calculated by the effective mass approximation are referred to as *deep centers*. They will be discussed in greater detail in the following sections.

Given an impurity atom in a host semiconductor, it is not easy to predict whether it will form a shallow or a deep center. The following guidelines have been found to be helpful, although not foolproof.

- If the "core" (the atom minus the outer valence electrons) of the impurity atom resembles the core of the host atom (allowing, of course, for the difference of one nuclear charge), the impurity levels tend to be shallow. For example, consider Ge_{Ga} in GaAs. The core of the Ge atom is almost identical to the core of the Ga atom, so Ge_{Ga} is a shallow donor in GaAs.
- If the impurity atom induces a strongly localized potential, such as a strain field around the impurity atom, the result is most likely a deep center. The part of a defect potential which is localized within one unit cell is known as a **central cell correction**. Central cell corrections violate the assumption that the defect potential is slowly varying in space in the effective mass approximation and therefore may result in deep centers. In some cases a defect may have both shallow and deep bound states. For example, electrons with *s*-symmetry envelope wave functions are more likely to behave like deep centers than electrons with *p*-symmetry envelope functions. The reason is that envelope functions with *s*-symmetry have nonzero probability densities at the origin (or the defect) and therefore are more sensitive to central cell corrections.

The above guidelines are not infallible principles. The problem is that the properties of defects in semiconductors depend on a number of factors, including the charge state of the defect, the band structure, and so on. An illustration of the difficulty in predicting whether an impurity will be shallow or deep is the case of Si in GaAs. While Si_{Ga} in GaAs is a perfect example of a hydrogenic donor, Si can also be incorporated as a deep donor known as the **DX center**, with completely different properties in GaAlAs alloys containing more than about 25 % of Al [4.5] or in pure GaAs under pressure [4.6]. It is now generally accepted that while the lowest electronic levels of Si_{Ga} in GaAs are hydrogenic there is an excited deep state which is resonant with the conduction band. Alloying with Al or applying hydrostatic pressure changes the conduction band structure and lowers the resonant deep state into the bandgap. When this deep state becomes lower in energy than the shallow levels a **shallow-to-deep instability** occurs. It has been suggested [4.7, 8] that this resonant level of Si in GaAs behaves like a deep center because of a large lattice relaxation associated with the impurity. The predicted configuration of this lattice distortion has not yet been confirmed experimentally. However, the theoretical models are now believed to be correct based on the verfications of their other predictions [See Appendix on DX Center on Web Page: http://Pauline.Berkeley.edu/textbook/DX-Center.pdf].

4.2.3 Donors Associated with Anisotropic Conduction Bands

Some of the most important semiconductors, such as Si, Ge, GaP, and even diamond, have their lowest conduction band minima near the zone boundaries. In these semiconductors the conduction band effective mass is strongly anisotropic. In addition, the conduction valleys are degenerate as a result of their symmetry. For example, we see in Fig. 2.10 that there are six conduction band minima in Si occurring in the six equivalent [100] directions about 85 % out towards the zone boundary. The impurity central cell corrections produce an interaction between the six degenerate valleys known as the **valley–orbit coupling**. These complications require a modification of the effective mass apoximation discussed in the last section. We will consider the donors in Si as an example but the technique can also be applied to donors in Ge with slight modifications.

Let us first neglect the valley–orbit coupling between the six equivalent conduction band valleys. The electron effective mass for each valley in Si can be written as a second-rank tensor

$$\begin{vmatrix} m_\mathrm{l} & 0 & 0 \\ 0 & m_\mathrm{t} & 0 \\ 0 & 0 & m_\mathrm{t} \end{vmatrix},$$

where m_l and m_t are, respectively, the effective masses longitudinal and transverse to the [100] axis (these masses have to be permuted when considering the equivalent [010] and [001] directions). For Si the masses have been determined by means of cyclotron resonance experiments to be $0.916m_0$ and $0.190m_0$, respectively. With this effective mass tensor the wave equation (4.22) for the envelope wave function $\Phi_j(\boldsymbol{R})$ (where the integer j now labels the valley along the [100] axis) becomes

$$\left[-\left(\frac{\hbar^2}{2}\right)\left(\frac{2\boldsymbol{\nabla}_\mathrm{t}^2}{m_t}+\frac{\boldsymbol{\nabla}_\mathrm{l}^2}{m_\mathrm{l}}\right)-|e|V_\mathrm{S}\right]\Phi_j(\boldsymbol{R})\approx[E-E_\mathrm{c}(\boldsymbol{k}_0)]\Phi_j(\boldsymbol{R}), \qquad (4.33)$$

where ∇_t and ∇_l are, respectively, the components of the operator $\boldsymbol{\nabla}$ projected along directions transverse and longitudinal to the [100] axis and $\boldsymbol{k}_0$ is the location of the conduction band minimum along the [100] axis in reciprocal lattice space. Equation (4.33) can be regarded as the wave equation for an "elliptically deformed hydrogen atom". If m_l is not too different from m_t, the solutions of this equation should be quite similar to those of the hydrogen atom and we can still label its eigenstates as $1S$, $2S$, and $2P$, etc. However, the lowering of the symmetry from spherical to cylindrical inherent in (4.33) means that states with the same principal quantum number N and angular momentum L but different magnetic quantum numbers m are no longer degenerate. It can be shown (for example, with time reversal symmetry arguments) that states with equal and opposite values of m remain degenerate. For example, $2P^1$ and $2P^{-1}$ are still degenerate but will have a different energy than $2P^0$. If m_l is much larger than m_t, (4.33) can be solved by using an adiabatic approximation to separate the longitudinal motion from the transverse one. More generally, (4.33) can be solved approximately by a variational technique. *Kohn* and *Luttinger* [4.9] solved (4.33) by using trial functions of the form

$$\Phi(x, y, z) = \exp[-a(y^2 + z^2) - bx^2]^{1/2}, \tag{4.34}$$

where the parameters a and b are varied to minimize the energy and the x axis is chosen along the direction ([100] in the present case) of the valley.

The results of a variational calculation performed by *Faulkner* [4.10] for several donors in Si are compared with the experimental results in Fig. 4.3. We see that the agreement between Faulkner's calculation and experiment is very good for the excited states but rather poor for the $1S$ ground state. This is expected since the excited states have larger orbits and are therefore less sensitive to central cell corrections. The theoretical $1S$ ground state energy can be greatly improved by including the *valley–orbit coupling*.

For simplicity, we shall treat the valley–orbit coupling among the $1S$ ground states of the six equivalent [100] valleys by perturbation theory and neglect their mixing with the excited states. As unperturbed wave functions, we construct six approximate wave functions similar to the donor wave function in (4.32):

$$x_j(\mathbf{r}) = \Phi_j(\mathbf{r})\psi_j(\mathbf{r}), \tag{4.35}$$

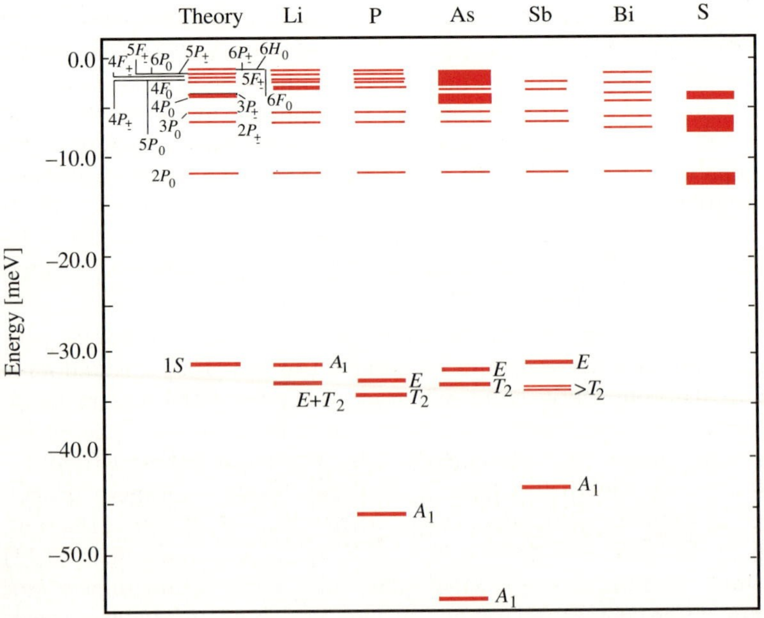

Fig. 4.3. Calculated and measured shallow donor energy levels in Si. (From [4.10])

where $\psi_j(\boldsymbol{r})$ is the Bloch function for the jth ($j = 1,\ldots,6$) conduction band minimum and $\Phi_j(\boldsymbol{r})$ the corresponding envelope function obtained by solving (4.33). For convenience, the six Bloch wave functions $\psi_j(\boldsymbol{r})$ are labeled X, $\overline{X}$, Y, $\overline{Y}$, Z and $\overline{Z}$ according to the direction of the corresponding conduction band minimum. If we assume that the donor in Si is substitutional (such as As_{Si}), the impurity potential has tetrahedral symmetry. The diagonalization of the perturbation Hamiltonian is simplified by symmetrizing the six functions according to the irreducible representations of the T_d group. Using the character table for the T_d group (Table 2.3) it can be shown (Problem 4.1) that these six Bloch functions form a singlet with A_1 symmetry, one doublet with E symmetry, and a triplet with T_2 symmetry. Since the impurity potential has only rotational and no translational symmetry, it is customary to use the point group notations for the symmetrized defect wave functions. The appropriate linear combinations are

$$A_1\colon (X + \overline{X} + Y + \overline{Y} + Z + \overline{Z})/\sqrt{6}; \tag{4.36a}$$

$$E\colon\ (X + \overline{X} - Y - \overline{Y})/2, \quad (2Z + 2\overline{Z} - X - \overline{X} - Y - \overline{Y})/\sqrt{12}; \tag{4.36b}$$

$$T_2\colon (X - \overline{X})/\sqrt{2}, \quad (Y - \overline{Y})/\sqrt{2}, \quad (Z - \overline{Z})/\sqrt{2}. \tag{4.36c}$$

In Fig. 4.3 we notice that the totally symmetric A_1 state is usually found experimentally to have the lowest energy. A plausible explanation is that the impurity potential is attractive and the A_1 state, like the s states in the hydrogen atom, has the highest probability of being near the origin, where attractive central cell corrections are generated. In addition, different donor species exhibit rather large **chemical shifts** in their ground state energies. This is because the impurity potential is usually not purely Coulombic near the core. There are corrections due to exchange and correlation effects between the donor electron and the core electrons. In addition, the screening of the Coulomb potential (via the dielectric constant ε_0) is reduced near the core. It is difficult to calculate these corrections from first principles. Once a realistic potential has been determined, the Hamiltonian for the envelope wave function can be diagonalized by numerical methods. Table 4.2 compares the experimental valley–orbit split energy levels for donors in Si with theoretical values computed by *Pantelides* and *Sah* [4.11]. The agreement between theory and experiment is very good for P and As but not so satisfactory for Sb.

Table 4.2. Experimental substitutional donor binding energies [meV] in Si compared with values computed numerically by solving the effective mass equation including valley–orbit interaction (from [4.11])

Level	P		As		Sb	
	Theory	Exp.	Theory	Exp.	Theory	Exp.
A_1	−44.3	−45.5	−53.1	−53.7	−31.7	−42.7
E	−30.5	−32.6	−29.6	−31.2	−28.5	−30.5
T_2	−31.3	−33.9	−29.8	−32.6	−27.8	−32.9

4.2.4 Acceptor Levels in Diamond- and Zinc-Blende-Type Semiconductors

The calculation of acceptor binding energies in tetrahedrally coordinated semiconductors presents a special challenge due to two factors. First, the valence bands are degenerate at the Brillouin zone center, and second, as a result of the degeneracy, the valence bands are warped (Sect. 2.6.2). Hence, it is not possible to define a simple effective mass tensor for the valence bands.

Several different approaches to solving the acceptor problem have appeared in the literature. Invariably they all involve numerical solutions of a wave equation for the envelope functions. Here we present a derivation of these wave equations without solving them. The purpose is to obtain some insight into the problem without going into the details of the numerical solutions. We will also try to simplify the calculation as much as possible. We start by assuming that the spin–orbit coupling is much larger than the acceptor binding energies so that only the heavy- and light-hole bands need to be considered. The wave equation for the four resultant envelope functions $\Phi_i(\boldsymbol{r})$ $(i = 1,\dots,4)$ can be cast in the form [4.2]

$$-\sum_j \sum_{\alpha\beta} D_{ij}^{\alpha\beta}\left(\frac{\partial}{\partial x_\alpha}\right)\left(\frac{\partial}{\partial x_\beta}\right)\Phi_j - \frac{e^2}{4\pi\varepsilon_0\varepsilon_0 r}\Phi_i = E\Phi_i. \tag{4.37}$$

$D_{ij}^{\alpha\beta}(\partial/\partial x_\alpha)(\partial/\partial x_\beta)$ is the 4×4 matrix operator obtained from the 4×4 matrix $\{H'_{ij}\}$ in Sect. 2.6.2 by using (4.11) to convert $\boldsymbol{k}$ into $(\partial/\partial\boldsymbol{r})$. In the absence of the Coulomb potential term, the solutions of (4.37) are four-fold degenerate (symmetry: Γ_8 in zinc-blende structure and Γ_8^+ in diamond structure). One could also have included in (4.37) the doubly degenerate, spin-orbit split bands (symmetries: Γ_7 in zinc-blende structure and Γ_7^+ in diamond structure) but except in the case of Si [4.12], their effect on acceptor levels is very small. The spin degeneracy is not lifted by the Coulomb potential so the solutions of (4.37) should remain at least doubly degenerate.

The approach adopted by *Kohn* and *Schechter* [4.13] was to expand $\Phi_i(\boldsymbol{r})$ in terms of radial wave functions and spherical harmonics as in the case of the hydrogen atom. For simplicity the expansion was truncated at an angular momentum (l) less than some value l_0, and then a variational technique was used to solve (4.37). With this approach *Kohn* and *Schechter* [4.13] obtained a value of 8.9 meV for the ground state acceptor binding energy in Ge as compared to an experimental value of between 10.2 to 11.2 meV. The disadvantage of this technique lies in the difficulty of improving its accuracy.

More recently, *Baldereschi* and *Lipari* [4.14, 15] have developed a different and more systematic approach to solving this problem. Their starting point is the Luttinger Hamiltonian in (2.70). This Hamiltonian has been constructed to reflect the cubic symmetry of the crystal. Noting that deviations from spherical symmetry in the warped heavy and light hole bands are small in most semiconductors, Baldereschi and Lipari rewrote the Luttinger Hamiltonian using **spherical tensors** instead of Cartesian tensors. The idea is that all the symmetry operations of a spherically symmetric potential form a group known as the **full rotational group**. The spherical harmonic functions form a complete orthonormal set of basis functions for the irreducible representations of this group. Symmetrizing the Luttinger Hamiltonian with spherical tensors is, therefore, the systematic way to decompose the Luttinger Hamiltonian into terms with spherical and cubic symmetries. The process involved is left as an exercise in Problems 4.2 and 4.3. The resultant "spherically symmetrized" Luttinger Hamiltonian obtained by Baldereschi and Lipari (H_{BL}) is given by

$$H_{\mathrm{BL}} = \frac{\gamma_1 \boldsymbol{p}^2}{2m_0} - \frac{3\gamma_3 + 2\gamma_2}{45m_0}\left(\mathcal{P}^{(2)} \cdot \boldsymbol{J}^{(2)}\right) + \frac{\gamma_3 - \gamma_2}{18m_0}\left\{\left[\mathcal{P}^{(2)} \times \boldsymbol{J}^{(2)}\right]^{(4)}_{-4} + \frac{\sqrt{70}}{5}\left[\mathcal{P}^{(2)} \times \boldsymbol{J}^{(2)}\right]^{(4)}_{0} + \left[\mathcal{P}^{(2)} \times \boldsymbol{J}^{(2)}\right]^{(4)}_{4}\right\}, \quad (4.38)$$

where $\mathcal{P}^{(2)}$ and $\boldsymbol{J}^{(2)}$ are second-order spherical tensors. Their definitions plus those of the tensor products $\mathcal{P}^{(2)} \cdot \boldsymbol{J}^{(2)}$ and $[\mathcal{P}^{(2)} \times \boldsymbol{J}^{(2)}]^{(4)}_i$ can be found in Problems 4.2 and 4.3.

Note that both the first and second terms in (4.38) are spherically symmetric. Only the third term proportional to $(\gamma_3 - \gamma_2)$ has lower, cubic symmetry and gives rise to warping of the valence bands. In most diamond- and zinc-blende-type semiconductors the spherical terms are much larger than the cubic one. This can be seen in Table 4.3, where the values of γ_1, γ_2, and γ_3 for a number of tetrahedrally bonded semiconductors are listed. As a result, the

Table 4.3. Values of the dimensionless valence band parameters γ_1, γ_2, γ_3, μ and δ for various semiconductors with the diamond and zinc-blende structures. Although most of these values have been reproduced from [4.14] many of the values in this reference have been changed to the most recent values[a,b]. In addition, the values of these parameters now agree with those computed from the parameters A, B and C in Table 2.4 using (2.71a) to (2.71c)

	γ_1	γ_2	γ_3	μ	δ
C	2.5	−0.1	0.63	0.27	0.29
Si	4.28	0.339	1.446	0.47	0.26
Ge	13.38	4.24	5.69	0.766	0.11
SiC[c]	2.8	0.51	0.67	0.433	0.488
GaN[d]	5.05	0.59	1.78	0.52	0.24
GaP	4.05	0.49	1.25	0.47	0.19
GaAs	6.9	2.2	2.9	0.75	0.1
GaSb	13.3	4.4	5.7	0.8	0.1
InP	5.15	0.95	1.62	0.523	0.13
InAs	20.4	8.3	9.1	0.861	0.039
InSb	36.41	16.24	17.34	0.928	0.03
ZnS	2.54	0.75	1.09	0.751	0.134
ZnSe	2.75	0.5			
ZnTe	3.8	0.72	1.3	0.562	0.153
CdTe	4.14	1.09	1.62	0.68	0.128

[a] Refer to Landolt-Börnstein Tables Vol. 22a. (Springer, Berlin Heidelberg 1987)
[b] H. Mayer and U. Rössler: Solid State Commun. **87**, 81 (1993)
[c] and [d] Values for the zinc-blende structure.

cubic term can be neglected in the first-order approximation and the heavy and light hole bands treated as spherical.

To calculate the acceptor binding energy, it is more convenient to simplify first the acceptor Hamiltonian:

$$H = H_{\mathrm{BL}} - \frac{|e|^2}{4\pi\varepsilon_0\varepsilon_0 r} \tag{4.39}$$

by

- defining two new parameters

$$\mu = 2(3\gamma_3 + 2\gamma_2)/5\gamma_1 \tag{4.40}$$

and

$$\delta = (\gamma_3 - \gamma_2)/\gamma_1, \tag{4.41}$$

where μ is a measure of the magnitude of the second spherical term in (4.38) while δ is proportional to the coefficient of the cubic term, and

- introducing **effective atomic units** in which both the *effective Bohr radius*:

$$a^* = \frac{\varepsilon_0\hbar^2\gamma_1}{m_0e^2} \cdot 4\pi\varepsilon_0 \tag{4.42}$$

and the *effective Rydberg*:

$$R = \frac{e^4 m_0}{2\hbar^2\varepsilon_0^2\gamma_1} \cdot \frac{1}{(4\pi\varepsilon_0)^2} \tag{4.43}$$

are set equal to unity [this implies $e^2 = 2\varepsilon_0$, $\hbar^2 = 2m_0/\gamma_1$].

With these simplifications, (4.39) becomes

$$H = \frac{\boldsymbol{p}^2}{\hbar^2} - \frac{2}{r} - \frac{\mu}{9\hbar^2}(\mathcal{P}^{(2)} \cdot \boldsymbol{J}^{(2)}) + \frac{\delta}{9\hbar^2}\left(\left[\mathcal{P}^{(2)} \times \boldsymbol{J}^{(2)}\right]^{(4)}_{-4} + \sqrt{\frac{70}{5}}\left[\mathcal{P}^{(2)} \times \boldsymbol{J}^{(2)}\right]^{(4)}_{0} + \left[\mathcal{P}^{(2)} \times \boldsymbol{J}^{(2)}\right]^{(4)}_{4}\right). \tag{4.44}$$

Except for the last term, which is proportional to δ, this Hamiltonian has full rotational symmetry. Hence it is important to compare the values of μ and δ for various semiconductors, shown in Table 4.3. We see that, *with the exception of Si, SiC and GaN*, the spherical parameter μ is at least four times larger than the cubic parameter δ.

We shall not attempt to diagonalize (4.44) since it is quite complicated, as one may expect. Instead we shall discuss the qualitative features of the solutions. We shall begin with the simplest case by neglecting most of the terms in (4.44). Then we shall introduce one additional term at a time and examine the consequence of each term using perturbation theory.

a) "One Spherical Band Approximation"

The simplest approximation of (4.44) one can make is to set the terms depending on μ and δ both equal to zero:

$$H = \frac{p^2}{\hbar^2} - \frac{2}{r}. \tag{4.45}$$

Within this approximation the heavy and light hole masses are equal (hence, they are treated as one band). Equation (4.45) becomes analogous to the hydrogenic donor, therefore its solutions are characterized by a principal quantum number N and an angular momentum L. The bound states will be labeled nS, nP, etc. and their energies are given by the Rydberg series (4.23), except that the Rydberg constant (4.24) is now defined by (4.43).

b) "Spherical Approximation"

Next, the term $\mathcal{P}^{(2)} \cdot \boldsymbol{J}^{(2)}$ is added as a perturbation to (4.45):

$$H = \frac{p^2}{\hbar^2} - \frac{2}{r} - \frac{\mu}{9\hbar^2}\left(\mathcal{P}^{(2)} \cdot \boldsymbol{J}^{(2)}\right). \tag{4.46}$$

The Hamiltonian still has spherical symmetry but now the heavy and light hole bands have different masses (Problem 4.3). The term $\mathcal{P}^{(2)} \cdot \boldsymbol{J}^{(2)}$ resembles the spin–orbit interaction if we note that the *"pseudo-angular momentum"* $\boldsymbol{J}$ plays the role of spin except that $J = 3/2$. Using this similarity, a *"pseudo-total angular momentum"* $\boldsymbol{F} = \boldsymbol{L} + \boldsymbol{J}$ can be defined. $\boldsymbol{F}$ is conserved just as the total angular momentum $\boldsymbol{L} + \boldsymbol{S}$ would be conserved if spin–orbit coupling were included (except that now $\boldsymbol{J}$ is the pseudo-angular momentum of the Bloch function while $\boldsymbol{L}$ represents the angular momentum of the impurity envelope function). As in atomic physics, the bound states of the acceptor can be labeled with the **spectroscopic notation**

$L = 0 : nS_{3/2}$;

$L = 1 : nP_{5/2},\ nP_{3/2}$, and $nP_{1/2}$;

$L = 2$, and so on.

In general, only terms with $L < 2$ are significant. Notice that there is only *one* $1S$ acceptor state. If we had made the assumption that the heavy and light hole bands can be treated as two separate spherical bands we would, instead, incorrectly obtain *two* $1S$ acceptor states.

In spite of the analogy with the spin–orbit coupling, the $\mathcal{P}^{(2)} \cdot \boldsymbol{J}^{(2)}$ term is not trivial to treat by perturbation theory because of the degeneracy of the heavy and light holes. Since $\mathcal{P}^{(2)}$ and $\boldsymbol{J}^{(2)}$ are second-order spherical tensors, the selection rules for the matrix elements of $\mathcal{P}^{(2)} \cdot \boldsymbol{J}^{(2)}$ are $\Delta F = 0$ and $\Delta L = 0$ or ± 2. For simplicity we will consider only these lowest energy states: $1S_{3/2}$, $2S_{3/2}$, $2P_{5/2}$, $2P_{3/2}$, and $2P_{1/2}$. Except for the $2P_{1/2}$ state, these states are coupled to higher energy levels by the $\mathcal{P}^{(2)} \cdot \boldsymbol{J}^{(2)}$ term. *Baldereschi* and *Lipari* [4.14] assumed that each of these states is coupled with no more than one other state. In this approximation, the envelope wave functions consist of linear combinations of at most two functions:

$$\begin{aligned} \Phi(S_{3/2}) &= f_0(r)\,|\,L = 0,\ J = 3/2,\ F = 3/2,\ F_z\rangle \\ &\quad + g_0(r)\,|\,L = 2,\ J = 3/2,\ F = 3/2,\ F_z\rangle; \end{aligned} \tag{4.47a}$$

$$\Phi(P_{1/2}) = f_1(r)\,|\,L = 1,\ J = 3/2,\ F = 1/2,\ F_z\rangle; \tag{4.47b}$$

$$\begin{aligned}\Phi(P_{3/2}) &= f_2(r)\,|\,L = 1,\ J = 3/2,\ F = 3/2,\ F_z\rangle\\ &\quad + g_2(r)\,|\,L = 3,\ J = 3/2,\ F = 3/2,\ F_z\rangle;\end{aligned} \tag{4.47c}$$

$$\begin{aligned}\Phi(P_{5/2}) &= f_3(r)\,|\,L = 1,\ J = 3/2,\ F = 5/2,\ F_z\rangle\\ &\quad + g_3(r)\,|\,L = 3,\ J = 3/2,\ F = 5/2,\ F_z\rangle.\end{aligned} \tag{4.47d}$$

Except for $\Phi(P_{1/2})$, substituting these wave functions into (4.46) produces two coupled differential equations for the radial functions f_i and g_i. These differential equations can be solved only approximately by numerical methods. However, the energy of the $2P_{1/2}$ state can be calculated exactly because its radial wave equation is similar to that of the p state in the hydrogen atom. Its eigenvalue can be shown to be equal to $[4(1+\mu)]^{-1}$. Noting that the effective mass of the light hole is given by $(1+\mu)^{-1}$ (Problem 4.4), we see that the energy of the $2P_{1/2}$ state is equal to that of the $n = 2$ level of a hydrogenic acceptor with the light hole mass.

The energies of the other three states have been calculated by *Baldereschi* and *Lipari* [4.14] using variational techniques and the values of the Kohn–Luttinger parameters listed in Table 4.3. Their results are shown in Table 4.4. Note that the values of some of the Kohn–Luttinger parameters have since been revised. The theoretical acceptor energy levels are compared with the experimental values available at the time of the calculation.

The agreement between the experimental and theoretical ground state energies of acceptors is quite good for most diamond- and zinc-blende-type semiconductors. The notable exceptions are Si, GaP, InP, and some II–VI semiconductors. As pointed out earlier, the spherical model is not expected to work well for Si because it has a relatively large cubic term ($\delta/\mu \approx 0.5$, see Table 4.3). The experimental acceptor value quoted by *Baldereschi* and *Lipari* [4.14]

Table 4.4. Comparison between the values of the lowest-energy bound states of acceptors for various semiconductors with the diamond or zinc-blende structures calculated by *Baldereschi* and *Lipari* [4.14, 16, 17] using (4.47) with available experimental values (values in italics represent measurements after 1973). All energies are in meV

	Experiment	$1S_{3/2}$	$2S_{3/2}$	$2P_{1/2}$	$2P_{3/2}$	$2P_{5/2}$
Si	*45*, 68.9	31.6	8.6	4.2	11.2	7.6
Ge	10.8	9.8	2.9	0.6	4.2	2.5
GaP	*57–64*	47.5	13.7	4.2	19.1	11.7
GaAs	*31*	25.6	7.6	1.6	11.1	6.5
GaSb	13–15	12.5	3.8	0.65	5.6	3.2
InP	*31*, 56.3	35.2	10.5	2	15.5	8.9
InAs	10–*20*	16.6	5.1	0.4	7.9	4.4
InSb	≈10	8.6	2.7	0.2	4.2	2.3
ZnS		175.6	52	11.7	75.1	44.1
ZnSe	114	110.1	33	6.1	48.6	28
ZnTe	*≈30*	77.7	23	5.1	33.4	19.6
CdTe	*≈30*	87.4	26.5	3.7	39.9	22.6

was for Al_{Si}. The more recent experimental value for B_{Si}, 45 meV, is in better agreement with the theoretical predictions. In the case of GaP and InP the experimental values quoted were for Zn_{Ga} and Cd_{In}, respectively. More recent values for the binding energies of Mg_{Ga} and Be_{Ga} in GaP are 56.6 and 59.9 meV, respectively. The corresponding energies for Mg_{In} and Be_{In} in InP are both 31 meV. These latest experimental values are in much better agreement with theory. The acceptors in ZnTe and CdTe, for which binding energies have become available recently, correspond to Li substituting for the cations. The understanding of impurities in these II–VI semiconductors is still relatively poor, in part due to inaccuracies in our knowledge of the Luttinger parameters. Lately, there has been much interest in shallow impurities in the larger bandgap II–VI semiconductors because of their use in blue and green lasers [4.18, 19]. This should lead to more precise determinations of acceptor binding energies.

c) Including the Cubic Term

When the cubic term is included, the wave functions have to be classified according to the irreducible representations of T_d. For example, an acceptor wave function formed from a $J = 3/2$ valence band with Γ_8 symmetry (double group notation) and an envelope function with S symmetry in the spherical model (or Γ_1 symmetry in T_d) has Γ_8 symmetry (the direct product of Γ_8 and Γ_1). Similarly, if the envelope function has P symmetry (or Γ_4), the symmetries of the acceptor wave functions will belong to the representations $\Gamma_8 \otimes \Gamma_4 = \Gamma_6 \oplus \Gamma_7 \oplus 2\Gamma_8$. It is easily shown that the doubly degenerate $P_{1/2}$ state will become the Γ_6^- state while the four-fold degenerate $P_{3/2}$ state becomes Γ_8^- in the case of Si and Ge. The six-fold degenerate $P_{5/2}$ state is reducible into a doubly degenerate Γ_7^- level plus a four-fold degenerate Γ_8^- level. From these symmetry considerations one concludes that the cubic term can

- only shift the energies of the $S_{3/2}$, $P_{1/2}$, and $P_{3/2}$ levels and
- split the degeneracy of the $P_{5/2}$ level.

Thus the $S_{3/2}$, $P_{1/2}$, and $P_{3/2}$ levels are not affected by the cubic term to first order in δ. When higher order terms in δ are included, the selection rules for the cubic term are $\Delta F_z = 0$ and ± 4 (the result of multiplying two second-order spherical tensors in the cubic term is a tensor of fourth order; see Problem 4.3). These selection rules greatly restrict the levels which can be coupled via the cubic term. As a result, the $1S_{3/2}$, $2S_{3/2}$, and $2P_{1/2}$ levels are basically unchanged by the cubic term. In summary, the "spherical model" in which the cubic term is neglected should be as good an approximation as models including the cubic term for calculating the lower energy levels of acceptors in tetrahedrally coordinated semiconductors.

The cubic term shifts the $P_{3/2}$ levels mainly by coupling them with the nearby $P_{5/2}$ levels. The corresponding shift in energy is typically less than 10 %. For example, in Si, where the largest shift is expected to occur, the cubic term changes the $2P_{3/2}$ level energy from 11.2 meV to 12.13 meV. The values of the $2P_{5/2}(\Gamma_7^-)$ and $2P_{5/2}(\Gamma_8^-)$ levels calculated numerically by *Baldereschi* and *Lipari* [4.15] by including the cubic term are listed in Table 4.5.

The measured excited-state energies of some common acceptors in Si and Ge are compared with the values calculated by Baldereschi and Lipari in Fig. 4.4. The agreement between theory and experiment is excellent for Ge but only fair for Si. The agreement is especially poor for the $1\Gamma_8^+$ ground state in Si. The calculated binding energy in Si is 44.4 meV while the experimental values vary from 45 meV for boron to 160 meV for indium. These discrepancies can be attributed, at least in part, to central cell corrections which become more important in Si. However, the calculation based on a 4×4 Luttinger Hamiltonian should break down when the calculated binding energy is larger than Δ_0. In this case, the full 6×6 Hamiltonian, including the Γ_7 bands split-off by Δ_0, must be solved [4.20]. It turns out that the spin–orbit splitting of the ground state of the acceptor levels can be much smaller than Δ [4.20, 21].

Table 4.5. Comparison between the theoretical values of the $2P_{5/2}$ bound states of acceptors calculated with and without the cubic term for various semiconductors with diamond and zinc-blende structures by *Baldereschi* and *Lipari* [4.14, 15]. All energies are in meV

	"Spherical model"	"Including cubic term"	
	$2P_{5/2}$	$2P_{5/2}(\Gamma_8^-, \Gamma_8)$	$2P_{5/2}(\Gamma_7^-, \Gamma_7)$
Si	7.6	8.51	5.86
Ge	2.5	2.71	2.04
GaP	11.7	13.04	9.42
GaAs	6.5	7.2	5.33
GaSb	3.2	3.59	2.61
InP	8.9	9.98	7.32
InAs	4.4	4.76	3.63
InSb	2.3	2.54	1.91
ZnS	44.1	49.56	35.57
ZnSe	28	31.47	22.68
ZnTe	19.6	22.32	15.36
CdTe	22.6	25.85	17.68

4.3 Deep Centers

We have shown in the last section that one characteristic of shallow impurity levels is that their electron wave functions typically extend over many primitive unit cells. As a result, those wave functions can be constructed from one Bloch function indexed by a single wave vector equal to that of the nearest band extremum, see (4.32). Deep centers, on the other hand, have localized wave functions which involve Bloch functions from several bands and over a large region of k-space. Thus, defects with highly localized potentials are expected to form deep centers. Such localized potentials can be caused by broken bonds, strain associated

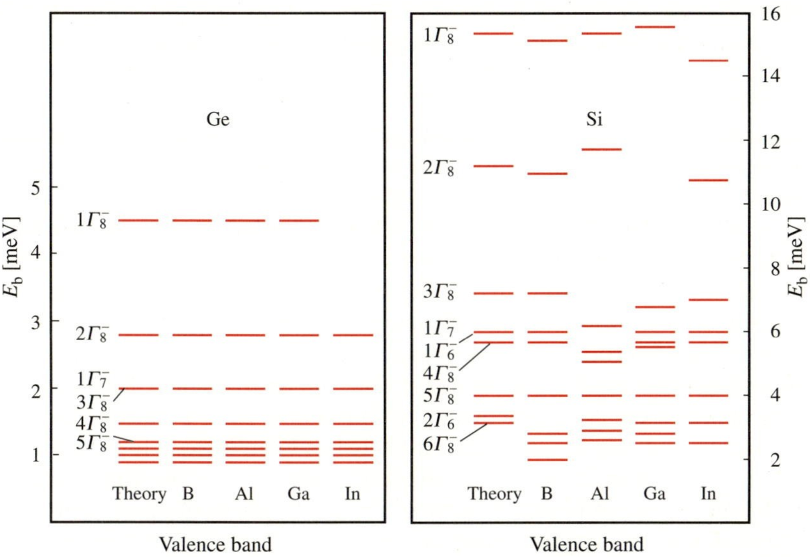

Fig. 4.4. Theoretical values of shallow acceptor energies in Si and Ge calculated with the *Baldereschi* and *Lipari Hamiltonian* [4.14] compared with experimental values. (From [4.22])

with displacement of atoms, and difference in electronegativity or core potentials between the impurity and host atoms. The localized nature of deep center potentials suggests that a tight-binding molecular orbital approach should be a better starting point for studying their electronic energies. Since a defect is imbedded in a semiconductor, it is necessary to consider also the interaction between the localized defect electrons and the Bloch electrons of the host. For example, when a deep center energy level (or *deep level* for short) overlaps with band states of the host, it becomes a **resonant state**.

To calculate the energies of deep levels one needs to know the defect potential and then find a way to solve the corresponding Schrödinger equation. It is very difficult to deduce the defect potential for deep centers because displacements of atoms (or **lattice relaxation**) can occur. Both the impurity atom and atoms surrounding it can be involved in the relaxation. The reason why lattice displacements are important in deep centers can be qualitatively seen from the following examples.

Suppose an impurity atom in a semiconductor has a choice of becoming either a shallow donor on a substitutional site or a deep level via a lattice distortion. Let us assume that, as a deep level, it is located deep in the bandgap at an energy E_0 (known as its **thermal ionization energy**) below the conduction band. On the other hand, as a shallow donor its energy will be near the conduction band edge. Thus the impurity atom can lower its electronic energy (by about E_0) by becoming a deep center. However, it may require a **lattice relaxation energy** E_d to produce the lattice distortion. If E_0 is larger than E_d it is energetically favorable for the impurity atom to distort spontaneously and become a deep center. It is necessary to know both E_0 and E_d in order to predict whether this impurity will be deep or shallow. Lattice displacements are responsible for the shallow-to-deep instability which converts shallow donors in GaAs to the DX center in AlGaAs alloy [4.7, 8]. In the case of the DX centers, E_0 is on the order of 0.4 eV for Si in an AlGaAs [4.23]. Actually, the electronic energy gained by DX centers via their lattice displacements is equal to $2E_0$ because of their peculiar property known as **negative** U. For further details on the DX center and on their negative properties, see the following Web Page: http://Pauline.berkeley.edu/textbook/DX-Center.pdf. Here U stands for the on-site repulsive Coulomb interaction between two electrons in the so-called *Hubbard model* [4.24]. For a deep center with a negative U, a second electron will actually be *attracted* to it even though it is occupied by one electron already as a result of lattice distortions. The resultant *negatively charged* state will be more stable than the neutral state. This negative U property makes it energetically even more favorable for a defect to undergo lattice relaxation and become a deep center. It has been known for some time that some large bandgap II–VI semiconductors, such as ZnS and ZnSe, cannot be doped p-type easily while others, such as ZnTe, cannot be doped n-type. This phenomenon is known as **self-compensation** [Ref. 4.4, p. 238]. While it is still not completely understood, recently it has been suggested that DX center formation by substitutional donors can explain why ZnTe cannot be doped n-type by common group III dopants like Al and Ga [4.25].

As another example of lattice relaxation in deep centers we consider a vacancy in a group-IV semiconductor. The removal of an atom results in a loss of four electrons in this case. This is equivalent to adding four positive charges to the otherwise neutral semiconductor. Obviously the electrons inside the semiconductor will respond by screening these positive charges. The response of the semiconductor is best understood in terms of a tight-binding model. A vacancy in a covalent semiconductor produces four unpaired **dangling bonds** as shown schematically in Fig. 4.5a. The four dangling bonds can

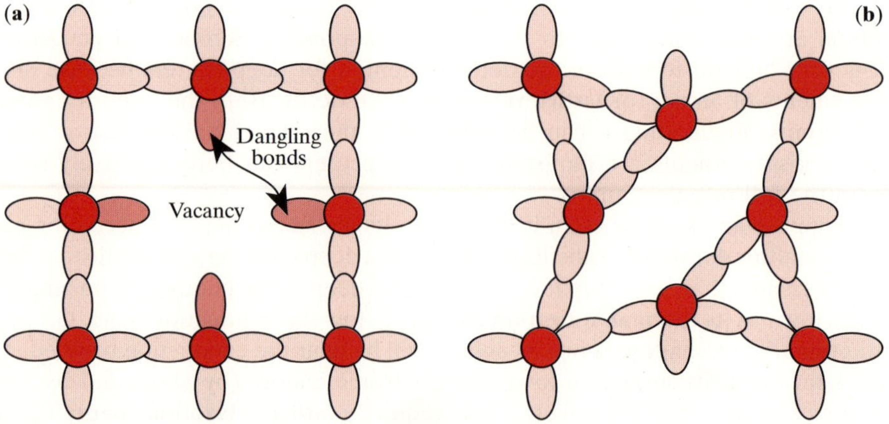

Fig. 4.5. Schematic diagram of a covalent semiconductor with (**a**) an unrelaxed vacancy involving four dangling bonds and (**b**) a relaxed vacancy with no dangling bonds

be "healed" by forming two pairs of bonds among neighboring dangling bonds. Since originally the distances between atoms with dangling bonds are larger than the bond lengths in the perfect crystal, atoms with dangling bonds have to move closer to each other in order to form new bonds, as shown in Fig. 4.5b. This displacement of neighboring atoms involves an elastic energy, which is compensated for by the lowering of the energy of the four electrons originally in the dangling bonds. Since each new bond can accommodate two electrons in the bonding state, the decrease in the energy of each electron in the dangling bond is of the order of the overlap parameters (V_{ss}, V_{pp}, and V_{sp}) discussed in Sect. 2.7.1. We will consider in more detail the electronic energy of a vacancy in diamond- and zinc-blende-type semiconductors in Sect. 4.3.2.

4.3.1 Green's Function Method for Calculating Defect Energy Levels

In this book we will avoid the difficult problem of determining deep center potentials. Instead we will study one method to solve for the deep center energies *if the potential is known*. There are now first-principles techniques for calculating deep center energies including lattice relaxation [4.7, 8], but these techniques are beyond the scope of this book. The method we will consider here is the **Green's function** approach to calculating the electronic structure of deep centers. Many books have been written on Green's functions. Discussions of their applications to the defect problem can be found in those by *Economou* [4.26] and *Lannoo* and *Bourgoin* [4.27]. We will apply this technique to study the systematics of defects, especially isoelectronic centers, in diamond- and zinc-blende-type semiconductors.

To understand the motivation for using Green's functions to study defects we will first introduce the concept of **density of states** (to be abbreviated as DOS). The DOS of a system is defined as the number of allowed states per unit energy lying in the energy range between E and $E + \delta E$. Suppose a particular system under study is described by a Hamiltonian H and its energy levels are denoted by E_k. These energy levels can be discrete or continuous. The particles in the system may be arranged periodically in space like atoms in a crystal, or be completely random as in an amorphous solid. The DOS of this system, $n(E)$, is equal to

$$n(E) = \sum_k \delta(E - E_k) \tag{4.48}$$

where $\delta(x)$ represents the **Dirac delta function**, [4.28] and k labels all possible states. $n(E)$ is an important quantity because it can be calculated theoretically and can also be measured experimentally. Next we will define the **resolvent** or **Green's function operator** G as

$$G = \lim_{\eta \to 0+} (E - H + \mathrm{i}\eta)^{-1}. \tag{4.49}$$

To understand the reason for defining G this way, we note that the expectation value of G for an eigenstate $|k\rangle$ of H is given by

$$\langle k | G | k \rangle = \lim_{\eta \to 0+} (E - E_k + i\eta)^{-1}. \tag{4.50}$$

Using a well-known property of the delta function, the right hand side of (4.50) can be rewritten as

$$\lim_{\eta \to 0+} (E - E_k + i\eta)^{-1} = \mathcal{P}(E - E_k)^{-1} - i\pi\delta(E - E_k), \tag{4.51}$$

where $\mathcal{P}[E - E_k]^{-1}$, known as the **Cauchy principal value** of $[E - E_k]^{-1}$, is defined by

$$\int_{-\infty}^{+\infty} f(E')\mathcal{P}[E - E']^{-1} dE' = \mathcal{P} \int_{-\infty}^{+\infty} \frac{f(E')}{E - E'} dE'$$
$$= \lim_{\varepsilon \to 0} \left(\int_{-\infty}^{E-\varepsilon} \frac{f(E')}{E - E'} dE' + \int_{E+\varepsilon}^{+\infty} \frac{f(E')}{E - E'} dE' \right). \tag{4.52a}$$

In some sense we can regard the function $\mathcal{P}[x]^{-1}$ as having the property

$$\mathcal{P}[x]^{-1} = \begin{cases} 0 & \text{when } x = 0, \\ x^{-1} & \text{when } x \neq 0. \end{cases} \tag{4.52b}$$

From (4.48) and (4.51) we obtain

$$n(E) = -\frac{1}{\pi} \sum_k \mathrm{Im}\{\langle k | G | k \rangle\} = -\frac{1}{\pi} \mathrm{Im}\{\mathrm{Tr} G\}, \tag{4.53}$$

where $\mathrm{Im}\{A\}$ stands for the imaginary component of a complex quantity A, and $\mathrm{Tr}\{G\}$ respresents the trace of the matrix $\langle k | G | k' \rangle$. Thus (4.53) establishes the relation between the operator G and $n(E)$. We should note that the matrix element of G defined in (4.50) is a complex function of E. Its real part is always related to its imaginary part in a particular way. For example, let us denote $\langle k | G | k \rangle$ by $g(E - E_k)$ [see (4.50)] and define

$$F(E) = \int_{-\infty}^{+\infty} g(E - E') f(E') dE', \tag{4.54}$$

where $f(x)$ is any given function. Substituting (4.50) into (4.54) we obtain

$$F(E) = \mathcal{P} \left(\int_{-\infty}^{+\infty} \frac{f(E')}{E - E'} dE' \right) + i\pi f(E). \tag{4.55}$$

Noting that $\mathrm{Im}\{F\} = \pi f(E)$ in (4.55), the real part of $F(E)$ can be rewritten as

$$\mathrm{Re}\{F(E)\} = \frac{1}{\pi} \mathcal{P} \left(\int_{-\infty}^{+\infty} \frac{\mathrm{Im}\{F(E')\}}{E - E'} dE' \right). \tag{4.56}$$

This equation expressing the real part of the function $F(E)$ in terms of its imaginary part is one of two equations known as the **Kramers–Kronig relations** or the **dispersion relations**. It is also called a **Hilbert transform**. In Chap. 6 we will utilize these important relations extensively in studying the linear optical response function (which will be defined in Chap. 6). Obviously the function $g(E - E_k)$ satisfies the Kramers–Kronig relations.

Another expression for $n(E)$ which is useful for defect problems is

$$n(E) = \frac{1}{\pi}\frac{d}{dE}\mathrm{Im}\Big\{\log\big(\det G\big)\Big\} \tag{4.57}$$

where $\det G$ represents the determinant of the matrix $\langle k \,|\, G \,|\, k' \rangle$.

Let us express the Schrödinger equation for the defect electron as:

$$(H_0 + V)\Phi = E\Phi, \tag{4.58}$$

where H_0 is the Hamiltonian of the perfect crystal, V is the defect potential, and E is the defect energy. As mention before, we will assume that V is known. Furthermore, we will assume that the wave functions Φ_0 and eigenvalues E_0 of the "unperturbed" equation

$$(E_0 - H_0)\Phi_0 = 0. \tag{4.59}$$

are known. Equation (4.58) can be rewritten as

$$\Phi = [E - H_0]^{-1}V\Phi. \tag{4.60}$$

We will now define the "unperturbed" Green's Function operator G_0 using (4.49):

$$G_0 = \lim_{\eta\to 0+}[E - H_0 + \mathrm{i}\eta]^{-1}, \tag{4.61}$$

so that Φ can be expressed as

$$\Phi = G_0 V\Phi \tag{4.62}$$

One should remember that G_0 is a function of the defect energy E.

There are two possible types of solutions for (4.58).

a) Bound State Solutions

If the defect electron energy E does not overlap with the eigenvalues E_0 of the unperturbed wave equation, that the eigenstate Φ of the defect electron has to be a bound state. Since E does not overlap with E_0 no extended state can exist at this energy. The defect electron wave function has to be localized around the defect. To determine E we can expand in terms of an orthonormal set of basis functions ψ_j:

$$\Phi = \sum_j a_j \psi_j. \tag{4.63}$$

Substituting this expression into (4.62) we obtain a system of homogeneous linear equations in the unknown a_j's:

$$\sum_j (I - G_0 V)_{ij} a_j = 0, \qquad i = 1, 2, \ldots, \tag{4.64}$$

where I is the unit matrix; (4.64) has nonvanishing solutions only if its determinant is zero. The defect energies E can thus be obtained, in principle, by solving the equation

$$\det[I - G_0 V] = |I - (E - H_0)^{-1} V| = 0. \tag{4.65}$$

For a general potential V (4.65) is very complicated and difficult to solve. For point defects the potential V is localized and (4.65) can be simplified by choosing a suitable set of localized basis functions ψ_j such that there will be only a small number of nonzero matrix elements V_{ij}. Applications of this approach will be studied in Sect. 4.3.2.

b) Resonant State Solutions

In the second case, the defect energy overlaps with the eigenvalues of the unperturbed states and a resonant state is formed. For $E = E_0$ the general solution Φ can be expressed as:

$$\Phi = \Phi_0 + G_0 V \Phi \tag{4.66}$$

The defect will modify the DOS $n_0(E)$ of the unperturbed state. Our goal will be to calculate $\delta n(E)$, the modification in $n(E)$ caused by the defect. For a highly localized defect potential it will be useful to define a **local density of states** since we expect $\delta(E)$ to be significant only in the vicinity of the defect. We will first introduce a matrix element of the operator G_0 defined by

$$G_0(\boldsymbol{r}, \boldsymbol{r}', E) = \langle \boldsymbol{r} | G_0 | \boldsymbol{r}' \rangle = \lim_{\eta \to 0+} \sum_{\boldsymbol{k}} \langle \boldsymbol{r} | \boldsymbol{k} \rangle \langle \boldsymbol{k} | \boldsymbol{r}' \rangle (E - E_{\boldsymbol{k}} + \mathrm{i}\eta)^{-1}. \tag{4.67}$$

Using (4.51) we can show that

$$\mathrm{Im}\{G_0(\boldsymbol{r}, \boldsymbol{r}, E)\} = -\pi \sum_{\boldsymbol{k}} |\langle \boldsymbol{r} | \boldsymbol{k} \rangle|^2 \delta(E - E_{\boldsymbol{k}}). \tag{4.68}$$

Now $|\langle \boldsymbol{r} | \boldsymbol{k} \rangle|^2$ equals the probability of finding an electron with wavevector $\boldsymbol{k}$ at the point $\boldsymbol{r}$ while $\sum \delta(E - E_{\boldsymbol{k}})$ is the DOS in the absence of defects [i.e., $n_0(E)$]. When these two terms are multiplied together, the resultant term

$$n_0(\boldsymbol{r}, E) = \sum_{\boldsymbol{k}} |\langle \boldsymbol{r} | \boldsymbol{k} \rangle|^2 \delta(E - E_{\boldsymbol{k}}) \tag{4.69}$$

has the meaning of an unperturbed local DOS. With the use of (4.68) $n_0(\boldsymbol{r}, E)$ can be related to G_0 by

$$n_0(\boldsymbol{r}, E) = -(1/\pi)\mathrm{Im}\{G_0(\boldsymbol{r}, \boldsymbol{r}, E)\}. \tag{4.70}$$

The total density of unperturbed states $n_0(E)$ is given by

$$n_0(E) = \sum_{\boldsymbol{r}} n_0(\boldsymbol{r}, E) = -\frac{1}{\pi}\mathrm{Im}\left\{\sum_{\boldsymbol{r}} G_0(\boldsymbol{r}, \boldsymbol{r}, E)\right\} = -\frac{1}{\pi}\mathrm{Im}\{\mathrm{Tr}\, G_0\}. \tag{4.71}$$

The local density of states $n(\boldsymbol{r}, E)$ in the presence of a defect can similarly be defined as

$$n(\boldsymbol{r}, E) = -\frac{1}{\pi}\mathrm{Im}\left\{\sum_{r} G(\boldsymbol{r}, \boldsymbol{r}, E)\right\}, \tag{4.72}$$

where $G(\boldsymbol{r}, \boldsymbol{r}', E)$ is given by

$$G(\boldsymbol{r}, \boldsymbol{r}', E) = \lim_{\eta \to 0+} \sum_{\boldsymbol{k}} \langle \boldsymbol{r} \,|\, \boldsymbol{k}\rangle \, \langle \boldsymbol{k} \,|\, \boldsymbol{r}'\rangle \, (E - H_0 - V + \mathrm{i}\eta)^{-1}. \tag{4.73}$$

For computational purposes it is more convenient to use (4.57) to define $n_0(\boldsymbol{r}, E)$ and $n(\boldsymbol{r}, E)$ so that the change in the local DOS induced by a defect can be expressed as

$$\begin{aligned} \delta n(\boldsymbol{r}, E) &= n(\boldsymbol{r}, E) - n_0(\boldsymbol{r}, E) \\ &= \frac{1}{\pi}\frac{d}{dE}\left(\mathrm{Im}\left\{\log\frac{\det G(\boldsymbol{r}, \boldsymbol{r}, E)}{\det G_0(\boldsymbol{r}, \boldsymbol{r}, E)}\right\}\right). \end{aligned} \tag{4.74}$$

Equation (4.74) can be simplified by expressing G in terms of G_0:

$$\begin{aligned} G &= \mathrm{Lim}(E - H_0 - V + \mathrm{i}\eta)^{-1} \\ &= \mathrm{Lim}(E - H_0 + \mathrm{i}\eta)^{-1}\left[1 + V(E - H_0 - V + \mathrm{i}\eta)^{-1}\right] \\ &= G_0 + G_0 V G. \end{aligned} \tag{4.75}$$

In the above equations it is understood that Lim stands for the limit $\eta \to 0+$. Equation (4.75) is known as the **Dyson equation**. Using this equation G can be expressed as $G = (1 - G_0 V)^{-1} G_0$ and det G as

$$\det G = \det(I - G_0 V)^{-1} \;\det G_0. \tag{4.76}$$

Substituting (4.76) for det$[G]$ into (4.74) we obtain an expression for $\delta n(\boldsymbol{r}, E)$ in terms of G_0 and V only:

$$\delta n(\boldsymbol{r}, E) = -\frac{1}{\pi}\frac{d}{dE}\Big(\mathrm{Im}\{\log(\det[I - G_0(\boldsymbol{r}, \boldsymbol{r}, E)V])\}\Big). \tag{4.77}$$

Thus, for both bound state and resonant state solutions, the change in the electron density of states induced by a defect potential V is determined by $\det[I - G_0 V]$.

4.3.2 An Application of the Green's Function Method: Linear Combination of Atomic Orbitals

We shall now apply the Green's function method to calculate the bound state energies of a deep center. Linear combinations of atomic orbitals will be used as the basis functions ψ_j. They are assumed to be given by the Löwdin orbitals $\phi_{ms}(\boldsymbol{r} - \boldsymbol{r}_{js})$ introduced in Sect. 2.7.2. In diamond- and zinc-blende-type semiconductors they can be further restricted to include only the four sp^3 orbitals. For illustration purposes we shall consider only one type of orbital (say an s-symmetry orbital) from each atom at the lattice vectors $\boldsymbol{r}_{js}$. The atoms are divided into two groups $\boldsymbol{A}$ and $\boldsymbol{B}$ such that the subspace $\boldsymbol{B}$ contains the location of all the atoms for which the defect potential V has zero matrix elements. In this simple case the defect wave function Φ in (4.63) can be expressed in terms of the two sets of basis functions $\{\psi_{\boldsymbol{A}}\}$ and $\{\psi_{\boldsymbol{B}}\}$ indexed by $\boldsymbol{A}$ and $\boldsymbol{B}$:

$$\Phi = \sum a_{\boldsymbol{A}}\psi_{\boldsymbol{A}} + a_{\boldsymbol{B}}\psi_{\boldsymbol{B}}. \tag{4.78}$$

The summation in (4.78) is over the subspaces $\boldsymbol{A}$ and $\boldsymbol{B}$ and the coefficients $\{a_{\boldsymbol{A}}\}$ and $\{a_{\boldsymbol{B}}\}$ are arrays. The matrices G_0 and V can be represented as

$$G_0 = \begin{pmatrix} G_{0AA} & G_{0AB} \\ G_{0BA} & G_{0BB} \end{pmatrix} \quad \text{and} \quad V = \begin{pmatrix} V_{AA} & 0 \\ 0 & 0 \end{pmatrix}, \tag{4.79}$$

where G_{0AA}, G_{0AB}, V_{AA}, etc. represent matrices of G_0 and V calculated for the two subspaces. Substituting (4.78) into (4.64) and (4.65) we obtain

$$\{a_{\boldsymbol{A}}\} = G_{0AA}V_{AA}\{a_{\boldsymbol{A}}\}, \tag{4.80a}$$

$$\{a_{\boldsymbol{B}}\} = G_{0BA}V_{AA}\{a_{\boldsymbol{A}}\}, \tag{4.80b}$$

and

$$\det\begin{pmatrix} I - G_{0AA}V_{AA} & 0 \\ -G_{0BA}V_{AA} & I \end{pmatrix} = \det(I - G_{0AA}V_{AA}) = 0. \tag{4.81}$$

In addition Φ has to satisfy the normalization condition

$$\langle\Phi|\Phi\rangle = 1. \tag{4.82}$$

For a very localized deep center, the subspace $\boldsymbol{A}$, representing the spatial extent of the defect, is very small. Hence the determinant in (4.81) and the matrix equations in (4.80) can be solved easily even with a personal computer. In particular, the defect energy E can be determined from (4.81) since the matrix element G_{0AA} is a function of E. We will show later how to evaluate G_{0AA} from the density of states of the perfect crystal.

Equation (4.82) ensures that the total probability of finding the electron in subspaces $\boldsymbol{A}$ plus $\boldsymbol{B}$ is unity. A more convenient form for (4.80a) can be obtained by writing $G_0^+G_0$ formally as

$$G_0^+G_0 = [E - H_0]^{-2} = -\frac{dG_0}{dE}, \tag{4.83}$$

where G_0^+ is the adjoint of G_0 defined as

$$G_0^+ = \lim_{\eta \to 0+} (E - H_0 - \mathrm{i}\eta)^{-1} \tag{4.84}$$

Substituting (4.83) and (4.59) into the normalization condition

$$\langle \Phi | \Phi \rangle = \langle \Phi | V^+ G_0^+ G_0 V | \Phi \rangle = 1 \tag{4.85}$$

an equation containing only a_A is obtained:

$$-a_{\boldsymbol{A}}^+ V_{\boldsymbol{AA}} \left(\frac{dG_{0\boldsymbol{AA}}}{dE} \right) V_{\boldsymbol{AA}} a_{\boldsymbol{A}} = 1. \tag{4.86}$$

Since (4.86) involves only a closed subspace $\boldsymbol{A}$, it can be solved numerically provided $\boldsymbol{A}$ is kept small. This will be the case for a localized deep center. Once a_A is known, $a_{\boldsymbol{B}}$ can be determined from (4.80b).

We shall now consider the extreme case where $\boldsymbol{A}$ contains only the defect site and four sp^3 orbitals. These orbitals are the same sp^3 orbitals as discussed already in Sect. 2.6 and denoted by S, X, Y, and Z. This approximation is reasonable for a deep substitutional impurity or a vacancy in the diamond- and zinc-blende-type semiconductors. From symmetry considerations, it is not difficult to see that S transforms according to the irreducible representation $A_1(\Gamma_1)$ of the T_d group while X, Y, and Z belong to the triply degenerate $T_2(\Gamma_4)$ representation. Using these four basis functions the determinants of G_0 and V can be written in the simplified form

$$\begin{vmatrix} G_{0A} & 0 & 0 & 0 \\ 0 & G_{0T} & 0 & 0 \\ 0 & 0 & G_{0T} & 0 \\ 0 & 0 & 0 & G_{0T} \end{vmatrix} \qquad \begin{vmatrix} V_A & 0 & 0 & 0 \\ 0 & V_T & 0 & 0 \\ 0 & 0 & V_T & 0 \\ 0 & 0 & 0 & V_T \end{vmatrix}$$

where the matrix elements G_{0A} and V_A represent the matrix elements $\langle S | G_0 | S \rangle$ and $\langle S | V | S \rangle$, respectively. G_{0T} and V_T are defined similarly using the T_2 triplet. Substituting these determinants into (4.81) we obtain two implicit equations

$$G_{0A}(E_A)V_A = 1 \quad \text{and} \quad G_{0T}(E_T)V_T = 1, \tag{4.87}$$

from which the defect energies E_A and E_T (for the deep levels with symmetries A_1 and T_2 respectively) can be determined.

Hjalmarson et al. [4.29] have studied the dependence of the deep level energies on the defect potential V in a large number of tetrahedrally bonded semiconductors. To understand their results, we shall consider G_{0A} as a function of E. According to the definition in (4.50), G_{0A} is equal to

$$G_{0A} = \langle S | G_{0A} | S \rangle = \sum_{\boldsymbol{k}} \lim_{\eta \to 0+} [E - E_S(\boldsymbol{k}) + \mathrm{i}\eta]^{-1}, \tag{4.88}$$

where $E_S(\boldsymbol{k})$ and $\boldsymbol{k}$ denote, respectively, the energy and wave vector of electrons in the bands which contain atomic orbitals with s-symmetry. Since we have assumed that the deep level with A_1 symmetry is a bound state, there are no energies $E_S(\boldsymbol{k})$ for which $E = E_S(\boldsymbol{k})$. As a result, (4.88) can be rewritten as

$$G_{0A}(E) = \sum_{\boldsymbol{k}} \mathcal{P}[E - E_S(\boldsymbol{k})]^{-1} \tag{4.89}$$

$$= \mathcal{P}\left(\int \frac{N_{\rm d}(E_S)}{E - E_S} dE_S\right). \tag{4.90}$$

In (4.90) we have transformed the summation over the electron wave vector $\boldsymbol{k}$ into an integration over electron energy by introducing a density of states $N_{\rm d}(E_S)$. Notice that $N_{\rm d}(E_S)$ is the DOS for electrons with s-symmetry only. This is sometimes referred to as a **partial** or **projected density of states**.

Equation (4.90) can also be derived from the Kramers–Kronig relations by regarding $G_{0A}(E)$ as a complex function of E. In general $\mathrm{Re}\{G_{0A}(E)\}$ is related to the integral of $\mathrm{Im}\{G_{0A}(E)\}$ over all E via (4.56). $\mathrm{Im}\{G_{0A}(E)\}$ turns out to be equal to $-\pi N_{\rm d}(E_S)$ [see (4.51)]. To visualize the function $G_{0A}(E)$ we first make a crude approximation to the partial DOS of the s-symmetry valence (bonding) and conduction (antibonding) bands in a typical zinc-blende-type semiconductor such as GaAs. This is shown schematically in Fig. 4.6a. From this partial DOS we can calculate $G_{0A}(E)$ using (4.90). We can also

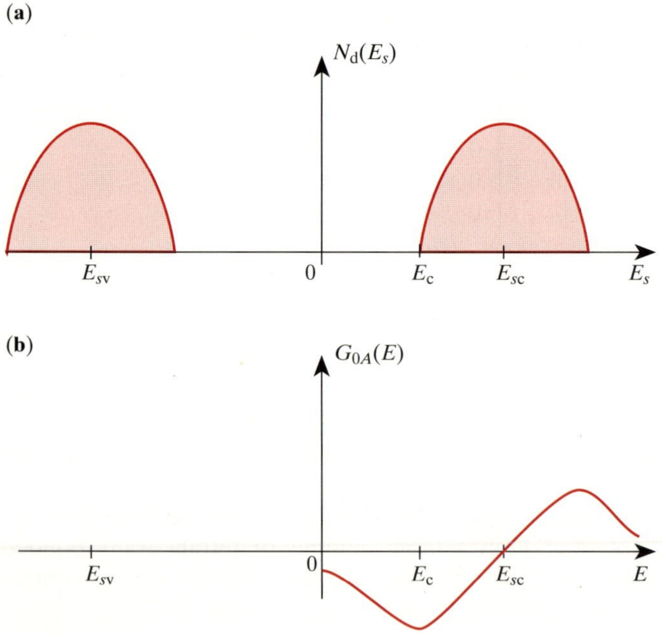

Fig. 4.6. Schematic plots of (**a**) the density of states for the valence band (centered at energy E_{sv}) and conduction band (centered at energy E_{sc}) derived from the s-symmetry atomic orbitals in a zinc-blende-type semiconductor and (**b**) the function $G_{0A}(E)$ obtained from the density of states in (**a**) using (4.89). The energy at the middle of the gap has been chosen as the zero in the energy scale. $E_{\rm c}$ denotes the bottom of the conduction band

"guess" the shape of $G_{0A}(E)$ by noticing that there are "correlations" between the shapes of the real and imaginary parts of a complex function satisfying the Kramers–Kronig relations. Such "correlations" are illustrated with an example in Problem 4.5. The $G_{0A}(E)$ "guesstimated" by using such correlations is sketched in Fig. 4.6b. For E lying within the energy gap ($0 < E < E_c$) $G_{0A}(E)$ is approximately a linear function of E (note that the origin of energies is assumed to be in the middle of the gap). As a result of this linear relation the solution of (4.81) produces an almost hyperbolic dependence of the trap energy E_A on V_A:

$$E_A \propto (1/V_A). \tag{4.91}$$

Figure 4.7 shows the energies of the A_1-symmetry deep impurity levels in several diamond- and zinc-blende-type semiconductors obtained by *Hjalmarson* et al. [4.29]. These authors first fitted the known band structures of various semiconductors with a nearest-neighbor tight-binding model. The overlap parameters obtained from this fitting procedure were used to determine the partial DOS of *s*-symmetry. The defect potential V_A was taken to be equal to the difference between the *s* atomic orbital energies of the impurity and the host atom it replaces. The deep level energies for various substitutional impurities were calculated with (4.87). The relevant impurities are listed in Fig. 4.7 above the upper horizontal axis in the order of their defect potentials V_A. Notice the nearly hyperbolic dependence of the deep level energy E_A on V_A as predicted by (4.91). The intercepts of the quasi-hyperbolas with

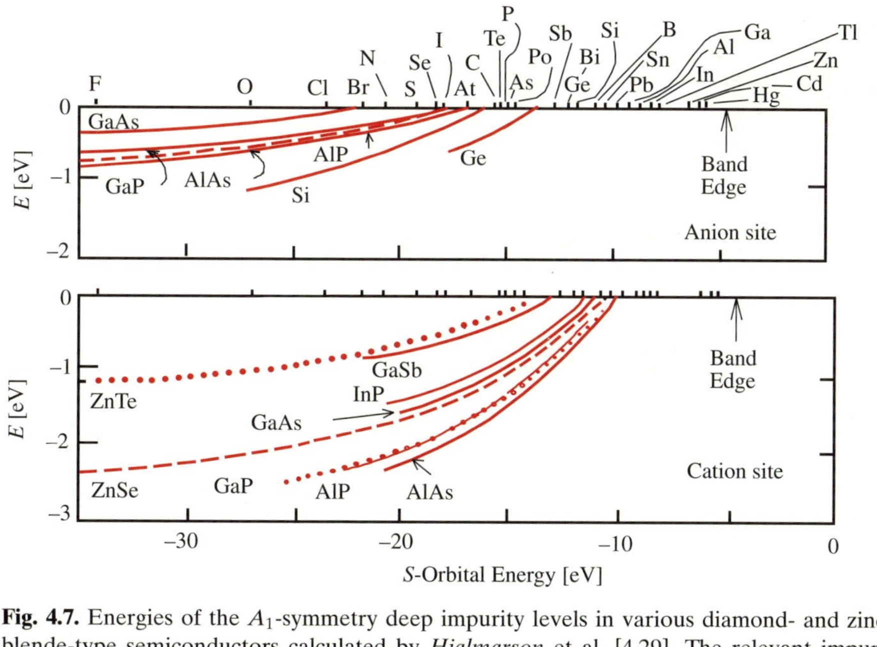

Fig. 4.7. Energies of the A_1-symmetry deep impurity levels in various diamond- and zinc-blende-type semiconductors calculated by *Hjalmarson* et al. [4.29]. The relevant impurities are listed above the upper horizontal axis in the order of their defect potentials

the band edges (the $E = 0$ axis being one of them) determine the thresholds for either an attractive or repulsive potential. For some semiconductors, such as Si and Ge, the quasi-hyperbolas terminate at the point where $-E$ equals the bandgap. Beyond that point, a resonant state rather than a bound state will be formed. For quasi-hyperbolas which do not terminate at the band edges, they asymptotically approach the energies $E_A(\pm\infty)$ when V_A approaches $\pm\infty$. At these energies, unless dG_{0AA}/dE in (4.86) vanishes also, $a_A = 0$, i.e., no electron or hole can be trapped at these energies. Thus the asymptotes of E_A are the dangling-bond or ideal vacancy energies. These energies are determined by the host semiconductor.

From the above discussion we can draw some qualitative conclusions about deep level energies. In general, the defect energy is not very sensitive to V when V is large. The reason is that E is given by the solution of $G_0(E) = 1/V$, which varies slowly with V when V is much larger than zero. This explains why impurities whose atomic energies differ by more than 10 eV often give rise to deep levels with defect energies differing by less than 1 eV. For infinitely large values of V, the defect energy approaches that of an ideal vacancy. Thus the energy of a vacancy can be considered as the lower bound for a deep level energy. Based on these considerations a substitutional deep center can be regarded as a vacancy in the lowest order of approximation. A deep center wave function is predominantly determined by the host crystal rather than by the impurity. One serious limitation of this model is its neglect of lattice relaxations. Since most deep centers usually induce some lattice distortion, this model cannot predict deep center energies quantitatively. However, its use lies in predicting their chemical trends. For example, from Fig. 4.7 we can predict that As on Ga sites and O on As sites both form deep donors in GaAs, in agreement with experiment. Another application of this model is the calculation of the *variation* of deep center energies with changes in the host band structure. Such changes can be induced by alloying or by hydrostatic pressure. The next section is devoted to one such application in order to understand substitutional nitrogen in GaAsP alloys. The model has been recently used to estimate the spin–orbit splitting of acceptor ground states, which is much smaller than Δ_0 for diamond and silicon [4.20, 21].

4.3.3 Another Application of the Green's Function Method: Nitrogen in GaP and GaAsP Alloys

Nitrogen substituting for P in GaP is an example of an isoelectronic or isovalent center because N has the same valence as P. In this case the isoelectronic centers produce localized defect levels because of the strength of V. In many other cases, isoelectronic impurities such as phosphorus replacing As or aluminum replacing Ga in GaAs do not produce localized states. Instead they generate resonant states which overlap with the band structure and hybridize with the Bloch states. When present in sufficient concentration they can be considered as forming a random alloy such as $Ga_xAl_{1-x}As$. The band structure of these alloys can be calculated by assuming the crystal to be perfect except for having an average effective or *virtual crystal potential.* Not surprisingly this approach is known as the **virtual crystal approximation**. For N_P in GaP, N can attract an electron because N is much more electronegative than P. (These electrons can be introduced by n-type doping or by optical excitation.) The electronegativity of N and P, as defined by Phillips, are respectively 3.0 and 1.64 [4.30]. Once N has bound an electron and become negatively charged (to be denoted by N^-), it attracts a hole just like an acceptor. As a result, N in GaP is known as an **isoelectronic acceptor**. Similarly Bi (whose electronegativity of 1.24 is less than that of P) is an **isoelectronic donor** in GaP. Isoelectronic impurities appear electrically neutral except within the immediate vicinity of the impurity. Their potentials are therefore always short-ranged. The highly localized nature of isoelectronic impurity potentials causes them to behave as deep centers in spite of the fact that they may have very small binding energies. For example, the binding energy of an electron to N_P in GaP is only about 9 meV [4.31].

The very short-ranged potential of N in GaP is responsible for its application in the fabrication of light-emitting diodes (LEDs). Once N has cap-

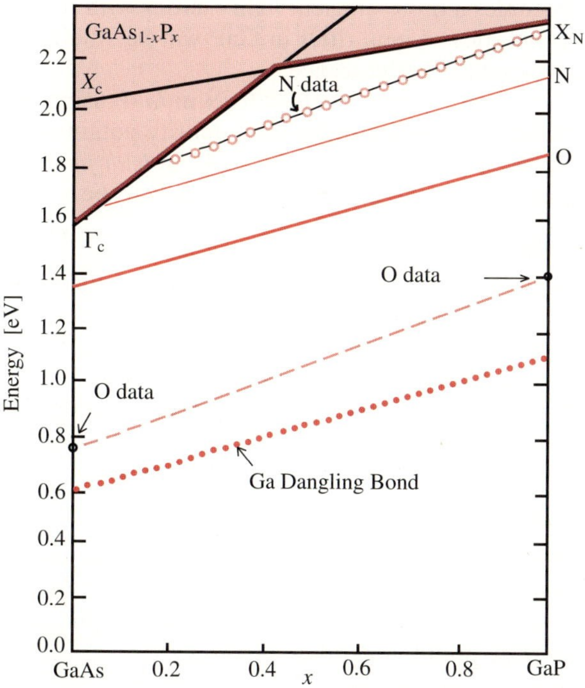

Fig. 4.8. The energy of an A_1-symmetry deep level for N substituting an anion (labeled "N data") in $GaAs_{1-x}P$ alloy as a function of the alloy composition x. The *open circles* are experimentally measured energies of an exciton bound to an isolated N center while the line labeled "N" represents the theoretical values. The point X_N represents an exciton bound to a neutral nitrogen. The lines labeled "O data" and "Ga dangling bond" are not relevant to this chapter. (From [4.29]). Γ_c and X_c denote, respectively, conduction band minima at the Γ and X points of the Brillouin zone

tured an electron, its Coulomb potential dominates over the short-range potential. This Coulomb potential is screened by the dielectric function of the semiconductor as usual and hence N^- behaves like a hydrogenic acceptor with a binding energy of 11 meV [4.32]. An alternative way of looking at a hole bound to form N^- is to regard it as an electron–hole pair bound to a neutral N (denoted by the point X_N in Fig. 4.8). An electron–hole pair is also known as an **exciton**. An exciton bound to an impurity like N in GaP via an attractive potential is known as a **bound exciton**. Excitons and bound excitons are very important in determining the optical properties of semiconductors and will be studied in more detail in Chaps. 6 and 7. At this point we simply note that normally GaP is not an efficient light emitter because it has an indirect bandgap. In order to conserve the wave vector during optical transitions, radiative recombination of an exciton in an indirect-bandgap semiconductor requires the cooperation of phonons (Sect. 6.2.6). As a result, indirect-bandgap semiconductors have smaller light emission probabilities than direct-bandgap semiconductors. However, excitons bound to N in GaP have a relatively high probability of radiative recombination because the localized N centers break the translational invariance of the crystal and thereby relax the wave vector conservation requirement in optical transitions. As a result, GaP becomes a more efficient light emitter with the introduction of N. The optical properties of impurities in GaP will be studied in more detail in Chap. 7 when we discuss luminescence.

We now calculate the binding energy of an electron to N in GaP using the Green's function method. First we have to determine the impurity potential V. Since V has a very short range we will simply assume *V to be a delta function in real space*. More precisely, we assume the matrix element of V to be

$$\langle n, \boldsymbol{R} \,|\, V \,|\, n', \boldsymbol{R}' \rangle = U_0 \delta_{n,n'} \delta_{\boldsymbol{R},\boldsymbol{R}'} \delta_{n,0} \delta_{\boldsymbol{R},\boldsymbol{O}}, \tag{4.92}$$

where n and n' are band indices (with the lowest conduction band numbered arbitrarily as 0); $\boldsymbol{R}$ and $\boldsymbol{R}'$ are lattice vectors (with N_P assumed to be the origin); and $|\, n, \boldsymbol{R} \rangle$ denotes the Wannier function $a_{n\boldsymbol{R}}$ defined in (4.4a). When defined in this way V is called the **Slater–Koster interaction potential**. Experimentally it is known that N in GaP forms only one bound state inside the energy gap. Using the result of Sect. 4.3.1 the energy E of this bound state can be obtained by solving (4.65). Substituting V into (4.65) we obtain

$$1 - U_0 \langle 0, \boldsymbol{0} \,|\, (E - H_0)^{-1} \,|\, 0, \boldsymbol{0} \rangle = 0. \tag{4.93}$$

There are two ways to solve (4.93). One approach is to use a simple tight-binding model with the minimum number of parameters, as discussed in the last section. Another method is to use a more exact band structure obtained by methods such as the empirical pseudopotentials method (Sect. 2.5). We shall briefly describe the results of both approaches for comparison.

a) Tight-Binding Method

Using this method *Hjalmarson* et al. [4.29] have calculated the energy of an electron bound to N in $GaAs_{1-x}P_x$ alloys as a function of the alloy composition x. Their results are shown in Fig. 4.8. According to their calculation N forms a resonant state above the bottom of the conduction band in GaAs. As the concentration of P in the alloy is increased, this resonant level gradually moves into the bandgap and becomes a bound state. The important thing to note is that the location of the lowest conduction minimum in the reciprocal space changes from the zone center to the X point at $x \approx 0.5$. As a result, the slope of the lowest conduction band minimum versus x in Fig. 4.8 changes abruptly at $x \approx 0.5$. On the other hand, the deep level energy varies smoothly with x. This is consistent with what we expect for a deep level. The energy of a highly localized center should be determined by the entire band and not by the lowest band minimum only, as in case of the shallow centers. The experimentally measured energies of an electron to N in the GaAsP alloys are shown as open circles in Fig. 4.8. While the calculated energies of the N deep level are lower than the experimental values, the measured dependence of the deep level on alloy concentration is in very good agreement with theory. In addition, the resonant nature of the N level in GaAs has been verified by *Wolford* et al. [4.33]. Instead of varying the concentration of P, Wolford et al. applied hydrostatic pressure to lower the conduction band minimum at X below the band minimum at Γ. At pressures above 30 kbar (or 3 GPa), the N deep level moved into the bandgap as in the case of increasing P concentration. Using this method, excitons bound to N in GaAs have been studied in great detail by photoluminescence experiments, see Fig. 4.9.

b) Empirical "Energy Moment" Approach

In this approach, we change the basis function in (4.93) from the Wannier functions back to Bloch functions $|n, \boldsymbol{k}\rangle$. The resultant secular equation becomes

$$1 + \frac{U_0}{N} \sum_{\boldsymbol{k}} \langle 0, \boldsymbol{k} \,|\, (E_{0,\boldsymbol{k}} - E)^{-1} \,|\, 0, \boldsymbol{k} \rangle = 0, \tag{4.94}$$

where $n = 0$ denotes the lowest conduction band. Equation (4.94) can be solved numerically. In particular, we recognize that the expression $\langle 0, \mathbf{0} \,|\, (E - H_0)^{-1} \,|\, 0, \mathbf{0} \rangle$ in (4.93) is related to the local DOS by the Kramers–Kronig relations. So E can be computed from the DOS of the host semiconductor in a relatively straightforward manner. Alternatively, we can obtain an analytic expression for E by introducing approximations such as expanding $(E_{0,\boldsymbol{k}} - E)^{-1}$ into a series:

$$\begin{aligned}(E_{0,\boldsymbol{k}} - E)^{-1} &= E_{0,\boldsymbol{k}}^{-1} + E_{0,\boldsymbol{k}}^{-1} \left(\frac{E}{E_{0,\boldsymbol{k}} - E} \right) \\ &= E_{0,\boldsymbol{k}}^{-1} \left\{ 1 + \frac{E}{E_{0,\boldsymbol{k}}} \left[1 + \frac{E}{E_{0,\boldsymbol{k}}} + \left(\frac{E}{E_{0,\boldsymbol{k}}} \right)^2 + \cdots \right] \right\}.\end{aligned} \tag{4.95}$$

Substituting (4.95) into (4.94) we obtain

$$\begin{aligned}0 = 1 &+ \frac{U_0}{N} \left\{ \sum E_{0,\boldsymbol{k}}^{-1} \right\} + E \frac{U_0}{N} \left\{ \sum E_{0,\boldsymbol{k}}^{-2} \right\} \\ &+ E^2 \frac{U_0}{N} \left\{ \sum E_{0,\boldsymbol{k}}^{-3} \right\} + \ldots .\end{aligned} \tag{4.96}$$

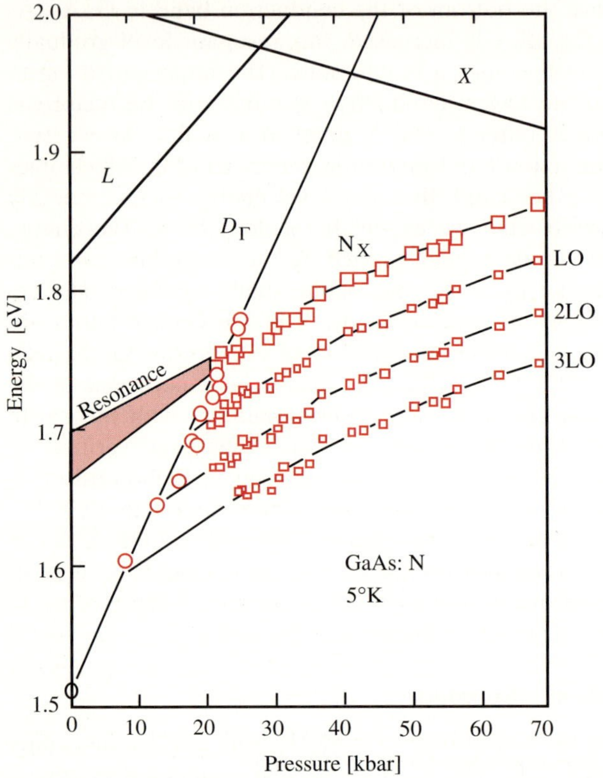

Fig. 4.9. The energy of excitons bound to isoelectronic substitutional N deep centers (denoted as N_X and indicated by *large open squares*) in GaAs as a function of hydrostatic pressure. The data points (indicated by *small open squares*) labeled LO, 2LO, etc., denote the LO phonon replicas of the main N_X peaks. The *open circles* labeled D_Γ denote the energy of an exciton bound to a neutral shallow donor as a function of pressure. The *solid lines* labeled L and X denote the higher conduction band minima at the L and X points of the Brillouin zone. (From [4.33])

Equation (4.96) can be expressed in a more compact form by defining "moments" of the band $n = 0$:

$$\langle E_0^{-1} \rangle = \left\{ \sum E_{0,\boldsymbol{k}}^{-1} \right\} / N, \tag{4.97a}$$

$$\langle E_0^{-2} \rangle = \left\{ \sum E_{0,\boldsymbol{k}}^{-2} \right\} / N, \tag{4.97b}$$

and so on. These moments can be calculated from the band structure. With this notation (4.96) becomes

$$0 = 1 + (U_0 \langle E_0^{-1} \rangle) + E(U_0 \langle E_0^{-2} \rangle) + E^2(U_0 \langle E_0^{-3} \rangle) + \dots \, . \tag{4.98}$$

In the case of N in GaP it is known that E satisfies the condition $E \langle E_0^{-1} \rangle \ll 1$, hence we can neglect the term in (4.98) that depends on E^2 and higher terms.

This gives the approximate solution for E as

$$E \approx -\left[1 + (U_0\langle E_0^{-1}\rangle)\right] / U_0\langle E_0^{-2}\rangle. \tag{4.99}$$

Since we have assumed that the potential V of N in GaP is attractive, U_0 is negative. Without loss of generality we can choose the energy of the bottom of the conduction band as zero. Since $\langle E_0^{-2}\rangle$ is positive, in order that E be a solution of (4.99) lying in the gap, E has to be negative. Hence $1 + (U_0\langle E_0^{-1}\rangle)$ has to be negative also. This imposes the condition on U_0 that $| U_0\langle E_0^{-1}\rangle | \geq 1$.

Faulkner [4.31] used the conduction band structure of GaP and the experimental value of the binding energy of an electron to N to derive a value for U_0 based on the above model. Unfortunately, the value of U_0 obtained does not explain the other experimental results such as the binding energy of an exciton to pairs of nitrogen atoms [4.34] and the properties of bound excitons in GaAsP alloys. Furthermore, the Slater–Koster potential allows the existence of only one bound state and therefore cannot explain the presence of excited states. Some of these shortcomings can be traced to the fact that the expansion in (4.95) is a good approximation only when the defect energy E is quite different from the band energies. In other words, the binding energy of the deep center has to be large. This condition is not satisfied by N in GaP. In spite of these shortcomings, this very simple model does provide a framework for understanding qualitatively some of the properties of isoelectronic impurities in semiconductors. It has become the starting point for later models proposed to overcome some of its shortcomings. In one model proposed by *Hsu* et al. [4.35], a long-range potential due to strain was added to the Slater-Koster potential. The energies of excited states of excitons bound to N and of excitons bound to nitrogen pairs were satisfactorily explained by choosing this long-range potential appropriately.

4.3.4 Final Note on Deep Centers

Our quantitative discussion of deep centers so far has neglected lattice distortions (i.e., relaxation). This topic is beyond the introductory nature of the present text. However, several deep defects, known as EL2 [4.36] and DX centers [4.23], have been found to play important roles in semiconductor devices based on III–V and perhaps even II–VI compound semiconductors. These defects exhibit interesting properties such as metastability, persistent photoconductivity, and negative U. Often these properties of deep centers can be understood only when lattice relaxations are included. Readers interested in these topics should refer to one of the review articles listed in the references, e. g. [4.23, 36] and to the Appendix on the DX Centers to be found under the Web Page http://Pauline.Berkeley.edu/textbook/DX-Center.pdf.

PROBLEMS

4.1 *Symmetrization of the Conduction Band Wave Functions in Si for Donor Binding Energy Calculations*

a) There are six conduction band minima in Si. They occur along the six equivalent [100] directions inside the Brillouin zone. Use the notation X, Y, and Z to represent the wave function for the minima occurring in the [100], [010], and [001] directions, respectively, ($\overline{X}$, $\overline{Y}$, and $\overline{Z}$ for those along [$\overline{1}$00], [0$\overline{1}$0], [00$\overline{1}$]). Apply the symmetry operations of the point group T_d to these six wave functions. Show that the characters of the corresponding representation are

$\{E\}$	$\{3C_2\}$	$\{6S_4\}$	$\{6\sigma\}$	$\{8C_3\}$
6	2	0	2	0

b) Using Table 2.3, show that these six wave functions form a reducible representation which can be reduced to

$$A_1 \oplus E \oplus T_2.$$

c) By examining the basis functions in Table 2.3, show that the six wave functions when written as the following linear combinations correspond to the irreducible representations:

A_1: $(X + \overline{X} + Y + \overline{Y} + Z + \overline{Z})/\sqrt{6}$;
E: $(X + \overline{X} - Y - \overline{Y})/2$, $(2Z + 2\overline{Z} - X - \overline{X} - Y - \overline{Y})/\sqrt{12}$;
T_2: $(X - \overline{X})/\sqrt{2}$, $(Y - \overline{Y})/\sqrt{2}$, $(Z - \overline{Z})/\sqrt{2}$.

4.2 *Full Rotation Group and Spherical Tensors*

Consider a particle moving in a spherically symmetric potential. A rotation about an axis in the three-dimensional space is clearly a symmetry operation. The set of all rotations can be shown to form an infinite group known as the **full rotation group** (see for example books on group theory listed for Chap. 2). All rotations through the same angle (irrespective of the axis of rotation) can be shown to belong to the same class. Rotations by different angles belong to different classes: there are an infinite number of classes. Hence there are also an infinite number of irreducible representations. In quantum mechanics we learn that the angular momentum l is a good quantum number for particles moving in spherically symmetric potentials. Their eigenfunctions are the spherical harmonic functions $\{Y_{lm}\}$ where $m = -l, -l+1, \ldots, -1, 0, 1, \ldots, l$. For integral values of l, these functions turn out to be the basis functions for the irreducible representations of the full rotation group. Hence an irreducible representation with $\{Y_{lm}\}$ as basis functions has dimension $2l + 1$. Suppose we choose a polar coordinate system (r, Θ, ϕ) with the z axis as the polar axis.

a) Show that the effect of a clockwise rotation of the coordinate axes by an angle α about the z axis on $Y_{lm}(\Theta, \phi)$ is given by

$$P_\alpha Y_{lm}(\Theta, \phi) = \exp(-im\alpha)\, Y_{lm}(\Theta, \phi).$$

b) Show that the representation matrix of such a rotation is a diagonal matrix with diagonal elements $\exp(-\mathrm{i}m\alpha)$, where $m = l, l-1, \ldots, 0, \ldots, -l+1, -l$. Hence show that the character for such a rotation is given by

$$\chi_l(P_\alpha) = \frac{\sin\left[\left(l+\frac{1}{2}\right)\alpha\right]}{\sin(\alpha/2)}.$$

The functions $\{Y_{lm}\}$ form a spherical tensor of order l and dimension $2l+1$. For the purpose of this chapter, we need to know only $\{Y_{lm}\}$ for $l = 0$, 1, and 2. For $l = 0$, $\{Y_{lm}\}$ is a scalar or any function that is invariant under all rotations. The $\{Y_{1m}\}$ for $l = 1$ are three functions defined by

$$Y_{10} = \left(\frac{3}{4\pi}\right)^{1/2} \frac{z}{r};$$

$$Y_{1\pm} = \left(\frac{3}{8\pi}\right)^{1/2} \frac{x \pm \mathrm{i}y}{r}.$$

The spherical harmonic functions for $l = 2$ are

$$\begin{aligned} Y_{20} &= \left(\frac{5}{16\pi}\right)^{1/2} \frac{3z^2 - r^2}{r^2}; \\ Y_{2\pm 1} &= \pm\left(\frac{15}{8\pi}\right)^{1/2} \frac{(x \pm \mathrm{i}y)z}{r^2}; \\ Y_{2\pm 2} &= \left(\frac{15}{32\pi}\right)^{1/2} \frac{x^2 - y^2 \pm 2\mathrm{i}xy}{r^2}. \end{aligned}$$

c) Suppose $\{T_{ij}\}$, where i and j stand for x, y, and z, is now a second-rank *Cartesian* tensor containing nine elements. These nine elements form a *reducible* representation of the full rotation group.

Show that the second-rank Cartesian tensor can be reduced to the following three spherical tensors of order 0, 1, and 2, respectively:

$$\begin{aligned} T_0^{(0)} &= T_{11} + T_{22} + T_{33} \\ T_0^{(1)} &= T_{12} - T_{21} \\ T_{\pm}^{(1)} &= \mp(1/\sqrt{2})\left[T_{23} - T_{32} \pm \mathrm{i}(T_{31} - T_{13})\right] \\ T_0^{(2)} &= (3/2)^{1/2} T_{33} \\ T_{\pm 1}^{(2)} &= \mp(T_{13} \pm \mathrm{i}T_{23}) \\ T_{\pm 2}^{(2)} &= (1/2)(T_{11} - T_{22} \pm 2\mathrm{i}T_{12}). \end{aligned}$$

4.3 *Luttinger Hamiltonian in Terms of Spherical Tensors*

Two spherical tensors can be multiplied together to generate another tensor as with Cartesian tensor. For example, the **scalar product** of two tensor $\{T_q^{(k)}\}$

and $\{U_q^{(k)}\}$ both of order k can be defined as

$$\left(T^{(k)} \cdot U^{(k)}\right) = \sum_q (-1)^q T_q^{(k)} U_{-q}^{(k)}.$$

The tensor product $[\boldsymbol{T}^{(a)} \times \boldsymbol{U}^{(b)}]_d^{(c)}$ of two spherical tensors of orders a and b, respectively, are spherical tensors of order c defined by

$$\left[\boldsymbol{T}^{(a)} \times \boldsymbol{U}^{(b)}\right]_d^{(c)} = (-1)^{a-b+d}(2c+1)^{1/2} \sum_{i,j} \begin{bmatrix} a & b & c \\ i & j & -d \end{bmatrix} \boldsymbol{T}_i^{(a)} \boldsymbol{U}_j^{(b)},$$

where $\begin{bmatrix} a & b & c \\ i & j & -d \end{bmatrix}$ is the $3-j$ symbol used in the addition of two angular momenta in quantum mechanics (see, for example, Appendix C in [4.37]). Using these definitions show that

$$\begin{aligned} P_x^2 J_x^2 + P_y^2 J_y^2 + P_z^2 J_z^2 \;=\; & \left(\frac{1}{3}\right) P^2 J^2 + \left(\frac{2}{45}\right) \left(\boldsymbol{P}^{(2)} \cdot \boldsymbol{J}^{(2)}\right) \\ & + \left(\frac{1}{18}\right) \Bigg\{ \left[\boldsymbol{P}^{(2)} \times \boldsymbol{J}^{(2)}\right]_{-4}^{(4)} \\ & + \frac{(70)^{1/2}}{5} \left[\boldsymbol{P}^{(2)} \times \boldsymbol{J}^{(2)}\right]_0^{(4)} + \left[\boldsymbol{P}^{(2)} \times \boldsymbol{J}^{(2)}\right]_4^{(4)} \Bigg\}, \end{aligned}$$

$$\begin{aligned} & \left[\{P_x P_y\}\{J_x J_y\} + \text{ cyclic permutations}\right] \;= \\ & \left(\frac{1}{30}\right) \left(\boldsymbol{P}^{(2)} \cdot \boldsymbol{J}^{(2)}\right) - \left(\frac{1}{36}\right) \Bigg\{ \left[\boldsymbol{P}^{(2)} \times \boldsymbol{J}^{(2)}\right]_{-4}^{(4)} + \\ & \frac{(70)^{1/2}}{5} \left[\boldsymbol{P}^{(2)} \times \boldsymbol{J}^{(2)}\right]_0^{(4)} + \left[\boldsymbol{P}^{(2)} \times \boldsymbol{J}^{(2)}\right]_4^{(4)} \Bigg\}. \end{aligned}$$

4.4 *Average Light and Heavy Hole Masses*

In Sect. 2.6.2 we have already attempted to approximate the warped heavy and light hole bands with spherical bands by defining averaged effective masses with (2.69a) and (2.69b):

$$\frac{1}{m_{\mathrm{hh}}^*} = \frac{1}{\hbar^2} \left[-2A + 2B \left(1 + \frac{|C|^2}{10B^2} \right) \right] \tag{2.69a}$$

$$\frac{1}{m_{\mathrm{lh}}^*} = \frac{1}{\hbar^2} \left[-2A - 2B \left(1 + \frac{|C|^2}{10B^2} \right) \right]. \tag{2.69b}$$

In this chapter we have presented a different approach. Using the method of *Baldereschi* and *Lipari* [4.14] we can express the Luttinger Hamiltonian in terms of irreducible representations of the full rotational group:

$$\begin{aligned} H_{\mathrm{BL}} \;=\; & \frac{\boldsymbol{p}^2}{\hbar^2} - \frac{\mu}{9\hbar^2} \left(\mathcal{P}^{(2)} \cdot \boldsymbol{J}^{(2)}\right) + \delta\Big(\left[\mathcal{P}^{(2)} \times \boldsymbol{J}^{(2)}\right]_{-4}^{(4)} \\ & + \frac{\sqrt{70}}{5} \left[\mathcal{P}^{(2)} \times \boldsymbol{J}^{(2)}\right]_0^{(4)} + \left[\mathcal{P}^{(2)} \times \boldsymbol{J}^{(2)}\right]_4^{(4)} \Big). \end{aligned}$$

The hole bands can be "sphericalized" by setting the cubic term, proportional to δ in H_{BL}, equal to zero. Show that, in this spherical approximation, the energy dispersions of the heavy- and light-hole bands are given by

$$E_{(\pm)} = -\frac{\hbar^2\gamma_1}{2m_0}(1 \pm \mu)k^2$$

and that the "sphericalized" heavy- and light-hole masses are given by

$$\frac{1}{m^*_{\mathrm{hh}}} = \frac{1}{\hbar^2}\left\{-2A + \frac{4}{5}B\left[1 + \frac{3}{2}\left(1 + \frac{4|C|^2}{9B^2}\right)^{1/2}\right]\right\}$$

$$\frac{1}{m^*_{\mathrm{lh}}} = \frac{1}{\hbar^2}\left\{-2A - \frac{4}{5}B\left[1 + \frac{3}{2}\left(1 + \frac{4|C|^2}{9B^2}\right)^{1/2}\right]\right\}.$$

Notice that in the limit $C = 0$ the above masses become identical to those in (2.69a) and (2.69b).

4.5 *Kramers–Kronig Relations*
Given that the imaginary part of a function $F(E)$ is a *Lorentzian*:

$$-\mathrm{Im}\{F(E)\} = \frac{1}{(E - E')^2 + \Gamma^2},$$

use the Kramers-Kronig relation in (4.56) to show that the real part of $F(E)$ is given by

$$\mathrm{Re}\{F(E)\} = \left(\frac{1}{\Gamma}\right)\frac{E - E'}{(E - E')^2 + \Gamma^2}.$$

An alternative, easier way to solve this problem "by inspection" is to note that if $F(E)$ is equal to the analytic function

$$F(E) = \left(\frac{1}{\Gamma}\right)\frac{1}{(E - E') + \mathrm{i}\Gamma}$$

then the imaginary part of $F(E)$ is equal to the given Lorentzian.

Figure 4.10a,b shows schematically the functions $\mathrm{Re}\{F(E)\}$ and $-\mathrm{Im}\{F(E)\}$. These plots illustrate the "correlations" between functions which are related to each other by the Kramers–Kronig relations:

- If $-\mathrm{Im}\{F(E)\}$ is symmetric with respect to the vertical axis $y = E_0$, then $\mathrm{Re}\{F(E)\}$ is antisymmetric with respect to $y = E_0$.
- If $-\mathrm{Im}\{F(E)\}$ is always positive and large over only a limited range of values of E; $E_1 < E < E_2$, then $\mathrm{Re}\{F(E)\}$ will be negative for E well below E_1 but positive for E much larger than E_2.

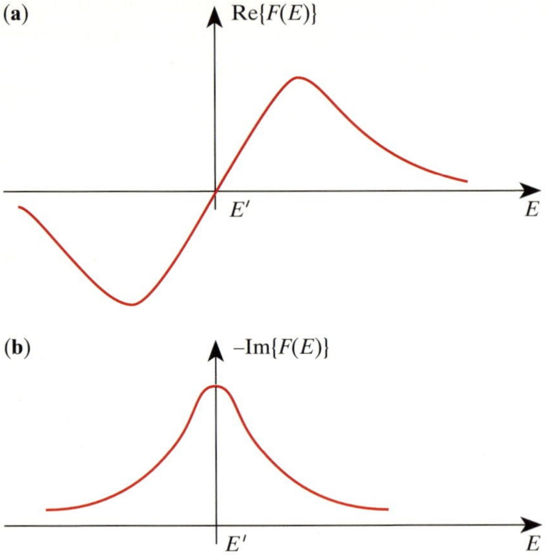

Fig. 4.10. Schematic plots of the functions (**a**) $\mathrm{Re}\{F(E)\}$ and (**b**) $-\mathrm{Im}\{F(E)\}$ defined in Problem 4.5

SUMMARY

This chapter dealt with the study of the electronic properties of defects in semiconductors because electrically active defects play an important role in the operation of many semiconductor devices. Since defects come in many different forms we restricted our discussions to *point defects* only. These are separated into *donors* and *acceptors* and further divided into *shallow* or *hydrogenic centers* and *deep centers*. For shallow centers we introduced the *effective mass approximation* for calculating their energies and wave functions. Properties of shallow centers were shown to be very similar to those of the hydrogen atom except for effective mass anisotropy and other corrections arising from the host crystal lattice. Hence energy levels of shallow centers are sometimes referred to as hydrogenic levels. Many defect centers cannot be understood within this approximation and they are referred to as *deep centers*. Properties of deep centers are often determined by potentials, localized within one unit cell, known as *central cell corrections*. These localized potentials are difficult to handle. We have, therefore, presented only a rather rudimentary *Green's functions* approach to calculating deep center energies. This approach was applied to explain the chemical trends of deep levels in tetrahedrally bonded semiconductors and to the special case of *isovalent* (substitutional nitrogen) impurities in GaAsP alloys. A very serious limitation of our approach is the neglect of *lattice relaxations*, which are often associated with deep centers.

5. Electrical Transport

CONTENTS

In Chap. 4 we studied electrons and holes located around defects. Since these electrons and holes are immobile they are known as **bound** *electrons and holes*, respectively. In contrast, electrons in the conduction band and holes in the valence band of a semiconductor can carry electrical current. Hence they are referred to as **free carriers**. In this chapter we will study the effect of an external electric field on free carriers in a semiconductor. The response of these carriers to an electric field depends on the field strength. We will first consider the case of weak electric fields, where the behavior of carriers can be described by **Ohm's law**. Under high electric fields, carriers in a semiconductor can acquire so much energy that their average kinetic energy becomes higher than that of the lattice. Such energetic electrons are known as **hot electrons**. It is very difficult to calculate their properties analytically, therefore our discussions of hot electrons will be qualitative.

5.1 Quasi-Classical Approach

Let $\boldsymbol{F}$ represent a weak external static electric field applied to a semiconductor. We can assume without loss of generality, that this semiconductor contains only free electrons (i. e., it is an n-*type* semiconductor). For simplicity, we will assume that the concentration of free electrons is low enough that we can neglect their interactions with each other (such as collision and screening effects). We will also neglect local field effects due to ionic charges, i. e. the field experienced by every free electron is assumed to be equal to the external applied field. Now let Φ be the electric potential associated with the applied field. The wave equation for the time evolution of an electron in a semiconductor under the influence of Φ is given by

$$(H_0 - e\Phi)\psi(\boldsymbol{r},t) = \mathrm{i}\hbar\frac{\partial\psi}{\partial t}, \tag{5.1}$$

where H_0 is the one-electron Hamiltonian (in the absence of external perturbations) that we have already studied in Sect. 2.1, e is the magnitude of the electronic charge and $\psi(\boldsymbol{r},t)$ is the electron wave function in the presence of the external field. As long as Φ is small and does not vary rapidly in space, we can use the effective mass approximation (Sect. 4.2.1) to solve (5.1). The approach is very similar to the way we handled the donor electron problem in Sect. 4.2. The difference is that, in the present case, we are interested in the nonstationary solutions, which produce a current in response to the applied field. This requires the expectation value of $e\boldsymbol{v}$ to be evaluated where $\boldsymbol{v}$ is the electron velocity operator. The current density operator $\boldsymbol{j}$ is then defined as

$$\boldsymbol{j} = ne\boldsymbol{v}, \tag{5.2}$$

where n is the electron density. As we saw in Sect. 4.2, solving the Schrödinger equation within the effective mass approximation is quite an involved process. Instead of the fully quantum mechanical approach, we will adopt here a quasi-classical approach [5.1]. We shall derive a classical equation of motion for our electron in the external field based on the effective mass approximation.

As we discussed in detail in Sect. 4.2.1, a wave equation for Bloch waves such as (5.1) can be replaced by an effective wave equation for the envelope functions $C(\boldsymbol{R},t)$. For simplicity we shall assume that our electron is in an isotropic and nondegenerate conduction band with an energy minimum at the zone center ($\boldsymbol{k} = 0$) and a dispersion given by

$$E_\mathrm{c}(\boldsymbol{k}) = E_\mathrm{c}(0) + \frac{\hbar^2 k^2}{2m^*}, \tag{5.3}$$

where m^* is the effective mass. These assumptions are valid for electrons in direct bandgap semiconductors such as GaAs and InP but not for Si and Ge. Within the effective mass approximation the wave equation for the envelope functions can be written as [see (4.22)]

$$\left[E_\mathrm{c}(0) - \left(\frac{\hbar}{2m^*}\right)\frac{\partial^2}{\partial \boldsymbol{R}^2} - e\Phi(\boldsymbol{R})\right] C(\boldsymbol{R},t) \approx \mathrm{i}\frac{\hbar\partial}{\partial t}C(\boldsymbol{R},t). \tag{5.4}$$

Instead of solving (5.4), we argue that the net effect of the crystal potential on the motion of this electron inside the semiconductor is to cause its mass to change from the value in free space to m^*. This suggests that, as a simple approximation, we can describe the motion of this electron in an external electric field by a classical equation of motion (see, for example, [5.1])

$$m^* \frac{d^2 \boldsymbol{r}}{dt^2} + \frac{m^*}{\tau} \left(\frac{d\boldsymbol{r}}{dt} \right) = -e\boldsymbol{F}, \tag{5.5}$$

where $\boldsymbol{r}$ is the position of the electron and τ is a phenomenological scattering time introduced to account for the scattering of the electron by impurities and phonons. Equation (5.5) is considered quasi-classical because the concept of an effective mass for the electron motion has been derived quantum mechanically.

Once we have established (5.5), the motion of any charge q can be derived via classical mechanics. For example, under the influence of $\boldsymbol{F}$ a stationary charge will accelerate. As its velocity increases, the retardation term $(m^*/\tau)(d\boldsymbol{r}/dt)$ will also increase. Eventually the retardation term will cancel the term due to $\boldsymbol{F}$ and a steady state in which the charge has no acceleration is attained. The steady-state velocity of the charge is known as its **drift velocity** $\boldsymbol{v}_{\mathrm{d}}$. It is obtained from (5.5) by setting the acceleration term $m^*(d^2\boldsymbol{r}/dt^2)$ to zero and denoting by q the electronic charge $-e$:

$$\boldsymbol{v}_{\mathrm{d}} = q\boldsymbol{F}\tau/m^*. \tag{5.6}$$

The current density $\boldsymbol{J}$ at steady state is related to $\boldsymbol{v}_{\mathrm{d}}$ by

$$\boldsymbol{J} = nq\boldsymbol{v}_{\mathrm{d}}. \tag{5.7}$$

Combining (5.6 and 7) we obtain an expression for the current density:

$$\boldsymbol{J} = nq^2\boldsymbol{F}\tau/m^*. \tag{5.8}$$

The second-rank **conductivity tensor** $\boldsymbol{\sigma}$ is defined in general by

$$\boldsymbol{J} = \boldsymbol{\sigma} \cdot \boldsymbol{F}. \tag{5.9}$$

For the case of an isotropic conduction band $\boldsymbol{\sigma}$ is a diagonal tensor with all diagonal elements given by

$$\sigma = nq^2\tau/m^*. \tag{5.10}$$

Since σ depends on q^2, the contributions to the conductivity of a semiconductor from electrons and holes always add. Semiconductors differ from metals in that their carrier densities can be varied widely by changing the temperature or the dopant concentration. It is therefore convenient to factor out the dependence of σ on n. This can be accomplished by defining a carrier **mobility** μ as

$$\boldsymbol{v}_{\mathrm{d}} = \mu\boldsymbol{F}. \tag{5.11}$$

Combining (5.6 and 11) we obtain

$$\mu = q\tau/m^*. \tag{5.12}$$

In a semiconductor containing both free electrons and free holes σ is given by

$$\sigma = e(n_{\mathrm{e}}\mu_{\mathrm{e}} + n_{\mathrm{h}}\mu_{\mathrm{h}}), \tag{5.13}$$

where the subscripts e and h refer to electrons and holes, respectively.

5.2 Carrier Mobility for a Nondegenerate Electron Gas

The expressions we derived in the previous section are valid when all the carriers have the same scattering time. We will now generalize these expressions to the case where the carriers are distributed in a band according to the Boltzmann distribution [to be defined later in (5.22)] and the scattering time depends on the carrier energy.

5.2.1 Relaxation Time Approximation

We define the **distribution function** $f_{\boldsymbol{k}}(\boldsymbol{r})$ of a carrier as the probability that a band state with energy $E_{\boldsymbol{k}}$ will be occupied by this carrier at a carrier temperature T. We assume that in the absence of an external field the carriers are at thermal equilibrium, so that $f_{\boldsymbol{k}}$ is equal to the **Fermi–Dirac distribution function**:

$$f_{\boldsymbol{k}}^0 = \frac{1}{\exp[(E_{\boldsymbol{k}} - \mu_{\mathrm{F}})/(k_{\mathrm{B}}T)] + 1}, \tag{5.14}$$

where μ_{F} is the chemical potential (also called Fermi energy when $T \approx 0$) and k_{B} the Boltzmann constant. The equation governing the variation of $f_{\boldsymbol{k}}$ in the presence of an external perturbation is known as the **Boltzmann equation**:

$$\frac{df_{\boldsymbol{k}}}{dt} = \left(\frac{\partial f_{\boldsymbol{k}}}{\partial t}\right)_{\mathrm{field}} + \left(\frac{\partial f_{\boldsymbol{k}}}{\partial t}\right)_{\mathrm{diff}} + \left(\frac{\partial f_{\boldsymbol{k}}}{\partial t}\right)_{\mathrm{scatt}}. \tag{5.15}$$

Equation (5.15) includes the effects on $f_{\boldsymbol{k}}$ due to the applied field, the diffusion of carriers, and the scattering of carriers by phonons, impurities, etc. For simplicity, we shall assume that the diffusion term is negligible and the applied field $\boldsymbol{F}$ is small enough that we can expand $f_{\boldsymbol{k}}$ about $f_{\boldsymbol{k}}^0$ as a function of $\boldsymbol{F}$:

$$f_{\boldsymbol{k}} = f_{\boldsymbol{k}}^0 + g_{\boldsymbol{k}}(\boldsymbol{F}). \tag{5.16}$$

With this approximation we can write $(\partial f_{\boldsymbol{k}}/\partial t)_{\mathrm{field}}$ as

$$\left(\frac{\partial f_{\boldsymbol{k}}}{\partial t}\right)_{\mathrm{field}} \approx \left(\frac{\partial f_{\boldsymbol{k}}^0}{\partial E_{\boldsymbol{k}}}\right)\left(\frac{dE_{\boldsymbol{k}}}{dt}\right) = \left(\frac{\partial f_{\boldsymbol{k}}^0}{\partial E_{\boldsymbol{k}}}\right) q\boldsymbol{v}_{\boldsymbol{k}} \cdot \boldsymbol{F}, \tag{5.17}$$

where $\boldsymbol{v}_{\boldsymbol{k}}$ is the velocity of carriers with wave vector $\boldsymbol{k}$. Within the **relaxation time approximation**, we assume that the net effect of the scattering processes is to cause $g_{\boldsymbol{k}}$ to relax with a time constant $\tau_{\boldsymbol{k}}$, so that

$$\left(\frac{\partial f_{\boldsymbol{k}}}{\partial t}\right)_{\mathrm{scatt}} \approx -\frac{g_{\boldsymbol{k}}}{\tau_{\boldsymbol{k}}}. \tag{5.18}$$

At steady state $df_{\boldsymbol{k}}/dt = 0$, and after substituting (5.17) and (5.18) into (5.15) we obtain

$$g_{\boldsymbol{k}} = \left(\frac{\partial f_{\boldsymbol{k}}^0}{\partial E_{\boldsymbol{k}}}\right) q\tau_{\boldsymbol{k}}\boldsymbol{v}_{\boldsymbol{k}} \cdot \boldsymbol{F}. \tag{5.19}$$

The corresponding generalized expression for the current density is now given by

$$\boldsymbol{j} = \int q f_{\boldsymbol{k}} \boldsymbol{v}_{\boldsymbol{k}} d\boldsymbol{k} = \int q g_{\boldsymbol{k}} \boldsymbol{v}_{\boldsymbol{k}} d\boldsymbol{k} \tag{5.20}$$

$$= q^2 \int \tau_{\boldsymbol{k}} \boldsymbol{v}_{\boldsymbol{k}} \left(\frac{\partial f_{\boldsymbol{k}}^0}{\partial E_{\boldsymbol{k}}} \right) (\boldsymbol{v}_{\boldsymbol{k}} \cdot \boldsymbol{F}) d\boldsymbol{k}, \tag{5.21}$$

since $\int q f_{\boldsymbol{k}}^0 \boldsymbol{v}_{\boldsymbol{k}} d\boldsymbol{k} = 0$. Using (5.21) the corresponding expressions for σ and μ can be easily obtained.

5.2.2 Nondegenerate Electron Gas in a Parabolic Band

As an example of how to apply (5.21) we will consider the simple case of a nondegenerate electron gas in a parabolic band with an isotropic effective mass m^* in a cubic crystal. In such crystal $\boldsymbol{j}$ and $\boldsymbol{v}_{\boldsymbol{k}}$ are parallel to $\boldsymbol{F}$, at least in the region where Ohm's law holds. The distribution function (5.14) for a nondegenerate electron gas can be approximated by the **Boltzmann distribution**:

$$f_{\boldsymbol{k}}^0 \propto \exp\left[- E_{\boldsymbol{k}}/(k_{\mathrm{B}} T) \right] \tag{5.22}$$

so that

$$\frac{\partial f_{\boldsymbol{k}}^0}{\partial E_{\boldsymbol{k}}} \propto - \frac{1}{k_{\mathrm{B}} T} \exp\left(\frac{-E_{\boldsymbol{k}}}{k_{\mathrm{B}} T} \right). \tag{5.23}$$

The integration over $\boldsymbol{k}$-space in (5.21) can be replaced by an integration over the energy $E_{\boldsymbol{k}}$ using the density of states (DOS) introduced in Sect. 4.3.1. For a parabolic band in three dimensions, the DOS $D(E)$ is given by (including spin degeneracy):

$$D(E) = \frac{1}{\pi^2} k^2 \frac{dk}{dE} = \frac{1}{2\pi^2} \left(\frac{2m^*}{\hbar^2} \right)^{3/2} E^{1/2}. \tag{5.24}$$

Substituting these results into (5.21) we can calculate j and hence σ:

$$\sigma = \left(\frac{q^2}{3\pi^2 m^* k_{\mathrm{B}} T} \right) \left(\frac{2m^*}{\hbar} \right)^{3/2} \int_0^\infty \tau(E) E^{3/2} \exp\left(- E/(k_{\mathrm{B}} T) \right) dE. \tag{5.25}$$

In analogy with (5.10) we can define an **average scattering time** $\langle \tau \rangle$ by

$$\sigma = \frac{1}{m^*} \int_0^\infty D(E) q^2 \langle \tau \rangle dE. \tag{5.26}$$

Comparing (5.25) and (5.26) we obtain

$$\langle \tau \rangle = \left(\frac{2}{3 k_{\mathrm{B}} T} \right) \frac{\int_0^\infty \tau(E) E^{3/2} \exp\left[- E/(k_{\mathrm{B}} T) \right] dE}{\int_0^\infty E^{1/2} \exp\left[- E/(k_{\mathrm{B}} T) \right] dE}. \tag{5.27}$$

Using $\langle \tau \rangle$ we can express the mobility for a nondegenerate electron gas as

$$\mu = q \langle \tau \rangle / m^*. \tag{5.28}$$

From the expression for $\langle\tau\rangle$ we notice that the mobility depends on the electron temperature T. In order to calculate this temperature dependence, it is necessary to know the dependence of the scattering mechanisms on electron energy.

5.2.3 Dependence of Scattering and Relaxation Times on Electron Energy

Carriers in a semiconductor are scattered by their interaction with the following excitations [Ref. 5.2, pp. 82–183]:

- phonons: both acoustic and optical,
- ionized impurities,
- neutral defects,
- surfaces and interfaces,
- other carriers (e. g., scattering between electrons and holes).

In order to calculate the relaxation time $\tau_{\boldsymbol{k}}$ to be used in (5.18), we have to first consider the effect of scattering on $f_{\boldsymbol{k}}$. Let us define $P(\boldsymbol{k},\boldsymbol{k}')$ as the probability per unit time that an electron with wave vector $\boldsymbol{k}$ will be scattered into another state $\boldsymbol{k}'$. Once this scattering rate is known the rate of change of $f_{\boldsymbol{k}}$ caused by scattering can be calculated with the equation

$$\left(\frac{\partial f_{\boldsymbol{k}}}{\partial t}\right)_{\text{scatt}} = \sum_{\boldsymbol{k}'\neq\boldsymbol{k}} \left[P(\boldsymbol{k}',\boldsymbol{k})f_{\boldsymbol{k}'}(1-f_{\boldsymbol{k}}) - P(\boldsymbol{k},\boldsymbol{k}')f_{\boldsymbol{k}}(1-f_{\boldsymbol{k}'})\right]. \qquad (5.29)$$

The first term inside the square brackets represents the rate at which an electron at $\boldsymbol{k}'$ will be scattered into the state at $\boldsymbol{k}$, while the second term is the rate for scattering out of the state $\boldsymbol{k}$ into $\boldsymbol{k}'$. The summation is over all processes which conserve both energy and wave vector. If we assume as before that the electron gas is nondegenerate, then $f_{\boldsymbol{k}}$ and $f_{\boldsymbol{k}'}$ are small and can be neglected compared with unity. Applying the **principle of detailed balance**, $P(\boldsymbol{k}',\boldsymbol{k})f^0_{\boldsymbol{k}'} = P(\boldsymbol{k},\boldsymbol{k}')f^0_{\boldsymbol{k}}$, see [5.3], (5.29) simplifies to

$$-\left(\frac{\partial f_{\boldsymbol{k}}}{\partial t}\right)_{\text{scatt}} = \sum_{\boldsymbol{k}'\neq\boldsymbol{k}} P(\boldsymbol{k},\boldsymbol{k}')\left[f_{\boldsymbol{k}} - (f_{\boldsymbol{k}'}f^0_{\boldsymbol{k}}/f^0_{\boldsymbol{k}'})\right]. \qquad (5.30)$$

In general, $(\partial f_{\boldsymbol{k}}/\partial t)_{\text{scatt}}$ *cannot* be expressed as $-(f_{\boldsymbol{k}} - f^0_{\boldsymbol{k}})/\tau_{\boldsymbol{k}}$ as assumed in the relaxation time approximation. Only with more assumptions do we obtain an expression of the form (Problem 5.2)

$$\left(\frac{\partial f_{\boldsymbol{k}}}{\partial t}\right)_{\text{scatt}} = -(f_{\boldsymbol{k}} - f^0_{\boldsymbol{k}})\left(\sum_{\boldsymbol{k}'\neq\boldsymbol{k}} P(\boldsymbol{k},\boldsymbol{k}')\right), \qquad (5.31)$$

from which a scattering time τ_{s} can be defined as

$$(1/\tau_{\text{s}}) = \sum_{\boldsymbol{k}'\neq\boldsymbol{k}} P(\boldsymbol{k},\boldsymbol{k}'), \qquad (5.32)$$

where the summation is over all the final states $\boldsymbol{k}'$ that satisfy both energy and momentum conservation. However, τ_s represents the residence time of the electron in state $\boldsymbol{k}$ before being scattered and is not the same as $\tau_{\boldsymbol{k}}$, which is equal to the time it takes a perturbed distribution to return to equilibrium. In order for thermal equilibrium to be achieved, carriers have to be scattered in and out of a state many times. When $P(\boldsymbol{k}, \boldsymbol{k}')$ is known, it is possible to calculate $\tau_{\boldsymbol{k}}$ numerically by following the time evolution of the distribution function of an electron gas using a Monte Carlo simulation technique [5.4, 5].

5.2.4 Momentum Relaxation Times

We shall now obtain analytical expressions for $\tau_{\boldsymbol{k}}$ by making some approximations. One approach is to equate $\tau_{\boldsymbol{k}}$ to a **momentum relaxation time** τ_m. We can argue that the most important effect of scattering on electron transport is the randomization of the electron velocity. The relevant quantity is a **momentum relaxation rate** defined by

$$\langle d\boldsymbol{k}/dt \rangle = \boldsymbol{k}(\tau_m)^{-1} = \sum_{\boldsymbol{k}' \neq \boldsymbol{k}} (\boldsymbol{k}' - \boldsymbol{k}) P(\boldsymbol{k}, \boldsymbol{k}'). \tag{5.33}$$

The scattering rate $P(\boldsymbol{k}, \boldsymbol{k}')$ can be calculated using Fermi's Golden Rule

$$P(\boldsymbol{k}, \boldsymbol{k}') = (2\pi/\hbar) |\langle \boldsymbol{k} \,|\, H_{\text{scatt}} \,|\, \boldsymbol{k}' \rangle|^2 \varrho_f, \tag{5.34}$$

where H_{scatt} is the Hamiltonian for the scattering process, and ϱ_f is the density of final states $\boldsymbol{k}'$ per unit volume of $\boldsymbol{k}$-space for scattering processes which conserve both energy and wave vector. Of the scattering processes listed earlier, the ones most effective in randomizing the electron momentum are those with impurities and with phonons. Scattering by the *static* potential of impurities is elastic. Scattering by *acoustic* phonons is nearly elastic (quasi-elastic) because of the small energy transfers involved; the average scattering angles are large. As we saw in Chap. 3, optical phonons in semiconductors have energies in the range of tens of meV, therefore scattering between electrons and optical phonons is inelastic.

We will now calculate the individual rates for these scattering processes. Afterwards these rates can be *added* together to calculate the total scattering rate whose inverse is a measure of the relaxation time.

a) Intraband Scattering by Acoustic Phonons

We will assume that an electron with initial energy E_k and wave vector $\boldsymbol{k}$ is scattered to another state with energy $E_{k'}$ and wave vector $\boldsymbol{k}'$ via the emission of an acoustic phonon with energy E_p and wave vector $\boldsymbol{q}$. Since the total energy and wave vector of the electron and phonon have to be conserved in the scattering in a periodic lattice, the initial and final electron energies and wave vectors are related by the equations (for the case of phonon emission)

$$E_{k'} - E_k = E_p \quad \text{and} \quad \boldsymbol{k}' - \boldsymbol{k} = \boldsymbol{q}. \tag{5.35}$$

For an acoustic phonon with a small $\boldsymbol{q}$, E_p is related to q by

$$E_p = \hbar v_s q, \tag{5.36}$$

where v_s is the phonon velocity. For simplicity we shall assume v_s to be isotropic.

We shall further assume that the electron is in a parabolic band with effective mass m^* and it is scattered by acoustic phonons within the same band (this is known as **intraband scattering**). Since the scattering process conserves energy and wave vector, the allowed values of $\boldsymbol{q}$ are obtained by combining (5.35) and (5.36) into

$$(\hbar^2/2m^*)(k^2 - |\boldsymbol{k} - \boldsymbol{q}|^2) = \hbar v_s q \tag{5.37}$$

and solving for $\boldsymbol{q}$. The final electronic states, after emission of an acoustic phonon, are shown schematically in Fig. 5.1a. From this picture it is clear that the allowed values of q lie between a minimum (q_{min}) and a maximum (q_{max}). For $k > mv_s/\hbar$, q_{min} is zero while q_{max} is reached when $\boldsymbol{k}'$ is diagonally opposite to $\boldsymbol{k}$, i. e., when the electron is scattered by 180° (backscattering). From (5.37) q_{max} can easily be calculated to be

$$q_{max} = 2k - (2mv_s/\hbar). \tag{5.38}$$

The energy lost by the electron in emitting this phonon is

$$E_k - E_{k'} = \hbar v_s q_{max} = 2\hbar v_s k - 2mv_s^2. \tag{5.39}$$

To estimate the order of magnitude of these quantities, we will assume the following values of the parameters involved: $m^* = 0.1m_0$ (m_0 is the free electron mass), $v_s = 10^6$ cm/s, and $E_k = 25$ meV (roughly corresponding to room temperature times k_B). For this electron $k = 2.6 \times 10^6\ \text{cm}^{-1}$, $q_{max} = 5 \times 10^6\ \text{cm}^{-1} \approx 2k$, $k' = 2.4 \times 10^6\ \text{cm}^{-1}$ ($\boldsymbol{k}' \approx -\boldsymbol{k}$) and $E_k - E_{k'} = 3.3$ meV. In emitting an acoustic phonon with wave vector $\boldsymbol{q}_{max}$, the electron completely

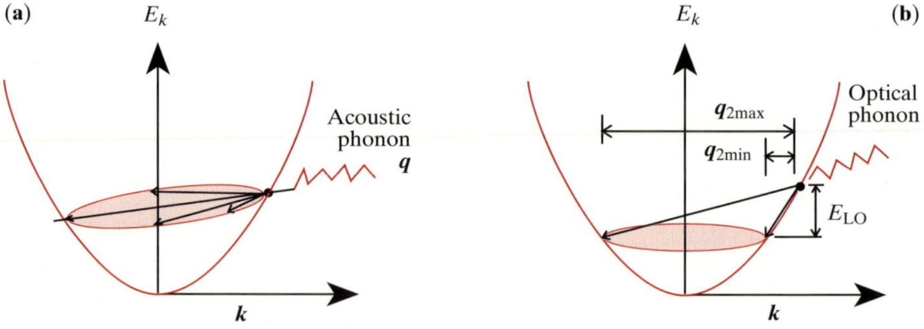

Fig. 5.1. Schematic diagrams for the scattering of an electron in a parabolic band by emission of (**a**) an acoustic phonon and (**b**) a longitudinal optical (LO) phonon showing the final electronic states and also the range of phonon wave vectors allowed by wave vector conservation

reverses its direction but its energy changes by only about 13 %. Thus scattering between electrons and acoustic phonons is nearly elastic (or quasi-elastic) and the main effect of these collisions is the relaxation of electron momentum.

Using (5.34) we shall now calculate the probability $P_{\mathrm{LA}}(\boldsymbol{k},\boldsymbol{q})$ that an electron in a parabolic and nondegenerate band will emit a LA phonon of wave vector $\boldsymbol{q}$. First we shall consider only the *deformation potential mechanism* and use the electron LA–phonon interaction Hamiltonian $H_{\mathrm{e-LA}}$ defined in (3.21 and 22). The scattering matrix element in (5.34) will now be written as $|\langle \boldsymbol{k}, N_{\boldsymbol{q}} \,|\, H_{\mathrm{e-LA}} \,|\, \boldsymbol{k}', N'_{\boldsymbol{q}}\rangle|^2$, where $|\,\boldsymbol{k}, N_{\boldsymbol{q}}\rangle$ represents the initial state with the electron in state $\boldsymbol{k}$ and the occupation number (for a definition see Sect. 3.3.1, p. 126) of LA phonons with wave vector $\boldsymbol{q}$ equal to $N_{\boldsymbol{q}}$. Similarly, $|\,\boldsymbol{k}', N'_{\boldsymbol{q}}\rangle$ is the final state, where the phonon occupation number is changed to $N'_{\boldsymbol{q}}$ and the electron is scattered to state $\boldsymbol{k}'$. We will be interested in one-phonon scattering only, i. e., $N'_{\boldsymbol{q}}$ differs from $N_{\boldsymbol{q}}$ by ± 1. Since our goal here is to calculate the temperature dependence of the electron mobility with (5.27), we are concerned only with the dependence of P_{LA} on electron energy and temperature. As discussed already in Sect. 3.3.1, we can express $|\langle \boldsymbol{k}, N_{\boldsymbol{q}} \,|\, H_{\mathrm{e-LA}} \,|\, \boldsymbol{k}', N'_{\boldsymbol{q}}\rangle|^2$ as

$$|\langle \boldsymbol{k}, N_{\boldsymbol{q}} \,|\, H_{\mathrm{e-LA}} \,|\, \boldsymbol{k}', N'_{\boldsymbol{q}}\rangle|^2 \propto q\left(N_{\boldsymbol{q}} + \tfrac{1}{2} \pm \tfrac{1}{2}\right), \qquad (5.40)$$

where the $+(-)$ sign in (5.40) corresponds to emission (absorption) of a phonon by the electron. As also shown in Sect. 3.3.1, $N_{\boldsymbol{q}}$ at room temperature can be approximated by $k_{\mathrm{B}}T/(\hbar v_{\mathrm{s}} q) \gg 1$, so that $|\langle \boldsymbol{k}, N_{\boldsymbol{q}} \,|\, H_{\mathrm{e-LA}} \,|\, \boldsymbol{k}', N'_{\boldsymbol{q}}\rangle|^2$ is proportional to $N_{\boldsymbol{q}} \gg 1$ for both phonon emission and absorption. As a result, we can deduce the following dependence of P_{LA} on T and E_k

$$P_{\mathrm{LA}}(\boldsymbol{k}) = \sum_{q} P_{\mathrm{LA}}(\boldsymbol{k},\boldsymbol{q}) \propto \int q N_q \delta[E_k - E_{k'} - \hbar v_{\mathrm{s}} q] d\boldsymbol{q}. \qquad (5.41)$$

Using polar coordinates, (5.41) can be expressed as

$$P_{\mathrm{LA}}(\boldsymbol{k}) \propto \int_0^{q_{\max}} q^3 (T/q) dq \int_0^{\pi} \delta\left\{\left[\hbar^2 q/(2m)\right](k\cos\Theta + q)\right\} d(\cos\Theta) \qquad (5.42a)$$

$$\approx T q_{\max}^2 / k, \qquad (5.42b)$$

where Θ is the angle between $\boldsymbol{k}$ and $\boldsymbol{q}$. Since $q_{\max}$ is approximately equal to $2k$, P_{LA} is roughly proportional to kT. Thus the P_{LA} produced by the deformation potential mechanism depends on the electron energy $E_{\boldsymbol{k}}$ and temperature as (see Problem 5.3)

$$P_{\mathrm{LA}} \propto T (E_k)^{1/2}. \qquad (5.43)$$

In noncentrosymmetric crystals, carriers can be scattered by both LA and TA phonons via the piezoelectric interaction (Sect. 3.3.3). Again we are interested mainly in the energy dependence of the corresponding scattering matrix element: $|\langle \boldsymbol{k}, N_{\boldsymbol{q}} \,|\, H_{\mathrm{pe}} \,|\, \boldsymbol{k}', N'_{\boldsymbol{q}}\rangle|^2$. From (3.30) and (3.22) we obtain

$$|\langle \boldsymbol{k}, N_{\boldsymbol{q}} \,|\, H_{\mathrm{pe}} \,|\, \boldsymbol{k}', N'_{\boldsymbol{q}}\rangle|^2 \propto (1/q)\left(N_{\boldsymbol{q}} + \tfrac{1}{2} \pm \tfrac{1}{2}\right). \qquad (5.44)$$

Unlike $|\langle \boldsymbol{k}, N_{\boldsymbol{q}} \,|\, H_{\mathrm{e-LA}} \,|\, \boldsymbol{k}', N'_{\boldsymbol{q}}\rangle|^2$, the constant of proportionality in (5.44) depends, in general, on the direction of $\boldsymbol{q}$, so we cannot simply substitute (5.44) into (5.41) to calculate the electron–phonon scattering rate due to the piezoelectric interaction. However, by comparing the dependence on q in (5.44 and 40) it is clear that the piezoelectric interaction is more important for small-q phonons. Since phonons with small q are less effective in relaxing carrier momentum, we may argue that the piezoelectric interaction is less important than the deformation potential interaction in momentum relaxation. This is only true if the electromechanical constants are small, such as for semiconductors with low ionicity. In more ionic crystals, such as II–VI compound semiconductors, the piezoelectric interaction tends to dominate over the deformation potential interaction [5.6]. If we can assume the constant of proportionality in (5.44) to be independent of the direction of $\boldsymbol{q}$, then it is straightforward to show (Problem 5.4) that the piezoelectric electron–acoustic-phonon scattering rate P_{pe} is proportional to

$$P_{\mathrm{pe}} \propto T(E_k)^{-1/2}. \tag{5.45}$$

This expression is obviously not valid for very low energy electrons. When E_k is small, the wave vector q of the acoustic phonons involved in the scattering will be small also. Usually, in the case under consideration, corresponding to (5.45), free carriers are also present in the semiconductor, and therefore the macroscopic piezoelectric field associated with these long-wavelength phonons will be screened by the free carriers. For a nondengenerate electron gas this screening effect can be included by introducing a **screening wave vector** q_0, which is defined as the reciprocal of the **Debye screening length** λ_{D}, see [Refs. 5.1, p. 151; 5.2, p. 179; 6.10, p. 497]

$$q_0^2 = \frac{1}{\lambda_{\mathrm{D}}^2} = \frac{4\pi N e^2}{4\pi\varepsilon_0\varepsilon_{\mathrm{s}} k_{\mathrm{B}} T}, \tag{5.46}$$

where ε_{s} is the static dielectric constant. The result is that the matrix element $|\langle \boldsymbol{k}, N_{\boldsymbol{q}} \,|\, H_{\mathrm{pe}} \,|\, \boldsymbol{k}', N'_{\boldsymbol{q}}\rangle|^2$ in (5.44) should be replaced by

$$|\langle \boldsymbol{k}, N_{\boldsymbol{q}} \,|\, H_{\mathrm{pe}} \,|\, \boldsymbol{k}', N'_{\boldsymbol{q}}\rangle|^2 \propto \left(\frac{q^3}{(q^2+q_0^2)^2}\right)\left(N_{\boldsymbol{q}} + \tfrac{1}{2} \pm \tfrac{1}{2}\right). \tag{5.47}$$

Notice that (5.44) is recovered if $q_0 = 0$. When the screening effect is included in (5.47), i. e., $q_0 \neq 0$, $|\langle \boldsymbol{k}, N_{\boldsymbol{q}} \,|\, H_{\mathrm{pe}} \,|\, \boldsymbol{k}', N'_{\boldsymbol{q}}\rangle|^2$ approaches zero as $q \to 0$. Similarly P_{pe} goes to zero, rather than diverging like in (5.45), as the electron energy decreases to zero. Figure 5.2 shows qualitatively the dependence of P_{pe} on electron energy.

From (5.41) we see that carriers are more likely to be scattered by LA phonons with large q via the deformation potential interaction. Since these phonons are more effective in randomizing the carrier momentum, acoustic phonon scattering (via the deformation potential interaction) is the dominant mechanism for momentum relaxation of carriers at room or lower temperatures in most semiconductors except for the very ionic ones. The acoustic

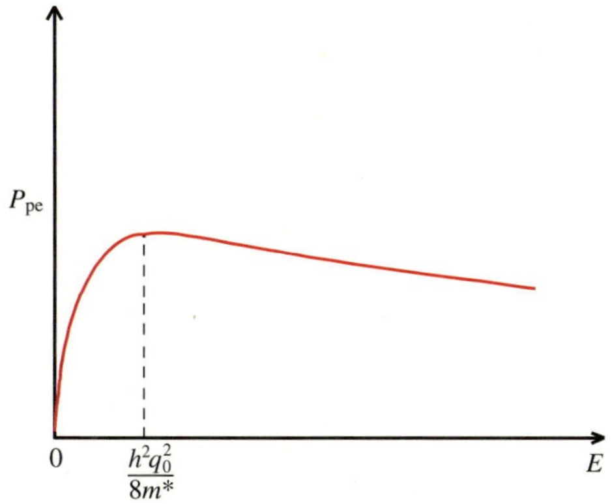

Fig. 5.2. Sketch of the dependence on the energy of an electron of the rate of scattering (P_{pe}) by acoustic phonons via the piezoelectric electron–phonon interaction [Ref. 5.2, Fig. 3.20]

phonon scattering time for conduction electrons in GaAs (via deformation potential interaction only) has been calculated by *Conwell* and *Vassel* [5.8]. Their results are shown in Fig. 5.3. Notice that this scattering time is of the order of several picoseconds (one picosecond is equal to 10^{-12} s and abbreviated as ps) in GaAs (Problem 5.3), and notice also how it decreases with increasing electron energy as predicted by (5.43).

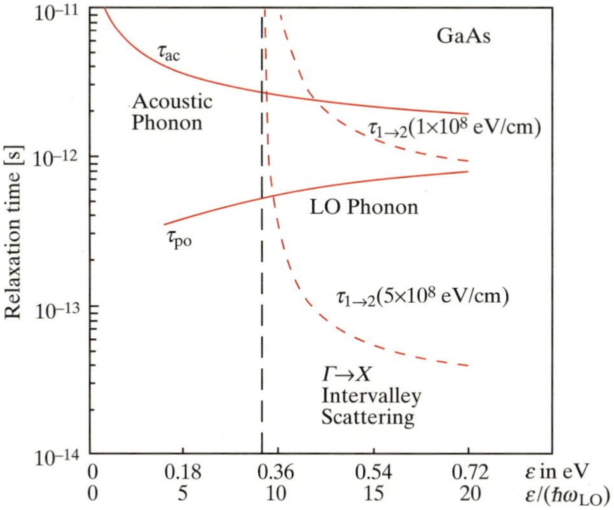

Fig. 5.3. Momentum scattering rates of a conduction electron in GaAs as a function of electron energy. Scattering by: small wave vector LA phonons (τ_{ac}) via the deformation potential interaction; small wave vector optical phonons (τ_{po}) via the Fröhlich interaction and via zone-edge phonons from Γ to the X valleys ($\tau_{1\to2}$) calculated by *Conwell* and *Vassel* [5.8]. Notice that the deformation potential for the Γ to X intervalley scattering has been assumed to be either 1×10^8 or 5×10^8 eV/cm. These values are smaller than the now accepted value of 10^9 eV/cm [5.9]

b) Intraband Scattering by Polar Optical Phonons

Optical phonons in semiconductors typically have energies of the order of tens of meV (Chap. 3). Hence at low temperatures ($T < 100$ K) most electrons do not have sufficient energy to emit optical phonons. In addition, the thermal occupation number N_q of optical phonons is very low, and consequently the probability of an electron absorbing an optical phonon is small also. Thus, optical phonon scattering processes are negligible at low temperatures. On the other hand, at room temperature, where there are sufficient high-energy electrons to emit optical phonons, they tend to dominate over acoustic phonon scattering. This is particularly true in polar semiconductors, where the Fröhlich electron–phonon interaction (Sect. 3.3.5 and Fig. 5.3) can be very strong. The distribution of final electronic states after optical phonon scattering is shown schematically in Fig. 5.1b. While scattering by acoustic phonons relaxes mainly the electron momentum, scattering by optical phonons contributes to both momentum and energy relaxation.

The LO phonon scattering probability (P_{LO}) corresponding to $P_{\mathrm{LA}}(\boldsymbol{k})$ in (5.42) can be shown to be [Ref. 5.2, p. 115, 5.10]

$$P_{\mathrm{LO}} \propto \int_0^\infty \left(\frac{q^2}{(q^2+q_0^2)^2}\right) q^2 dq \int_0^\pi d(\cos\Theta)\{N_{\mathrm{LO}}\delta[E_{k'}-(E_k+E_{\mathrm{LO}})] + (N_{\mathrm{LO}}+1)\delta[E_{k'}-(E_k-E_{\mathrm{LO}})]\}, \quad (5.48)$$

where N_{LO} and E_{LO} are, respectively, the LO phonon occupation number and energy. For simplicity we have assumed that the LO phonon is dispersionless, which is usually a good approximation since only phonons with $q \lesssim q_0$ contribute significantly to (5.48). The term inside the first set of parentheses comes from the Fröhlich matrix element, including screening. The terms proportional to N_{LO} and $N_{\mathrm{LO}}+1$ are identified with phonon absorption and emission, respectively. As a result of wavevector conservation in (5.35), (5.48) can be expressed as

$$P_{\mathrm{LO}} \propto \int_0^\infty \left(\frac{q^4}{(q^2+q_0^2)^2}\right) dq \int_0^\pi \left\{ N_{\mathrm{LO}}\delta\left[\left(\frac{\hbar^2 q}{2m}\right)(k\cos\Theta + q) - E_{\mathrm{LO}}\right] + (N_{\mathrm{LO}}+1)\delta\left[\left(\frac{\hbar^2 q}{2m}\right)(-k\cos\Theta+q)+E_{\mathrm{LO}}\right]\right\} d(\cos\Theta). \quad (5.49)$$

Integrating over Θ results in the following expression for P_{LO}:

$$P_{\mathrm{LO}} \propto N_{\mathrm{LO}} \int_{q_{1\,\mathrm{min}}}^{q_{1\,\mathrm{max}}} \left(\frac{q^3}{(q^2+q_0^2)^2}\right) dq + (N_{\mathrm{LO}}+1) \int_{q_{2\,\mathrm{min}}}^{q_{2\,\mathrm{max}}} \left(\frac{q^3}{(q^2+q_0^2)^2}\right) dq, \quad (5.50)$$

where $q_{i\,\mathrm{max}}$ and $q_{i\,\mathrm{min}}$ are, respectively, the maximum and minumum values of the LO phonon wave vector for phonon absorption ($i = 1$) and phonon emission ($i = 2$) (Problem 5.5). From Fig. 5.1b one can easily identify the electron final states corresponding to the minumum and maximum values of q. Since $q_{i\,\min}$ is nonzero for optical phonons, the screening wave vector in (5.50) is not as important as for piezoelectric acoustic phonons, except in the case of highly doped semiconductors. If we neglect q_0 in (5.50), P_{LO} decreases as $1/q$ and scattering by small-q LO phonons is more likely than by large-q LO phonons. In contrast to the case of acoustic phonon scattering, scattering between electrons and LO phonons tends to relax the electron energy rather than its momentum.

The momentum relaxation time of an electron due to LO phonon scattering can be deduced from (5.50) using (5.33). The result is [Ref. 5.2, p. 118]

$$\begin{aligned}\left(\frac{1}{\tau_{\mathrm{m}}}\right) \propto\; & N_{\mathrm{LO}}\left(\frac{E_k+E_{\mathrm{LO}}}{E_k}\right)^{1/2}+(N_{\mathrm{LO}}+1)\left(\frac{E_k-E_{\mathrm{LO}}}{E_k}\right)^{1/2}\\ &+\left(\frac{E_{\mathrm{LO}}}{E_k}\right)\left[-N_{\mathrm{LO}}\sinh^{-1}\left(\frac{E_k}{E_{\mathrm{LO}}}\right)^{1/2}\right.\\ &\left.+(N_{\mathrm{LO}}+1)\sinh^{-1}\left(\frac{E_k-E_{\mathrm{LO}}}{E_{\mathrm{LO}}}\right)^{1/2}\right]. \qquad (5.51)\end{aligned}$$

A plot of the relaxation time for electrons in GaAs due to LO phonon scattering (via the Fröhlich interaction) is shown in Fig. 5.3 under the label τ_{po}. Typically the relaxation time due to scattering by LO phonons is less than 1 ps.

c) Intervalley Scattering

The role of intervalley scattering in electron relaxation is different in direct and indirect bandgap semiconductors. In direct bandgap semiconductors, such as GaAs and InP, intervalley scattering is important only for electrons with sufficient energy to scatter into the higher conduction band valleys. Since these valleys are several tenths of an eV above the conduction band minimum at zone center, in these semiconductors *intervalley* scattering is important only for their hot electron transport properties. This will be discussed in more detail in Sect. 5.4. The situation is quite different in indirect bandgap semiconductors, such as Si and Ge. In these materials the electrons are located in conduction band minima which are not at zone center and are degenerate. In addition to *intraband* scattering by phonons, electrons can be scattered from one degenerate valley to another via intervalley scattering. In both materials the latter scattering processes turn out to be more important than the intraband processes in relaxing the momentum and energy of conduction electrons. In this section we shall consider only intervalley scattering of electrons in indirect bandgap semiconductors.

As an example, we discuss the case of Si, where the conduction band minima occur along the six equivalent [100] directions (Δ_1) at about 0.83 of the distance from zone center to the zone edge. Electrons in one of the minima (say in the [100] direction) can be scattered either into the valley along the [$\bar{1}$00] direction or into one of the four equivalent [0,±1,0] and [0,0,±1] valleys (Fig. 5.4a). The former process is known as a **g-process** and the phonon involved is known as a **g-phonon** while the latter processes are called

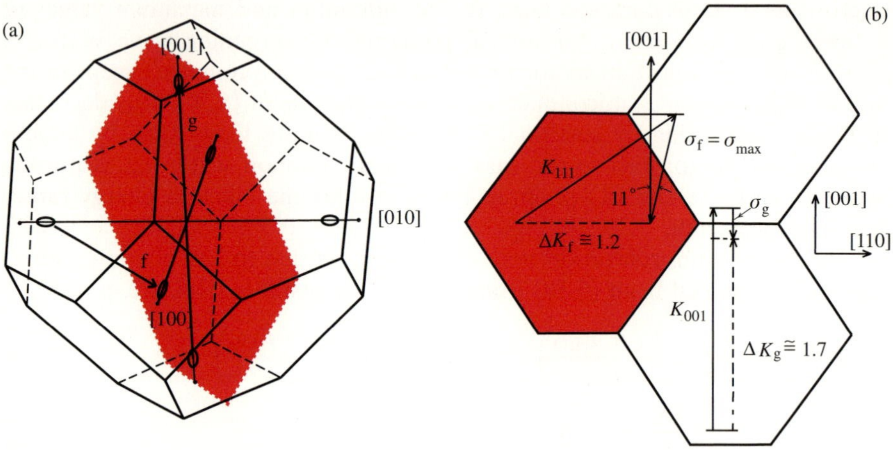

Fig. 5.4a,b. Schematic diagram of the intervalley scattering processes for electrons in the conduction band minimum of Si showing the phonon wave vectors involved in the g- and f-processes [Refs. 5.2 (Fig. 3.16), 5.11]. ΔK is given in units of $(2\pi/a_0)$.

f-processes. For both processes the electron valley after scattering can lie in the same Brillouin zone or in an adjacent one (Fig. 5.4b). The former process is known as a **normal process**, the latter as an **umklapp process** [Ref. 5.7, p. 146]. As shown in Fig. 5.4b, the wave vector $\Delta \boldsymbol{K}_g$ (note that, in order to show both the g- and f-phonons in this figure, the g-phonon is now chosen along the [001] direction) of the phonon mediating the normal g-process is about 1.7 times the Brillouin zone length along the X axis. On the other hand, the phonon wave vector ($\boldsymbol{\sigma}_g$) of the umklapp g-process is only ≈ 0.34 of the Brillouin zone edge (measured from the zone center) in the $[00\bar{1}]$ direction. The reciprocal lattice vector involved in the g umklapp process, denoted by $\boldsymbol{K}_{001}$, is also along the X direction. The f-phonon $\Delta \boldsymbol{K}_f$ is approximately equal to 1.2 times $(2\pi/a_0)(1, \pm 1, 0)$, a_0 being the size of the unit cube in Si. The wave vector lies outside the first Brillouin zone. Hence both g- and f-processes require phonons with wave vectors outside the first Brillouin zone. Thus in the reduced zone scheme only umklapp processes are allowed. Combined with a reciprocal lattice vector $\boldsymbol{K}_{111} = (2\pi/a_0)(1, 1, 1)$, the resultant umklapp wave vector $\boldsymbol{\sigma}_f$ is about 11° off the [001] direction (Σ_1 symmetry) with length almost exactly equal to that of the zone-boundary value along that direction (Fig. 5.4b). The symmetries and energies of intervalley phonon modes allowed by group theory in Si are (see Fig. 3.1 and [5.2], p. 110)

g-process: Δ_2' (LO), 63 meV

f-processes: Σ_1 (LA, TO; 45 and 57 meV; an average of 54 eV was used in [5.2], p. 110).

The corresponding selection rules for intervalley scattering in III–V compounds have been derived by *Birman* et al. [5.12].

In order to fit the temperature dependence of the mobility of Si it was found necessary to include a contribution to intervalley scattering from a phonon of 16 meV energy [5.11]. According to the phonon dispersion curves of Si, this 16 meV phonon (Fig. 3.1; 16 meV $\Leftrightarrow$ 130 cm^{-1} $\Leftrightarrow$ 3.9 THz) can be attributed to an LA mode with wave vector about 0.3 times the zone boundary, corresponding to the g umklap process of Fig. 5.4. This process is strictly speaking forbidden since the corresponding matrix element has $\Delta_1 \otimes \Delta_1 \otimes \Delta_{2'} = \Delta_{2'}$ symmetry and therefore vanishes. The presence of such processes, required by the first of the resistivity vs T curve has been explained by expanding the electron–phonon interaction as a function of phonon wave vector beyond the lowest (zeroth) order term [5.13]. Except for the higher order electron–phonon interactions, the calculation of the intervalley scattering rates is similar, in principle, to that for acoustic phonons in direct bandgap semiconductors. In practice the calculation is complicated by the anisotropy in the electron mass in Si. Assuming for the sake of simplicity an isotropic mass, we find that the intervalley scattering rate due to the zeroth-order electron–phonon interaction is given by

$$(1/\tau_{\mathrm{iv}}) \propto N_q(E_k + E_{\mathrm{p}})^{1/2} + (N_q + 1)(E_k - E_{\mathrm{p}})^{1/2}U(E_k - E_{\mathrm{p}}), \tag{5.52a}$$

where E_{p} is the phonon energy and $U(x)$ is the step function:

$$U(x) = \begin{cases} 0 & \text{for } x < 0, \\ 1 & \text{for } x \geq 0. \end{cases} \tag{5.52b}$$

Readers interested in the contribution to the scattering rate from the first-order electron–phonon interaction should consult either [5.13] or [Ref. 5.2, p. 110].

d) Scattering by Impurities

Typically a semiconductor contains defects such as impurities and dislocations. Carriers are scattered elastically by these defects, the details of the scattering mechanism depending on the defect involved. Here we will concern ourselves with scattering by the most common kind of defects, namely, charged impurities. Free carriers are produced in semiconductors by ionization of shallow impurities (except in the intrinsic case where they are produced by thermal ionization across the gap). As a result, free carriers will, in principle, always be scattered by the ionized impurities they leave behind. A way to avoid this will be presented in the next section.

Our approach to calculating impurity scattering rates will be different from the approach for phonon scattering. Phonons are quantized lattice waves with well-defined wave vectors and therefore they scatter electrons from one Bloch state to another. Impurity potentials are localized in space. Hence they do not scatter electrons into well-defined Bloch states. This problem can be treated quantum mechanically using scattering theory and what is referred to as the **Brooks–Herring approach** [Refs. 5.2, p. 143; 5.14]. In this approach the impurity potential is approximated by a screened Coulomb potential. The screening can be either Debye [with the screening length of (5.46)] or *Thomas–Fermi* [Ref. 5.7, p. 266]. The scattering cross section is calculated within the Born approximation. The quantum mechanical results obtained by *Brooks* [5.14] are not too different from the classical results obtained earlier by *Conwell* and *Weisskopf* [5.15] so we shall first consider the **Conwell–Weisskopf approach**.

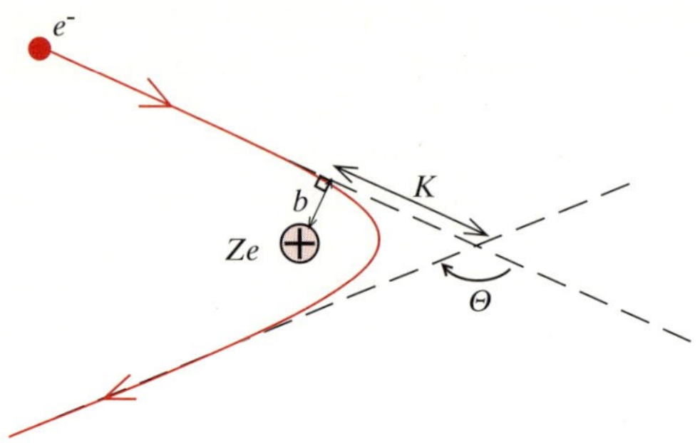

Fig. 5.5. Coulomb scattering of an electron by a positively charged ion. The impact parameter b and the length K are discussed in the text. [Ref. 5.2, Fig. 4.1]

Conwell–Weisskopf Approach. In this approach an electron is assumed to be scattered classically via Coulomb interaction by an impurity ion with charge $+Ze$. The corresponding scattering cross section is calculated in exactly the same way as for the Rutherford scattering of α particles [5.16]. The scattering geometry of this problem is shown schematically in Fig. 5.5. The *scattering cross section* σ as a function of the scattering angle Θ is given by

$$\sigma(\Theta)d\Omega = 2\pi b\, d\,|\,b\,| \tag{5.53}$$

where $d\Omega$ is an element of solid angle, b is known as the **impact parameter**. and $d\,|\,b\,|$ is the change in $|\,b\,|$ required to cover the solid angle $d\Omega$. The impact parameter and the solid angle $d\Omega$ are related to the scattering angle by

$$b = K\cot(\Theta/2); \quad d\Omega = 2\pi\sin\Theta d\Theta \tag{5.54}$$

where K is a characteristic distance defined by

$$K = \frac{Ze^2}{mv_k^2} \tag{5.55}$$

and v_k is the velocity of an electron with energy $E_k = mv_k^2/2$. If $d\Omega$ and $d\,|\,b\,|$ are expressed in terms of $d\Theta$, (5.53) can be simplified to

$$\sigma(\Theta) = \left(\frac{K}{2\sin^2(\Theta/2)}\right)^2 . \tag{5.56}$$

The well-known dependence of σ on the electron velocity to the power of -4 in Rutherford scattering is contained in the term K^2. The scattering rate R (per unit time) of particles traveling with velocity v by N scattering centers per unit volume, each with *scattering cross section* σ, is given by

$$R = N\sigma v. \tag{5.57}$$

Since scattering by impurities relaxes the momenta of carriers, but not their energy, we can define a momentum relaxation time τ_i due to impurity scattering by

$$1/\tau_\mathrm{i} = N_\mathrm{i} v_k \int \sigma(\Theta)(1 - \cos\Theta) 2\pi \sin\Theta\, d\Theta, \tag{5.58}$$

where N_i is the concentration of ionized impurities. Within the integrand, the term $(1 - \cos\Theta)$ is the fractional change in the electron momentum due to scattering event and the term $2\pi\sin\Theta\, d\Theta$ represents integration of the solid angle Ω. In principle, Θ has to be integrated from 0 to π. The divergence of (5.56) at $\Theta = 0$ makes the integral in (5.58) diverge. This problem can be avoided by arguing that b cannot be larger than a maximum value $b_{\max}$ equal to half of the average separation between the ionized impurities:

$$b_{\max} = \tfrac{1}{2} N_i^{-1/3}. \tag{5.59}$$

As a result the minimum value of Θ allowable in (5.58) is

$$\Theta_m = 2\cot^{-1}(b_{\max}/K). \tag{5.60}$$

Integrating (5.58) with the condition (5.60) we obtain the following expression for $1/\tau_{\mathrm{i}}$:

$$1/\tau_{\mathrm{i}} = -4\pi N_{\mathrm{i}} v_k K^2 \ln[\sin(\Theta_m/2)] \tag{5.61}$$

or, in terms of the electron energy E_k,

$$\frac{1}{\tau_{\mathrm{i}}} = 2\pi N_{\mathrm{i}} \left(\frac{2E_k}{m}\right)^{1/2} \left(\frac{Ze^2}{2E_k}\right)^2 \ln\left[1 + \left(\frac{E_k}{N_{\mathrm{i}}^{1/3} Ze^2}\right)^2\right]. \tag{5.62}$$

The scattering rate of electrons by ionized impurities is independent of temperature and depends on the electron energy approximately like $E_k^{-3/2}$ due to the dependence of the scattering cross section on the particle velocity in (5.56). As a result, ionized impurity scattering tends to become dominant at low temperature where the electron energies are small and, moreover, phonon scattering freezes out.

Brooks–Herring Approach. As mentioned earlier, in the Brooks–Herring approach the scattering rate of electrons by ionized impurities is calculated quantum mechanically. If the effect of screening of the impurity potentials by free carriers is taken care of by introducing a screening wave vector q_0 [which is the reciprocal of the Debye screening length] defined in (5.46), then (5.56) must be replaced by

$$\sigma(\Theta) = \left[\frac{K}{2\sin^2\dfrac{\Theta}{2} + \left(\dfrac{q_0}{2k}\right)^2}\right]^2. \tag{5.63}$$

In this way the divergence in $\sigma(\Theta)$ at $\Theta = 0$ in (5.56) is automatically avoided without introducing, in a somewhat artificial way, the maximum impact parameter $b_{\max}$.

The electron mobility (assuming that the only scattering process is by ionized impurities) calculated by the Brooks–Herring approach (BH for short) is compared with that calculated by the Conwell-Weisskopf approach (curves labeled CW) in Fig. 5.6 as a function of carrier density for a hypothetical uncompensated semiconductor. Except for high carrier concentrations the two results are almost identical. However, for very high carrier concentrations even the Brooks–Herring approach breaks down. The question of screening at high carrier concentrations has been studied by many researchers and a detailed account can be found in [Ref. 5.2, pp. 145–152].

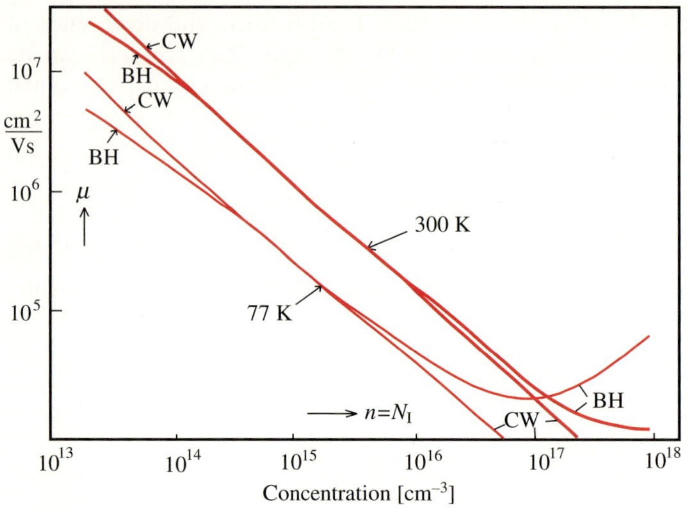

Fig. 5.6. Mobility of electrons calculated by considering only ionized impurity scattering as a function of impurity concentration. The curves labeled CW have been calculated classically by *Conwell* and *Weisskopf* [5.15]. The curves labeled BH have been calculated quantum mechanically using the Brooks–Herring approach [5.14]

5.2.5 Temperature Dependence of Mobilities

We are now in a position to discuss the temperature dependence of carrier mobilities in a nondegenerate semiconductor with parabolic bands. From the scattering rates of an electron in a band we first calculate the electron lifetime $\tau(E_k)$ by taking the reciprocal of the total scattering rate. Next we substitute $\tau(E_k)$ into (5.27) to obtain $\langle\tau\rangle$. Since the different scattering mechanisms have different dependences on electron energy and temperature, they result in different temperature dependences of the mobility. By comparing the measured temperature dependence of the mobility with theory one can determine the contributions from the different scattering mechanisms. To facilitate this comparison we note that if $\tau(E_k)$ can be expressed as a function of E_k and T as being proportional to E_k^n and T^m then

$$\int E^{3/2}\exp(-E/k_B T)\tau(E)dE \propto T^{m+n+5/2} \tag{5.64}$$

and

$$\langle\tau\rangle \propto T^{m+n}. \tag{5.65}$$

Using the energy and temperature dependence of the scattering rate obtained in the previous section we can conclude that

- $\mu \propto T^{-3/2}$ for acoustic phonon (deformation potential) scattering;
- $\mu \propto T^{-1/2}$ for acoustic phonon (piezoelectric) scattering;
- $\mu \propto T^{3/2}$ for ionized impurity scattering.

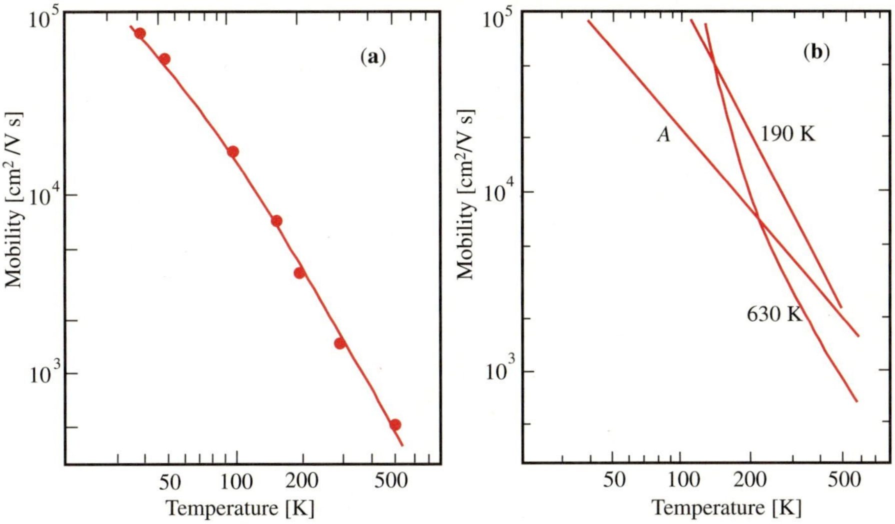

Fig. 5.7a,b. Temperature dependence of mobility in n-type Si. The *solid curve* in (**a**) represents the experimental results. The *curves* in (**b**) are the relative contributions to the mobility from scattering by different kinds of phonons. Curve *A* represents the contribution from *intravalley* acoustic phonon scattering while the other two curves represent contributions from *intervalley* scattering by phonons whose energies correspond to temperatures of 190 K (or 16 meV) and 630 K (54 meV). The *filled circles* in (**a**) display a fit to the experimental curve using the theoretical curves in (**b**) [5.13]

Figure 5.7 shows the temperature dependence of the mobility in intrinsic n-type Si. The experimental results [solid curve in (a)] can be explained by a combination of intravalley scattering by acoustic phonons and intervalley scattering by two phonons of energies 16 meV (129 cm^{-1}, TA) and 54 meV (436 cm^{-1}, LO). The electron–phonon interaction for the LO intervalley scattering is symmetry allowed for phonons along Δ while that for the TA phonon is forbidden. *Ferry* [5.13] attributed the nonvanishing value of the latter to contributions from phonons close to but not exactly along Δ.

Figure 5.8 shows the mobility in n-type Si for various donor concentrations. The shape of the experimental curves at high donor concentrations can be explained by the dominance of ionized impurity scattering at low temperatures, as sketched in the inset. The experimental temperature dependence of mobility in n-type GaAs is compared with theory in Fig. 5.9. Again the experimental results can be understood in terms of ionized impurity scattering at low temperatures and phonon scattering at higher temperatures. At room temperature, the scattering of electrons in GaAs is dominated by the polar LO phonon processes.

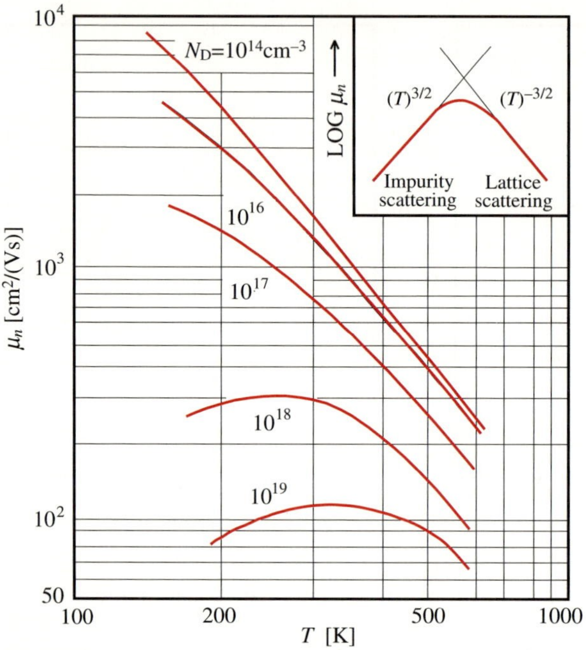

Fig. 5.8. Temperature dependence of mobilities in n-type Si for a series of samples with different electron concentrations. The inset sketches the temperature dependence due to lattice and impurity scattering [5.17]

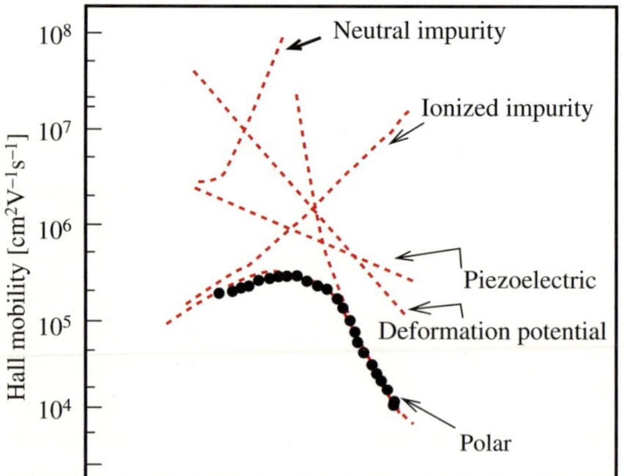

Fig. 5.9. Temperature dependence of mobility in n-type GaAs determined by Hall measurements (*points*) by *Stillman* et al. [5.18]. The dashed curves are the corresponding contributions from various scattering mechanisms calculated by *Fletcher* and *Butcher* [5.19] (from [5.19])

5.3 Modulation Doping

From the temperature dependence of mobilities in Si and GaAs we conclude that scattering by ionized impurities ultimately limits the carrier mobility at low temperatures. This limitation can be circumvented by using the method of **modulation doping** proposed by *Störmer* et al. [5.20].

The idea behind modulation doping is illustrated in Fig. 5.10. Two materials with almost identical lattice constants but different bandgaps are grown on top of each other to form a **heterojunction** (see Chap. 9). One example of a well-behaved heterojunction used commonly in the fabrication of semiconductor laser diodes is GaAs/$Al_xGa_{1-x}As$ (other examples will be discussed in Chap. 9). The lattice constants of these two semiconductors differ by less than 1 %. The bandgap of AlGaAs with less than 40 % of Al is direct and larger than that of GaAs. The difference between their bandgaps is divided in an approximately 60/40 split between the conduction and valence bands (for further discussions see Chap. 9). The results are very abrupt discontinuities, known as **band offsets**, in their energy bands at the interface, as shown schematically in Fig. 5.10. If the material with the larger bandgap (AlGaAs) is then doped with shallow donors, the Fermi level is shifted from the middle of the bandgap of AlGaAs to the donor level. In order to maintain a constant chemical potential throughout the two materials, electrons will flow from AlGaAs to GaAs. This causes the band edges to bend at the interface as shown in Fig. 5.10, a phenomenon known as **band bending**. We shall show in Sect. 8.3.3 that band bending also occurs near the surface of semiconductors as a result of the existence of surface states.

Due to band bending the electrons in GaAs are now confined by an approximately triangular potential near the interface and form a *two-dimensional (2D) electron gas*. These 2D electrons are physically separated from the ion-

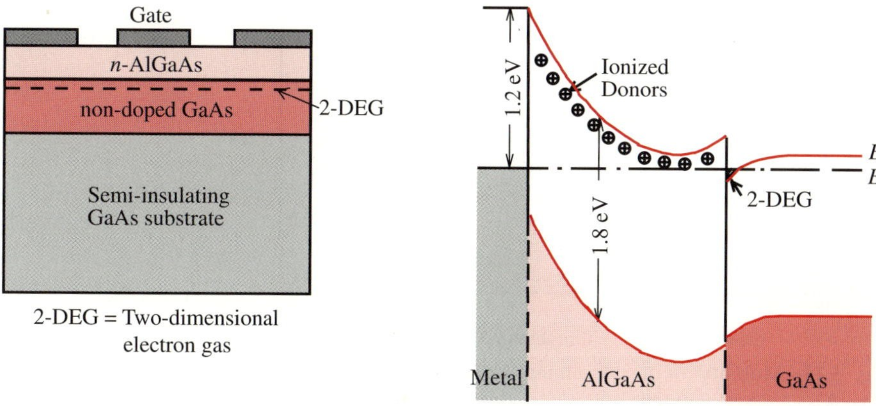

Fig. 5.10. The structure and band diagram of a modulation-doped heterojunction between GaAs and n-AlGaAs. E_c and E_F represent, respectively, the conduction band edge and the Fermi energy

ized impurities in AlGaAs, hence they are only weakly scattered by the charged impurities. This method constitutes the *modulation doping* technique mentioned above [5.20]. If scattering by interface defects can be avoided, the mobility of the 2D electron gas in a modulation-doped sample can approach the theoretical limit set by phonon scattering in the absence of impurity scattering. Using this method carrier mobilities exceeding 10^6 cm^2/(Vs) have been achieved in GaAs. Figure 5.11 shows the temperature dependence of the mobility of a 2D electron gas at a GaAs/$Al_{0.3}Ga_{0.7}As$ heterojunction. Notice that unlike the mobility depicted in Fig. 5.9, the mobility in Fig. 5.11 does not decrease when the temperature is lowered towards zero, as would be expected if scattering by ionized impurities were present. However, some residual scattering by the potential of the impurities located inside the $Al_{0.3}Ga_{0.7}As$ is still present; it is labeled "remote impurities" in Fig. 5.11. It approaches a small constant value at low temperatures so that modulation doping improves the carrier mobility when phonon scattering is frozen. Since electron scattering in these samples is dominated by phonons (except at very low temperatures), it is possible to determine the absolute volume deformation potentials of LA

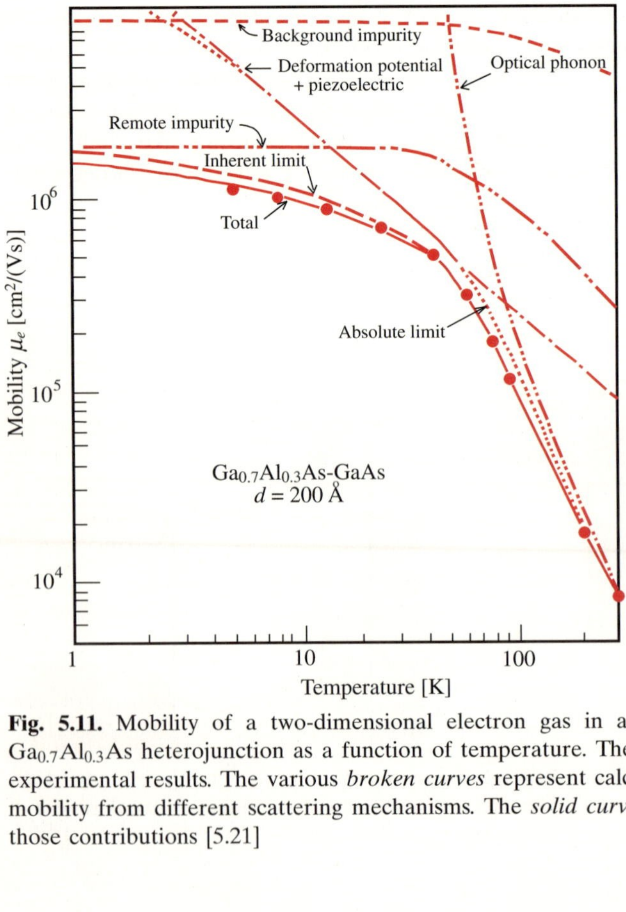

Fig. 5.11. Mobility of a two-dimensional electron gas in a modulation-doped GaAs/$Ga_{0.7}Al_{0.3}As$ heterojunction as a function of temperature. The closed circles indicate the experimental results. The various *broken curves* represent calculated contributions to the mobility from different scattering mechanisms. The *solid curve* represents the sum of all those contributions [5.21]

phonons quite accurately from the temperature dependence of the electron mobility [5.21].

Modulation doping is now utilized in the fabrication of field-effect transistors with very high mobility. These transistors are known either as **MODFETs** (which stands for *modulation-doped field-effect transistors*) or **HEMTs** (*high electron mobility transistors*) [Ref. 5.22, p. 698].

5.4 High-Field Transport and Hot Carrier Effects

The formalism developed in Sect. 5.1 for calculating the carrier drift velocity leads to Ohm's law and is valid only at low electric fields. In most semiconductors we find that Ohm's law breaks down at electric fields exceeding 10^4 V/cm. In this section we shall study the effect of high electric fields on carrier distributions and also other transport phenomena which can occur under high electric fields. As we pointed out in the introduction, these high-field effects can only be calculated numerically [5.4, 5.23] and therefore our discussions will necessarily be qualitative.

The main difficulty in these calculations results from the very fast rates at which carriers gain energy under the high electric field. When these rates are larger than those for energy loss to the lattice, the carriers are no longer at thermal equilibrium with the lattice. There are two possible scenarios for these nonequilibrium situations. In one case, carriers are in thermal equilbrium among themselves but not with the phonons. In this situation carriers can be said to be in **quasi thermal equilibrium**. Their distribution can still be characterized by a Fermi–Dirac distribution, albeit with a temperature different from the *sample temperature* (defined as that of the phonons which are in equilibrium with the thermal sink to which the sample is attached). Usually these carriers have higher temperatures than the lattice, hence they are known as **hot carriers**. In the second scenario, the carriers cannot be described by an equilibrium distribution and therefore do not have a well-defined temperature. In the literature one may also find that these nonthermal equilibrium carriers are loosely referred to as hot carriers. More precisely these carriers should be called **nonequilibrium carriers**. Hot-carrier effects are important in the operation of many semiconductors devices such as *laser diodes*, *Gunn oscillators*, and *short-channel field-effect transistors* [Ref. 5.22, pp. 698–720]. The hot carriers in these devices are generated electrically by a high field or by injection through a barrier. Hot carriers can also be produced optically by high intensity photon beams such as in *laser annealing* [5.24].

What conditions determine whether a carrier distribution is an equilibrium one or not? The answer depends on the magnitude of the various time scales which characterize the interaction among the carriers and their interaction with the lattice relative to the carrier lifetime. Let us define the time it takes a nonequilibrium carrier distribution to relax to equilibrium as the **thermal-**

ization time. Processes contributing to thermalization are carrier–carrier and carrier–phonon interactions. As shown in Fig. 5.3, carrier–phonon interaction times can range from 0.1 ps (for polar optical phonons and for phonons in intervalley scattering) to tens of picoseconds (for acoustic phonons). Carrier–carrier interaction times depend strongly on carrier density. This has been measured optically in GaAs. At high densities ($>10^{18}$ cm^{-3}) carriers thermalize in times as short as *femtoseconds* (equal to 10^{-15} s and abbreviated as fs) [5.25]. Thus the thermalization time is determined by carrier–carrier interaction at high carrier densities. At low densities, it is of the order of the shortest carrier–phonon interaction time. Often carriers have a finite lifetime because they can be trapped by defects. If both electrons and holes are present, the carrier lifetime is limited by **recombination** (Chap. 7). In samples with a very high density of defects (such as amorphous semiconductors) carrier lifetimes can be picoseconds or less. Since the carrier lifetime determines the amount of time carriers have to thermalize, a distribution is a nonequilibrium one when the carrier lifetime is shorter than the thermalization time. A transient nonequilibrium situation can also be created by perturbing a carrier distribution with a disturbance which lasts for less than the thermalization time.

The properties of hot carriers can be different from those of equilibrium carriers. One example of this difference is the dependence of the drift velocity on electric field. Figure 5.12 shows the drift velocity in Si and GaAs as a function of electric field. At fields below 10^3 V/cm, the carriers obey Ohm's law, namely, the drift velocity increases linearly with the electric field. At higher fields the carrier velocity increases sublinearly with field and saturates at a velocity of about 10^7 cm/s. This leveling off of the carrier drift velocity at high field is known as **velocity saturation**. n-Type GaAs shows a more complicated behavior in that its velocity has a maximum *above* the saturation velocity. This phenomenon is known as **velocity overshoot** (Fig. 5.12) and is found usually only in a few n-type semiconductors such as GaAs, InP, and InGaAs. For

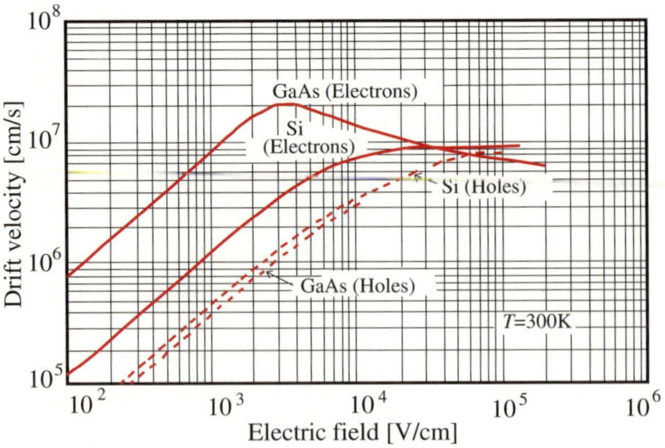

Fig. 5.12. Dependence of drift velocity on electric field for electrons and holes in Si and GaAs [5.17]. Notice the velocity overshoot for electrons in GaAs

electric fields between 3×10^3 V/cm and 2×10^5 V/cm, the velocity of electrons in GaAs *decreases* with increasing electric field. This phenomenon is known as **negative differential resistivity**. We shall consider these high field behaviors of carriers separately.

5.4.1 Velocity Saturation

The Boltzmann equation (5.15) is difficult to solve under high-field conditions, for which the carrier distribution can be a nonequilibrium one. One simplifying approach is to expand the carrier distribution as a function of carrier velocity in a Taylor series. This leads to a field-dependent mobility of the form

$$\mu(F) = \mu(0) + \beta F^2 + \dots . \tag{5.66}$$

Carriers are sometimes referred to as **warm carriers** when only the terms up to βF^2 are important [Ref. 5.26, p. 102]. As mentioned in Sect. 5.1, hot carrier properties are usually calculated numerically using a Monte Carlo simulation method; a discussion of these calculations is beyond the scope of this book [5.27]. Instead, we will present a highly simplified explanation of why the drift velocity in most semiconductors saturates at more or less the same value of 10^7 cm/s.

Within the quasi-classical approach that we adopted in Sect. 5.1, carriers in a semiconductor are regarded as free particles with effective mass m^*. Their average energy $\langle E \rangle$ can be defined as $\langle E \rangle = m^* v_\mathrm{d}^2/2$, where v_d is their drift velocity. The saturation in v_d under a high electric field therefore implies that the average energy of the carriers no longer increases with electric field at high field strengths. This can be understood if there is an energy loss mechanism that becomes dominant at large $\langle E \rangle$. Scattering with optical phonons is the most obvious candidate. A simple-minded picture of what happens is as follows. At low fields, carriers are scattered elastically by acoustic phonons and these processes lead to momentum relaxation of the carriers. The carrier distribution is essentially a "drifted" equilibrium distribution (Problem 5.2). At intermediate fields, $\langle E \rangle$ becomes large enough for some of the carriers in the high-energy tail of the distribution to start to scatter *inelastically* with optical phonons. This distorts the carrier distribution and causes $\langle E \rangle$ to increase sublinearly with the field. At still higher fields, more of the carriers are scattered inelastically, until energy relaxation processes dominate. When the rate at which carriers gain energy from the field is balanced exactly by the rate of energy loss via optical phonon emission, $\langle E \rangle$ (and also v_d) becomes independent of electric field.

The saturation velocity v_s can thus be deduced from the energy-loss rate equation

$$\frac{d\langle E \rangle}{dt} = eFv_\mathrm{s} - \frac{E_\mathrm{op}}{\tau_\mathrm{e}}, \tag{5.67}$$

where E_op is the optical phonon energy and τ_e is the energy relaxation time. A corresponding rate equation for momentum relaxation is

$$\frac{d(m^* v_s)}{dt} = eF - \frac{m^* v_s}{\tau_m}, \tag{5.68}$$

where τ_m is the momentum relaxation time. Assuming that scattering by optical phonons is the dominant process at high fields, both τ_e and τ_m are equal to the optical phonon scattering time τ_{op}. In the steady state, $d\langle E\rangle/dt = 0$ and $d(m^* v_s)/dt = 0$. Therefore the solution for v_s from (5.67) and (5.68) is simply

$$v_s = (E_{op}/m^*)^{1/2}. \tag{5.69}$$

Notice that τ_{op} is absent in the expression for v_s. For most tetrahedrally bonded semiconductors, the optical phonon energy is of the order of 40 meV and the carrier effective mass m^* is of the order of $0.1m_0$. Substituting these values into (5.69) we obtain $v_s \approx 2 \times 10^7$ cm/s. This explains both the constancy and order of magnitude of the experimental saturation velocities in many semiconductors.

5.4.2 Negative Differential Resistance

Electrical resistance is normally associated with dissipation of electrical energy in the form of heat in a conductor. Thus a negative electrical resistance suggests that electrical energy can be "created" in such a medium. However, it should be noted that a negative *differential* resistance (often abbreviated as NDR) is only a negative AC resistance. This means that NDR can be used in designing an AC amplifier only. When we study electronics we learn that an amplifier coupled with a properly designed positive feedback circuit can be made into an oscillator. Thus one important application of materials exhibiting NDR is in the construction of high-frequency (typically microwave frequencies) oscillators. The **Esaki tunnel diode**[1] is an example of a device exhibiting NDR [Ref. 5.22, pp. 641–643]. More recently **resonant tunneling diodes** (see Chap. 9 for further discussions) have also been shown to exhibit NDR [5.28]. The principles behind the NDR in these devices are different from those of n-type GaAs under a high electric field.

To understand the NDR in GaAs we have to refer to the conduction band structure shown in Fig. 2.14. While the lowest conduction band minimum in GaAs occurs at the Brillouin zone center, there are also conduction band minima at the L points, which lie about 0.3 eV higher in energy. The effective mass of electrons in these L valleys is not isotropic. For motion along the axis of the valleys, the longitudinal mass is $1.9m_0$ while the transverse mass is $0.075m_0$ [5.29]. These masses are much larger than the effective mass (to be denoted by m^*_Γ) of $0.067m_0$ for electrons in the Γ valley. On the basis of (5.12) we expect the mobility of electrons in the Γ valley (to be denoted by μ_Γ) to be larger than that of electrons in the L valleys (denoted by μ_L). At low electric fields all the electrons are in the Γ valley and the electron mobility is high

[1] Leo Esaki shared the 1973 Nobel Prize in Physics for his discovery of the phenomenon of electron tunneling in the diodes named after him.

because of the small m_Γ^*. As the field is increased, electrons gain energy until some of them have sufficient energy to transfer via intervalley scattering to the L valleys. This intervalley scattering now competes with intravalley relaxation via scattering by optical phonons. In GaAs the time it takes an electron to emit a LO phonon via the Fröhlich interaction (Fig. 5.3) is about 200 fs. The corresponding Γ to L intervalley scattering time at room temperature is less than 100 fs [5.30, 31]. However, the time it takes the electron to return to the Γ valley is of the order of picoseconds because the density of states of the Γ valley is much smaller than that of L valleys. As a result, for high enough electrics fields, we expect that a significant fraction of the electrons will be excited into the L valleys and the electron conductivity will become

$$\sigma = e(N_\Gamma \mu_\Gamma + N_L \mu_L), \tag{5.70}$$

where N_Γ and N_L are, respectively, the number of electrons in the Γ and L valleys. Since μ_L is smaller than μ_Γ, the conductivity decreases with increasing field, leading to a negative differential resistance. The field dependence of the electron drift velocity in GaAs deduced from the above picture is shown in Fig. 5.13. At even higher fields ($E > E_b$ in Fig. 5.13) there will be more electrons in the L valleys than in the Γ valley because the former have larger density of states and the intervalley transfer of electron stops. Now the electrons in the L valleys are accelerated by the external field and their velocities increase linearly with field again as shown schematically in Fig. 5.13. The threshold field E_c at which the drift velocity begins to decrease is commonly referred to as the **critical field**. It should be noted that at one time it was thought that the NDR in GaAs was caused by transfer of electrons to the X valleys. After it was demonstrated that the L valleys were lower in energy than the X valleys [5.32], transfer to the L valleys became accepted as the mechanism responsible for NDR in GaAs.

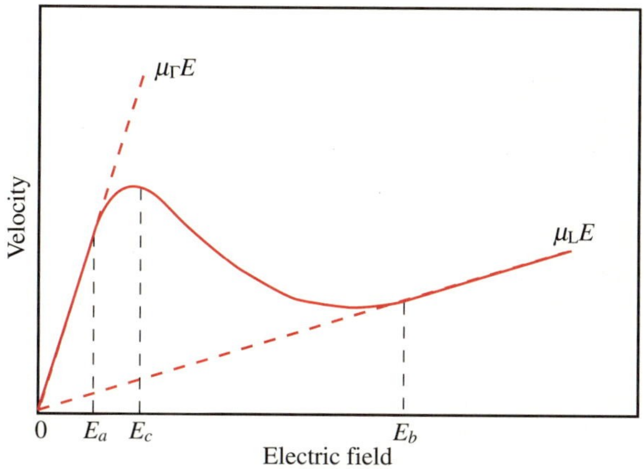

Fig. 5.13. Sketch of the dependence of the drift velocity of electrons in GaAs on electric field based on the qualitative model in Sect. 5.4.2

In order to observe NDR it is necessary to satisfy the following conditions. There must be higher energy valleys to which carriers can be excited under high electric field. The mobility of carriers in these higher energy valleys should be much smaller than in the lower energy valleys. The separation of the higher energy valleys from the lower energy valleys should be much larger than $k_B T$, where T is the device operation temperature, in order that the higher valleys are not populated thermally. However, this separation should not be larger than the bandgap. Otherwise, before the carriers have acquired enough energy from the field to transfer to the higher valleys they can already excite carriers from the valence band into the conduction band via **impact ionization** [Ref. 5.22, pp. 322–384]. These conditions are satisfied by the conduction bands of GaAs, InP [5.26], and the ternary alloy $In_xGa_{1-x}As$ ($x < 0.5$).

5.4.3 Gunn Effect

In 1963 *J. B. Gunn* [5.33, 34] discovered that when a thin sample (thickness of the order 10 μm) of n-type GaAs is subjected to a high voltage (such that the electric field exceeds the critical field E_c in GaAs), the current through the sample spontaneously breaks up into oscillations at microwave frequencies as shown in Fig. 5.14. The frequency of oscillation is inversely proportional to the length of the sample across which the field is applied. As will be shown below, the frequency of oscillation turns out to be equal to the saturation velocity v_s divided by the sample length. For sample lengths of the order of 10 μm and

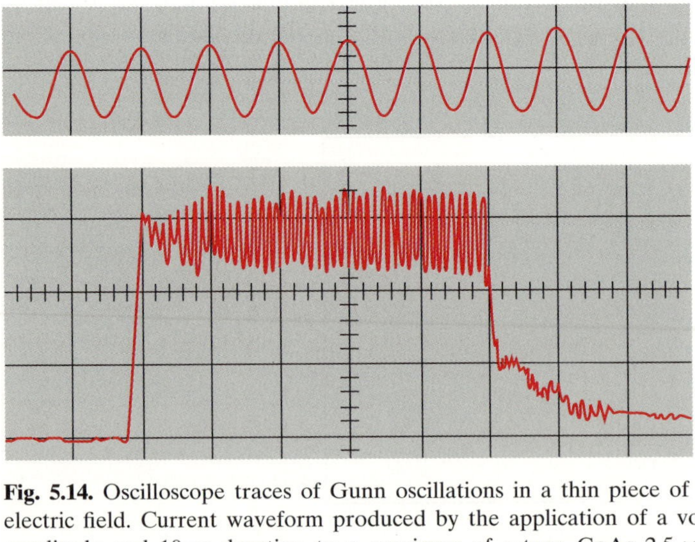

Fig. 5.14. Oscilloscope traces of Gunn oscillations in a thin piece of GaAs under a high electric field. Current waveform produced by the application of a voltage pulse of 16 V amplitude and 10 ns duration to a specimen of n-type GaAs 2.5×10^{-3} cm in length. The frequency of the oscillating component is 4.5 GHz. *Lower trace*: 2 ns/cm horizontally, 0.23 A/cm vertically. *Upper trace*: expanded view of lower trace [5.33]

v_s equal to 10^7 cm/s, the oscillation frequency is of the order of 10^{10} Hz, i. e. 10 gigahertz[2]. Electromagnetic waves of such high frequencies are known as **microwaves**. Hence the obvious application of the Gunn effect is the fabrication of microwave generators known as **Gunn diodes**.

The Gunn effect is one example of how NDR can lead to high-frequency oscillations. To understand this effect qualitatively [5.35], we will assume that the field dependence of the drift velocity in n-type GaAs has the simple form shown schematically in Fig. 5.15a. Suppose a constant high voltage is applied to the sample so that carriers drift from the left to right as shown schematically in Fig. 5.15b. We assume that the electric field is maintained at a value slightly below the threshold field E_c. Due to fluctuations in the electric field at finite temperatures, a small region labeled D in Fig. 5.15b has a field slightly above E_c at time $t = 0$. The carriers on both sides of D have now higher drift velocities than carriers inside D. As a result, carriers will pile up on the left hand side in D while the carrier density will drop on its right hand side. This charge pile-up in D at $t > 0$ leads to an increase in electric field inside D and a decreasing field outside, as shown in Fig. 5.15b. Because of the NDR for

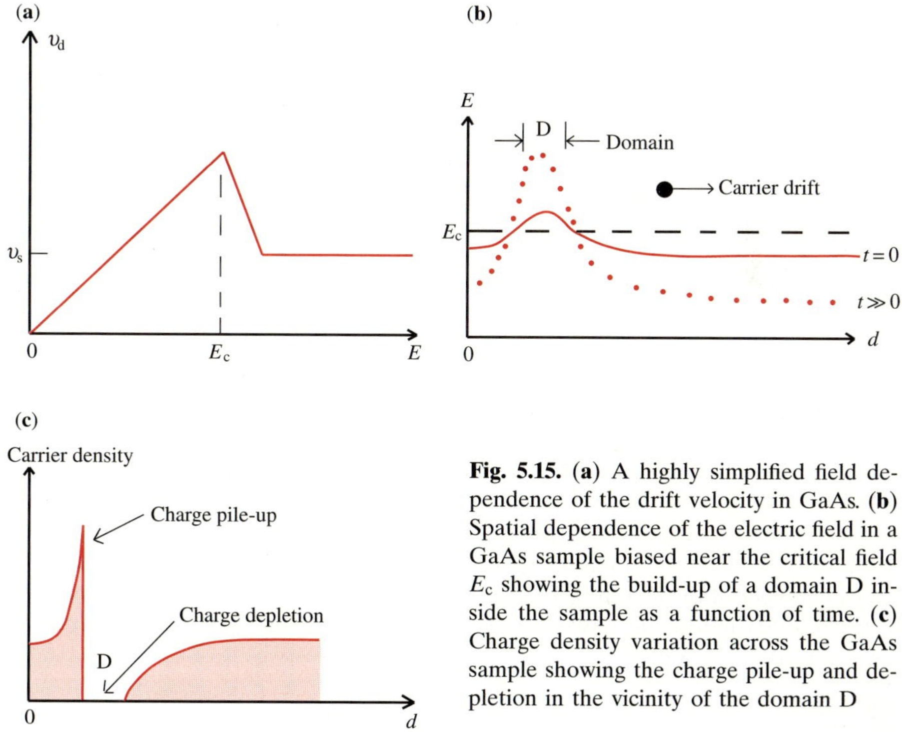

Fig. 5.15. (a) A highly simplified field dependence of the drift velocity in GaAs. **(b)** Spatial dependence of the electric field in a GaAs sample biased near the critical field E_c showing the build-up of a domain D inside the sample as a function of time. **(c)** Charge density variation across the GaAs sample showing the charge pile-up and depletion in the vicinity of the domain D

[2] The unit of frequency is hertz (Hz), named after Heinrich Hertz who produced and detected radio waves in 1888. One hertz is equal to one cycle per second.

fields larger than E_c, the increase in the field inside D leads to further slowing down of electrons inside D and hence more charge pile-up. This process, once started, will continue until most of the applied field is across D, as shown in the dotted curve in Fig. 5.15b for $t \gg 0$. Figure 5.15c displays the charge distribution along the length of the sample. The region D where the electric field is high is known as a **domain**. Only one domain can exist inside the sample at one time since most of the applied voltage will be across this domain. The most likely place for domains to be formed is the cathode, since the field fluctuations tend to be largest there. Under the influence of the applied voltage this domain will drift across the sample with the saturation velocity until it reaches the anode, thus giving rise to a periodic oscillation in current. The frequency of this oscillation is equal to v_s divided by the length of the sample. As a result of this oscillatory current, electromagnetic waves are radiated from the sample. From this simple description it is clear that Gunn oscillators are very efficient, yet miniature, microwave generators.

5.5 Magneto-Transport and the Hall Effect

We conclude this chapter by discussing the electric current induced in a sample in the presence of both electric and magnetic fields. As we pointed out in Sect. 5.1, in a cubic crystal the second-rank conductivity tensor $\boldsymbol{\sigma}$ can usually be represented by a diagonal matrix. This is not true when a magnetic field is present. In this case the conductivity tensor contains off-diagonal elements that are linearly dependent on the magnetic field. In this section we will derive this magneto-conductivity tensor and use the result to study an important phenomenon known as the Hall effect. Our approach in this section will be classical, leaving a quantum mechanical treatment to Chap. 9, where we shall consider the quantum Hall effect in two-dimensional electron gases.

5.5.1 Magneto-Conductivity Tensor

We shall first assume that the sample is an infinite, cubic and nonmagnetic crystal. Without loss of generality we can suppose that a magnetic field B_z is applied to the sample along the z axis, while an electric field $\boldsymbol{F}$ is applied along any arbitrary direction. To calculate the resultant current we use the quasi-classical approach adopted in Sect. 5.1. In the presence of both electric and magnetic fields, the equation of motion for the electrons (5.4) is replaced by the **Lorentz equation**:*

$$m^* \frac{d^2\boldsymbol{r}}{dt^2} + \frac{m^*}{\tau}\frac{d\boldsymbol{r}}{dt} = (-e)\left[\boldsymbol{F} + (\boldsymbol{v} \times \boldsymbol{B}/c)\right], \qquad (5.71)$$

* The equations in this section are transformed into SI units by deleting the velocity of light c.

where c is the speed of light in vacuum. In (5.71) we have assumed that m^* and τ are isotropic. For generalizations see [5.36, 37]. Under steady-state conditions, $d\boldsymbol{v}/dt = d^2\boldsymbol{r}/dt^2 = 0$, we obtain

$$(m^*/\tau)\boldsymbol{v}_\mathrm{d} = (-e)\left[\boldsymbol{F} + (\boldsymbol{v}_\mathrm{d} \times \boldsymbol{B}/c)\right] \tag{5.72}$$

for the electron drift velocity $\boldsymbol{v}_\mathrm{d}$. In terms of its components along the x, y, and z axes, the three components of this equation can be written as

$$(m^*/\tau)v_{\mathrm{d},x} = (-e)\left[F_x + (v_{\mathrm{d},y}B_z/c)\right], \tag{5.73a}$$

$$(m^*/\tau)v_{\mathrm{d},y} = (-e)\left[F_y - (v_{\mathrm{d},x}B_z/c)\right], \tag{5.73b}$$

$$(m^*/\tau)v_{\mathrm{d},z} = (-e)F_z. \tag{5.73c}$$

By multiplying each of the above equations by the electron density n and charge $(-e)$ we obtain the corresponding equations for the current density $\boldsymbol{j} = n(-e)\boldsymbol{v}_\mathrm{d}$:

$$j_x = (ne^2\tau/m^*)F_x - (eB_z/m^*c)\tau j_y, \tag{5.74a}$$

$$j_y = (ne^2\tau/m^*)F_y + (eB_z/m^*c)\tau j_x, \tag{5.74b}$$

$$j_z = (ne^2\tau/m^*)F_z. \tag{5.74c}$$

It is convenient at this point to introduce the definitions

$$\sigma_0 = ne^2\tau/m^*, \tag{5.75}$$

which can be recognized as the zero-field conductivity, and

$$\omega_\mathrm{c} = eB_z/(m^*c), \tag{5.76}$$

which is the classical **cyclotron frequency** of the electron in the presence of the magnetic field B_z. Using these definitions, (5.74) can be simplified to

$$j_x = \sigma_0 F_x - \omega_\mathrm{c}\tau j_y, \tag{5.77a}$$

$$j_y = \sigma_0 F_y + \omega_\mathrm{c}\tau j_x, \tag{5.77b}$$

$$j_z = \sigma_0 F_z. \tag{5.77c}$$

Solving (5.77), we obtain the three components of the current density:

$$j_x = \frac{1}{1+(\omega_\mathrm{c}\tau)^2}\sigma_0(F_x - \omega_\mathrm{c}\tau F_y), \tag{5.78a}$$

$$j_y = \frac{1}{1+(\omega_\mathrm{c}\tau)^2}\sigma_0(F_y + \omega_\mathrm{c}\tau F_x), \tag{5.78b}$$

$$j_z = \sigma_0 F_z. \tag{5.78c}$$

Based on (5.78) we can define a generalized **magneto-conductivity tensor $\boldsymbol{\sigma}(B)$** for the electrons as

$$\boldsymbol{\sigma} = \frac{\sigma_0}{1+(\omega_\mathrm{c}\tau)^2}\begin{pmatrix} 1 & -\omega_\mathrm{c}\tau & 0 \\ \omega_\mathrm{c}\tau & 1 & 0 \\ 0 & 0 & 1+(\omega_\mathrm{c}\tau)^2 \end{pmatrix}. \tag{5.79}$$

Notice that (5.79) contains the sum of a diagonal and an antisymmetric tensor. The sign of the off-diagonal elements depends on the sign of the charge.

From (5.79) we conclude that the effects of the magnetic field on the charge transport are twofold. (1) The conductivity perpendicular to the magnetic field is decreased by the factor $[1+(\omega_c\tau)^2]^{-1}$. The corresponding increase in the sample resistance induced by a magnetic field is known as **magnetoresistance** (for small values of B it is proportional to B^2). (2) The magnetic field also generates a current transverse to the applied electric field, resulting in off-diagonal elements in the conductivity tensor. These are linearly proportional to the magnetic field whereas the diagonal elements are quadratic in the magnetic field. The off-diagonal elements give rise to the Hall effect to be discussed in the next section.

5.5.2 Hall Effect

Let us consider a sample in the form of a rectangular bar oriented with its longest axis along the x axis, as shown in Fig. 5.16a. The electric field $\boldsymbol{F}$ is now applied along the x axis while the magnetic field $\boldsymbol{B}$ is still along the z axis. According to Lorentz's law, when electrons start to drift along the x axis under the influence of the electric field, they also experience a force in the y direction. This results in a current in the y direction although there is no applied electric field along that direction. One typical experimental configuration involves a closed current loop in the x direction while leaving an open circuit in

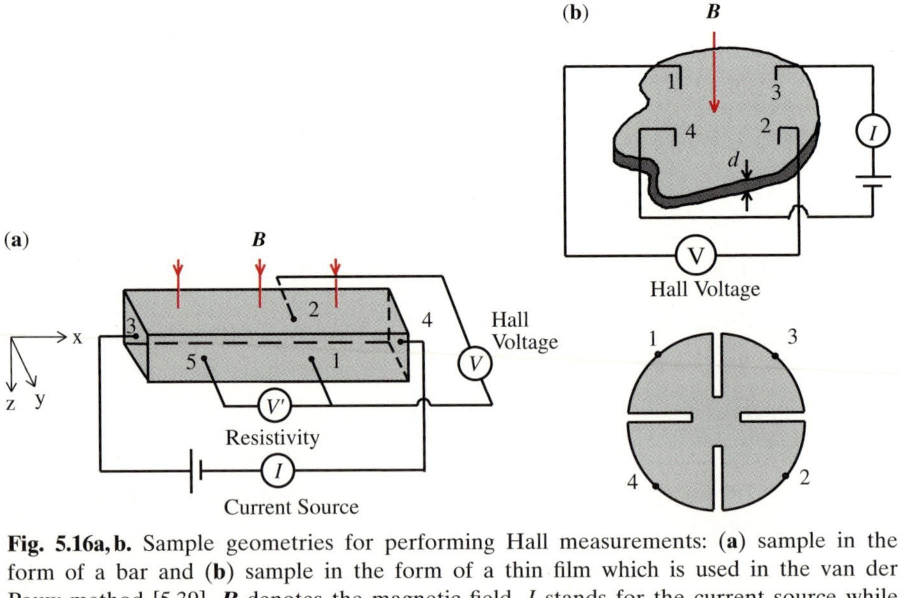

Fig. 5.16a,b. Sample geometries for performing Hall measurements: (**a**) sample in the form of a bar and (**b**) sample in the form of a thin film which is used in the van der Pauw method [5.39]. $\boldsymbol{B}$ denotes the magnetic field. I stands for the current source while V represents the meter for measuring the Hall voltage

the y direction, as shown in Fig. 5.16a. As a result of this open circuit the current density j_y must be zero. From (5.78b) we see that an electric field F_y is induced by the presence of the $\boldsymbol{B}$ field. This phenomenon is known as the **Hall effect** [5.38] after E.H. Hall (1855–1938), who discovered it at Johns Hopkins University in 1879 and then became a professor at Harvard [from 1881 to 1921].

A simple physical picture of what happens is as follows. The magnetic field causes the charges to drift in the y direction. As a result, charges pile up on the two opposite sample surfaces perpendicular to the y axis and create an electric field F_y, which cancels the effect of the Lorentz force. Under the steady-state condition $j_y = 0$, the induced field F_y is equal to

$$F_y = -\omega_c \tau j_x / \sigma_0 \tag{5.80a}$$

while the current measured in the x direction is given by

$$j_x = \sigma_0 F_x . \tag{5.80b}$$

The measured quantity in this experiment is F_y while the externally controlled parameters are j_x and $\boldsymbol{B}$. Therefore one defines the **Hall coefficient** R_H as the ratio

$$R_H = F_y / (j_x B_z). \tag{5.81}$$

Combining (5.80a) and (5.80b), we find that R_H is equal to[3]

$$R_H = -\frac{\omega_c \tau}{\sigma_0 B_z} = -\frac{1}{nec}. \tag{5.82}$$

Notice that the sign of R_H depends on the sign of the charge. While the Hall coefficient in (5.82) is negative (since we have assumed the charges are electrons), it can be easily shown that R_H becomes positive for holes. Thus we see that the Hall effect is an important technique for determining both the concentration and the sign of charged carriers in a sample. This technique is not limited to semiconductors only but is also used extensively in the study of metals. In compensated semiconductor samples, where both electrons and holes are present, R_H can be shown to be given by (Problem 5.6)

$$R_H = \frac{N_p - b^2 N_n}{ec(bN_n + N_p)^2}, \tag{5.83}$$

where N_n and N_p are the concentrations of the negative and positive charges, respectively, and b is the ratio of their mobilities: μ_n/μ_p. Corrections to (5.82) for the case of anisotropic masses and τ have been given by *Herring* and *Vogt* [5.37].

5.5.3 Hall Coefficient for Thin Film Samples (van der Pauw Method)

One limitation of the Hall effect measurement described in the previous section is the requirement that the sample be in the shape of a rectangular bar. As discussed in Sect. 1.2, samples are often grown in the form of thin epitaxial

[3] To convert to SI units delete c.

films on some insulating substrate. The extension of the Hall technique to such thin films was developed by *van der Pauw* [5.39]. Two common geometries for the van der Pauw method of measuring the Hall coefficient and resistivity in a thin sample are shown in Fig. 5.16b. This method is particularly convenient for a disk of irregular shape. The current is fed through the contacts 3 and 4 while the Hall voltage is measured across the contacts 1 and 2. The "clover" shape in Fig. 5.16b has the advantage of keeping the current flow away from the Hall voltage contacts. To minimize the error in the measurement of the Hall voltage due to the fact that the current flow may not be perpendicular to the line joining the contacts 1 and 2, one usually measures the voltage both with the magnetic field $V_{12}(\pm B)$ and without the field $V_{12}(0)$. Van der Pauw showed that the Hall coefficient is given by

$$R_{\mathrm{H}} = \frac{[V_{12}(B) - V_{12}(0)]d}{I_{34}B} = \frac{[V_{12}(B) - V_{12}(-B)]d}{2I_{34}B}, \tag{5.84}$$

where d is the thickness of the film, B is the magnetic field, and I_{34} is the current flowing from contact 3 to contact 4.

The sample resistivity ϱ can also be measured with the van der Pauw method. In this case two adjacent contacts such as 2 and 3 (I_{23}) are used as current contacts while the two remaining contacts are used for measuring the voltage drop (V_{41}). The resultant resistance is defined as $R_{41,23}$:

$$R_{41,23} = |V_{41}|/I_{23}. \tag{5.85}$$

Another measurement is then made in which current is instead sent through the contacts 1 and 3 and the voltage is measured across the contacts 2 and 4. From the resulting resistance $R_{24,13}$, together with $R_{41,23}$, ϱ can be calculated with the expression

$$\varrho = \frac{\pi d(R_{24,13} + R_{41,23})f}{2\ln 2} \tag{5.86}$$

where f is a factor that depends on the ratio $R_{24,13}/R_{41,23}$; f is equal to 1 when this ratio is exactly 1 [Ref. 5.26, p. 63]. When this ratio is equal to 10, f decreases to 0.7. Usually a large value for this ratio is undesirable and suggests that either the contacts are bad or that the sample is inhomogeneously doped.

5.5.4 Hall Effect for a Distribution of Electron Energies

So far we have assumed that all the charged carriers have the same properties. We shall now consider a collection of electrons with a range of energies E and a distribution function $f(E)$. We denote the average of any electron property $a(E)$ by $\langle a \rangle$:

$$\langle a \rangle = \int a(E)f(E)dE \Big/ \int f(E)dE. \tag{5.87}$$

Using this definition, (5.78) can be rewritten as

$$\langle j_x \rangle = \alpha F_x - \gamma B_z F_y \tag{5.88a}$$

$$\langle j_y \rangle = \alpha F_y + \gamma B_z F_x \tag{5.88b}$$

$$\langle j_z \rangle = \langle \sigma_0 \rangle F_x, \tag{5.88c}$$

where*

$$\alpha = \frac{ne^2}{m^*} \left\langle \frac{\tau}{1 + (\omega_c \tau)^2} \right\rangle \tag{5.89a}$$

$$\gamma = \frac{ne^3}{m^{*2}c} \left\langle \frac{\tau^2}{1 + (\omega_c \tau)^2} \right\rangle \tag{5.89b}$$

In the limit of a weak magnetic field or, when $(\omega_c \tau)^2 \ll 1$, we can approximate $1 + (\omega_c \tau)^2$ by one and thus write:

$$\alpha \simeq \frac{ne^2}{m^*} \langle \tau \rangle \quad \text{and} \quad \gamma \simeq \frac{ne^3 \langle \tau^2 \rangle}{m^{*2}c} \tag{5.90}$$

Within this approximation the Hall coefficient for a distribution of electrons can be expressed as[4]

$$R_H = \frac{\langle \tau^2 \rangle}{(-nec)\langle \tau \rangle^2} = -\frac{r_H}{nec} \tag{5.91}$$

The factor $r_H = \langle \tau^2 \rangle / \langle \tau \rangle^2$ is called the **Hall factor**. Its magnitude depends on the scattering mechanisms that contribute to τ and is usually of the order of 1 [Ref. 5.26, p. 57]. In the limit of strong magnetic fields, or for very pure samples, when $(\omega_c \tau)^2 \gg 1$, (5.91) remains valid with $r_H = 1$ (Problem 5.7). In principle we can determine the carrier mobility by measuring R_H and σ_0 and using (5.75 and 82) to obtain

$$\mu = R_H \sigma_0, \tag{5.92}$$

but in practice the carriers usually have a distribution of energies, so that the mobility calculated from (5.92) is not the same as the mobility μ defined by (5.12). Instead the mobility defined by (5.92) is referred to as the **Hall mobility** μ_H and is related to μ by

$$\mu_H = r_H \mu. \tag{5.93}$$

PROBLEMS

5.1 *Drifted Carrier Distributions*

Using (5.19) show that, within the relaxation time approximation, the carrier distribution in the presence of the field $\boldsymbol{F}$ can be approximated by

$$f_{\boldsymbol{k}} = f^0_{\boldsymbol{k}}(E_{\boldsymbol{k}} + q\tau_{\boldsymbol{k}} \boldsymbol{v}_{\boldsymbol{k}} \cdot \boldsymbol{F}).$$

for a small external field. This means that the effect of the electric field is to shift the entire distribution (without distortion) by the energy $q\tau_{\boldsymbol{k}} \boldsymbol{v}_{\boldsymbol{k}} \cdot \boldsymbol{F}$. This is just the energy gained by a charge q with velocity $\boldsymbol{v}_{\boldsymbol{k}}$ in the field $\boldsymbol{F}$ during a time $\tau_{\boldsymbol{k}}$.

[4] remove c to convert to SI units.

5.2 *a) Drifted Maxwell–Boltzmann Distributions*
Suppose the thermal equilibrium electron distribution f_k^0 in (5.14) is approximated by a Boltzmann distribution function

$$f_k^0 = A \exp\left[-E_k/(k_B T)\right].$$

For free carriers located in a spherical band with effective mass m^* the electron energy is given by $E_k = (1/2)m^* v_k^2$. The resultant f_k^0 is known as a **Maxwell–Boltzmann distribution**. Using the result of Problem 5.1, show that the carrier distribution in the presence of a weak external electric field $\boldsymbol{F}$ can be approximated by

$$f_k \approx A \exp\left(-\frac{(\boldsymbol{v}_k + \boldsymbol{v}_\mathrm{d})^2}{2m^* k_B T}\right),$$

where $\boldsymbol{v}_\mathrm{d}$ is the drift velocity defined in (5.6). The interpretation of this result is that the external field causes the carrier velocities to increase uniformly by an amount equal to $\boldsymbol{v}_d$ while leaving the distribution function unchanged. Consequently the resultant distribution is known as a **drifted Maxwell–Boltzmann distribution.**

b) Relaxation Time Approximation
The purpose of this problem is to show how the relaxation time approximation (5.18) can be obtained starting from (5.30). We will make the following assumptions:

1) The electronic band is isotropic, with effective mass m^*.
2) Scattering is completely elastic (so that $E_{k'} = E_k$ and $f_k^0 = f_{k'}^0$).
3) The applied field is weak so that (5.19) is valid.

a) With these assumptions, show that

$$f_k - f_{k'} = (f_k - f_k^0)\left[1 - (\boldsymbol{v}_{k'} \cdot \boldsymbol{F})/(\boldsymbol{v}_k \cdot \boldsymbol{F})\right].$$

Substitute this result into (5.30) and show that

$$\tau_k^{-1} = \sum_{k'} P(\boldsymbol{k}, \boldsymbol{k}')\left[1 - (\boldsymbol{v}_{k'} \cdot \boldsymbol{F})/(\boldsymbol{v}_k \cdot \boldsymbol{F})\right].$$

b) To simplify the above expression we choose (without loss of generality) coordinate axes such that $\boldsymbol{k}$ is parallel to the z-axis and $\boldsymbol{F}$ lies in the yz-plane. Let θ and θ' be, respectively, the angles between $\boldsymbol{k}$ and $\boldsymbol{F}$ and between $\boldsymbol{k}'$ and $\boldsymbol{F}$. Let the polar coordinates of $\boldsymbol{k}'$ in this system be (k', α, β). Show that

$$(\boldsymbol{v}_{k'} \cdot \boldsymbol{F})/(\boldsymbol{v}_k \cdot \boldsymbol{F}) = \tan\theta \, \sin\alpha \, \sin\beta + \cos\alpha.$$

c) Assume that the elastic scattering probability $P(\boldsymbol{k}, \boldsymbol{k}')$ depends on α but not on β. Show that

$$\tau_k^{-1} = \sum_{k'} P(\boldsymbol{k}, \boldsymbol{k}')(1 - \cos\alpha).$$

The physical interpretation of this result is that the relaxation time in elastic scattering is dominated by large angle scattering (i. e., processes with $\alpha \simeq \pi$).

5.3 *Intravalley Scattering by LA Phonons*
a) Using (5.42) show that the contribution of deformation potential interaction to the intravalley LA phonon scattering time (τ_{ac}) of an electron in a nondegenerate band with isotropic effective mass m^* and energy E is given by

$$\frac{1}{\tau_{ac}} = \frac{\sqrt{2}(a_c)^2(m^*)^{3/2}k_B T(E)^{1/2}}{\pi\hbar^4\varrho v_s^2},$$

where a_c is the volume deformation potential for the electron, T is the temperature, k_B is the Boltzmann constant, ϱ is the crystal density, and v_s is the LA phonon velocity.

b) Assume that the parameters in (a) have the following values, appropriate to GaAs: $m^* = 0.067$; $E = 0.36$ meV; $a_c = 6$ eV; $\varrho = 5.31$ g/cm^3, and $v_s = 5.22 \times 10^5$ cm/s. Show that $\tau_{ac} = 7 \times 10^{-12}$ s (7 ps) at $T = 300$ K.

5.4 *Piezoelectric Acoustic Phonon Scattering Rate*
Assume that the constant of proportionality in (5.44) is independent of angle. Substitute the result into (5.41) and perform the integration corresponding to that in (5.42). Show that instead of a scattering probability proportional to $T(E_k)^{1/2}$ as in (5.43), the probability for piezoelectric scattering is proportional to $T(E_k)^{-1/2}$. See p. 202 for discussions on how to remove the divergence in the scattering probability when $E_k \to 0$.

5.5 *Rate of Scattering by LO Phonons for Electrons in a Parabolic Band*
a) Consider a polar semiconductor with a dispersionless LO phonon energy $\hbar\omega_{LO}$. The Fröhlich electron–LO-phonon interaction Hamiltonian is given by (3.36). For an electron in a nondegenerate conduction band with isotropic effective mass m^*, the electron wave vector and energy above the band minimum are denoted by $\boldsymbol{k}$ and $E_{\boldsymbol{k}}$, respectively. Show that for $E_{\boldsymbol{k}} > \hbar\omega_{LO}$ the minimum and maximum phonon wave vectors, $q_{2\,\mathrm{min}}$ and $q_{2\,\mathrm{max}}$, of the phonon *emitted* by the electron are given by

$$q_{2\,\mathrm{min}} = k[1 - f(E_{\boldsymbol{k}})]$$

and

$$q_{2\,\mathrm{max}} = k[1 + f(E_{\boldsymbol{k}})]$$

where

$$f(E_{\boldsymbol{k}}) = \left[1 - (\hbar\omega_{LO}/E_{\boldsymbol{k}})\right]^{1/2}.$$

b) Show that the corresponding $q_{1\,\mathrm{min}}$ and $q_{1\,\mathrm{max}}$ for *absorption* of one LO phonon are given by

$$q_{1\,\mathrm{min}} = k[f'(E_{\boldsymbol{k}}) - 1]$$

and

$$q_{1\,\mathrm{max}} = k[f'(E_{\boldsymbol{k}}) + 1]$$

where

$$f'(E_{\boldsymbol{k}}) = \left[1 + (\hbar\omega_{\mathrm{LO}}/E_{\boldsymbol{k}})\right]^{1/2}.$$

In the special case that $k = 0$,

$$q_{1\,\mathrm{min}} = q_{1\,\mathrm{max}} = [2m^*\omega_{\mathrm{LO}}/\hbar]^{1/2}.$$

c) Substituting the above results into (5.50), show that the momentum relaxation rate by LO phonon scattering is given by (5.51) in the limit $q_0 = 0$.

d) Assume that a conduction band electron in GaAs has $E_{\boldsymbol{k}}/(\hbar\omega_{\mathrm{LO}}) = 4$. Use the *Fröhlich* interaction in (3.36) to calculate the scattering rate of this electron by LO phonons. Some materials parameters for GaAs are: density $\varrho = 5.31\ \mathrm{g/cm^3}$; $m^* = 0.067m_0$; $\varepsilon_0 = 12.5$; $\varepsilon_\infty = 10.9$; and $\hbar\omega_{\mathrm{LO}} = 36\ \mathrm{meV}$.

5.6 *Hall Coefficient for Samples Containing both Electrons and Holes*
a) Show that, for a sample containing N_{n} electrons and N_{p} holes per unit volume with the corresponding mobilities μ_{n} and μ_{p}, the equations for the current density in the presence of applied electric field $\boldsymbol{F}$ and magnetic field B_z are

$$j_x = (\alpha_{\mathrm{n}} + \alpha_{\mathrm{p}})F_x - (\beta_{\mathrm{n}} + \beta_{\mathrm{p}})B_z F_y,$$

$$j_y = (\alpha_{\mathrm{n}} + \alpha_{\mathrm{p}})F_y + (\beta_{\mathrm{n}} + \beta_{\mathrm{p}})B_z F_x,$$

$$j_z = \sigma_0 F_x,$$

where $\alpha_i = N_i e \mu_i$ and $\beta_i = -\alpha_i \mu_i / c$ (delete c for SI units) for $i = \mathrm{n}$ and p.

b) Assume that $j_y = 0$, as in a conventional Hall effect measurement, and derive F_y. Calculate the Hall coefficient $R_{\mathrm{H}} = F_y/(j_x B_z)$.

5.7 *Hall Factor in the Limit of Strong and Weak Magnetic Fields*
Show that the Hall coefficient for electrons with a distribution of energies and scattering times τ is given by:

$$R_{\mathrm{H}} = -\frac{\gamma}{\alpha^2}\left[1 + \frac{\gamma^2 B_z^2}{\alpha^2}\right]^{-1}$$

where α and γ are defined in (5.89a,b).

SUMMARY

In this chapter we have discussed the transport of charges in semiconductors under the influence of external fields. We have used the *effective mass approximation* to treat the *free carriers* as having classical charge and renormalized masses. We first considered the case of weak fields in which the field does not distort the carrier distribution but causes the entire distribution to move with a *drift velocity*. The drift velocity is determined by the length of time, known as the *scattering time*, over which the carriers can accelerate in the field before they are scattered. We also defined *mobility* as the constant of proportionality between drift velocity and electric field. We calculated the scattering rates for carriers scattered by *acoustic phonons, optical phonons, and ionized impurities*. Using these scattering rates we deduced the *temperature dependence of the carrier mobilities*. Based on this temperature dependence we introduced *modulation doping* as a way to minimize scattering by ionized impurities at low temperatures. We discussed qualitatively the behavior of carriers under high electric fields. We showed that these *hot carriers* do not obey Ohm's law. Instead, their drift velocities at high fields saturate at a constant value known as the saturation velocity. We showed that the saturation velocity is about 10^7 cm/s in most semiconductors as a result of energy and momentum relaxation of carriers by scattering with optical phonons. In a few n-type semiconductors, such as GaAs, the drift velocity can *overshoot* the saturation velocity and exhibit *negative differential resistance*. This is the result of these semiconductors having secondary conduction band valleys whose energies are of the order of 0.1 eV above the lowest conduction band minimum. The existence of negative differential resistance leads to spontaneous current oscillations at microwave frequencies when thin samples are subjected to high electric fields, a phenomenon known as the *Gunn effect*. Under the combined influence of an electric and magnetic field, the transport of carriers in a semiconductor is described by an antisymmetric second rank *magneto-conductivity tensor*. One important application of this tensor is in explaining the *Hall effect*. The *Hall coefficient* provides the most direct way to determine the sign and concentration of charged carriers in a sample.

In this chapter we have discussed the transport of charges in semiconductors under the influence of external fields. We have used the [illegible]

6. Optical Properties I

CONTENTS

The fundamental energy gaps of most semiconductors span the energy range from zero to about 6 eV. Photons of sufficient energy can excite electrons from the filled valence bands to the empty conduction bands. As a result, the optical spectra of semiconductors provide a rich source of information on their electronic properties. In many semiconductors, photons can also interact with lattice vibrations and with electrons localized on defects, thus making optical techniques also useful for studying these excitations. Their optical properties are the basis of many important applications of semiconductors, such as lasers, light emitting diodes, and photodetectors.

Figure 6.1 shows schematically some of the optical processes which can occur when a medium is illuminated by light. At the surface of the medium,

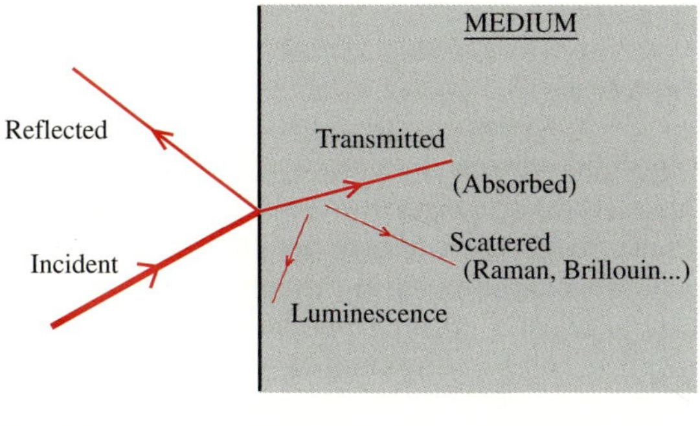

Fig. 6.1. Schematic diagram showing the linear optical processes that occur at the surface and in the interior of a medium. The incident beam is assumed to arrive at the surface of the medium from vacuum (or air)

a fraction of the incident light is reflected and the rest transmitted. Inside the medium some of the radiation may be absorbed or scattered while the remainder passes through the sample. Some of the absorbed electromagnetic waves may be dissipated as heat or reemitted at a different frequency. The latter process is known as **photoluminescence**. Electromagnetic waves are scattered by inhomogeneities inside the medium. These inhomogeneities may be static or dynamic. An example of a dynamic fluctuation is the density fluctuation associated with an acoustic wave. Scattering of light by acoustic waves is usually referred to as **Brillouin scattering** [6.1a]. (This phenomenon was independently discovered by *Mandelstam* [6.1b].) Scattering of light by other elementary excitations, such as optical phonons or plasmons, is known as **Raman scattering**. (Sir C.V. Raman received the Nobel Prize in Physics in 1930 for discovering, in 1928 in Calcutta, the effect named after him.) In general, the strongest optical processes are **reflection** and **absorption** because they involve the lowest order of interaction between electromagnetic waves and elementary excitations inside the medium. Light scattering involves two such interactions (since there is incident radiation and scattered radiation) hence it tends to be weaker. In this book we shall not consider nonlinear optical processes (such as sum and difference frequency generations), which involve higher-order optical interactions. Readers interested in these topics should consult [6.2–4]. Because there are many different ways that photons can interact with excitations inside semiconductors, we shall divide the discussion on the optical properties of semiconductors into two chapters. In this chapter, the fundamental optical properties of semiconductors will be presented. Chapter 7 will discuss the more specialized topics such as photoluminescence and light scattering.

6.1 Macroscopic Electrodynamics

In a dielectric medium an external sinusoidal electromagnetic wave with electric field vector $\boldsymbol{E}(\boldsymbol{r}, t) = \boldsymbol{E}(q, \omega)\sin(\boldsymbol{q}\cdot\boldsymbol{r} - \omega t)$, where $\boldsymbol{q}$ is the wavevector and ω is the frequency, will induce a polarization vector $\boldsymbol{P}$, which is related to the applied field via a second-rank tensor:

$$P_i(\boldsymbol{r}', t') = \int \chi_{ij}(\boldsymbol{r}, \boldsymbol{r}', t, t')E_j(\boldsymbol{r}, t)d\boldsymbol{r}\, dt. \tag{6.1}$$

χ_{ij} is known as the **electric susceptibility tensor**. To simplify the notation, summation over the repeated index j in (6.1) is automatically implied. Time is, of course, homogeneous in the absence of time-dependent perturbations. It is often assumed that space is also homogeneous to avoid complications such as *local field corrections* (avoided by averaging all microscopic quantities over a unit cell).[1] With this assumption, (6.1) can be simplified to

[1] In recent years there has been considerable interest in the propagation of electromagnetic waves through periodic structures composed of macro- or microscopic elements such as semiconductor spheres. These samples are called photonic crystals [6.5].

$$P_i(\boldsymbol{r}',t') = \varepsilon_0 \int \chi_{ij}(|\boldsymbol{r}-\boldsymbol{r}'|,|t-t'|)E_j(\boldsymbol{r},t)d\boldsymbol{r}\,dt. \tag{6.2}$$

From the *convolution theorem*, (6.2) can be expressed in terms of the Fourier transforms of $\boldsymbol{P}$, χ, and $\boldsymbol{E}$ (see Sect. 4.2.1 for a definition of Fourier transforms):

$$P_i(\boldsymbol{q},\omega) = \varepsilon_0\chi_{ij}(\boldsymbol{q},\omega)E_j(\boldsymbol{q},\omega). \tag{6.3}$$

In principle, all the linear optical properties of the medium are determined by the complex electric susceptibility tensor $\chi_{ij}(\boldsymbol{q},\omega)$. Note that while $\boldsymbol{P}(\boldsymbol{r}',t')$, $\chi_{ij}(\boldsymbol{r},\boldsymbol{r}',t,t')$, and $\boldsymbol{E}(\boldsymbol{r},t)$ are all real, their Fourier transforms can be complex. The fact that $\chi_{ij}(\boldsymbol{r},\boldsymbol{r}',t,t')$ is real imples that $\chi_{ij}(\boldsymbol{q},\omega) = \chi_{ij}^*(-\boldsymbol{q},-\omega)$.

For comparison with experiments it is often more convenient to define another complex, second-rank tensor, known as the **dielectric tensor**. This tensor $\varepsilon_{ij}(\boldsymbol{q},\omega)$ is defined by

$$D_i(\boldsymbol{q},\omega) = \varepsilon_0\varepsilon_{ij}(\boldsymbol{q},\omega)E_j(\boldsymbol{q},\omega) \tag{6.4}$$

where $\boldsymbol{D}(\boldsymbol{q},\omega)$ is the Fourier transform of the electric displacement vector $\boldsymbol{D}(\boldsymbol{r},t)$ defined by $\boldsymbol{D}(\boldsymbol{r},t) = \boldsymbol{E}(\boldsymbol{r},t) + 4\pi\boldsymbol{P}(\boldsymbol{r},t)$. It follows from their definitions that $\chi_{ij}(\boldsymbol{q},\omega)$ and $\varepsilon(\boldsymbol{q},\omega)$ are related by

$$\varepsilon_{ij}(\boldsymbol{q},\omega) = 1 + 4\pi\chi_{ij}(\boldsymbol{q},\omega) = 1 + \chi_{ij}(\boldsymbol{q},\omega) \text{ in SI units.} \tag{6.5}$$

The real and imaginary parts of the dielectric tensor will be denoted by $\boldsymbol{\varepsilon}_{\mathrm{r}}(\boldsymbol{q},\omega)$ and $\boldsymbol{\varepsilon}_{\mathrm{i}}(\boldsymbol{q},\omega)$, respectively. We will now state, without proof, some of the properties of $\boldsymbol{\varepsilon}(\boldsymbol{q},\omega)$ [6.6]:

$$\boldsymbol{\varepsilon}(-\boldsymbol{q},-\omega) = \boldsymbol{\varepsilon}^*(\boldsymbol{q},\omega), \tag{6.6}$$

$$\varepsilon_{ij}(\boldsymbol{q},\omega) = \varepsilon_{ji}(-\boldsymbol{q},\omega). \tag{6.7}$$

Equation (6.6) follows from the fact that $\boldsymbol{\varepsilon}(\boldsymbol{r},t)$ has to be a real function of space and time. Equation (6.7) is one example of a general property of all so-called kinetic coefficients known as the **Onsager relations**.

In most of the cases that we shall study, the wavelength of light is much larger than the lattice constants or other relevant dimensions (such as the exciton radius to be discussed in Sect. 6.3) of the semiconductor crystals. As a result, the magnitude of the photon wavevector $\boldsymbol{q}$ can be assumed to be zero. Unless noted otherwise, we shall therefore assume that to be the case and abbreviate the dielectric tensor as $\boldsymbol{\varepsilon}(\omega)$. If we assume that $\boldsymbol{\varepsilon}(\boldsymbol{q},\omega)$ is independent of $\boldsymbol{q}$ whatever the value of $\boldsymbol{q}$ (although in fact this is only true for small $\boldsymbol{q}$'s) and calculate its Fourier transform (see also Sect. 4.2.1), we find that $\boldsymbol{\varepsilon}(\boldsymbol{r},\omega)$ is proportional to (the Dirac δ-function) $\delta(\boldsymbol{r})$. This means that the response to $\boldsymbol{E}(\boldsymbol{r})$ is **local**, i.e., $\boldsymbol{D}(\boldsymbol{r})$ depends only on the field applied at the point $\boldsymbol{r}$. On the other hand, if $\boldsymbol{\varepsilon}$ is taken to depend on $\boldsymbol{q}$, its Fourier transform depends on $\boldsymbol{r}-\boldsymbol{r}'$ and the response is **nonlocal**. The variation of $\boldsymbol{\varepsilon}$ with $\boldsymbol{q}$ is called **spatial**

dispersion. In most cases of interest in semiconductor optics this variation is rather small [$\lesssim 10^{-5}$ times the values of $\varepsilon_{ii}(0,\omega)$]. Nevertheless it has been observed in semiconductors such as Ge and GaAs [6.7, 8]. Readers interested in this topic should consult the book by *Agranovich* and *Ginzburg* [6.9].

The rest of our discussions on the optical properties of semiconductors will be concerned mostly with the calculation of the dielectric tensor and its properties. In either isotropic media or cubic crystals the dielectric tensor $\boldsymbol{\varepsilon}(\omega)$ has only three identical diagonal elements. In cases where the tensor nature of $\boldsymbol{\varepsilon}(\omega)$ is not important we shall replace $\varepsilon_{ij}(\omega)$ by a scalar function $\varepsilon(\omega)$ known as the **dielectric function**. This will be assumed to be true for the rest of our discussions unless stated otherwise.

Among the semiconductors we have studied so far, only the II–VI compounds with the wurtzite structure, e.g., CdS and ZnO (Problem 3.7), are not cubic. These crystals are said to be **uniaxial** since they contain a special axis known as the **optical axis** (which coincides with the c-axis in the wurtzite structure). When plane electromagnetic waves propagate along this axis they have the same velocity independent of their polarization direction. Along other crystallographic axes, the light velocity varies with the polarization, giving rise to the phenomenon known as **birefringence**. Crystals with tetragonal or hexagonal symmetry (e.g. wurtzite or chalcopyrite structures) are uniaxial. Those with lower symmetries have two directions along which plane electromagnetic waves travel with the same velocity regardless of their polarization. They are said to be *biaxial* (e.g. orthorhombic GeS, GeSe).

The macroscopic optical properties of an isotropic medium can also be characterized by a **complex refractive index** $\tilde{n}$. The real part n of $\tilde{n}$ is usually also referred to as the **refractive index**. The imaginary part $\varkappa$ is known as the **extinction index or coefficient**. The normal-incidence **reflection coefficient** or **reflectance** $\mathcal{R}$ of a semi-infinite isotropic medium in vacuum is given by [6.10]

$$\mathcal{R} = |(\tilde{n}-1)/(\tilde{n}+1)|^2. \tag{6.8}$$

When light is absorbed in passing through a medium from a point $\boldsymbol{r}_1$ to another point $\boldsymbol{r}_2$, the **absorption coefficient** α of the medium is defined by

$$I(\boldsymbol{r}_2) = I(\boldsymbol{r}_1)\exp(-\alpha|\boldsymbol{r}_2-\boldsymbol{r}_1|), \tag{6.9}$$

where $I(\boldsymbol{r})$ denotes the intensity at $\boldsymbol{r}$. The absorption coefficient is related to $\varkappa$ by

$$\alpha = 4\pi\varkappa/\lambda_0, \tag{6.10}$$

where λ_0 is the wavelength of the light *in vacuum*. The complex refractive index $\tilde{n}$ is related to $\varepsilon(\omega)$ by

$$\varepsilon(\omega) = (\tilde{n})^2. \tag{6.11}$$

6.1.1 Digression: Units for the Frequency of Electromagnetic Waves

At this point it is important to make a digression to present the various units of frequency of electromagnetic waves found in the literature. In principle the frequency of light (to be denoted by ν) is given in *Hertz* (Hz), i.e., cycles per second. The angular frequency ω is related to ν by $\omega = 2\pi\nu$ (radians per second). In light scattering spectroscopy and infrared spectroscopy one often encounters another unit, known as *wavenumber*, which is defined as the reciprocal of the wavelength λ. A wavenumber of 1 cm^{-1} corresponds to a wavelength of 1 cm or a frequency of 3×10^{10} Hz (i.e., 30 GHz). When we study quantum processes involving electromagnetic waves, we quantize the energy of electromagnetic waves into photons. The energy of a photon is equal to $\hbar\omega$, where $\hbar$ is *Planck's constant.* Photon energies are often expressed in electron volts (or eV) or sometimes in units of an equivalent temperature T (in Kelvin). The photon energy is then equal to $k_B T$, where k_B is Boltzmann's constant. The conversion factors between eV and the various units of frequency are 1 eV $\leftrightarrow$ 8065.5 cm^{-1} $\leftrightarrow$ 2.418×10^{14} Hz $\leftrightarrow$ $\lambda = 1.2398$ µm $\leftrightarrow$ 11600 K or 1 cm^{-1} $\leftrightarrow$ 0.12398 meV.

6.1.2 Experimental Determination of Optical Functions

There are several ways to determine the optical functions of a semiconductor as a function of photon energy. The appropriate method depends on whether the photon energy is above or below the bandgap [6.11].

If the photon energy is below the electronic bandgap and well above any phonon energies, the sample absorption coefficient is either zero or very small. The relevant optical function is then the refractive index (which is real). This can be measured by several different methods. One very accurate method is to fabricate the material into a prism and to measure the angle of minimum deviation of a beam of light passing through this prism. Another method is to polish the sample into a thin slab with parallel surfaces and to measure the interference fringes in the sample transmission or reflection spectra.

When the photon energy is increased from below to above the bandgap, typically the semiconductor absorption coefficient increases rapidly to values as large as or larger than 10^4 cm^{-1}. As a result the sample becomes opaque for photon energies higher than the bandgap unless its thickness is very small. Since the intensity of light transmitted through a sample decreases exponentially with thickness according to (6.9), it is necessary to thin the sample down to about α^{-1} in order to detect easily the transmitted radiation. The quantity α^{-1} is therefore known as the **optical penetration depth**. Since α tends to increase rapidly as a function of photon energy above the bandgap, a series of progressively thinner samples is necessary in order to measure α over a wide photon energy range.

An alternative and more popular method for determining the complex dielectric function of strongly absorbing samples is to use reflection measurements. The major drawback of these measurements is their sensitivity to the sample surface quality. As pointed out above, the penetration depth of light into a sample is equal to α^{-1} and for semiconductors α is typically of the order of 10^4–10^6 cm^{-1} above the absorption edge. For such large absorption coefficients, light will probe only a thin layer, about 1 μm or less thick, at the top of the sample. As a result, the reflectance will be very sensitive to the presence of surface contaminants such as oxides or even air pollutants. Unless great care is taken to achieve an atomically clean surface in an *ultra-high vacuum* (see Chap. 8 for a definition of ultra-high vacuum and discussions on surface properties of semiconductors), the measured reflectance is that of a "composite", consisting of a surface contaminant layer and the bulk sample. Some authors have labeled the dielectric functions deduced from the reflectance of such contaminated samples as "**pseudodielectric functions**" [6.12]. As an example of the sensitivity of the dielectric function to sample surface quality, we compare in Fig. 6.2 the "pseudodielectric function" of an oxidized GaAs surface (oxide thickness 10 Å) with that of GaAs with an abrupt and clean surface. Notice that the largest difference in ε between these surfaces occurs around 4.8 eV photon energy. In this energy range the value of $\varkappa$ is about 4 [6.12], with a corresponding penetration depth of about 50 Å. Obviously, even an oxide layer of only 10 Å will make a difference in the reflectivity.

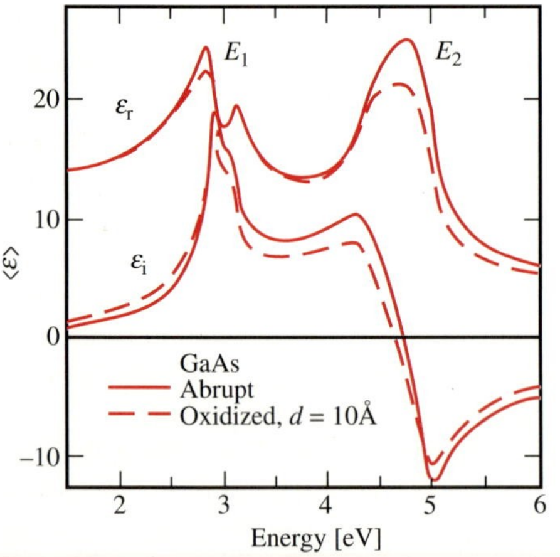

Fig. 6.2. The real and imaginary parts of the dielectric function of GaAs measured by ellipsometry. The *solid curve*, labeled "abrupt", was obtained for an atomically clean surface, the *broken* one for a surface covered by an oxide layer. [6.12]

Except for this surface sensitivity, determining the dielectric function from reflection measurements is quite straightforward. It involves irradiating the sample at either normal or oblique incidence. In oblique incidence techniques, the reflectance $\mathcal{R}_s$ and $\mathcal{R}_p$ of the s- and p-polarized components of the incident light are measured (components of the incident radiation perpendicular and parallel to the plane of incidence are labeled, respectively, as s- and p-polarized). These reflectances are related to the complex refractive index by the Fresnel formulae [6.10]

$$\mathcal{R}_s = |r_s|^2 = \left| \frac{\cos\phi - (\tilde{n}^2 - \sin^2\phi)^{1/2}}{\cos\phi + (\tilde{n}^2 - \sin^2\phi)^{1/2}} \right|^2 \tag{6.12a}$$

and

$$\mathcal{R}_p = |r_p|^2 = \left| \frac{\tilde{n}^2\cos\phi - (\tilde{n}^2 - \sin^2\phi)^{1/2}}{\tilde{n}^2\cos\phi + (\tilde{n}^2 - \sin^2\phi)^{1/2}} \right|^2 , \tag{6.12b}$$

where r_s and r_p are the complex *reflectivity* for s- and p-polarized light, respectively, and ϕ is the angle of incidence. The complex refractive index can be deduced by measuring both $\mathcal{R}_s$ and $\mathcal{R}_p$ at a fixed ϕ.

An oblique angle of incidence technique which has become very popular in the past decade is **ellipsometry**. This name derives from the fact that when linearly polarized light that is neither s- nor p-polarized is incident on a medium at an oblique angle, the reflected light is elliptically polarized (Fig. 6.3). The ratio (σ) of the complex reflectivities r_p/r_s can be determined by measuring the orientation and the ratio of the axes of the polarization ellipse corresponding to the reflected light. The complex dielectric function can be determined from σ and ϕ using the expression (Problem 6.1)

$$\varepsilon = \sin^2\phi + \sin^2\phi\tan^2\phi\left(\frac{1-\sigma}{1+\sigma}\right)^2 . \tag{6.13}$$

Figure 6.3 shows schematically the principal components of an ellipsometer [6.13]. The light source can be either a laser or a broad-band source such as a *xenon gas discharge* or a *quartz-halogen lamp*. In the case of a broad-band

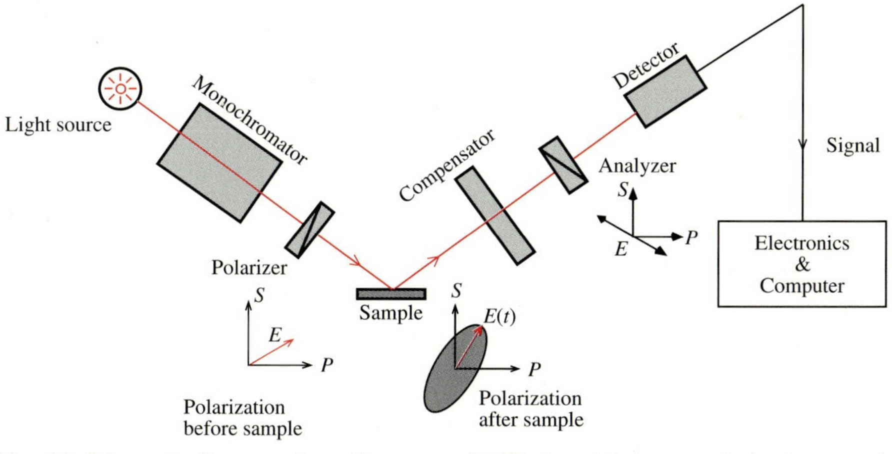

Fig. 6.3. Schematic diagram of an ellipsometer [6.13]. P and S denote polarizations parallel or perpendicular to the plane of incidence, respectively

source the light is passed through a *monochromator* to select a narrow band of frequencies. Ellipsometry performed over a wide range of photon frequencies is known as *spectroscopic ellipsometry*. The light leaving the monochromator is passed through a linear polarizer. After reflection from the sample surface the light experiences a relative *phase shift* θ between its s and p components and becomes elliptically polarized. There are several variations in the method to detect the ellipticity of the reflected light. In Fig. 6.3 a *compensator* introduces another relative phase shift $-\theta$ which exactly cancels the ellipticity induced by the reflection and the light becomes linearly polarized again. This null condition can be easily detected by passing the light through an *analyzer*, which consists of another linear polarizer oriented to block out the light after the compensator.

Despite the increasing popularity of ellipsometers, perhaps the most common and also the simplest technique for determining the optical constants is to measure the normal-incidence reflectance. Figure 6.4 shows the construction of the apparatus used by *Philipp* and *Ehrenreich* [6.14] to measure the normal-incidence reflectance $\mathcal{R}$ of semiconductors from about 1 eV to 20 eV. Modern versions of this setup are basically the same except for improvements in the light source, gratings and detectors. Instead of gas-discharge lamps *synchrotron radiation* [6.15, 16] is preferred nowadays as the source of high intensity and broad-band radiation extending from the infrared to the X-ray region. *Holographic gratings* have mostly replaced mechanically ruled gratings since much more uniform and closely spaced grooves can be generated by interference effects. In the area of detectors, the biggest improvement has been the appearance of multichannel detectors such as CCDs (which stands for *charge-coupled devices*). These are solid-state detectors that are sensitive from the near infrared (wavelengths of about 1 μm) to the ultraviolet. They can also be used in the vacuum ultraviolet and soft-x-ray region together with a scintillator and image-intensifiers. They allow the entire optical spectrum, covering a wide range of wavelengths, to be recorded electronically in one exposure. Finally, the electronic signal from these detectors can be processed by desktop computers to display the optical constants in real time with time resolution of milliseconds.

In principle, it is necessary to measure both the normal-incidence reflectance $\mathcal{R}$ and the absorption coefficient α in order to determine the complex refractive index and hence the dielectric function $\varepsilon(\omega)$. In practice it is sufficient to simply measure $\mathcal{R}$ over a wide range of photon frequencies and then deduce the absorption coefficients using the *Kramers–Kronig relations* (*KKRs*) or *dispersion relations*, which were introduced in Sect. 4.3.1.

6.1.3 Kramers–Kronig Relations

If we assume that the field strength of the incident radiation is weak enough that the induced polarization is linearly dependent on the electric field, both $\chi(\omega)$ and $\varepsilon(\omega)$ describe the linear responses of a medium to an external field.

It can be shown that *linear response functions* such as ε or χ satisfies the KKRs

$$\varepsilon_r(\omega) - 1 = \frac{2}{\pi}\mathcal{P}\int_0^\infty \frac{\omega'\varepsilon_i(\omega')d\omega'}{\omega'^2 - \omega^2} \tag{6.14}$$

and

$$\varepsilon_i(\omega) = -\frac{2\omega}{\pi}\mathcal{P}\int_0^\infty \frac{\varepsilon_r(\omega')d\omega'}{\omega'^2 - \omega^2}, \tag{6.15}$$

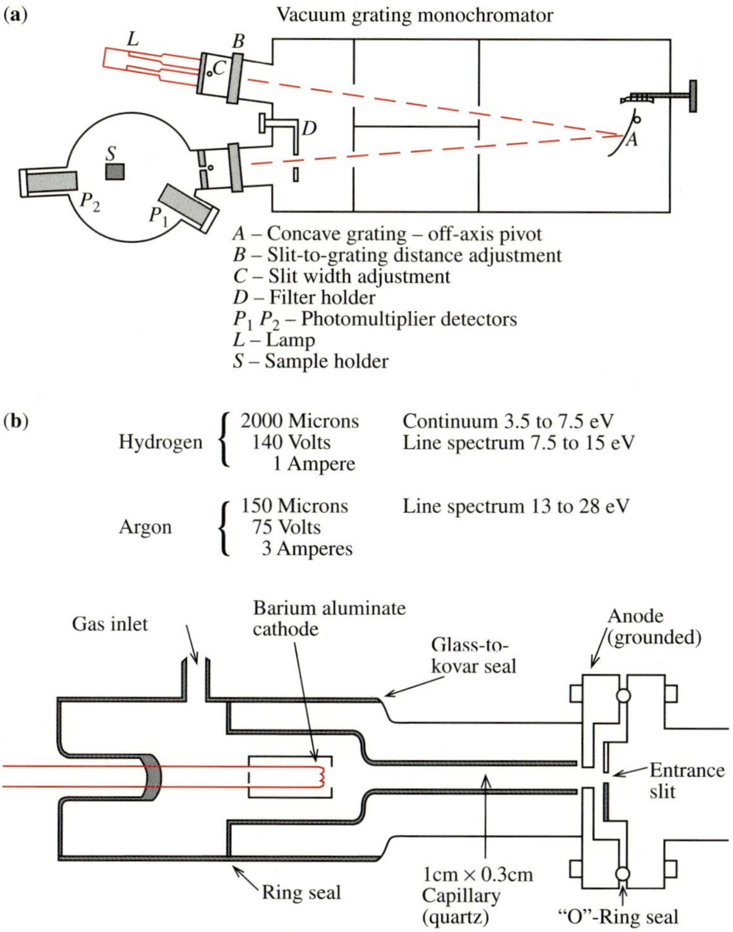

Fig. 6.4. (**a**) The vacuum reflectometer used by *Philipp* and *Ehrenreich* [6.14] to measure the normal incidence reflectance of semiconductors from about 1 to 20 eV. (**b**) Detailed construction of the gas discharge lamp they used

where $\mathcal{P}$ means the principal value of the integral. The proof of this relation (see, for example, [6.10] and Problem 6.2) is based on *the principle of causality* (i. e., a response to an applied field such as the polarization cannot precede the applied field). Although the refractive index is not a response function, we can derive a KKR for $\tilde{n}$ based on the fact that it has the same analytic properties as ε. For example, $\tilde{n}$ approaches unity as ω approaches infinity and, if relativistic causality holds (i. e., the response cannot propagate faster than the speed of light), $\tilde{n}$ is analytic in the upper half plane of the complex variable ω. As a result, a set of KKRs for $\tilde{n}$ can be obtained from (6.14 and 15) by replacing ε_r and ε_i by n and $\varkappa$, respectively.

Since it is easier to measure the normal-incidence reflectance than either ε or n of a bulk sample in most experiments, it is desirable to derive a KKR for the reflectance. Such a relation can be obtained by constructing a complex function, known as complex *reflectivity*, with analytic properties similar to ε. Let us define such a function $\tilde{r}$ as

$$\tilde{r} = \frac{(\tilde{n}-1)}{(\tilde{n}+1)} = \varrho \exp(\mathrm{i}\theta). \tag{6.16}$$

Consider a *contour integral* of the function

$$f(\omega') = \left(\frac{1+\omega'\omega}{1+\omega'^2}\right) \ln \tilde{r}(\omega')/(\omega'-\omega) \tag{6.17}$$

over the contour C shown in Fig. 6.5. The function f is constructed so that, in contrast to $\ln \tilde{r}$, it approaches zero as ω' approaches infinity. Furthermore, the only poles of $f(\omega')$ are i and ω and the residue of f at ω gives $\ln \tilde{r}(\omega)$. From the Cauchy theorem the contour integral of f is given by the sum of the residues at i and ω. The result can be written as

$$\oint_C f(\omega')d\omega' = 2\pi\mathrm{i}\left[\ln \tilde{r}(\omega) - \frac{1}{2}\ln \tilde{r}(\mathrm{i})\right]. \tag{6.18}$$

Taking the real part of both sides of (6.18) we obtain (note that the principal part of the integral in (6.19) is equal to the contour integral in (6.18) minus half of the residue at $\omega' = \omega$, see Fig. 6.5):

$$\mathcal{P}\int_{-\infty}^{\infty} \frac{(1+\omega'\omega)\ln \varrho(\omega')d\omega'}{(1+\omega'^2)(\omega'-\omega)} = -\pi\theta \tag{6.19}$$

[the fact that $\ln \tilde{r}(\mathrm{i})$ is real, needed for this derivation, can be surmised from (6.48 and 49) below]. By simplifying this equation further, we can express θ in terms of an integral of $\ln \varrho(\omega)$:

$$\theta = -\frac{2\omega}{\pi}\mathcal{P}\int_0^{\infty} \frac{\ln \varrho(\omega')d\omega'}{(\omega'^2-\omega^2)}. \tag{6.20}$$

Using (6.20), the complex reflectivity, refractive index, and dielectric function can all be deduced from measurement of the reflectance over a wide frequency range. Since it is difficult to measure the reflectance from zero to infinite fre-

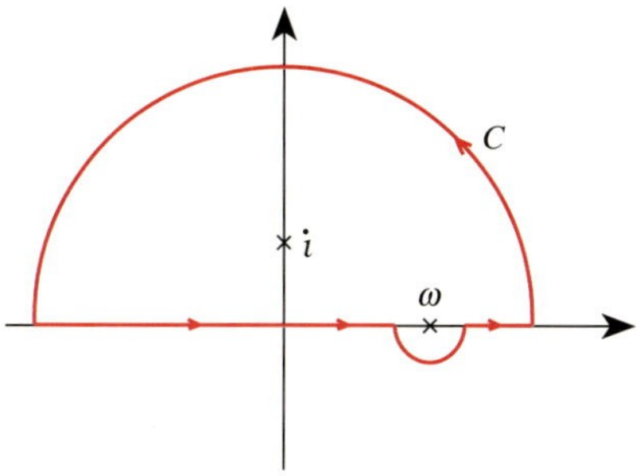

Fig. 6.5. Contour C for integrating the function f in (6.17) to obtain the KKR for the reflectivity

quency, it becomes necessary to use extrapolations to both low and high frequencies. At low frequency in an undoped sample $\mathcal{R}$ can be approximated by a constant. At high enough frequencies the semiconductor can be approximated by a free electron gas with the dielectric constant

$$\varepsilon = 1 - (\omega_{\mathrm{p}}/\omega)^2, \tag{6.21}$$

where the **plasma frequency** ω_{p} (see Problem 6.3 for its definition and derivation) is determined by the density of valence electrons only. The core electrons are so tightly bound that their contributions to the dielectric function can usually be neglected.

6.2 The Dielectric Function

6.2.1 Experimental Results

Some typical reflectance spectra of group IV and III–V semiconductors measured by *Philipp* and *Ehrenreich* [6.14] are shown in Figs. 6.6a–8a. The corresponding real and imaginary parts of the dielectric function and the imaginary part of the so-called energy loss function $(1/\varepsilon)$ deduced from the reflectance spectra using the KKRs are shown in Figs. 6.6b–8b. The dielectric function of GaAs displayed in Fig. 6.8b compares well with that measured by ellipsometry over a smaller photon energy range, shown in Fig. 6.2.

Notice that both the reflectance spectra and the dielectric functions in Si, Ge, and GaAs show considerable structure in the form of peaks and *shoulders*. These structures arise from optical transitions from the filled valence bands to the empty conduction bands. That such structures occur in the optical transitions between valence bands and conduction bands in crystalline semiconductors should not be surprising (they are, however, smoothed out by disorder in **amorphous semiconductors**). In Sect. 2.3.4 we studied the selection rules for

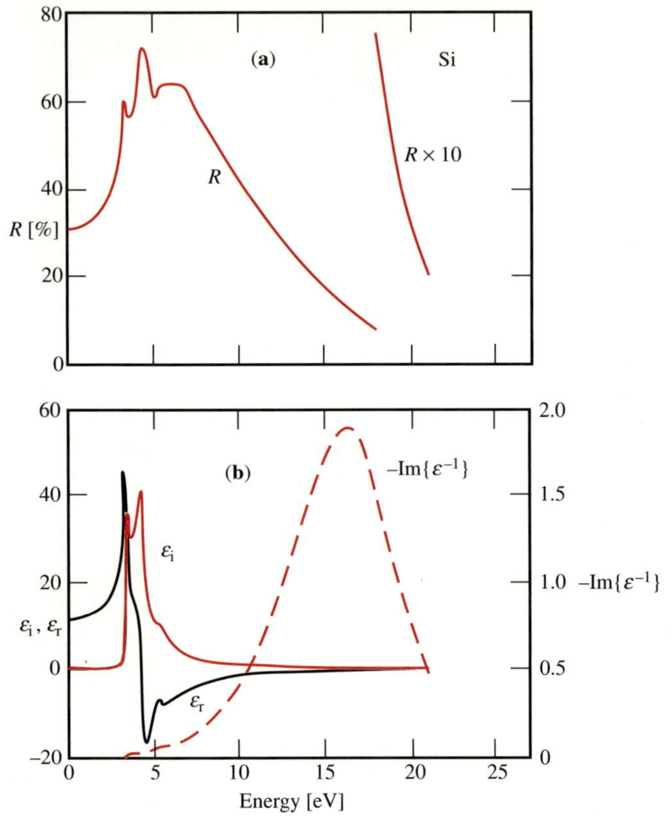

Fig. 6.6. (**a**) Reflectance curve of Si measured at room temperature. (**b**) The real (ε_r) and imaginary (ε_i) parts of the dielectric function and the imaginary part of $(-1/\varepsilon)$ (known as the energy loss function) of Si deduced from the reflectivity curve in (**a**) using the Kramers–Kronig relation [6.14]. Notice that the peak of $\text{Im}\{-1/\varepsilon\}$, occurs at the plasma energy of the valence electrons (Problem 6.3)

the matrix elements of the electric dipole operator between two given electron wavefunctions in zinc-blende-type crystals. In the following sections we will discuss the relation between the band structure of a semiconductor and its optical spectra based on a microscopic theory of the dielectric function.

6.2.2 Microscopic Theory of the Dielectric Function

We will use a *semi-classical approach* to derive the Hamiltonian describing the interaction between an external electromagnetic field and Bloch electrons inside a semiconductor. In this approach the electromagnetic field is treated classically while the electrons are described by quantum mechanical (Bloch)

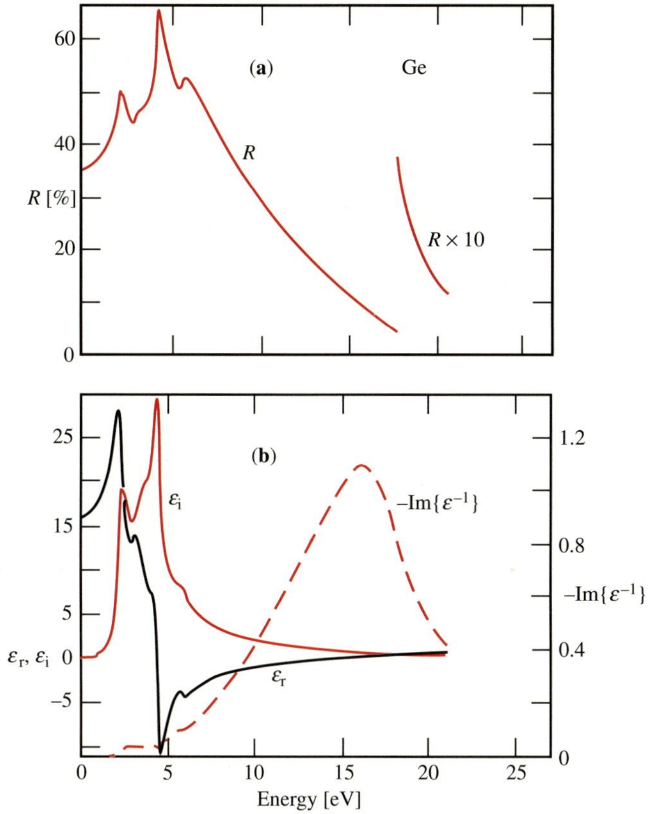

Fig. 6.7a,b. Curves for Ge similar to those for Si in Fig. 6.6

wave functions. Although this approach may not be regarded as being as rigorous as a fully quantum mechanical treatment in which the electromagnetic waves are quantized into photons (e. g., [6.17]), it has the advantage of being simpler and easier to understand. This approach generates the same results as the quantum mechanical treatment, including even spontaneous emission (to be discussed in Chap. 7).

We start with the unperturbed one-electron Hamiltonian introduced already in (2.4):

$$\mathcal{H}_0 = p^2/2m + V(\boldsymbol{r}). \tag{6.22}$$

To describe the electromagnetic fields we introduce a vector potential $\boldsymbol{A}(\boldsymbol{r}, t)$ and a scalar potential $\Phi(\boldsymbol{r}, t)$. Because of **gauge invariance**, the choice of these potentials is not unique. For simplicity, we will choose the *Coulomb gauge* [6.10], in which

$$\Phi = 0 \text{ and } \boldsymbol{\nabla} \cdot \boldsymbol{A} = 0. \tag{6.23}$$

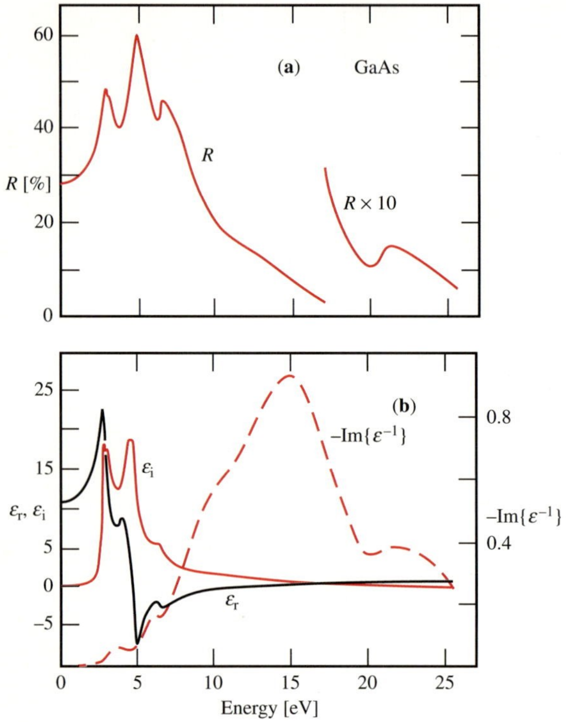

Fig. 6.8a,b. Curves for GaAs similar to those for Si in Fig. 6.6

In this gauge the electric and magnetic fields $(\boldsymbol{E}, \boldsymbol{B})$ are given by[2]

$$\boldsymbol{E} = -\frac{1}{c}\frac{\partial \boldsymbol{A}}{\partial t} \text{ and } \boldsymbol{B} = \nabla \times \boldsymbol{A}, \tag{6.24}$$

where c is the velocity of light. The classical Hamiltonian of a charge Q in the presence of an external magnetic field can be obtained from the free-particle Hamiltonian by replacing the momentum $\boldsymbol{P}$ by $\boldsymbol{P}-(Q\boldsymbol{A}/c)$, where $\boldsymbol{P}$ is the momentum conjugate to the position vector [6.10, p. 409]. Correspondingly, we obtain the quantum mechanical Hamiltonian describing the motion of a charge $-e$ in an external electromagnetic field by replacing the electron momentum operator $\boldsymbol{p}$ in (6.22) by $\boldsymbol{p} + (e\boldsymbol{A}/c)$:

$$\mathcal{H} = \frac{1}{2m}[\boldsymbol{p} + (e\boldsymbol{A}/c)]^2 + V(\boldsymbol{r}). \tag{6.25}$$

The term $[\boldsymbol{p} + (e\boldsymbol{A}/c)]^2/2m$ can be expanded [keeping in mind that $\boldsymbol{p}$ is an operator which does not commute with $\boldsymbol{A}(\boldsymbol{r})$] as

$$\frac{1}{2m}\left(\boldsymbol{p} + \frac{e\boldsymbol{A}}{c}\right)^2 = \frac{p^2}{2m} + \frac{e}{2mc}\boldsymbol{A}\cdot\boldsymbol{p} + \frac{e}{2mc}\boldsymbol{p}\cdot\boldsymbol{A} + \frac{e^2A^2}{2mc^2}. \tag{6.26}$$

[2] In (6.24, 25, 26, 28, 29, 31) delete c for SI units.

Using the definition of $\boldsymbol{p}$ as the operator $(\hbar/\mathrm{i})\nabla$ we can express the term $\boldsymbol{p}\cdot\boldsymbol{A}$ as

$$(\boldsymbol{p}\cdot\boldsymbol{A})f(r) = \boldsymbol{A}\cdot\left(\frac{\hbar}{\mathrm{i}}\nabla f\right) + \left(\frac{\hbar}{\mathrm{i}}\nabla\cdot\boldsymbol{A}\right)f. \tag{6.27}$$

From (6.23), $\nabla\cdot\boldsymbol{A} = 0$, and therefore $[e/(2mc)]\boldsymbol{p}\cdot\boldsymbol{A} = [e/(2mc)]\boldsymbol{A}\cdot\boldsymbol{p}$. For the purpose of calculating linear optical properties we can also neglect the term $e^2A^2/(2mc^2)$, which depends quadratically on the field. Under this assumption we can approximate $\mathcal{H}$ by

$$\mathcal{H} = \mathcal{H}_0 + \frac{e}{mc}\boldsymbol{A}\cdot\boldsymbol{p}. \tag{6.28}$$

Compared with the unperturbed Hamiltonian $\mathcal{H}_0$, the extra term $[e/(mc)]\boldsymbol{A}\cdot\boldsymbol{p}$ describes the interaction between the radiation and a Bloch electron. As a result, this term will be referred to as the **electron–radiation interaction Hamiltonian** $\mathcal{H}_{\mathrm{eR}}$:

$$\mathcal{H}_{\mathrm{eR}} = \frac{e}{mc}\boldsymbol{A}\cdot\boldsymbol{p}. \tag{6.29}$$

Note that the form of $\mathcal{H}_{\mathrm{eR}}$ depends on the gauge we choose. Another form of $\mathcal{H}_{\mathrm{eR}}$ commonly found in the literature is

$$\mathcal{H}_{\mathrm{eR}} = (-e)\boldsymbol{r}\cdot\boldsymbol{E}. \tag{6.30}$$

Equation (6.30) can be shown to be equivalent to (6.29) in the limit that the wavevector $\boldsymbol{q}$ of the electromagnetic wave is small [(6.30) corresponds to the **electric dipole approximation**; (6.29) is more general]. Both forms of $\mathcal{H}_{\mathrm{eR}}$ neglect the term quadratic in the field. In the case of (6.30), the interaction between the electrons and the electromagnetic field via the Lorentz force has been neglected. Since this force depends on $\boldsymbol{v}\times\boldsymbol{B}$ and since the velocity $\boldsymbol{v}$ varies with $\boldsymbol{E}$, this term has a quadratic dependence on the applied field. The advantage of using (6.29) for semiconductors is that the matrix elements of the electron momentum enter directly into the $\boldsymbol{k}\cdot\boldsymbol{p}$ method of band structure calculation. For example, the matrix element between the lowest Γ_{1c} conduction band and the top Γ_4 valence band in tetrahedrally bonded semiconductors can be determined from the conduction band effective mass using (2.44). Other matrix elements are related to the dispersion of the valence bands as shown in (2.63–65).

There are several ways to calculate the dielectric function of a semiconductor from $\mathcal{H}_{\mathrm{eR}}$. Again, we will take the simplest approach. We first assume that $\boldsymbol{A}$ is weak enough that we can apply time-dependent perturbation theory (in the form of the Fermi Golden Rule) to calculate the transition probability per unit volume R for an electron in the valence band state $|\mathrm{v}\rangle$ (with energy E_{v} and wavevector $\boldsymbol{k}_{\mathrm{v}}$) to the conduction band $|\mathrm{c}\rangle$ (with corresponding energy E_{c} and wavevector $\boldsymbol{k}_{\mathrm{c}}$). To do this we need to evaluate the matrix element $|\langle\mathrm{c}|\mathcal{H}_{\mathrm{eR}}|\mathrm{v}\rangle|^2$:

$$|\langle\mathrm{c}|\mathcal{H}_{\mathrm{eR}}|\mathrm{v}\rangle|^2 = (e/mc)^2|\langle\mathrm{c}|\boldsymbol{A}\cdot\boldsymbol{p}|\mathrm{v}\rangle|^2. \tag{6.31}$$

We will now write the vector potential $\boldsymbol{A}$ as $A\hat{\boldsymbol{e}}$, where $\hat{\boldsymbol{e}}$ is a unit vector parallel to $\boldsymbol{A}$. In terms of the amplitude of the incident electric field $\boldsymbol{E}(\boldsymbol{q},\omega)$, the amplitude of $\boldsymbol{A}$ can be written as

$$A = -\frac{E}{2q}\{\exp[\mathrm{i}(\boldsymbol{q}\cdot\boldsymbol{r}-\omega t)]+\mathrm{c.c.}\}, \tag{6.32}$$

where c.c. stands for complex conjugate. The calculation of the matrix element $\langle c|\boldsymbol{A}\cdot\boldsymbol{p}|v\rangle$ involves integration over space. The integration over time of the term $\exp[\mathrm{i}(-\omega t)]$ in (6.32) and the corresponding factors in the electron Bloch functions leads formally to

$$\int \exp(\mathrm{i}E_c t/\hbar)\exp[\mathrm{i}(-\omega t)]\exp(-\mathrm{i}E_v t/\hbar)dt \propto \delta(E_c(\boldsymbol{k}_c)-E_v(\boldsymbol{k}_v)-\hbar\omega), \tag{6.33}$$

i. e., the delta function found in the Fermi Golden Rule. This result means that the electron in the valence band absorbs the photon energy and is then excited into the conduction band. Hence this term in (6.32) describes an absorption process. Similarly, the matrix element of the complex conjugate: $\langle c|\exp(\mathrm{i}\omega t)|v\rangle$, gives rise to $\delta(E_c(\boldsymbol{k}_c)-E_v(\boldsymbol{k}_v)+\hbar\omega)$. This term is nonzero when an electron which is initially in the conduction band emits a photon and ends up in the valence band. Since this emission process occurs now in the presence of an external field, this term describes a **stimulated emission** process. In other words, the two terms in (6.32) describe, respectively, absorption and emission of photons by electrons in a semiconductor under the influence of an external electromagnetic field. Notice that the magnitudes of the matrix elements describing both processes are equal. The stimulated emission term, represented as c.c. in (6.33), will be discussed in more detail in Chap. 7 but will be neglected in this chapter.

Writing the Bloch functions [see (2.6)] for the electrons in the conduction and valence bands, respectively, as

$$|c\rangle = u_{c,\boldsymbol{k}_c}(\boldsymbol{r})\exp[\mathrm{i}(\boldsymbol{k}_c\cdot\boldsymbol{r})] \tag{6.34a}$$

and

$$|v\rangle = u_{v,\boldsymbol{k}_v}(\boldsymbol{r})\exp[\mathrm{i}(\boldsymbol{k}_v\cdot\boldsymbol{r})] \tag{6.34b}$$

and using the expression for A in (6.32), we obtain

$$|\langle c|\boldsymbol{A}\cdot\boldsymbol{p}|v\rangle|^2 = \frac{|E|^2}{4q^2}\left|\int u^*_{c,\boldsymbol{k}_c}\exp[\mathrm{i}(\boldsymbol{q}-\boldsymbol{k}_c)\cdot\boldsymbol{r}](\hat{\boldsymbol{e}}\cdot\boldsymbol{p})u_{v,\boldsymbol{k}_v}\exp(\mathrm{i}\boldsymbol{k}_v\cdot\boldsymbol{r})dr\right|^2. \tag{6.35}$$

Operating with $\boldsymbol{p}$ on $u_{v,\boldsymbol{k}_v}\exp(\mathrm{i}\boldsymbol{k}_v\cdot\boldsymbol{r})$ yields two terms:

$$\boldsymbol{p}u_{v,\boldsymbol{k}_v}\exp(\mathrm{i}\boldsymbol{k}_v\cdot\boldsymbol{r}) = \exp(\mathrm{i}\boldsymbol{k}_v\cdot\boldsymbol{r})\boldsymbol{p}u_{v,\boldsymbol{k}_v} + \hbar\boldsymbol{k}_v u_{v,\boldsymbol{k}_v}\exp(\mathrm{i}\boldsymbol{k}_v\cdot\boldsymbol{r}). \tag{6.36}$$

The integral of the second term in (6.36) multiplied by $u^*_{c,\boldsymbol{k}_c}$, vanishes because $u_{c,\boldsymbol{k}_c}$ and $u_{v,\boldsymbol{k}_v}$ are orthogonal. We can split the corresponding integral of the first term

$$\int u^*_{c,\boldsymbol{k}_c}\exp[\mathrm{i}(\boldsymbol{q}-\boldsymbol{k}_c+\boldsymbol{k}_v)\cdot\boldsymbol{r}]\boldsymbol{p}u_{v,\boldsymbol{k}_v}dr$$

into two parts by writing $\boldsymbol{r}=\boldsymbol{R}_j+\boldsymbol{r}'$, where $\boldsymbol{r}'$ lies within one unit cell and $\boldsymbol{R}_j$ is a lattice vector. Because of the periodicity of the functions $u_{c,\boldsymbol{k}_c}$ and $u_{v,\boldsymbol{k}_v}$, we find

$$\int u^*_{c,k_c} \exp[i(q - k_c + k_v) \cdot r] p u_{v,k_v} dr$$
$$= \left(\sum_j \exp[i(q - k_c + k_v) \cdot R_j] \right) \int_{\substack{\text{unit}\\\text{cell}}} u^*_{c,k_c} \exp[i(q - k_c + k_v) \cdot r'] p u_{v,k_v} dr'. \quad (6.37)$$

The summation of $\exp[i(q - k_c + k_v) \cdot R_i]$ over all the lattice vectors R_j results in a delta function $\delta(q - k_c + k_v)$. This term ensures that wavevector is conserved in the absorption process:

$$q + k_v = k_c. \quad (6.38)$$

Equation (6.38) is a consequence of the translation symmetry of the crystal and therefore must be satisfied for all processes in a perfect crystal. It is, however, relaxed in amorphous semiconductors (see Appendix by J. Tauc on p. 566).

Using (6.38), the integral over the unit cell in (6.37) simplifies to

$$\int_{\substack{\text{unit}\\\text{cell}}} u^*_{c,k_c} \exp[i(q - k_c + k_v) \cdot r] p u_{v,k_v} dr' = \int_{\substack{\text{unit}\\\text{cell}}} u^*_{c,k_v+q} p u_{v,k_v} dr'. \quad (6.39)$$

This expression can be further simplified if we assume that q is much smaller than the size of the Brillouin zone, a condition usually satisfied by visible photons, whose wavelengths are of the order of 500 nm. For small q the wavefunction u_{c,k_v+q} can be expanded into a Taylor series in q:

$$u_{c,k_v+q} = u_{c,k_v} + q \cdot \nabla_k u_{c,k_v} + \dots . \quad (6.40)$$

When q is small enough that all the q-dependent terms in (6.40) can be neglected, the matrix element $|\langle c|\hat{e} \cdot p|v\rangle|^2$ is given by

$$|\langle c|\hat{e} \cdot p|v\rangle|^2 = \left(\int_{\substack{\text{unit}\\\text{cell}}} u^*_{c,k} (\hat{e} \cdot p) u_{v,k} dr' \right)^2. \quad (6.41a)$$

This approximation is known as the **electric dipole approximation** [it can be shown to be equivalent to using (6.30)] and the corresponding matrix element in (6.41a) as the *electric dipole transition matrix element*. Notice that the electric dipole approximation is equivalent to expanding the term $\exp(iq \cdot r)$ in (6.32) into a Taylor series: $1 + i(q \cdot r) + \dots$ and neglecting all the q-dependent terms. In this case we have $k_v = k_c$, and the transitions are said to be **vertical** or **direct**.

If the electric dipole matrix element is zero, the optical transition is determined by the $q \cdot \nabla_k u_{k_v}$ term in (6.41). The matrix element

$$|\langle c|\hat{e} \cdot p|v\rangle|^2 = \left(\int_{\substack{\text{unit}\\\text{cell}}} q \cdot (\nabla_k u^*_{c,k}) (\hat{e} \cdot p) u_{v,k} dr' \right)^2 \quad (6.41b)$$

gives rise to **electric quadrupole** and **magnetic dipole transitions**. These higher order optical transitions can also be considered as arising from the $\mathrm{i}(\boldsymbol{q}\cdot\boldsymbol{r})$ term in the Taylor expansion of $\exp(\mathrm{i}\boldsymbol{q}\cdot\boldsymbol{r})$. Compared to the electric dipole transitions they are reduced in strength by a factor of (lattice constant/wavelength of light)^2 [6.7].

From now on we shall restrict ourselves to electric dipole transitions unless stated otherwise. To simplify the notation we drop the subscript v or c in the electron wavevectors $\boldsymbol{k}_{\mathrm{v}}$ and $\boldsymbol{k}_{\mathrm{c}}$, since they are the same. In most cases the momentum matrix element in (6.41a) is not strongly dependent on $\boldsymbol{k}$ [6.18] so we shall replace it by the constant $|P_{\mathrm{cv}}|^2$. Equation (6.31) can then be simplified to read

$$|\langle c|\mathcal{H}_{\mathrm{eR}}|v\rangle|^2 = (e/mc)^2|A|^2|P_{\mathrm{cv}}|^2. \tag{6.42}$$

In using (6.42) one must remember that we have defined $\boldsymbol{A}$ in (6.32) as a sum of $\exp[\mathrm{i}(\boldsymbol{q}\cdot\boldsymbol{r}-\omega t)]$ and its complex conjugate $\exp[-\mathrm{i}(\boldsymbol{q}\cdot\boldsymbol{r}-\omega t)]$ in order that $\boldsymbol{A}$ be a real function of space and time. However, of these two terms only the one containing $\exp(-\mathrm{i}\omega t)$ yields the absorption process. Its complex conjugate, which gives rise to stimulated emission, has been completely disregarded in the present discussion.

The electric dipole transition probability R for photon absorption per unit time obtained by substituting (6.42) and (6.32) into the Fermi Golden Rule:

$$R = (2\pi/\hbar)\sum_{\boldsymbol{k}_{\mathrm{c}},\boldsymbol{k}_{\mathrm{v}}}|\langle c|\mathcal{H}_{\mathrm{eR}}|v\rangle|^2\delta(E_{\mathrm{c}}(\boldsymbol{k}_{\mathrm{c}})-E_{\mathrm{v}}(\boldsymbol{k}_{\mathrm{v}})-\hbar\omega) \tag{6.43a}$$

is thus given by

$$R = \frac{2\pi}{\hbar}\left(\frac{e}{m\omega}\right)^2\left|\frac{E(\omega)}{2}\right|^2\sum_{k}|P_{\mathrm{cv}}|^2\delta(E_{\mathrm{c}}(\boldsymbol{k})-E_{\mathrm{v}}(\boldsymbol{k})-\hbar\omega). \tag{6.43b}$$

If we restrict the summation to those $\boldsymbol{k}$'s allowed per *unit volume* of crystal then (6.43b) gives the absorption transition rate *per unit volume* of the crystal. The power lost by the field due to absorption in *unit volume* of the medium is simply the transition probability per unit volume multiplied by the energy in each photon:

$$\text{Power loss} = R\hbar\omega. \tag{6.44}$$

This power loss from the field can also be expressed in terms of either α or ε_{i} of the medium by noting that the rate of decrease in the energy of the incident beam per unit volume is given by $-dI/dt$, where I is the intensity of the incident beam:

$$-\frac{dI}{dt} = -\left(\frac{dI}{dx}\right)\left(\frac{dx}{dt}\right) = \frac{c}{n}\alpha I \tag{6.45}$$

$$= \frac{\varepsilon_{\mathrm{i}}\omega I}{n^2}. \tag{6.46}$$

The energy density I can be related to the field amplitude by

$$I = \frac{n^2}{8\pi}|E(\omega)|^2. \tag{6.47}$$

Equating $-dI/dt$ with the expression for the power loss per unit volume of the field in (6.44) we obtain

$$\varepsilon_i(\omega) = \frac{1}{4\pi\varepsilon_0}\left(\frac{2\pi e}{m\omega}\right)^2 \sum_{\boldsymbol{k}} |P_{cv}|^2 \delta(E_c(\boldsymbol{k}) - E_v(\boldsymbol{k}) - \hbar\omega). \tag{6.48}$$

By using the KKRs we can then obtain the expression for ε_r:

$$\varepsilon_r(\omega) = 1 + \frac{4\pi e^2}{4\pi\varepsilon_0 m}\left[\sum_{\boldsymbol{k}}\left(\frac{2}{m\hbar\omega_{cv}}\right)\frac{|P_{cv}|^2}{\omega_{cv}^2 - \omega^2}\right] \tag{6.49}$$

where $\hbar\omega_{cv} = E_c(\boldsymbol{k}) - E_v(\boldsymbol{k})$. We have purposely written (6.49) in a form similar to the dielectric function for a collection of classical, charged, harmonic oscillators with frequencies ω_i (Problem 6.4):

$$\varepsilon_r(\omega) = 1 + \frac{4\pi e^2}{4\pi\varepsilon_0 m}\left(\sum_i \frac{N_i}{\omega_i^2 - \omega^2}\right), \tag{6.50}$$

where N_i is the number of oscillators per unit volume with frequencies ω_i. Comparing the two expressions for ε_r in (6.49) and (6.50), we see that the dimensionless quantity

$$f_{cv} = \frac{2|P_{cv}|^2}{m\hbar\omega_{cv}} \tag{6.51}$$

is essentially the "number" of oscillators with frequency ω_{cv}. Therefore f_{cv} is known as the **oscillator strength** of the optical transition.[3]

6.2.3 Joint Density of States and Van Hove Singularities

Note that in (6.48) most of the dispersion in ε_i comes from the summation over the delta function $\delta(E_c(\boldsymbol{k}) - E_v(\boldsymbol{k}) - \hbar\omega)$. This summation can be converted into an integration over energy by defining a **joint density of states** for the (doubly degenerate) conduction and valence bands (see Sect. 4.3.1 for the definition of density of states):

$$D_j(E_{cv}) = \frac{1}{4\pi^3}\int \frac{dS_k}{|\nabla_{\boldsymbol{k}}(E_{cv})|}, \tag{6.52}$$

where E_{cv} is the abbreviation for $E_c - E_v$, and S_k is the constant energy surface defined by $E_{cv}(\boldsymbol{k})$ = const. We have assumed that both the conduction and valence bands are doubly degenerate (as a result of spin), which is strictly

[3] The summation in (6.49) is performed over the $\boldsymbol{k}$ vectors allowed *per unit volume* of crystal.

valid for centrosymmetric crystals but not for zinc-blende-type crystals [6.19]. Since $\mathcal{H}_{eR}$ does not involve electron spin, the spin state of the electron does not change in an optical transition (provided there is no spin–orbit coupling). With (6.52) we can make the following replacement in (6.4 and 49):

$$\sum_{\boldsymbol{k}} \rightarrow \int D_{\mathrm{j}}(E_{\mathrm{cv}}) dE_{\mathrm{cv}} \tag{6.53}$$

It has been pointed out by *Van Hove* [6.20] that the density of states of electron and phonon bands possesses singularities at points where $|\nabla_{\boldsymbol{k}}(E)|$ vanishes [see (6.52)]. These points are known as **critical points** and the corresponding singularities in the density of states are known as **Van Hove singularities**. Assuming that $\boldsymbol{k} = 0$ is a critical point in three-dimensional space, $E(\boldsymbol{k})$ can be expanded as a function of $\boldsymbol{k}$ about the critical point:

$$E(\boldsymbol{k}) = E(0) + \alpha_1 k_1^2 + \alpha_2 k_2^2 + \alpha_3 k_3^2 + \dots \tag{6.54}$$

Van Hove singularities are classified according to the number of negative coefficients α_i in (6.54). In three-dimensional space there are four kinds of Van Hove singularities, labeled M_0, M_1, M_2, and M_3 *critical points*. For example, a M_0 critical point has no negative α_i's and therefore represents a minimum in the band separation E_{cv}. M_1 and M_2 are known as *saddle points*, since the plots of their energies versus wavevector resemble a saddle. An M_3 critical point represents a maximum in the interband separation. It has been shown extensively in the literature (see for example [6.11]) that the density of states in the vicinity of an M_0 critical point of doubly degenerate valence and conduction bands is

$$D_{\mathrm{j}} = \begin{cases} (2\pi^2\alpha^{3/2})^{-1}(E - E_0)^{1/2}, & E > E_0 \\ 0, & E < E_0 \end{cases} ; \ \alpha^3 = \alpha_1\alpha_2\alpha_3 \tag{6.55}$$

The dependences on E of the densities of states in the vicinity of Van Hove singularities are listed in Table 6.1 for one-dimensional to three-dimensional $\boldsymbol{k}$-space. The corresponding Van Hove singularities in ε_{i} are sketched in Fig. 6.9.

6.2.4 Van Hove Singularities in ε_{i}

The real and imaginary parts of the dielectric function can be calculated readily from the band structure of a semiconductor using (6.49) and (6.48), respectively. A detailed comparison of the theoretical and experimental curves is one of the most stringent tests of the accuracy of a band structure calculation. From such a comparison it is possible to identify the optical transitions in reciprocal space which give rise to the structures in the experimental dielectric functions. In Figs. 6.10 and 11 we show comparisons of the theoretical and experimental dielectric function curves for Si and GaAs. The theoretical curves have been calculated from the band structures obtained by the empirical pseudopotential method. Figure 6.12 shows a similar comparison for the imaginary part of the dielectric function in Ge [6.25] except that the band structure was

Table 6.1. Van Hove singularities in one, two, and three dimensions and the corresponding density of states D_j. C stands for an energy-independent constant

	Type	D_j	
		$E < E_0$	$E > E_0$
Three dimensions	M_0	0	$(E - E_0)^{1/2}$
	M_1	$C - (E_0 - E)^{1/2}$	C
	M_2	C	$C - (E - E_0)^{1/2}$
	M_3	$(E_0 - E)^{1/2}$	0
Two dimensions	M_0	0	C
	M_1	$-\ln(E_0 - E)$	$-\ln(E - E_0)$
	M_2	C	0
One dimension	M_0	0	$(E - E_0)^{-1/2}$
	M_1	$(E_0 - E)^{-1/2}$	0

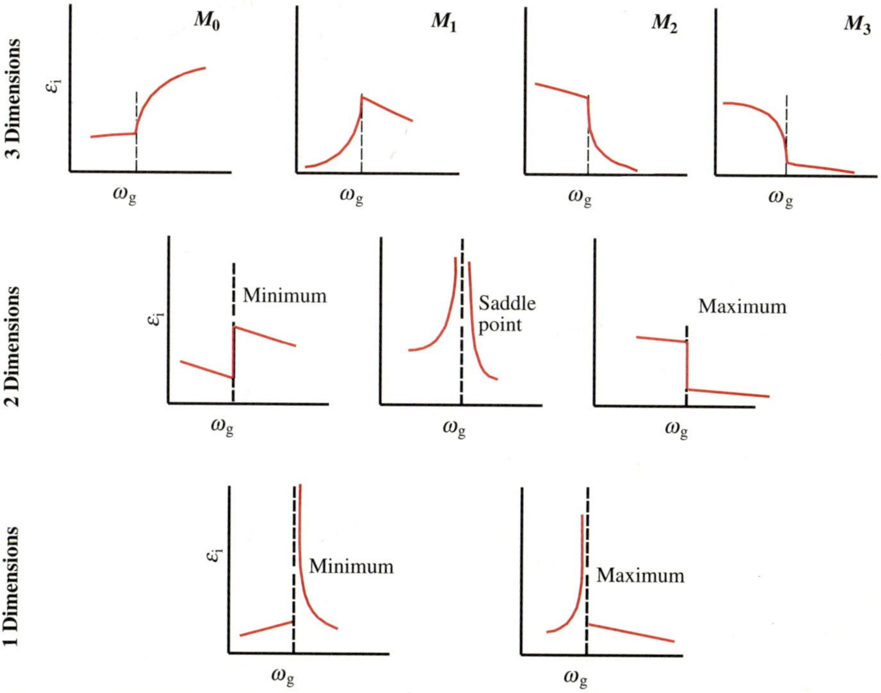

Fig. 6.9. Schematic dependence of the imaginary part of the dielectric constant (ε_i) on frequency near Van Hove singularities (i.e., interband critical points) in one, two, and three dimensions

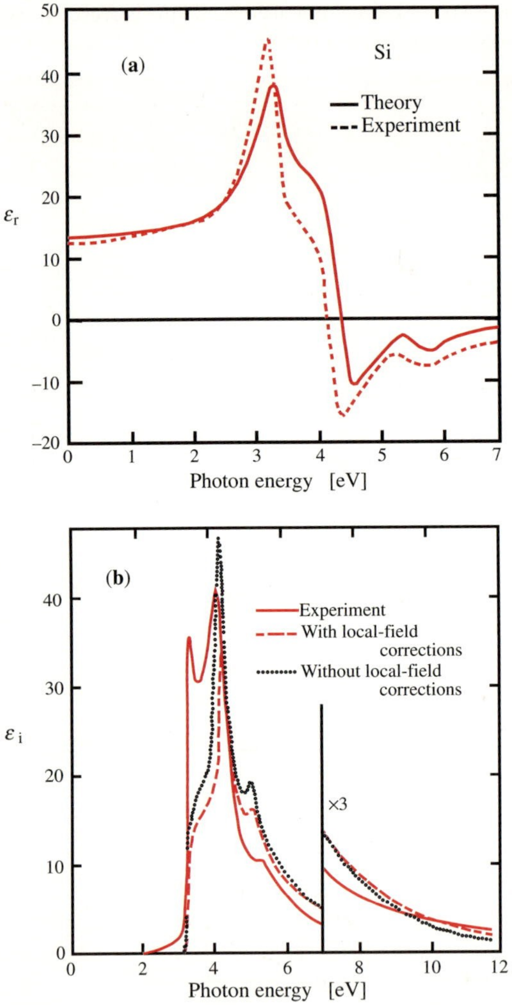

Fig. 6.10a,b. A comparison between the experimental and calculated dielectric function of Si: (**a**) real part and (**b**) imaginary part. In (**b**) the results of two theoretical calculations, one including and the other excluding local-field corrections, are presented [6.21]. For more recent theoretical results see [6.22, 23]. The local field corrections are due to changes in the higher Fourier components of the crystal (pseudo)-potential induced by the electromagnetic field. Note that, according to (**b**), their effect is small; they do not improve agreement with experimental data. The sharp peak observed at 3.3 eV has been attributed to excitonic effects (see Sect. 6.3.3 and [6.24]

calculated with the $\boldsymbol{k} \cdot \boldsymbol{p}$ method. The agreement between theory and experiment is quite good in all three cases because some of the input parameters for the band calculations were actually determined by fitting structures observed in the experimental curves. This limitation of the empirical methods of band structure calculations has been overcome in more recent *ab initio* pseudopotential calculations as discussed in Sect. 2.5.2.

There are many similarities in the overall shape of the imaginary part of the dielectric functions in Figs. 6.10–12. This is true not only for the three semiconductors shown here, but also for most of the tetrahedrally bonded semiconductors in the group IV, III–V and II–VI families. The main difference between their dielectric functions lies in the energies of the transitions. A close examination of the ε_i of these semiconductors shows that they all possess the following features:

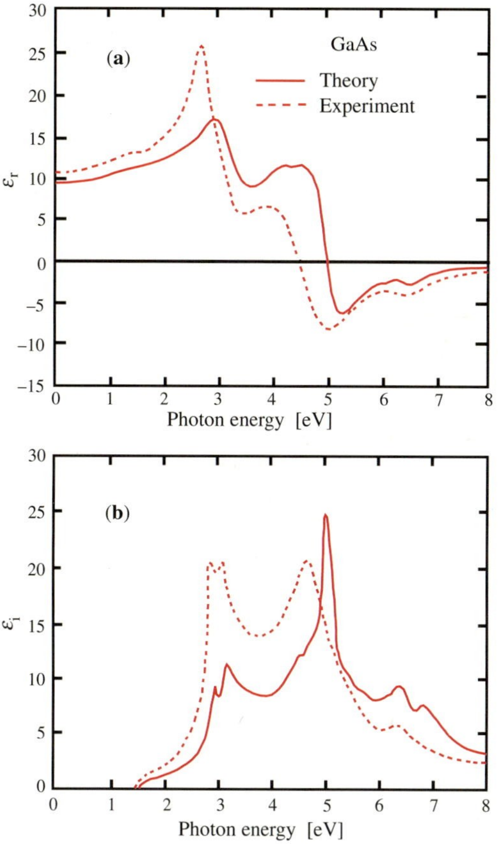

Fig. 6.11a,b. Results for GaAs similar to those presented in Fig. 6.10 for Si (the theoretical results do not include local field corrections) [2.8] p. 105

- An onset due to transitions between the absolute valence band maximum and the conduction band minimum. This is known as the **fundamental absorption edge**. The strength of this absorption edge depends on whether the valence band maximum and the conduction band minimum occur at the same point in the Brillouin zone. Since transitions between bands with the same wavevector are labeled as direct, semiconductors whose fundamental absorption edge involves a direct transition are said to have a *direct absorption edge*. Otherwise the absorption edge is said to be *indirect*. Diamond, Si, SiC, Ge, AlAs, AlSb, and GaP have indirect absorption edges, while GaN, GaAs, GaSb, InP, InAs, InSb and all the II–VI semiconductors have direct absorption edges. Within the scheme presented so far, optical transitions across an indirect bandgap are not allowed by the wavevector conservation condition. As we shall show in the next section, optical transitions between two bands with different wavevectors (knows as **indirect transitions**) are possible with the involvement of phonons, although they are orders of magnitude weaker than direct transitions. Hence they can be observed only when their energy is below that of all the direct transitions. ε_i becomes appreciable usually at the onset of the lowest-energy direct

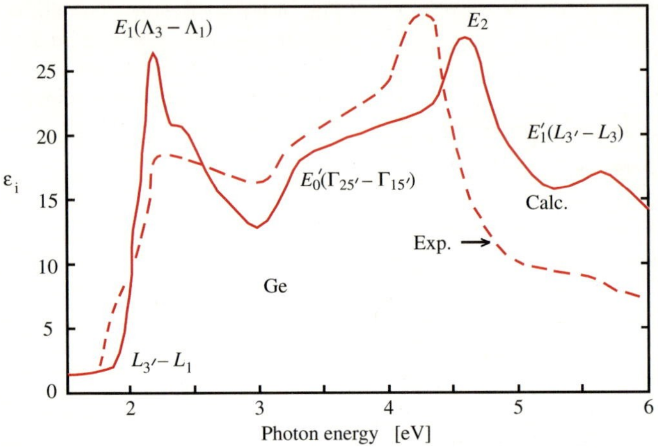

Fig. 6.12. A comparison between the experimental and calculated imaginary part of the dielectric function in Ge. The theoretical curve was calculated from a band structure obtained by the $\boldsymbol{k} \cdot \boldsymbol{p}$ method without spin-orbit interaction [6.25]

transition. In the diamond and zinc-blende-type semiconductors this transition usually occurs at the center of the Brillouin zone between the Γ_{4v} valence band and the Γ_{1c} conduction band and is usually referred to in the literature as the E_0 *transition*. Whenever the valence band has a sizable spin–orbit interaction, for example in semiconductors containing heavy elements such as In, As, and Sb, this transition is split by spin–orbit coupling into two transitions. The higher energy Γ_{7v}–Γ_{6c} transition involving the split-off valence band is labeled the $E_0 + \Delta_0$ *transition*.

- Above the fundamental absorption edge, ε_i typically rises to an asymmetric peak related to transitions occurring along the eight equivalent [111] directions of the Brillouin zone (a direction and all equivalent ones are represented as $\langle 111 \rangle$). If the spin–orbit coupling is small (such as in Si and GaP), only one peak is observed and the transition is known as the E_1 *transition*. Band structures suggest that these transitions involve M_1-type critical points in their joint density of states. [Since the negative longitudinal mass of these transitions (i.e., $(1/\alpha_1)$, with α_1 defined in (6.54)) is very large, the E_1 critical points are often modeled by a two-dimensional M_0 critical point.] When the spin–orbit interaction in the valence bands is large, the E_1 transitions are split into the E_1 and $E_1 + \Delta_1$ *transitions*. Using the $\boldsymbol{k} \cdot \boldsymbol{p}$ method, the spin–orbit splitting in the valence band along the $\langle 111 \rangle$ directions Δ_1 can be shown to be approximately 2/3 of the spin–orbit splitting Δ_0 at zone center [6.11]. This "*two-thirds rule*" provides a consistency check on the identification of the E_1 transitions. Table 6.2 lists Δ_0, Δ_1 and the ratio Δ_0/Δ_1 in a number of tetrahedrally coordinated semiconductors. Except for InP and GaN, this rule is well obeyed by all the semiconductors in the list (see Problem 6.20).
- ε_i reaches a strong absolute maximum known as the E_2 *peak*. This peak contains contributions from transitions occurring over a large region of the

Table 6.2. The valence band spin–orbit splitting at zone center (Δ_0) and in the ⟨111⟩ directions (Δ_1) and their ratio in several tetrahedrally coordinated semiconductors (list compiled from [6.19]). GaN values were calculated with the LCAO-LDA method in [6.26]. The experimental value of Δ_0 for GaN is 17 ± 1 meV [6.26]

Semiconductor	Δ_0 [eV]	Δ_1 [eV]	Δ_0/Δ_1
Si	0.044	0.03	1.47
Ge	0.296	0.187	1.58
GaN	0.019	0.032	0.59
GaAs	0.341	0.220	1.55
InP	0.108	0.133	0.81
InAs	0.38	0.267	1.42
InSb	0.803	0.495	1.62
ZnSe	0.432	0.27	1.59
CdTe	0.949	0.62	1.53

Brillouin zone close to the edges in the ⟨100⟩ and ⟨110⟩ directions [6.21]. Some of these transitions are associated with M_2 critical points.

- Superimposed on these features are weaker structures labeled *E_0' and E_1' transitions*. These involve transitions between the valence bands and higher conduction bands at the zone center and along the ⟨111⟩ directions, respectively.

The above system for labeling the interband optical transitions was proposed by *Cardona* [6.11]. Transitions occurring at the zone center, along the ⟨111⟩ directions and along the ⟨100⟩ directions are denoted by subscripts 0, 1, and 2, respectively. As an illustration, these optical transitions are indicated by arrows in the band structure of Ge in Fig. 6.13. This band structure has been calculated by a "full-zone $\boldsymbol{k}\cdot\boldsymbol{p}$ method", which is an extension of the $\boldsymbol{k}\cdot\boldsymbol{p}$ method discussed in Sect. 2.6 in which the electron wavevector $\boldsymbol{k}$ is no longer restricted to be near a critical point but can extend all the way to the zone edge [6.18]. The overall agreement between theory and experiment in Figs. 6.10–12 supports the identifications of the structures in the optical spectra in terms of critical points in the density of states. It should be pointed out that many of the remaining disagreements between theory and experiment in Figs. 6.10–12 have since been removed in more recent calculations. The present "state of the art" is represented by *ab initio* calculations which include Coulomb interaction between the excited electron and the hole left behind. This interaction enhances the E_1 transitions while weakening E_1', thus correcting deficiencies of Figs. 6.10, 11 [6.22, 23].

Table 6.3 lists the experimental energies of structures in the optical spectra measured at low temperatures for a number of semiconductors. In principle, by comparing these energies with critical point energies in the calculated joint density of states, the types of Van Hove singularities responsible for these structures can be deduced. In practice, the higher energy transitions are often found to contain contributions from several critical points of different types. Only the E_0 and E_1 transitions can be attributed definitively to three-dimensional critical points with M_0 and M_1 (or two-dimensional M_0) types of Van Hove singularities.

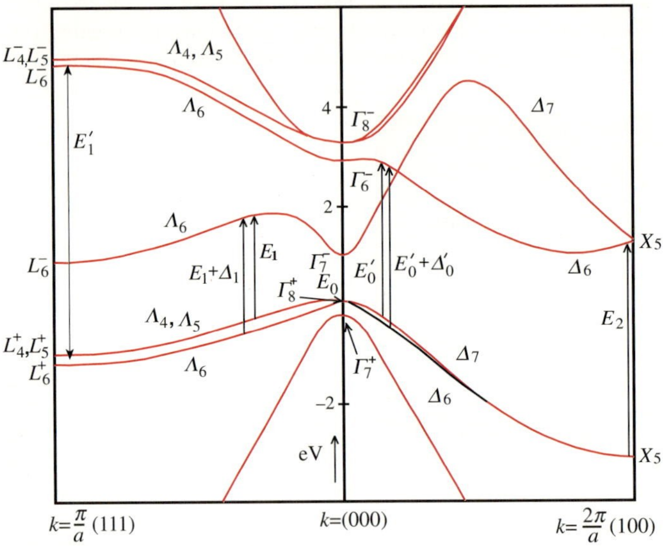

Fig. 6.13. The band structure of Ge showing the various direct transitions responsible for the structures in the imaginary part of the dielectric function shown in Fig. 6.12. The transitions giving rise to the various structures in the dielectric function are identified

Table 6.3. The measured energies [eV] of the prominent structures in the optical spectra of some diamond and zinc-blende-type semiconductors. All energies are low temperature values except that of the E_0 transition in Si, which was measured at room temperature. Compiled from data listed in [6.27], [6.28] and [6.29].

Transition	Si	Ge	GaAs	InP	GaP	GaN
E_0	4.185	0.898	1.5192	1.4236	2.869	3.302[a]
$E_0 + \Delta_0$	4.229	1.184	1.859	1.532	2.949	3.319[a]
E_1	3.45	2.222	3.017	3.287	3.780	7.03
$E_1 + \Delta_1$	—	2.41	3.245	3.423	3.835	
E_0'	3.378	3.206	4.488	4.70	4.72	
$E_0' + \Delta_0'$	—	3.39	4.659	5.17	4.88	
E_2	4.330	4.49	5.110	5.05	5.22	7.63
E_1'	5.50	5.65	6.63		6.8	

[a] Grown on MgO.

6.2.5 Direct Absorption Edges

We shall now consider in more detail the optical transitions at the fundamental absorption edge, since many semiconductor optoelectronic devices, such as lasers and photodetectors, involve these transitions. As pointed out in the preceding section, there are direct and indirect absorption edges. In the case of a direct absorption edge, ε_i can be calculated from (6.48). Let m_c and m_υ denote, respectively, the effective masses of the conduction and valence bands (assumed to be spherical for simplicity) and E_g the direct energy gap. The energy difference E_{cv} in the vicinity of the energy gap can be expanded as

$$E_{cv} = E_g + (\hbar^2/2\mu)k^2, \tag{6.56}$$

where μ is the effective mass defined by $\mu^{-1} = m_c^{-1} + m_v^{-1}$. Using (6.52) the joint density of states D_j can be calculated to be

$$D_j = \begin{cases} [2^{1/2}\mu^{3/2}/(\pi^2\hbar^3)](E_{cv} - E_g)^{1/2} & \text{for } E_{cv} > E_g, \\ 0 & \text{for } E_{cv} < E_g. \end{cases} \tag{6.57}$$

Substituting this result into (6.48) we obtain ε_i near E_g as

$$\varepsilon_i(\omega) = \begin{cases} Ax^{-2}(x-1)^{1/2} & \text{for } x > 1, \\ 0 & \text{for } x < 1, \end{cases} \tag{6.58a}$$

where

$$A = \frac{2e^2(2\mu)^{3/2}}{m^2\hbar}|P_{cv}|^2 E_g^{-3/2}$$

and

$$x = \hbar\omega/E_g. \tag{6.58b}$$

Using the KKRs, ε_r near E_g can be shown to have the form (Problem 6.2).

$$\varepsilon_r(\omega) = \begin{cases} \text{const} + Ax^{-2}[2 - (1+x)^{1/2}] & \text{for } x > 1, \\ \text{const} + Ax^{-2}[2 - (1+x)^{1/2} - (1-x)^{1/2}] & \text{for } x < 1, \end{cases} \tag{6.59}$$

where the constant term is determined by contributions from transitions above the fundamental absorption edge. According to (6.58) a plot of ε_i^2, or the square of the absorption coefficient, as a function of the photon energy should be a straight line. The energy gap is given by the intercept of the line with the x axis and either μ or $|P_{cv}|^2$ can be determined from its slope. An example of such a plot is given in Fig. 6.14 for PbS. An equivalent semilogarithmic plot for InSb is given in Fig. 6.15. A fit to the experimental values of $\varepsilon_r(\omega)$ in PbS with the expression (6.59) is shown in Fig. 6.14 (b).

6.2.6 Indirect Absorption Edges

If the lowest energy gap is indirect, a photon can excite an electron from the valence band to the conduction band with the assistance of a phonon. The wavevector difference between the electrons in the two bands is supplied by the phonon. If the phonon energy and wavevector are denoted by E_p and $\boldsymbol{Q}$, the energy- and wavevector-conservation conditions in the optical process are represented by

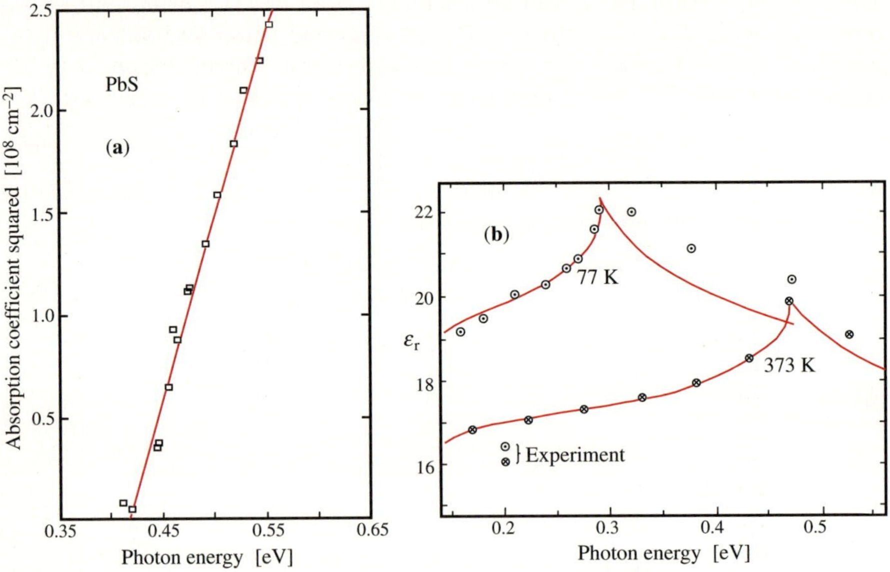

Fig. 6.14. (**a**) Plot of the square of the absorption coefficient of PbS as a function of photon energy showing the linear behavior discussed in the text. The intercept with the x-axis defines the direct energy gap. Reproduced from [6.30]. (**b**) Fits (*curves*) to the experimental values of the real part of the dielectric function of PbS (*data points*) measured at 77 and 373 K with the expression in (6.59). Reproduced from [6.144].

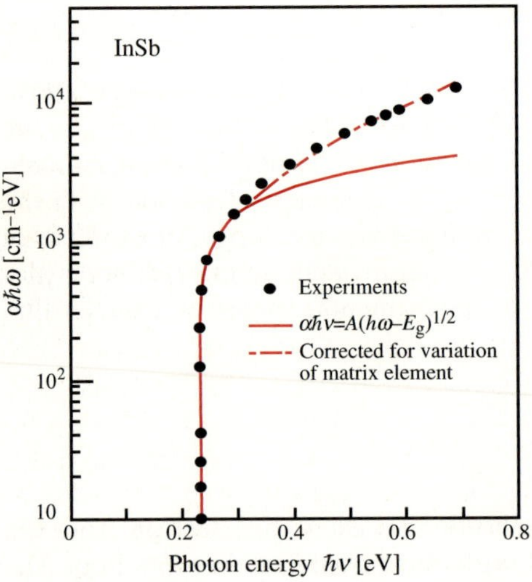

Fig. 6.15. Semilogarithmic plot of the absorption coefficient of InSb at 5 K as a function of photon energy. The *filled circles* represent experimental results from [6.31]. The *curves* have been calculated using various models. The intercept with the x-axis gives the direct bandgap of InSb [6.32]

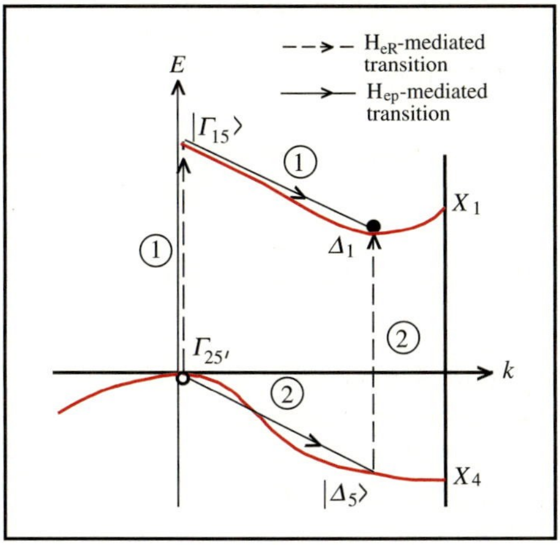

Fig. 6.16. Schematic band structure of Si as an indirect-bandgap semiconductor showing the phonon-assisted transitions (labeled **1** and **2**) which contribute to the indirect absorption edge. $|\Gamma_{15}\rangle$ and $|\Delta_5\rangle$ represent intermediate states

$$\hbar\omega = E_{\mathrm{cv}} \pm E_{\mathrm{p}} \text{ and } \boldsymbol{k}_{\mathrm{c}} - \boldsymbol{k}_{\mathrm{v}} = +\boldsymbol{Q}, \tag{6.60}$$

where + and − correspond, respectively, to emission and absorption of a phonon. Processes involving several phonons are in principle also possible but usually with much smaller probabilities.

As a specific example of an indirect absorption edge we will consider Si. The band structure of Si in the vicinity of the indirect gap $\Delta_1 - \Gamma_{25'}$ is shown schematically in Fig. 6.16. The absorption processes near this energy gap now consist of two steps. One of these steps involves the electron–photon interaction $\mathcal{H}_{\mathrm{eR}}$ while the other step involves the electron–phonon interaction $\mathcal{H}_{\mathrm{ep}}$. Both interactions will be assumed to be weak enough for perturbation theory to be valid. The calculation of the optical transition probability can therefore be performed by using second order perturbation theory. A systematic way to calculate this probability makes use of **Feynman diagrams**. We shall defer the discussion of this technique until Chap. 7, when we apply it to study Raman scattering. Using Feynman diagrams to calculate the transition probability at an indirect absorption edge is left for Problem 7.7. In this chapter we shall point out, without proof, several possible processes which contribute to the indirect absorption edge. In one of them an electron is first excited via a *virtual* transition (i. e., a transition which does not conserve energy – although such transitions have to conserve wavevector as a result of the translation symmetry of the crystal) from the valence band to an intermediate state $|\mathrm{i}\rangle$ by absorbing the incident photon. A second virtual transition takes the electron from $|\mathrm{i}\rangle$ to the Δ_1 conduction band state via absorption or emission of a phonon. In the final state there is an electron in the Δ_1 conduction band, a hole in the $\Gamma_{25'}$ valence band state and a phonon has been either created or annihilated. This process is shown schematically for Si in Fig. 6.16 by arrows labeled 1. A second possible phonon-assisted indirect optical transition is also shown in the same figure by arrows labeled 2. These processes are similar except that for process 2 the intermediate states occur at Δ.

There are other possible processes which contribute to absorption at an indirect bandgap. In principle these terms have to be combined with the contributions from the above two in calculating the transition probability. To understand why the above two terms have been singled out, we shall examine their contribution to the transition probability R_{ind} using an extension of Fermi's Golden Rule to second order perturbation (see Sect. 7.2.4 for further details):

$$R_{\text{ind}} = \frac{2\pi}{\hbar} \sum_{\boldsymbol{k}_c, \boldsymbol{k}_v} \left| \sum_{\text{i}} \frac{\langle \text{f}|\mathcal{H}_{\text{ep}}|\text{i}\rangle \langle \text{i}|\mathcal{H}_{\text{eR}}|0\rangle}{E_{\text{i}0} - \hbar\omega} \right|^2 \delta(E_{\text{c}}(\boldsymbol{k}_{\text{c}}) - E_{\text{v}}(\boldsymbol{k}_{\text{v}}) - \hbar\omega \pm E_{\text{p}}); \tag{6.61}$$

$|0\rangle$ represents the initial state of the system with a filled valence band and an empty conduction band and phonon occupation number N_{p}. In the final state $|\text{f}\rangle$ an electron has been excited into the Δ_1 conduction band, a hole has been created at the $\Gamma_{25'}$ valence band, and N_p has changed by one. For the two processes in Fig. 6.16, the intermediate state $|\text{i}\rangle$ involves either an electron excited into the Δ_1 conduction band (shown by arrow 2) or a hole created at the $\Gamma_{25'}$ valence band (shown by arrow 1). In principle, it is necessary to permute the time order in which $\mathcal{H}_{\text{eR}}$ and $\mathcal{H}_{\text{ep}}$ occur and also to sum over all possible intermediate states $|\text{i}\rangle$ in calculating R_{ind}. However, processes in which the phonons excite an electron across the gap make a negligible contribution because of the energy denominator in (6.61). Similarly, intermediate states for which $E_{\text{i}0} \gg \hbar\omega$ are unimportant. As a result, the two processes shown in Fig. 6.16 are usually the most important ones.

In many semiconductors, the matrix elements which appear in (6.61) can be assumed to be constant in the vicinity of the indirect bandgap. Therefore the photon energy dependence of R_{ind} can be obtained by summing over the delta function in (6.61). By converting the summations over $\boldsymbol{k}_{\text{c}}$ and $\boldsymbol{k}_{\text{v}}$ to integrations over the conduction and valence band energies E_{c} and E_{v}, respectively, via their density of states $D_{\text{v}}(E_{\text{v}})$ and $D_{\text{c}}(E_{\text{c}})$ we obtain

$$R_{\text{ind}} \propto \iint D_{\text{v}}(E_{\text{v}}) D_{\text{c}}(E_{\text{c}}) \delta(E_{\text{c}} - E_{\text{v}} - \hbar\omega \pm E_{\text{p}}) dE_{\text{c}} dE_{\text{v}}. \tag{6.62}$$

Assuming that the bands are parabolic and three-dimensional, we find

$$D_{\text{v}} \propto \begin{cases} (-E_{\text{v}})^{1/2} & \text{for } E_{\text{v}} < 0, \\ 0 & \text{for } E_{\text{v}} > 0, \end{cases} \tag{6.63}$$

and

$$D_{\text{c}} \propto \begin{cases} (E_{\text{c}} - E_{\text{ig}})^{1/2} & \text{for } E_{\text{c}} > E_{\text{ig}}, \\ 0 & \text{for } E_{\text{c}} < E_{\text{ig}}. \end{cases} \tag{6.64}$$

The zero of the energy scale has been taken at the top of the valence band and E_{ig} is the indirect energy gap. Substituting D_{v} and D_{c} into (6.62) and integrating over E_{v} one obtains

$$R_{\text{ind}} \propto \int_{E_{\text{ig}}}^{\hbar\omega \mp E_{\text{p}} - E_{\text{ig}}} (E_{\text{c}} - E_{\text{ig}})^{1/2} (\hbar\omega \mp E_{\text{p}} - E_{\text{c}})^{1/2} dE_{\text{c}}. \tag{6.65}$$

By changing the variable to

$$x = \frac{E_c - E_{ig}}{\hbar\omega \pm E_p - E_{ig}},$$

(6.65) can be expressed as

$$R_{ind} \propto (\hbar\omega \mp E_p - E_{ig})^2 \int_0^1 x^{1/2}(1-x)^{1/2} dx. \tag{6.66}$$

On performing the integral in (6.66) we can conclude that, in the vicinity of an indirect bandgap, ε_i depends on the photon energy as

$$\varepsilon_i(\omega) \propto \begin{cases} (\hbar\omega \mp E_p - E_{ig})^2 & \text{for } \hbar\omega \geq E_{ig} \pm E_p, \\ 0 & \text{otherwise.} \end{cases} \tag{6.67}$$

Thus an indirect energy gap can, in principle, be distinguished from a direct one by the different dependence of their absorption coefficients on photon energy. In addition, every indirect energy gap gives rise to two absorption edges, at $E_{ig} + E_p$ and $E_{ig} - E_p$, for each phonon E_p that can mediate the indirect transition. The edge at $E_{ig} - E_p$ corresponds to phonon absorption. The electron–phonon matrix element $|\langle f|\mathcal{H}_{ep}|i\rangle|^2$ in (6.61) is proportional to N_p, where N_p is the phonon occupation number (Sect. 3.3.1). This absorption edge is therefore present only at high temperatures and disappears at temperatures too low for such phonons to be thermally excited. On the other hand, the higher energy edge at $E_{ig} + E_p$ involves phonon emission and hence is proportional to $(1 + N_p)$. It is present at both high and low temperatures. Identification of these two edges by their different temperature dependence enables not only E_{ig} but also E_p to be determined.

Figures 6.17–19 show the absorption edges of three indirect bandgap semiconductor: Si, Ge, and GaP. In these materials several phonons can participate in indirect transitions, giving rise to a number of absorption thresholds. The temperature dependence of these edges in Ge and GaP is shown in Figs. 6.18 and 19. Notice that at low temperature the shape of these indirect absorption edges deviates from the proportionality to the square of the photon energy predicted by (6.67). Instead, their shape resembles more the square root dependence given in (6.58) for direct gaps. The explanation for this deviation is that *exciton effects* modify the shape of these indirect absorption edges at low temperatures. These effects will be discussed in Sect. 6.3.

6.2.7 "Forbidden" Direct Absorption Edges

The last case we shall consider involves a direct-bandgap semiconductor where an electric dipole transition between the top valence band and the lowest conduction minimum is forbidden by a selection rule. One example of a semicon-

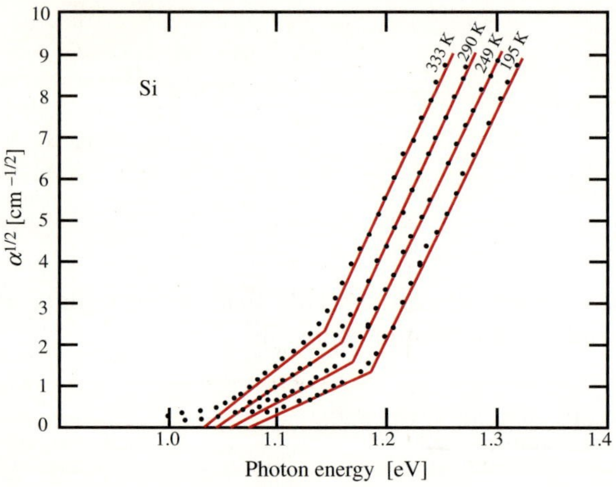

Fig. 6.17. Plots of the square root of the absorption coefficients of Si versus photon energy at several temperatures. The two segments of a straight line drawn through the experimental points represent the two contributions due to phonon absorption and emission [6.33]

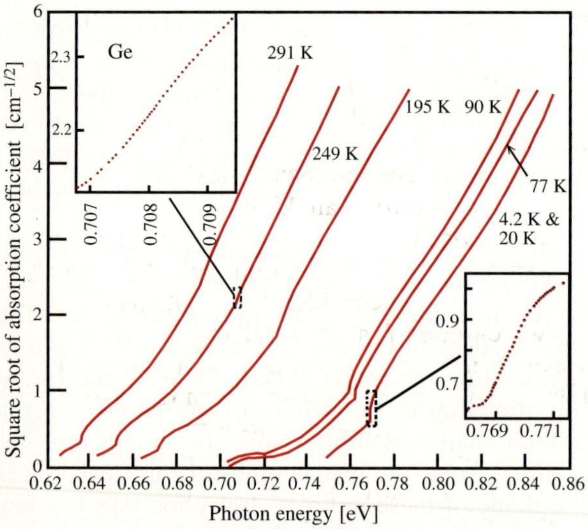

Fig. 6.18. Plots of the square root of the absorption coefficients of Ge versus photon energy at several temperatures. The two *insets* compare the exciton-induced abruptness of the absorption edge due to phonon emission at high and low temperatures [6.34]

ductor with such a "forbidden" direct absorption edge is Cu_2O, whose crystal structure and zone-center phonon properties were studied in Problem 3.1. This crystal is centrosymmetric. Both the conduction and valence band extrema oc-

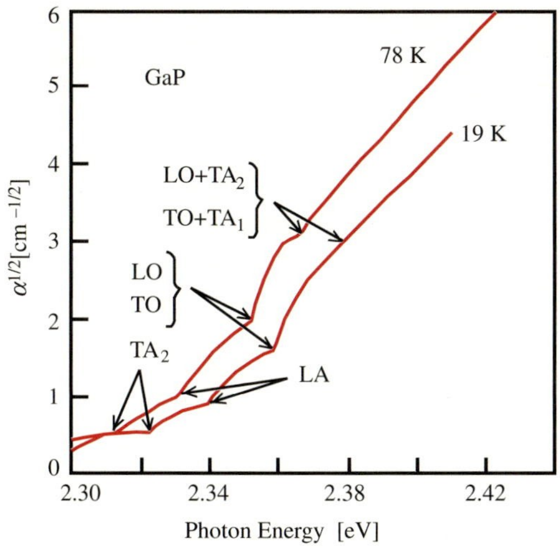

Fig. 6.19. Plots of the square root of the absorption coefficients of GaP versus photon energy at two different temperatures. The labels denote the exciton-enhanced absorption thresholds associated with the emission of various phonon modes. Note the square-root singularities at the onset of the various phonon-aided processes. These square-roots are typical of indirect excitonic absorption (e.g. without k conservation) [6.35]

cur at the zone center and have even parity [6.36]. The electron momentum operator $\boldsymbol{p}$ in $\mathcal{H}_{eR}$ has odd parity under inversion. As a result of the matrix-element theorem discussed in Sect. 2.3.3, electric dipole transitions are allowed only between states of different parity. Hence they are forbidden by the parity selection rule at the absorption edge of Cu_2O.

In the case of optical transitions involving isolated atoms or molecules, the electronic density of states has discrete peaks. When the electric dipole matrix element vanishes one may have to consider higher order optical transitions (such as quadrupole transitions). For interband transitions the density of states is a continuum. When $|P_{vc}|^2$ is zero at an M_0 critical point, this does not imply that there will be no electric dipole absorption edge; the density of states is zero anyway at an M_0 critical point. The vanishing of $|P_{vc}|^2$ exactly at the critical point just means that one has to consider the possibility that $|P_{vc}|^2$ does not vanish for electron wavevectors $\boldsymbol{k}$ slightly off the critical point. In general, one finds $|P_{vc}|^2$ to have a nonzero $\boldsymbol{k}$-dependent component. For example, in the case of Cu_2O one can use the $\boldsymbol{k}\cdot\boldsymbol{p}$ expansion in (2.37) to express either the conduction or valence band wavefunctions as a function of k. This expansion introduces some mixture of odd-parity wavefunctions, giving rise to parity-allowed electric-dipole transition matrix elements. Phenomenologically one can expand $|P_{vc}|^2$ as a Taylor series in $\boldsymbol{k}$. Assuming that the critical point occurs at $\boldsymbol{k} = 0$ and $|P_{vc}(0)|^2 = 0$, we find

$$|P_{vc}(\boldsymbol{k})|^2 = |dP_{vc}/dk|^2 k^2 + 0(k^4). \tag{6.68}$$

When we substitute (6.68) into (6.48) the optical transition matrix element introduces a term proportional to k^2. Since the joint density of states at a direct gap is proportional to k in three dimensions (6.57), $\varepsilon_i(\omega)$ becomes proportional to k^3. Hence for a direct and "forbidden" bandgap semiconductor $\varepsilon_i(\omega)$ depends on photon energy as

$$\varepsilon_{\mathrm{i}}(\omega) \propto \begin{cases} (\hbar\omega - E_{\mathrm{g}})^{3/2} & \text{for } \hbar\omega > E_{\mathrm{g}} \\ 0 & \text{for } \hbar\omega < E_{\mathrm{g}}. \end{cases} \tag{6.69}$$

Compared to direct but "allowed" absorption edges, a "forbidden" edge increases more slowly with energy in the vicinity of the bandgap. As a result, "forbidden" absorption edges are difficult to identify except when there are strong exciton effects which change the smooth absorption edges into sharp peaks. These effects will be studied in the next section.

6.3 Excitons

The approach we have adopted so far to view optical absorption processes is that an incident radiation field excites an electron–hole pair inside the semiconductor. The properties of the electron and the hole are both described by the band structure within the one-electron approximation. In this section we shall go beyond this approximation and consider the effects of **electron–electron interaction** on the absorption spectra.

To simplify the calculation we shall make the following assumptions. We shall include only the Coulombic part of the electron–electron interaction neglecting both exchange and correlation terms. Furthermore, the interaction between the excited electron in the conduction band and those left behind in the now almost filled valence band will be replaced by an electron–hole interaction. Attraction between the electron and the hole causes their motion to be *correlated* and the resultant electron–hole pair is known as an **exciton**. Typically excitons have been studied in two limiting cases. For strong electron–hole attraction, as in ionic crystals, the electron and the hole are tightly bound to each other within the same or nearest-neighbor unit cells. These excitons are known as **Frenkel excitons**. In most semiconductors, the Coulomb interaction is strongly screened by the valence electrons via the large dielectric constant. As a result, electrons and holes are only weakly bound. Such excitons are known as **Wannier–Mott excitons** [6.37, 38] or simply as Wannier excitons. In this book we shall be concerned with Wannier excitons only [6.39–41].

The properties of Wannier excitons can be calculated with the effective mass approximation introduced in Sect. 4.2. Within this approximation, the electron and the hole are considered as two particles moving with the effective masses of the conduction and valence bands, respectively. Donors and acceptors studied in Chap. 4 can be regarded as "excitons" in which one of the particles has an infinite effective mass. Since the difference in effective mass between the electron and the hole in a semiconductor is not as large as that between the electron and the proton, excitons are more analogous to *positro-*

nium, an electron–positron pair. As a result of the Coulomb interaction between the electron and hole, the potential acting on an electron (or a hole) in a crystal is not translationally invariant.

As in any two-particle system, the exciton motion can be decomposed into two parts: a *center-of-mass* (CM) *motion* and a *relative motion* of the two particles about the CM. With this decomposition, the potential acting on the exciton CM still has translational invariance since the Coulomb interaction depends only on the *relative coordinate* of the electron and hole. Within the effective mass approximation, the exciton CM behaves like a free particle with mass $M = m_e + m_h$ (where m_e and m_h are, respectively, the electron and hole effective masses). The relative motion of the electron and the hole in the exciton is similar to that of the electron and the proton inside the hydrogen atom. There are bound states and continuum states. The bound states are quantized with principal quantum numbers $n = 1, 2, 3$, etc., and orbital angular momentum $l = 0, \hbar, 2\hbar$, etc. In the continuum states, excitons can be considered to be ionized into free electrons and free holes but their wavefunctions are still modified by their Coulomb attraction.

In the literature, one often finds schematic diagrams in which the exciton energy levels are shown superimposed on a one-electron energy band structure. Strictly speaking this is incorrect. Since the exciton is a two-particle state, its energy levels cannot be represented by one electron energy levels. To clarify this point further, we compare the energies of an electron–hole pair in a one-electron energy band diagram and in a two-particle energy diagram in Fig. 6.20. In the one-electron picture the ground state of the semiconductor is represented by a filled valence band and an empty conduction band. Since there are no electron–hole pairs in the ground state, this state is represented by the origin in the two-particle picture. In the one-particle picture, the excited state is represented by an electron in the conduction band (with wavevector $\boldsymbol{k}_e$) and a hole in the valence band (wavevector $\boldsymbol{k}_h = -\boldsymbol{k}_v$). This excited state corresponds to an exciton in the two-particle state. We note that optical excitation is not the only mechanism for creating such excited states. An energetic electron, for example, can also create an exciton. In the two-particle picture, the exciton wavevector $\boldsymbol{K}$ is given by $\boldsymbol{k}_e + \boldsymbol{k}_h$. As pointed out earlier, the potential for the exciton CM motion is translationally invariant even when the electron is attracted to the hole. $\boldsymbol{K}$ is therefore a good quantum number. The kinctic energy E_{ke} of the exciton is related to $\boldsymbol{K}$ by the free-particle expression [to be derived later, see (6.78)]

$$E_{ke} = \frac{\hbar^2 K^2}{2M}. \tag{6.70}$$

Thus exciton levels can be represented by parabolas in the two-particle energy diagram in Fig. 6.20b. We shall show later that the exciton state wavefunction with CM wavevector $\boldsymbol{K}$ is a linear combination of many electron–hole pair wavefunctions with wavevectors $\boldsymbol{k}_e$ and $\boldsymbol{k}_h$ satisfying the condition $\boldsymbol{K} = \boldsymbol{k}_e + \boldsymbol{k}_h$. Thus the one-particle picture in Fig. 6.20a is not correct even for the exciton continuum states.

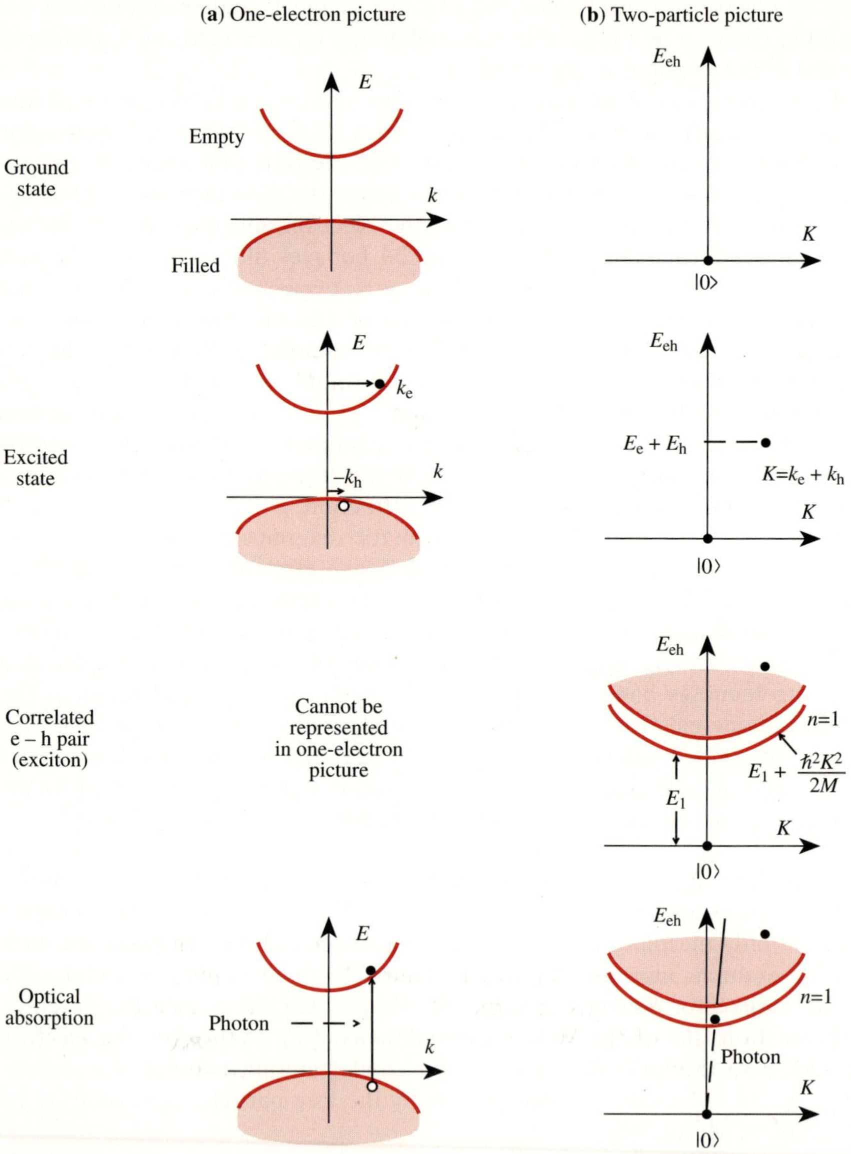

Fig. 6.20. Comparison between the energy levels of the ground state and excited states of a semiconductor in a one-electron band picture (**a**) and in a two-particle picture (**b**). Also, schematic diagrams showing processes in which a photon is absorbed while producing an electron-hole pair

Figure 6.20b shows schematically an optical transition in the two-particle energy diagram. Since the CM motion is translationally invariant, wavevector conservation applies only to the exciton wavevector and not to those of the individual electrons or holes. To conserve both energy and wavevector during absorption, this process can occur only at the intersection between the radiation (or photon) and the exciton dispersion curves, as shown in Fig. 6.20b. At this point the electron–radiation (or more precisely the *photon–exciton*) interaction couples the exciton and radiation to form a mixed mechanical-electromagnetic wave known as an *exciton-polariton* [6.42, 43]. These will be studied after we have considered excitonic effects at different kinds of critical points.

6.3.1 Exciton Effect at M_0 Critical Points

We now consider quantitatively the effect of Coulomb attraction on the motion of electrons and holes in the vicinity of an M_0 critical point of a direct bandgap semiconductor in three dimensions. We shall assume the conduction band to be spherical with energy

$$E_e(\boldsymbol{k}_e) = E_g + \frac{\hbar^2 k_e^2}{2m_e}, \tag{6.71}$$

where E_g is the bandgap, and the corresponding hole energy to be given by

$$E_h(\boldsymbol{k}_h) = \frac{\hbar^2 k_h^2}{2m_h}. \tag{6.72}$$

Let the Bloch functions for the electron and the hole be represented by $\psi_{\boldsymbol{k}_e}(\boldsymbol{r}_e)$ and $\psi_{\boldsymbol{k}_h}(\boldsymbol{r}_h)$, respectively. As in Sect. 4.2 we assume that the Coulomb interaction between electron and hole is weak due to screening by the valence electrons, so that the effective mass approximation is valid. We write the exciton wavefunction Ψ as a linear combination of the electron and hole wavefunctions:

$$\Psi(\boldsymbol{r}_e, \boldsymbol{r}_h) = \sum_{\boldsymbol{k}_e, \boldsymbol{k}_h} C(\boldsymbol{k}_e, \boldsymbol{k}_h) \psi_{\boldsymbol{k}_e}(\boldsymbol{r}_e) \psi_{\boldsymbol{k}_h}(\boldsymbol{r}_h). \tag{6.73}$$

Similar to the case of the donor electron in Sect. 4.2, the electron and hole in an exciton are localized relative to their CM, so it is more convenient to express their wavefunctions in terms of Wannier functions rather than Bloch functions. In terms of the Wannier functions $a_{\boldsymbol{R}_e}(\boldsymbol{r}_e)$ and $a_{\boldsymbol{R}_h}(\boldsymbol{r}_h)$ for electron and hole, respectively, the exciton wavefunction can be written as

$$\Psi(\boldsymbol{r}_e, \boldsymbol{r}_h) = N^{-1/2} \sum_{\boldsymbol{R}_e, \boldsymbol{R}_h} \Phi(\boldsymbol{R}_e, \boldsymbol{R}_h) a_{\boldsymbol{R}_e}(\boldsymbol{r}_e) a_{\boldsymbol{R}_h}(\boldsymbol{r}_h), \tag{6.74}$$

where $\Phi(\boldsymbol{R}_e, \boldsymbol{R}_h)$ is the **exciton envelope wavefunction**. The wave equation for $\Phi(\boldsymbol{R}_e, \boldsymbol{R}_h)$ analogous to (4.22) is [6.44]

$$\left[-\left(\frac{\hbar^2}{2m_e}\right) \nabla^2_{\boldsymbol{R}_e} - \left(\frac{\hbar^2}{2m_h}\right) \nabla^2_{\boldsymbol{R}_h} - \frac{e^2}{4\pi\varepsilon_0\varepsilon_0 |\boldsymbol{R}_e - \boldsymbol{R}_h|} \right] \Phi(\boldsymbol{R}_e, \boldsymbol{R}_h) = E\Phi(\boldsymbol{R}_e, \boldsymbol{R}_h), \tag{6.75}$$

where ε_0 is the zero-frequency dielectric constant of the semiconductor. Equation (6.75) can be solved in the same way as in the case of the hydrogen atom.

One expresses $\boldsymbol{R}_\mathrm{e}$ and $\boldsymbol{R}_\mathrm{h}$ in terms of two new coordinates: a center-of-mass coordinate $\boldsymbol{R}$ and a relative coordinate $\boldsymbol{r}$ defined by

$$\boldsymbol{R} = \frac{m_\mathrm{e}\boldsymbol{R}_\mathrm{e} + m_\mathrm{h}\boldsymbol{R}_\mathrm{h}}{m_\mathrm{e} + m_\mathrm{h}} \text{ and } \boldsymbol{r} = \boldsymbol{R}_\mathrm{e} - \boldsymbol{R}_\mathrm{h}. \tag{6.76}$$

The equation of motion for the CM is now decoupled from that for the relative motion because the Coulomb interaction term does not involve $\boldsymbol{R}$. The two resultant equations are

$$\left(-\frac{\hbar^2}{2M}\right)\nabla^2_{\boldsymbol{R}}\psi(\boldsymbol{R}) = E_R\psi(\boldsymbol{R}), \tag{6.77a}$$

$$\left(-\frac{\hbar^2}{2\mu}\nabla^2_{\boldsymbol{r}} - \frac{e^2}{4\pi\varepsilon_0\varepsilon_0 r}\right)\phi(\boldsymbol{r}) = E_r\phi(\boldsymbol{r}), \tag{6.77b}$$

where μ, the **reduced mass** of the exciton, is defined by

$$\frac{1}{\mu} = \frac{1}{m_\mathrm{e}} + \frac{1}{m_\mathrm{h}}. \tag{6.77c}$$

The total energy of the exciton E is simply the sum of E_R and E_r. The solutions of (6.77a and b) can be obtained readily. Equation (6.77a) describes a free particle whose eigenfunction and energy are given by

$$\psi_K(\boldsymbol{R}) = (N)^{-1/2}\exp(\mathrm{i}\boldsymbol{K}\cdot\boldsymbol{R}) \text{ and } E_R = \frac{\hbar^2K^2}{2M}. \tag{6.78}$$

E_R represents the kinetic energy of the CM motion and is therefore the same as E_ke in (6.70).

Equation (6.77b) is similar to the equation describing the motion of the donor electron discussed in Sect. 4.2.2. As in the hydrogen atom, its wavefunctions and energies can be indexed by three quantum numbers: a principal quantum number n, the angular momentum quantum number l and the magnetic quantum number m. The wavefunction ϕ can be expressed in polar coordinates (r, θ, ϕ) as:

$$\phi_{nlm}(\boldsymbol{r}) = R_\mathrm{nl}(r)Y_{lm}(\theta, \phi), \tag{6.79}$$

where $R_{nl}(r)$ can be expressed in terms of the associated Laguerre polynomials and $Y_{lm}(\theta, \phi)$ are the *spherical harmonic* functions. These functions are tabulated in many quantum mechanics textbooks and therefore will not be reproduced here. For isotropic effective masses E_r depends on n only and is given by

$$E_r(n) = E_r(\infty) - \frac{R^*}{n^2}, \tag{6.80}$$

where $E_r(\infty)$ is the minimum energy of the continuum states, i. e., the energy gap E_g in (6.69), and R^* is the Rydberg constant for the exciton defined as

$$R^* = \frac{\mu e^4}{2\hbar^2 (4\pi\varepsilon_0)^2 \varepsilon_0^2} = \left(\frac{\mu}{m\varepsilon_0^2}\right) \times 13.6\ \text{eV}. \quad (6.81)$$

If the hole mass is much heavier than the electron mass, as in the case of most tetrahedrally bonded direct gap semiconductors, the reduced mass μ is close to the electron effective mass and hence the exciton Rydberg constant should be comparable to the donor binding energy. Furthermore, the exciton Bohr radius is also comparable to the donor electron Bohr radius.

Combining the above results for the relative motion and CM motion of the exciton, we obtain the following envelope wavefunctions and energies for the exciton:

$$\Phi_{nlm}(\boldsymbol{R}, \boldsymbol{r}) = (1/\sqrt{N}) \exp(\mathrm{i}\boldsymbol{K} \cdot \boldsymbol{R}) R_{nl}(r) Y_{lm}(\theta, \phi), \quad (6.82)$$

$$E_{nlm} = E_{\mathrm{g}} + \frac{\hbar^2 K^2}{2M} - \frac{R^*}{n^2}. \quad (6.83)$$

The energy spectrum of a Wannier exciton is shown schematically in Fig. 6.20b and in greater detail in Fig. 6.21.

The above model of excitons based on electrons and holes with spherically symmetric parabolic dispersion is useful for understanding qualitatively exciton effects on optical spectra. However, it is not accurate enough for quantitative interpretation of experimental spectra in diamond- and zinc-blende-type semiconductors. As we discussed in Chap. 2, the valence band structure in these families of materials is complicated by degeneracies and warping. Of the various attempts to calculate excitonic effects based on realistic band structures, we shall mention the one by *Baldereschi* and *Lipari* [6.45]. They calculated the exciton binding energies by using a "spherical effective hamiltonian"

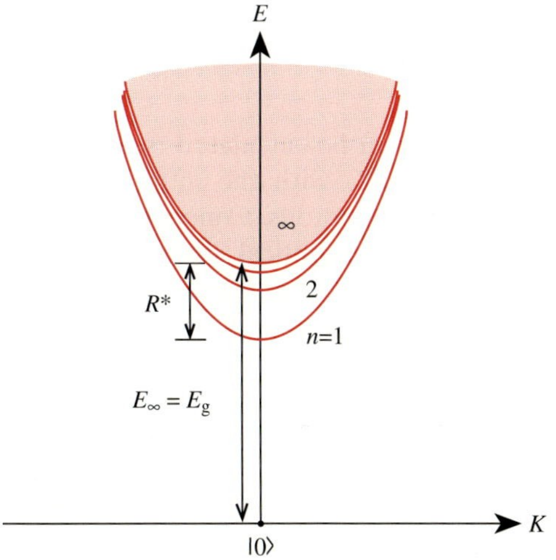

Fig. 6.21. The energy states of a Wannier exciton showing both its bound states $n = 1$ to 3 and the continuum states. E_{g} is the bandgap and R^* the exciton binding energy

Table 6.4. Exciton binding energy (R^*) and Bohr radius (a_0) in some direct bandgap semiconductors. The three semiconductors labeled (W) have the wurtzite crystal structure while the others have the zinc-blende structure. Experimental values of R^* and a_0 are from [Ref. 6.48, p. 155, 6.49]. The theoretical values of R^* are from [6.49]

Semiconductor	R^* [meV]	R^* (theory) [meV]	a_0 [Å]
GaN[a]	27	32	24
GaAs	4.9	4.4	112
InP	5.1	5.14	113
CdTe	11	10.71	12.2
ZnTe	13	11.21	11.5
ZnSe	19.9	22.87	10.7
ZnS	29	38.02	10.22
ZnO (W)	59		
CdSe (W)	15		
CdS (W)	27		

[a] for the A-exciton of wurtzite GaN, see [6.45]

to treat the holes similar to the one they proposed for hydrogenic acceptors (Sect. 4.2.4) [6.47]. Table 6.4 lists the experimentally determined exciton binding energies and Bohr radii in a number of direct bandgap semiconductors. The binding energies are compared with the theoretical values obtained in [6.45].

While the agreement between theory and experiment is quite good for R^* in the smaller-bandgap semiconductors, there are significant discrepancies for the more ionic materials. We mention that *Altarelli* and *Lipari* [6.50] have calculated the exciton dispersion in semiconductors with degenerate valence bands and showed that there are striking deviations from the parabolic dependence of (6.83). However, a detailed discussion of this calculation is beyond the scope of this book.

6.3.2 Absorption Spectra of Excitons

In principle, the absorption spectra of excitons can be calculated from the exciton energies and wavefunctions in (6.83 and 82) with the introduction of an interaction Hamiltonian between excitons and photons. Conceptually, however, optical absorption by excitons is different from optical absorption in the one-electron picture. In the two-particle picture, optical absorption is the conversion of a photon into an exciton; conservation of energy and wavevector requires that this process must occur at the point where the photon dispersion curve (broken line in Fig. 6.22b) intersects the exciton dispersion curves. At these intersections the photon and exciton are degenerate. When an exciton–photon interaction (even a very weak one) is introduced, the resultant eigenstates are linear combinations of the photon and exciton wavefunctions. Such a "coupled state" of an exciton with a photon is known as an **exciton-polariton**.

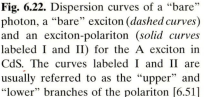

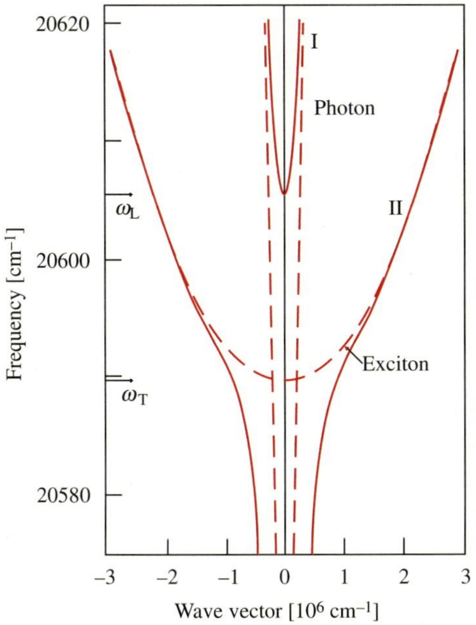

Fig. 6.22. Dispersion curves of a "bare" photon, a "bare" exciton (*dashed curves*) and an exciton-polariton (*solid curves* labeled I and II) for the A exciton in CdS. The curves labeled I and II are usually referred to as the "upper" and "lower" branches of the polariton [6.51]

Its dispersion relations (solid curves labeled I and II) are different from those of the uncoupled or "bare" photon and exciton as shown in Fig. 6.22 (drawn for the so-called A exciton in CdS).

In general, "polariton" is the name given to any coupled electromagnetic and polarization wave traveling inside a medium. The polarization wave in the present case is associated with the electric dipole moments of the excitons (assumed to be nonzero). As excitons travel in the medium they radiate electromagnetic waves. In turn, the electromagnetic waves can excite excitons. In principle, there is no way to separate the exciton wave from the electromagnetic wave. Thus introducing an exciton–photon interaction does not necessarily mean that energy will be lost by photons inside the medium. In this polariton picture energy is converted from photons to excitons and vice versa. Suppose the sample is a thin slab and light is incident on the sample from the left. Outside the sample there is only an electromagnetic field associated with the photons. As it enters the sample the electromagnetic wave is converted into a polariton wave (Fig. 6.23). Unless there are other interactions that can scatter the polaritons inside the sample, they will travel unattenuated to the sample surface on the right. On exiting the slab from the right surface, the polaritons are reconverted into photons with no loss except for those polaritons reflected back at the surface. *Thus no optical absorption has occurred inside the medium.* In order for absorption to occur (that is, for energy to be dissipated from the photon field that enters the sample) polaritons have to be scattered inelastically, e. g., by phonons. After inelastic scattering some polaritons will eventually exit the sample and appear in the form of emission (or

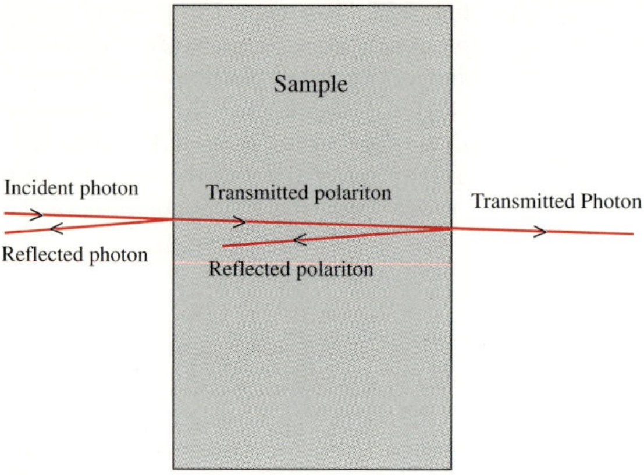

Fig. 6.23. A schematic diagram showing the transmission of photons through a semiconductor slab via propagation of polaritons inside the sample

luminescence, see Chap. 7 for further details) at a *different* photon energy. The relative probability of re-emission and relaxation via phonons depends on the polariton branch. High energy polaritons from branch I in Fig. 6.22 have large photon components in their wavefunction (and therefore are said to be photon-like); they have little interaction with phonons and are more likely to escape from the medium. However, once they are scattered elastically by defects into branch II, with a large exciton component in their wavefunctions (or exciton-like), they lose their energy efficiently via scattering with phonons or by nonradiative recombination. It is predominantly through the latter that energy in polaritons becomes dissipated inside a medium, resulting in optical absorption [6.52].

It is rather complicated to calculate the optical absorption using the above exciton-polariton picture since it is necessary to introduce energy dissipation processes for polaritons via phonon scattering. One way to avoid this difficulty is to assume that, as a result of scattering between excitons and phonons, the exciton damping constant is larger than the exciton–photon interaction. In this approximation one can replace polaritons by the "bare" photons and excitons. Whenever a photon is converted into an exciton it will lose its energy completely inside the medium via exciton damping processes. As a result, the rate of dissipation of energy from the photon field is completely determined by the rate of conversion of photons into excitons. Within this approximation we can use Fermi's Golden Rule to calculate the optical transition probability per unit volume for converting a photon into an exciton. Similar to (6.43) we obtain

$$R = (2\pi/\hbar)\sum_{\mathrm{f}} |\langle \mathrm{f}|\mathcal{H}_{\mathrm{xR}}|0\rangle|^2 \delta(E_{\mathrm{f}}(\boldsymbol{K}) - E_0 - \hbar\omega), \tag{6.84}$$

where $|0\rangle$ is the initial (ground) state with no excitons, $|\mathrm{f}\rangle$ the final state, where an exciton with energy E_{f} and wavevector $\boldsymbol{K}$ has been excited optically, and H_{xR} the exciton–photon interaction. Because of wavevector conservation, $\boldsymbol{K}$

should be equal to the photon wavevector, which is negligible, i. e., $\boldsymbol{K} = \boldsymbol{k}_{\mathrm{e}} - \boldsymbol{k}_{\mathrm{v}} \approx 0$. From now on we will therefore denote both $\boldsymbol{k}_{\mathrm{v}}$ and $\boldsymbol{k}_{\mathrm{e}}$ by $\boldsymbol{k}$.

The contributions to the imaginary part of the dielectric function (ε_{i}) due to exciton absorption consists of two parts: one arising from the bound states and the other from the continuum.

For the discrete bound states, we can express the exciton wavefunction in terms of the envelope wavefunctions given in (6.82). The optical matrix element can be shown to be ([6.44] and Problem 6.7):

$$\langle \mathrm{f}|\mathcal{H}_{\mathrm{xR}}|0\rangle = \sum_{r,k}(1/\sqrt{N})\mathrm{e}^{\mathrm{i}\boldsymbol{k}\cdot\boldsymbol{r}}\phi_{nlm}(\boldsymbol{r})\langle\psi_{\boldsymbol{k}}(\boldsymbol{r}_{\mathrm{e}})\psi_{-\boldsymbol{k}}(\boldsymbol{r}_{\mathrm{h}})|\mathcal{H}_{\mathrm{xR}}|0\rangle \tag{6.85a}$$

$$= \sum_{r,k}(1/\sqrt{N})\mathrm{e}^{\mathrm{i}\boldsymbol{k}\cdot\boldsymbol{r}}\phi_{nlm}(\boldsymbol{r})\langle\psi_{\boldsymbol{k}}^{\mathrm{c}}|\mathcal{H}_{\mathrm{eR}}^{\mathrm{c}}|\psi_{\boldsymbol{k}}^{\mathrm{v}}\rangle. \tag{6.85b}$$

If we assume that the matrix element $\langle\psi_{\boldsymbol{k}}^{\mathrm{c}}|\mathcal{H}_{\mathrm{eR}}^{\mathrm{c}}|\psi_{\boldsymbol{k}}^{\mathrm{v}}\rangle$ is independent of $\boldsymbol{k}$, the summation of $\exp(\mathrm{i}\boldsymbol{k}\cdot\boldsymbol{r})$ over $\boldsymbol{k}$ in (6.85) results in the delta function $\delta(\boldsymbol{r})$. Hence the summation over $\boldsymbol{r}$ can be easily performed, giving

$$|\langle \mathrm{f}|\mathcal{H}_{\mathrm{xR}}|0\rangle|^2 = N\,|\phi_{nlm}(0)|^2\;|\langle\psi_{\boldsymbol{k}}^{\mathrm{c}}|\mathcal{H}_{\mathrm{eR}}|\psi_{\boldsymbol{k}}^{\mathrm{v}}\rangle|^2. \tag{6.86}$$

When $\boldsymbol{r} = 0$, we have $\boldsymbol{R}_{\mathrm{e}} = \boldsymbol{R}_{\mathrm{h}}$ and therefore $|\phi_{nlm}(0)|^2$ represents the probability of finding the electron and hole within the same primitive cell. The physical interpretation of (6.86) is that the probability of exciting an exciton optically is proportional to the overlap of the electron and hole wavefunctions. Since $|\phi_{nlm}(0)|^2$ is nonzero only for $l = 0$, only excitons with s symmetry can be optically excited. Using the hydrogen atom wavefunctions one can show that ε_{i} for the bound states is equal to (**in atomic units with** $m_0 = e = \hbar = 1$)

$$\varepsilon_{\mathrm{i}}(\hbar\omega) = \frac{8\pi\,|P|^2\,\mu^3}{\omega^2(4\pi\varepsilon_0)^3\varepsilon_0^3}\sum_1^\infty \frac{1}{n^3}\delta(\omega-\omega_n), \tag{6.87}$$

where $|P|^2 = |\langle\psi_{\boldsymbol{k}}^{\mathrm{c}}|\hat{\boldsymbol{e}}\cdot\boldsymbol{p}|\psi_{\boldsymbol{k}}^{\mathrm{v}}\rangle|^2$, see (6.41), and a factor of 2 has been included to take into account spin degeneracy. Thus the oscillator strength of the bound states with quantum number n decreases as n^{-3} while their binding energy decreases as n^{-2}. In the limit $n \to \infty$ the discrete peaks of the bound states merge into a quasi continuum with density of states given by [6.44]

$$\frac{dn}{d\omega} = \frac{n^3(4\pi\varepsilon_0)^2\varepsilon_0^2}{\mu}. \tag{6.88}$$

Substituting (6.88) into (6.87) we find that, as $\hbar\omega$ approaches the bandgap, $\varepsilon_{\mathrm{i}}(\hbar\omega)$ approaches

$$\varepsilon_{\mathrm{i}}(\hbar\omega) \simeq \frac{8\pi\,|P|^2\,\mu^2}{\omega_{\mathrm{g}}^2 4\pi\varepsilon_0\varepsilon_0}. \tag{6.89}$$

For the continuum states the exciton wavefunctions can be expressed in terms of *confluent hypergeometric functions* [6.44, 53]. The corresponding contribution of exciton absorption to the imaginary part of the dielectric constant is given by ([6.44] and Problem 6.7)

$$\varepsilon_{\mathrm{i}} = \frac{2\,|P|^2\,(2\mu)^{3/2}(\omega-\omega_{\mathrm{g}})^{1/2}\tau \mathrm{e}^{\tau}}{\omega^2 \sinh\tau}, \tag{6.90}$$

where τ is defined (**in atomic units**) as.

$$\tau = \pi \left| \frac{R^*}{\omega - \omega_{\mathrm{g}}} \right|^{1/2} \tag{6.91}$$

In the limit $\omega \to \omega_{\mathrm{g}}$ one finds (Problem 6.7) that ε_{i} approaches (6.89), and therefore it varies smoothly from the discrete bound states to the continuum, a physically rather appealing result. Figure 6.24 shows schematically the exciton absorption coefficient, including contributions from both the bound and continuum states. The broken curve displays the corresponding absorption coefficient when the exciton effect is neglected. We note that the exciton effect enhances the absorption coefficient both above and below the bandgap. Instead of decreasing to zero at the bandgap, the absorption coefficient approaches a constant as in the case of an M_0 critical point in two dimensions. For a comparison of *Elliott*'s theory with experimental results, we show in Fig. 6.25 the absorption spectra of GaAs near the bandgap at different temperatures. Since the binding energy of the exciton in GaAs is about 5 meV (Table 6.4), only a broadened $n = 1$ bound state is observed at 21 K. Notice that excitons in GaAs should become *thermally ionized* at room temperature; however, the GaAs absorption edge is still modified by excitonic effects.

Excitonic effects also modify the shape of the absorption edge in indirect bandgap semiconductors such as Ge and GaP at low temperatures, as we pointed out in Sect. 6.2.6. Instead of rederiving the absorption coefficient in this case, we present the following argument. As shown in Fig. 6.21, the exciton energy levels consist of series of parabolas centered at $\boldsymbol{K} = 0$ in direct

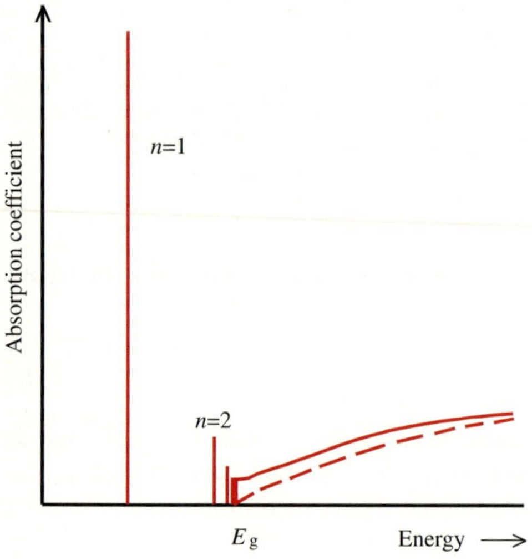

Fig. 6.24. Comparison between the absorption spectra in the vicinity of the bandgap of a direct-gap semiconductor with (*solid lines*) and without (*broken curve*) exciton effects

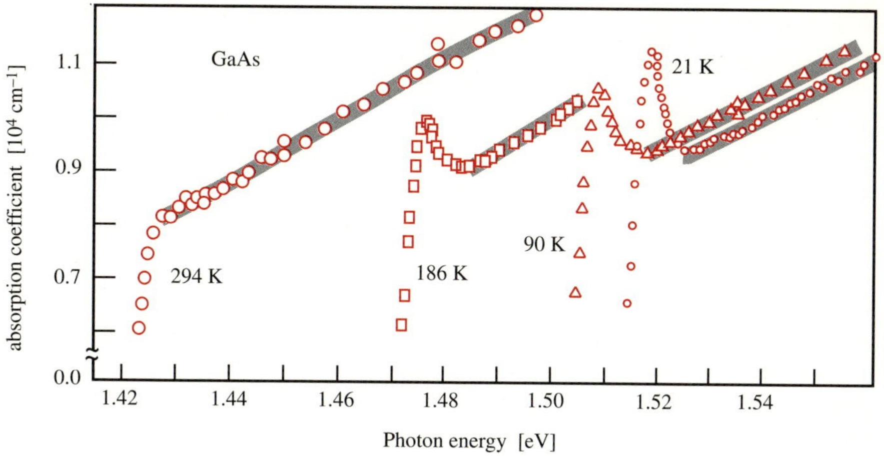

Fig. 6.25. Excitonic absorption spectra of GaAs near its bandgap for several sample temperatures. The *gray lines* drawn through the 21, 90 and 294 K data points represent fits with (6.90) [6.54]

bandgap semiconductors. In an indirect bandgap semiconductor, such as Ge, the conduction band minima are at L and the valence band maximum is at the zone center. We therefore expect an exciton created from these band extrema to form parabolas centered at the wavevectors $\boldsymbol{k}_0 = \boldsymbol{k}_L$, where $\boldsymbol{k}_L$ is the wavevector of the electron at L at Ge. In the direct gap materials, excitons can be excited directly by photons only at the point where the photon and exciton dispersion curves intersect (Fig. 6.22). This wavevector conservation condition is relaxed in indirect gap materials by the participation of phonons. As a result, photons can excite excitons at any point on the parabolas with nearly the same probability. Thus the absorption coefficient is proportional to the density of final states, which has the same shape as the absorption edge at direct bandgap semiconductors for uncorrelated electron–hole pairs (i. e., without excitonic effects). Hence the absorption edge in indirect bandgap semiconductors including excitonic effects has the same dependence on energy as in a direct bandgap semiconductor without excitonic effects [represented by (6.58a)].

So far the best example of an excitonic Rydberg series has been found in the semiconductor Cu_2O. As pointed out in Sect. 6.2.6, the conduction and valence band extrema in Cu_2O have the same parity under inversion. Optical transitions between these two bands are electric-dipole forbidden but possible as magnetic-dipole or electric-quadrupole transitions. Their optical transition probability is thus proportional to the k-dependent term in the matrix element $|\langle c, \boldsymbol{k}|\boldsymbol{k}\cdot p|v, \boldsymbol{k}\rangle|^2$. Using (6.68) one can show that this transition probability depends on the derivative $|d\Phi_{nlm}(0)/dr|^2$ of the excitonic envelope function evaluated at $\boldsymbol{r} = 0$ [6.44]. This leads to the conclusion that exciton bound states with p symmetry are weakly electric-dipole active at a "forbidden" direct bandgap. The absorption coefficient is proportional to

$$\left| \frac{d\Phi_{nlm}(0)}{dr} \right|^2 \propto \frac{n^2 - 1}{n^5} \tag{6.92}$$

and is identically zero for the $n = 1$ level (since this is the only level without a p state). Figure 6.26 shows the weakly allowed excitonic absorption peaks in Cu_2O involving the so-called yellow exciton series $2p$, $3p$, $4p$, etc., measured by *Baumeister* [6.55]. The observed peaks are fitted very well with the Rydberg series:

$$E_n = (2.166 - 0.097/n^2)\ \text{eV}\ (n = 2, 3, \ldots). \tag{6.93}$$

Separately, a very sharp and weak exciton peak associated with the $1s$ exciton has been observed in Cu_2O at 2.033 eV. It is excited via magnetic-dipole and electric-quadrupole transitions [6.56]. From the continuum threshold energy of 2.166 eV in (6.93) one obtains a binding energy of 0.133 eV for the $1s$ state, while the higher exciton states in the same series can be fitted with the smaller Rydberg energy of 0.097 eV. The reason for this difference is the central cell correction discussed already in Sect. 4.2.2 in connection with hydrogenic impurities. A discussion of central cell effects in excitons is beyond the scope of this book.

6.3.3 Exciton Effect at M_1 Critical Points or Hyperbolic Excitons

In Table 6.1 we showed that the Van Hove singularity in the density of states at an M_1 critical point has a shape described by $C - (E_0 - E)^{1/2}$ (where C is a constant and E_0 is the energy of the critical point) for $E < E_0$ and is equal to C for $E > E_0$. The corresponding shape of ε_i in the vicinity of such a critical

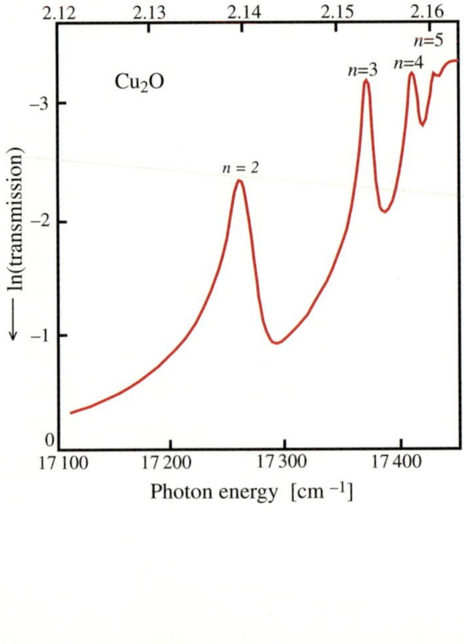

Fig. 6.26. The low-temperature absorption spectrum of Cu_2O showing the excitonic p series associated with its "dipole-forbidden" band edge [6.55]

point in the joint density of states is sketched in Fig. 6.9. The characteristic feature of ε_i is that it rises sharply as E approaches E_0 from below and decreases slowly for E above E_0. In Sect. 6.2.4 we mentioned that the E_1 and $E_1 + \Delta_1$ transitions in the optical spectra of diamond- and zinc-blende-type semiconductors are attributed to M_1-type critical points based on band structure calculations. However, experimentally it was found that the shape of the E_1 transitions deviates significantly from that expected for M_1 critical points. Figure 6.27 shows the imaginary part of the dielectric constant (ε_i) in the region of the E_1 transitions in two II–VI semiconductors: CdTe and ZnTe. The dashed curve was calculated from the pseudopotential band structures by *Walter* et al. [6.58] without exciton effects. It shows the asymmetric shape expected for the three-dimensional M_1-type critical points. However, the experimental curves

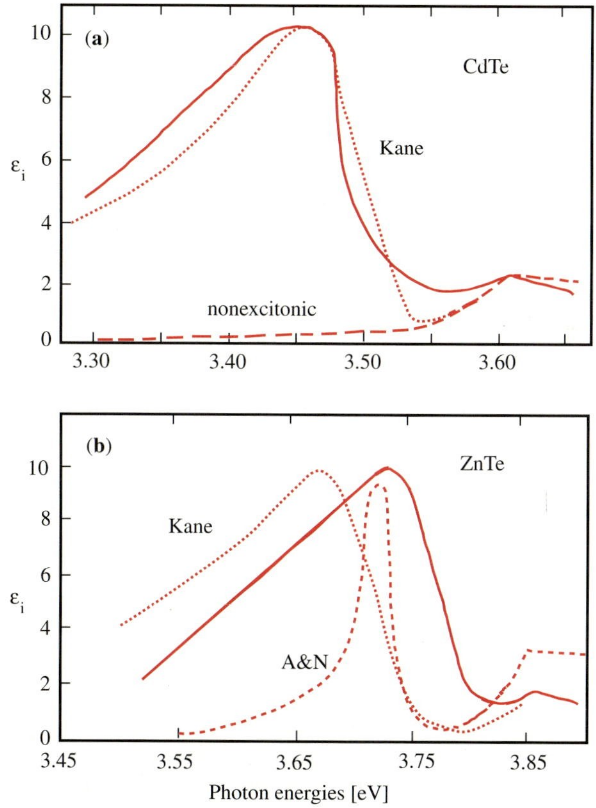

Fig. 6.27a, b. Imaginary part of the dielectric constant (ε_i) in the region of the E_1 transitions for (**a**) CdTe and (**b**) ZnTe. The red solid curves represent the experimental results of *Petroff* and *Balkanski* [6.57]. The dashed curve labeled "nonexcitonic" is calculated from the band structure obtained by *Walter* et al. [6.58] without excitonic effects. Dotted curves labeled *Kane* and A&N have been calculated, respectively, by *Kane* [6.59] and *Antoci* and *Nardelli* [6.60], both including excitonic effects

measured by *Petroff* and *Balkanski* [6.57] show asymmetries in the opposite direction, i. e., the peaks are sharper above the critical point rather than below. In addition, a weaker structure appears at energies higher than the main strong peak in both spectra. These discrepancies between the experiment and the theory based on a one-electron band calculation have been explained by excitonic effects at M_1 critical points. Such excitons occurring at saddle points are known as **hyperbolic excitons** (since the constant energy surfaces near M_1 critical points are hyperboloids).

Kane [6.59] calculated the lineshape of ε_i near an M_1 saddle point ω_1 by solving the wave equation for the relative motion of the electron and hole:

$$\left(\frac{P_1^2}{2\mu_1} + \frac{P_2^2}{2\mu_2} + \frac{P_3^2}{2\mu_3} - \frac{e^2}{4\pi\varepsilon_0\varepsilon_0 r}\right)\Phi(\boldsymbol{r}) = E\Phi(\boldsymbol{r}), \quad (6.94)$$

where μ_1, $\mu_2 > 0$ and $\mu_3 < 0$. In the extreme case when $|\mu_3| \to \infty$ (6.94) reduces to the exciton equation of motion for an M_0 critical point in two dimensions. Analytic solutions in this case are known [6.61]. Quasi-two-dimensional excitons are known to exist in layered-type semiconductor such as GaSe (Sect. 7.2.7). More recently they have been found to be important for the optical properties of quantum wells (Chap. 9). Their bound state energies (indexed by the quantum number n) E_{2D} are given by the series

$$E_{2D}(n) = E_{2D}(\infty) - R^*/(n + \tfrac{1}{2})^2 \text{ for } n = 0, 1, 2, \ldots, \quad (6.95)$$

where the effective Rydberg R^* is the same one as defined in (6.81) for three-dimensional excitons. The corresponding oscillator strengths, analogous to those in (6.87) but for the two-dimensional case, are proportional to $(n+\frac{1}{2})^{-3}$. The ratio f_0/f_1 of the oscillator strengths of the $n = 0$ peak to the $n = 1$ peaks is given by $(1/2)^{-3}/[(3/2)^{-3}] = 27$. The important feature of the two-dimensional M_0 exciton is the dominance of the $n = 0$ peak. For finite $|\mu_3|$ but $|\mu_3| \gg \mu_1, \mu_2$, *Kane* solved (6.94) using the adiabatic approximation for the heavier mass direction. The numerical solutions show that the effects of a finite $|\mu_3|$ are to decrease the binding energy and to broaden asymmetrically the $n = 0$ peak in the corresponding two-dimensional exciton. His results for the cases of $m_3 = \mu_3/\mu_1 = -320$, -40, and -5 are shown in Fig. 6.28. *Phillips* [6.62] has interpreted this asymmetric broadening in terms of a **Fano interference** [6.63] between a discrete state and a continuum. When μ_3 is negative, there is a continuum of allowed states with energy below the M_1 critical point. The $n = 0$ state in the "two-dimensional" exciton can decay into these continuum states and becomes a resonant state (Sect. 4.3). The lineshape of the $n = 0$ state is similar to the so-called auto-ionizing states found in atomic spectra. *Kane*'s theory explained quantitatively the absorption spectra in CdTe. (Fig. 6.27a) but not in ZnTe [6.57] where the theoretical peak position is lower in energy than the experimental one although their lineshapes agree qualitatively. Later *Antoci* and *Nardelli* [6.60] showed that the theoretical peak position in ZnTe can be made to agree better with experiment by slightly modifying the energy bands along the ⟨111⟩ directions. Their results are also compared with experiment in Fig. 6.27b.

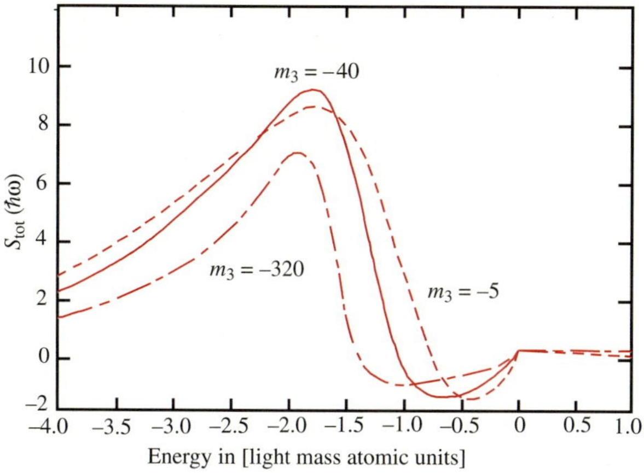

Fig. 6.28. Contribution to the imaginary part of the dielectric constant (ε_i) in the vicinity of an M_1 critical point (with energy chosen at the origin) calculated numerically by *Kane* [6.59] for the cases $m_3 = \mu_3/\mu_1 = -320$, -40, and -5. The $m_3 = -40$ and -5 curves have been multiplied by factors of 5 and 20, respectively, in order to be displayed on the same vertical scale as the $m_3 = -320$ curve. The light mass atomic unit of energy is $2\mu_1/\varepsilon_0$ Ry

In recent years more elaborate many-body calculations of exciton effects on the whole dielectric function have largely corrected the discrepancies between theory and experiment shown in Figs. 6.10–12 [6.64, 65]. For an example of such calculations and a comparison with the measured UV spectrum of GaN see Sect. 6.7.

6.3.4 Exciton Effect at M_3 Critical Points

We have found so far that excitonic effects tend to enhance the oscillator strength at M_0 and M_1 critical points. Since the total oscillator strength is proportional to the total number of valence electrons (see the sum rule in Problem 6.5) it should be "conserved" in some way. In other words, the gain in oscillator strength at M_0 and M_1 critical points induced by the excitonic interaction must be compensated by losses elsewhere. Optical transitions are suppressed at M_2 and M_3 critical points [6.66]. The equation for the relative motion [i. e., the analog of (6.77b)] for electron–hole pairs at M_3 critical points is almost identical to that for M_0 critical points except that the reduced mass is negative. This is equivalent to keeping the reduced mass positive but changing the sign of the Coulomb attraction term in (6.77b). This means that at an M_3 critical point the electron and hole can be regarded as having normal positive masses but repelling rather than attracting each other via the Coulomb interaction. The result is that no bound states are formed. The repulsion keeps them apart and therefore the optical transition probability is suppressed. The solutions of the wave equation can be shown to be similar to the continuum solutions of (6.77b). Now $\varepsilon_i(\omega < \omega_3)$, where ω_3 is the frequency of the M_3 critical point, is given (**in atomic units,** $e = m_0 = \hbar = 1$) by [6.66]

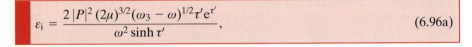

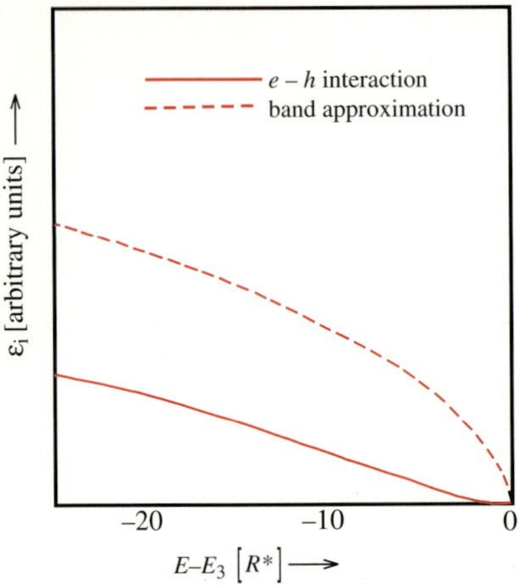

Fig. 6.29. Comparison between the lineshapes of the imaginary part of the dielectric constant in the vicinity of an M_3 critical point calculated with and without exciton effects [6.66]. The energy units are exciton Rydbergs

$$\varepsilon_{\mathrm{i}} = \frac{2\,|P|^2\,(2\mu)^{3/2}(\omega_3-\omega)^{1/2}\tau' \mathrm{e}^{\tau'}}{\omega^2 \sinh \tau'}, \tag{6.96a}$$

where τ' is defined as

$$\tau' = -\pi \left| \frac{R^*}{\omega_3 - \omega} \right|^{1/2} . \tag{6.96b}$$

The resultant shape of ε_{i} is shown in Fig. 6.29. The strong suppression of the singularity in the optical transition strength is presumably the reason why no M_3 critical points have been positively identified in optical spectra.

6.4 Phonon-Polaritons and Lattice Absorption

In Chap. 3 we discussed how phonons in crystals containing more than one atom per primitive cell are classified into acoustic and optical phonons. As their names imply, optical phonons can interact with electromagnetic radiation. We shall study this interaction in this section.

Since phonons are quantized simple harmonic oscillators, we shall begin by reviewing the response of a collection of identical and charged simple harmonic oscillators (SHO) to a radiation field in the form of the plane wave

$$\boldsymbol{E}(\boldsymbol{r},t) = \boldsymbol{E}_0 \exp \mathrm{i}(\boldsymbol{k}\cdot\boldsymbol{r} - \omega t). \tag{6.97}$$

We shall assume that these SHO are isotropic and uniformly distributed in the entire space (in order to avoid problems related to the presence of surfaces). The mass and charge of the SHOs are M and Q, respectively. The natural vibrational frequency of each SHO is ω_{T}. In response to the applied field the

SHO are displaced from their equilibrium positions by the vector $\boldsymbol{u}$. The equation of motion of the SHOs is

$$M(d^2\boldsymbol{u}/dt^2) = -M\omega_{\mathrm{T}}^2\boldsymbol{u} + Q\boldsymbol{E}. \tag{6.98}$$

In the steady state the solutions to (6.98) can be expressed as

$$\boldsymbol{u} = \boldsymbol{u}_0 \exp[\mathrm{i}(\boldsymbol{k}\cdot\boldsymbol{r} - \omega t)]. \tag{6.99}$$

Substituting this into (6.98) we obtain the solution for $\boldsymbol{u}_0$:

$$\boldsymbol{u}_0 = \frac{Q\boldsymbol{E}_0}{M(\omega_{\mathrm{T}}^2 - \omega^2)}. \tag{6.100}$$

Since these SHOs are charged and they are all displaced by the same amount $\boldsymbol{u}$, they produce a macroscopic polarization $\boldsymbol{P}$ oscillating also at frequency ω:

$$\boldsymbol{P} = NQ\boldsymbol{u}, \tag{6.101}$$

where N is the density of the SHOs. The electric displacement vector $\boldsymbol{D}$ of the medium is given by

$$\boldsymbol{D} = \boldsymbol{E} + 4\pi\boldsymbol{P} = \varepsilon\boldsymbol{E}, \quad \boldsymbol{D} = \varepsilon_0\boldsymbol{E} + \boldsymbol{P} = \varepsilon_0\varepsilon\boldsymbol{E} \text{ (SI units)} \tag{6.102}$$

where ε is the dielectric function of the isotropic medium. Substituting (6.100 and 101) into (6.102) we obtain ε:

$$\varepsilon = 1 + \frac{4\pi NQ^2}{M(\omega_{\mathrm{T}}^2 - \omega^2)} = 1 + \frac{NQ^2}{\varepsilon_0 M\,(\omega_{\mathrm{T}}^2 - \omega^2)} \text{ (SI units)} \tag{6.103}$$

When (6.103) is generalized to a collection of SHOs with different resonance frequencies ω_{i} we obtain (6.50). Equation (6.103) can also be expressed in terms of the refractive index n of the medium:

$$n^2 = 1 + \sum_i \frac{4\pi N_i Q^2}{M(\omega_i^2 - \omega^2)} = 1 + \sum_i \frac{N_i Q^2}{\varepsilon_0 M(\omega_i^2 - \omega^2)} \text{ (SI units)}. \tag{6.104}$$

Equation (6.104) has been used to explain the anomalous dispersion in the refractive indices of gases in the vicinity of absorption lines and is known as **Sellmeier's equation** [6.67].

Before proceeding further with (6.103) we shall include also the contribution of the valence electrons to the total dielectric function ε of the medium. To distinguish the valence electron contribution from that of SHOs, we shall denote one by ε_{e} and the other by ε_{l} (e and l stand, respectively, for electrons and lattice). We shall assume that the bandgap $E_{\mathrm{g}} \gg \hbar\omega$, so that the radiation field appears to be static to the electrons and $\varepsilon_{\mathrm{e}}(\omega)$ can be approximated by $\varepsilon_{\mathrm{e}}(0)$. On the other hand if $\omega \gg \omega_{\mathrm{T}}$ the SHOs cannot follow the electric field and they no longer contribute to the total dielectric function and therefore $\varepsilon_{\mathrm{l}} \to 1$ in (6.103). Thus for $(E_{\mathrm{g}}/\hbar) \gg \omega \gg \omega_{\mathrm{T}}$ we have the total $\varepsilon \approx \varepsilon_{\mathrm{e}}(0)$. It is usual to designate $\varepsilon_{\mathrm{e}}(0)$ as ε_∞ and call it the high frequency dielectric constant since it is the dielectric constant at a frequency much higher that the vibrational frequencies but below electronic excitation energies. When ε_∞ is included in (6.103), $\varepsilon = \varepsilon_l + \varepsilon_e$ becomes

$$\varepsilon(\omega) = \varepsilon_\infty + \frac{4\pi NQ^2}{M(\omega_T^2 - \omega^2)} = \varepsilon_\infty + \frac{NQ^2}{\varepsilon_0 M(\omega_T^2 - \omega^2)} \text{ (SI units)} \tag{6.105}$$

provided $\omega \ll (E_g/\hbar)$.

As there are no excess charges in the medium, the electric displacement $\boldsymbol{D}$ satisfies the Gauss equation

$$\operatorname{div}\boldsymbol{D} = 0 \tag{6.106}$$

or equivalently

$$\varepsilon(\boldsymbol{k}\cdot\boldsymbol{E}_0) = 0. \tag{6.107}$$

This equation is fulfilled when either $\varepsilon = 0$ or $(\boldsymbol{k}\cdot\boldsymbol{E}_0) = 0$. We shall consider these two cases separately.

Case 1 (Transverse Field): $(\boldsymbol{k}\cdot\boldsymbol{E}_0) = 0$

If $\boldsymbol{E}_0$ is also zero we obtain the trivial situation $\boldsymbol{u} = \boldsymbol{P} = \boldsymbol{D} = 0$. If $\boldsymbol{E}_0$ is nonzero then $(\boldsymbol{k}\cdot\boldsymbol{E}_0) = 0$ implies that $\boldsymbol{E}_0$ is perpendicular to the propagation direction. In other words, the electric field has to be transverse. For such transverse fields the response of the SHOs is described by the dielectric function in (6.105). In particular, ε diverges when ω approaches ω_T. As a result, ω_T represents the resonance frequency of the medium when a transverse vibration is excited (the **transverse resonance frequency** in short).

Case 2 (Longitudinal Field): $\boldsymbol{E}_0 \parallel \boldsymbol{k}$ and $\varepsilon = 0$

When the electric field is longitudinal (so that $\boldsymbol{E}_0\cdot\boldsymbol{k} \neq 0$), ε has to vanish in order that (6.107) be satisfied. From (6.105) we see that this can occur at frequencies ω_L defined by $\varepsilon(\omega_L) = 0$. Solving (6.105) we obtain

$$\omega_L^2 = \omega_T^2 + \frac{4\pi NQ^2}{M\varepsilon_\infty} = \omega_T^2 + \frac{NQ^2}{\varepsilon_0\varepsilon_\infty M} \text{ (SI units).} \tag{6.108}$$

To understand what happens at ω_L, we note that (6.102) can be written as $\boldsymbol{E} = (1/\varepsilon)\boldsymbol{D}$. When $\varepsilon = 0$, $\boldsymbol{E}$ is not necessary zero even for $\boldsymbol{D} = 0$. In fact the longitudinal electric field at ω_L, to be denoted as $\boldsymbol{E}_L$, is given by

$$\boldsymbol{E}_L = -(4\pi/\varepsilon_\infty)\boldsymbol{P} = -(4\pi/\varepsilon_\infty)NQ\boldsymbol{u} = -(1/\varepsilon_0\varepsilon_\infty)NQ\boldsymbol{u} \text{ (SI units)} \tag{6.109}$$

(when $\boldsymbol{D} = 0$). This implies that no external charges are required to generate an electric field when the SHOs are oscillating at ω_L. Instead, the longitudinal electric field $\boldsymbol{E}_L$ is produced by the polarization induced by the oscillations. From (6.109) we see that $\boldsymbol{E}_L$ is a macroscopic field (since $\boldsymbol{P}$ is a macroscopic quantity). Furthermore, this field points in the opposite direction to the polarization and therefore it contributes an additional restoring force to the longitudinal oscillation. This explains why ω_L, the longitudinal resonance frequency, in (6.108) is always larger than the transverse frequency ω_T. It should be emphasized that E_L should not be confused with the macroscopic electric field in the parallel-plate capacitor shown in Fig. 3.3. While the planes of positive and negative charges in a solid appear to resemble parallel-plate capacitors, the fields they produce are microscopic. At equilibrium, in the absence of an LO phonon, these *microscopic* fields when summed over many unit cells will produce a zero *macroscopic* field. Otherwise a macroscopic polarization would exist even without an external field as in *ferroelectric* materials. When an LO phonon is excited in the medium, the relative displacements of the charges result in many induced dipole moments. They sum to a macroscopic polarization, which gives rise to the longitudinal field $\boldsymbol{E}_L$.

The charge Q and mass M of the SHOs are microscopic properties that are sometimes difficult to measure. On the other hand quantitities such as ω_T, ω_L, and ε_∞ can be determined experimentally. Often it is convenient to introduce another quantity $\varepsilon_0 = \varepsilon(0)$, the **low frequency dielectric constant**, so that the dielectric function can be expressed in terms of measurable quantities (Problem 6.4):

$$\varepsilon(\omega) = \varepsilon_\infty \left(1 + \frac{\omega_L^2 - \omega_T^2}{\omega_T^2 - \omega^2}\right) = \varepsilon_\infty \frac{\omega_L^2 - \omega^2}{\omega_T^2 - \omega^2} \tag{6.110a}$$

or

$$\varepsilon(\omega) = \varepsilon_\infty + \frac{\varepsilon_0 - \varepsilon_\infty}{1 - (\omega^2/\omega_T^2)}. \tag{6.110b}$$

From these results one can derive the **Lyddane–Sachs–Teller (LST) relation** (Problem 6.4):

$$\frac{\varepsilon_0}{\varepsilon_\infty} = \frac{\omega_L^2}{\omega_T^2}. \tag{6.111}$$

6.4.1 Phonon-Polaritons

The above discussion of the interaction between electromagnetic fields and charged harmonic oscillators neglects the radiation produced by the oscillating macroscopic polarization $\boldsymbol{P}$. As a result, the above theory predicts that transverse and longitudinal resonance occur at ω_T and ω_L, respectively, even for zero wavevector. Since whether a wave is transverse of longitudinal depends on the direction of its displacement relative to its wavevector, one expects however, that the transverse and longitudinal vibrations should become degenerate in the limit that the wavevector approaches zero. After all, there is no way to distinguish between the two when the wavevector is zero. This shortcoming is rectified by taking *retardation effects* into consideration [6.68].

A complete description of the interaction between electromagnetic waves and charges must invoke Maxwell's equations. Equation (6.106) is only one of them. The remaining three equations (in the absence of any current in the medium) are

$$\operatorname{div}\boldsymbol{B} = 0, \tag{6.112a}$$

$$\operatorname{curl}\boldsymbol{H} = (1/c)(\partial\boldsymbol{D}/\partial t), \tag{6.112b}$$

$$\operatorname{curl}\boldsymbol{E} = (-1/c)(\partial\boldsymbol{B}/\partial t), \tag{6.112c}$$

(set $c = 1$ for SI units)
where $\boldsymbol{B}$ and $\boldsymbol{H}$ are, respectively, the magnetic induction and the magnetic field. Since we have limited ourselves to considering only nonmagnetic semiconductors in this book, we can use the approximation $\boldsymbol{B} = \boldsymbol{H}$ (for SI units $B = \mu_0 H$, where μ_0 is the permeability of vacuum). It is well known, of course, that Maxwell's equations can be combined to generate two wave equations (one for $\boldsymbol{E}$ and one for $\boldsymbol{H}$) that describe the propagation of electromagnetic fields [6.10].

Since electromagnetic waves are transverse (in an infinite medium) they couple to transverse excitations such as TO phonons but not to LO phonons. We can therefore limit our considerations from now on to Case 1 only. Equation (6.112a) implies that the $\boldsymbol{H}$ field is also perpendicular to $\boldsymbol{k}$. As in the case of $\boldsymbol{E}$, we shall assume that $\boldsymbol{H}$ is represented by a plane wave:

$$\boldsymbol{H}(\boldsymbol{r},t) = \boldsymbol{H}_0 \exp[\mathrm{i}(\boldsymbol{k}\cdot\boldsymbol{r} - \omega t)]. \tag{6.113}$$

By substituting $\boldsymbol{E}$ and $\boldsymbol{H}$ from (6.97 and 113) into (6.112b and c), respectively, we obtain two linear homogeneous equations in $\boldsymbol{E}_0$ and $\boldsymbol{H}_0$. In order that these two equations have nontrivial solutions we require that the characteristic determinant be zero. This condition can be expressed as

$$k^2 = (\omega^2/c^2)\varepsilon, \tag{6.114a}$$

which we recognize as the dispersion of a transverse electromagnetic wave inside a nonmagnetic medium with dielectric constant ε. Substituting into (6.114a) the expression for ε in (6.111) we obtain the dispersion relation

$$k^2 = \frac{\omega^2}{c^2}\left(\varepsilon_\infty + \frac{\varepsilon_0 - \varepsilon_\infty}{1 - (\omega^2/\omega_\mathrm{L}^2)}\right). \tag{6.114b}$$

Figure 6.30 (solid curves) shows a plot of the solutions of (6.114b). If we rewrite (6.114b) as a quadratic equation in ω^2, we see that for every value of k there are two solutions for ω^2, in other words there are two "branches" to the dispersion curve, similar to the exciton-polariton dispersion sketched in Fig. 6.22. It is easy to show that as $k \to 0$, one solution of (6.114b) approaches $\omega^2 = c^2k^2/\varepsilon_0$ (known as the "lower branch") while the other one ("upper branch") approaches the constant value ω_L. Thus the frequency of the transverse oscillations in the limit of zero wavevector (for the upper branch) becomes degenerate with the longitudinal oscillation frequency, as expected because of the cubic symmetry. On the other hand, when $k \to \infty$ the dispersion of the upper branch is given by $\omega^2 = c^2k^2/\varepsilon_\infty$ while the lower branch

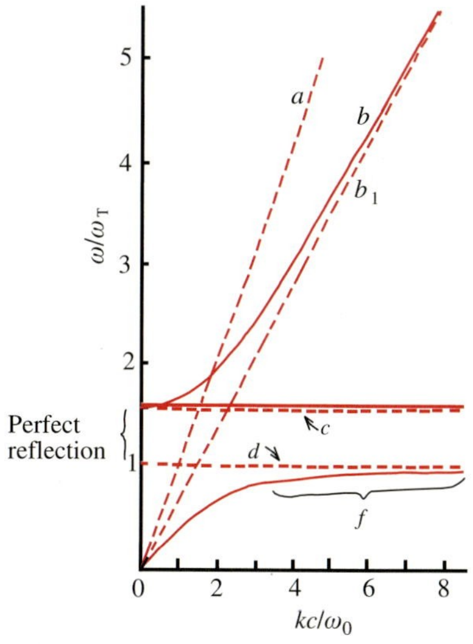

Fig. 6.30. Schematic diagram of the dispersion curves of an uncoupled light wave and lattice vibrations and of their coupled optical wave (called a phonon-polariton) in a polar crystal [6.68]. *a*: light in vacuo; *b*: photon-phonon coupled mode (upper polariton); b_1: photon dispersion in the medium but without coupling to the phonons; *c*, *d*: longitudinal and transverse uncoupled lattice vibrations, respectively; *f*: transverse phonons coupled to the photons (lower polaritons)

approaches ω_T. Since the longitudinal oscillations cannot couple to the transverse electromagnetic wave, they have no dispersion, a fact which is represented in Fig. 6.30 by the horizontal straight line passing through ω_L.

Although the above results have been derived for SHOs distributed uniformly in space, they can be shown [6.68] to be valid also for an ionic (or partly ionic) crystal containing two atoms per unit cell provided we make the following substitutions. The displacement of the SHO is replaced by the relative displacement of the two ions in the primitive unit cell or

$$M\boldsymbol{u} \rightarrow \mu(\boldsymbol{u}_+ - \boldsymbol{u}_-), \tag{6.115}$$

where $\boldsymbol{u}_+$ and $\boldsymbol{u}_-$ are, the displacements of the positive and the negative ions and μ is the reduced mass of the two ions A and B with masses m_A and m_B, respectively ($\mu^{-1} = m_A^{-1} + m_B^{-1}$). The charge Q should be replaced by an *effective ionic charge* e^* on the ions (one positive and the other negative). The meaning of e^* will be discussed in 6.4.4. The transverse and longitudinal oscillation frequencies are now identified with the frequencies of the TO and LO phonons, respectively. The resultant transverse wave whose dispersion is described by (6.114b) in crystals is known as a **phonon-polariton**. Its dispersion curve can be understood in terms of coupled vibrations. In the absence of the TO phonons the dispersion of the electromagnetic oscillation in Fig. 6.30 is given by the straight line $\omega = ck/\varepsilon_\infty^{1/2}$. If the TO phonon (which is assumed to be dispersionless) does not couple to radiation it will be represented in the same figure by a horizontal straight line passing through ω_T. At the wavevector where these two lines intersect (i. e., the modes are degenerate) they can couple to each other. The reason is that electromagnetic waves can excite the TO phonons while the oscillating charges can radiate electromagnetic waves. The two kinds of waves cannot be separated. As a result of this coupling the frequencies of the two modes are altered: one is raised while the other is lowered. In other words, the two modes seem to "repel" each other. This is sometimes also referred to as **level anticrossing**. Notice that for $\omega \ll \omega_T$ the TO phonon also contributes to the dielectric function. As a result, the polariton dispersion for small ω is given by $(ck/\omega) = \varepsilon_0^{1/2}$. In Chap. 7 we shall describe in further detail the measurement of the TO and LO phonon frequencies in semiconductors near $k = 0$ and of the phonon-polariton dispersion in GaP using light scattering techniques (Fig. 7.26).

6.4.2 Lattice Absorption and Reflection

Once the contribution to the dielectric constant from lattice vibrations is known, it is straightforward to calculate the corresponding optical properties of the sample, such as absorption coefficients, refractive index and reflectance, using the KKRs and (6.10, 11, and 8). For example, the imaginary part of the dielectric function is obtained from (6.105) with the KKR (Problem 6.4). The resultant expression consists of a delta function centered at the TO phonon frequency. This is therefore also the lineshape of the absorption spectrum of an infrared-active optical phonon. Equation (6.110) shows that ε is negative between the TO and LO phonon frequencies and the complex refractive index pure imaginary. The reflectance of the sample from (6.8) is therefore 1 between ω_L and ω_T. This means that no incident light can penetrate the sample. As we shall see later, the reflectivity at ω_T is reduced from 100 % when damping of the TO phonon is included in the calculation. Still, the reflectivity remains highest at ω_T. As a result the frequency of the TO phonon is known as the **reststrahlen** (after a German word meaning residual rays) **frequency**.

For better agreement with experiment it is necessary to introduce a *damping constant* γ for the TO phonons [6.70]. The result is that the equation of motion (6.98) becomes

$$M(d^2\boldsymbol{u}/dt^2) - M\gamma(d\boldsymbol{u}/dt) = -M\omega_T^2\boldsymbol{u} + Q\boldsymbol{E} \tag{6.116}$$

and the corresponding dielectric constant in (6.103) is complex:

$$\varepsilon(\omega) = \varepsilon_\infty + \frac{4\pi NQ^2}{4\pi\varepsilon_0 M(\omega_T^2 - \omega^2 - i\omega\gamma)} \text{ (SI units).} \tag{6.117a}$$

$\varepsilon(\omega)$ can also be expressed in terms of ε_∞ and ε_0 as in (6.110b):

$$\varepsilon(\omega) = \varepsilon_\infty + \frac{\varepsilon_0 - \varepsilon_\infty}{[1 - (\omega^2/\omega_T^2)] - i(\omega\gamma/\omega_T^2)}. \tag{6.117b}$$

A plot of the real and imaginary parts of $\varepsilon(\omega)$ for $\gamma/\omega = 0.05$ is shown in Fig. 6.31a. The reflectivity coefficients calculated from (6.117b) are displayed for several values of γ/ω in Fig. 6.31b. Figure 6.32 exhibits the measured lattice reflection spectra in several zinc-blende-type semiconductors. They can all be fitted reasonably well by curves calculated from (6.117b) with (6.8) using the phonon frequency and the damping constant as the only adjustable parame-

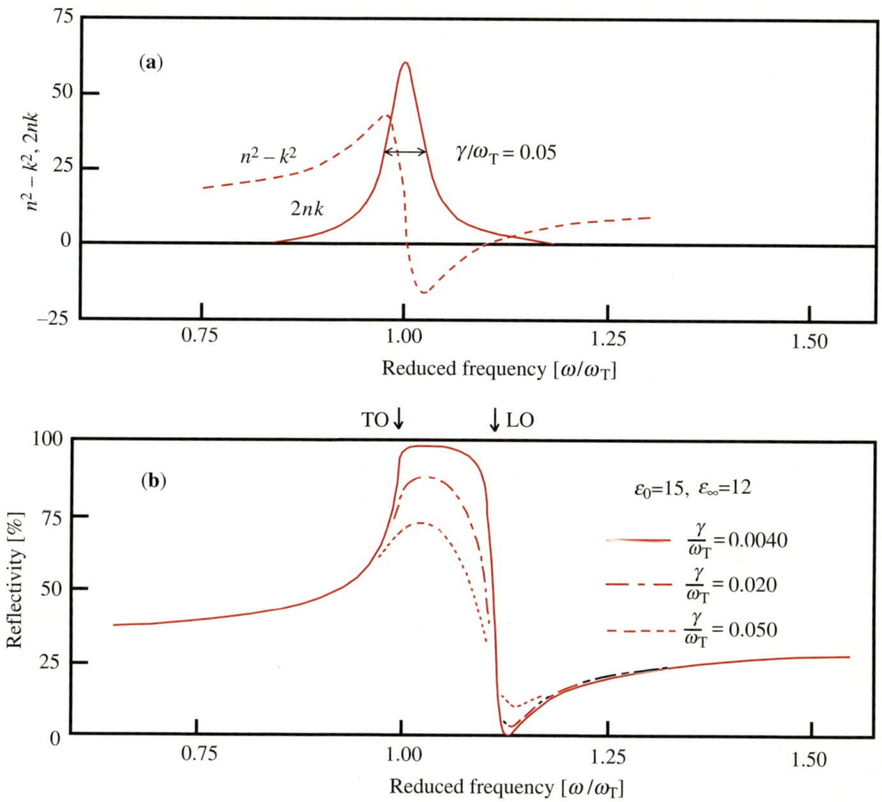

Fig. 6.31. Plot of (**a**) the real and imaginary parts of the complex dielectric constant and (**b**) the reflectivity coefficients calculated from (6.117b). The vertical arrows indicate the frequencies of the TO and LO phonons. Note the deep minimum in the reflectivity which corresponds to $\varepsilon_r \approx 1$ [6.69]

ters. The parameters obtained in this way are listed in Table 6.5. The phonon frequencies can be measured independently by other techniques such as Raman scattering (Chap. 7), while γ can be deduced from time-resolved Raman scattering since it is the inverse of the phonon lifetime [6.70, 73]. For calculations of phonon linewidths based on the electronic band structure see [6.70]. These widths can be considered as the imaginary part of an anharmonic correction to the energy (self-energy). Its real part is an anharmonic correction to the phonon frequency [6.71].

6.4.3 Multiphonon Lattice Absorption

Although the optical phonons in covalent semiconductors, such as Si and Ge, are not infrared active (Sect. 2.3.4) these semiconductors also absorb infrared

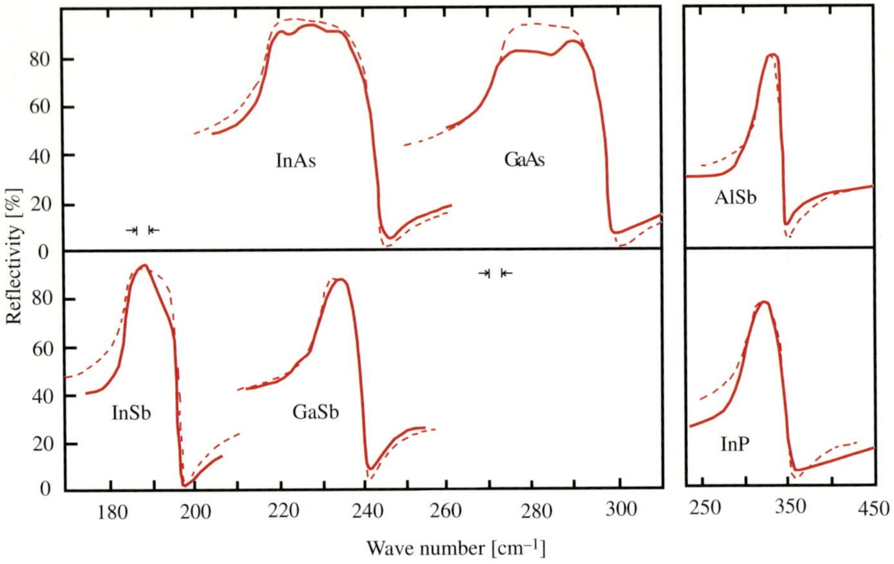

Fig. 6.32. Comparison between experimental (*solid curves*) lattice reflection spectra in several zinc-blende-type semiconductors with those calculated from (6.117b) using (6.8) (*broken curves*). The TO and LO phonon frequencies and the corresponding damping constants were adjusted to fit the experimental spectra. The spectra on the left-hand side were measured at liquid helium temperature; those on the right are room temperature spectra [6.69]

Table 6.5. The TO (ω_T) and LO phonon (ω_L) frequencies and the ratio of the damping constant (γ) to ω_T determined from lattice reflection spectra in several zinc-blende-type semiconductors [6.69] and from Raman scattering [6.72]

Semiconductor	Temperature [K]	ω_T [cm^{-1}]	ω_L [cm^{-1}]	γ/ω_T
InSb	4.2	184.7	197.2	<0.01
	300	179.1	190.4	0.016
InAs	4.2	218.9	243.3	<0.01
InP	300	307.2	347.5	0.01
GaSb	4.2	230.5	240.3	<0.01
GaAs	4.2	273.3	297.3	<0.01
	296	268.2	291.5	0.007
GaP	300	366.3	401.9	0.003
GaN	300	555	740	–
AlSb	300	318.8	339.6	0.0059
CdTe	1.2	145	170	–
ZnSe	80	211	257	0.01

radiation. Figure 6.33a shows their infrared absorption spectra measured by *Collins* and *Fan* [6.74]. The corresponding spectra in Si measured as a function of temperature by *Johnson* [6.75] are shown in Fig. 6.33b. A mechanism

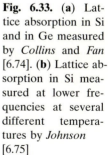

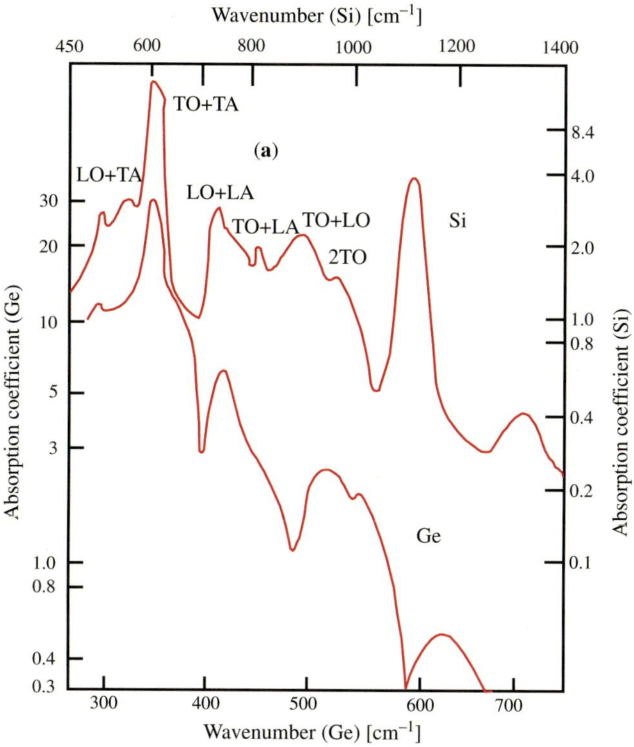

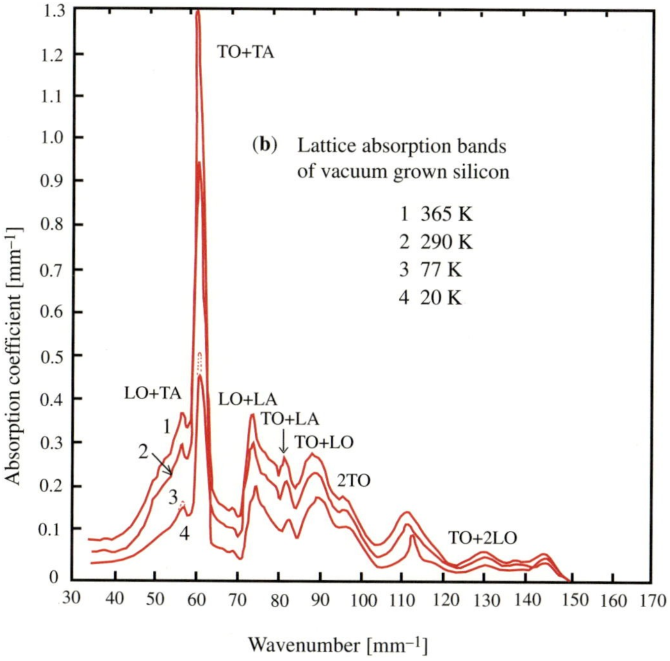

Fig. 6.33. (**a**) Lattice absorption in Si and in Ge measured by *Collins* and *Fan* [6.74]. (**b**) Lattice absorption in Si measured at lower frequencies at several different temperatures by *Johnson* [6.75]

for these absorption processes has been proposed by *Lax* and *Burstein* [6.76]. In general, we can regard the polarization $\boldsymbol{P}$ in a medium as a function of the TO phonon displacements $\boldsymbol{u}$. When we expand $\boldsymbol{P}$ as a function of $\boldsymbol{u}$ in zinc-blende-type semiconductors we find the lowest order nonzero term to be proportional to $\boldsymbol{u}$ as in (6.101) (crystals containing a constant term independent of $\boldsymbol{u}$ are known as ferroelectrics). In crystals with the diamond structure this first-order term vanishes because of the parity selection rule (the optical phonons are even, hence they cannot produce on odd parity polarization vector). As a result one has to include the second-order terms. Nonzero second-order terms in phonon displacements give rise to the infrared absorption in Si and Ge. According to *Lax* and *Burstein* [6.76], these terms can be visualized as the joint effect of two vibrations. The first one breaks the inversion symmetry and induces charges on the two atoms in the primitive cell. The second vibration causes these charges to oscillate and generate an electric-dipole moment which couples to the electromagnetic wave. The combinations of two-phonon modes contributing to the electric-dipole moment in the diamond structure can be determined by using group theory [6.77]. Many of these two-phonon modes involve one optical phonon plus an acoustic one, both at zone edges.

When two phonons participate in the lattice absorption, conservation of energy and wavevector requires that

$$\boldsymbol{q}_1 + \boldsymbol{q}_2 = \boldsymbol{k} \approx 0, \tag{6.118a}$$

where $\boldsymbol{q}_1$ and $\boldsymbol{q}_2$ are the wavevectors of the two phonons and $\boldsymbol{k}$ is the photon wavevector, and

$$\omega_1 + \omega_2 = \omega, \tag{6.118b}$$

where ω_1, ω_2, and ω are the corresponding phonon and photon frequencies. Unlike the one-phonon absorption process, the phonon wavevectors are thus no longer restricted to the zone center. This means that the two-phonon absorption spectra can be a broad continuum determined by the two-phonon density of states (DOS). Thus one can try to identify the structures in the two-phonon spectra in Fig. 6.33 with critical points in the phonon DOS. The energies of the critical points deduced in this way can be compared with phonon dispersion curves determined by neutron scattering. Table 6.6 shows the identifications of the peaks in the infra-red absorption spectra of Si based on the assumption that these peaks involve sums and differences of four zone-edge phonon frequencies in the TA, LA, TO and LO phonon branches:

$$\omega_{\text{TA}} = 127.4\ \text{cm}^{-1};\ \omega_{\text{LA}} = 333.9\ \text{cm}^{-1};$$

$$\omega_{\text{TO}} = 482.3\ \text{cm}^{-1};\ \text{and}\ \omega_{\text{LO}} = 413.8\ \text{cm}^{-1}.$$

Notice that here the TO and LO phonons are not degenerate since these are zone-edge phonons rather than zone-center ones. These multiphonon absorption peaks in Si have also been correlated with critical points in the phonon DOS (at points L, X, W and Σ in the Brillouin zone) [6.79]. Multiphonon absorption spectra have been observed in zinc-blende-type semiconductors. Be-

Table 6.6. List of the peak frequencies in the multiphonon absorption spectra of Si and their identification in terms of combinations of four zone-edge phonons [6.78]

Observed peak frequency [cm^{-1}]	Identification	Calculated peak frequency [cm^{-1}]
371.8	TO−TA	354.9
566.2	LO+TA	541.2
609.8	TO+TA	607.9
739.7	LO+LA	747.7
818.7	TO+LA	816.2
896.1	TO+LO	896.1
963.9	2TO	964.6
1301.9	TO+2LO	1309.9

sides the Lax–Burstein mechanism, an additional mechanism is present in this case. Since one-phonon absorption is allowed, the incident photon can virtually excite the zone-center TO phonon. This TO phonon then decays via anharmonic interactions into two phonons. These results have been reviewed by *Spitzer* [6.78] and will not be repeated here.

Amorphous Ge and Si, in which the translational symmetry is lifted, exhibit rather strong infrared absorption [6.80].

6.4.4 Dynamic Effective Ionic Charges in Heteropolar Semiconductors

Using (6.108) and (6.111) we can express e^* in terms of experimentally measurable quantities:

$$e^* = \left(4\pi\varepsilon_0 \frac{\mu(\varepsilon_0 - \varepsilon_\infty)}{4\pi N}\right)^{1/2} \omega_{\mathrm{T}}. \quad (6.119)$$

Since this effective charge is associated with the absorption induced by TO phonons it is referred to as the **transverse** or **Born effective charge**. Using (6.119) we obtain an expression for the macroscopic longitudinal electric field $\boldsymbol{E}_{\mathrm{L}}$ generated by LO phonons:

$$E_{\mathrm{L}} = [4\pi\mu N\omega_{\mathrm{T}}^2(\varepsilon_0 - \varepsilon_\infty)/4\pi\varepsilon_0\varepsilon_\infty^2]^{1/2}(u_+ - u_-), \quad (6.120a)$$

which can also be written as

$$E_{\mathrm{L}} = -[4\pi\mu N\omega_{\mathrm{L}}^2(\varepsilon_\infty^{-1} - \varepsilon_0^{-1})(4\pi\varepsilon_0)^{-1}]^{1/2}(u_+ - u_-). \quad (6.120b)$$

Equation (6.120b) has already been introduced in (3.32). Sometimes it is convenient to express (6.120b) as

$$E_{\mathrm{L}} = -4\pi N e_{\mathrm{L}}^*(u_+ - u_-) \quad (6.121)$$

by introducing another effective ionic charge $e^*_{\rm L}$, which is known as the **longitudinal** or **Callan effective charge**. Both e^* and $e^*_{\rm L}$ are examples of effective charges determined from lattice vibrations. They are referred to as **dynamic effective charges**, as distinct from the **static effective charges** which result from the static transfer of electrons from the cation to the anion when an ionic crystal is formed. The static effective charge is related to the **ionicity** or **polarity** of the crystal [6.81, 82]. However, since ionicity is a somewhat qualitative concept, several functional relationships have been proposed in the literature to relate the ionicity to static and dynamic charges. We reproduce here the one proposed by *Harrison* [6.82] for the transverse charge:

$$e^*_{\rm H} = Z - 4 + (\alpha_{\rm P}/3)(20 - 8\alpha^2_{\rm P}), \tag{6.122}$$

where Z is the cation core charge (three and two for III–V and II–VI compounds, respectively) and $\alpha_{\rm P}$ is the **polarity** defined by *Harrison* [6.82]. The latter can be calculated within the LCAO scheme. If the bonding is assumed to be dominated by the p orbitals, then $\alpha_{\rm P}$ is defined by

$$\alpha_{\rm P} = \frac{E^{\rm c}_p - E^{\rm a}_p}{[(E^{\rm c}_p - E^{\rm a}_p)^2 + 4V^2_{\rm xx}]^{1/2}} \tag{6.123}$$

where $E^{\rm c}_p$, $E^{\rm a}_p$, and $V_{\rm xx}$ are, respectively, the atomic p electron energies of the cation and anion and their overlap parameter defined in (2.80c) (Sect. 2.7.2). Notice that $\alpha_{\rm P} = 0$ when the cation and anion are identical (such as in a homopolar crystal).

In Table 6.7 we display the experimental values of e^* and $e^*_{\rm L}$ together with those of $\alpha_{\rm P}$ and the corresponding values of $e^*_{\rm H}$ obtained with (6.122). Note the nonmonotonic dependence of e^* and $e^*_{\rm H}$ on $\alpha_{\rm P}$ (compare, for example InSb with CdTe).

The transverse effective charge can also be estimated from pseudopotential form factors such as those in Table 2.21. This determination, however, requires a knowledge of the wavefunctions and an integration over the entire Brillouin zone, similar to that needed for the calculation of the dielectric function. The corresponding expressions have been given by *Vogl* [6.86] and by *Sanjurjo* et al. [6.85]. The latter group of researchers have shown that the rather forbidding complete expression for the effective charge $e^*_{\rm PS}$ in terms of pseudopotential form factors can be reduced to the following simple form:

$$e^*_{\rm PS} = Z - 4 + \frac{8P_i}{1 + P^2_i} \tag{6.124}$$

where Z is the cation core charge as in (6.122) and $P_i = v^{\rm a}_3/v^{\rm s}_3$ with $v^{\rm s}_3$ and $v^{\rm a}_3$ being the symmetric and antisymmetric pseudopotential form factors defined in Sect. 2.5.1. Table 6.7 shows some of the values of $e^*_{\rm PS}$ in zinc-blende-type semiconductors obtained with (6.124) by *Sanjurjo* et al. [6.85] and also values calculated by de *Gironcoli* et al. [6.84] using the *ab initio* pseudopotential method. Using the same method the latter group and also *King-Smith* and *Vanderbilt* [6.87] have calculated, in addition, the piezoelectric constants (see also Sect. 3.3.3), which are also related to $\alpha_{\rm P}$. Equation (6.124) yields a poor

Table 6.7. Experimental values of the transverse (e^*) and longitudinal (e^*_{L}) effective charges in zinc-blende-type semiconductors [6.69, 83]. Provided for comparison with these experimental values are theoretical transverse charges e^*_{H} estimated from the values of α_{P} using (6.126) (both values obtained from [6.82]) and e^*_{PS} deduced from the semiempirical pseudopotential form factors. All effective charges are in units of the electronic charge

Semiconductor	e^*	e^*_{L}	α_{P}	e^*_{H}	e^*_{PS}
InSb	2.5	0.16	0.53	2.11	2.0[a]
InAs	2.6	0.22	0.55	2.24	2.3[a]
InP	2.5	0.26	0.58	2.35	2.4[a]
GaSb	1.8	0.13	0.45	1.74	1.6[a]
GaAs	2.2	0.2	0.48	1.92	2.0[a]
GaP	2.0	0.24	0.51	2.43	2.0[a]
GaN	2.7[c]	0.52[c]	0.60	2.43	2.9[b]
AlSb	1.9	0.19	0.45	1.78	1.8[a]
AlAs	2.3	0.26	0.48	1.91	2.1[a]
AlP	2.28	0.31	0.51	2.03	2.2[a]
AlN	2.75	0.59	0.58	2.36	2.9[b]
ZnSe	2.03	0.34	0.75	1.88	2.8[b]
ZnTe	2.00	0.27	0.74	1.86	2.5[b]
CdTe	2.35	0.32	0.78	1.94	1.6[b]

[a] From [6.84].
[b] Calculated with (6.124) using the pseudopotential form factors from [6.85].
[c] Obtained from available parameters with (6.119)

approximation in cases where $P_i \ll 1$. For the II–VI compounds and large-bandgap III–V compounds (such as GaN and AlN) it gives reasonable results. However, together with (6.122), it has proven to be useful for estimating the effect of external perturbations (such as hydrostatic and uniaxial stress) on the effective charge [6.88].

6.5 Absorption Associated with Extrinsic Electrons

In Chap. 4 we studied the electronic properties of impurities in semiconductors. The corresponding extrinsic electrons can also interact with electromagnetic waves. In fact, infrared absorption and related optical spectroscopies are the main techniques for determining the impurity energy levels discussed in Chap. 4. In this section we shall concentrate on absorption associated with donors and acceptors in diamond- and zinc-blende-type semiconductors only. The optical properties of these impurity electrons depend on the sample temperature. At low temperatures carriers are trapped on the bound states of the impurities. At sufficiently high temperature the impurities are ionized and their carriers become free, as in metals. Free-carrier absorption is particularly important in small-bandgap semiconductors such InSb, lead chalcogenides (such as PbTe) and some II–VI compounds (e. g. HgCdTe) since their impurity levels are so shallow that they are completely ionized at relatively low temperatures. We shall consider the optical properties of extrinsic electrons in these two regimes separately.

6.5.1 Free-Carrier Absorption in Doped Semiconductors

The electrical and optical properties of free carriers in simple metals, such as the alkali metals, have been covered extensively in many textbooks on solid-state physics (e. g. [6.89, 90]) on the basis of the **Drude model**. We shall use this model also, since free carriers introduced into semiconductors by doping behave in many ways like those in simple metals. One important difference between the two is that the carrier concentration in a semiconductor can be changed. Since the dopant concentration is typically less than 10^{20} cm^{-3} (except in very special cases) the plasma frequencies of carriers in semiconductors (Problem 6.3) are usually in the infrared range whereas they are in the visible or ultraviolet for metals.

To obtain the free-carrier contribution to the dielectric function of a semiconductor within the Drude model, we start with the corresponding expression (6.117a) for the TO phonon. In the same spirit as in Problem 6.4 we can obtain the "Drude free-carrier expression" by setting the phonon frequency ω_{T} in (6.117a) to zero:

$$\varepsilon(\omega) = \varepsilon_\infty - \frac{4\pi N_{\mathrm{c}} e^2}{4\pi\varepsilon_0 m^*(\omega^2 +)\mathrm{i}\omega\gamma_{\mathrm{c}}} \quad \text{(SI units)}. \tag{6.125}$$

In (6.125), N_{c}, e, and m^* are, respectively, the density, charge and effective mass of the free carriers. γ_{c}^{-1} now represents their scattering time and is related to the phenomenological scattering time τ introduced in Sect. 5.1. We can decompose (6.125) into a real and an imaginary part:

$$\varepsilon_{\mathrm{r}}(\omega) = \varepsilon_\infty \left(1 - \frac{\omega_{\mathrm{p}}^2}{\omega^2 + \gamma_{\mathrm{c}}^2}\right), \tag{6.126a}$$

$$\varepsilon_{\mathrm{i}}(\omega) = \frac{\varepsilon_\infty \omega_{\mathrm{p}}^2 \gamma_{\mathrm{c}}}{\omega(\omega^2 + \gamma_{\mathrm{c}}^2)}, \tag{6.126b}$$

where

$$\omega_{\mathrm{p}}^2 = 4\pi N_{\mathrm{c}} e^2/(4\pi\varepsilon_0 m^* \varepsilon_\infty) \tag{6.126c}$$

and ω_{p} is the plasma frequency of the free carriers *screened by the dielectric constant* ε_∞. Notice that $\varepsilon_{\mathrm{i}}(\omega)$, the imaginary part of the dielectric function is proportional to γ_c, and hence the absorption coefficient is too. The reason for this is well known in metal optics. A free-carrier absorption process is the annihilation of a photon with the excitation of a carrier from a filled state below the Fermi energy E_{F} to an empty state above it.This is shown schematically in Fig. 6.34. As in the case of excitons (Sect. 6.4), both energy and wavevector have to be conserved in this process. Equivalently, this process can occur only at the point where the photon dispersion curve in Fig. 6.34 (with its origin at 1) intersects the free carrier dispersion. As seen from Fig. 6.34, this does not happen, because of the large slope of the photon curve. Wavevector can be conserved if the transition is accompanied by scattering with a phonon or an impurity (represented by the horizontal arrow in Fig. 6.34). By using the definition of the absorption coefficient in (6.10) and the expression in (6.126b) we obtain the free-carrier absorption coefficient α_c:

$$\alpha_{\mathrm{c}}(\omega) = \frac{\varepsilon_\infty \omega_{\mathrm{p}}^2 \gamma_{\mathrm{c}}}{n_{\mathrm{r}} c(\omega^2 + \gamma_{\mathrm{c}}^2)}, \tag{6.127}$$

where n_{r} is the real part of the refractive index and c the speed of light. At low frequencies. (i. e. $\omega \ll \gamma_{\mathrm{c}}$) α_{c} can be written as

$$\alpha_{\mathrm{c}} = \frac{4\pi N_{\mathrm{c}} e^2}{4\pi\varepsilon_0 n_{\mathrm{r}} c m^* \gamma_{\mathrm{c}}}. \tag{6.128}$$

Comparing α_{c} with the corresponding expression for the electrical conductivity σ in (5.10) we can rewrite (6.128) as

$$\alpha_{\mathrm{c}} = \frac{4\pi\sigma}{n_{\mathrm{r}} c} \tag{6.129}$$

provided we can equate $(\gamma_{\mathrm{c}})^{-1}$ to the scattering time τ. As we saw in Chap. 5, the conductivity for a sample at nonzero temperatures is determined by the

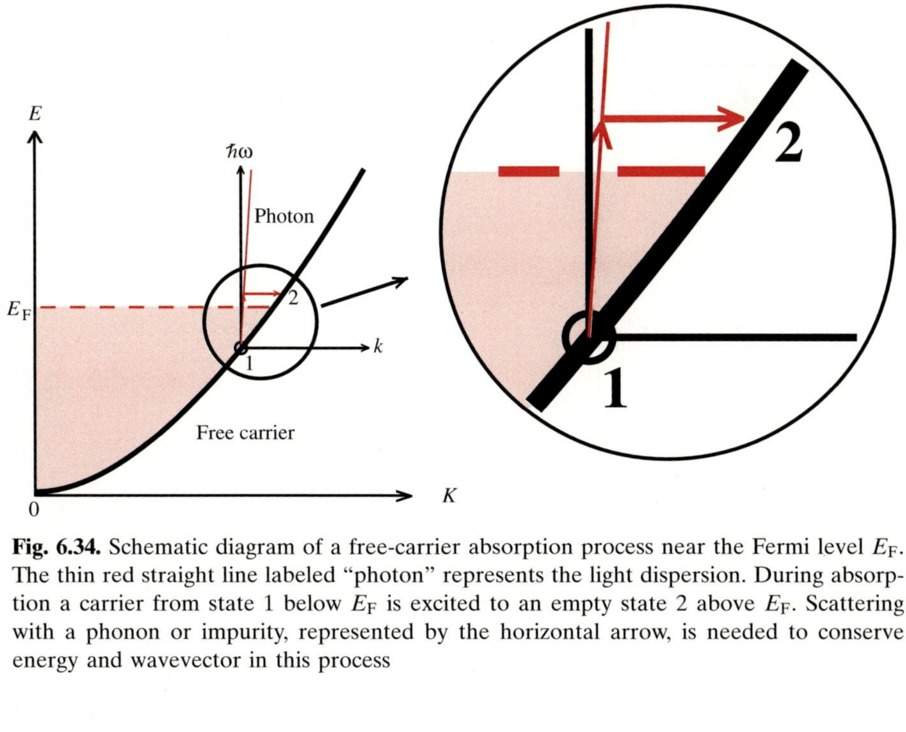

Fig. 6.34. Schematic diagram of a free-carrier absorption process near the Fermi level E_{F}. The thin red straight line labeled "photon" represents the light dispersion. During absorption a carrier from state 1 below E_{F} is excited to an empty state 2 above E_{F}. Scattering with a phonon or impurity, represented by the horizontal arrow, is needed to conserve energy and wavevector in this process

average scattering time $\langle\tau\rangle$ defined by (5.27), whereas the scattering processes which determine τ_c involve a band of carriers around E_F with width equal to $\hbar\omega$ or $k_B T$, whichever is smaller. Due to the dependence of the scattering probabilities on carrier energy (Sect. 5.2.3) $\gamma_c \neq (\langle\tau\rangle)^{-1}$ and (6.129) is only an approximation. However, the various scattering processes which contribute to γ_c are similar to those for $\langle\tau\rangle$, such as scattering with acoustic phonons, optical phonons, and ionized impurities. We shall not repeat these calculations here but instead refer interested readers to the review article by *Fan* [6.91] for details. Thus, in conjunction with electrical transport measurements, free-carrier absorption is a very useful technique for studying the scattering mechanisms of carriers in semiconductors.

Figure 6.35 shows a log–log plot of the free-carrier absorption coefficient in n-type InAs at room temperature for electron concentrations varying between 2.8×10^{16} and 3.9×10^{18} cm^{-3} [6.92]. The straight line drawn through the data points suggests that α_c can be fitted with the simple expression

$$\alpha_c \propto \lambda^p, \tag{6.130}$$

where λ is the wavelength of the infrared radiation. The exponent p is equal to 3 in InAs. Other zinc-blende-type semiconductors have also been found to obey (6.130), except that the value of p varies between 2 and 3 [6.91]. Equation (6.130) can be understood if we assume that $\omega \gg \gamma_c$ in (6.126b) and the refractive index n_r is approximately independent of ω (e. g., if the valence electron contribution ε_∞ to n_r is much larger than the free-carrier contribution). With these assumption α_c can be written as

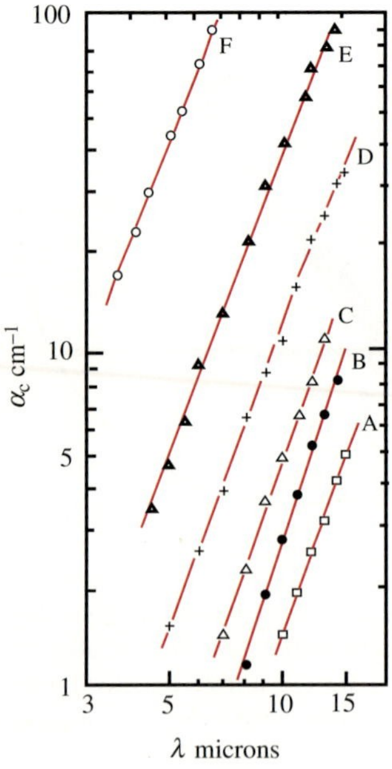

Fig. 6.35. Free-carrier absorption in n-type InAs at room temperature for six different carrier concentrations (in units of 10^{17} cm^{-3}: A: 0.28; B: 0.85; C: 1.4; D: 2.5; E: 7.8; and F: 39 [6.92]

$$\alpha_c = \frac{\varepsilon_i \omega}{n_r c} \propto (\omega^2 + \gamma_c^2)^{-1} \propto \lambda^2. \tag{6.131}$$

Deviations of p from 2 can be explained by the dependence of γ_c on ω. The fact that the scattering mechanisms which make the dominant contribution to γ_c vary with the semiconductor can be invoked to explain the variation in p for different materials [6.91].

In the optical transition depicted in Fig. 6.34 the initial and final states of the electron lie within the same band. The resultant free-carrier absorption is therefore called *intraband*. *Interband* free-carrier absorption can occur if the band containing the carriers is separated from another empty band by an amount smaller than the bandgap. The inset of Fig. 6.36 shows schematically two such transitions between the spin–orbit-split (so) hole band and the heavy

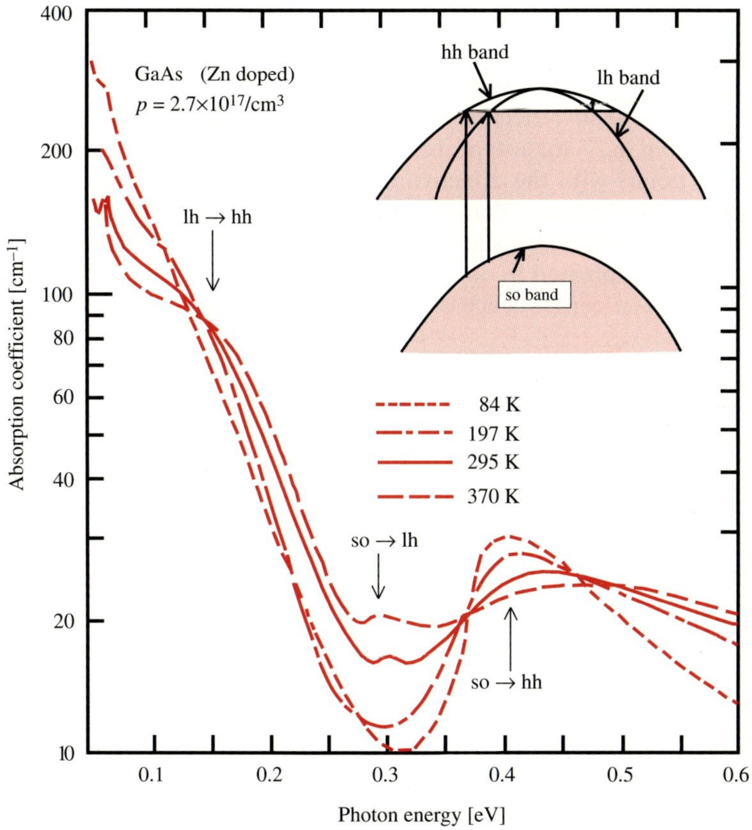

Fig. 6.36. The infrared absorption spectra of p-type GaAs (hole concentration: 2.7×10^{17} cm^{-3}) at four different temperatures. The peaks at 0.15, 0.31 and 0.42 eV are identified, respectively, with the transitions lh→hh, so→lh, and so→hh (hh, lh and so stand for the heavy hole, light hole, and spin–orbit split hole bands, respectively). The inset shows schematically transitions between the spin–orbit split (so) hole band and the heavy hole (hh) and light hole (lh) bands and also between the latter two [6.93]

hole (hh) and light hole (lh) bands in p-doped diamond- and zinc-blende-type semiconductors. The corresponding infrared absorption spectra due to intervalence transitions in p-type GaAs are shown in Fig. 6.36 as a function of temperature [6.93]. Three broad peaks at 0.15, 0.31 and 0.42 eV can be identified and have been attributed to the lh→hh, so→lh and so→hh transitions, respectively. From such intervalence band absorption measurements, the spin–orbit coupling and the ratios of effective masses of the top valence bands in semiconductors such as InSb, InAs, GaAs, and GaSb have been determined [6.91].

According to the Drude model the reflectivity of metals drops very steeply in the range $0 \lessapprox \varepsilon_r \lessapprox 1$, i. e. at the frequency slightly above the plasma frequency [6.89]. If $\varepsilon_i \equiv 0$ the reflectivity drops to zero when $\varepsilon_r = 1$. This decrease in reflectivity is known as the **plasma edge** or **Drude edge**. In most real metals the reflectivity does not vanish when $\varepsilon_r = 1$ because the plasma frequency is usually high enough for some interband transitions to contribute to ε_i. In doped semiconductors the plasma frequency is often below the bandgap so that the only contribution to ε_i comes from scattering of the free carriers and is given by (6.126b). As a result, doped semiconductors can exhibit very sharp plasma edges in their reflectivity curves. An excellent example of these edges is shown in Fig. 6.37 for n-type InSb [6.94]. The solid curves are fits to the experimental points with the effective mass m^* as the adjustable parameter. *Spitzer* and *Fan* [6.94] found that m^* in InSb increases from $0.023m_0$ (free electron mass) at low carrier concentration to $0.041m_0$ at high concentrations. This behavior can be explained by the large *nonparabolicity* of the conduction band of InSb, which is a consequence of its small bandgap (see Problem 6.15).

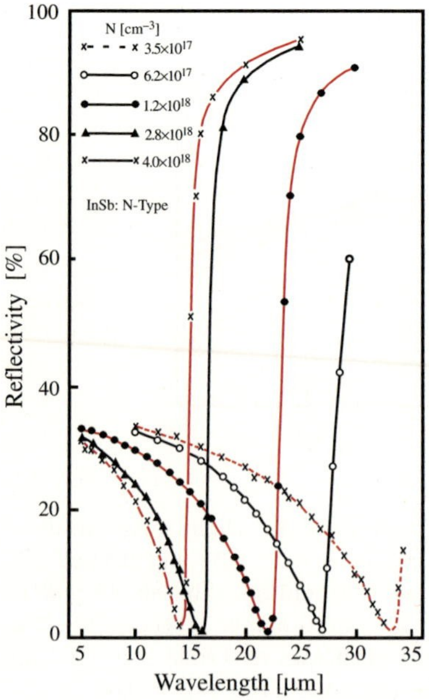

Fig. 6.37. Plasma edges observed in the room temperature reflectivity spectra of n-type InSb with carrier concentration N varying between 3.5×10^{17} cm^{-3} and 4.0×10^{18} cm^{-3} [6.94]. The solid curves are fits to the experimental points using the reflectivity calculated with (6.8) and (6.126b)

6.5.2 Absorption by Carriers Bound to Shallow Donors and Acceptors

In Chap. 4 we showed that shallow donors and acceptors in diamond- and zinc-blende-type semiconductors behave somewhat like "hydrogen atoms embedded in a solid". It is well known that a hydrogen atom can absorb electromagnetic radiation via electronic transitions between its quantized levels. These transitions give rise to series of sharp absorption lines known as the Lyman, Balmer, Paschen, etc., series in the spectra of atomic hydrogen [6.95]. By analogy with the hydrogen atom we expect that the electron in a donor atom or the hole on an acceptor can also be excited optically from one bound state to another. In addition, we expect these transitions to obey selection rules similar to those in the hydrogen atom, e. g., electric-dipole transitions are allowed between states with s and p symmetries (i. e., when the difference in the angular momentum quantum number Δl equals one) but forbidden between states of the same symmetry.

Figure 6.38a shows the absorption spectrum of phosphorus donors in Si at liquid helium temperature measured by *Jagannath* et al. [6.96]. The concentration of donors is around 1.2×10^{14} cm^{-3}. Notice that the absorption peaks are very sharp because of the discrete nature of the energy levels involved. These peaks are assigned to transitions originating from the $1s(A_1)$ ground state (Sect. 4.2.3) of the P donors in Si to their excited levels labeled $2p_0$, $2p_\pm$, etc. As pointed out in Sect. 4.2.3, the degeneracy of the three p levels with magnetic quantum numbers $m = 0, \pm 1$ in a hydrogen atom are split in the donor atoms in Si because of the anisotropic effective mass tensor associated with the lowest conduction band. In addition to the $1s(A_1)$ ground state, donors in Si have two other slightly higher energy states with symmetries $1s(E)$ and $1s(T_2)$ as a result of the valley–orbit coupling (Fig. 4.3). These states are not occupied at liquid helium temperature. They become thermally populated and contribute to infrared absorption at higher temperatures. Transitions from these higher energy $1s$ states have been observed in Si:P by *Aggarwal* and *Ramdas* [6.97] as shown in Fig. 6.38b. The energies of the bound states in shallow donors in Si, determined with great precision from such infrared absorption spectra, have already been compared with theoretical calculations in Fig. 4.3.

In spite of its sensitivity and resolution, infrared absorption spectroscopy of shallow impurities has its limitations. The oscillator strength of a transition decreases as the quantum number n of the final state increases and hence higher energy bound states are more difficult to observe. There is another technique which is even more sensitive than infrared absorption in measuring shallow impurity energy levels in semiconductors. This remarkable technique, known as **photothermal ionization spectroscopy** (PTIS), can detect impurity concentrations as low as 10^8 cm^{-3}! It was first reported by *Lifshits* and *Nad* [6.98]. The basic process in this technique can be described as **phonon-assisted photoconductivity**. Photoconductivity is the name for electric conductivity induced by shining light on a sample [6.99]. For example, an intrinsic semiconductor at low temperature may have a very low conductivity because there are

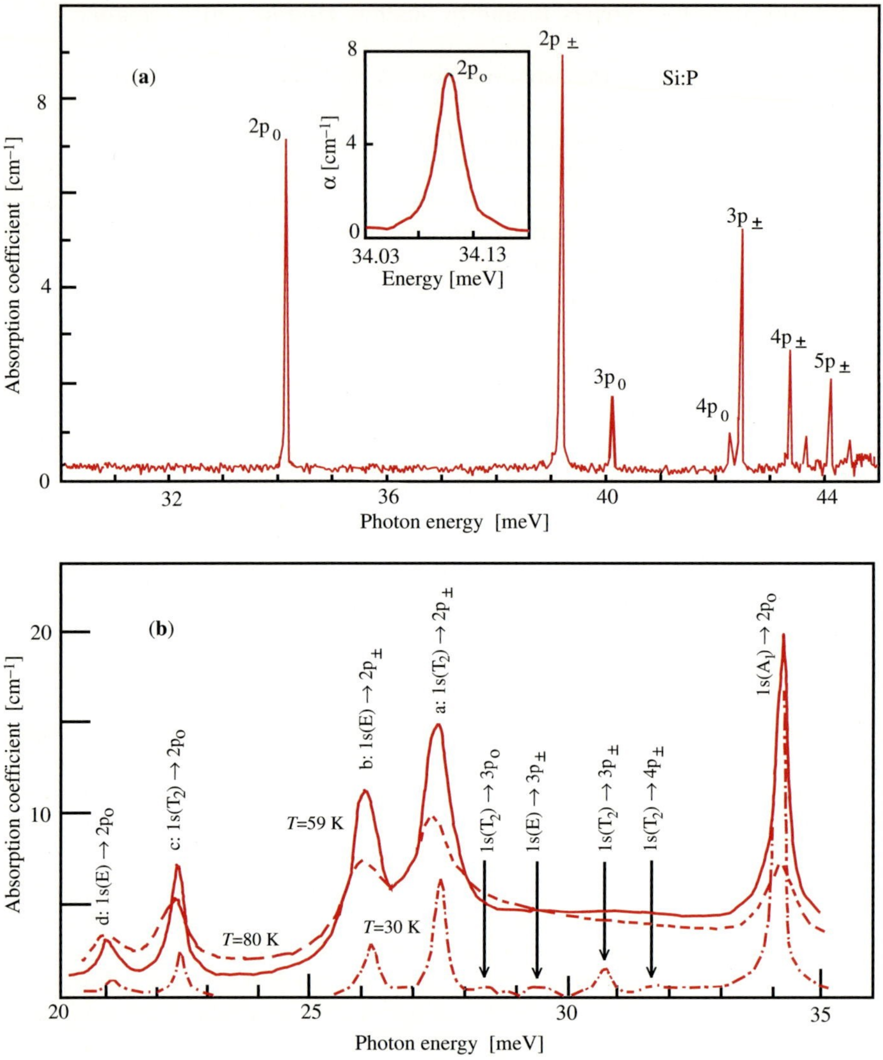

Fig. 6.38a,b. Absorption spectrum of phosphorus donors in Si. (**a**) Sample at liquid helium temperature containing around 1.2×10^{14} cm^{-3} of P. The inset shows the $2p_0$ line on an expanded horizontal scale [6.96]. (**b**) Between 30 and 80 K in a sample containing 5.2×10^{15} cm^{-3} of P [6.97]

very few thermally, excited carriers. When it is illuminated by light with photon energy larger than its bandgap, free carriers (both electrons and holes) are excited. The resultant increase in conductivity is known as **intrinsic** *photoconductivity*. In a doped semiconductor at low temperature (when all the carriers are frozen on the shallow impurities) **extrinsic** *photoconductivity* can occur when the photon energy of the incident light is sufficient to ionize the impurities. Thus impurity ionization energies can be measured with photoconductivity. This technique does not work for the excited bound states of the impurities unless they are occupied. However, these states may be so close to the continuum that once carriers from the ground state have been excited optically into them they can be ionized by absorbing a phonon. The phonons required can be thermally created by maintaining the sample at a temperature T such that $k_B T$ is slightly larger than the ionization energy of the excited state to be studied. A schematic diagram of the photothermal ionization process is shown in the inset of Fig. 6.39. The PTIS spectrum of $\sim 2 \times 10^{14}$ cm^{-3} P donors in Si is shown in the same figure. A comparison between Figs. 6.38a and 39 shows that the strengths of the transitions to the higher excited states ($n = 3$ to 6) in PTIS are enhanced relative to the $n = 2$ and 3 levels. A PTIS spectrum in ultra-pure p-type Ge measured by *Haller* and *Hansen* [6.100] at 8 K is shown in Fig. 6.40. This sample contains a net acceptor concentration of only 10^{10} cm^{-3}! The predominant group III acceptor is Al. The concentrations of B and Ga are about 20 times lower. The observed peaks have been identified by comparing them with the theoretical acceptor levels calculated by *Baldereschi* and *Lipari* [6.47] (Sect. 4.2.4). Table 6.8 summarizes the binding energies for various excited states of group III shallow acceptors in Ge as determined by PTIS. For comparison the corresponding energies calculated within the model of *Baldereschi* and *Lipari* [6.47] are shown under the column labeled "Theory". The group theoretical notations in parentheses are based on the cubic point group of the crystal. Since PTIS can mea-

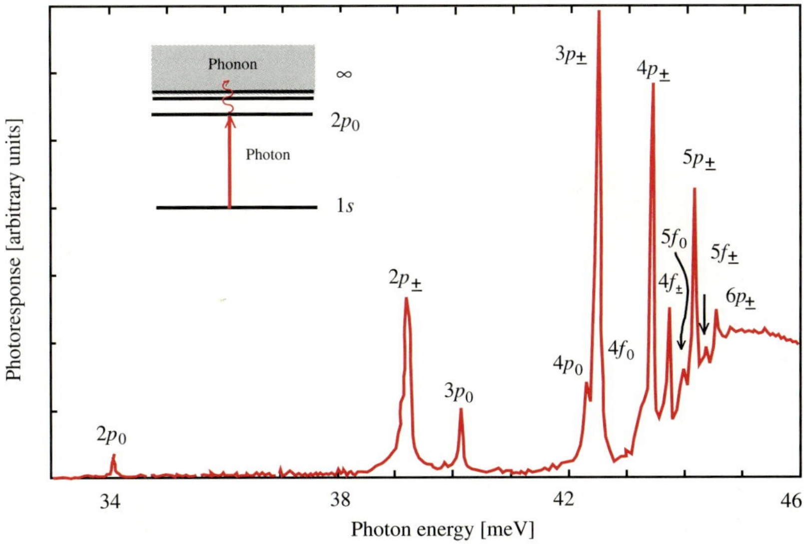

Fig. 6.39. Photothermal ionization spectrum of phosphorus-doped Si measured by *Jagannath* et al. [6.96]. The inset shows schematically the photothermal ionization process for a donor atom

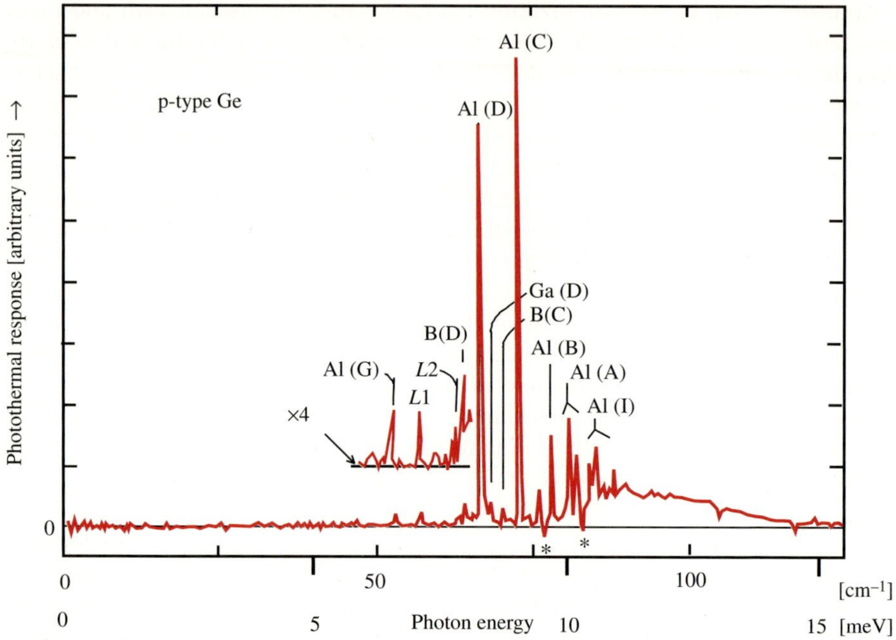

Fig. 6.40. Photothermal ionization spectrum of ultra-pure p-type Ge measured at 8 K [6.100]. For the meaning of the labels A, B, C, etc. in parentheses see Table 6.8. The peaks labeled L_1, L_2, and * are artifacts ("ghosts") produced by the Fourier transform infrared spectrometer

Table 6.8. Binding energies [meV] of group III acceptors in Ge as determined by PTIS. To convert the transition energies measured by PTIS into binding energies the energy of the $2\Gamma_8^-$ state (peak D) has been assumed to be equal to the theoretical value [6.47] of 2.88 meV [6.101]

	Binding energy [meV]					
Peak	B	Al	Ga	In	Tl	Theory
G	4.61	4.65	4.58	4.57	4.52	4.58 ($1\Gamma_8^-$)
E	3.27		3.3	3.54	3.57	
D	2.88	2.88	2.88	2.88	2.88	2.88 ($2\Gamma_8^-$)
C	2.14	2.13	2.13	2.10	2.13	2.13 ($1\Gamma_7^-$)
						2.11 ($3\Gamma_8^-$)
a	1.76			1.76	1.78	
B	1.49	1.48	1.48	1.48	1.5	1.48 ($4\Gamma_8^-$)
						1.22 ($5\Gamma_8^-$)
A*	1.16	1.13	1.15	1.15	1.14	1.14 ($2\Gamma_7^-$)
						1.13 ($6\Gamma_8^-$)
A′	1.03	1.00	1.01	1.00	1.01	

sure only the energy difference between an excited state and the ground level, the binding energies in Table 6.8 have been obtained by assuming that the $2\Gamma_8^-$ level (peak D) has the calculated binding energy of 2.88 meV. From this table we see that the energy separations of shallow acceptor levels in Ge are in very good agreement with theory. Also, the chemical shifts between different acceptors due to central cell corrections are largest for the $1\Gamma_8^-$ level.

6.6 Modulation Spectroscopy

The spectra of the dielectric function of semiconductors above the fundamental absorption edge are rather broad (Figs. 6.11 and 12). They could be described as broad bands with some superimposed Van Hove singularities (i. e., interband critical points). The reason for this can be found in the rather weak nature of Van Hove singularities, especially those of the three-dimensional variety. According to Table 6.1 they are of the form $(E - E_i)^{1/2}$, i. e., the optical functions remain finite for $E = E_i$ although their *derivatives* with respect to either E or E_i diverge. This suggests measuring directly one of these derivatives instead of the dielectric function: the background should largely disappear and sharp peaks should appear at the interband critical points (CPs). This is accomplished with any of the many **modulation spectroscopy** techniques to be described in this section. Most of our precise knowledge of CP energies stems from reflectance modulation measurements.

We note that the extreme accuracy with which $\varepsilon(\omega)$ can be measured with present day ellipsometric methods enables one to numerically obtain derivatives of $\varepsilon(\omega)$ (up to third order). This fact produced a certain shift of emphasis from modulation spectroscopy to spectral ellipsometry in the decade 1975–1985. Since 1985, however, there has been a revival in the use of modulation spectroscopy techniques in connection with semiconductor superlattices and quantum wells ([6.102], see also Chap. 9).

The use of modulation techniques for studying interband CPs can probably be traced to *Frova* and *Handler* [6.103] and *Seraphin* and *Hess* [6.104], who measured so-called **electrotransmission** and **electroreflectance** spectra, where the modulation was produced by an applied ac electric field. Essential for the development of these techniques was the commercial availability of *lock-in amplifiers*, which are devices that select a small ac signal synchronous with the modulation (acting as reference) while any other signal, including noise, is rejected. The measured signal is normalized to the average intensity of the transmitted or the reflected beam and to the amplitude of the modulating agent. In this manner the logarithmic derivative of the transmittance or reflectance with respect to the modulation agent is obtained while the intensity I_0 of the incident light, including its fluctuation, is eliminated. Typical modulation amplitudes are of the order of 10^{-4}–10^{-5} of the incident beam intensity; noise levels can be kept below 10^{-6}. For reviews of modulation spectroscopy see [6.11, 105, 106].

The power of reflectance modulation techniques is illustrated in Fig. 6.41 for GaAs. Note the sharp structures at the E_0, $E_0 + \Delta_0$, E_1, $E_1 + \Delta_1$, E'_0 and E_2 critical points which appear in the photon energy (i. e., frequency) derivative. These structures are well-reproduced in the theoretical spectra (broken curve in Fig. 6.41c) calculated from the band structure using the pseudopotential method [6.21]. Notice that the experimental E_1 and $E_1 + \Delta_1$ peaks are much sharper than the corresponding theoretical ones because of excitonic effects discussed in Sect. 6.3.3. A transmittance modulation spectrum yields directly the corresponding derivative of the absorption coefficient α (except for minor reflection corrections). The relationship between the reflectance mod-

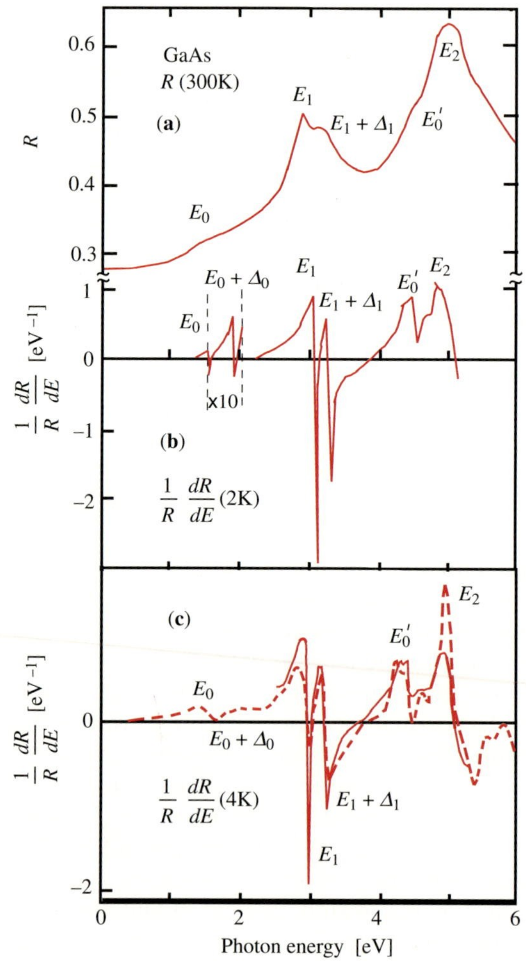

Fig. 6.41. Reflectance and frequency modulated reflectance spectra of GaAs. (**a**) Room temperature reflectance spectrum of GaAs measured by *Philipp* and *Ehrenreich* [6.14], the same as that in Fig. 6.8 (**b**) $(1/R)(dR/dE)$ spectrum calculated by *Sell* and *Stokowski* [6.107] by taking the derivative with respect to photon energy E of a 2 K reflectance curve. (**c**) The solid curve is the $(1/R)(dR/dE)$ spectrum measured by *Zucca* and *Shen* [6.108] with a wavelength modulation spectrometer. The broken curve is a theoretical curve calculated from a pseudopotential band structure. The spectra in (**a**) and (**b**) are adapted from [6.105]. The spectra in (**c**) are adapted from [6.21]

ulation and the corresponding modulation of real and imaginary parts of the dielectric function is more complex. Since the modulation amplitude is small, we can always linearize this relation by introducing the so-called **Seraphin coefficients** β_r and β_i:

$$\frac{\Delta\mathcal{R}}{\mathcal{R}} = \beta_r \Delta\varepsilon_r + \beta_i \Delta\varepsilon_i \tag{6.132a}$$

with

$$\beta_r = \frac{\partial \ln \mathcal{R}}{\partial \varepsilon_r} \text{ and } \beta_i = \frac{\partial \ln \mathcal{R}}{\partial \varepsilon_i}. \tag{6.132b}$$

The coefficients β_r and β_i can be obtained by differentiation of (6.8) using (6.11) (Problem 6.16). The spectral dependence of β_r and β_i on frequency is sketched in Fig. 6.42 for germanium [6.109]. Note that around the fundamental edge only β_r is significant. Therefore in this region the shape of the reflectance modulation spectra corresponds to the modulation of ε_r. Around E_1 and $E_1 + \Delta_1$ both $\Delta\varepsilon_r$ and $\Delta\varepsilon_i$ contribute equally to the spectrum. Around E_0', however, $\Delta\varepsilon_i$ dominates, while at higher frequencies (E_2, E_1'), $\Delta\varepsilon_r$ dominates again with a sign reversal (note that β_r is negative above 3.5 eV). Obviously $\Delta\varepsilon_r$ and $\Delta\varepsilon_i$ can be separated by using the Kramers–Kronig relations. In this case, extrapolations beyond the spectral range are not important (Problem 6.16c).

Conceptually, the simplest modulation spectra are those in which the light frequency is modulated. They just yield the derivative of ε_r and ε_i with respect to E, i. e., for a critical point $(E - E_i)^{1/2}$ we obtain a spectrum looking like $(E - E_i)^{-1/2}$. Other types of modulation spectra (with so-called external modulation) are obtained by sinusoidally modulating an external agent which affects E_i, such as the temperature, an applied stress, or an electric field. In

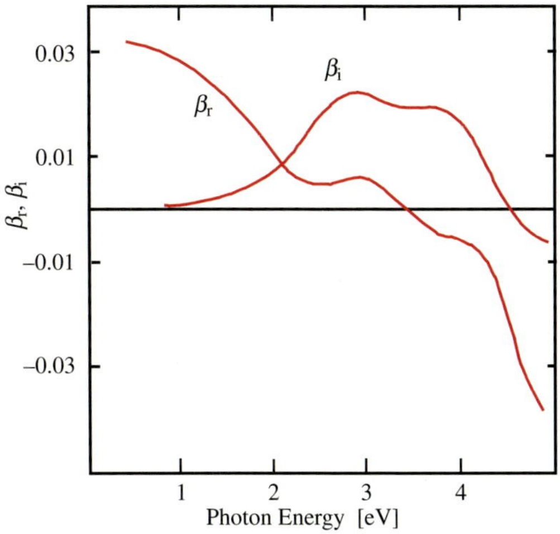

Fig. 6.42. Spectral dependence of the Seraphin coefficients β_r and β_i calculated for Ge at room temperature [6.109]

many cases such spectra are easier to obtain than their frequency-modulated counterparts. Moreover, they may lead to additional information on the coupling of the external perturbation to the electronic system. These agents not only change E_i but also modify the prefactor of $(E-E_i)^{1/2}$ by changing the oscillator strength and the reduced masses (6.51 and 56). Moreover, it becomes imperative to consider the **lifetime broadening** of the critical points produced by interaction with phonons and other scatterers. Here we shall represent such broadening by the energy Γ and write, for instance in the case of a three-dimensional M_0 CP, the density of states above E_i as

$$D(E) \propto \mathrm{Re}\{(E - E_i + \mathrm{i}\Gamma)^{1/2}\} \tag{6.133}$$

(The corresponding expressions for $\varepsilon_r(E)$ and $\varepsilon_i(E)$ near M_0, and also other types of CPs including broadening are given in [6.11].) An external perturbation will, in general, modulate not only E_i but also Γ (the latter effect is particularly strong in the case of temperature modulation).

Among the various external perturbation agents we must also distinguish between those which preserve translational invariance (e. g., temperature, strain) and those which destroy it along one or more directions (electric field, magnetic field). Only in the former cases can we assume that the modulated signal is proportional to the derivative of (6.133) with respect to E_i or Γ. Since (6.133) is the consequence of translational symmetry in all three directions, the application of a uniform electric field $\mathscr{E}$, which destroys the translational invariance along its direction z (because the perturbation Hamiltonian is proportional to $-e\mathscr{E}z$) makes (6.133) meaningless. This case will be treated in Sect. 6.6.3.

Using dimensionless energy units

$$W = \frac{E - E_0}{\Gamma}, \tag{6.134a}$$

(6.133) can be rewritten as

$$D(W) \propto \Gamma^{1/2}\mathrm{Re}\{(W + \mathrm{i})^{1/2}\} = \Gamma^{1/2}\left(\frac{W}{2} + \frac{1}{2}(W^2 + 1)^{1/2}\right). \tag{6.134b}$$

By means of (6.134a, b, 48, and 49) we obtain

$$\begin{aligned} \frac{d\varepsilon_r}{dE} &= -\frac{d\varepsilon_r}{dE_0} \propto \frac{1}{2}\Gamma^{-1/2}F(-W), \\ \frac{d\varepsilon_i}{dE} &= -\frac{d\varepsilon_i}{dE_0} \propto \frac{1}{2}\Gamma^{-1/2}F(W), \end{aligned} \tag{6.135a}$$

where the function $F(W)$, shown in Fig. 6.43, is given by

$$F(W) = (W^2 + 1)^{-1/2}\left[(W^2 + 1)^{1/2} + W\right]^{1/2}. \tag{6.135b}$$

Likewise we can calculate the derivatives of ε_r and ε_i with respect to the broadening energy Γ. We find at a three-dimensional (3D) M_0 CP

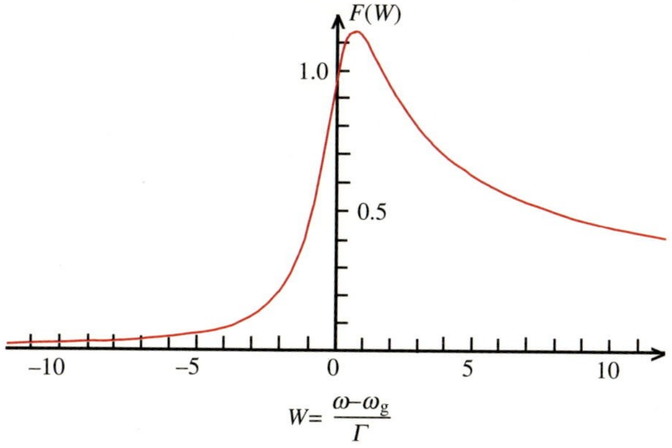

Fig. 6.43. Universal function $F(W)$ used to represent the derivatives of ε_r and ε_i with respect to E, E_0 and Γ at three-dimensional critical points

$$\frac{d\varepsilon_r}{d\Gamma} \propto -\frac{1}{2}\Gamma^{-1/2}F(W), \tag{6.135c}$$

$$\frac{d\varepsilon_i}{d\Gamma} \propto \frac{1}{2}\Gamma^{-1/2}F(-W). \tag{6.135d}$$

The expressions equivalent to (6.135) for other types of CPs can be easily derived by the reader (Problem 6.17).

We conclude this introduction by mentioning that phonons in semiconductors also modulate their dielectric function at the corresponding vibrational frequency. This modulation is responsible for the phenomenon of **Raman scattering** [to be discussed in Sect. 7.2, see (7.36)].

6.6.1 Frequency Modulated Reflectance and Thermoreflectance

A frequency modulation spectrometer is constructed by allowing the frequency of the light coming out of the exit slit of the monochromator to have a small modulation sinusoidally dependent on time. The corresponding change in the reflected light, synchronous with the frequency modulation, is detected with a lock-in amplifier. **Thermoreflectance** is measured by subjecting the sample to a periodic variation in temperature produced by an alternating current or a periodically chopped laser beam.

An increase in temperature preserves the average symmetry of a crystal. However, in principle, both the E_i and Γ of a critical point are changed. At E_0 and E_1 the gaps decrease with increasing temperature at a rate of about $-4.5\cdot 10^{-4}$ eV K^{-1} in most zinc-blende-type semiconductors. Γ increases at a rate of about $1.5\cdot 10^{-4}$ eV K^{-1} [6.110], see Fig. 6.44. Thus the thermoreflectance spectra are expected to be dominated by the shift in the gap, i. e., by (6.135a) for M_0 3D CPs or their equivalents in the case of other types of CPs. The frequency-modulated spectra of GaAs are shown in Fig. 6.41. They have been obtained by two different methods: the spectrum in Fig. 6.41b was computed from a reflectance spectrum of GaAs measured at 2 K, those in Fig. 6.41c were measured directly with a wavelength-modulated spectrometer. The two different approaches give very similar results. The corresponding thermoreflectance spectra at 80 K and 300 K are shown in Fig. 6.45. The frequency modulated spectrum of the E_1 and $E_1 + \Delta_1$ CPs is nearly the same as the thermo-reflectance spectrum at 80 K shown in Fig. 6.45. This confirms the fact

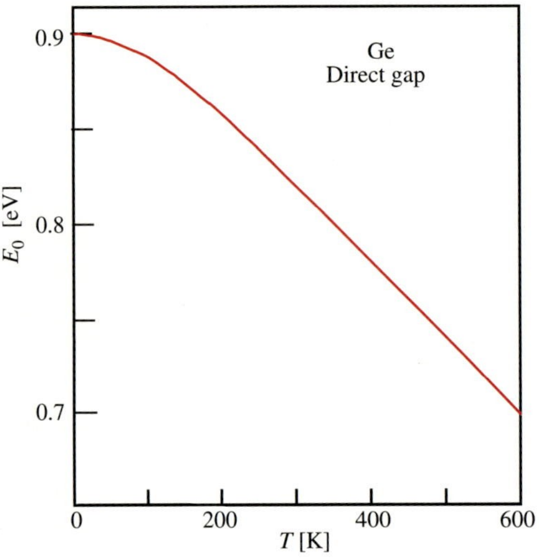

Fig. 6.44. Temperature dependence of the direct gap (E_0) of germanium [6.111]

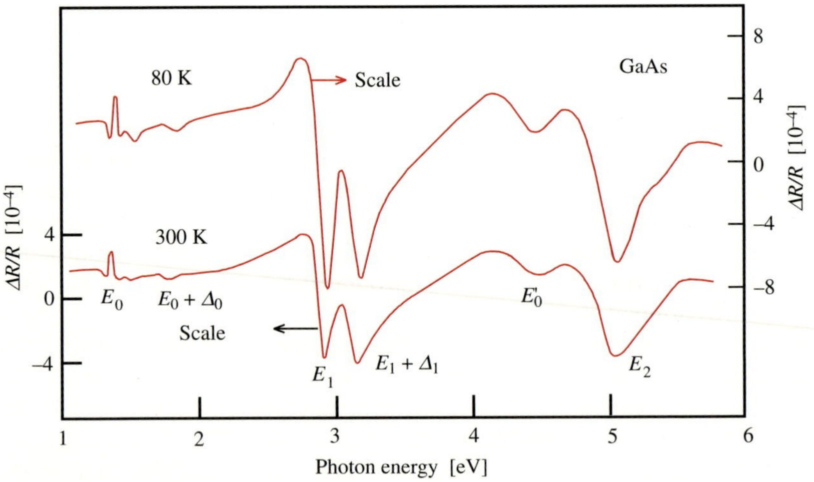

Fig. 6.45. Thermoreflectance spectrum of GaAs at 80 and 300 K as reported by *Matatagui* et al. [6.112]. Note the structures at the critical points E_0, $E_0 + \Delta_0$, E_1, $E_1 + \Delta_1$, E_0', and E_2

that, at least for these CPs, the thermoreflectance spectrum is mainly produced by the gap shift and not by the increase of Γ with increasing temperature. This conclusion, however, should not be regarded as general.

6.6.2 Piezoreflectance

Piezoreflectance spectra are usually obtained by applying a periodically modulated uniaxial stress to the sample. This can be accomplished in a number of ways, for example by coupling the cone of a loudspeaker to the sample with a pushrod. In another method, more common these days, the sample is mounted on a piezoelectric transducer in the manner shown in Fig. 6.46 or in another similar way.

Results obtained for the E_0 gap of CdTe samples oriented along either [100] or [111] are shown in Fig. 6.47. These spectra are determined mainly by the derivative of the E_0 exciton energy with respect to the strain. The strongly anisotropic nature of these spectra (note the difference in sign between $\hat{e} \parallel \chi$ and $\hat{e} \perp \chi$, χ represents here the strain direction) is due to the fact that the strain splits the energies of the hole in the exciton, as discussed in Sect. 3.3.2, with a splitting linear in the magnitude of the strain χ and proportional to the deformation potentials of the top of the valence band, b for [100] strain and d for [111] strain. At the same time the strain has a hydrostatic component which leads to the uniform shift of both split components by an amount pro-

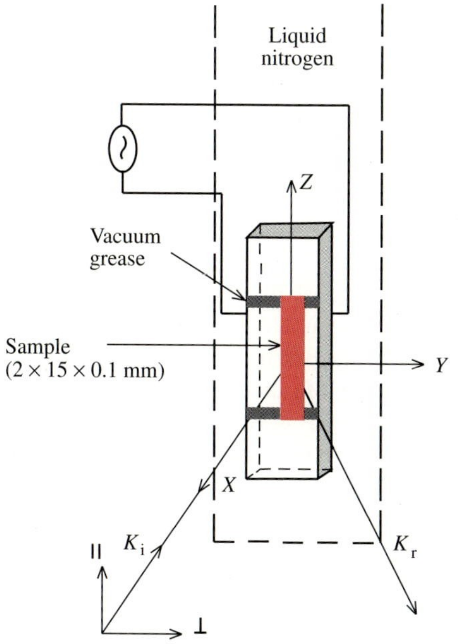

Fig. 6.46. Details of the piezoelectric transducer modulator arrangement used by *Gavini* and *Cardona* [6.113] for piezoreflectance measurements. The *black-shaded* parallellopiped represents a lead zirconate-titanate (PZT) transducer. The ends of a thin (100 μm) sample (*shaded red*) are glued to the transducer with vacuum grease (*shaded black*) which freezes at low temperatures, thus giving a strong bond to the transducer

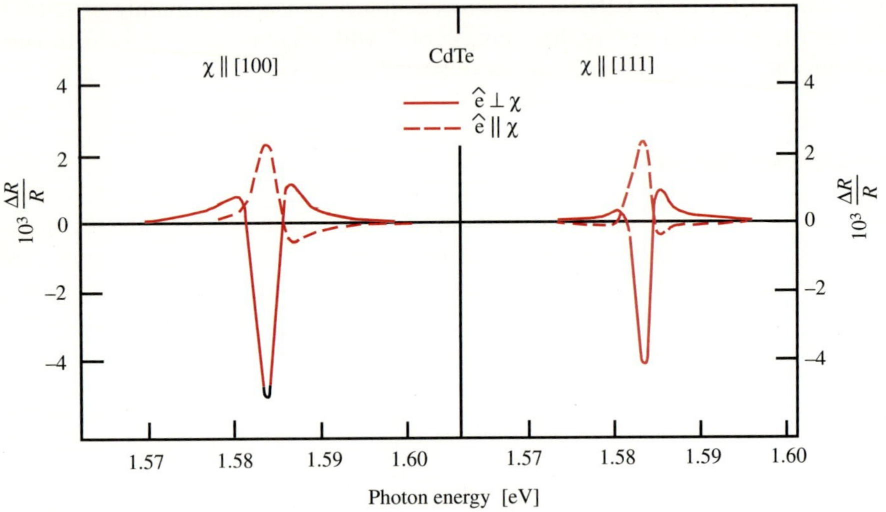

Fig. 6.47. Piezoreflectance spectra of CdTe [6.113] at 80 K for two directions of the applied stress $\boldsymbol{\chi}$, taken with polarizations $\hat{\boldsymbol{e}}$ parallel and perpendicular to $\boldsymbol{\chi}$

portional to a, where a is the volume deformation potential defined in (3.20). Using the wavefunctions shown in (3.20) we can see that the $(\frac{3}{2}, \frac{3}{2})$ split exciton couples *only* to the $\hat{\boldsymbol{e}} \perp \chi$ polarization while the $(\frac{3}{2}, \frac{1}{2})$ exciton couples *mainly* to $\hat{\boldsymbol{e}} \parallel \chi$. This is the reason for the opposite signs of the modulated signal exhibited for these two polarizations in Fig. 6.47. From the ratios of strengths of these two signals one can obtain the corresponding ratio of uniaxial to hydrostatic deformation potentials (b/a and d/a) (Problem 6.18). In cases in which a is known (as it was for CdTe when the work of Fig. 6.47 was performed) one can determine both b and d from the anisotropy of the signals in Fig. 6.47. One thus finds for CdTe [6.113] $b = -1.1$ eV and $d = -5.45$ eV, in rather good agreement with other measurements and calculations [6.29].

6.6.3 Electroreflectance (Franz–Keldysh Effect)

As mentioned above, even a uniform electric field of magnitude $\mathscr{E}$ (the notation $\mathscr{E}$ is used here for electric field to distinguish it from energy E) lifts the translational symmetry along its direction. We discuss next the theory of the dielectric function of a semiconductor around an M_0 critical point under the presence of $\mathscr{E}$. We assume, for simplicity, that the CP is isotropic (a restriction which can be easily lifted, the reader is encouraged to do it as an exercise), with reduced mass μ.

The applied uniform field $\mathscr{E}$ along z can be represented by a contribution to the Hamiltonian $\mathscr{H}_{\mathscr{E}} = -e\mathscr{E}z$ which, obviously, is not translationally invariant along z. The effect of $\mathscr{H}_{\mathscr{E}}$ on an electron–hole pair, however, depends only on the relative separation of electron and hole, $\boldsymbol{r}$, and not on the center-of-mass coordinate $\boldsymbol{R}$ [Sect. 6.3.1; (6.76)]. As in the case of excitons, the center-of-mass motion is not affected by the uniform electric field, leading to an equation equivalent to (6.77a). The solutions of this equation are plane waves and only those with $K = 0$ contribute to the optical absorption. Hence we can neglect the center-of-mass motion in the calculation of $\varepsilon(\omega, \mathscr{E})$. The equation for the relative motion, equivalent to (6.77b), becomes

$$\left(-\frac{\hbar^2}{2\mu}\nabla_{\boldsymbol{r}}^2 - e\mathscr{E}z - E_{\boldsymbol{r}}\right)\phi(\boldsymbol{r}) = 0, \tag{6.136a}$$

where e is the (negative) charge of the electron. Equation (6.136a) can be separated into an equation for the components of $\boldsymbol{r}$ perpendicular to $\mathscr{E}$, not involving $\mathscr{E}$, and another for the z component

$$\left(-\frac{\hbar^2}{2\mu}\frac{d^2}{dz^2} - e\mathscr{E}z - E_z\right)\phi(z) = 0. \tag{6.136b}$$

The solution $\phi(z)$ of (6.136b) must be multiplied by the plane-wave solution of the corresponding x, y equation

$$\phi(x, y) = \frac{1}{\sqrt{N}}\exp[-\mathrm{i}(k_x x + k_y y)], \tag{6.136c}$$

where N is the appropriate normalization constant and the energy E_z must be added to the kinetic energy in the x, y plane in order to obtain the total "relative coordinate" energy $E_{\boldsymbol{r}}$:

$$E_{\boldsymbol{r}} = E_0 + \frac{\hbar^2(k_x^2 + k_y^2)}{2\mu} + E_z. \tag{6.136d}$$

Equation (6.136b) can be written in the simple form

$$\frac{d^2\phi(\xi)}{d\xi^2} = -\xi\phi(\xi) \tag{6.137a}$$

by introducing the dimensionless reduced variable

$$\xi = \frac{E_z}{\Theta} - z\left(2\frac{\mu|e|}{\hbar^2}\mathscr{E}\right)^{1/3}, \tag{6.137b}$$

where the so-called **electrooptical energy** Θ is given by

$$\Theta = \left(\frac{e^2\mathscr{E}^2\hbar^2}{2\mu}\right)^{1/3}. \tag{6.137c}$$

The solutions of (6.137a) which fulfill the appropriate regularity conditions for $z \to \pm\infty$ can be written in terms of the **Airy function** $\mathrm{Ai}(\xi)$ [6.114] as

$$\phi_{E_z}(\xi) = \frac{(|e|\mathscr{E})^{1/2}}{\Theta}\mathrm{Ai}(\xi), \tag{6.138}$$

where the prefactor $(|e|\mathscr{E})^{1/2}/\Theta$ guarantees orthonormality with respect to the continuous variable E_z.

The dipole matrix element needed for the calculation of $\varepsilon_i(\omega, \mathscr{E})$ with (6.48) is given by the equivalent of (6.86)

$$|P_{vc}|^2_{E_z} = NP^2|\phi_{E_z}(0)|^2. \tag{6.139}$$

Equation (6.48) must be integrated over E_z and summed over all possible values of k_x and k_y. The latter is accomplished by introducing the density of states, which, instead of the three-dimensional (6.55), for two dimensions and at an M_0 CP is (including the spin degeneracy)

$$\begin{aligned} D(E_{x,y} - E_0) &= \frac{\mu}{\pi\hbar^2} \text{ for } E_{x,y} > E_0, \\ &= 0 \quad \text{for } E_{x,y} < E_0 \end{aligned} \tag{6.140}$$

The corresponding $\varepsilon_i(\omega, \mathscr{E})$ thus becomes

$$\varepsilon_i(\omega, \mathscr{E}) = \frac{2\pi e^2 (2\mu)^{3/2} P^2 \Theta^{1/2}}{m^2 \hbar E^2} \int_{\xi_0}^{-\infty} \mathrm{Ai}^2(\xi) d\xi \tag{6.141}$$

with

$$\xi_0 = \frac{E_0 - E}{\Theta} \text{ and } E = \hbar\omega.$$

The integral in (6.141) can be easily performed by using a standard integral representation for Ai(ξ) [Ref. 6.11; App. I]. We find

$$\int_{\xi_0}^{\infty} \mathrm{Ai}^2(\xi) d\xi = \xi \mathrm{Ai}^2(\xi) - \mathrm{Ai}'^2(\xi) \tag{6.142}$$

Actually, the quantity of interest in a modulation spectroscopy experiment is not $\varepsilon_i(\omega, \mathscr{E})$ but rather the change introduced in ε_i by $\mathscr{E}$. This can be written as

$$\Delta\varepsilon_i(\omega, \mathscr{E}) = \varepsilon_i(\omega, \mathscr{E}) - \varepsilon_i(\omega, 0) = \frac{2e^2 (2\mu)^{3/2} P^2 \Theta^{1/2}}{m^2 \hbar E^2} F\left(\frac{E_0 - E}{\Theta}\right) \tag{6.143a}$$

where

$$F(\xi) = \pi\left[\mathrm{Ai}'^2(\xi) - \xi \mathrm{Ai}^2(\xi)\right] - (-\xi)^{1/2} \mathrm{H}(-\xi) \tag{6.143b}$$

and H represents the Heaviside step function (which is zero [one] for negative [positive] argument).

The real part of $\Delta\varepsilon(\omega, \mathscr{E})$ can be obtained by replacing $F(\xi)$ in (6.143) by its Kramers–Kronig transform:

$$G(\xi) = \pi\left[\mathrm{Ai}'(\xi)\mathrm{Bi}'(\xi) - \xi \mathrm{Ai}(\xi)\mathrm{Bi}(\xi)\right] + \xi^{1/2} H(\xi), \tag{6.143c}$$

where $\mathrm{Bi}(\xi)$ is the **modified Airy function**, which diverges for $\xi \to \infty$. For details of the derivation above see [6.11] and the original derivation of $\varepsilon_i(E, \mathscr{E})$ by *Thamarlingham* [6.115].

The functions $F(\xi)$ and $G(\xi)$ are plotted in Fig. 6.48. The inset shows a schematic plot of the imaginary part of the dielectric function ε_i in the vicinity of a 3D M_0 CP, both with and without a field $\mathscr{E}$. The effect of an electric field on an optical absorption edge was first discussed independently by *Franz* [6.117] and *Keldysh* [6.118] and hence it is known as the **Franz–Keldysh effect**. The oscillations in $\varepsilon_i(\omega)$ above the band gap are known as Franz–Keldysh oscillations. In the presence of $\mathscr{E}$, ε_i is no longer zero below the bandgap E_0 but decreases exponentially. This can be understood simply by the fact that the term $e\mathscr{E}z$ in (6.136b) tilts the bands spatially. As a result, the bandgap vanishes. The conduction band energy can always be lowered enough to overlap with the valence band at different points of real space by making z large enough. However, the electron wishing to make a transition from the valence band to the conduction band has to *tunnel* through a distance dependent on $\mathscr{E}$. For photon energy $\hbar\omega < E_0$ the absorption can be regarded as **photon-assisted tunneling** from the valence to the conduction band (see Sect. 9.5).

So far we have neglected lifetime broadening of E_0. Also, very often the modulating field is produced at a surface depletion layer (8.22 and 23) and is, therefore, strongly nonuniform. Both effects can be taken care of by introducing Airy functions of a complex variable. This, and also the treatment of other

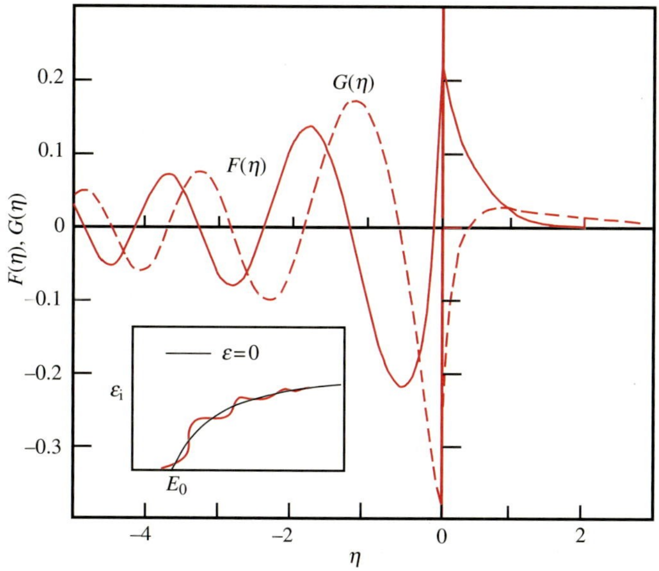

Fig. 6.48. Three-dimensional electrooptic functions $F(\eta)$ (*solid line*) and $G(\eta)$ (*dashed line*) according to [6.116]. The inset shows the imaginary part of the dielectric function with (*red*) and without (*black*) applied electric field

types of CPs, is beyond the scope of this book (interested readers are referred to [6.11, 119]).

Equation (6.143a), in connection with Fig. 6.48 indicates that around an M_0 critical point the modulation spectrum ($\Delta\mathcal{R}/\mathcal{R}$) should exhibit a sharp peak at the gap energy E_0, decay rapidly below E_0, and oscillate above E_0. It can be shown, using standard asymptotic expansions for the Airy functions [6.114] that the oscillatory behavior, in the limit of large $|\xi|$, has the form [6.120]

$$\frac{\Delta\mathcal{R}}{\mathcal{R}} \propto |\xi|^{-1} \cos\left(\frac{2}{3}\xi^{3/2} - \frac{\pi}{2}\right). \tag{6.144}$$

Labeling the maxima and minima consecutively by an index n we find from (6.144 and 137c)

$$(E_n - E_0)^{3/2} = An\mathcal{E} + B, \tag{6.145}$$

where A depends only on $\mu (A \propto \mu^{1/3})$ and physical constants. The larger $\mathcal{E}$, the more widely spaced the oscillations become. A fit of the energies E_n of maxima and minima with (6.145) allows μ to be determined provided $\mathcal{E}$ is known [6.121].

As an example of the Franz–Keldysh oscillations we show in Fig. 6.49 the spectrum obtained for a GaAs layer (100 nm thick) by modulating the field between 0.9 and 1.1×10^5 V/cm in the configuration displayed in the inset. A large number of oscillations are observed, with some evidence of a beating be-

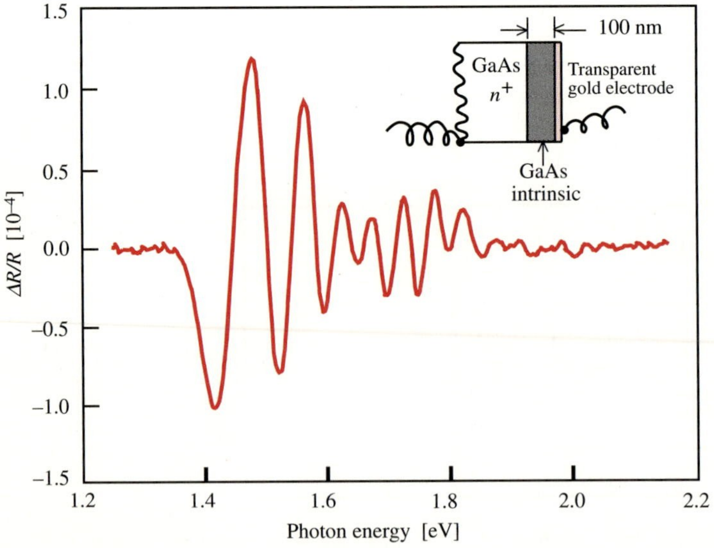

Fig. 6.49. Electroreflectance signal observed for strongly uniform modulating fields produced in an intrinsic GaAs layer (*see inset*) by a modulating voltage (1 ± 0.1 V) applied between the n^+ (*heavily doped*) substrate and a thin transparent gold film [6.122]

tween oscillations, which corresponds to heavy-hole to electron and light-hole to electron transitions [6.121].

We discuss next the so-called low-field regime, in which the broadening parameter Γ is larger than the electrooptic energy Θ. In this case, which can be described exactly by the theory given above modified to include Γ (i. e., by using Airy functions of imaginary argument), the oscillations shown in Figs. 6.48 and 49 should be largely washed out and, instead, a peak should remain at E_0.

It was suggested by Aspnes that in the limit $\Gamma \gg \Theta$ the spectral shape of either $\Delta\mathcal{R}$ or $\Delta\varepsilon$ corresponds to the **third derivative** of the primary spectra with respect to E. This important and by no means obvious fact is vividly illustrated in Fig. 6.50 for the E_1 and $E_1 + \Delta_1$ CPs of germanium. In this figure the primary $\varepsilon_r(E)$ and $\varepsilon_i(E)$ spectra of germanium obtained ellipsometrically at 300 K are shown together with their first, second, and third derivatives with respect to E. At the bottom of the figure, the corresponding $\Delta\varepsilon_r$ and $\Delta\varepsilon_i$ low-field electroreflectance spectra are displayed [6.105]. The close correspondence between the latter and the third derivative spectra (except for a trivial sign reversal) is striking confirmation of Aspnes' third-derivative approximation to the low field limit. Although this approximation was derived rigorously by Aspnes, we give below a *heuristic* treatment which reveals the underlying physical insight.

Let us approximate (6.49) in the neighborhood of a CP as

$$\varepsilon_r - 1 = CE^{-2}\sum_k \frac{1}{E_c - E_v - E}, \tag{6.146a}$$

where C involves the dipole matrix element P and the gap E_0 and is assumed to be constant near the gap. The contribution of a given transition to ε_r for $E \neq E_c - E_v$ is said to arise from **virtual transitions**. According to the time–energy uncertainty principle these transitions last a time τ given by

$$\tau = \frac{\hbar}{E_c - E_v - E}. \tag{6.146b}$$

In the presence of a field $\mathcal{E}$ a conduction band electron (energy E_c) is accelerated by the field, moving, during time τ, a distance z:

$$z = -\frac{e}{2}\frac{\mathcal{E}}{m_c}\tau^2 = -\frac{1}{2}\frac{e\mathcal{E}}{m_c}\frac{\hbar^2}{(E_c - E_v - E)^2}. \tag{6.146c}$$

This displacement results, via the Hamiltonian $ez\mathcal{E}$, in a change in energy with respect to the unperturbed E_c:

$$\Delta E_c = \frac{e^2\mathcal{E}^2}{2m_c}\frac{\hbar^2}{(E_c - E_v - E)^2} \tag{6.147a}$$

and, correspondingly, for an electron at the top of the valence band,

$$\Delta E_v = -\frac{e^2\mathcal{E}^2}{2m_v}\frac{\hbar^2}{(E_c - E_v - E)^2}. \tag{6.147b}$$

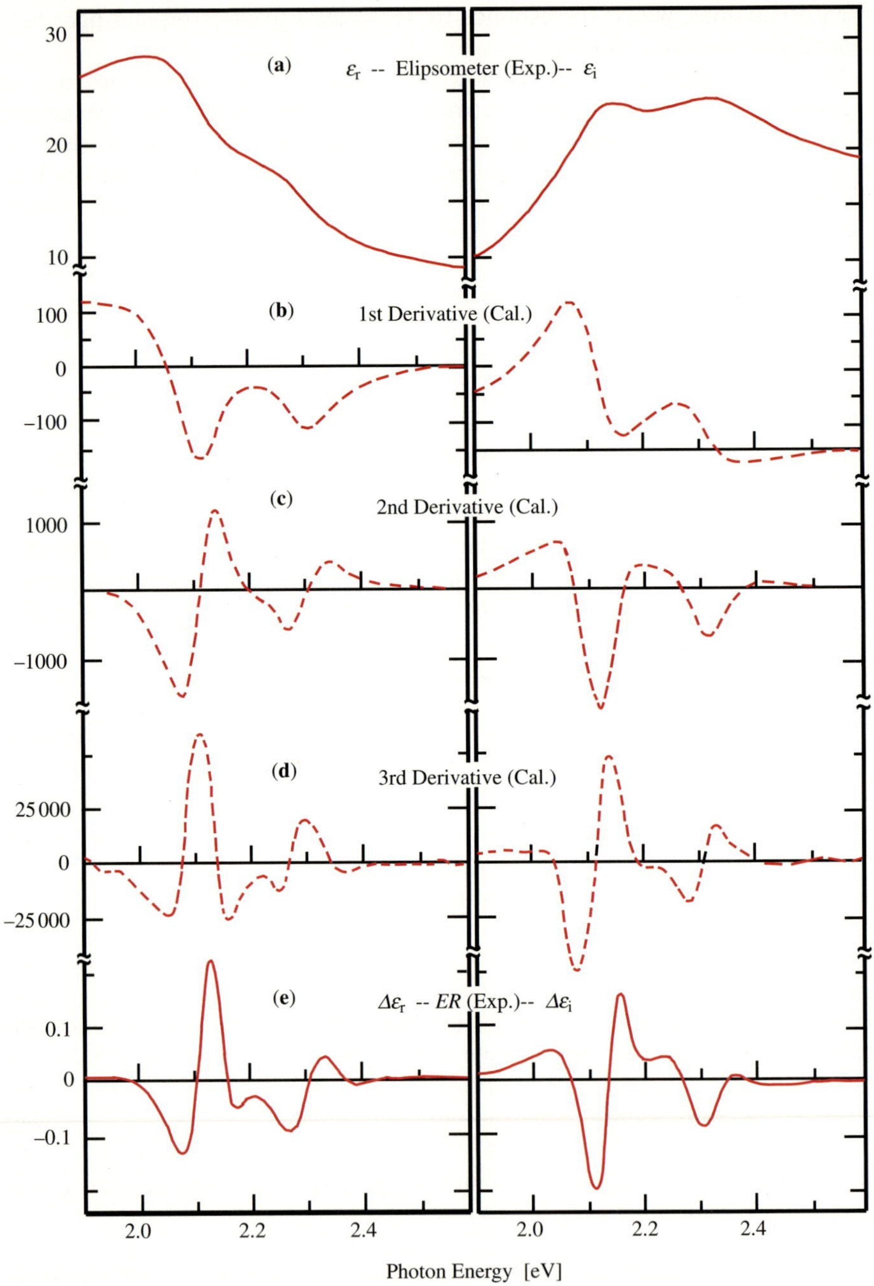

Fig. 6.50. Real and imaginary parts of (**a**) ε; (**b**) $E^{-2}(d/dE)(E^2\varepsilon)$, in eV^{-1}; (**c**) $E^{-2}(d^2/dE^2)(E^2\varepsilon)$, in eV^{-2}; (**d**) $E^{-2}(d^3/dE^3)(E^2\varepsilon)$, in eV^{-3}; (**e**) experimental field-induced change, $\Delta\varepsilon$, from ER measurement at $\mathscr{E} = 38\,\mathrm{kV\,cm^{-1}}$ in germanium. The spectra in (**b–d**) (*dashed curves*) were evaluated by numerically differentiating the experimental ellipsometric data shown in (**a**) [6.105]

Equation (6.147a and b) result in a shift of $(E_c - E_v)$:

$$\Delta(E_c - E_v) = \frac{e^2\mathcal{E}^2}{2\mu} \frac{\hbar^2}{(E_c - E_v - E)^2}. \tag{6.147c}$$

We now differentiate (6.146a) in order to calculate the effect of this shift on ε_r:

$$\begin{aligned}\Delta\varepsilon_r &= \frac{d\varepsilon_r}{d(E_c - E_v)} \cdot \Delta(E_c - E_v) = -C\frac{e^2\mathcal{E}^2}{2\mu E^2} \sum_k \frac{\hbar^2}{(E_c - E_v - E)^4} \\ &= \frac{\hbar^2 e^2\mathcal{E}^2}{12\mu E^2} \frac{\partial^3 E^2 \varepsilon_r}{\partial E^3} = \frac{1}{6E^2}\Theta^3 \frac{\partial^3 E^2 \varepsilon_r}{\partial E^3}.\end{aligned} \tag{6.147d}$$

Because of the analytic properties of ε, a similar expression holds for ε_i. We thus reach the surprisingly simple low-field result

$$\Delta\varepsilon = \frac{1}{6E^2}\Theta^3 \frac{\partial^3 E^2 \varepsilon}{\partial E^3}. \tag{6.148}$$

The rigorous derivation given in [6.105] yields the same expression except for a numerical factor of 1/2. Note that, since Θ^3 is proportional to $\mathcal{E}^2$, $\Delta\varepsilon$ also depends quadratically on $\mathcal{E}$, a physically meaningful result since (6.148) also applies to materials with a center of inversion and hence there can be no effect linear in $\mathcal{E}$ (Problem 6.13). For zinc-blende-type crystals, however, there is also a **linear electric effect** (Problem 6.14), which will not be discussed here any further. The interested reader should look at [6.123].

For a discussion of the effects of an electric field on excitonic absorption see [6.124].

Electroreflectance, usually of the low-field variety, is commonly used for the characterization of semiconductors, particular of mixed crystal samples (e. g., Ge_xSi_{1-x}, $Ga_xAl_{1-x}As$). We show in Fig. 6.51 a survey of the dependence on Ge concentration of the various interband critical points in the Ge_xSi_{1-x} system. These data were obtained [6.125] with the electrolytic method of electroreflectance, which is particularly easy to implement: the field is applied by immersing the sample into water with some salt (e. g., NaCl) and polarizing it with respect to the electrolyte. Modulation of the polarization voltage results in electroreflectance modulation [6.126].

6.6.4 Photoreflectance

In a photoreflectance measurement the reflectivity is modulated by a periodically chopped light (usually a laser beam) which is incident, at a different angle, on the same sample spot as the monochromatized probing beam. Care must be taken in order to avoid the detector registering some of the modulating beam (a small fraction of which is always diffusely scattered along the reflected probing beam). Photoreflectance measurements have recently become

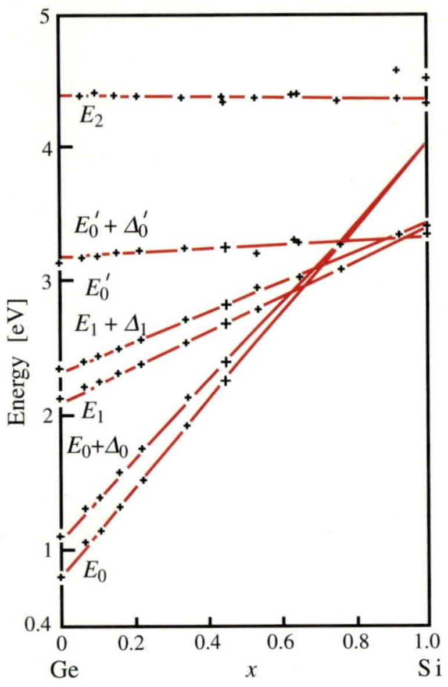

Fig. 6.51. Variation with composition of the energies of the E_0, $E_0 + \Delta_0$, E_1, $E_1 + \Delta_1$, E_0', and E_2 electroreflectance peaks observed in the Ge-Si alloy system at room temperature by the electrolyte method [6.125]

very important for the investigation of semiconductor microstructures [6.106]. The signal is usually sublinear with respect to the power W_L of the modulating laser (often $\propto W^{1/3}$), hence it is preferable to use low laser power (≤ 1 mW) in order to enhance the ratio of signal to spurious scattered light.

Several mechanisms have recently been suggested as contributing to photoreflectance and may be simultaneously operating in a given particular case. The photoreflectance is often due to the screening of the depletion layer field (Sect. 8.3.3) by the carriers generated when the modulating laser beam is absorbed. In this case, the technique is equivalent to electroreflectance (it is sometimes called **contactless electroreflectance**). For this purpose it is convenient to choose the doping such that the depletion layer thickness matches the penetration depth of the light at the frequency of interest.

For undoped to moderately doped semiconductors the carriers freeze out at low temperatures and the depletion layer becomes infinitely thick while the corresponding field tends to zero. In this case no significant field modulation can take place. Optical features involving strongly exciton modified CPs can, however, be modulated by the laser beam through screening of the exciton interaction by the induced carriers (for a theoretical treatment of this effect see [6.127]).

An example of the latter is shown in Fig. 6.52 [6.128] for the direct gap E_0 of three germanium crystals of different (stable) isotopic compositions. The dependence of the gap on isotopic mass shown in this figure is produced by gap renormalization through electron–phonon interaction and is closely related to the decrease of E_0 with increasing temperature [6.129]. Since the spectra in Fig. 6.52 were measured at very low temperatures (2 K) the renormalization is connected with the zero point vibrational amplitude of the phonons (Problem 6.19).

Figure 6.53 displays a photoreflectance spectrum which, in view of the large number of Franz–Keldysh-like oscillations (Figs. 6.49 and 50), must be attributed to space-charge-layer field modulation and is thus equivalent to electroreflectance. According to (6.145) the index which labels the extrema should be proportional to $(E - E_0)^{3/2}$ (except for a small offset). This proportionality is demonstrated in the inset of Fig. 6.53 [6.130].

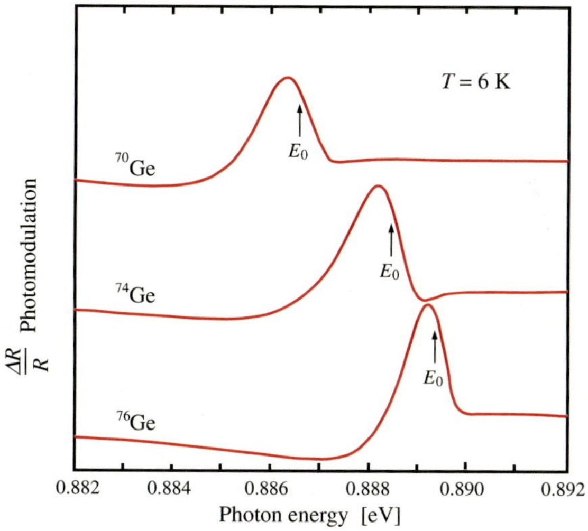

Fig. 6.52. Photomodulated reflectivity showing the E_0 direct gap of single crystals of nearly isotopically pure ^{70}Ge, ^{74}Ge, and ^{76}Ge at $T = 6$ K [6.128]. Note the remarkable dependence of E_0 on isotopic composition

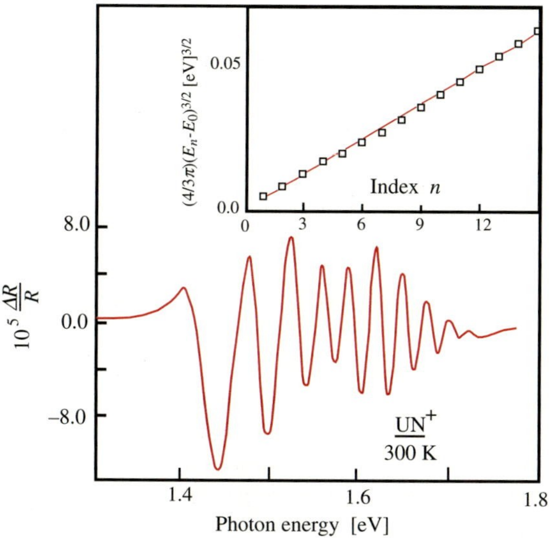

Fig. 6.53. Photoreflectance spectrum of a GaAs sample at room temperature obtained with a 633 nm laser for a modulating power between 3 μW cm^{-2} and 2 μW cm. The inset shows a plot of $(4/3\pi)(E_n - E_0)^{3/2}$ as a function of the index n which labels the extrema [6.130]

6.6.5 Reflectance Difference Spectroscopy

We conclude this chapter by discussing a technique which has recently become very powerful as an *in situ* diagnostic method for vapor-phase epitaxial growth [6.131, 132]. The technique is based on measuring the difference in normal-incidence reflectance for two different linear polarization directions. The first such measurements were performed by rotating the sample surface about an axis perpendicular to it [6.133, 134] while the probing linearly polarized light was reflected at normal incidence on the sample surface. A lock-in amplifier detected the signal synchronous with the sample rotation. (hence the name **rotoreflectance** used in [6.133]). A more recent variation of this technique, which has become the standard one these days, consists of flipping the linear polarization between two perpendicular directions with the help of a *photoelastic modulator* [6.131]. The technique is nowadays known as **reflectance difference spectroscopy** (RDS).

We show in Fig. 6.54 the RDS spectrum measured for a [110] silicon surface covered by the standard oxide layer (~2 nm) and after chemical stripping of the oxide. The differences exemplify the strong sensitivity of the technique to surface conditions, which can be used advantageously for the *in situ* investigation of epitaxial growth.

Several mechanisms can be invoked to explain the RDS phenomenon. The simplest one is of bulk origin and applies, in principle, to the case of Fig. 6.54. It is related to the $\boldsymbol{k}$-dependence of $\varepsilon(\omega, \boldsymbol{k})$, i. e., to the phenomenon of spatial dispersion mentioned in Sect. 6.1. This mechanism is effective for a [110] sur-

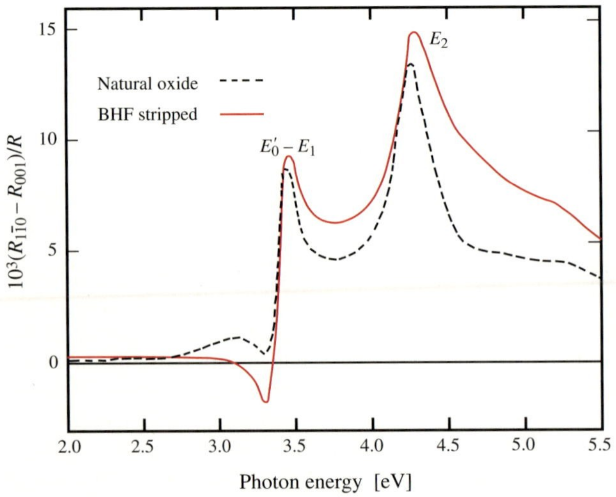

Fig. 6.54. Reflectance difference spectra of a silicon [110] surface covered with natural oxide and after stripping off the oxide with buffered HF. Note the sharp structure at the $E_0' - E_1$ and E_2 critical points [6.134]

face because of the symmetry of $\varepsilon(\omega)$. The symmetry of the system *crystal plus light* (compare with the concept of polaritons) is lower than cubic, in fact only those symmetry operations of the crystal which also preserve the $\boldsymbol{k}$ vector are symmetry operations of the whole. The resulting symmetry is orthorhombic (C_{2v} point group) and therefore the reflectivity for $[1\bar{1}0]$ polarization should be different from that for [001] polarization [6.7]. This contribution to the RDS is, however, an order of magnitude smaller than that shown in Fig. 6.54. In order to explain the magnitude of the observed effect surface modification of the effect of local fields on excitons is often invoked [6.134, 135].

Of particular interest is the case of [001] surfaces, which are most commonly employed for molecular beam epitaxy (MBE). For the sake of the discussion we place, as crystal growers do, the Ga atom at the center of the coordinate system and As at $(a/4)(111)$ (note that the convention used in Chap. 2, which leads to *positive* antisymmetrical pseudopotential form factors, is the opposite one). In this case at a [001] surface terminated by Ga the topmost As–Ga bonds are along [110], while if the surface is As-terminated the topmost Ga–As bonds are along $[1\bar{1}0]$ (readers should provide themselves with a model of the crystal and examine this fact). Hence optical surface anisotropy results, which can be measured by RDS. This anisotropy is different depending on whether the surface is Ga- or As-terminated, a fact which allows *in situ* monitoring of the growth. For a theory of the effect see [6.136]. For reviews of the applications of RDS to epitaxial growth see [6.132, 137].

6.7 Addendum (Third Edition): Dielectric Function

As already mentioned in connection with Fig. 6.10, considerable progress was made in the late 1990's concerning *ab initio* calculations of the dielectric function of semiconductors [6.22, 23]. Because of the topical importance of GaN, we present here an example of such state-of-the-art calculationis and a comparison with ellipsometric measurements of $\varepsilon_2(\omega)$ for the wurtzite modification of this semiconductor. Since the wurtzite structure is optically uniaxial, two sets of dielectric functions are needed to describe the optical behavior of GaN: one for $\boldsymbol{E} \perp \boldsymbol{c}$ and the other for $\boldsymbol{E} \parallel \boldsymbol{c}$. In Fig. 6.55 we show $\varepsilon_2(\omega)$ for $\boldsymbol{E} \perp \boldsymbol{c}$ measured ellipsometrically for GaN using synchrotron radiation as the light source. The agreement between the measured spectrum and the *ab initio* calculations is excellent [6.138]. The measured epitaxial thin film, oriented perpendicular to the c-axis, did not allow measurements in the $\boldsymbol{E} \parallel \boldsymbol{c}$ polarization configuration. The calculations indicate that the E_2 peak, split from E_1 by the hexagonal crystal field, should not appear for $\boldsymbol{E} \parallel \boldsymbol{c}$. This has recently been confirmed experimentally for a bulk GaN crystal [6.139]. The calculations of Fig. 6.55 start with an *ab initio* pseudopotential and include the so-called quasiparticle self-energy correction between the excited electron and the hole left behind plus excitonic interaction between the quasiparticles.

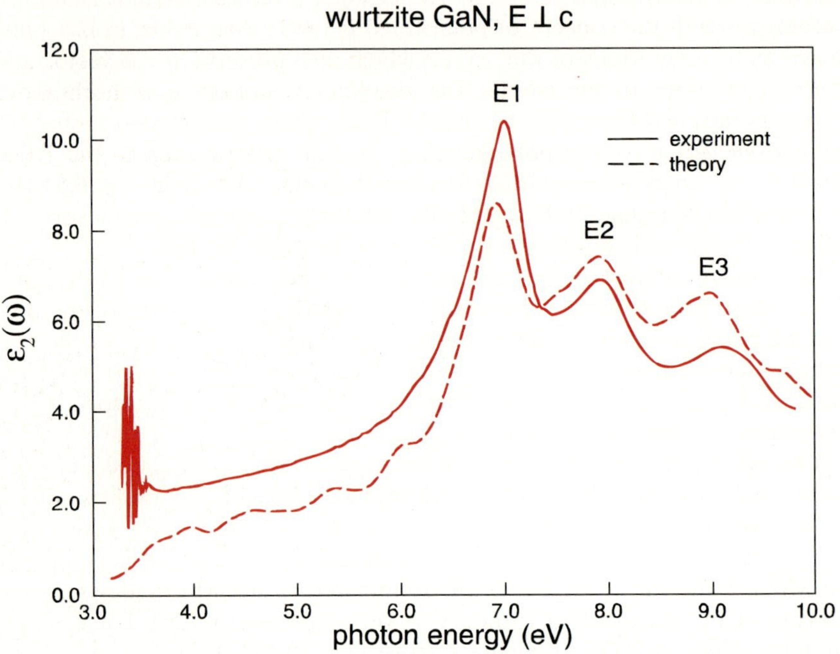

Fig. 6.55. Dielectric function $\varepsilon_2(\omega)$ of wurtzite GaN for $\boldsymbol{E} \perp \boldsymbol{c}$ as measured by ellipsometry using UV synchrotron radiation, compared with *ab initio* calculations [6.138]

PROBLEMS

6.1 *Ellipsometry*
Derive (6.13) from (6.12a and b).

6.2 *Kramers–Kronig Relations*
a) Show that if a linear response function, such as the linear electric susceptibility $\chi(\omega)$ or the dielectric function $\varepsilon(\omega) - 1$, satisfies the following two conditions: (1) it is analytic in the upper half of the complex ω-plane and (2) it approaches zero sufficiently fast as ω approaches infinity, then it satisfies the KKRs given in (6.14 and 15).
b) Apply the KKR (6.14) to the ε_i in (6.58) to derive the expression for ε_r given in (6.59). (The contour of integration for this proof can be found in [6.11, p. 20].) Derive the corresponding expressions for ε_i and ε_r assuming that the oscillator strength f_{vc} in (6.51) is constant instead of the matrix element $|P_{vc}|^2$ being constant.
c) Show that the dielectric function obtained in (b) satisfies the conditions listed in (a).

6.3 *Plasma Frequency and Plasmons*

a) Show that, when damping is neglected, the dielectric constant for a three-dimensional free electron gas of density N immersed in a uniform background of equal and opposite positive charges (known as a **plasma**) in three dimensions is given by

$$\varepsilon(\omega) = 1 - \frac{4\pi N e^2}{4\pi\varepsilon_0 m\omega^2}, \tag{6.149}$$

where m is the electron mass.

b) Show that $\varepsilon(\omega) = 0$ when $\omega = \omega_\mathrm{p} = (4\pi N e^2/4\pi\varepsilon_0 m)^{1/2}$. ω_p is known as the **plasma frequency** of the free electron gas.

c) Show that when $\varepsilon(\omega) = 0$ it is not necessary to apply an external field to the free electron gas in order to produce an internal electric field. Calculate this internal field E in terms of the oscillation amplitude of the electrons. Hint: $E = D$(external field)$/\varepsilon$. When $\varepsilon(\omega_\mathrm{p}) = 0$, E is not necessarily zero even when $D = 0$. The electrons can be excited to oscillate at ω_p with the Coulomb attraction between the electrons and the positive background providing the restoring forces. Such oscillations of a free electron gas are known as **plasma oscillations**. As in the case of a simple harmonic oscillator, the energy of plasma oscillations can be quantized into units of $\hbar\omega_\mathrm{p}$. The quantized entities are known as **plasmons**. Calculate the plasma frequency of the valence electrons in diamond, silicon (see caption of Fig. 6.6) and GaAs.

d) A traveling wave (known as a **plasma wave**) can also be excited in a free electron gas. Show that plasma waves are longitudinal waves (i. e., the displacements of the electrons are parallel to the direction of propagation) and they produce an oscillating macroscopic longitudinal electric field.

6.4 *Dielectric Function of a Collection of Charged Harmonic Oscillators*

a) Derive the expressions (6.110a and b) for the dielectric functions of a collection of charged simple harmonic oscillators starting from (6.105 and 108). Show that (6.108) reduces to the equation of a free electron gas when $\omega_\mathrm{T} = 0$ and $\varepsilon_\infty = 1$.

b) Derive the Lyddane–Sachs–Teller relation (6.111) from (6.110a).

c) Use the KKRs [or (4.55)] to show that the imaginary part of the dielectric function and hence the absorption spectrum of such oscillators consist of a delta function at ω_T.

6.5 *Sum Rules*

Define the jth moment M_j of the imaginary part of the dielectric function (ε_i) of a semiconductor as

$$M_j = \int_0^\infty \varepsilon_\mathrm{i}\omega^j d\omega \tag{6.150}$$

a) Use the KKRs to show that M_1 is related to the total charge density N in the semiconductor by

$$M_1 = 2\pi^2 e^2 N/m. \tag{6.151}$$

b) Show that M_{-1} is related to the low-frequency dielectric constant $\varepsilon_r(0)$ by

$$M_{-1} = (\pi/2)[\varepsilon_r(0) - 1]. \tag{6.152}$$

6.6 *Van Hove Singularities*
Verify the form of the Van Hove singularities in the density of states given in Table 6.1.

6.7 *Absorption Spectrum of Wannier Excitons at Direct Bandgaps*
a) By using (6.72 and 82) derive the exciton–photon interaction matrix element $|\langle f|H_{xR}|0\rangle|^2$ in (6.85).
b) Looking up the form of the associated Laguerre polynomials $R_{nl}(r)$ in a quantum mechanics textbook, show that $|\langle f|H_{xR}|0\rangle|^2$ varies with the principal quantum number n as n^{-3}.
c) Show that the probability of finding an exciton in the continuum state at the origin $|\phi_E(\mathbf{0})|^2$ is given by (in atomic units with $m_0 = e = \hbar = 1$)

$$|\phi_E(\mathbf{0})|^2 = \frac{\tau e^{\tau}}{NV_0 \sinh \tau} \tag{6.153}$$

where τ is defined as

$$\tau = \pi|R^*/(\omega - \omega_g)|^{1/2}. \tag{6.154}$$

N is the number of unit cells in the crystal, V_0 is the volume of each unit cell, and R^* the exciton Rydberg. Use this result to derive (6.90).
d) Show that, in the limit $\omega \to \omega_g$ (or $\tau \to \infty$), ε_i approaches a constant value:

$$\varepsilon_i \to \frac{8\pi|P|^2\mu^2}{4\pi\varepsilon_0\omega_g^2\varepsilon_0}. \tag{6.155}$$

e) Show that in the limit $\omega \to \infty$ (or $\xi \to 0$) $|\phi_E(\mathbf{0})|^2 \to \mathbf{1}$ and ε_i approaches the value obtained in the absence of an exciton effect. Show, however, that the exciton effect modifies the absorption edge even for photon energies well above the band edge.

6.8 *Low Frequency Dielectric Constant and the "Average Gap"*
From Problem 6.5b we see that the zero-frequency dielectric constant $\varepsilon_r(0)$ is determined by contributions from optical transitions throughout the entire frequency range from zero to infinity. Penn has proposed a simple two-band model with an "average bandgap" E_g (known as the **Penn gap**) to account for $\varepsilon_r(0)$ in a semiconductor [6.140]. In this model $\varepsilon_r(0)$ is given by

$$\varepsilon_r(0) = 1 + (E_p/E_g)^2, \tag{6.156}$$

where E_p is the plasma energy of the valence electrons. It was pointed out in [6.141] that, because the oscillator strength at the E_2 transition is the strongest in many tetrahedrally coordinated semiconductors, a good approximation to the Penn gap is the energy of the E_2 transition. Calculate the

Table 6.9. Comparison between the Penn gap [eV] calculated from the experimental values of $\varepsilon_r(\infty)$ and the E_2 transition energies [eV] for a few representative semiconductors [6.142]

	Si	Ge	GaAs	InP	GaP
$\varepsilon_r(\infty)$	12.0	16.0	10.9	9.6	9.1
Penn gap	4.8	4.3	5.2	5.2	5.75
E_2	4.44	4.49	5.11	5.05	5.21

energy of the Penn gap from the experimental values of $\varepsilon_r(0)$ in Table 6.9 and compare with that of the E_2 transitions obtained from Table 6.3.

Note that "zero frequency" in this context means a frequency low compared to interband transitions but higher than phonon frequencies. The value of $\varepsilon_r(0)$ listed in Table 6.9 is sometimes also referred to as the "high frequency" dielectric constant $\varepsilon_r(\infty)$ because it is measured at a frequency much higher than the phonon frequencies.

6.9 *Exciton-Polariton Dispersion Curve*

An exciton-polariton is a propagating mode in a dielectric medium in which the electromagnetic wave is coupled with the polarization wave of excitons. To obtain the polariton dispersion curve depicted in Fig. 6.22 we shall start with the usual electromagnetic wave dispersion $\omega^2 = c^2k^2/\varepsilon$ (6.114a), where ω and k are, respectively, the frequency and wavevector of the electromagnetic wave, c is the speed of light in vacuum, and ε is the dielectric function of the medium. We shall regard the excitons as a collection of identical, charged simple harmonic oscillators with mass M, charge q, and (transverse) resonance frequencies ω_{ex}. Using the result for infrared-active optical phonons in Sect. 6.4, the dielectric function of the medium is given by (6.103)

$$\varepsilon(\omega) = 1 + \frac{4\pi N q^2}{4\pi\varepsilon_0 M(\omega_{\text{ex}}^2 - \omega^2)}.$$

To take into account the fact that excitons are propagating polarization waves with dispersion, replace ω_{ex} in the above expression by the exciton dispersion in (6.83)

$$\omega_{\text{ex}}(k) = \omega_{\text{ex}} + \frac{\hbar k^2}{2M},$$

where k is now the exciton wavevector. The result is a *spatially dispersive* dielectric constant $\varepsilon(k, \omega)$.

a) Obtain the exciton-polariton dispersion curve from $\varepsilon(k, \omega)$.

b) Calculate the **longitudinal exciton frequency** ω_{L}, i. e., the frequency for which $\varepsilon(\omega_{\text{L}}) = 0$ at $k = 0$. The frequency $\omega_{\text{ex}}(0)$ is known as the **transverse exciton frequency** and is denoted by ω_{T} in Fig. 6.22.

c) Sketch the polariton dispersion curves by substituting different values of ω into the dispersion relation, such as $\omega < \omega_{\text{ex}}(0)$, $\omega = \omega_{\text{ex}}(0)$, $\omega_{\text{L}} > \omega > \omega_{\text{ex}}(0)$, $\omega = \omega_{\text{L}}$, and $\omega > \omega_{\text{L}}$, and solving $\omega^2 = c^2k^2/\varepsilon(\omega, k)$ for k.

6.10 *Coupled Plasmon–LO Phonon Modes*
In a doped zinc-blende-type semiconductor, there are two possible longitudinal resonances; one corresponds to the plasma oscillation of the free carriers (as discussed in Problem 6.3), while the other is the LO phonon. These two longitudinal oscillations couple with each other and exhibit the phenomenon of "level anticrossing" discussed in Sect. 6.4. Assume that the free carriers (when "uncoupled" to the LO phonon) have the plasma frequency ω_p defined in Problem 6.3 while the corresponding LO phonon frequency in the absence of free carriers is ω_L.
a) Using the results in Sect. 6.4 and Problem 6.3, show that the total dielectric function of the semiconductor in the presence of the free carriers is given by

$$\varepsilon(\omega) = \varepsilon_\infty \left(1 + \frac{\omega_L^2 - \omega_T^2}{\omega_T^2 - \omega^2} - \frac{\omega_p^2}{\omega^2}\right). \qquad (6.157)$$

b) Obtain the two new longitudinal oscillation frequencies ω_{L^+} and ω_{L^-} by solving the equation $\varepsilon(\omega) = 0$. These longitudinal oscillations are known as the **coupled plasmon–LO phonon modes**. Show that, as a function of ω_p, the two solutions never cross each other (i. e., they anti-cross). The higher frequency branch (L^+) approaches ω_L for $\omega_p \ll \omega_L$ while the lower branch (L^-) approaches ω_p. For $\omega_p \gg \omega_L$ the L^+ branch approaches ω_p while the L^- branch approaches ω_T, the TO phonon frequency. This means that at high free carrier density (when $\omega_p \gg \omega_L$) the free carriers completely screen the extra Coulombic restoring force induced by the phonon displacement (Sect. 6.4) so that the transverse and longitudinal oscillations occur at the same frequency ω_T.

6.11 *Surface Plasmons and Phonons*
a) Show that at a flat vacuum-solid interface the Laplace equation $\nabla^2\phi = 0$ has solutions which propagate along the interface and decay exponentially away from that interface when the dielectric function of the solid medium $\varepsilon(\omega)$ is equal to -1. Such waves are known as **surface waves**. Assuming that ε of a metallic medium is given by (6.149) (Problem 6.3), express the frequency of the surface wave (known as the **surface plasmon** frequency) in terms of the bulk plasmon frequency ω_p. You can neglect the propagation (i. e., retardation) of the surface wave. Under what conditions is this approximation valid?
b) Deduce the effects of retardation on the surface wave in (a) from (6.12b) by imposing the condition that the reflectance $\mathcal{R}_p \to \infty$, i. e., a reflected electromagnetic field is produced without an incident one. This means that a self-sustaining resonant oscillation occurs (similar to what happens in the bulk when $\varepsilon = 0$). See [6.143] for hints.
c) For a dielectric with ε given by (6.110a), the condition that $\varepsilon = -1$ is satisfied at a frequency between the TO and LO phonon frequencies. Calculate the frequency of this "**surface phonon**" (neglecting retardation).

6.12 *Modeling the E_1 and $E_1 + \Delta_1$ Transitions by Two-Dimensional M_0 Critical Points*

In Sect. 6.2.4 it was pointed out that the conduction and valence bands in many diamond- and zinc-blende-type semiconductors are nearly parallel along the $\langle 111 \rangle$ directions. As a result, the critical points E_1 and $E_1 + \Delta_1$ and their joint density of states can be modeled by two-dimensional M_0 critical points.
a) Calculate the effective mass of these critical points using the $\boldsymbol{k} \cdot \boldsymbol{p}$ theory (the corresponding matrix element of $\boldsymbol{p}$ is $\approx 2\pi\hbar/a_0$).
b) Calculate their contributions to ε_i and ε_r. See [6.144, 7.102] for hints.

6.13 *Effects of Uniaxial Strain on Optical Phonons and on Effective Charges*

a) Use symmetry arguments to deduce the effects of uniaxial strains along the [100] and [111] axes on the zone-center optical phonons in diamond-type semiconductors like Ge.
b) Repeat these arguments for the zone-center TO and LO phonons in GaAs. From these results discuss the effects of uniaxial strains on the effective charge e^*. See [6.145] for hints.

6.14 *Electrooptic Tensors*

The effect of a dc electric field E on the dielectric function can be expressed as

$$\delta\varepsilon_{ij} = \delta\varepsilon_{ijk} E_k + \delta\varepsilon_{ijkl} E_k E_l, \tag{6.158}$$

where the third-rank tensor $\delta\varepsilon_{ijk}$ and the fourth-rank tensor $\delta\varepsilon_{ijkl}$ are known as the electrooptic tensors. The latter describes the **Kerr effect**, the former the **Pockels effect**.
a) Show that $\delta\varepsilon_{ijk}$ is identically zero in a centrosymmetric material (assuming electric dipole transitions only).
b) Use symmetry arguments to determine the linearly independent and nonzero coefficients of these electrooptic tensors for diamond- and zinc-blende-type crystals. See [6.146] for hints.
c) Use the results of (a) and (b) to calculate the change in ε induced by an electric field along the [001], [111], and [110] directions of Si and of GaAs.

6.15 *Effect of Band Nonparabolicity on Plasma Frequency*

In small-bandgap semiconductors such as InSb and InAs, the conduction band dispersion is nonparabolic. As a result, the electron effective mass to be used in calculating the free carrier plasma frequency in (6.126c) depends on the carrier concentration. The effective mass obtained by fitting the plasma reflection is known as the **optical mass** m^*_{op} to distinguish it from the effective mass determined by other techniques, such as cyclotron resonance. The effect of conduction band nonparabolicity on the plasma frequency can be estimated using the $\boldsymbol{k} \cdot \boldsymbol{p}$ method.
a) Write down the matrix corresponding to the $\boldsymbol{k} \cdot \boldsymbol{p}$ Hamiltonian around the Brillouin zone center (Γ point) of zinc-blende-type semiconductors. Include in your basis functions only the lowest s-like conduction band and the top p-like valence bands (neglect spin–orbit coupling). Diagonalize this 4×4 matrix and

expand the energy of the conduction band, $E_c(k)$, to k^4 in the following form:

$$E_c(k) = \frac{\hbar^2 k^2}{2m^*}(1 + Nk^2). \tag{6.159}$$

Determine the nonparabolicity constant N.

b) Repeat the calculation in (a) including the spin–orbit interaction.

c) Define the optical mass as

$$\frac{1}{m^*_{\rm op}} = \frac{1}{\hbar^2}\left(\frac{1}{k}\frac{\partial E_c}{\partial k}\right)_{E_F}, \tag{6.160}$$

where the subscript E_F signifies that (6.160) must be evaluated at the Fermi energy for a degenerate semiconductor. Calculate and plot the optical mass of electrons in InSb and their plasma frequency as a function of carrier concentration, N_e, in the range $10^{17} \leq N_e \leq 10^{19}$ cm^{-3}. Compare the plasma frequencies you obtained with the data in Fig. 6.37 and the calculated optical masses with the experimental values in [6.91].

6.16 *Seraphin Coefficients*

a) Derive an expression for the Seraphin coefficients in (6.132b) as a function of ε_r and ε_i.

b) Using the experimental data in Fig. 6.10 sketch the frequency dependence of these coefficients for Si. Note that for small photon energies β_i is negligible while β_r and β_i are comparable around the E_0' and E_1 critical points.

c) Discuss how to obtain $\Delta\varepsilon_r$ and $\Delta\varepsilon_i$ from a reflectance modulation spectrum by using the KKRs. Why are the extrapolations outside the experimental range unimportant?

6.17 *Modulation of Critical Points*

Derive analytic expressions similar to those in (6.135a) for the derivatives $d\varepsilon_r/dE$ and $d\varepsilon_i/dE$ near all the other types of critical points displayed in Table 6.1. Sketch the spectral dependence of your results.

6.18 *Piezoreflectance*

Calculate the ratio of the peaks shown in Fig. 6.46 for light polarized parallel to the stress axis ($\hat{\boldsymbol{e}} \parallel \chi$) and perpendicular to the stress axis ($\hat{\boldsymbol{e}} \perp \chi$) as a function of the ratio of shear to hydrostatic deformation potentials b/a (for $\chi \parallel [100]$) and d/a (for $\chi \parallel [111]$).

6.19 *Temperature Dependence and Isotopic Shift of Bandgaps*

a) Show that the temperature (T) dependence of an interband gap energy E_g can be written as

$$E_g(T) - E_g(0) = A\left(\frac{2}{\exp[\hbar\Omega/(K_B T)] - 1} + 1\right), \tag{6.161}$$

where A is a temperature-independent constant, k_B is the Boltzmann constant, and $\hbar\Omega$ represents an average phonon energy. Hint: the term inside the paren-

thesis in (6.161) represents the ensemble-averaged square of the phonon displacement. (See also [6.144].)
b) Show that $\Delta E_g(T) = E_g(T) - E_g(0)$ becomes linear in T in the limit of $k_B T \gg \hbar\Omega$ as shown in Fig. 6.54.
c) For small T, $\Delta E_g(T)$ can also be written as

$$\Delta E_g(T) = \left(\frac{\partial E_g}{\partial V}\right)_T \left(\frac{dV}{dT}\right)_P \Delta T + \left(\frac{\partial E_g}{\partial T}\right)_V \Delta T, \tag{6.162}$$

where the first term describes the change in E_g caused by thermal expansion. Its sign can be positive or negative. The second term is the result of electron–phonon interaction. Its sign is usually negative. Estimate the contribution of these effects to $E_g(0)$ by extrapolating $E_g(T)$ to $T = 0$ using its linear dependence at large T. The resultant energy is known as the **renormalization** of the bandgap at $T = 0$ by electron–phonon interaction. Determine this energy for the E_0 gap of Ge from Fig. 6.44.
d) The result in (c) can be used to estimate the dependence of bandgap on isotopic mass. Since the bonding between atoms is not affected by the isotopic mass, the average phonon energy $\hbar\Omega$ in solids with two identical atoms per unit cell, like Ge, can be assumed to depend on atomic mass M as $M^{-1/2}$. Calculate the difference in the E_0 bandgap energies between the following isotopes: ^{70}Ge, ^{74}Ge, and ^{76}Ge. Compare your results with those in Fig. 6.52.

6.20 *Third-Order Nonlinear Optical Susceptibility in Ge*
In general this book has concentrated on linear optical properties and avoided nonlinear optical phenomena. However, some nonlinear optical properties of solids can be deduced from the effect of electric field E on their linear optical properties. The third-order nonlinear optical susceptibility $\chi^{(3)}$ is defined by (see also Problem 6.14):

$$\chi_{ij}(E) = \chi_{ij}^{(1)} + \chi_{ijk}^{(2)} E_k + \chi_{ijkl}^{(3)} E_k E_l, \tag{6.163}$$

where $\chi_{ij}^{(1)}$ is the field-independent linear electric susceptibility tensor introduced in (6.1). $\chi_{ijk}^{(2)}$ is the second-order nonlinear susceptibility tensor and is related to nonlinear optical effects such as second harmonic generation [6.148]. $\chi^{(2)}$ is identically zero in centrosymmetric crystals if only electric-dipole transitions are considered. The third-order nonlinear susceptibility $\chi^{(3)}$ is responsible for nonlinear phenomena such as two-photon absorption and third-harmonic generation. It is related to the fourth-rank electrooptic tensor $\delta\varepsilon_{ijkl}$ in Problem 6.14 via the expression

$$\delta\varepsilon_{ijkl} = 4\pi\chi_{ijkl}^{(3)}. \tag{6.164}$$

In general $\delta\varepsilon_{ijkl}$ depends on the photon frequency ω. For $\omega \approx 0$ (i. e., ω much less than the E_0 bandgap but ω much greater than the TO phonon frequency) the diagonal element $\delta\varepsilon_{iiii}(0)$ in Ge can be calculated under the assumption that its low-frequency dielectric constant $\varepsilon_r(0)$ is dominated by the contribution of interband transition between a parabolic valence band and a parabolic conduction band separated by the E_0 gap with both bands extending to infinity (why is this assumption reasonable in this exercise but not in Problem 6.8?). Hint: Start by expressing $\varepsilon_r(0)$ in terms of the E_0 gap with (6.49). Next differentiate $\varepsilon_r(0)$ with respect to the bandgap and use (6.147c) [6.149].

6.21 *The ratio Δ_0/Δ_1*

Δ_0/Δ_1 is about 3/2 for most materials listed in Table 6.2. Notice, however, that there are two exceptions. What might be the reasons for this anormalous behavior? Suggest other zinc-blende-type materials that may exhibit similar anormalous ratios.

6.22 *Chirality*

Discuss the symmetry of the phonons of the trigonal elemental semiconductors selenium and tellurium. Their structure can appear in two different space groups: D_3^4 and D_3^6. Show that two crystals with different space groups are mirror images of each other. This property is known as *chirality* (after the Greek $\chi\iota\rho\varepsilon$: hand).

a) Discuss possible effects of chirality on the optical response of the crystals.

b) Find out how many of the 6 optical vibrational modes at Γ are ir-active (i.e., $e_T \neq 0$) and explain qualitatively the origin of the e_T's. For hints see [2.5] and [3.8].

SUMMARY

Chapters 6 and 7 are devoted to the study of the optical properties of semiconductors. In this chapter we have discussed those phenomena involving only one photon frequency. In processes like *absorption* and *reflection* an incident electromagnetic wave illuminates the sample and the frequency of the wave is unchanged by its interaction with the sample. In the following chapter we shall discuss phenomena in which the frequency of the incident wave is altered by the sample. The optical properties of the sample studied in this chapter can be completely described by its complex *dielectric function*. A microscopic theory of this function shows that photons interact mainly with the electrons in semiconductors by exciting interband and intraband transitions. *Interband transitions* from the valence bands to the conduction bands produce peaks and shoulders in the optical spectra which can be attributed to *Van Hove singularities* in the valence–conduction band *joint density of states*. These structures can be greatly enhanced by using the technique of *modulation spectroscopy*, in which the derivatives of some optical response function with respect to either frequency or an external modulation (such as electric and stress fields) are measured. These optical measurements have provided an extremely sensitive test of existing electronic band structure calculations. Occasionally, disagreements between experimental and theoretical spectral peak positions and lineshapes have been found. These can be explained by the *excitonic effect* as a result of the Coulomb interaction between excited electrons and holes in the semiconductor. *Intraband electronic transitions* occur in doped semiconductors and their contribution to the optical properties can be obtained by using the *Drude model* proposed for free electrons in simple metals.

Transitions between the discrete levels of impurities in semiconductors can also contribute to absorption of photons in the infrared. Although these *extrinsic absorption* processes are much waker than those involving intrinsic electronic transitions, they can give rise to extremely sharp peaks and have been a very useful and highly sensitive probe of the electronic energy levels of impurities. Finally, in polar semiconductors, such as those with the zincblende crystal structure, photons can be absorbed and reflected as a result of interaction with optical phonons. The reflectivity becomes particularly high for photons with frequency between the TO and LO phonon frequencies, giving rise to a phenomenon known for a long time as *reststrahlen*. The coupling between infrared-active optic phonons and electromagnetic waves can be so strong that they cannot be separated inside the medium. Instead, they should be regarded as coupled waves or quasiparticles known as *phonon-polaritons*.

7. Optical Properties II

CONTENTS

In Chap. 6 we studied the interactions between a semiconductor and an electromagnetic field in which the semiconductor is brought from the ground state to an excited state via absorption of photons from the applied radiation. In this chapter we study other important optical phenomena in semiconductors. These optical processes involve emission of radiation from the sample. One of these processes is **luminescence**, while the other is **inelastic scattering** of light (also known as **Raman** or **Brillouin scattering**; their difference will be discussed later in this chapter). In a typical luminescence process electrons in the sample are excited electrically or optically. After some energy loss (relaxation) the excited electrons return to the ground state while emitting light. In a Raman or Brillouin process light is scattered by fluctuations inside the sample. One important difference between these two processes is that luminescence involves real excitation of electrons, while in light scattering typically virtual excitations of electrons are sufficient. We shall study the physical mechanisms of these processes in this chapter. We begin with a discussion of luminescence.

7.1 Emission Spectroscopies

In order for a sample to emit radiation it has to be energized by some external means. One possible way is to excite the sample by injection of electrons and holes via an external current, leading to **electroluminescence**. Another common method is by absorption of photons of energy higher than that of the bandgap, the resulting process, in which photons of energy lower than the exciting photons are radiated, is known as **photoluminescence**. The production of radiation by heating the sample is known as **thermoluminescence**, while the process of inducing light emission by electron bombardment is also called

electroluminescence or **cathodoluminescence**. In this chapter we shall be more concerned with the emission process than with how the sample is excited.

In Sect. 6.2.2 we pointed out that absorption and emission are related to each other. They are described by two terms which are complex conjugates of each other in the interaction Hamiltonian $\mathcal{H}_{eR}$ in (6.29). In an absorption process energy is removed from an incident electromagnetic wave while electron–hole pairs are created. Emission is the inverse of this process, i. e., an electron–hole pair inside the medium is destroyed with the emission of electromagnetic radiation (the pair is said to undergo **radiative recombination**). In the semiclassical approach we adopted in Sect. 6.2.2, an electromagnetic wave has to be present in order for emission to occur (since $\mathcal{H}_{eR}$ involves its vector potential $\boldsymbol{A}$). Such an emission process is known as **stimulated emission**. The probability of the stimulated emission process is proportional to the strength of the field, just as for the absorption process. From our everyday experience we know that a lamp can emit light in the absence of an external radiation field. Therefore the same must be true for an excited semiconductor containing electron–hole pairs. Radiation produced this way without an external field is known as **spontaneous emission**. A rigorous description of such processes requires a quantum mechanical treatment of the electromagnetic radiation (readers interested in such a treatment should refer to [7.1]). One has to quantize the electromagnetic waves into photons in a way very similar to the quantization of lattice vibrations into phonons (both photons and phonons are *bosons*). The probability of creating a photon with energy $\hbar\omega$ (where ω is the angular frequency of the electromagnetic wave) is proportional to $1 + N_p$, where N_p is the photon occupation number and is also given by the Bose–Einstein distribution function introduced in Sect. 3.3.1. Clearly this probability is nonzero even when N_p is zero. In this case one can understand the emission as induced by the **zero-point amplitude** of the photons. In the case of simple harmonic oscillators, the existence of such zero-point motion is well known in quantum mechanics and is explained by the uncertainty principle, namely, that the position and momentum of the vibrating particle cannot be zero simultaneously. The zero-point motion term in $(1 + N_p)$ is independent of N_p and induces spontaneous emission, whereas the term proportional to N_p gives rise to stimulated emission. Interestingly, the relation between absorption, spontaneous emission, and stimulated emission was first proposed by *Einstein* [7.2, 3] without invoking quantum mechanics. This is the approach we shall follow.

Einstein denoted the rates for absorption and stimulated emission per unit electromagnetic energy density (within the frequency interval between ν and $\nu + \Delta\nu$) due to the transition of an electron between a level n and another level m by the coefficient B_{nm}. The rate for spontaneous emission of radiation due to transition from level n to m was denoted by $\boldsymbol{A}_{nm}$. These rates are now commonly referred to as **Einstein's *A* and *B* coefficients** (Problem 7.1). On the basis of the principle of detailed balance (Sect. 5.2.3), *Einstein* [7.2, 3] showed that the coefficients A_{nm}, B_{nm}, and B_{mn} for two nondegenerate levels in a medium with refractive index n_r are related by

$$B_{nm} = B_{mn} \tag{7.1a}$$

and

$$A_{nm} = \frac{8\pi h\nu^3 n_r^3}{c^3} B_{nm}, \tag{7.1b}$$

where h is Planck's constant ($h = 2\pi\hbar$), ν the photon frequency [related to ω by $\nu = \omega/(2\pi)$], and c the velocity of light in vacuum. The term $8\pi h\nu^3 n_r^3/c^3$ is equal to the product of $h\nu$ and the density of electromagnetic modes with frequency between ν and $\nu + \Delta\nu$ *inside* the medium. The total emission rate R_{nm} of radiation from level n to level m for a system in thermal equilibrium at the temperature T can be written as

$$R_{nm} = A_{nm} + B_{nm}\varrho_e(\nu), \tag{7.2}$$

where $\varrho_e(\nu)$, the **photon energy density**, is defined as the energy density of photons with frequency between ν and $\nu + \Delta\nu$. It can be easily shown to be $8\pi h\nu^3 n_r^3/c^3$ multiplied by the photon occupation number N_p. Using (7.1), (7.2) can be rewritten as

$$R_{nm} = A_{nm}[1 + (B_{nm}/A_{nm})\varrho_e(\nu)] = A_{nm}(1 + N_p). \tag{7.3}$$

Equation (7.3) contains two terms. The second term, proportional to N_p, can be identified with stimulated emission, while the first one corresponds to spontaneous emission. The expression (7.3), obtained by Einstein using classical physics, is in complete agreement with the quantum mechanical result. While stimulated emission is important for understanding semiconductor **laser diodes**, we shall limit ourselves to considering only spontaneous emission in this book. However, from (7.3) it is clear that one can obtain the stimulated emission rate by simply multiplying the spontaneous emission rate by the occupation number N_p of photons present.

We shall now assume that the photon density $\varrho(\nu)$ (equal to $\varrho_e(\nu)$ divided by the photon energy $h\nu$) in the frequency interval $\Delta\nu$, is small, so that we can neglect the stimulated emission term. Furthermore, instead of discrete levels we shall consider a conduction band (c) and a valence band (v). The emission rate for the transition from the conduction band to the valence band is given by

$$R_{cv} = A_{cv} f_c (1 - f_v), \tag{7.4}$$

where f_c and f_v are the electronic occupancies in the conduction and valence bands, respectively. The term $1 - f_v$ gives the probability that the corresponding valence band states are empty in order that the transitions satisfy Pauli's exclusion principle. The corresponding absorption rate for the inverse transition (valence to conduction band) per photon is denoted by P_{vc}. At thermal equilibrium, the principle of detailed balance (Sect. 5.2.3) requires

$$P_{vc}(\nu)\varrho(\nu) = R_{vc}(\nu). \tag{7.5}$$

This relation between emission and absorption is known as the **van Roosbroek–Shockley relation** [7.4a]. The absorption rate can thus be related to the absorption coefficient α defined in (6.10) by

$$P_{vc} = \alpha c/n_r. \tag{7.6}$$

Combining (7.5 and 6) and using for $\varrho(\nu)$ the Planck distribution at the temperature T we obtain a relation between the emission probability and the absorption coefficient:

$$R_{vc}(\nu) = \frac{\alpha(\nu)8\pi\nu^2 n_r^2}{c^2[\exp(h\nu/k_B T) - 1]}. \tag{7.7}$$

This relation applies to intrinsic semiconductors only. In heavily doped semiconductors the emission spectra are modified by additional effects, such as the introduction of bandtail states within the gap, *bandgap shrinkage* (or *renormalization*) and filling up of the near-bandgap states. Some of these effects will be considered in more detail in Sect. 7.1.2. The band filling effect causes a blue shift of the absorption edge known as the **Burstein–Moss shift** [7.5, 6]. These effects are important for devices such as laser diodes, and readers interested in this aspect should refer to [7.7]. Notice that (7.7) describes the emission spectrum inside the sample. The spectrum measured externally outside the sample will be different from the internal spectrum as a result of reabsorption of the emitted photons as they propagate from the interior to the surface of the sample.

Alternatively, (6.10) can be used to relate R_{vc} to the imaginary part of the refractive index $\varkappa(\nu)$. Figure 7.1 shows the $\varkappa(\nu)$ curve for Ge and the corresponding $R_{vc}(\nu)$ curve computed from (7.7). The $\varkappa(\nu)$ curve shows two absorp-

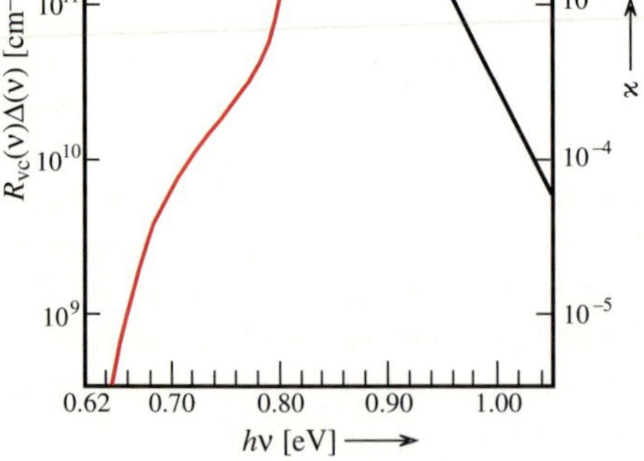

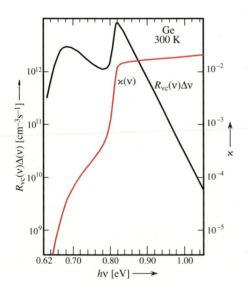

Fig. 7.1. Plots of the photon energy dependence of the imaginary part of the refractive index $\varkappa(\nu)$ and the emission rate $R_{vc}(\nu)$ of Ge at $T = 300$ K, multiplied by $\Delta\nu = k_B T/h$ (*for historical reasons*). (From [7.4b])

tion shoulders at 0.65 eV and 0.82 eV. They can be identified, respectively, as the indirect (see also Fig. 6.18) and direct absorption edges of Ge. While the direct absorption edge is much stronger than the indirect one, the heights of the corresponding emission peaks are comparable. The explanation lies in the exponential factor $\exp(h\nu/k_B T)$ in the denominator of (7.7). At thermal equilibrium the carrier population decreases exponentially with energy and therefore emissions are strongest from the lowest energy states. As a result *luminescence is a sensitive probe of low-lying energy levels*, such as defect levels inside the gap, provided electrons and holes can recombine radiatively at these levels. Such defects are known as **radiative recombination centers**, otherwise they are referred to as **nonradiative traps**. Although (7.5) and (7.7) were derived for transitions between conduction and valence bands, they should apply, in principle, to transitions between any two states in a system *at thermal equilibrium*. The amount of emission produced by a body under thermal equilibrium at room temperature is very small, hence most experiments are carried out under nonequilibrium conditions. The creation of such conditions and the detection of the resultant spontaneous emission from the sample is the essence of luminescence experiments.

In a luminescence experiment one excites initially a nonequilibrium distribution of electron–hole (to be abbreviated as e-h) pairs in a semiconductor. In most cases the electrons and holes will thermalize among themselves and reach quasi-thermal equilibrium (Sect. 5.3) in a time short compared to the time it takes for electrons and holes to recombine. Often these electrons and holes have different quasi-equilibrium distributions. In the final step the e-h pairs recombine radiatively, producing the spontaneous emission. Thus a luminescence process involves three separate steps:

- *Excitation:* Electron–hole pairs have to be excited by an external source of energy.
- *Thermalization:* The excited e-h pairs relax towards quasi-thermal-equilibrium distributions.
- *Recombination:* The thermalized e-h pairs recombine radiatively to produce the emission.

In special circumstances [7.8] emission from incompletely thermalized e-h pairs can be observed; it is referred to as **hot luminescence**. Because luminescence is produced by (either fully or partially) thermalized e-h pairs, the emitted photons have no correlation with the excitation process. We shall see later in this chapter that this is one important distinction between light scattering and photoluminescence. While the frequency of the scattered photon follows that of the incident photon, the energy of the emitted photon, in the case of thermalized luminescence, depends only on the band structure and energy levels of the sample. This is one useful way to distinguish between experimental peaks arising from these two processes. Figure 7.2 shows the position of all the observed peaks in a light scattering experiment involving Cu_2O when excited with photons in the vicinity of its "green" excitonic series [7.9]. (There are four exciton series in Cu_2O, labeled yellow, green, blue, and indigo according

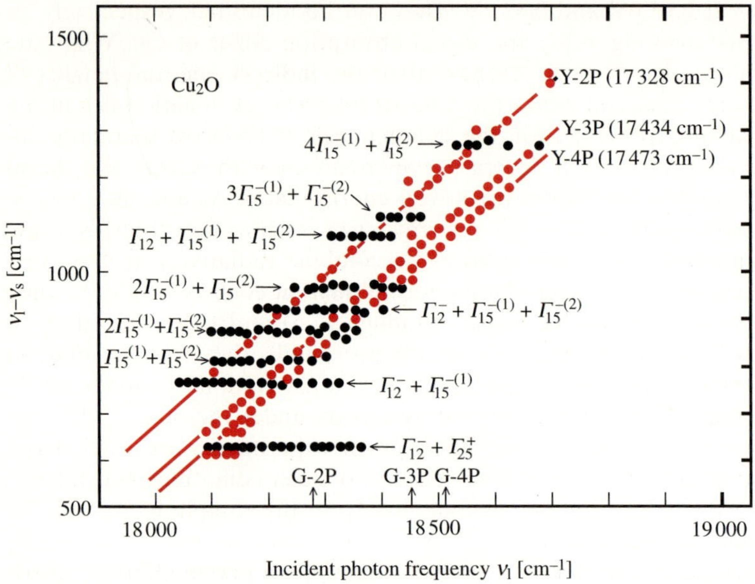

Fig. 7.2. Peaks observed in the emission spectra of Cu_2O when excited with photons in the vicinity of the green exciton series (labeled G−2P to G−4P). The difference between the exciting frequency ν_l and the peak frequency ν_s is plotted vs. ν_l. The *straight red lines* labeled Y−2P to Y−4P represent the expected dependence of the yellow exciton luminescence peaks on the incident photon energy ν_l. The *horizontal arrows* indentify the various multiphonon Raman peaks which are enhanced as a result of resonance of the incident and scattered photon energies with the green and yellow excitons respectively. (From [7.9])

to their photon frequencies, Sect. 6.3.2.) Some of these peaks are photoluminescence peaks associated with recombination of the "yellow" excitonic series. However, many are the results of light scattering. These Raman peaks appear only when the incident photon is resonant with the exciton peaks (a process known as *resonant Raman scattering* to be discussed later in this chapter). They are characterized by constant **Raman frequencies** (defined as the difference $\nu_l - \nu_s$ between the incident laser frequency ν_l and the scattered photon frequency ν_s). As a result they fall on horizontal lines in Fig. 7.2. On the other hand the luminescence peaks have emission frequencies ν_s independent of the laser frequency ν_l. Thus their frequency difference from ν_l increases linearly with ν_l, following the solid red lines in Fig. 7.2.

We shall consider in more detail the various radiative recombination processes found commonly in semiconductors. Unless specified differently we shall assume that the e-h pairs have been excited optically, i. e., the experimental technique is photoluminescence.

7.1.1 Band-to-Band Transitions

In a perfect semiconductor e-h pairs will thermalize and accumulate at the conduction and valence band extrema, where they tend to recombine. If this semiconductor has a direct bandgap and electric dipole transitions are allowed, the e-h pairs will recombine radiatively with a high probability. As a result, high quality direct-bandgap semiconductors, such as GaAs, are strong emitters of bandgap radiation. They are important materials for lasers and **light emitting diodes** (LEDs). In indirect bandgap semiconductors, such as Si and Ge, e-h pairs can recombine radiatively only via phonon-assisted transitions. Since the probability of these transitions is smaller than for competing nonradiative processes, these materials are not efficient emitters. The indirect bandgap semiconductor GaP is an exception. In GaP (Sect. 4.3.3) the radiative transition can be enhanced by localizing the e-h pair at defects such as isovalent nitrogen. There is much ongoing effort to make Si a more efficient emitter of light by fabricating Si into the form of nanometer-size crystallites known as *nanocrystals*. It is argued that by physically confining electrons and holes one can enhance their radiative recombination rate. One such technique involves the use of electrolysis [7.10] to produce a spongy form of Si known as *porous* Si. Unlike bulk Si, porous Si has been shown to produce efficient visible photoluminescence and electroluminescence [7.11, 12]. The reasons for this increased emission efficiency in porous Si are, however, still controversial [7.12, 13].

Band-to-band transitions involve the recombination of free electrons and free holes. Let us define τ_r as the **radiative recombination time** of one electron and one hole. If the free electron and hole concentrations are, respectively, n_e and n_h, then the rate of emission of photons by their recombination is given by $n_e n_h/\tau_r$ *assuming that* τ_r *is the same for all possible choices of recombining pairs*. For a thermalized distribution of free electrons and holes, the radiative recombination time depends on the electron and hole energies and therefore changes with the photon energy. In general τ_r should then be replaced by an *averaged radiative recombination time* $\langle\tau_r\rangle$. The averaging procedure depends also on whether wavevector is conserved in the recombination processes. While wavevector is expected to be conserved in recombination in perfect crystals, it has been found not to be conserved when a high density of e-h pairs is excited. In electroluminescence, where it is possible to inject one extra minority carrier into a semiconductor containing an equilibrium distribution of electrons and holes, it is usual to define a **minority carrier radiative lifetime** τ_{rad} as the time for this extra carrier to be annihilated radiatively by the majority carriers. For the case of an extra electron injected into a p-type sample, this time is given by

$$\frac{1}{\tau_{rad}} = \frac{n_h}{\langle\tau_r\rangle}. \tag{7.8}$$

In an intrinsic semiconductors n_h ($= n_e$) is given by the concentration of thermally excited holes (usually denoted by p_i, see [Ref. 7.14, p. 206]). For large-

Table 7.1. Minority carrier radiative lifetime in several tetrahedrally bonded semiconductors at room temperature. From [Ref. 7.15, p. 111]

	τ_{rad}	
Semiconductor	Intrinsic	10^{17} cm^{-3} majority carriers
Si	4.6 h	2.5 ms
Ge	0.61 s	0.15 ms
GaP		3.0 ms
GaAs	2.8 μs	0.04 μs
InAs	15 μs	0.24 μs
InSb	0.62 μs	0.12 μs

bandgap semiconductors p_i $(= n_i)$ is very small and hence τ_{rad} tends to be large, ranging from hours in Si to microseconds in smaller bandgap semiconductors like InSb (Table 7.1). τ_{rad} is always very large for indirect bandgap semiconductors, such as Si and Ge, since their electric-dipole indirect transition probabilities are smaller. In doped semiconductors the majority carrier concentrations are often much higher than the intrinsic carrier concentrations. As a result the minority carrier radiative lifetime depends on the doping concentration. In Table 7.1 we show also these lifetimes in several semiconductors when doped with 10^{17} cm^{-3} majority carriers.

There is a limit as to how much we can decrease τ_{rad} by increasing the majority carrier concentration. At most, we can make this concentration equal to the entire band population. For GaAs this minimum τ_{rad} is about 0.31 ns [Ref. 7.15, p. 133]. Note that this limit is not valid for the stimulated emission lifetime since the stimulated emission rate depends also on the photon density. It is not unusual to have stimulated radiative lifetimes of less than 0.1 ps.

In photoluminescence (commonly abbreviated as PL) experiments one always excites equal numbers of electrons and holes. Since the intrinsic carrier concentrations n_i and p_i are usually very low, it is relatively easy to excite optically in an intrinsic semiconductor enough carriers that $n_e = n_h \gg n_i = p_i$ and n_i and p_i become negligible. If wavevector conservation is not necessary (which may happen because of defects or phonons), the radiative recombination rate $(1/\tau_{rad})$ for each optically excited carrier is equal to $n/\langle\tau_r\rangle$, where $n = n_e = n_h$. Even if wavevector is conserved, the radiative lifetime will still depend on the intensity of the excitation light. In addition to radiative decay processes, the photoexcited e-h pairs can also recombine nonradiatively. The **total decay rate** $(1/\tau_{tot})$ of the photoexcited population of e-h pairs is given by

$$(1/\tau_{tot}) = (1/\tau_{rad}) + (1/\tau_{nonrad}), \tag{7.9}$$

where $1/\tau_{nonrad}$ is the **nonradiative recombination rate**. In nonradiative processes the energy of the e-h pair is dissipated as heat via excitation of phonons.

If τ_{tot} is much longer than the electron–phonon interaction times, the electrons and holes can reach quasi-thermal-equilibrium separately with the phonons.

To calculate the shape of the band-to-band PL spectra, we shall assume a direct bandgap semiconductor with gap E_g and joint density of states [see (6.57)]

$$D_j \propto (E - E_g)^{1/2} \tag{7.10}$$

Let f_e and f_h represent the quasi-equilibrium distribution functions for the electrons and holes, respectively. For low photoexcitation density f_e and f_h can be approximated by Boltzmann distributions:

$$f_e \text{ or } f_h \propto \exp[-E/(k_B T)]. \tag{7.11}$$

Substituting (7.10 and 11) into (7.4), we obtain the PL spectral shape

$$I_{PL}(\hbar\omega) \propto \begin{cases} (\hbar\omega - E_g)^{1/2} \exp[-(\hbar\omega - E_g)/(k_B T)] & \text{for } \hbar\omega > E_g, \\ 0 & \text{otherwise,} \end{cases} \tag{7.12}$$

where $\hbar\omega$ is the emitted photon energy (note the relationship between (7.12) and (7.7)). Figure 7.3 shows the experimental PL spectra in GaAs measured at room temperature under a pressure of 29.4 kbar [7.16]. The theoretical curve is a plot of (7.12) with $T = 373$ K, which is higher than room temperature because of heating of the small sample by the laser beam. For samples under

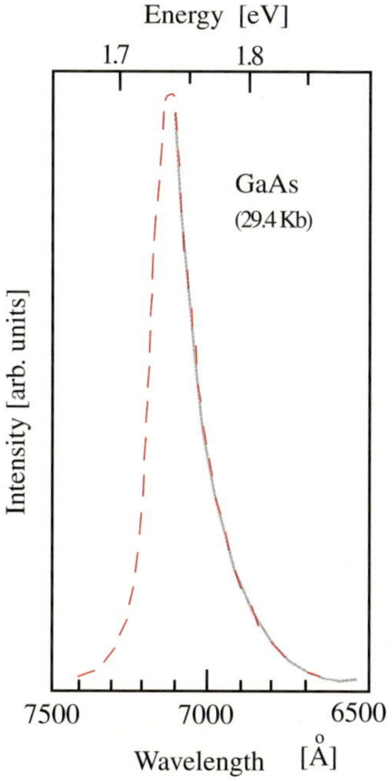

Fig. 7.3. Photoluminescence spectrum due to band-to-band transition in GaAs measured (*broken line*) at room temperature and a pressure of 29.4 kbar. The theoretical curve (*solid line*) is a plot of the expression (7.12), approximately proportional to $\exp[-(\hbar\omega - E_g)/(k_B T)]$, with $T = 373$ K. (From [7.16])

high intensity excitation the electron–electron and hole–hole interactions can become stronger than the carrier–phonon interactions. In such cases it is quite common for the electrons and holes to attain a temperature much higher than that of the lattice. Photoluminescence is one of the few methods capable of measuring this carrier temperature directly.

7.1.2 Free-to-Bound Transitions

Band-to-band transitions tend to dominate at higher temperatures where all the shallow impurities are ionized. At sufficiently low temperatures, carriers are frozen on impurities. For example, consider a PL experiment on a p-type sample containing N_A acceptors per unit volume. At low photoexcitation the density n_e of free electrons created in the conduction band is much smaller than N_A. These free electrons can recombine radiatively (and sometimes also nonradiatively) with the holes trapped on the acceptors. Such transitions, involving a free carrier (an electron in this case) and a charge (a hole in this case) bound to an impurity, are known as **free-to-bound transitions**. The emitted photon energy in this example is given by $E_g - E_A$, where E_A is the shallow acceptor binding energy. Thus, emission due to free-to-bound transitions is a simple way of measuring impurity binding energies. Figure 7.4 shows the electroluminescence specra of p-type GaAs at 4.2 K for various dopant concentrations. The spectra for a sample with N_A equal to 3.7×10^{17} and 1.9×10^{18} cm^{-3} can be assigned to free-to-bound transitions. As the acceptor concentration is increased these spectra exhibit several interesting changes. They first broaden because of changes in the acceptor level density of states. As the acceptors become closer to one another, their wavefunctions begin to overlap. This results in broadening of the acceptor levels into an **impurity band** (similar to the broadening of atomic levels into bands in solids discussed in

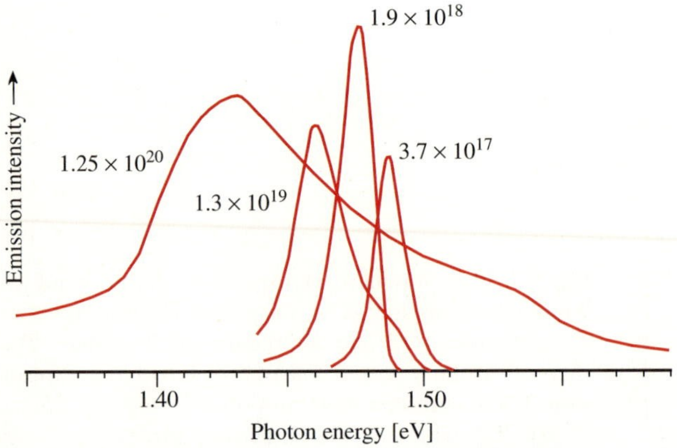

Fig. 7.4. Electroluminescence of p-type (Zn-doped) GaAs at 4.2K for increasing dopant concentrations in units of cm^{-3}. (From [Ref. 7.15, p. 136])

Sect. 2.7). When the band is so broad that it overlaps with the valence band, holes are no longer localized on the acceptors and become free carriers. This transformation of carriers from a localized state to a delocalized one is known as a **Mott transition** [Ref. 7.14, p. 268]. (Actually the Mott transition often occurs before the overlap takes place: the impurity levels broaden into bands, which are half-filled because of the spin degeneracy. When the broadening is larger than the additional energy required to put two electrons in the same impurity level of a given atom, as opposed to putting them on separate atoms, the material becomes conducting.) In addition to this broadening of the acceptor density of states, the carrier distribution function also becomes *degenerate* as their concentration increases. This manifests itself as a deviation from the exponential dependence $\exp[-\hbar\omega/(k_B T)]$ in the higher density spectra in Fig. 7.4. The highest density spectrum clearly resembles the Fermi–Dirac distribution more than the Boltzmann one. Finally the peak in the emission spectra red-shifts with increase in dopant concentration. This is a many-body effect known as bandgap renormalization [7.17, 18]. A detailed discussion of this effect is beyond the scope of this book.

Wavevector conservation in free-to-bound transitions is relaxed since the translational symmetry of the crystal is broken by the defects. Hence an electron in the conduction band can recombine with a hole on an acceptor regardless of its wavevector. The radiative recombination rates for free-to-bound transitions in direct-bandgap zinc-blende-type semiconductors have been calculated by *Dumke* [7.19] and compared with those of band-to-band transitions. His result for $(\tau_{e-A})^{-1}$, the rate of conduction band to acceptor transition, is

$$\left(\frac{1}{\tau_{e-A}}\right) = \frac{64\sqrt{2}\pi n_r e^2\hbar^2\omega\,|P_{cv}|^2 N_A}{c^3 m^2 (m_h E_A)^{3/2}}, \tag{7.13}$$

where m_h is an average hole mass and P_{cv} is an averaged electron momentum matrix element between the conduction and valence bands and N_A the acceptor concentration. For GaAs (7.13) becomes

$$\left(\frac{1}{\tau_{e-A}}\right) = 0.43 \times 10^{-9} N_A\,\mathrm{cm^3/s}. \tag{7.14}$$

This equation predicts that electrons in GaAs samples containing 10^{18} shallow acceptors per cubic centimeter will have a radiative lifetime of about 2 ns as a result of recombination with bound holes. This time is comparable to the radiative lifetime due to band-to-band transitions. Thus we expect that at low temperatures (when $k_B T \ll E_A$) electron-acceptor recombination will dominate the PL spectra in p-type GaAs. At higher temperatures, as more holes are excited into the valence band, both free-to-bound and band-to-band emission peaks will be observed. Finally, at high enough temperatures the spectrum will be dominated by band-to-band emission. The electron-to-acceptor PL intensity (I_{e-A}) should vary with temperature as $1 - \exp[-E_A/(k_B T)]$. The acceptor ionization energy can therefore be determined from the slope of a plot of $\ln[1 - (I_{e-A}/I_0)]$ versus $1/(k_B T)$, where I_0 is the emission intensity at $T = 0$ K. Such plots are known as **Arrhenius plots**.

7.1.3 Donor–Acceptor Pair Transitions

Quite often a semiconductor may contain both donors and acceptors. Such semiconductors are said to be **compensated** because, under equilibrium conditions, some of the electrons from the donors will be captured (or compensated) by the acceptors. As a result, a compensated sample contains both ionized donors (D^+) and acceptors (A^-).[1] By optical excitation, electrons and holes can be created in the conduction and valence bands, respectively. These carriers can then be trapped at the D^+ and A^- sites to produce neutral D^0 and A^0 centers. In returning to equilibrium some of the electrons on the neutral donors will recombine radiatively with holes on the neutral acceptors. This process is known as a **donor–acceptor pair transition** (or DAP transition). It can be represented by the reaction

$$D^0 + A^0 \rightarrow \hbar\omega + D^+ + A^- . \tag{7.15}$$

At first sight one may expect the photon emitted in a DAP transition to have the energy

$$\hbar\omega = E_g - E_A - E_D , \tag{7.16}$$

where E_g is the bandgap energy and E_D and E_A are the donor and acceptor binding energies, respectively. The problem with (7.16) is that it neglects the Coulomb interaction between the ionized donors and acceptors. Suppose the distance between the D^+ and A^- is R, then this Coulomb energy is equal to $-e^2/(4\pi\varepsilon_0\varepsilon_0 R)$ (provided R is much larger than the lattice constant), where ε_0 is the static dielectric constant. The energy of the emitted photon in a DAP transition should then be given by

$$\hbar\omega = E_g - E_A - E_D + e^2/(4\pi\varepsilon_0\varepsilon_0 R). \tag{7.17}$$

The emitted photon energy is *increased* by the amount $e^2/(4\pi\varepsilon_0\varepsilon R)$ because the energy of the final state in (7.15) is lowered by the Coulomb attraction. Notice that, in the case of excitonic absorption, the external photon creates a pair of positive and negative charges. The Coulomb attraction between these charges *lowers* the energy of the photon required to excite them. In the present case, the energy of the initial state is shared in the final state between the emitted photon and a pair of positive and negative charges. Any decrease in the energy of the charge pair by Coulomb attraction ends up in the emitted photon energy. In both cases a Coulomb interaction appears in the final state only and therefore this interaction is referred to as a **final state interaction**. In principle, there should also be an **initial state interaction** between the neutral donor and acceptors. This interaction is similar to the *van der Waals interaction* between two neutral atoms [7.20, 21]. Unlike the interaction between atoms, the separations between donors and acceptors are not continuously variable but are, instead, determined by the crystal parameters (to be

[1] Is is implicitly assumed that there are more donors than acceptors.

discussed in more detail below). For distant pairs we expect the van der Waals interaction to be completely negligible. For close pairs this interaction is still rather weak and will be neglected in the lowest order approximation.

a) Spectral Lineshapes

There is an important difference between the Coulomb interaction in excitons and in DA pairs. While the electron and hole separations in excitons are determined by quantum mechanics (via the solution of the Schrödinger equation), the separation R between the ionized impurities is determined by the crystal structure and the lattice constants. Since the values of R are discrete, the DAP transitions produce a series of sharp peaks converging towards the photon energy $E_g - E_A - E_D$ (corresponding to $R = \infty$). The best and most carefully studied examples of DAP transitions are found in GaP. Because this is a binary compound there should be two different ways to distribute substitutional donors and acceptors on its sublattices.

In *type* I DAP *spectra* both donors and acceptors are located on the same sublattice. For example, pairs such as S_P–Si_P, Se_P–Si_P or Si_{Ga}–Zn_{Ga} produce type I spectra.

In *type* II DAP *spectra* the donors and acceptors occupy different sublattices, such as S_P–Zn_{Ga} or O_P–Cd_{Ga}.

Since the lattice constant of GaP is known, one can calculate the *relative* number of DA pairs for a given separation R by assuming that the donors and acceptors are randomly distributed. Figure 7.5 shows the calculated distributions for both type I and II spectra in GaP. The horizontal scale is given in terms of m, the shell number for the neighboring pairs. This can be translated into the energy of the emitted photon by adding to $e^2/(4\pi\varepsilon_0\varepsilon_0 R_m)$ the appropriate energy $E_g - E_A - E_D$. Figure 7.6 shows the type I DAP spectra in GaP due to S_P–Si_P and Te_P–Si_P pairs measured at 1.6 K by *Thomas* et al.

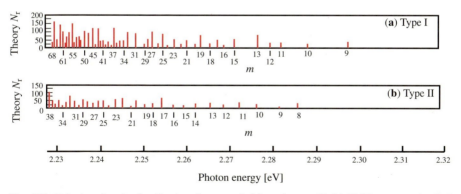

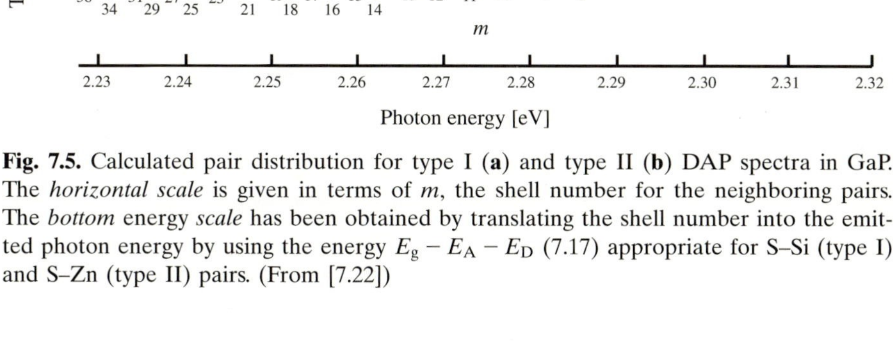

Fig. 7.5. Calculated pair distribution for type I (**a**) and type II (**b**) DAP spectra in GaP. The *horizontal scale* is given in terms of m, the shell number for the neighboring pairs. The *bottom* energy *scale* has been obtained by translating the shell number into the emitted photon energy by using the energy $E_g - E_A - E_D$ (7.17) appropriate for S–Si (type I) and S–Zn (type II) pairs. (From [7.22])

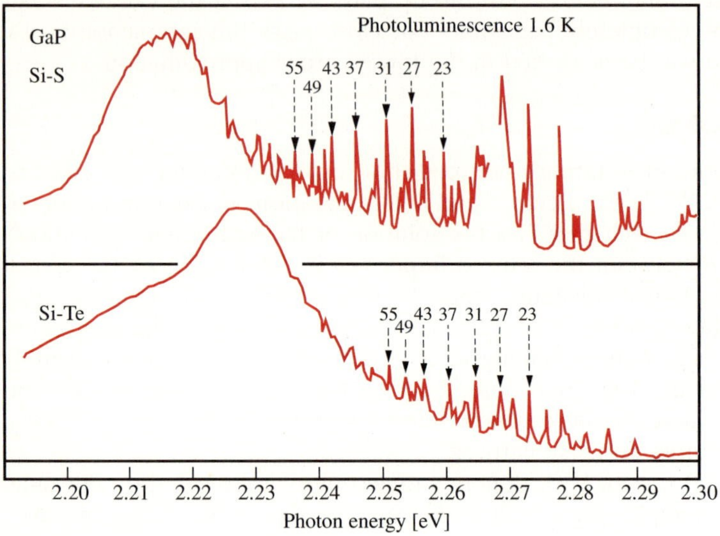

Fig. 7.6. DAP recombination spectra in GaP containing S–Si and Te–Si (type I) pairs measured at 1.6 K. The integers above the discrete peaks are the shell numbers of the pairs which have been identified by comparison with theoretical plots similar to those in Fig. 7.5. (From [7.22])

[7.22]. The numbers above the sharp peaks in the S-Si spectrum represent the shell numbers determined with the help of Fig. 7.5a. Figure 7.7 shows a type II spectrum in GaP due to S_P–Mg_{Ga} measured also at 1.6 K by *Dean* et al. [7.23]. The richness of information contained in the DAP spectra becomes obvious from these figures. In particular, one can determine the energy $E_A + E_D$ (taking the known low temperature values of the indirect bandgap

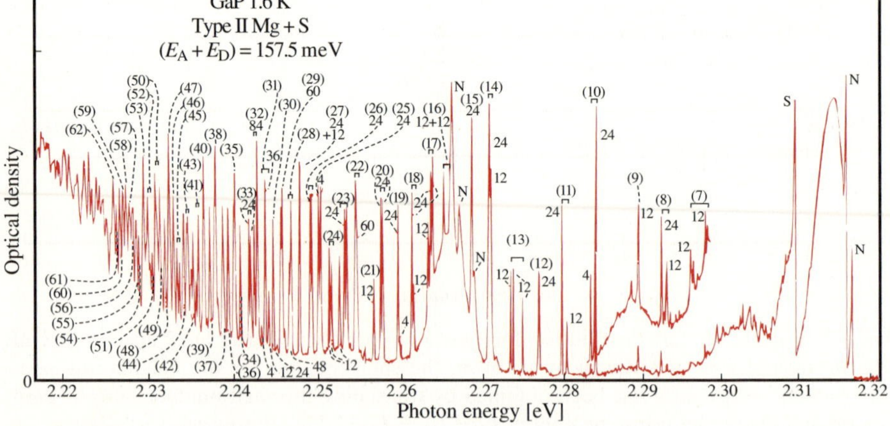

Fig. 7.7. Type II DAP spectra in GaP due to S_P–Mg_{Ga} pairs, measured at 1.6 K [7.23]

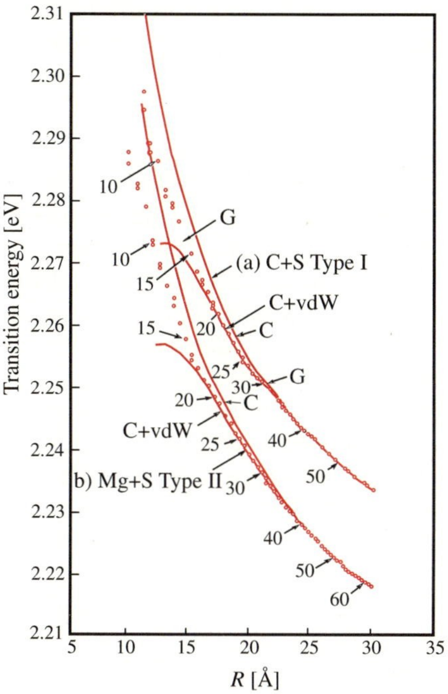

Fig. 7.8. Fit of experimental type I (S_P–C_C) and type II (S_P–Mg_{Ga}) DAP recombination peak energies in GaP with (7.17). The curves labeled C were fitted with (7.17). The curves labeled C+vdW were fitted to (7.17) including a van der Waals correction to the initial state energies. G denotes gaps in the Type I spectra while the numbers indicate the shell number. (From [7.23])

energy of GaP to be 2.339 ± 0.001 eV and ε equal to 10.75) with great precision by fitting the large number of observed peaks in the DAP spectra to (7.17). Figure 7.8 shows the fit of the type I (S_P–C_P) and type II (S_P–Mg_{Ga}) spectra in GaP. The curves labeled C were fitted directly to (7.17). The curves labeled C+vdW were fitted to (7.17) including a van der Waals correction to the initial state energies. These theoretical curves show that the van der Waals correction is significant only for pair separation less the 20 Å and also that the van der Waals approach over-corrects for the interaction between the neutral donor and acceptor. From these fits one can determine very accurately the difference in the binding energies of the two acceptors C and Mg to be $(E_A)_{Mg} - (E_A)_C = 5.6 \pm 0.3$ meV. Taking for the binding energy of the shallow donor S_P the known value of 104.2 ± 0.3 meV, we can determine the acceptor binding energies of Mg and C to be, respectively, 51.5 ± 1 meV and 48 ± 1 meV in GaP.

b) Temporal Evolution

The intensity of the DAP transitions and its temporal dependence also exhibits interesting properties [7.24]. Since the electron and the hole are spatially separated, their radiative recombination probability $(\tau_{DA})^{-1}$ depends on the overlap of their wavefunctions. This overlap depends exponentially on their separation as $\exp[-2(R/a_{D,A})]$, where $a_{D,A}$ is the larger of the Bohr radii of the donor and acceptor. From the discussions on donors and acceptors in Sects. 4.2.2

and 4.2.4, we find that the donor Bohr radius is usually larger in the tetrahedrally bonded semiconductors. Hence we shall assume that $a_{D,A} \approx a_D$ and

$$(1/\tau_{DA}) \propto \exp[-2(R/a_D)]. \qquad (7.18)$$

The calculation of the intensity distribution of DAP spectra is complicated because one needs to know the distributions of the excited donors and acceptors. These depend on the respective concentrations and on whether the impurities are distributed randomly or there is preferential pairing. They also depend on the level of excitation. For low levels of photoexcitation only a fraction of the donors and acceptors are excited and hence there will only be recombination from distant pairs. At high enough intensity, however, all donors and acceptors are excited (this is known as **saturation**) and therefore closer pairs also contribute to be recombination spectra. Thus one characteristic of DAP recombination is that the emission spectra *shift to higher energy* as the intensity of the excitation light is increased. This trend is just opposite to the effect of heating induced by higher light intensity because the bandgap in most tetrahedrally bonded semiconductors decreases with increasing temperature [6.147, 7.17], see also Fig. 6.44.

Another interesting characteristic of DAP recombination is its temporal dependence after excitation by a short pulse. If we assume that all the donors and acceptors are excited, the rate of recombination will be faster for the closer pairs with smaller R because of (7.18). These recombination peaks will also have higher photon energies according to (7.17). As a result, the DAP spectrum peaks initially at higher photon energies. As recombination depletes the number of neutral donors and acceptors their average separation increases and the recombination spectrum peak shifts towards lower energy. Hence the time decay of DAP spectra is non-exponential and the emission peaks redshift as time evolves. An example of the temporal behavior of DAP spectra in GaP is shown in Fig. 7.9.

The theoretical curves in Fig. 7.9 were calculated by *Thomas* et al. [7.24] with the following model. The concentration of acceptors was assumed to be larger than that of donors ($N_A > N_D$) and the impurities randomly distributed. Only emission from distant pairs was considered. Since the energy spacings between the emission peaks are very small, the acceptors were assumed to be arranged in spherical shells around the donor. At time $t = 0$ all the acceptors are assumed to be excited so that the number of neutral acceptors $N_A^0(0)$ was equal to N_A. After time t, $N_A^0(t)$ will have decreased because of recombination with neutral donors. The fraction of acceptors in a shell of radius R is proportional to R^2. Similarly, let $\langle Q(t)\rangle$ be the average probability for the electron to be on the donor at time t. Since the fraction of donors on a shell of radius R is also proportional to R^2, the total number of donor–acceptor pairs with distance R from each other at time t is proportional to $N_A^0(t)\langle Q(t)\rangle R^4$. If we define E' as the energy of the emission measured with respect to $E_g - E_A - E_D$, then E' is proportional to $1/R$ because

$$E' = \hbar\omega - (E_g - E_A - E_D) = e^2/(4\pi\varepsilon_0\varepsilon_0 R). \qquad (7.19)$$

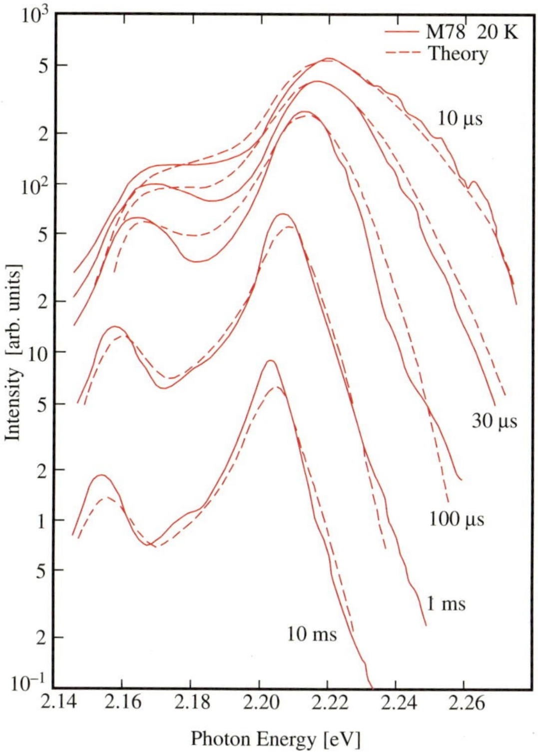

Fig. 7.9. Comparison between the measured time-resolved DAP spectra of GaP:S,C (*solid line*) and theory (*dashed line*). (From [7.24])

The number of donor–acceptor pairs contributing to the emission intensity $I_{\mathrm{DA}}(E',t)$ at time t is therefore proportional to $N_{\mathrm{A}}^{0}(t)\langle Q(t)\rangle(1/E')^4$. The probability that these DAP will recombine radiatively is given by (7.18), hence $I_{\mathrm{DA}}(E',t)$ is proportional to

$$I_{\mathrm{DA}}(E',t) \propto N_{\mathrm{A}}^{0}(t)\langle Q(t)\rangle(1/E')^4 \exp[-2e^2/(4\pi\varepsilon_0\varepsilon_0 E' a_{\mathrm{D,A}})]. \tag{7.20}$$

Assuming that the rate of decrease in the population of neutral acceptors is due entirely to recombination with neutral donors we can write [using (7.18)]

$$\begin{aligned} N_{\mathrm{A}}^{0}(t) &= N_{\mathrm{A}} \exp(-t/\tau_{\mathrm{DA}}) \\ &= N_{\mathrm{A}} \exp\left\{-(t/\tau_{\mathrm{DA}}^{0})\exp[-2e^2/(4\pi\varepsilon_0\varepsilon_0 E' a_{\mathrm{D,A}})]\right\}, \end{aligned} \tag{7.21}$$

where $1/\tau_{\mathrm{DA}}^{0}$ is the recombination rate for a DAP with $R = 0$. Combining (7.20) and (7.21) we obtain the following rather complicated expression for $I_{\mathrm{DA}}(E',t)$:

$$I_{\mathrm{DA}}(E',t) \propto N_{\mathrm{A}}\langle Q(t)\rangle\left(\frac{1}{E'}\right)^4 \exp\left[-\frac{2e^2}{4\pi\varepsilon_0\varepsilon_0 E' a_{\mathrm{D,A}}} - \left(\frac{t}{\tau_{\mathrm{DA}}^{0}}\right)\exp\left(-\frac{2e^2}{4\pi\varepsilon_0\varepsilon_0 E' a_{\mathrm{D,A}}}\right)\right]. \tag{7.22}$$

Figure 7.9 shows that the experimental spectra agree quite well with the predictions of (7.22).

7.1.4 Excitons and Bound Excitons

In photoluminescence experiments on high purity and high quality semiconductors at low temperatures, we expect the photoexcited electrons and holes to be attracted to each other by Coulomb interaction and to form excitons. As a result, the emission spectra should be dominated by radiative annihilation of excitons producing the so-called free exciton peak. When the sample contains a small number of donors or acceptors in their neutral state (a common occurrence at low temperature) the excitons will be attracted to these impurities via van der Waals interaction. Since this attraction lowers the exciton energy, neutral impurities are very efficient at trapping excitons to form **bound excitons** at low temperature. Figure 7.10 shows the low temperature PL spectra of GaAs measured by *Sell* et al. [7.25]. The peak labeled (D^0, X) is due to recombination of an exciton bound to a neutral donor, while the peaks in the inset are attributed to free excitons. We shall now consider these two types of emission peaks in more detail.

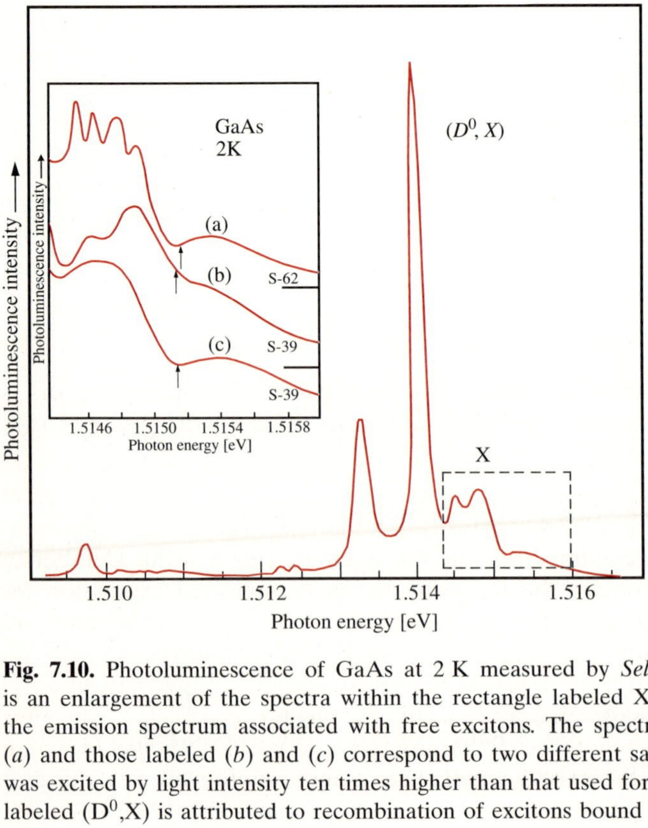

Fig. 7.10. Photoluminescence of GaAs at 2 K measured by *Sell* et al. [7.25]. The *inset* is an enlargement of the spectra within the rectangle labeled X. It contains the part of the emission spectrum associated with free excitons. The spectrum in the inset labeled (*a*) and those labeled (*b*) and (*c*) correspond to two different samples. The spectrum (*c*) was excited by light intensity ten times higher than that used for spectrum (*b*). The peak labeled (D^0,X) is attributed to recombination of excitons bound to neutral donors

a) Free-Exciton Emission

In principle, we should consider radiative recombination of excitons in terms of exciton-polaritons (Sect. 6.3.2) or polaritons for brevity. To understand the importance of the polariton approach we shall first neglect this effect and examine how well the theoretical results agree with experiment. Within this approximation, the emission process is simply a radiative decay of excitons into photons. Since wavevector has to be conserved in this process only excitons with wavevector $\boldsymbol{k}$ equal to the photon wavevector (i. e., $\boldsymbol{k} \approx 0$) can convert to photons. The emission spectra should be essentialy a delta function at the energy of the exciton ground state when damping is neglected. When exciton lifetime broadening is included, the emission spectrum becomes a Lorentzian. This conclusion disagrees with the experimental results in most high quality samples at low temperature. Often the observed free exciton emission spectra have an asymmetric lineshape quite different from a Lorentzian. As an illustration, we show in Fig. 7.11 excitonic emission spectra in four typical semiconductors with bandgaps ranging from around 1.5 eV (GaAs) to over 3 eV (CuCl) [7.26]. In no case does the emission spectrum resemble a Lorentzian. Instead, all the spectra exhibit an asymmetrical peak plus a higher energy shoulder.

Toyozawa [7.27] first pointed out that luminescence spectra in semiconductors at low temperature should be interpreted in terms of polaritons. Within this picture, PL involves the conversion of external photons, on entering a medium, into excitonic polaritons. These polaritons relax towards lower energies by scattering with phonons via their exciton component. Their photon part has a very weak interaction with phonons. This relaxation process randomizes their distribution. Some of the polaritons will be scattered backwards to emerge from the sample as luminescence photons. Since the polariton dispersion curve shown in Fig. 6.22 has no energy minimum where relaxation processes normally terminate, there is no *a priori* reason to assume that polaritons will attain thermal equilibrium via scattering processes. Thus one may expect the polariton emission spectra to show no peaks at all! Even if there is a peak, its width can be much larger than predicted by the sample temperature. *Toyozawa* [7.27] pointed out that polaritons could accumulate at a "*bottleneck*" near the *transverse exciton energy* ($E_T = \hbar\omega_T$ in Fig. 6.22, see also Problem 6.9) where their lifetimes are longest. The lower polaritons above this bottleneck possess a large exciton component [7.28] and therefore have short lifetimes as a result of strong scattering by phonons. These phonon scattering rates decrease as polaritons become more photon-like. On the other hand, once their energies decrease below E_T polaritons are short-lived again because they now have high *group velocities* (defined as $d\omega/dk$, i. e., by the slopes of the polariton dispersion curves such as those in Fig. 6.22) and can easily escape from the sample as photons. Thus the polariton distribution function can have a peak near its bottleneck. The emission spectrum represents the product of this distribution function and the transmission coefficient of polaritons at the sample surface. Unfortunately, polariton transmission coefficients cannot be calculated simply by using Maxwell's boundary conditions [7.29] only. Since

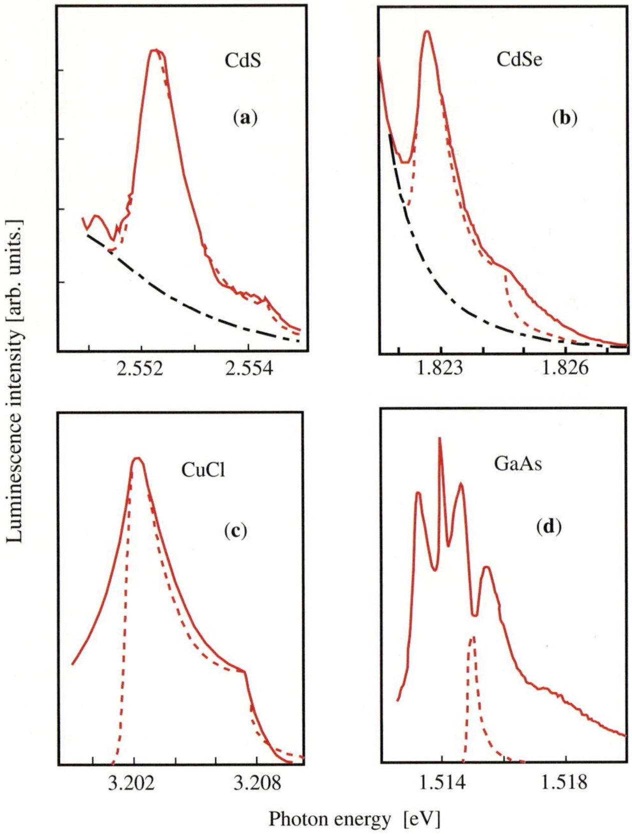

Fig. 7.11a–d. Comparison between experimental and calculated polariton emission spectra in four semiconductors. The *solid curves* are experimental spectra while the red *dashed curves* were calculated [7.26] using a two-branch polariton model with the Pekar ABC. The *black dashed-dotted curves* represent a bound exciton background

excitons are involved, it is necessary to introduce **additional boundary conditions** (known as ABCs) to describe the behavior of excitons near the sample surface. There have been many theoretical treatments of the ABC problem (e. g. [7.30–33]) and a detailed description of these theories is beyond the scope of this book. The whole question of which ABC to choose for a particular sample is still unresolved since it will presumably depend on the details and quality of the sample surface [7.34].

Polariton luminescence spectra have been computed by *Askary* and *Yu* [7.26] using two different types of ABCs. Figure 7.12a shows the calculated polariton distributions in the lower branch (abbreviated as LB in the figure) of CdS for the two ABCs. A large peak occurs at the bottleneck near the transverse exciton energy as predicted by *Toyozawa*. The corresponding PL spectra including the upper branch (labeled UB) and the transmission coeffi-

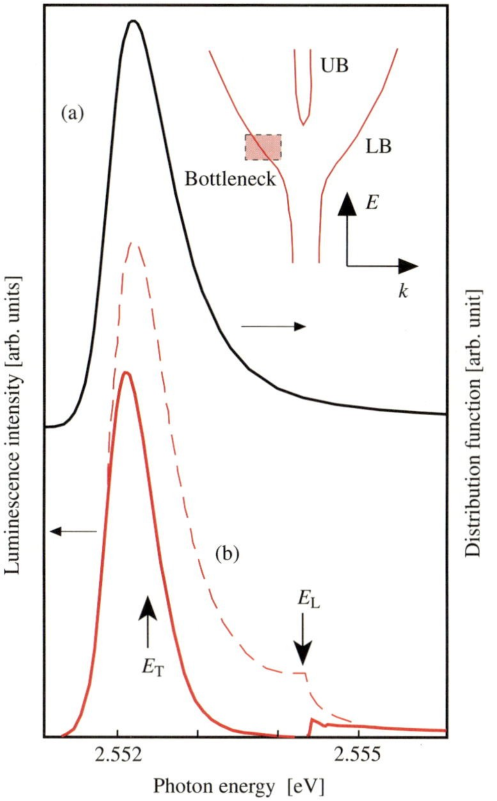

Fig. 7.12. Theoretical polariton population (*a*) in the lower polariton branch and (*b*) polariton luminescence spectra calculated with two different ABCs. The broken curves correspond to the Pekar ABC [7.30]. The inset shows schematically the two polariton branches included in the model. The material parameters used in the calculation are appropriate for the A exciton in CdS (From [7.26])

cients are shown in Fig. 7.12b. The PL peak can be explained by the large polariton population at the bottleneck and is relatively insensitive to the ABCs. The higher energy shoulder visible in the experimental spectra in Fig. 7.11 can now be identified with a change in the transmission coefficients near the longitudinal exciton energy ($E_L = \hbar\omega_L$ in Fig. 6.22, see also Problem 6.9) and is very sensitive to the ABC chosen. The theoretical PL spectra calculated by *Askary* and *Yu* using the ABC proposed by *Pekar* [7.30] are compared with the experimental spectra in Fig. 7.11. Note that a background due to emission associated with defects and possibly phonon emission has been added to some of the theoretical spectra to achieve quantitative agreement with experiment. The agreement between theory and experiment in Fig. 7.11 is quite good except for GaAs. Instead of a peak at E_T the experimental spectra exhibited a dip. More recent experiments [7.35] showed that the minimum in GaAs was caused by scattering with impurities such as donors. Since the polariton group velocity has a minimum near E_T, its scattering probability with defects is maximum at this energy. This dip in the PL spectrum is absent in purer GaAs epi-layers. For recent work on polaritons in GaAs and ABC's see [7.34].

b) Bound-Exciton Emission

In Fig. 7.10 we have already seen that the emission spectrum of GaAs at 2 K is dominated by a strong peak occurring at an energy slightly below the free exciton energy. This sharp peak was identified with recombination of an exciton bound to a neutral donor atom, usually denoted by (D^0X). In addition to neutral donors, an exciton can also bind to a neutral acceptor forming the complex (A^0X). Figure 7.13 shows the bound exciton emission spectrum of the wurtzite-type crystal CdS at low temperature [7.36]. The very sharp peaks labeled I_1 and I_2 correspond to recombination of (A^0X) and (D^0X), respectively. The nature of both excitons was established by their splittings under a magnetic field (from which the g-values of the mobile particles can be determined). The many phonon sidebands of these bound excitons present in the emission spectrum suggest that the interaction between the bound charges and phonons is enhanced. These bound excitons can be considered as analogs of the hydrogen molecule H_2 except for the different binding energies. Bound excitons have smaller binding energies because the hole mass is much smaller than that of the proton. The binding energy of the hydrogen molecule is known to be equal to 4.75 eV [7.37]. Hence the ratio

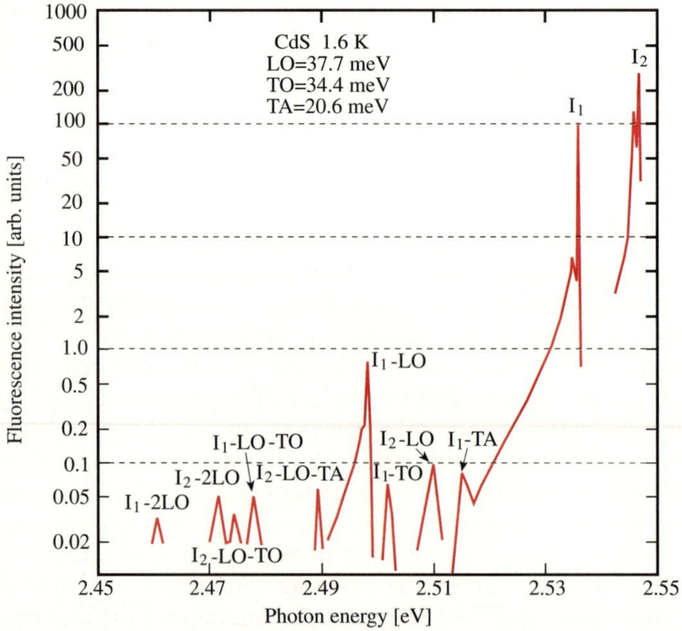

Fig. 7.13. Emission spectrum of a high quality CdS crystal at 1.6 K showing the zero-phonon bound exciton recombination peaks I_1 and I_2 and their phonon sidebands. The energies of the transverse optical (TO), longitudinal optical (LO) and zone-edge transverse acoustic (TA) phonons deduced from the data are listed in the figure. (From [7.36])

of the binding energy of H_2 to that of a single electron in the hydrogen atom is 4.75 eV/13.6 eV = 0.35. The ratio of the binding energy of bound excitons to that of a free exciton will depend on the ratio (r) of the hole effective mass (m_h^*) to the electron effective mass (m_e^*). This dependence has been estimated by *Hopfield* [7.38] and the results are shown in Fig. 7.14. In general, as r decreases the bound exciton binding energies also decrease.

Excitons can also bind to ionized impurities, forming bound excitons denoted as (D^+X) and (A^-X). The former can be regarded as an analog of the hydrogen molecule ion $H_2{}^+$. The ratio of the $H_2{}^+$ binding energy (2.6 eV) to that of the hydrogen atom is 0.2. At first sight it seems that (D^+X) can also be regarded as a hole bound to a neutral donor, to be denoted as (D^0h). The more appropriate picture depends on which state has a larger binding energy. Let us define E_I as the energy required to remove both electron and hole from (D^+X) or (D^0h) leaving behind the ion D^+. In the (D^+X) picture E_I is given by the sum of $E_{(D^+X)}$, the binding energies of (D^+X), and the ionization energy of the exciton E_X:

$$E_I = E_{(D^+X)} + E_X. \tag{7.23}$$

In the other picture E_I is equal to

$$E_I = E_{(D^0h)} + E_D, \tag{7.24}$$

where $E_{(D^0X)}$ and E_D are, respectively, the binding energies of (D^0h) and that of the electron to the donor ion. Equating (7.23) and (7.24) we obtain

$$E_{(D^+X)} + E_X = E_{(D^0h)} + E_D. \tag{7.25}$$

For a semiconductor with dielectric constant ε, $E_D = 13.6m_e^*/(m_0\varepsilon^2)$ eV while $E_X = 13.6\mu/(m_0\varepsilon^2)$ eV. Since $m_e^* > \mu$ we find from (7.25) that $E_{(D^+X)} > E_{(D^0h)}$

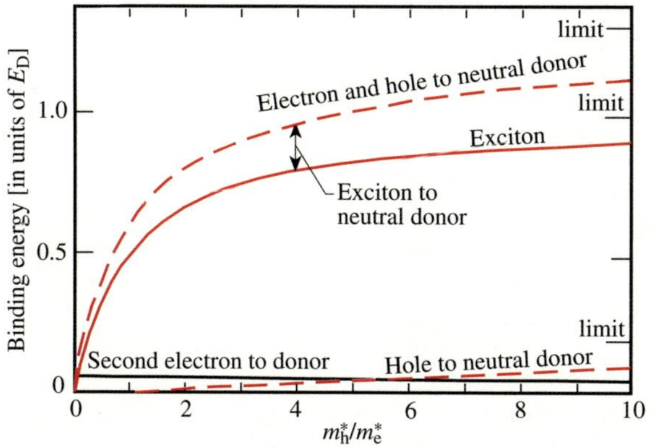

Fig. 7.14. The binding energies of various exciton complexes relative to the donor binding energy shown as a function of the mass ratio (m_h^*/m_e^*). (From [7.38])

and therefore the bound exciton (D^+X) represents the correct way to describe this complex.

Hopfield [7.38] had estimated that the (D^+X) would not be bound for $r < 1.4$. As a result, excitons cannot be bound to both ionized donors and acceptors in the same material. When r is larger than 1.4, (D^+X) will be stable but, since $(1/r)$ is less than 1.4, (A^-X) wil be unstable. In most semiconductors $r > 1$ and therefore no PL peaks have been attributed to (A^-X). Figure 7.15 shows the (D^+X) recombination peak (labeled I_3) in the emission spectra in CdS at low temperature. In order to increase the population of ionized donors at low temperature the sample has to be illuminated with infrared light to photoionize the donors. The label I.R. in Fig. 7.15 denotes the voltage applied to the infrared source. Notice how the intensity of the I_3 peak increases with the intensity of the infrared source at the expense of the I_2 (D^0X) bound exciton peak.

Finally one may ask if there is an analog of the hydrogen atom ion H^- containing two electrons moving around a proton. The binding energy of the second electron in H^- is only 0.75 eV [7.39]. The impurity analogs to H^- are an electron bound to a neutral donor (D^0e) or D^- and a hole bound to a neutral acceptor (A^0h) or A^+. Obviously these must be very weakly bound states. The D^- state has been observed in n-type Si [7.40] with a thermal ionization energy of the order of 1 meV. More recently the binding energy of D^- has been found to become enhanced in two dimensions, especially under high magnetic fields [7.41].

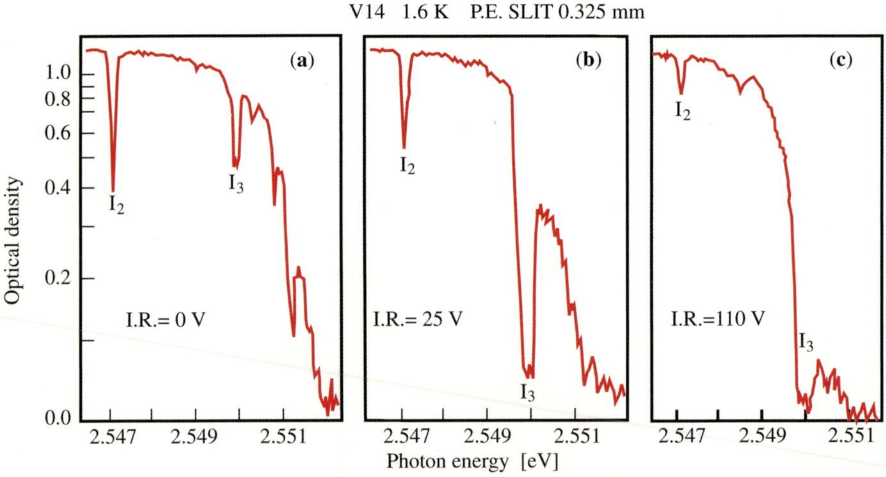

Fig. 7.15a–c. Emission spectrum of a high quality CdS crystal at 1.6 K showing the zero-phonon bound exciton recombination peaks I_2 and I_3 as a function of the IR radiation intensity. The infrared radiation ionizes shallow donors with the result that excitons bound to neutral donors (I_2) are quenched while excitons bound to ionized donors (I_3) are enhanced. I.R. stands for the voltage applied to the infrared source. Thus the intensity of the infrared radiation increases from (**a**) to (**c**). (From [7.36])

In the case of bound excitons (D^+X) and (A^-X) an electron–hole pair is bound to immobile charged impurities. In the negatively charged hydrogen ion H^- two electrons are bound to a heavy (albeit mobile) proton. From these considerations one may ask whether it will be possible to have a "three-carrier" complex consisting of either two electrons and one hole or two holes and one electron. These complexes, known as **trions**, can be considered as charged excitons. Trions were first proposed by Lampert in 1958 [7.42] as analogs of the positively charged hydrogen molecules H_2^+ which contain two positive charges and one electron. Lampert suggested that a trion of two holes and one electron be abbreviated as X_2^+. Nowadays trions are usually considered as being closer to a positively charged exciton and therefore are abbreviated as X^+ and X^-. From the discussion on the binding energy of bound excitons in the last section it is obvious that the binding energy of trions would be even smaller than those of excitons bound to charged impurities. Under normal circumstances it would be extremely difficult to observe trions. Indeed trions were not observed in semiconductors until recently when two developments made this possible. The first development is the fabrication of thin layers of semiconductors sandwiched between semiconductor layers of larger band gap to form quantum wells (see p. 5 and also Chap. 9). The confinement of electrons and holes within such quantum wells to form two-dimensional excitons increases their binding energy [see (6.95)] when compared to three-dimensional excitons. The other development is the ability to modulation-dope (see Sect. 5.3) the quantum wells so that excitons can be bound to free carriers without competition from charged impurities. Trions (of the X^- variety) were experimentally observed in modulation doped quantum wells of CdTe by Finkelstein et al. [7.43]. Since then both X^- and X^+ have been found in GaAs quantum wells and in other II-VI semiconductor quantum wells. Further details can be found in a review paper by Cox et al. [7.44].

For a review of luminescence in gallium nitride see [1.1].

7.1.5 Luminescence Excitation Spectroscopy

With the availability of continuously tunable lasers, such as those based on liquid dyes [7.45], color centers in alkali halides [7.46], or sapphire doped with titanium (abbreviated as Ti: sapphire laser) [7.47, 48], a new kind of emission spectroscopy has become possible. In this technique the spectrometer is set to detect emission of a particular photon energy from the sample. The intensity of this emission is then recorded as a function of the excitation photon energy. This technique is known as **photoluminescence excitation spectroscopy** or PLE. It has become very popular for studying thin epilayers grown on opaque bulk substrates. It is often assumed that the PLE spectrum is roughly equivalent to the absorption spectrum of the sample. Since it is difficult to remove the substrate from an epilayer so as to be able to perform absorption measurements, PLE has become accepted as a simple alternative. In this section we shall examine the conditions under which this assumption may be valid.

As we discussed in Sect. 7.1.1, the photoexcited e-h pairs relax during photoluminescence towards lower energy states and reach quasi-thermal equilibrium with the lattice. We expect the e-h pairs to "forget" how they were excited during this relaxation. Therefore the emission intensity should not necessarily have any correlation with the absorption coefficient. To analyze this question more quantitatively, however, let us write the relation between the emission intensity I_{em} and the excitation intensity I_{ex} as

$$I_{em} = P_{abs} P_{rel} P_{em} I_{ex}. \tag{7.26}$$

In (7.26) P_{abs}, P_{rel}, and P_{em} denote, respectively, the probability of the incident photon being absorbed by the sample, the probability that the photoexcited e-h pairs will relax to the emitting state, and their probability of radiative recombination after relaxation. While P_{em} can be assumed to be a constant in a PLE experiment P_{rel} depends strongly on the e-h pair energies. In defect-free semiconductors, excited e-h pairs relax predominantly via electron–phonon interaction. However, when many defects are present the majority of e-h pairs are trapped by defects and recombine nonradiatively. The probabilities for both processes depend on the electron energy. As a result, it is usually not possible to correlate I_{em} with P_{abs} without some knowledge of the relative magnitudes of the defect trapping rate and the electron–phonon relaxation rate. One exceptional case is a semiconductor in which nonradiative recombination is negligible compared with scattering by electron–phonon interaction. Examples of such high quality materials are quantum wells (QWs, see Chaps. 1 and 9 for more details) based on GaAs and related III–V semiconductors. As we have seen in Chaps. 3 and 5, energetic electrons in these semiconductors relax predominantly by scattering with LO phonons (via the Fröhlich interaction) and acoustic phonons (via the piezoelectric or deformation potential interaction). These scattering processes occur on picosecond and subpicosecond time scales, which are significantly shorter than the radiative lifetime (Sect. 7.1.1). In high quality QW samples the nonradiative lifetimes are often also long enough that P_{rel} can be almost unity and independent of e-h energy. Thus in these samples at low temperature, one finds a good correspondence between the PLE and absorption spectra.

As an illustration, we show in Fig. 7.16a the absorption spectra in a 20 nm GaAs/GaAlAs QW measured at low temperature [7.49]. Notice the existence of several sharp peaks which have been identified as excitonic transitions associated with the quantized levels in the well (see Chap. 9). They are labeled according to the quantum number $n = 1, 2, 3$ and 4. Some of these transitions, such as $n = 1$ and 2, exhibit a doublet structure due to the splitting of the heavy and light hole valence bands caused by the confinement potential (more details on this effect can be found in Chap. 9). Figure 7.16b shows the PLE of a different but equally high quality QW consisting of a wide GaAs well (width 10 nm) separated by a AlGaAs barrier from a narrower well (width 5 nm) [7.50]. The broken curves are the PL spectra of the wide (upper panel) and narrow wells (lower panel). When the spectrometer was set to detect the PL

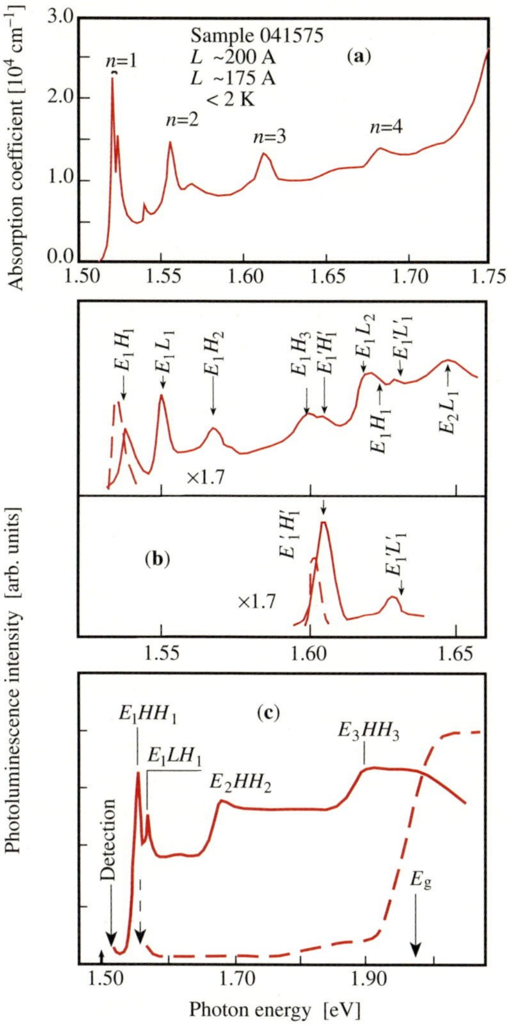

Fig. 7.16a–c. Comparison between absorption and PLE spectra in GaAs/AlGaAs quantum well (QW) samples measured at low temperature. (**a**) Absorption spectra in a 20 nm well showing structures due to excitonic absorption peaks (labeled $n = 1, 2$, etc.) [7.49]. (**b**) PLE spectra from a sample containing two QWs with widths of 5 and 10 nm (shown in *lower* and *upper panels* respectively). The *broken curves* show the PL from these two wells. The primes on E and H label structures belonging to the 5 nm well [7.50]. (**c**) PLE spectra from a p-doped well. The broken curve was obtained by setting the spectrometer to admit only excitonic emission (indicated by the *broken vertical arrow*). The *vertical arrow* labeled E_g marks the energy gap of the AlGaAs barrier layer. The *solid curve* is the PLE spectrum obtained when the spectrometer was set at the energy of the electron-to-neutral-acceptor emission (indicated by the *solid vertical arrow labeled "Detection"*) [7.51]

from one of the two wells, the PLE spectra (solid curves) in Fig. 7.16b were obtained. In the upper panel the spectrometer was set to record only the PL from the wider well. All the prominent structures in the PLE spectrum can be identified with optical transitions to excited (excitonic) states in this well. The lower panel shows the PLE (solid curve) spectrum of the narrow well. There is good overall resemblance in shape between these PLE spectra and the absorption spectrum in Fig. 7.16a even though they were measured in QW of different widths. However, this is not always the case. Figure 7.16c shows the PLE of another GaAs/AlGaAs QW sample which contains shallow acceptors. The broken curve represents the PLE when the spectrometer was set to detect only the lowest excitonic emission (indicated by the vertical broken arrow). In this case no structure due to excited excitonic states was observed. The rising edge in the vicinity of the arrow labeled E_g was assigned to the bandgap of the AlGaAs barrier. On the other hand, when emission produced by the capture of electrons at neutral acceptors was detected (energy indicated by the solid arrow labeled "Detection"), the PLE spectrum (solid curve) looked like the absorption spectrum in Fig. 7.16a. The explanation for two completely different PLE spectra in the same sample lies in their different P_{rel}. In this doped sample excitons are efficiently trapped by neutral acceptors. As a result P_{rel} is ≈ 1 for relaxation into bound excitons only. Hence, while the PLE of the bound exciton emission resembles the absorption spectrum the PLE of the free exciton does not.

Another sample of the application of PLE can be found in *inhomogeneously broadened* systems. As an example we shall consider a typical donor–acceptor pair recombination spectrum which contains many overlapping peaks associated with different pair separations. The width of the emission from a pair with a well-defined separation is known as its **homogeneous linewidth**, and often can be very narrow. However, due to the distribution of many defect pairs with slightly different separations, many of these sharp peaks overlap slightly and form a much broader band whose width is referred to as the **inhomogeneous linewidth**. This is not the only cause of inhomogeneous broadening. For example, emissions from gas molecules are broadened by the Doppler effect associated with the random motion of these molecules. Multiple quantum wells with a distribution of well widths are other examples of systems whose emission spectra are inhomogeneously broadened.

Figure 7.17a shows the low-temperature DAP bands (peaks labeled P and Q) of ZnSe containing two shallow acceptors: Li and Na [7.52]. These peaks are featureless and have widths of the order of several meV. From these spectra alone it would not be possible to identify the acceptors or to determine their energy levels. *Tews* et al. [7.52] overcame this problem by using selective excitation and PLE spectroscopy. When they set the spectrometer to admit a narrow band of photon energies (< 1 meV) centered at 2.705 eV located on the high-energy side of the Li-related Q band, they found that the PLE spectrum (Fig. 7.17b) showed considerable structure with one peak (labeled Li: $2P_{3/2}$) as narrow as 0.5 meV. They assigned it, together with another peak labeled Li: $2S_{3/2}$, to excitation of the neutral acceptor into its excited states

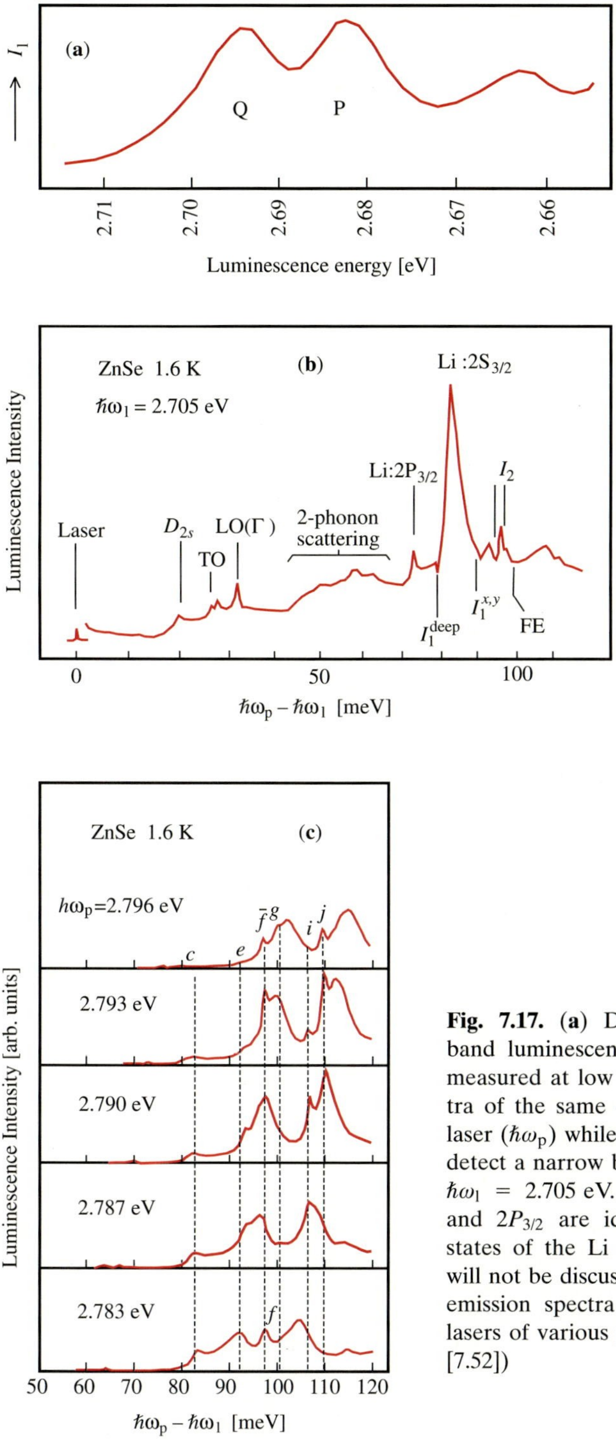

Fig. 7.17. (**a**) Donor–acceptor pair (DAP) band luminescence spectrum in ZnSe:Na,Li measured at low temperature. (**b**) PLE spectra of the same obtained by scanning a dye laser ($\hbar\omega_p$) while the spectrometer was set to detect a narrow band of photons with energy $\hbar\omega_1 = 2.705$ eV. The peaks labeled Li:$2S_{3/2}$ and $2P_{3/2}$ are identified as due to excited states of the Li acceptors. The other peaks will not be discussed in this chapter. (**c**) DAP emission spectra selectively excited by dye lasers of various photon energies $\hbar\omega_p$. (From [7.52])

(the remaining peaks in this PLE spectrum are not of interest here). These peaks are sharp because the detector selected out emission from only a small subset of acceptors. These acceptors satisfy the condition that their separations R from the donor are determined by the emission energy via (7.17). When the incident photon excites these acceptors to their $2S_{3/2}$ and $2P_{3/2}$ states (see Sect. 4.2.4 for the acceptor energy levels in zinc-blende-type semiconductors) the pair emission produced by transitions from their $1S_{3/2}$ ground state to the donor becomes enhanced. This process is shown schematically in Fig. 7.18. Since only pairs with a selected separation are enhanced, the DAP band exhibits a sharp peak superimposed on a broader background due to the non-selectively excited pairs. These DAP emission spectra under selective excitations are shown in Fig. 7.17c. Notice that in Fig 7.17 the bands of the PL

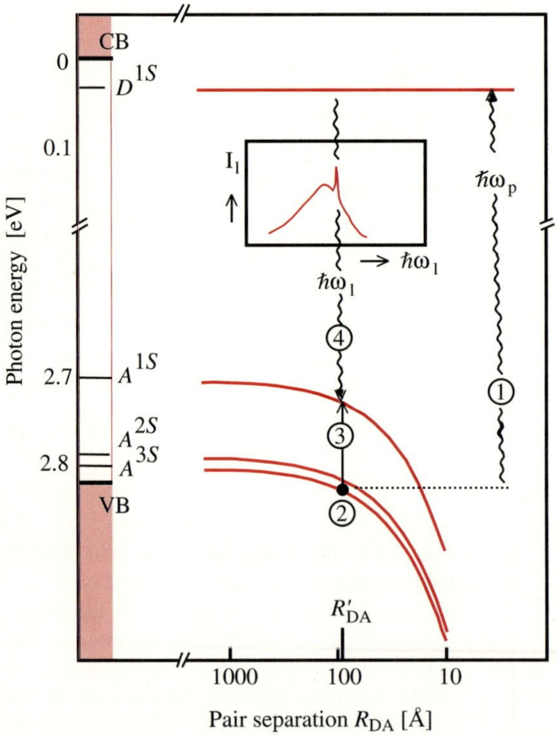

Fig. 7.18. Schematic diagram showing the selective excitation process responsible for the sharp emission peaks in the DAP spectra in Fig. 7.17c. The pump photon of energy $\hbar\omega_p$ (labeled *1*) excites resonantly an electron from the 3S excited state of a neutral acceptor (labeled *2*) to the 1S ground state of a donor a distance R'_{DA} away. Afterwards the acceptor hole relaxes to the 1S ground state (labeled *3*) before recombining with the electron on the donor at distance R'_{DA}, emitting the photon $\hbar\omega_l$ (labeled *4*). The *inset* shows how this selectively excited DAP emission forms a sharp peak superimposed on the broad background due to DAP emission from many other pairs with different pair separation

spectra are plotted as a function of the difference between the laser photon energy $\hbar\omega_p$ and the emission photon energy $\hbar\omega_l$. A number of sharp peaks labeled c, e, f, g, i and j are observed depending on the excitation energy. Their positions correspond to the separation between the ground and excited states of the neutral acceptors. The peaks c and e are assigned to Li acceptors while the remaining peaks are identified with Na acceptors. Their widths are now determined by homogeneous broadening only. From these selectively excited DAP spectra, *Tews* et al. [7.52] have determined accurately the binding energy and excited state energies of Na and Li acceptors in ZnSe.

7.2 Light Scattering Spectroscopies

Although most of the light traveling through a medium is either transmitted or absorbed following the standard laws of reflection and refraction (which obtain from $\boldsymbol{k}$-conservation), a very tiny fraction is scattered, in all directions, by inhomogeneities inside the medium. These inhomogeneities may be static or dynamic. Defects such as dislocations in a crystal are static scatterers and scatter the light elastically (i. e., without frequency change). Fluctuations in the density of the medium that are associated with atomic vibrations are examples of dynamic scatterers. Other examples of scattering mechanisms in semiconductors are fluctuations in the charge or spin density. Inelastic scattering of light by acoustic waves was first proposed theoretically by *Brillouin* [7.53] and later independently by *Mandelstam* [7.54]. Inelastic scattering of light by molecular vibrations was first reported by *Raman* [7.55]. In 1930 *Raman* was awarded the Nobel prize for his discovery of Raman scattering. Today Raman scattering and resonant Raman scattering have become standard spectroscopic tools in the study of semiconductors. In this section we shall first present a macroscopic theory of Raman scattering by phonons in solids. This is followed by a microscopic theory and a discussion of resonant Raman scattering. The rest of the section is devoted to discussions of Brillouin scattering by acoustic modes and *resonant Brillouin scattering* by exciton-polaritons.

7.2.1 Macroscopic Theory of Inelastic Light Scattering by Phonons

Consider an infinite medium with electric susceptibility χ. As shown in Sect. 6.1, the electrical susceptibility should be a second rank tensor in general. For the time being we shall assume the medium to be isotropic so that χ can be represented by a scalar. When a sinusoidal plane electromagnetic field described by

$$\boldsymbol{F}(\boldsymbol{r},t) = \boldsymbol{F}_i(\boldsymbol{k}_i,\omega_i)\cos(\boldsymbol{k}_i\cdot\boldsymbol{r}-\omega_i t) \tag{7.27}$$

is present in this medium, a sinusoidal polarization $\boldsymbol{P}(\boldsymbol{r},t)$ will be induced:

$$\boldsymbol{P}(\boldsymbol{r},t) = \boldsymbol{P}(\boldsymbol{k}_\mathrm{i},\omega_\mathrm{i})\cos(\boldsymbol{k}_\mathrm{i}\cdot\boldsymbol{r}-\omega_\mathrm{i}t). \tag{7.28}$$

Its frequency and wavevector are the same as those of the incident radiation while its amplitude is given by

$$\boldsymbol{P}(\boldsymbol{k}_\mathrm{i},\omega_\mathrm{i}) = \chi(\boldsymbol{k}_\mathrm{i},\omega_\mathrm{i})\boldsymbol{F}_\mathrm{i}(\boldsymbol{k}_\mathrm{i},\omega_\mathrm{i}). \tag{7.29}$$

If the medium is at a finite temperature there are fluctuations in χ due to thermally excited atomic vibrations. We have seen in Chap. 3 that the normal modes of atomic vibrations in a crystalline semiconductor are quantized into phonons. The atomic displacements $\boldsymbol{Q}(\boldsymbol{r},t)$ associated with a phonon can be expressed as plane waves:

$$\boldsymbol{Q}(\boldsymbol{r},t) = \boldsymbol{Q}(\boldsymbol{q},\omega_0)\cos(\boldsymbol{q}\cdot\boldsymbol{r}-\omega_0 t) \tag{7.30}$$

with wavevector $\boldsymbol{q}$ and frequency ω_0. These atomic vibrations will modify χ. We assume that the characteristic electronic frequencies which determine χ are much larger than ω_0, hence χ can be taken to be a function of $\boldsymbol{Q}$. This is known as the quasi-static or adiabatic approximation. Normally the amplitudes of these vibrations at room temperature are small compared to the lattice constant and we can expand χ as a Taylor series in $\boldsymbol{Q}(\boldsymbol{r},t)$:

$$\chi(\boldsymbol{k}_\mathrm{i},\omega_\mathrm{i},\boldsymbol{Q}) = \chi_0(\boldsymbol{k}_\mathrm{i},\omega_\mathrm{i}) + (\partial\chi/\partial\boldsymbol{Q})_0\boldsymbol{Q}(\boldsymbol{r},t) + \dots, \tag{7.31}$$

where χ_0 denotes the electric susceptibility of the medium with no fluctuations. The second term in (7.31) represents an oscillating susceptibility induced by the lattice wave $\boldsymbol{Q}(\boldsymbol{r},t)$. Substituting (7.31) into (7.29) we can express the polarization $\boldsymbol{P}(\boldsymbol{r},t,\boldsymbol{Q})$ of the medium in the presence of atomic vibrations as

$$\boldsymbol{P}(\boldsymbol{r},t,\boldsymbol{Q}) = \boldsymbol{P}_0(\boldsymbol{r},t) + \boldsymbol{P}_\mathrm{ind}(\boldsymbol{r},t,\boldsymbol{Q}), \tag{7.32}$$

where

$$\boldsymbol{P}_0(\boldsymbol{r},t) = \chi_0(\boldsymbol{k}_\mathrm{i},\omega_\mathrm{i})\boldsymbol{F}_\mathrm{i}(\boldsymbol{k}_\mathrm{i},\omega_\mathrm{i})\cos(\boldsymbol{k}_\mathrm{i}\cdot\boldsymbol{r}-\omega_\mathrm{i}t) \tag{7.33}$$

is a polarization vibrating in phase with the incident radiation and

$$\boldsymbol{P}_\mathrm{ind}(\boldsymbol{r},t,\boldsymbol{Q}) = (\partial\chi/\partial\boldsymbol{Q})_0\boldsymbol{Q}(\boldsymbol{r},t)\boldsymbol{F}_\mathrm{i}(\boldsymbol{k}_\mathrm{i},\omega_\mathrm{i})\cos(\boldsymbol{k}_\mathrm{i}\cdot\boldsymbol{r}-\omega_\mathrm{i}t) \tag{7.34}$$

is a polarization wave induced by the phonon (or other similar fluctuation). Polarization waves can also be induced indirectly by longitudinal optical (LO) phonons via their macroscopic electric fields (Sect. 6.4). For the time being we shall neglect this effect.

To determine the frequency and wavevector of $\boldsymbol{P}_\mathrm{ind}$ we rewrite $\boldsymbol{P}_\mathrm{ind}(\boldsymbol{r},t,\boldsymbol{Q})$ as

$$\begin{aligned}\boldsymbol{P}_\mathrm{ind}(\boldsymbol{r},t,\boldsymbol{Q}) &= (\partial\chi/\partial\boldsymbol{Q})_0\boldsymbol{Q}(\boldsymbol{q},\omega_0)\cos(\boldsymbol{q}\cdot\boldsymbol{r}-\omega_0 t)\\ &\quad\times\boldsymbol{F}_i(\boldsymbol{k}_\mathrm{i},\omega_i)\cos(\boldsymbol{k}_\mathrm{i}\cdot\boldsymbol{r}-\omega_\mathrm{i}t)\end{aligned} \tag{7.35a}$$

$$\begin{aligned}&= \tfrac{1}{2}(\partial\chi/\partial\boldsymbol{Q})_0\boldsymbol{Q}(\boldsymbol{q},\omega_0)\boldsymbol{F}_\mathrm{i}(\boldsymbol{k}_\mathrm{i},\omega_\mathrm{i}t)\\ &\quad\times\{\cos[(\boldsymbol{k}_\mathrm{i}+\boldsymbol{q})\cdot\boldsymbol{r}-(\omega_\mathrm{i}+\omega_0)t]\\ &\quad+\cos[(\boldsymbol{k}_\mathrm{i}-\boldsymbol{q})\cdot\boldsymbol{r}-(\omega_\mathrm{i}-\omega_0)t]\}\,.\end{aligned} \tag{7.35b}$$

$\boldsymbol{P}_{\text{ind}}$ consists of two sinusoidal waves: a *Stokes* shifted wave with wavevector $\boldsymbol{k}_{\text{S}} = (\boldsymbol{k}_{\text{i}} - \boldsymbol{q})$ and frequency $\omega_{\text{S}} = (\omega_{\text{i}} - \omega_0)$ and an *anti-Stokes* shifted wave with wavevector $\boldsymbol{k}_{\text{AS}} = (\boldsymbol{k}_{\text{i}} + \boldsymbol{q})$ and frequency $\omega_{\text{AS}} = (\omega_i + \omega_0)$.

The radiation produced by these two induced polarization waves is known, respectively, as **Stokes scattered** and **anti-Stokes scattered** light. Since the phonon frequency is equal to the difference between the incident photon frequency ω_{i} and the scattered photon frequency ω_{s}, this difference is referred to as the **Raman frequency** or **Raman shift** (one also speaks of **Stokes** and **anti-Stokes shifts**). Raman spectra are usually plots of the intensity of the scattered radiation versus the Raman frequency.

When compared to nonlinear optical spectroscopy, light scattering can be regarded as a kind of *parametric process*, since it involves periodically changing a parameter (namely, the electrical susceptibility) of the medium. However, the change induced is bilinear in the phonon displacement and the electric field and therefore light scattering is not a nonlinear optical process as are optical parametric processes.[2] The phonon modulation of the susceptibility at frequency ω_0 generates sidebands at frequencies $\omega_{\text{i}} \pm \omega_0$ to the incident radiation at frequency ω_{i}. In this respect light scattering resembles *frequency modulation* (FM) in radio transmission. The incident radiation plays the role of the *carrier wave*.

Notice that *both frequency and wavevector are conserved in the above scattering processes*. As a result of wavevector conservation, the wavevector q of phonons studied by one-phonon Raman scattering must be smaller than twice the photon wavevector. Assuming that visible lasers are used to excite Raman scattering in a sample with refractive index about 3, q is of the order of $10^6\ \text{cm}^{-1}$. This value is about 1/100 of the size of the Brillouin zone in a semiconductor. Hence *one-phonon Raman scattering probes only zone-center phonons*. In such experiments q can usually be assumed to be zero.

The expansion in (7.31) can be easily extended to second or even higher orders in the phonon displacements. The second-order terms give rise to induced polarizations whose frequencies are shifted from the laser frequency by the amount $\pm\omega_{\text{a}} \pm \omega_{\text{b}}$ (where $\omega_{\text{a}} > \omega_{\text{b}}$ are the frequencies of the two phonons involved). These induced polarizations give rise to **two-phonon Raman scattering**. For two different phonons, peaks with Raman frequencies $\omega_{\text{a}} + \omega_{\text{b}}$ and $\omega_{\text{a}} - \omega_{\text{b}}$ are referred to as the **combination** and **difference modes**, respectively. If the two phonons are identical, the resultant two-phonon Raman peak is called an **overtone**. Wavevector conservation in two-phonon Raman scattering is satisfied when $\boldsymbol{q}_{\text{a}} \pm \boldsymbol{q}_{\text{b}} \approx 0$, where $\boldsymbol{q}_{\text{a}}$ and $\boldsymbol{q}_{\text{b}}$ are the wavevectors of the two phonons a and b, respectively. In overtone scattering this condition implies $\boldsymbol{q}_{\text{a}} = -\boldsymbol{q}_{\text{b}}$, i. e., the two phonons have equal and opposite wavevectors. Thus in two-phonon Raman scattering there is no restriction on the magnitudes of the *individual* phonon wavevectors as there is in one-phonon scattering (only their sum must be near zero). Hence the *overtone* Raman spectrum, after di-

---

[2] Stimulated Raman scattering (see pp. 258 and 395) however, is a nonlinear optical process.

viding the Raman frequency by two, is a measure of the phonon density of states, although modified by a factor dependent on the phonon occupancy and the scattering efficiency.

For Raman scattering with highly monochromatic x-rays, one cannot assume that $q \approx 0$. In fact, scattering wavevectors sweeping the whole BZ can be obtained by varying the angle between the $\boldsymbol{k}$'s of incident and scattered photons. See [3.7] and [3.28b].

7.2.2 Raman Tensor and Selection Rules

The intensity of the scattered radiation can be calculated from the time-averaged power radiated by the induced polarizations $\boldsymbol{P}_{\text{ind}}$ into unit solid angle. Since the induced polarizations for Stokes and anti-Stokes scattering differ only in their frequencies and wavevectors, we will restrict ourselves to Stokes scattering. This intensity will depend on the polarization of the scattered radiation, $\boldsymbol{e}_{\text{s}}$, as $|\boldsymbol{P}_{\text{ind}} \cdot \boldsymbol{e}_{\text{s}}|^2$. If we denote the polarization of the incident radiation as $\boldsymbol{e}_{\text{i}}$, the scattered intensity I_{s} calculated from (7.35) is proportional to

$$I_{\text{s}} \propto |\boldsymbol{e}_{\text{i}} \cdot (\partial\chi/\partial\boldsymbol{Q})_0 \boldsymbol{Q}(\omega_0) \cdot \boldsymbol{e}_{\text{s}}|^2. \tag{7.36}$$

In (7.36) we have approximated $\boldsymbol{q}$ by zero for one-phonon scattering and allowed for the possibility of χ being complex. Notice that the scattered intensity is proportional to the vibration amplitude $\boldsymbol{Q}$ squared. In other words, there will be no Stokes scattering if no atomic vibration is present. This result is a consequence of our classical treatment. Once we quantize the vibrational modes into phonons, in Stokes scattering, where a phonon is excited in the medium by the incident radiation, the intensity becomes proportional to $(N_q + 1)$, where N_q is the phonon occupancy. (the summand 1 in $N_q + 1$ corresponds to the zero-point motion mentioned, in connection with photons, in Sect. 7.1). Similarly, the anti-Stokes intensity will be proportional to N_q and vanish at low temperatures.

Let us assume that $\boldsymbol{Q}$ is the vector displacement of a given atom induced by the phonon so that $(\partial\chi/\partial\boldsymbol{Q})$ is a third-rank tensor with complex components. By introducing a unit vector $\hat{\boldsymbol{Q}} = \boldsymbol{Q}/|\boldsymbol{Q}|$ parallel to the phonon displacement we can define a complex second rank tensor $\mathfrak{R}$ as

$$\mathfrak{R} = (\partial\chi/\partial\boldsymbol{Q})_0 \hat{\boldsymbol{Q}}(\omega_0) \tag{7.37}$$

such that I_{s} is proportional to

$$I_{\text{s}} \propto |\boldsymbol{e}_{\text{i}} \cdot \mathfrak{R} \cdot \boldsymbol{e}_{\text{s}}|^2. \tag{7.38}$$

$\mathfrak{R}$ is known as the **Raman tensor**. In general $\mathfrak{R}$ is obtained by a contraction of $\boldsymbol{Q}$ and the derivative of χ with respect to $\boldsymbol{Q}$, and therefore it is a second-rank tensor with complex components like χ.

By measuring the dependence of the scattered intensity on the incident and scattered polarizations one can deduce the symmetry of the Raman tensor and hence the symmetry of the corresponding Raman-active phonon. Thus *Raman scattering can be used to determine both the frequency and symmetry of a zone-center phonon mode.* By means of two-phonon Raman scattering, phonon densities of states can also be estimated. Obviously, Raman scattering is a very powerful tool for studying vibrational modes in a medium. Later in this chapter we shall show that, in addition to studying atomic vibrations, resonant Raman scattering can be used to study interband electronic transitions, excitons, and even electron–phonon interactions. It is also useful to study magnetic excitations [7.56]. Thus Raman scattering is truly one of the most versatile spectroscopic techniques for studying not only semiconductors but also other condensed media.

At first sight, the Raman tensor as defined in (7.37) appears to be a symmetric second-rank tensor, since the susceptibility is a symmetric tensor. This is only exactly correct if we can neglect the slight difference in frequency between the incident and scattered radiation. We shall come back to this point later in this section. Within this approximation, antisymmetric components in the Raman tensor can be introduced only by magnetic fields [7.56]. Since most semiconductors are nonmagnetic we can usually assume the Raman tensor in semiconductors to be symmetric. Additional requirements are often imposed on Raman tensors as a result of the symmetries of the medium and of the vibrational modes involved in the scattering. The result of these symmetry requirements is that the scattered radiation vanishes for certain choices of the polarizations $\boldsymbol{e}_{\mathrm{i}}$ and $\boldsymbol{e}_{\mathrm{s}}$ and scattering geometries. These so-called **Raman selection rules** are very useful for determining the symmetry of Raman-active phonons.

The simplest example of Raman selection rules can be found in centrosymmetric crystals. In these crystals phonons can be classified as having even or odd parity under inversion. Since the crystal is invariant under inversion, its tensor properties, such as $(\partial\chi/\partial\boldsymbol{Q})$, should remain unchanged under the same operation. On the other hand, the phonon displacement vector $\boldsymbol{Q}$ of an odd-parity phonon changes sign under inversion, implying that $(\partial\chi/\partial\boldsymbol{Q})$ changes sign. Hence *the Raman tensor of odd-parity phonons in centrosymmetric crystals (within the approximation that the phonon wavevector is zero) must vanish.* As we have seen in Sect. 6.4, these odd-parity phonons can be infrared active while the even-parity phonons cannot. Thus infrared absorption and Raman scattering are complementary in centrosymmetric crystals. In some crystals there are phonon modes which are neither infrared nor Raman active. These phonons are said to be *silent*.

As another example we shall consider Raman selection rules in the zinc-blende-type semiconductor GaAs. Its zone-center optical phonon has symmetry Γ_4 (also called Γ_{15}) as discussed in Sect. 3.1. This is a triply degenerate representation whose three components can be denoted as X, Y, and Z. In this particular case we can regard these three components as equal to the projections of the relative displacement of the two atoms in the unit cell along the crystallographic axes. As we showed in Chap. 3 a third-rank tensor in the zinc-blende crystal, such as the piezoelectric or the electromechanical tensor, has only one linearly independent and nonzero component, namely, the component with indices xyz and its cyclic permutations, such as yzx, zxy, etc. Thus the third-rank tensor $\partial\chi/\partial\boldsymbol{Q}$ has only one linearly independent component, which we shall denote by d. The nonzero components of the corresponding Raman tensor are dependent on the phonon displacement. For an optical phonon polarized along the x direction, its Raman tensor $\mathcal{R}(X)$ will have only two nonzero components: $\mathcal{R}_{yz}(X) = \mathcal{R}_{zy}(X) = d$. We can represent $\mathcal{R}(X)$ as a 3×3 matrix:

$$\mathcal{R}(X) = \begin{bmatrix} 0 & 0 & 0 \\ 0 & 0 & d \\ 0 & d & 0 \end{bmatrix}. \tag{7.39a}$$

Using similar arguments we can derive the Raman tensors for the equivalent optical phonons polarized along the y and z axes as

$$\mathcal{R}(Y) = \begin{bmatrix} 0 & 0 & d \\ 0 & 0 & 0 \\ d & 0 & 0 \end{bmatrix} \quad \text{and} \quad \mathcal{R}(Z) = \begin{bmatrix} 0 & d & 0 \\ d & 0 & 0 \\ 0 & 0 & 0 \end{bmatrix}. \tag{7.39b}$$

We should keep in mind that the zone-center optical phonon in GaAs is split into a doubly degenerate transverse optical (TO) mode and a longitudinal optical (LO) mode for $q \neq 0$. The Raman tensor elements for these two phonons are different because the LO mode can scatter light via its macroscopic longitudinal electric field (Sect. 6.4). In order to distinguish them we shall use d_{TO} and d_{LO} in their respective Raman tensors.

Using the Raman tensors defined in (7.39) we can now derive the selection rules for Raman scattering in GaAs. Since these selection rules are dependent on the scattering geometry, we shall introduce a notation for describing scattering geometries which can be specified by four vectors: $\boldsymbol{k}_{\mathrm{i}}$ and $\boldsymbol{k}_{\mathrm{s}}$ (the directions of the incident and scattered photons, respectively) and $\boldsymbol{e}_{\mathrm{i}}$ and $\boldsymbol{e}_{\mathrm{s}}$ (the polarizations of the incident and scattered photons, respectively). These four vectors define the scattering configurations usually represented as $\boldsymbol{k}_{\mathrm{i}}(\boldsymbol{e}_{\mathrm{i}}, \boldsymbol{e}_{\mathrm{s}})\boldsymbol{k}_{\mathrm{s}}$.[3]

EXAMPLE: Raman Selection Rule for Backscattering from the (100) Surface of a GaAs Crystal

Since GaAs is opaque to the usual visible laser light (bandgap 1.52 eV at 4 K, see Table 6.3) the simplest scattering geometry is the backscattering one, i. e., $\boldsymbol{k}_{\mathrm{i}}$ and $\boldsymbol{k}_{\mathrm{s}}$ are antiparallel to each other. In order to conserve wavevector, the $\boldsymbol{q}$ of the phonon must be along the [100] direction also for backscattering from a (100) surface. The polarization of a TO phonon must be perpendicular to $\boldsymbol{q}$ (or the x-axis) and therefore its Raman tensor is a linear combination of $\mathcal{R}(Y)$ and $\mathcal{R}(Z)$. The nonzero components of both tensors in (7.39b) dictate that either $\boldsymbol{e}_{\mathrm{i}}$ or $\boldsymbol{e}_{\mathrm{s}}$ must have a projection along the x-axis. If $\boldsymbol{k}_{\mathrm{i}}$ and $\boldsymbol{k}_{\mathrm{s}}$ are both parallel to the x-axis, then $\boldsymbol{e}_{\mathrm{i}}$ and $\boldsymbol{e}_{\mathrm{s}}$ are both perpendicular to

[3] This notation is due to S.P.S. Porto, a Brazilian pioneer of light scattering in semiconductors

Table 7.2. Raman selection rules for backscattering geometries in zinc-blende-type crystals. d_{TO} and d_{LO} denote the non-zero Raman tensor elements for the TO and LO phonons, respectively. y' and z' denote the [011] and [0$\bar{1}$1] axes, while x'', y'' and z'' denote the set of three mutually perpendicular [111], [1$\bar{1}$0] and [11$\bar{2}$] axes (see Problem 7.4)

Scattering geometry	Selection rule	
	TO phonon	LO phonon
$x(y,y)\bar{x}$; $x(z,z)\bar{x}$	0	0
$x(y,z)\bar{x}$; $x(z,y)\bar{x}$	0	$\|d_{\mathrm{LO}}\|^2$
$x(y',z')\bar{x}$; $x(z',y')\bar{x}$	0	0
$x(y',y')\bar{x}$; $x(z',z')\bar{x}$	0	$\|d_{\mathrm{LO}}\|^2$
$y'(x,x)\bar{y}'$	0	0
$y'(z',x)\bar{y}'$	$\|d_{\mathrm{TO}}\|^2$	0
$y'(z',z')\bar{y}'$	$\|d_{\mathrm{TO}}\|^2$	0
$x''(z'',z'')\bar{x}''$	$(2/3)\|d_{\mathrm{TO}}\|^2$	$(1/3)\|d_{\mathrm{LO}}\|^2$
$x''(z'',y'')\bar{x}''$	$(2/3)\|d_{\mathrm{TO}}\|^2$	0

the x-axis, and therefore Raman scattering by the TO phonon is forbidden in this backscattering geometry. For the LO phonon the situation is different since its $\boldsymbol{q}$ is along the x-axis. Its Raman tensor is given by $\mathcal{R}(X)$ instead. For the scattering geometries $x(y,z)\bar{x}$ or $x(z,y)\bar{x}$ the corresponding scattered intensity is proportional to $|d_{\mathrm{LO}}|^2$. On the other hand, the LO phonon is forbidden in the geometries $x(y,y)\bar{x}$ and $x(z,z)\bar{x}$. One can also show that the LO phonon is forbidden in the geometries $x(y',z')\bar{x}$ and $x(z',y')\bar{x}$ but allowed in the geometries $x(y',y')\bar{x}$ and $x(z',z')\bar{x}$, where y' and z' denote the [011] and [0$\bar{1}$1] axes, respectively. These and other additional Raman selection rules for backscattering from zinc-blende-type crystals are summarized in Table 7.2. The derivations of these selection rules and the corresponding ones for other scattering geometries are left as an exercise (Problem 7.4).

The general problem of deriving the symmetry of Raman tensors in crystals can be solved with the help of group theory. Again we shall use simple examples as illustrations. The rigorous derivations and discussions can be found in articles by *Loudon* [7.57, 58]. Let us consider again a cubic crystal. We have already shown in Sect. 3.3.1 that a symmetric second rank tensor, like the strain tensor, can be decomposed into three components transforming according to the irreducible representations Γ_1, Γ_3 and Γ_4. Similarly, we expect that the Raman tensor in a zinc-blende crystal can be decomposed into three (irreducible) tensors: $\mathcal{R}(\Gamma_1)$, $\mathcal{R}(\Gamma_3)$ and $\mathcal{R}(\Gamma_4)$. The tensors corresponding to the three irreducible components of Γ_4 have been given in (7.39) already. The Raman tensors belonging to the other two irreducible representations can be derived in the same way as for the strain tensors $e_{ij}(\Gamma_1)$ and $e_{ij}(\Gamma_3)$ in Sect. 3.3.1:

$$\mathcal{R}(\Gamma_1) = \begin{bmatrix} a & 0 & 0 \\ 0 & a & 0 \\ 0 & 0 & a \end{bmatrix} \tag{7.40}$$

$$\mathfrak{R}(\Gamma_3) = \begin{bmatrix} b & 0 & 0 \\ 0 & b & 0 \\ 0 & 0 & -2b \end{bmatrix}; \ \sqrt{3}\begin{bmatrix} b & 0 & 0 \\ 0 & -b & 0 \\ 0 & 0 & 0 \end{bmatrix}. \tag{7.41}$$

While there are no *optical* phonons belonging to these irreducible representations in crystals such as GaAs, these tensor components can be found in two-phonon Raman spectra (to be shown for Si in the next section).

The quantity often measured in a scattering experiment is the **scattering efficiency** η. This can be defined as the ratio of the energy of electromagnetic waves scattered per unit time divided by the energy of incident electromagnetic modes crossing the scattering area per unit time. Using the expression (see, e. g., [7.29]) for the power radiated by the induced dipole in (7.35) we can derive the following expression for η [7.59]:

$$\eta = (\omega_s/c)^4 VL|\boldsymbol{e}_i \cdot (\partial\chi/\partial\boldsymbol{Q})_0 \boldsymbol{Q}(\omega_0) \cdot \boldsymbol{e}_s|^2. \tag{7.42}$$

In (7.42) L is the **scattering length**. If the sample is transparent to the incident light, L is equal to the thickness of the sample along the path of the incident light. Otherwise L is equal to $(\alpha_i + \alpha_s)^{-1}$, where α is the absorption coefficient. V, equal to AL (A being the area of the incident beam), is the volume of the sample producing the scattered radiation. Sometimes the efficiency is defined per unit scattering length L and denoted as S. Notice that η depends on the fourth power of ω_s. Thus short-wavelength light is scattered more efficiently than long-wavelength radiation. This important property of light scattering is responsible for the blue color of the sky on a sunny day and the red sunset. The scattering process in these cases is known as **Rayleigh scattering** and results either from entropy fluctuations or, more commonly these days, from pollutants.

One deficiency of the classical treatment of light scattering presented so far is that radiation fields are not explicitly quantized into photons. Raman scattering should be regarded as the inelastic scattering of *photons* by quantized excitations in a medium. The efficiency of scattering of particles is usually defined in terms of a **scattering cross section** σ. Let the *flux* of the incident photon beam be N_i photons *per unit area* (A). If N_s is the total number of particles integrated over all directions (or 4π solid angle) and all scattered frequencies, then σ is defined by[4]

$$N_s = N_i\sigma. \tag{7.43}$$

Clearly σ has the dimensions of an area. In experiments one usually collects only photons scattered into a cone (with solid angle $\delta\Omega$) pointing in a specific direction and within a scattered frequency range $\delta\omega_s$ centered on ω_s. In such cases the ratio between the number of scattered photons and the incident pho-

[4] The cross section for scattering of a particle by another particle is uniquely defined. In the case of a solid, however, it depends on the volume under consideration, i.e., a primitive cell, a crystallographic unit cell, 1 cm^3, etc.

ton flux is known as the **differential scattering cross section** $d^2\sigma/d\Omega\, d\omega_s$. It is related to the scattering efficiency in (7.42) by

$$\frac{d^2\eta}{d\Omega\, d\omega_s} = \left(\frac{\omega_i}{\omega_s A}\right)\left(\frac{d^2\sigma}{d\Omega\, d\omega_s}\right) \tag{7.44}$$

We stated earlier that the Raman tensor is symmetric with respect to interchange between its two subindices because the electric susceptibility tensor $\chi(\omega)$, from which it is derived, has such symmetry in the limit where the photon wave vectors are negligible [see (6.1)]. This is not strictly correct. $\chi(\omega)$ depends on the photon frequency ω only, while the Raman tensor involves two slightly different frequencies: ω_i and ω_s. *The Raman tensor is symmetric only when we neglect the small difference between* ω_i *and* ω_s. We can derive the correct result, without neglecting this difference, by using **time-reversal symmetry**.

Let us assume that a beam of N_i photons with frequency ω_i and polarization $\boldsymbol{e}_i$ is incident on a unit area of a medium. The discussion in the rest of this section will refer to a unit volume of sample and per unit time, unless otherwise stated. This beam is Stokes scattered in the medium as shown in Fig. 7.19a. the polarization of the scattered beam is $\boldsymbol{e}_s$ and its frequency ω_s. The total number of spontaneously scattered photons (i.e., in spontaneous emission discussed in Sect. 7.1; detailed discussion of this point will be postponed until Sect. 7.2.4) will be proportional to $|\boldsymbol{e}_i \cdot \mathcal{R}(\omega_i, \omega_s) \cdot \boldsymbol{e}_s|^2 N_i$. Let us denote the cross section corresponding to $|\boldsymbol{e}_i \cdot \mathcal{R}(\omega_i, \omega_s) \cdot \boldsymbol{e}_s|^2$ as $\sigma(\omega_i, \omega_s)$. Thus the number of scattered photons N_s is equal to $\sigma(\omega_i, \omega_s) N_i$ as long as the number of scattered photons is small and *stimulated* scattering (as in stimulated emission) can be neglected. Notice that the scattering cross section is a scalar quantity defined in terms of photons. However, it can be calculated from the tensor $\mathcal{R}$ defined for macroscopic electromagnetic fields. The arguments in $\sigma(\omega_i, \omega_s)$ serve as reminders of the tensor components in $\mathcal{R}$ from which σ can be derived. Let us assume that the incident photon flux is such that only one scattered photon is produced (Fig. 7.19a). This is equivalent to having

$$N_i = 1/\sigma(\omega_i, \omega_s). \tag{7.45a}$$

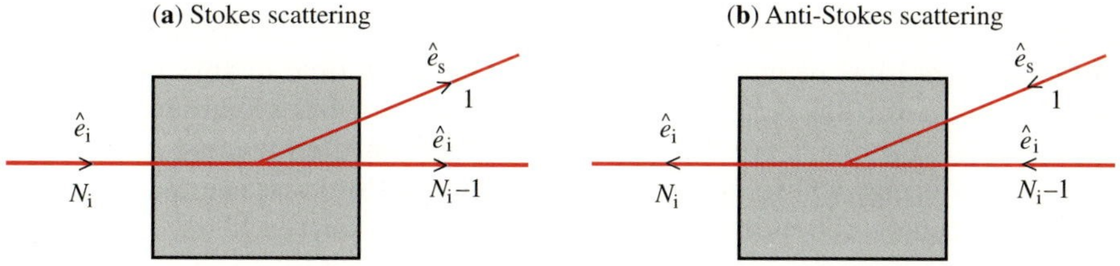

Fig. 7.19. Schematic diagram of (**a**) a Stokes Raman scattering process in a medium and (**b**) its time-reversed anti-Stokes process

The flux of the unscattered beam is therefore $N_i - 1$. In Fig. 7.19b we reverse the direction of time so that the outgoing beams in Fig. 7.19a now become the incoming beams and vice versa. We denote by N_s the total number of photons emerging from unit area of the medium with frequency ω_i. This beam contains $N_i - 1$ photons, which are the unscattered photons. We can neglect loss of photons from this beam due to scattering since it requires at least N_i photons to produce one scattered photon. In addition to these unscattered photons there is an anti-Stokes scattered photon from the single incoming photon with polarization $\boldsymbol{e}_s$ and frequency ω_s. Using the same notation as before for Stokes scattering, the anti-Stokes scattering cross section will be denoted by $\sigma_A(\omega_s, \omega_i)$. Since there are now $N_i - 1$ photons present with the anti-Stokes frequency ω_i we cannot neglect the contribution from stimulated emission. The probability of stimulated emission is proportional to one plus the number of photons present (Sect. 7.1). Therefore the anti-Stokes scattered photon flux is given by $N_i \sigma_A(\omega_s, \omega_i)$. Hence we obtain

$$N_s = (N_i - 1) + N_i \sigma_A(\omega_s, \omega_i). \tag{7.45b}$$

Time-reversal symmetry requires that the number of photons N_s emerging from the medium after time reversal (Fig. 7.19b) be equal to the number of in-coming photons N_i before reversal (Fig. 7.19a). Therefore

$$N_i = (N_i - 1) + N_i \sigma_A(\omega_s, \omega_i) \tag{7.46a}$$

or

$$1 = N_i \sigma_A(\omega_s, \omega_i). \tag{7.46b}$$

Since we have chosen N_i to be $1/\sigma(\omega_s, \omega_i)$ in (7.45a), we obtain from (7.46b)

$$\sigma(\omega_i, \omega_s) = \sigma_A(\omega_s, \omega_i). \tag{7.47}$$

From (7.47) we can go back to $\mathcal{R}$ to show that the *Stokes* Raman tensor element for incident photon frequency ω_i and incident and scattered photons polarizations equal to $\boldsymbol{e}_i$ and $\boldsymbol{e}_s$, respectively, is equal to the corresponding *anti-Stokes* tensor element for incident photon frequency ω_s and incident and scattered photon polarizations equal to $\boldsymbol{e}_s$ and $\boldsymbol{e}_i$, respectively. If we neglect the difference between ω_i and ω_s, there is no distinction between Stokes and anti-Stokes scattering and the Raman tensor is *symmetric* with respect to an interchange between $\boldsymbol{e}_i$ and $\boldsymbol{e}_s$. The equality between the Stokes scattering cross section $\sigma(\omega_i, \omega_s)$ and the corresponding anti-Stokes scattering cross section $\sigma_{AS}(\omega_s, \omega_i)$ has been tested directly in GaAs multiple quantum wells (see Sect. 9.2.4 for further description of the optical properties of these quantum wells) under resonance condition (see Sect. 7.2.8 on the meaning of resonant Raman scattering) where both the Stokes and anti-Stoke scattering cross sections are strongly dependent on the incident photon energy [7.60].

7.2.3 Experimental Determination of Raman Spectra

a) Experimental Techniques

The measurement of a Raman spectrum requires at least the following equipment:

- a source of collimated and monochromatic light;
- an efficient optical system to collect the weak scattered radiation;
- a spectrometer to analyze the spectral content of the scattered radiation and
- a highly sensitive detector for the scattered radiation.

Since Raman efficiencies are typically very small (in some cases as small as 10^{-12}), every component in this system has to be optimized. We shall now consider these components individually.

Light Source

In the days before the advent of lasers, the light source was typically a high power discharge lamp. Discrete emission lines of a gas or vapor (typically mercury vapor) were used. In those days only transparent samples could be studied because of their larger scattering lengths. Since many common semiconductors are opaque, Raman studies of semiconductors became feasible only after the advent of lasers. High power pulsed lasers, such as the ruby laser which appeared first, made it possible to observe *stimulated Raman scattering* (see, for example, [7.61]). However, they are not well suited for studying spontaneous Raman scattering for which a continuous wave (cw) laser of high time-averaged power is preferred. As a result the cw He-Ne laser (wavelength $\lambda = 632.8$ nm) was the first laser to be used in Raman scattering. But soon it was replaced by the Nd:YAG, Ar^+ and Kr^+ ion lasers. The latter two produce several high power (> 1 W in a single line) discrete emission lines covering the red (647 nm), yellow (564 nm), green (514 nm), blue (488 nm) and violet (458 nm) regions of the visible spectrum. With these high average power cw lasers it became feasible to obtain not only one-phonon Raman spectra in semiconductors but also their two-phonon spectra. With continuously tunable cw lasers based on dyes (1 μm $\geq \lambda \geq$ 450 nm), color-centers in ionic crystals (3 μm $\geq \lambda \geq$ 1 μm) and more recently Ti-doped sapphire (1 μm $\geq \lambda \geq$ 700 nm) it became possible to perform *Raman excitation spectroscopies*, i. e., resonant Raman scattering. In analogy to the luminescence excitation spectroscopy discussed in Sect. 7.1.5, in resonant Raman spectroscopy one monitors the Raman efficiency as a function of the excitation laser wavelength. The physics involved in this kind of spectroscopy will be discussed in Sect. 7.2.7.

Spectrometers

In most Raman experiments on semiconductors the signal is 4–6 orders of magnitude weaker than the elastically scattered laser light. At the same time the difference in frequency between the Raman signal and the laser light is only about 1% of the laser frequency. In order to observe this weak sideband in the vicinity of the strong laser light, the spectrometer must satisfy several stringent conditions. First it must have good *spectral resolving power*. Modern Raman spectrometers typically have resolving power $(\lambda/\Delta\lambda) > 10^4$, which can be obtained easily with diffraction gratings. It is, however, important that these gratings do not produce "ghosts" and "satellites", which can be confused with Raman signals. Modern holographic gratings (Sect. 6.1.2) have practically eliminated this problem. A Raman spectrometer must also have an excellent *stray light rejection ratio*. This is defined as the ratio of the background stray light (light at all wavelengths other than the nominal one specified by the spectrometer) to the signal. Stray light is produced by imperfections in the optics (such as mirrors and gratings) and by scattering of light off walls and dust particles inside the spectrometer. Most spectrometers have a rejection ratio of 10^{-4}–10^{-6}. As a result, the background stray light can be orders of magnitude larger than the Raman signal. This situation can be solved by: (a) making the sample surface as smooth as possible to minimize the elastically scattered laser light; (b) using a "notch filter", which will block out the laser light; (c) putting two or more spectrometers in series. A properly designed *double monochromator* can have rejection ratios as small as 10^{-14}, equal to the product of the ratios for the two monochromators. This rejection ratio is adequate for Raman studies in most semiconductors. Nowadays *triple spectrometers* have become popular for use with **multichannel detectors** to be described next. In these spectrometers two monochromators are put "back-to-back" for use as a notch filter. The *third monochromator* provides all the dispersion required for separating the Raman signal from the laser light.

Detector and Photon-Counting Electronics

Raman recorded the weak inelastically scattered light in his pioneering experiment in 1928 by using photographic plates. These detectors actually have many of the desirable characteristics of modern systems. They have the sensitivity to detect individual photons. They are *multichannel detectors* in that they can measure many different wavelengths at the same time. Finally, they can integrate the signal over long periods of time, from hours to even days. They also have one big advantage compared to modern detector systems: they are inexpensive! However, they also have some serious drawbacks: Their outputs are not linear in the light intensity and it is also cumbersome to convert the recorded signal into digital form for analysis. The first major advance in photoelectric recording (see Chap. 8 for discussions of the photoelectric effect) of Raman spectra was the introduction of *photon counting* methods [7.62]. Instead of integrating all the photocurrent pulses arriving at the photomultiplier tube anode as the signal, a *discriminator* selects and counts only those pulses with large enough amplitude to have originated at the photocathode. The background pulses (noise) remaining in such systems are those generated by thermionic emission of electrons at the photocathode. This can be minimized by cooling the entire photomultiplier tube to about −20C (via thermoelectric coolers) or to liquid nitrogen temperature. One of the most popular photomultipliers for Raman scattering has a GaAs photocathode cooled to −20C. When coupled to properly designed counting electronics, such a detector system has a background noise (or *dark counts*) of a few counts per second and a *dynamic range* of 10^6.

The above detector system has one major disadvantage compared with the photographic plate. It counts the total number of photons emerging from the spectrometer without spatially resolving the positions (and hence the wavelengths) of the photons. As a result, the Raman spectrum is obtained only after scanning the spectrometer output over a wavelength range containing the Raman peak. Recently several multichannel detection systems have become available commercially. These systems are based on either *charge-coupled devices* (CCDs) or *position-sensitive imaging photomultiplier tubes*. These detectors have been reviewed by *Chang* and *Long* [7.63] and by *Tsang* [7.64]. The CCD detector is essentially the same as a modern television camera. Its sensitivity can be enhanced by adding an image-intensifier tube. This tube consists of a photocathode as in a photomultiplier tube. The photoelectrons generated at the cathode are multiplied by a factor of 10^6–10^7 through a *microchannel plate*. This is essentially a honeycomb consisting of many tiny glass tubes whose interior walls are coated with a secondary electron emitter. Just one such glass tube with an enlarged entrance in the shape of a funnel is known as a *channeltron* (see Fig. 8.9 for a sketch). A high voltage is applied between the entrance and exit ends of each glass tube. When an electron enters the tube and hits the secondary emitter wall it generates several additional electrons. These will, in turn, produce more secondary electrons when they impact the glass wall. Thus an "avalanche" of secondary electrons is created as they travel down the narrow tube. A phosphor at the exit end of the microchannel plate converts the electron pulses back into a brighter image.

An imaging photomultiplier tube [7.65, 66] (also known as a *Mepsicron*) has essentially the same construction as an image-intensifier tube except that the phosphor is replaced by an anode with four output leads (Fig. 7.20). When an amplified electron pulse hits the anode, it generates four electrical output pulses from these four anode leads. Depending on the position of the electron pulse on the anode, these four output pulses emerge at different times. A timing circuit measures the arrival time delays between these four pulses. An analog computer calculates the position of the original electron pulse at the anode based on these time delays. This detector has all the advantages of a photomultiplier tube plus a much lower dark count. Since each pixel is equal to one "channel" and the area of a pixel is much smaller than a photomultiplier tube, its dark current per channel is also much smaller. The dark current of a cooled imaging tube can be as low as 0.01 counts/second per channel. The major disadvantage of this detector is its finite lifetime. Every time a photoelectron is amplified by the microchannel plate positive ions are emitted in the channel plates and accelerated towards the photocathode. As a result, such detectors are rated to have a total lifetime of 10^{13} photoelectrons/pixel. Obviously, such detectors should be used only for extremely weak signals. For obvious reasons

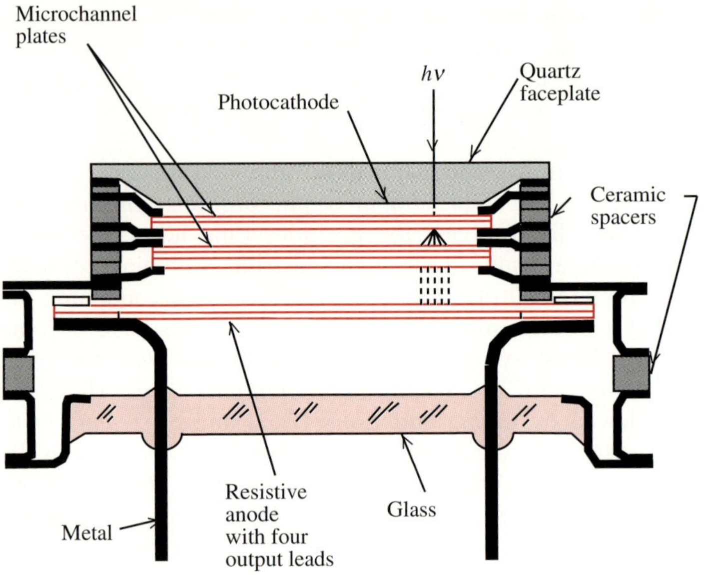

Fig. 7.20. Schematic diagram of the construction of a position-sensitive imaging photomultiplier tube

the dynamical range of the mepsicrons is small, typically $< 10^5$ electrons/pixel. The relative merits of the CCD and the Mepsicron multichannel detection systems have been compared by *Tsang* [7.64].

b) Experimental Phonon Raman Spectra in Semiconductors

One-Phonon Raman Spectra

Figure 7.21 shows Raman spectra of several group III–V semiconductors (GaAs, InP, AlSb) measured by *Mooradian* and *Wright* [7.67] using a Nd:YAG laser (1.06 μm wavelength) as the excitation source in a 90° scattering geometry (note that these semiconductors are transparent to this wavelength). In this geometry both the TO and LO phonons are allowed by the selection rules discussed in Sect. 7.2.2. Figure 7.22 shows the Raman spectra of Si obtained by *Temple* and *Hathaway* [7.68] in backscattering geometry but with several different polarization configurations. These configurations allowed them to extract components of the Raman tensor with different symmetries. Notice that there is only one very strong one-phonon peak at 519 cm^{-1} (at 305 K), corresponding to the zone-center optical phonons in Si (the TO and LO phonons are degenerate at zone center in diamond-type crystals as pointed out in Sect. 3.1). In agreement with selection rules, this peak appears only in scattering configurations where the $\Gamma_{25'}$ components of the Raman tensor are allowed. It appears rather weakly in the "forbidden" configurations; it is easy to figure out possible reasons (e. g., imperfect polarizers).

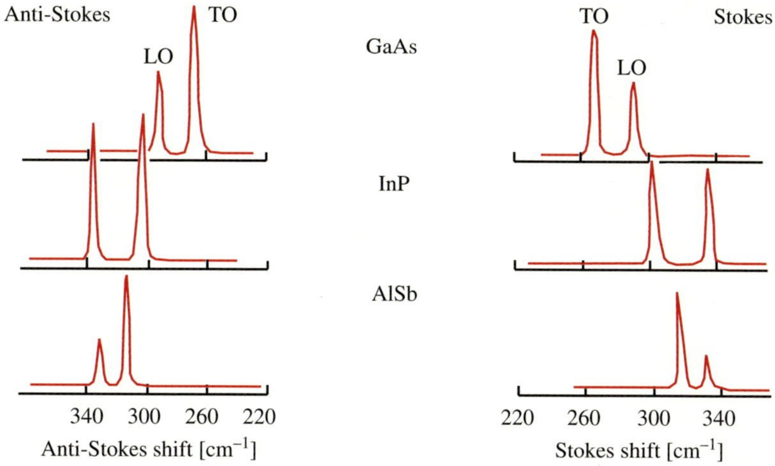

Fig. 7.21. Raman spectra of three zinc-blende-type semiconductors showing the TO and LO phonons in both Stokes and anti-Stokes scattering. Note that the vertical scales are not the same for all spectra. (From [7.67])

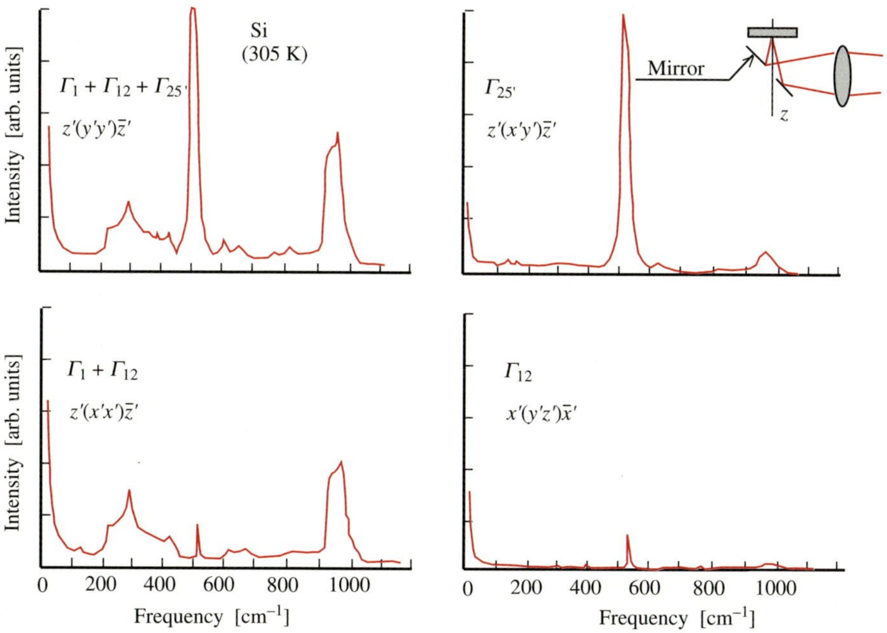

Fig. 7.22. First (peak at 520 cm^{-1}) and second order Raman spectra of Si obtained in the scattering geometry shown in the *inset*. The notations for the scattering configuration in each spectrum are defined in Sect. 7.2.2. (From [7.68])

Two-Phonon Raman Spectra

In addition to the one-phonon peak, *Temple* and *Hathaway* also observed a number of weaker structures, which were identified with two-phonon Raman scattering. While the selection rules for two phonon-scattering in semiconductors are beyond the scope of this book, they have been studied by several researchers [7.69, 70]. In Si, components of the two-phonon Raman tensor with symmetries $\Gamma_{25'}$, Γ_{12} and Γ_1 are allowed. Notice that the two-phonon spectra show peaks and shoulders that are reminiscent of structures in the three-dimensional density of states associated with critical points. This is not surprising since, as we pointed out in Sect. 7.2.1, overtone two-phonon Raman spectra mimic the phonon density of states. The two-phonon Raman peaks in Si roughly fall into three groups (similar results are found in zinc-blende-type semiconductors although the phonon frequencies may be different). The broad low-energy peak in the range of 200–450 cm^{-1} in Fig. 7.22 is the result of overtone scattering from the acoustic phonons. The few bands near the one-phonon peak are combination modes involving one optical phonon and one acoustic phonon. Finally, the high-energy peak located between 900 and 1000 cm^{-1} is due to overtone scattering by two optical phonons. Figure 7.23 shows the two-phonon Raman spectrum in Ge obtained by *Weinstein* and

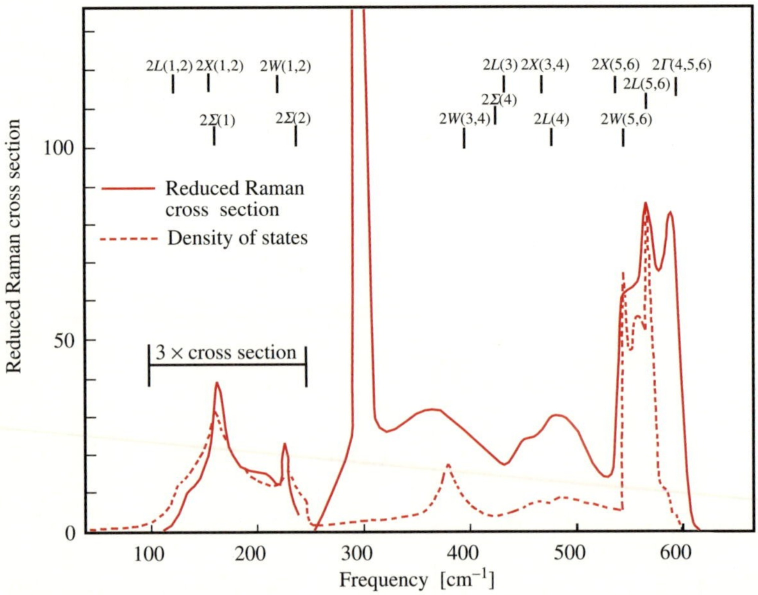

Fig. 7.23. "Reduced" two-phonon Raman spectrum of Ge compared with the density of two-phonon overtone states. The experimental two-phonon Raman spectrum containing the components with symmetry $\Gamma_1 + 4\Gamma_{12}$ has been divided by the factor $[N(\omega) + 1]^2$, where $N(\omega)$ is the Bose–Einstein occupation number of the phonon mode with frequency ω. (From [7.71]). Note that at low frequencies the reduced Raman cross section lies below the density of states (see Problem 7.15)

Cardona [7.71]. In order to compare the two-phonon Raman spectrum with the experimental phonon density of states (deduced from neutron scattering results), these researchers took the linear combination $\Gamma_1 + 4\Gamma_{12}$ of the two-phonon Raman spectra and then divided it by the factor $[N(\omega) + 1]^2$ [where $N(\omega)$ is the Bose–Einstein occupation number of the phonon mode with frequency ω] to eliminate the effect of phonon occupation number. The resultant "reduced" two-phonon spectrum is the solid curve in Fig. 7.23. This is compared with the two-phonon "overtone" density of states (broken curve) obtained from the one-phonon density of states curve based on the phonon dispersion curves of Ge [7.72] but with the phonon frequency doubled. Except for the strong, sharp one-phonon peak at around 300 cm^{-1} in the experimental spectrum (which could not be completely eliminated because of imperfect polarizers), the agreement between the two curves is quite good, especially in the two-acoustic-phonon part of the spectrum. The vertical bars and labels in this figure highlight the critical points in the Brillouin zone [numbers in parentheses denote the phonon branches, counting from lower (TA) to higher (TO) frequencies] which contribute to the structures in the two-phonon "overtone" density of states.

Raman Spectra of Semiconductor Monolayers

The development of high intensity laser sources for Raman spectroscopies has made it possible to measure the Raman spectra of opaque semiconductors. Still the thickness of sample probed is of the order of the optical penetration depth, which is typically larger than 100 nm. With the appearance of optical multichannel detectors it is now possible to measure the Raman spectra of monolayers of semiconductor materials. Figure 7.24c shows the Raman spec-

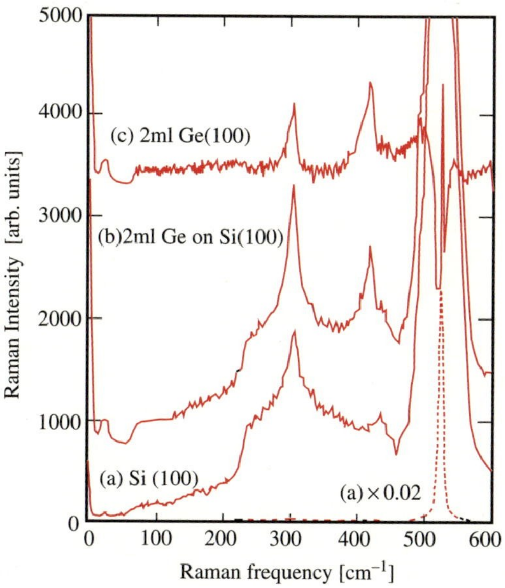

Fig. 7.24. Raman spectrum (*c*) of two monolayers of Ge deposited on a Si(100) substrate obtained by subtracting spectrum (*a*) from (*b*). (From [7.64])

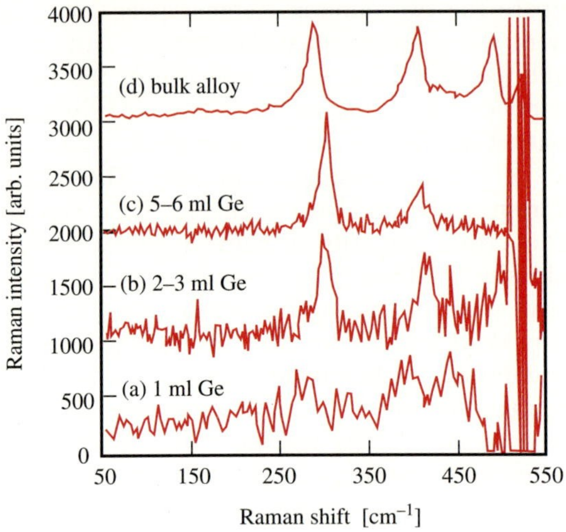

Fig. 7.25. Evolution of Raman spectra of monolayers of Ge deposited on a Si(100) substrate and then protected by 10 nm of Si as a function of Ge layer thickness. These spectra have been obtained in the same way as those in Fig. 7.24. Spectrum (*d*) is that of a bulk Ge-Si alloy. (From [7.64])

trum of two monolayers of Ge deposited on a Si(100) substrate. It is obtained by subtracting the spectrum of the Si substrate (Fig. 7.24a) from that of the Ge layer plus the Si substrate (Fig. 7.24b). Compared with the Raman spectrum of bulk Ge, the Ge monolayer shows one extra peak at 410 cm^{-1}. Figure 7.25 shows the thickness dependence of the Raman spectra of Ge monolayers on Si(100) substrates. For comparison, Fig. 7.25d shows the Raman spectrum of a bulk Ge-Si alloy. These spectra indicate that the 410 cm^{-1} peak is also present in the bulk Ge-Si alloy. It is strongest in the two-monolayer Ge film. As the Ge film thickness increases, the intensity of this peak decreases relative to that of the bulk Ge TO Raman peak at 300 cm^{-1}. All these properties are consistent with the interpretation that this 410 cm^{-1} peak is associated with the vibration of a Ge-Si bond that forms at the interface between the Ge monolayer and the Si substrate. These results emphasize the power of Raman spectroscopy for characterizing semiconductor systems.

Raman Spectra of Phonon-Polaritons

Usually the phonon wavevector observed by Raman scattering in semiconductors is too small to be used for mapping out the phonon dispersion over the entire Brillouin zone (Exception: the novel technique of Raman scattering with x-rays, see Fig. 3.6 and [3.28b]). Nevertheless, Raman scattering is a good technique for measuring dispersion of phonon-polaritons near the zone-center (Sect. 6.4). We showed in that section that the coupling between the electromagnetic field and the polarization wave generated by the atomic vibration resulted in polaritons with dispersion given by (6.115b). As shown in Fig. 6.30, polaritons exhibit considerable dispersion for $q < 10^5$ cm^{-1}, a region which is difficult to probe by neutron scattering but can be conveniently studied by forward Raman scattering. Figure 7.26a shows a comparison between the theoretical polariton dispersion in GaP calculated from (6.114b) and the

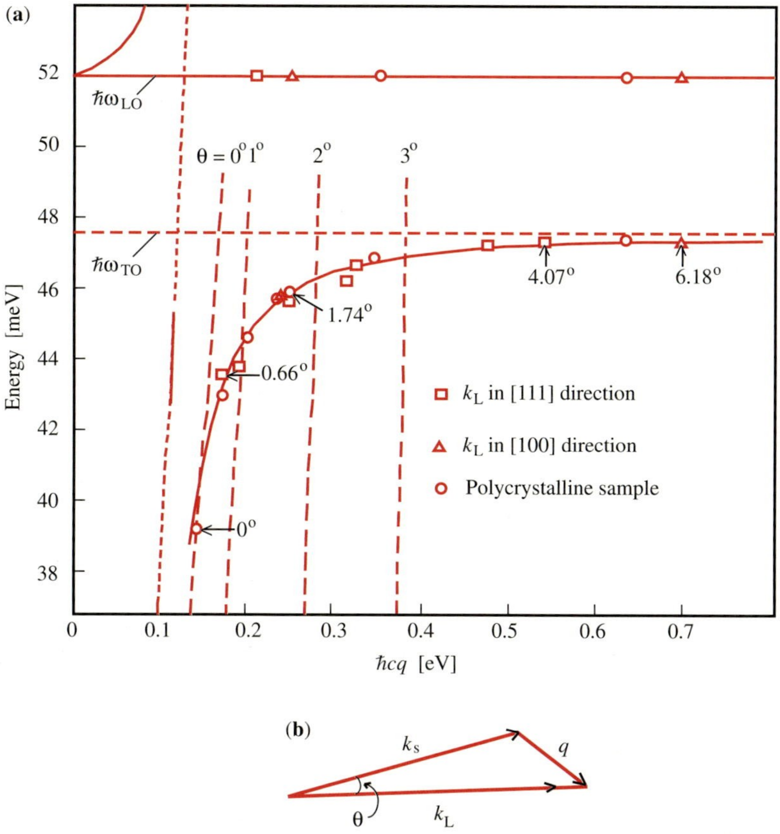

Fig. 7.26. (**a**) A comparison between the theoretical polariton dispersion (*solid curves*) in GaP with experimental points (□, △, ○) determined by *Henry* and *Hopfield* [7.73]. The numbers by the data points are values of the scattering angle θ defined by the scattering geometry shown in (**b**). The *broken lines* in (**a**) represent the variation of Raman frequency as a function of q for different values of θ. The wavevector q has been multiplied in the abscissa by $\hbar c$ so as to obtain units of energy (eV). A He-Ne laser line (1.96 eV) was used

experimental results obtained by *Henry* and *Hopfield* [7.73] using forward Raman scattering. Their scattering geometry is shown in Fig. 7.26b. The broken curves in Fig. 7.26a labeled with different values of θ show the variation of the calculated Raman frequency as a function of q for different values of the scattering angle θ (Problem 7.6). Although their scattering configuration allowed them to observe the lower polariton branch only, there is no question that the experimental polariton dispersion is in excellent agreement with theory.

7.2.4 Microscopic Theory of Raman Scattering

To describe microscopically inelastic scattering of light by phonons in a semiconductor, we have to specify the state of the three systems involved:

- incident and scattered photons with frequencies ω_i and ω_s, respectively;
- electrons in the semiconductor;
- the phonon involved in the scattering.

In the initial state $|\mathrm{i}\rangle$ (before scattering occurs) there are, respectively, $N(\omega_\mathrm{i})$ and $N(\omega_\mathrm{s})$ photons with the frequencies ω_i and ω_s. There are also N_q phonons present in the semiconductor (assumed to be at a nonzero temperature T) while the electrons are all in their ground states (i. e., the valence bands are completely filled and the conduction band empty). In the final state $|\mathrm{f}\rangle$, after Stokes Raman scattering, $N(\omega_\mathrm{i})$ has decreased by one while $N(\omega_\mathrm{s})$ and N_q have both increased by one. The electrons remain unchanged. At first sight it seems that this scattering process does not involve electrons and therefore it can be described by an interaction Hamiltonian involving photons and phonons only. The strength of this interaction, however, is very weak unless the photons and phonons have comparable frequency. While such direct (spontaneous) inelastic scattering of photons by phonons has been proposed theoretically [7.74], it has not been identified experimentally to our knowledge. The main experimental obstacle is, probably, the lack of laser sources and single-photon detectors in the far-infrared.

When visible photons are used to excite Raman scattering in a semiconductor, they couple by-and-large only to electrons via the electron–radiation interaction Hamiltonian $\mathcal{H}_{\mathrm{eR}}$ in (6.29). The scattering proceeds in three steps.

Step 1. The incident photon excites the semiconductor into an intermediate state $|a\rangle$ by creating an electron–hole pair (or exciton).

Step 2. This electron–hole pair is scattered into another state by emitting a phonon via the electron–phonon interaction Hamiltonian $\mathcal{H}_{\mathrm{e-ion}}$ (Sect. 3.3). This *intermediate state* will be denoted by $|b\rangle$.

Step 3. the electron–hole pair in $|b\rangle$ recombines radiatively with emission of the scattered photon.

Thus *electrons mediate the Raman scattering of phonons although they remain unchanged after the process*. Since the transitions involving the electrons are virtual they do not have to conserve energy, although they still have to conserve wavevectors.

We notice that spontaneous emission of the scattered photon is involved in step 3 and therefore what we have described is known as **spontaneous Raman scattering** as distinct from stimulated Raman scattering. In principle, such spontaneous scattering processes can be described rigorously only by quantizing the radiation fields. We will avoid this again by using the semiclassical approach described in Sect. 7.1. A vector potential $\boldsymbol{A}(\omega_\mathrm{s})$ associated with the scattered radiation field will be assumed to exist (quantum-mechanical zero-point amplitude) so that the Raman transition probabilities can be calculated. Afterwards this stimulated Raman scattering probability is converted into a spontaneous Raman scattering probability using the ratio between the Einstein A and B coefficients as in (7.1). In the rest of this section we shall write down only the stimulated scattering probabilities and leave the conversion into the corresponding efficiencies for spontaneous scattering to the reader.

7.2.5 A Detour into the World of Feynman Diagrams

As long as all the interactions in the above Raman scattering processes are weak, the scattering probability (for phonons) can be calculated with third-order perturbation theory. However, it is not a trivial matter to enumerate all the terms involved in such a third-order perturbation calculation. This is usually done, systematically, with the help of Feynman diagrams. It is beyond the scope of the present book to discuss them in detail. Instead we shall simply describe what they are and how to use them to derive the scattering probability. Readers can easily find further details in many books on quantum mechanics and many-body problems (e. g. [7.75–77]).

The rules for drawing Feynman diagrams are:

- Excitations such as photon, phonons and electron–hole pairs in Raman scattering are represented by lines (or **propagators**) as shown in Fig. 7.27. These propagators can be labeled with properties of the excitations such as their wavevectors, frequencies, and polarizations.
- The interaction between two excitations is represented by an intersection of their propagators. This intersection is known as a **vertex** and is sometimes highlighted by a symbol such as a filled circle or empty rectangle.
- Propagators are drawn with an arrow to indicate whether the corresponding excitations (quasiparticles) are created or annihilated in an interaction. Arrows pointing towards a vertex represent excitations which are annihilated. Those pointing away from the vertex are created.

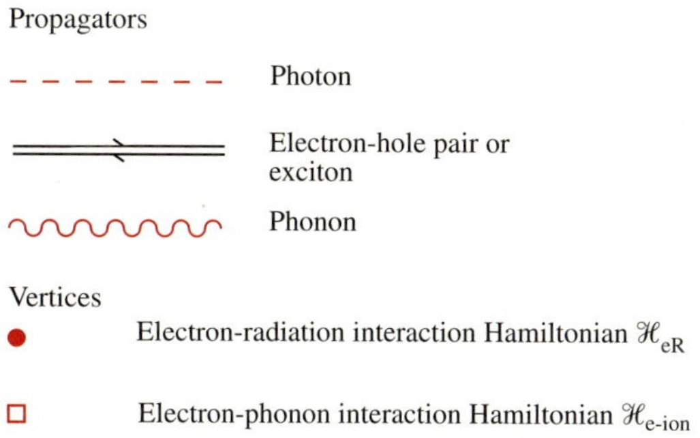

Fig. 7.27. Symbols used in drawing Feynman diagrams to represent Raman scattering

- When several interactions are involved they are always assumed to proceed sequentially from the left to the right as a function of time.
- Once a diagram has been drawn for a certain process, other possible processes are derived by permuting the time order in which the vertices occur in this diagram.

It should be noted that there may be slight differences among publications in the above rules for drawing Feynman diagrams.

We shall illustrate the application of Feynman diagrams by using them to represent the Raman scattering by phonons [7.77, 78]. The diagram for the Raman process described earlier in this section is shown in Fig. 7.28a. The other five possible permutations of the time order of the three vertices involved in this process are shown in Fig. 7.28b–f.

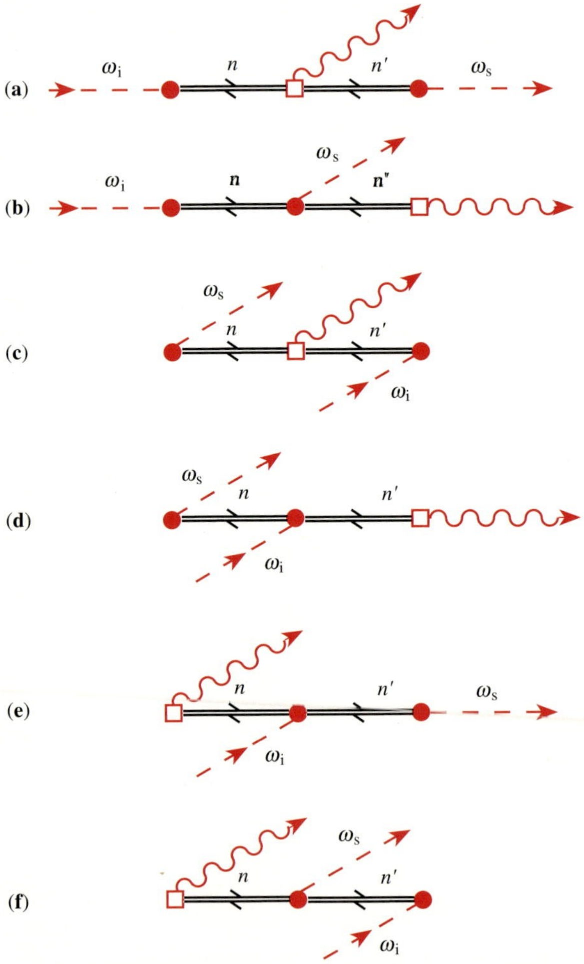

Fig. 7.28a–f. Feynman diagrams for the six scattering processes that contribute to one-phonon (Stokes) Raman scattering

After all the possible Feynman diagrams have been drawn, the next step is to translate them into terms in the perturbation expansion of the scattering probability. The probability for scattering a system from the initial state $|\mathrm{i}\rangle$ to the $|\mathrm{f}\rangle$ state can be derived, as usual, via the Fermi Golden Rule (Sect. 5.2.4). The resulting rules for translating a Feynman diagramm into a term in the perturbation theory series [7.77] are best illustrated by an example. Let us consider the diagram in Fig. 7.28a.

The first vertex introduces a term of the form

$$\sum_n \frac{\langle n|\mathcal{H}_{\mathrm{eR}}(\omega_\mathrm{i})|\mathrm{i}\rangle}{[\hbar\omega_\mathrm{i} - (E_n - E_\mathrm{i})]} \tag{7.48}$$

into the scattering probability. In this term $|\mathrm{i}\rangle$ is the initial state and E_i is its energy. $|n\rangle$ is an intermediate electronic state with energy E_n. The sign of $\hbar\omega_\mathrm{i}$ in the energy denominator depends on whether the quantum of energy $\hbar\omega_\mathrm{i}$ was absorbed (+ sign) or emitted (− sign). Notice the summation over all intermediate states $|n\rangle$ in (7.48). When there is a second vertex, as in Fig. 7.28a, (7.48) is multiplied by another similar term to become

$$\sum_{n,n'} \frac{\langle n'|\mathcal{H}_{\mathrm{e-ion}}(\omega_0)|n\rangle\ \langle n|\mathcal{H}_{\mathrm{eR}}(\omega_\mathrm{i})|\mathrm{i}\rangle}{[\hbar\omega_\mathrm{i} - (E_n - E_\mathrm{i})][\hbar\omega_\mathrm{i} - (E_n - E_\mathrm{i}) - \hbar\omega_0 - (E_{n'} - E_n)]}, \tag{7.49a}$$

where $|n'\rangle$ is another intermediate state. The sign of $\hbar\omega_0$ in the denominator is negative now because the quantum of energy (a phonon in this case) is emitted. Thus each vertex adds a term involving the matrix element of the interaction Hamiltonian to the numerator and an energy term to the denominator. Notice how the second energy denominator also involves the energy denominator for the first vertex. By simplifying the energy denominators (7.49a) can be rewritten as

$$\sum_{n,n'} \frac{\langle n'|\mathcal{H}_{\mathrm{e-ion}}(\omega_0)|n\rangle\ \langle n|\mathcal{H}_{\mathrm{eR}}(\omega_\mathrm{i})|\mathrm{i}\rangle}{[\hbar\omega_\mathrm{i} - (E_n - E_\mathrm{i})][\hbar\omega_\mathrm{i} - \hbar\omega_0 - (E_{n'} - E_\mathrm{i})]}. \tag{7.49b}$$

This process is continued until the last vertex in the diagram is reached. A diagram with n vertices will therefore produce a term containing n matrix elements in the numerator. In principle, there should also be n energy terms in the denominator. The last energy denominator, however, represents the overall energy conservation condition and is converted to a delta function. For example, the last energy term in the denominator corresponding to Fig. 7.28a can be written as

$$\begin{aligned}&[\hbar\omega_\mathrm{i} - (E_n - E_\mathrm{i}) - \hbar\omega_0 - (E_{n'} - E_n) - \hbar\omega_\mathrm{s} - (E_\mathrm{f} - E_{n'})]\\ &\quad = [\hbar\omega_\mathrm{i} - \hbar\omega_0 - \hbar\omega_\mathrm{s} - (E_\mathrm{i} - E_\mathrm{f})].\end{aligned}$$

We have pointed out that Raman scattering of visible photons by phonons is mediated by electrons. However, the electrons are unchanged after the scattering process so the final electronic state $|\mathrm{f}\rangle$ in this case should be identical to the initial state $|\mathrm{i}\rangle$. Hence the last term in the denominator is simply

$$[\hbar\omega_\mathrm{i} - \hbar\omega_0 - \hbar\omega_\mathrm{s}].$$

It should vanish because of the energy conservation condition of Raman scattering (Sect. 7.2.1). As a result it must be replaced by the delta function $\delta[\hbar\omega_i - \hbar\omega_0 - \hbar\omega_s]$ when writing down the scattering probability using the Golden Rule.

If diagram Fig. 7.28a is the only term contributing to the scattering process, the scattering probability as given by the Golden Rule will be

$$P_{ph}(\omega_s) = \left(\frac{2\pi}{\hbar}\right) \left| \sum_{n,n'} \frac{\langle i|\mathcal{H}_{eR}(\omega_s)|n'\rangle \langle n'|\mathcal{H}_{e-ion}|n\rangle \langle n|\mathcal{H}_{eR}(\omega_i)|i\rangle}{[\hbar\omega_i - (E_n - E_i)][\hbar\omega_i - \hbar\omega_0 - (E_{n'} - E_i)]} \right|^2 \times \delta[\hbar\omega_i - \hbar\omega_0 - \hbar\omega_s]. \quad (7.50a)$$

To obtain the scattering probability P_{ph} due to all six diagrams in Fig. 7.28, we have to sum first their individual contributions using the above rules and then square (Problem 7.8):

$$\begin{aligned} P_{ph}(\omega_s) = \left(\frac{2\pi}{\hbar}\right) \Bigg| \sum_{n,n'} & \frac{\langle i|\mathcal{H}_{eR}(\omega_i)|n\rangle \langle n|\mathcal{H}_{e-ion}|n'\rangle \langle n'|\mathcal{H}_{eR}(\omega_s)|i\rangle}{[\hbar\omega_i - (E_n - E_i)][\hbar\omega_i - \hbar\omega_0 - (E_{n'} - E_i)]} \\ & + \frac{\langle i|\mathcal{H}_{eR}(\omega_i)|n\rangle \langle n|\mathcal{H}_{eR}(\omega_s)|n'\rangle \langle n'|\mathcal{H}_{e-ion}|i\rangle}{[\hbar\omega_i - (E_n - E_i)][\hbar\omega_i - \hbar\omega_s - (E_{n'} - E_i)]} \\ & + \frac{\langle i|\mathcal{H}_{eR}(\omega_s)|n\rangle \langle n|\mathcal{H}_{e-ion}|n'\rangle \langle n'|\mathcal{H}_{eR}(\omega_i)|i\rangle}{[-\hbar\omega_s - (E_n - E_i)][-\hbar\omega_s - \hbar\omega_0 - (E_{n'} - E_i)]} \\ & + \frac{\langle i|\mathcal{H}_{eR}(\omega_s)|n\rangle \langle n|\mathcal{H}_{eR}(\omega_i)|n'\rangle \langle n'|\mathcal{H}_{e-ion}|i\rangle}{[-\hbar\omega_s - (E_n - E_i)][-\hbar\omega_s + \hbar\omega_i - (E_{n'} - E_i)]} \\ & + \frac{\langle i|\mathcal{H}_{e-ion}|n\rangle \langle n|\mathcal{H}_{eR}(\omega_i)|n'\rangle \langle n'|\mathcal{H}_{eR}(\omega_s)|i\rangle}{[-\hbar\omega_0 - (E_n - E_i)][-\hbar\omega_0 + \hbar\omega_i - (E_{n'} - E_i)]} \\ & + \frac{\langle i|\mathcal{H}_{e-ion}|n\rangle \langle n|\mathcal{H}_{eR}(\omega_s|n'\rangle \langle n'|\mathcal{H}_{eR}(\omega_i)|i\rangle}{[-\hbar\omega_0 - (E_n - E_i)][-\hbar\omega_0 - \hbar\omega_s - (E_{n'} - E_i)]} \Bigg|^2 \\ & \times \delta(\hbar\omega_i - \hbar\omega_s - \hbar\omega_0). \end{aligned} \quad (7.50b)$$

By substituting some typical values appropriate for semiconductors for the parameters in (7.50b), *Loudon* [7.57] estimated the Raman efficiency to be around 10^{-6}–10^{-7} [sterad cm]$^{-1}$. Usually (7.50b), in spite of its generality, is not too useful for calculating absolute Raman efficiencies because of the large number of unknown parameters involved (such as electron–phonon interaction matrix elements).

7.2.6 Brillouin Scattering

We have already pointed out in Sect. 7.2 that inelastic scattering of light by acoustic waves was first proposed by *Brillouin* [7.53]. As a result, this kind of light scattering spectroscopy is known as Brillouin scattering. In terms of physics, there is very little difference between Raman scattering and Brillouin scattering. In semiconductors the main difference between them arises from the difference in dispersion between optical and acoustic phonons. Except for infrared-active phonons (polaritons), optical phonon energies typically do not change much for a wavevector $\boldsymbol{q}$ varying between 0 and 10^6 cm^{-1} (which corresponds to backscattering), on the other hand the acoustic phonon dispersion is linear with q:

$$\omega_{\mathrm{ac}}(q) = v_{\mathrm{ac}} q, \tag{7.51}$$

where ω_{ac} and v_{ac} are, respectively, the acoustic phonon angular frequency and velocity. Substituting this phonon dispersion into the energy and wavevector conservation equations for light scattering in crystals, we obtain the **Brillouin frequency**, i. e., the acoustic phonon frequency ω_{ac} involved in the Brillouin scattering [7.79]

$$\omega_{\mathrm{ac}}(q) = (\omega_{\mathrm{i}} v_{\mathrm{ac}}/c)[(n_{\mathrm{i}} - n_{\mathrm{s}})^2 + 4 n_{\mathrm{i}} n_{\mathrm{s}} \sin^2(\theta/2)]^{1/2}, \tag{7.52}$$

where c is the velocity of light, n_{i} and n_{s} are the refractive indices of the medium at the incident and scattered photon frequencies, respectively, ($n_{\mathrm{i}} \simeq n_{\mathrm{s}}$) and θ is the scattering angle (inside the medium) as defined in Fig. 7.26. In (7.52) we have neglected terms of higher order in v_{ac}/c since they are usually too small to be observed experimentally. In most semiconductors the difference between n_{i} and n_{s} is negligible and (7.52) can be simplified to

$$\omega_{\mathrm{ac}}(q) \approx (2 n_{\mathrm{i}} \omega_{\mathrm{i}} v_{\mathrm{ac}}/c) \sin(\theta/2). \tag{7.53}$$

Thus one application of Brillouin scattering is to determine either v_{ac} or n_{i}.

The spectral shape of Brillouin scattering deduced from (7.53) is, in principle, a delta function for a well-defined scattering angle. In practice, the peak is broadened by experimental factors such as a nonzero collection angle and spectrometer resolution. In addition to these external factors, there are intrinsic broadening mechanisms, such as phonon lifetime and opacity of the sample. Let us neglect the external factors and assume that the dominant broadening mechanism is the damping of the acoustic phonon with damping constant equal to Γ_q. The Brillouin peak is broadened into a Lorentzian:

$$I_{\mathrm{s}}(\omega_{\mathrm{s}}) = \frac{\Gamma_q}{\pi[(\omega_{\mathrm{i}} - \omega_{\mathrm{s}} - \omega_{\mathrm{ac}})^2 + \Gamma_q^2]}. \tag{7.54}$$

Many semiconductors are strongly absorbing at laser frequencies such as those of the Ar and Kr ion lasers used in scattering experiments. This attenuation of the laser light introduces uncertainties in the photon wavevector inside the semiconductor. Suppose the refractive indices n_{i} and n_{s} are now complex:

$$n_{\mathrm{i}} = \eta_{\mathrm{i}} + \mathrm{i}\kappa_{\mathrm{i}}, \tag{7.55a}$$

$$n_{\mathrm{s}} = \eta_{\mathrm{s}} + \mathrm{i}\kappa_{\mathrm{s}}, \tag{7.55b}$$

where η and κ are the real and imaginary parts of n. The photon wavevector is now defined in terms of the real refractive index:

$$k' = \eta\omega/c, \tag{7.56a}$$

while the imaginary refractive index results in the photon wavevectors inside the semiconductor spreading by the amount

$$k'' = \kappa\omega/c. \tag{7.56b}$$

As a result, the Brillouin peak is also broadened into a Lorentzian [7.80]

$$I_s(q) = \frac{I_0\omega_s^2}{c^2[(q - k_i' - k_s')^2 + (k_i'' + k_s'')^2]}, \tag{7.57}$$

where I_0 is a material-dependent parameter. *Pine* and *Dresselhaus* [7.81] pointed out that when the scattering occurs very close to the sample surface, as in opaque material, the acoustic phonon consists of the incident wave plus a wave reflected from the sample surface. When the contributions from both waves are included the lineshape becomes an asymmetric Lorentzian. The lineshape derived by *Dervisch* and *Loudon* [7.80] is given by

$$I_s(q) = \left(\frac{\omega_s^2}{c^2}\right) \frac{4I_0q^2}{[(q - k_i' - k_s')^2 - (k_i'' + k_s'')^2] + 4(k_i' + k_s')^2(k_i'' + k_s'')^2}. \tag{7.58}$$

A slightly different expression has been obtained by *Pine* and *Dresselhaus* [7.81]. This asymmetric Lorentzian has been observed in the Brillouin spectra of several semiconductors, as will be shown in the next section.

7.2.7 Experimental Determination of Brillouin Spectra

a) Experimental Techniques

Because the frequency difference between incident and scattered light in Brillouin scattering (typically a few cm^{-1}) is much smaller than in Raman scattering, Brillouin spectra are usually analyzed by Fabry-Perot interferometers rather than grating spectrometers. Since many semiconductors are opaque to visible lasers, the amount of light elastically scattered at surface imperfections tends to be very large. One interferometer often does not have sufficient stray light rejection and resolution to separate the Brillouin peak from the strong elastic peak. As in Raman scattering, this problem can be solved by using several interferometers in tandem. One such tandem system is shown in Fig. 7.29a. It consists of a plane-parallel Fabry-Perot interferometer (labeled PPFP in the figure) followed by a confocal spherical one (labeled CSFP) with the former used as a pre-filter. It is located inside a pressure cell containing an inert gas. The separation between its two plane mirrors is tuned by changing the pressure inside the cell. The mirror spacing in the spherical Fabry-Perot is tuned via piezoelectric transducers (PZT).

The most important development in Brillouin scattering instrumentation was the invention of the multi-pass Fabry-Perot interferometer by *Sandercock* [7.84]. The arrangement of one such unit is shown in Fig. 7.29b. By sending the scattered beam through the same plane-parallel interferometer several times using retroreflectors, the problem of synchronization between tandem interferometer is avoided.

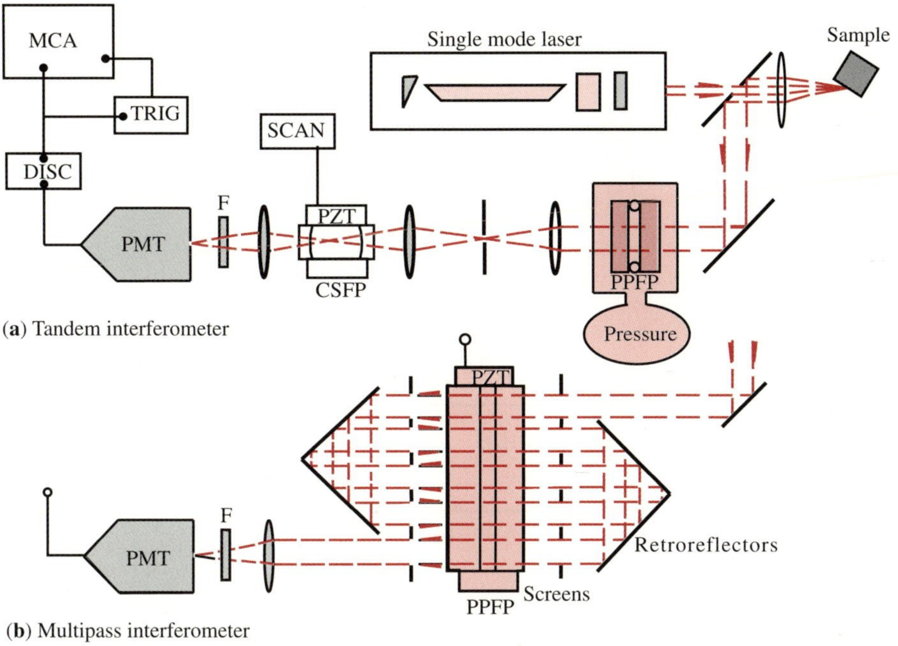

Fig. 7.29. (**a**) Schematic diagram of a Brillouin-experiment setup based on two Fabry–Perot interferometers in tandem. (**b**) Construction of a multipass Fabry–Perot interferometer designed by *Sandercock* [7.82–84]. (From [7.79])

b) Experimental Brillouin Spectra in Semiconductors

Figure 7.30 shows some of the Brillouin scattering spectra measured by *Sandercock* [7.82, 83] for opaque semiconductors like Si, Ge and GaAs. The incident laser (488 nm) was strongly absorbed in the cases of Ge and GaAs while the penetration depth was considerably longer for Si. The asymmetry in the Brillouin spectra of Ge and GaAs caused by opacity is quite apparent. From the known acoustic phonon velocities, *Sandercock* [7.82] was able to determine the complex refractive index of Ge.

7.2.8 Resonant Raman and Brillouin Scattering

Equation (7.42 and 7.50) show that additional information about the medium, besides phonon energies, can be deduced by Raman scattering. The determination of the phonon energies utilizes only the energy conservation condition of light scattering. The scattering cross section contains, at least in principle, information on electron–phonon interaction, electron–radiation interaction and the electron band structure. However, it is usually impossible to extract this information because of the summation over many intermediate states involved in (7.50). This becomes feasible if only one or a small number of intermediate states make the dominant contribution to (7.50). One way to achieve

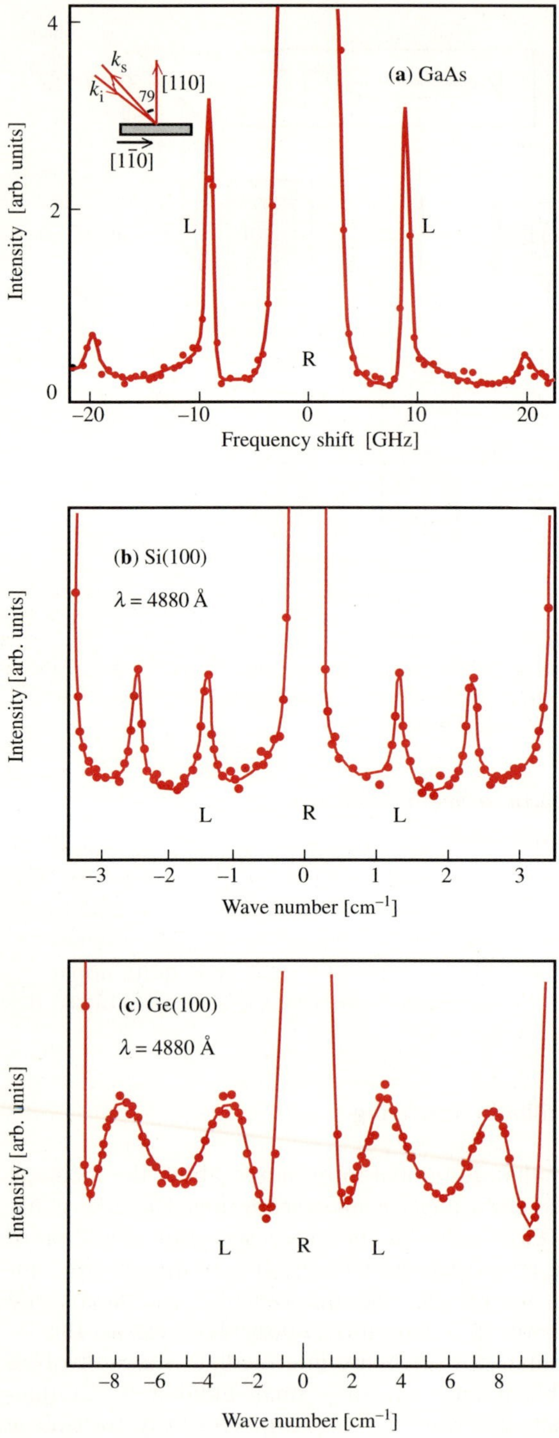

Fig. 7.30. Brillouin spectrum of (**a**) GaAs, (**b**) Si and (**c**) Ge measured by *Sandercock* [7.82, 83] using a multipass Fabry–Perot interferometer that he developed. The inset in (**a**) shows the scattering geometry. R denotes the elastic Rayleigh peak. L denotes the Brillouin peaks due to the longitudinal acoustic phonon

this is by tuning the incident laser to resonate with a strong electronic interband transition. The enhancement of the Raman cross section near an electronic resonance is known as **resonant Raman scattering**. Obviously, **resonant Brillouin scattering** is defined similarly. Such Raman and Brillouin excitation spectroscopy requires tunable lasers as the excitation sources. As pointed out in Sect. 7.2.3, several kinds of continuously tunable lasers, such as dye lasers, are now widely available. The photon energy dependence of the Raman (or Brillouin) cross section is known as the Raman (or Brillouin) excitation spectrum or resonant Raman (or Brillouin) profile.

Under resonance conditions, the contributions of the nonresonant terms of the scattering probability can be regarded as constant. In addition, of the six Feynman diagrams involving the resonant state, the one shown in Fig. 7.28a has the strongest contribution. To see this we assume the initial electronic state is the ground state $|0\rangle$ of the semiconductor with no electron–hole pairs excited, and its energy is taken to be zero. We shall denote the resonant intermediate state as $|a\rangle$ with energy E_a. The Raman scattering probability for a given phonon mode in the vicinity of E_a (after summing over ω_s to remove the delta function) can be approximated by

$$P_{\mathrm{ph}}(\omega_\mathrm{i}) \approx \left(\frac{2\pi}{\hbar}\right)\left|\frac{\langle 0|\mathcal{H}_{\mathrm{eR}}(\omega_\mathrm{i})|a\rangle\;\langle a|\mathcal{H}_{\mathrm{e-ion}}|a\rangle\;\langle a|\mathcal{H}_{\mathrm{eR}}(\omega_\mathrm{s})|0\rangle}{(E_a-\hbar\omega_\mathrm{i})(E_a-\hbar\omega_\mathrm{s})}+C\right|^2, \quad (7.59)$$

where C is a constant background. Notice that there are other terms in (7.50) which contain either $(E_a-\hbar\omega_\mathrm{i})$ or $(E_a-\hbar\omega_\mathrm{s})$ in their denominators. These terms will also exhibit enhancement when $\hbar\omega_\mathrm{i}$ is in the vicinity of E_a. However, the difference between $\hbar\omega_\mathrm{i}$ and $\hbar\omega_\mathrm{s}$ is equal to the phonon energy and is usually small compared with electronic energies. Whenever $(E_a-\hbar\omega_\mathrm{i})$ is small $(E_a-\hbar\omega_\mathrm{s})$ will also be small. Thus the term we include in (7.59) has "almost" two resonant denominators while the other terms contain at most one. [Except for special circumstances, these two denominators do not vanish simultaneously. The case $E_a = \hbar\omega_\mathrm{i}$ is referred to as an **incoming resonance** while $E_a = \hbar\omega_\mathrm{s}$ is an **outgoing resonance**.] We have therefore lumped these less resonant terms together with the nonresonant contributions in the constant C. It is important to note that the constant term C is added to the resonant term first and then the total sum is squared to obtain the scattering probability. This means that it is possible for the resonant term to interfere with C depending on the relative sign. While such interference effects are not uncommon [7.85, 86], we shall neglect them here. Within this approximation the constant term can be put outside the absolute square sign in (7.59).

When either the incident $(\hbar\omega_\mathrm{i})$ or the scattering photon energy $(\hbar\omega_\mathrm{s})$ is resonant with E_a, (7.59) diverges (the energy denominator vanishes). One way to avoid this unphysical situation is to assume that the intermediate state $|a\rangle$ has a finite lifetime τ_a due to radiative and nonradiative decay processes. As a result, E_a has to be replaced by a complex energy $E_a - \mathrm{i}\Gamma_a$, where Γ_a is the **damping constant** [7.1] related to τ_a by $\Gamma_a = \hbar/\tau_a$. If the resonant state E_a is a discrete state (such as a bound state of an exciton) and is well separated from other intermediate states, the Raman scattering probability in the vicinity of E_a can be written as[5]

$$P_{\text{ph}} \approx \left(\frac{2\pi}{\hbar}\right) \left| \frac{\langle 0|\mathcal{H}_{\text{eR}}(\omega_{\text{s}})|a\rangle \ \langle a|\mathcal{H}_{\text{e-ion}}|a\rangle \ \langle a|\mathcal{H}_{\text{eR}}(\omega_{\text{i}})|0\rangle}{(E_a - \hbar\omega_{\text{i}} - \text{i}\Gamma_a)(E_a - \hbar\omega_{\text{s}} - \text{i}\Gamma_a)} \right|^2 . \tag{7.60}$$

If the phonon involved in the Raman scattering has a nonnegligible damping constant Γ_0, it can also be included in (7.60) by replacing $\hbar\omega_{\text{s}}$ with $\hbar\omega_{\text{i}} - (\hbar\omega_0 - \text{i}\Gamma_0)$.

In general the behavior of Raman cross sections under resonance conditions depends on whether the intermediate states form a continuum or not. In the case of resonance with excitons, the results depend on the magnitude of the exciton oscillator strength and also on its damping constant. We shall consider some typical cases of resonant Raman and Brillouin scattering.

a) Resonant Raman Scattering in the Vicinity of Absorption Continua

For simplicity we shall consider resonant Raman scattering in the vicinity of a direct bandgap formed by spherical conduction and valence bands. In addition, we assume that (1) the wavevectors of the incident and scattered photons and of the phonon are all negligible and (2) the three matrix elements in (7.60) are constant and independent of wavevectors. With these simplifications the dependence of P_{ph} on the incident photon energy $\hbar\omega_{\text{i}}$ can be expressed as

$$P_{\text{ph}} \propto \left(\frac{1}{\hbar\omega_0}\right)^2 \left| \frac{1}{E_a - \hbar\omega_{\text{i}} - \text{i}\Gamma_a} - \frac{1}{E_a - \hbar\omega_{\text{s}} - \text{i}\Gamma_a} \right|^2 . \tag{7.61}$$

By using (4.51) to express $(E_a - \hbar\omega_{\text{i}} - \text{i}\Gamma_a)^{-1}$ in terms of real and imaginary parts and then comparing the results with the expressions (6.48 and 49) for the real and imaginary parts of the dielectric functions in the vicinity of a direct bandgap, we can express P_{ph} in terms of the complex dielectric function ε:

$$P_{\text{ph}} \propto \left(\frac{1}{\hbar\omega_0}\right)^2 |\varepsilon(\omega_{\text{i}}) - \varepsilon(\omega_{\text{s}})|^2 . \tag{7.62}$$

Martin [7.87b] has presented a more rigorous derivation of this result plus discussions of the effects of nonzero phonon wavevectors on resonant Raman scattering.

In the limit that the phonon energy approaches zero, (7.62) can be rewritten as

$$P_{\text{ph}} \propto \left| \frac{\partial \varepsilon}{\partial E} \right|^2 . \tag{7.63}$$

Since the phonon frequency in Brillouin scattering is quite small, we may expect (7.63) to be more valid for resonant Brillouin experiments. One way

[5] The reader may wonder why the ω_s^4 term of (7.42) does not appear in (7.60). For a solution to the puzzle see [7.87a].

to state the result of (7.63) is that resonant Raman and Brillouin scattering can be regarded simply as a form of modulation spectroscopy. The same conclusion can be reached by starting from the macroscopic theory of Raman scattering in Sect. 7.2.1. In this approach the atomic vibrations modulate the electric susceptibility χ (hence also the dielectric function) at the phonon frequency. When the damping of the electrons is larger than the phonon frequency ω_0 the resulting oscillation of the electrons at the phonon frequency is over-damped and the modulation appears to them to be static. As a result, the term $(\partial\chi/\partial Q)_0$ in the Raman tensor $\mathcal{R}$, see (7.37), can be regarded as a derivative of χ with respect to a *static* modulation Q. For example, if an optical phonon of amplitude Q changes the bandgap energy by an amount δE_g then $\mathcal{R}$ is proportional to $(\partial\chi/\partial E_g)$ for photon energy in the vicinity of E_g. Within this approximation the dispersion of the Raman intensity I_s is proportional to $|\partial\chi/\partial E_g|^2$ or $|\partial\varepsilon/\partial E|^2$. The tensor $\mathcal{R}$ for Brillouin scattering (the Brillouin tensor) is actually the derivative of χ with respect to strain (i. e., the so-called elasto-optic constants) contracted with the strain induced by the corresponding acoustic phonon (Problem 7.7).

Comparison of (7.63) with experimental results is complicated by the fact that the lowest energy bandgap in most diamond- and zinc-blende-type semiconductors involves two transitions (E_0 and $E_0 + \Delta_0$) split by spin–orbit coupling. The data points in Fig. 7.31 show the experimental dispersion in the

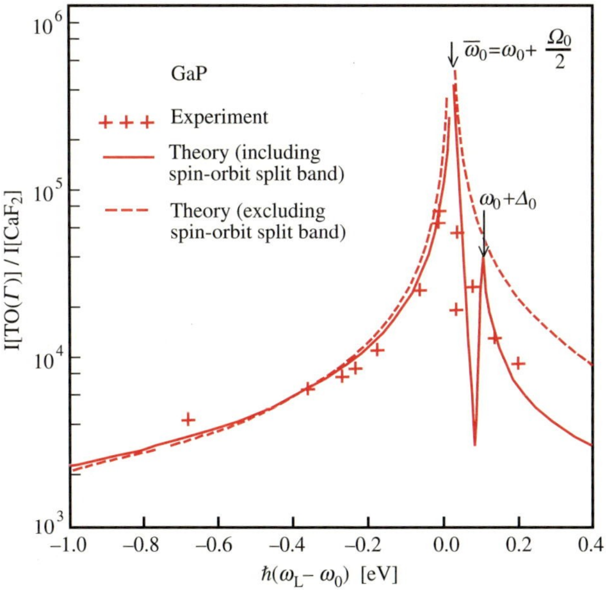

Fig. 7.31. The room temperature Raman cross section of the TO phonon in GaP plotted as a function of the incident photon frequency ω_L relative to the bandgap frequency ω_0. The *broken curve* contains the contribution from the E_0 transition only while the solid curve includes also that from the $E_0 + \Delta_0$ transition. In both curves the bandgap frequency was taken to be equal to the actual bandgap frequency ω_0 plus half of the phonon frequency (Ω_0). (From [7.88])

Raman cross section of the TO phonon in GaP at room temperature. The broken curve is a plot of the calculated dispersion in $|\partial\chi/\partial E|^2$ including only the contribution of the E_0 transition. The solid curve is calculated similarly but including the $E_0 + \Delta_0$ transitions. In the theoretical curves the bandgap frequency was taken to be equal to the actual bandgap frequency ω_0 plus half of the phonon frequency Ω_0. For more recent low temperature data on GaP where incoming and outgoing resonance can be separated see [7.89].

b) Resonance Raman Scattering at Excitons

Free Excitons

In Fig. 7.31 the enhancement in the Raman cross section at resonance is only two orders of magnitude relative to the nonresonant background. Furthermore, only a limited amount of information about the electron–phonon interaction and the energy band structure is obtained from these measurements. The importance of excitons as resonant intermediate states has been pointed out by *Birman* and co-workers [7.90, 91]. Because of their small damping constants at low temperatures, both free excitons and bound excitons have been shown to enhance the Raman cross section by several orders of magnitude. Such strong resonance effects have made possible the observation of new phenomena, such as wavevector-dependent electron–LO phonon interaction, electric-dipole forbidden transitions, higher order Raman scattering involving more than three phonons, and the determination of exciton dispersions. Many reviews [7.78, 92, 93] have been written on these topics. Here we limit ourselves to mentioning some of the representative results.

Using (7.60) we can write the Raman scattering probability of a phonon in the vicinity of an excitonic series indexed by the principal quantum number n (Sect. 6.3) as

$$P_{\text{ph}} \approx \left(\frac{2\pi}{\hbar}\right) \left| \sum_{n,n'} \frac{\langle 0|\mathcal{H}_{\text{eR}}(\omega_\text{i})|n\rangle\ \langle n|\mathcal{H}_{\text{e-ion}}|n'\rangle\ \langle n'|\mathcal{H}_{\text{eR}}(\omega_\text{s})|0\rangle}{(E_n - \hbar\omega_\text{i} - \text{i}\Gamma_n)(E_{n'} - \hbar\omega_\text{s} - \text{i}\Gamma_{n'})} \right|^2 . \tag{7.64}$$

In the special case where the $1s$ exciton state is the only important intermediate state, (7.64) simplifies further to

$$P_{\text{ph}} \approx \left(\frac{2\pi}{\hbar}\right) \left| \frac{\langle 0|H_{\text{eR}}(\omega_\text{i})|1\rangle\ \langle 1|H_{\text{e-ion}}|1\rangle\ \langle 1|H_{\text{eR}}(\omega_\text{s})|0\rangle}{(E_1 - \hbar\omega_\text{i} - \text{i}\Gamma_1)(E_1 - \hbar\omega_\text{s} - \text{i}\Gamma_1)} \right|^2 . \tag{7.65}$$

An example of resonant Raman scattering in the vicinity of such a single isolated exciton level is found in the layered-type semiconductor GaSe. The quasi-two-dimensional nature of the crystal structure enhances its exciton binding energy and oscillator strength. Figure 7.32 shows the enhancement in the Raman cross section of its 255 cm^{-1} LO phonon in the vicinity of its $n = 1$ exciton at 2.102 eV [7.94]. Notice that both the Stokes and the anti-Stokes Raman modes show two resonance peaks of equal strength as predicted by (7.65). One of these peaks occurs when the incident photon energy $\hbar\omega_\text{i}$ is resonant with the exciton (indicated by the arrow labeled X) and corresponds to the incoming resonance. The second one results from the resonance

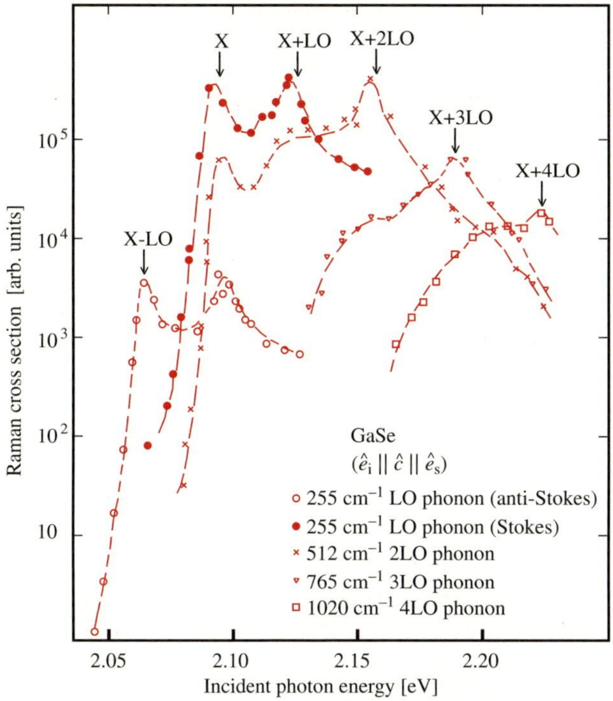

Fig. 7.32. Enhancements in the Raman cross sections of the 255 cm^{-1} LO phonon mode in GaSe plotted as a function of the incident photon energy for T = 80 K. In addition to the one-phonon mode in both Stokes and anti-Stokes scattering, the higher order multiphonon modes are also shown. The *broken curves* drawn through the data points are just to guide the eye. Notice that the higher order modes tend to show a much stronger outgoing resonance. (From [7.94])

of the scattered photon energy $\hbar\omega_s$ with the exciton and is the outgoing resonance. Notice that the outgoing resonance in the anti-Stokes mode occurs at one phonon energy below the exciton energy (indicated by X − LO). The reason why the observed enhancements are so strong is because the damping of the exciton is only about 3 meV. In Fig. 7.32 we find strong resonance not only in the one-phonon modes but also in the higher order modes. Notice that in these multiphonon resonant Raman profiles the outgoing resonance peaks tend to be stronger than the incoming ones. We shall return to this point later in this section.

Bound Excitons

Bound excitons (Sect. 7.1.4) which have "giant" oscillator strength [7.95] and long nonradiative lifetimes are also good candidates for studying resonant Raman scattering. An exciton which is strongly localized around an impurity is expected to interact strongly with the vibrational modes of that impurity. If these vibrational modes are also localized at an impurity, they are known as

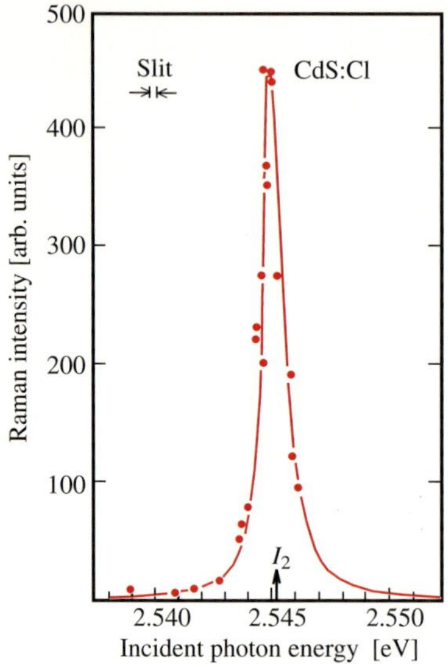

Fig. 7.33. Enhancements in the Raman intensity of the 116 cm^{-1} vibrational mode of Cl donors in CdS doped with 2×10^{17} cm^{-3} of Cl plotted as a function of the incident photon energy. The *arrow* labeled I_2 is the energy of excitons bound to the Cl donors determined from photoluminescence. The *solid curve* is a plot of (7.66) assuming the matrix elements are constant [7.96b]. Temperature equal to 2 K

local modes [7.96a]. Thus we expect the local modes of a given impurity to show specially strong enhancement at excitons bound to the same impurity. In this way resonant Raman scattering can be a useful technique for studying a small quantity of impurities in a host crystal. As an example of this application we show in Fig. 7.33 the very sharp and strong enhancement of the 116 cm^{-1} vibrational mode of shallow Cl donor impurities in CdS measured at 2K [7.96b]. This resonance occurs at 2.5453 eV, i.e., at the energy of excitons bound to shallow Cl donors (indicated by the arrow labeled I_2). The sample contained 2×10^{17} cm^{-3} of Cl and at this low concentration the Raman signal from the Cl vibrational modes should normally be too weak to be observed. The solid curve in Fig. 7.33 is a plot of (7.65) modified to *include the damping of the impurity vibrational mode*:

$$P_{\text{ph}} \approx \left(\frac{2\pi}{\hbar^2}\right) \left| \frac{\langle 0|\mathcal{H}_{\text{eR}}(\omega_{\text{i}})|b\rangle \; \langle b|\mathcal{H}_{\text{e-ion}}|b\rangle \; \langle b|\mathcal{H}_{\text{eR}}(\omega_{\text{s}})|0\rangle}{(E_{\text{b}} - \hbar\omega_{\text{i}} - \text{i}\Gamma_{\text{b}})(E_{\text{b}} - \hbar\omega_{\text{s}} - \text{i}\Gamma_0)} \right|^2 . \tag{7.66}$$

E_{b} and Γ_{b} are, respectively, the energy and damping of the bound exciton, and Γ_0 is the damping of the impurity vibrational mode. This modification is necessary because Γ_0 (equal to 2.8 meV) is larger than $10\Gamma_{\text{b}}$ (Γ_b being equal to 0.25 meV). As a result of the large Γ_0 the outgoing resonance is about 100 times weaker than the incoming resonance.

Electric-Dipole-Forbidden Excitons

One can take advantage of the extremely strong enhancement in Raman cross sections at resonance with sharp excitons to study weak electric-dipole-forbidden excitons. In Sect. 6.3.2 we pointed out that the 1S state of the lowest energy exciton series (also known as the yellow exciton series) in the *centrosymmetric* crystal Cu_2O is electric-dipole forbidden by the parity selection rule. This is also true for its higher *S*- and *D*-symmetry states in the same series. These levels can be optically excited via electric-quadrupole and magnetic-dipole transitions. However, they are very difficult to observe in absorption experiments because they are masked by the weakly electric-dipole-allowed *P*-symmetry transitions. The zone-center phonons in Cu_2O also have definite parity (Problem 3.1). Most of them have odd parity and therefore are not Raman active. *Washington* et al. [7.97] have taken advantage of these parity selection rules to observe enhancement in the Raman cross sections of *odd parity* phonons in Cu_2O at the forbidden excitons. They were able to determine accurately the energies of the normally forbidden *S* and *D* exciton series via resonant Raman scattering. Their experimental results for the 109 cm^{-1} Γ_{12}^- (also called $\Gamma_{12'}$ or Γ_3^-) mode are plotted in Fig. 7.34.

These results can be understood on the basis of (7.50a). First we shall assume that one of the exciton–photon interactions is an electric-dipole transition (to be denoted as $\mathcal{H}_{eR}^-$ since it has odd parity) while the other one is an electric-quadrupole or magnetic-dipole transition (to be denoted as $\mathcal{H}_{eR}^+$). Of

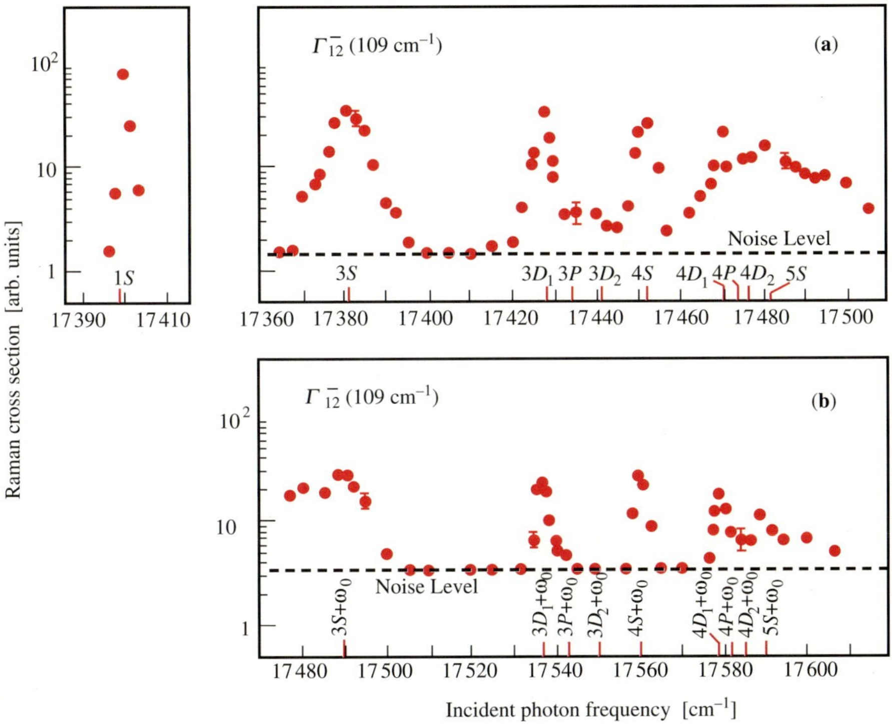

Fig. 7.34a,b. Enhancements in the Raman cross section of the odd-parity Γ_{12}^- (109 cm^{-1}) mode of Cu_2O plotted as a function of the incident photon energy in the region of the yellow excitonic series. (**a**) The peaks in the Raman excitation spectrum are identified as incoming resonances with the electric-dipole forbidden series of *S* and *D* symmetries (see red tics on the abscissa). (**b**) The outgoing resonance peaks are similarly identified. $T = 4$ K. (From [7.97])

the two intermediate exciton states n and n', again one will have even parity (for example yellow excitons in S and D states) while the other will be odd (i. e., P states of the yellow exciton series in Cu_2O). It should be remembered that these symmetries refer to the symmetry of the exciton envelope functions only. These envelope functions are multiplied by the Wannier functions for the electron and the hole to form the total wavefunction for the exciton, see (6.72). Since the binding energies of the yellow excitons of Cu_2O are small, one can use instead of Wannier functions the periodic Bloch functions of the corresponding band edges, [see (4.32)] which are both *even*: the absorption edge of Cu_2O is direct but forbidden (see Sect. 6.2.7). Correspondingly, transitions via $\mathcal{H}_{\mathrm{eR}}^{-}$ to excitonic states with an even envelope function will be (dipole) forbidden while those to states with odd envelope functions will be allowed, and vice versa for the quadrupole or magnetic dipole transitions induced by $\mathcal{H}_{\mathrm{eR}}^{+}$. With these results in mind, one can easily rewrite (7.50a) such that all the matrix elements in the numerator are nonzero:

$$P_{\mathrm{ph}} \approx \left(\frac{2\pi}{\hbar}\right)\left|\frac{\langle 0|\mathcal{H}_{\mathrm{eR}}^{-}(\omega_{\mathrm{i}})|n'(P)\rangle\langle n'(P)|\mathcal{H}_{\mathrm{e-ion}}|n(S,D)\rangle\langle n(S,D)|\mathcal{H}_{\mathrm{eR}}^{+}(\omega_{\mathrm{s}})|0\rangle}{(E_{n'}-\hbar\omega_{\mathrm{i}}-\mathrm{i}\Gamma_n)(E_n-\hbar\omega_{\mathrm{s}}-\mathrm{i}\Gamma_n)}\right.$$
$$\left. + \frac{\langle 0|\mathcal{H}_{\mathrm{eR}}^{+}(\omega_{\mathrm{i}})|n(S,D)\rangle\ \langle n(S,D)|\mathcal{H}_{\mathrm{e-ion}}|n'(P)\rangle\ \langle n'(P)|\mathcal{H}_{\mathrm{eR}}^{-}(\omega_{\mathrm{s}})|0\rangle}{(E_n-\hbar\omega_{\mathrm{i}}-\mathrm{i}\Gamma_n)(E_{n'}-\hbar\omega_{\mathrm{s}}-\mathrm{i}\Gamma_{n'})}\right|^2 \quad (7.67)$$

provided the electron–phonon interaction $\mathcal{H}_{\mathrm{e-ion}}$ that connects two states of opposite parity involves an odd parity phonon. Normally one expects the electric-quadrupole and magnetic-dipole transition matrix elements in (7.67) to be several orders of magnitude smaller than the electric-dipole matrix elements. In principle such "forbidden" (according to the Raman selection rules given in Sect. 7.2.2) phonon modes should be too weak to be observed. However, the small optical matrix elements result in small radiative decay probabilities and hence small damping constants provided nonradiative decay processes are negligible. At resonance the small damping constant in the denominators in (7.67) "over-compensates" the small matrix elements in the numerator, since the damping constants, when due only to radiative transitions, are proportional to the square of the optical matrix elements. This is particularly true for the 1S level, whose lifetime is about 1.5 ns and is determined by its decay into a lower energy *paraexciton* state [7.98].

Based on (7.67) one predicts that: (1) only the odd parity phonon modes are enhanced at the S and D excitonic states; (2) both odd and even parity phonons are enhanced at the P states; and (3) both incoming and outgoing resonances should be observed at the S, P and D states. Figure 7.34a shows that resonances at the 3P and 4P states are indeed observed, although they are weaker than the resonances at the S and D states. This is presumably a result of their larger damping constants since the P states have larger probability of radiative decay. The outgoing resonances were also observed by *Washington* et al. [7.97] as shown in Fig. 7.34b. In addition to the Γ_{12}^{-} mode, enhancements in other odd parity phonons were also observed in Cu_2O and found to be in good agreement with (7.67).

Wavevector Dependence of Exciton–LO Phonon Interaction

In principle one can determine the magnitude of electron–phonon interactions by measuring the resonant Raman profile. In practice this is rather difficult because it is necessary to know the absolute values of the corresponding Raman cross section. So far this has been accomplished only in a few cases [7.99–101]. It is much easier to obtain qualitative information or relative magnitudes of electron–phonon interactions via resonant Raman scattering. As an example of such an application, we shall study the wavevector dependence of the exciton–LO phonon (Fröhlich) interaction and the important role it plays in resonant Raman scattering.

In Sect. 3.3.5 we investigated the Fröhlich interaction for electrons which arises from the interaction between the electron and the macroscopic longitudinal electric field produced by an LO phonon. A similar interaction for holes can be derived by simply changing the sign of the charge. An exciton consists of an electron and a hole whose motions are correlated. We shall now write down, without derivation, the Fröhlich interaction Hamiltonian $\mathcal{H}_{\mathrm{F,X}}$ for an exciton derived from spherical electron and hole bands with effective masses m_{e} and m_{h}, respectively [7.27]. The resultant expression is easier to understand if we write $\mathcal{H}_{\mathrm{F,X}}$ in terms of creation and annihilation operators both for phonons [denoted by C_q^+ and C_q respectively, see also (3.22, 37)] and for excitons (a_K^+ and a_K respectively, $\boldsymbol{K}$ being the exciton wavevector):

$$\mathcal{H}_{\mathrm{F,X}} = (\mathrm{i}C_{\mathrm{F}}/q)[\exp(\mathrm{i}p_{\mathrm{h}}\boldsymbol{q}\cdot\boldsymbol{r}) - \exp(\mathrm{i}p_{\mathrm{e}}\boldsymbol{q}\cdot\boldsymbol{r})](a_{K+q}^+ a_K)(C_{-q}^+ + C_q), \tag{7.68}$$

where C_{F} is the same coefficient as defined in (3.39), $\boldsymbol{r}$ is the relative position of the electron and hole defined in (6.76) and the quantities p_{e} and p_{h} are defined as

$$p_{\mathrm{e}} = \frac{m_{\mathrm{e}}}{m_{\mathrm{e}} + m_{\mathrm{h}}} \quad \text{and} \quad p_{\mathrm{h}} = \frac{m_{\mathrm{h}}}{m_{\mathrm{e}} + m_{\mathrm{h}}}. \tag{7.69}$$

The term $a_{K+q}^+ a_K C_{-q}^+$ in (7.68) describes the scattering of an exciton from the state with wavevector $\boldsymbol{K}$ to state $\boldsymbol{K} + \boldsymbol{q}$ by emission of an LO phonon with wavevector $\boldsymbol{q}$. Similarly, the remaining term describes the scattering of the exciton with absorption of an LO phonon.

The matrix element of $\mathcal{H}_{\mathrm{F,X}}$ for the 1s state of the exciton with Bohr radius a_B can be shown to be given by (see [7.27, 92, 102])

$$|\langle 1s|\mathcal{H}_{\mathrm{F,X}}|1s\rangle| = \frac{C_{\mathrm{F}}}{q}\left(\frac{1}{[1+(p_{\mathrm{h}}a_{\mathrm{B}}q/2)^2]^2} - \frac{1}{[1+(p_{\mathrm{e}}a_{\mathrm{B}}q/2)^2]^2}\right). \tag{7.70a}$$

A plot of this matrix element (squared) is shown in the inset of Fig. 7.35 for $p_{\mathrm{e}} = 0.4$ and $p_{\mathrm{h}} = 0.6$. The important features of this plot are (1) $|\langle 1s|\mathcal{H}_{\mathrm{F,X}}|1s\rangle|^2$ vanishes for $q = 0$ and increases as q^2 for $qa_{\mathrm{B}} \ll 1$; (2) it reaches a maximum at $qa_{\mathrm{B}} \approx 2$; and (3) it decreases to zero again for large qa_{B}. This behavior can be understood with the following simple arguments. For $q = 0$ the macroscopic electric field of the LO phonon is uniform in space. Since the exciton is neutral, its energy cannot be changed by a uniform field.

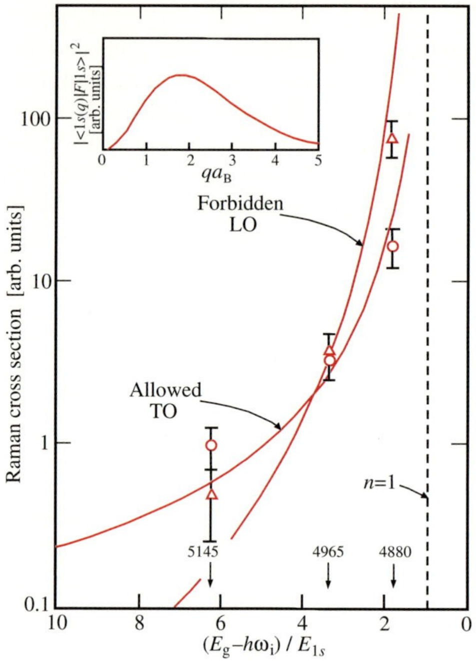

Fig. 7.35. Enhancements in the allowed TO and forbidden LO phonon Raman cross sections in CdS near its lowest energy exciton measured at 6 K. The *solid curves* represent theoretical calculations using the B exciton in CdS as the only intermediate state [7.103]. The *inset* shows a plot of the squared exciton–LO-phonon interaction matrix element $|\langle 1s|\mathcal{H}_{F,X}|1s\rangle|^2$ for $p_e = 0.4$ and $p_h = 0.6$ [see (7.70a)]

For small but nonvanishing values of q, the denominators in (7.70a) can be expanded in powers of q, to yield, to first order in q,

$$|\langle 1s|\mathcal{H}_{F,X}|1s\rangle| \simeq C_F q a_B \frac{m_e - m_h}{m_e + m_h}. \tag{7.70b}$$

Hence this matrix element is proportional to q, like the matrix element for optical quadrupole transitions. Notice that (7.70b), and also (7.70a), vanishes exactly for all values of q when $m_e = m_h$. This reflects the fact that the effect of the electric field on the electron is cancelled exactly by the corresponding effect on the hole. For $m_e \neq m_h$ the interaction between the exciton and the electric field will be largest when the wavelength of the field is of the order of the exciton Bohr radius or $qa_B \approx 1$. Finally, when q is very large the matrix element of the interaction Hamiltonian decreases as q^{-5}.

The wavevector dependence of the matrix element in (7.70) has a strong influence on the Raman selection rule for LO phonon scattering when excitons form resonant intermediate states. As we pointed out in Sect. 7.2.2, one-phonon Raman selection rules are usually derived based on the assumption that the phonon wavevector $\boldsymbol{q}$ is zero. This assumption makes it possible to deduce the nonzero and linearly independent elements of the Raman tensor $\mathfrak{R}$ using only the symmetry properties of the crystal. If $\boldsymbol{q}$ is nonzero the allowed symmetry operations are those that leave $\boldsymbol{q}$ also invariant, and hence the resultant Raman selection rule will now depend on the direction of $\boldsymbol{q}$. Breakdown of the Loudon Raman selection rules for LO phonons at resonance with excitons was observed by *Martin* and *Damen* [7.104] in CdS and was labeled "forbidden" scattering. It was found that the enhancement of the LO phonon in the "allowed" scattering configuration was weaker than those in the "forbidden" geometries. Furthermore, the enhancement was always strongest when the incident and scattering radiations were polarized parallel to each other, irrespective of the symmetry of the crystal. These results (Fig. 7.35) were explained in terms of the wavevector-dependent exciton LO-phonon interaction by *Martin* [7.103]. *Colwell* and *Klein* [7.105] tried to verify directly the wavevector dependence in the one-LO-phonon Raman scattering resonant at excitons in CdS by comparing the forward ($q \approx 0$) and backward scattering (q maximum) cross sections. They found nearly the same results for both cases and concluded that defect-induced scattering led to a breakdown of wavevector conservation, i. e., to a matrix element larger than that of (7.70a,b), nearly independent of the wavevector of the laser and scattered light. Asymmetry in forward and backward resonant Raman scattering of LO phonons was later observed in high quality CdS crystals by *Permogorov* and *Reznitsky* [7.106] and by *Yu* [7.92]. *Gross* et al. [7.107] adopted a different approach. They measured the ratio of the intensities between the "forbidden" one-LO phonon scattering in CdS and the wavevector-independent two LO phonons. The measured ratios were found to vary with excitation photon energy in accordance with the q^{-2} dependence in the squared exciton-LO phonon matrix element. Since then it has been shown that it is possible to choose scattering configurations such that the "allowed" LO phonon Raman tensor (due to the deformation potential interaction) will interfere either constructively or destructively with the wavevector-dependent "forbidden" tensor [7.86, 101]. More recent calculations show that resonant profiles and absolute scattering efficiencies can be reproduced provided the exciton continuum is included [7.100, 101].

Exciton-Mediated Multiphonon Resonant Raman Scattering

In Fig. 7.32 we found that, in resonant Raman scattering of the 255 cm^{-1} LO mode in GaSe in the vicinity of its lowest energy exciton state, not only does the one-phonon mode become strongly enhanced but the higher order multiphonon modes also exhibit large enhancements whose strengths decrease slowly with the scattering order. Such strong multi-LO-phonon scattering was first observed in CdS by *Leite* et al. [7.108] and *Klein* and *Porto* [7.109]. In CdS up to nine LO-modes have been observed (Fig. 7.36). At first sight it seems that such n-order Raman processes with $n \gg 1$ should be extremely weak since they should involve scattering of excitons by either n phonons simultaneously or by sequential applications of the one phonon interaction n times. If the first case applies, then the corresponding one-phonon interaction would be expected to be so strong that perturbation theory would break down. This is not consistent with other experimental results, such as the rather well behaved electron mobility that is determined by electron–phonon interaction. In the second case one would expect each successive application of exciton–phonon interaction to decrease the scattering cross section by at least one or-

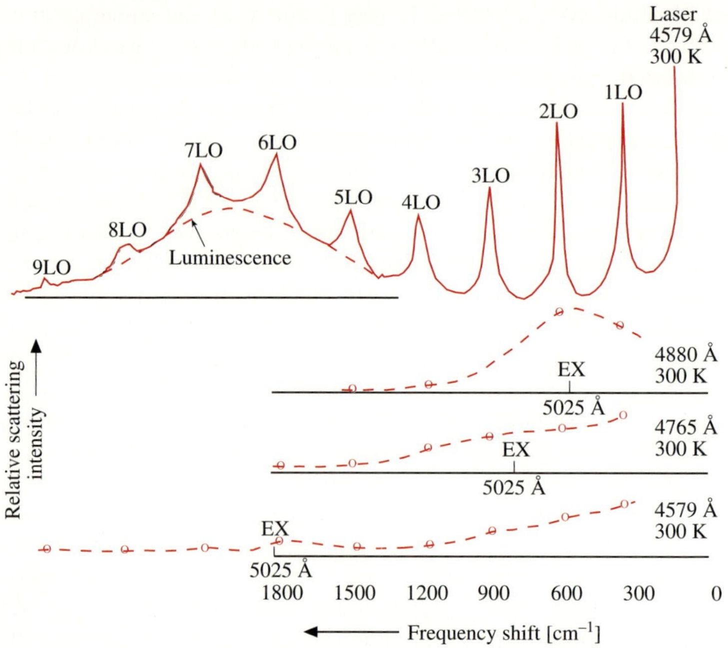

Fig. 7.36. Multi-LO-phonon Raman spectra in CdS excited at room temperature by the various laser lines of an Ar ion laser. The *broken lines* labeled *luminescence* represent the photoluminescence background from the sample. When excited off-resonance the luminescence is normally much stronger than the Raman signals. These spectra show that the Raman signal becomes enhanced when the scattered photons are at resonance with the exciton (out-going resonance). The *dashed curves* represent the scattered intensities after removing the luminescence and correcting for spectrometer response. (Adapted from [7.108]; similar results were obtained simultaneously by *Klein* and *Porto* [7.109])

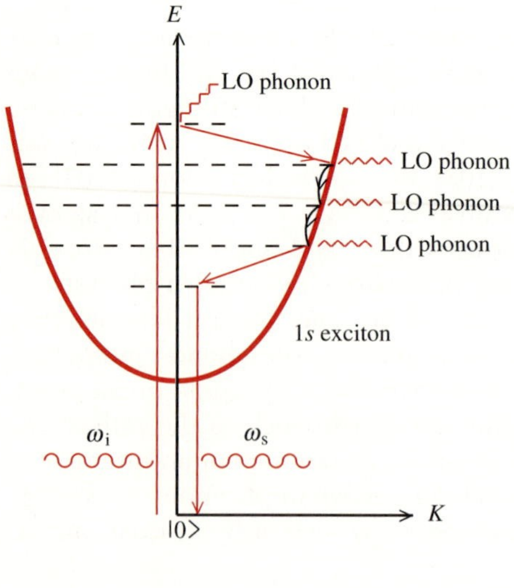

Fig. 7.37. Schematic diagram of the cascade model of multi-LO-phonon scattering with the 1s exciton as the resonant intermediate state. Notice that $E = \hbar^2K^2/(2M)$, where $M = m_e + m_h$ and K is the center-of-mass wavevector ($K = k_e + k_h$)

der of magnitude. These exciton-enhanced multiphonon Raman results have been explained in terms of the "cascade model" [7.110]. This model can be understood with the help of Fig. 7.37.

In this model the scattering process is decomposed into sequential steps: (1) absorption of the incident photon ($\hbar\omega_{\mathrm{i}}$) with excitation of a 1s exciton with energy $E_1(\boldsymbol{K}_1)$ and wavevector $\boldsymbol{K}_1$ and the emission of an LO phonon to conserve wavevector (but not necessarily energy); (2) relaxation of this exciton into lower energy states $E_1(\boldsymbol{K}_2)$ etc. with successive emission of LO phonons ("cascade") within the 1s exciton band; (3) radiative recombination of the exciton with emission of the scattered photon ($\hbar\omega_{\mathrm{s}}$) plus another LO phonon for wavevector (not energy) conservation. The scattering probability for the n-LO-phonon Raman processes can be written as

$$P_{\mathrm{ph}}^{(n)} \propto \alpha_1 \left(\frac{\tau_{\mathrm{rel}}}{\Gamma_1(\boldsymbol{K}_1)}\right)\left(\frac{\tau_{\mathrm{rel}}}{\Gamma_1(\boldsymbol{K}_2)}\right)\cdots\left(\frac{\tau_{\mathrm{rel}}}{\Gamma_1(\boldsymbol{K}_{n-2})}\right)\left(\frac{\tau_{\mathrm{rad}}}{\Gamma_1(\boldsymbol{K}_{n-1})}\right), \tag{7.71}$$

where α_1 is the absorption coefficient for LO-phonon-assisted creation of excitons, τ_{rel} is equal to the rate of relaxation of excitons via emission of one LO phonon, $\Gamma_1(\boldsymbol{K}_{\mathrm{i}})$ is the damping of the exciton state $E_1(\boldsymbol{K}_{\mathrm{i}})$, and τ_{rad} is the rate of LO-phonon-assisted radiative recombination of excitons. In this model, the enhancement of the nth order LO phonon relative to the $(n-1)$th mode depends on the relative magnitudes of τ_{rel} and $\Gamma_1(\boldsymbol{K}_{n-1})$. For polar semiconductors, in which exciton trapping by impurities is not as important as relaxation via LO phonon emission, the square of the matrix element (7.70a,b) appears both in the numerators and in the denominators of (7.71), and thus (7.71) becomes independent of the usually small electron–phonon coupling strength. Consequently the multiphonon cascade decays slowly with n, as shown in Fig. 7.36. Such cascade processes often terminate when the excitons no longer have sufficient energy to relax by LO phonon emission. When this occurs, $\hbar\omega_{\mathrm{s}} \approx E_1(\boldsymbol{K} \approx 0)$. Hence these multiphonon Raman processes tend to exhibit outgoing resonances at excitons. A rigorous formulation of the cascade model has since been presented by Zeyher [7.111], who concluded that the above argument is valid in three-dimensional crystals for $n \geq 4$. The argument (Fig. 7.37) considers only phonon-induced real transitions, for which the real parts of the denominators in the transition probabilities vanish [see for example (7.65)]. For $n \leq 3$ there is a significant contribution of virtual transitions, in which these denominators do not vanish.

It was pointed out by *Menéndez* and *Cardona* [7.112] that the same model may explain the much stronger outgoing resonances observed in even the one-LO mode if one invokes elastic scattering via impurities as a substitute for one of the LO phonons.

The fact that all the processes depicted in Fig. 7.37 are real (except for the first and last one away from resonance) suggests that the cascade model of resonant Raman scattering resembles the photoluminescence process we described in Sect. 7.1. Many authors have preferred to denote such processes as **hot luminescence** rather than resonant Raman scattering [7.113]. The distinction between these two kinds of processes has generated a fair amount of controversy [7.114]. A measurement of the *coherence* of the scattered photon for a well defined set of scattering phonons is necessary before one can distinguish these two types of processes. Since in most resonant Raman experiments such a measurement is not performed, the choice of label for the emission process becomes usually a case of semantics.

In closing this section we would like to mention recent calculations of resonant Raman scattering by two phonons in several III–V semiconductors that include not only the discrete exciton states but also the continuum. They show that the latter is essential for reproducing the experimentally determined scattering efficiencies [7.115].

Exciton Dispersion Determined by Resonant Raman Scattering

One consequence of the cascade model of multi-LO-phonon scattering (i. e., of the dominance of real transitions) is that the wavevectors of the LO phonons involved are determined by the exciton dispersion (since for real transitions the exciton relaxation processes have to satisfy both energy and wavevector conservation). In principle, the wavevector of the LO phonon emitted by an exciton will vary with the exciton energy and hence with $\hbar\omega_{\mathrm{i}}$. As a result, the corresponding LO-phonon frequencies will change with $\hbar\omega_{\mathrm{i}}$. This is usually not observed, mainly because most LO phonons have very small dispersion near the Brillouin zone center (since the Fröhlich interaction tends to favor small-$\boldsymbol{q}$ phonons; see, however, [7.116]). The opposite is true for longitudinal acoustic (LA) phonons. Acoustic phonons have *linear* dispersions (at least for small $\boldsymbol{q}$) and the deformation potential electron LA–phonon interaction $\mathcal{H}_{\mathrm{e-LA}}$ [see (3.21)] has an explicit linear dependence on q. At very low temperatures, when the LA phonon occupation number N_{LA} is much smaller than one, the term $N_{\mathrm{LA}} + 1$ in the probability of emitting an LA phonon is approximately unity. Thus the matrix element (squared) of $\mathcal{H}_{\mathrm{e-LA}}$ for phonon emission can be shown to be also proportional to q (Sect. 3.3.1). Unlike the Fröhlich interaction, the deformation potential interactions between the LA phonon with the electron and hole do not cancel each other. The net result is that the squared matrix element $|\langle 1s|\mathcal{H}_{\mathrm{X,LA}}|1s\rangle|^2$ of the exciton LA–phonon interaction Hamiltonian $\mathcal{H}_{\mathrm{X,LA}}$ dependends linearly on q. As a consequence, differences between LA and LO phonons appear: multiphonon resonant Raman peaks involving the former are broader, with a peak position occurring at the largest value of q allowed by the exciton dispersion. Furthermore, this peak frequency will change with photon energy as excitons with different energies are excited as intermediate states. Such dispersive multiphonon Raman modes involving acoustic phonons have been observed in Cu_2O at resonance with its extremely narrow 1s yellow exciton [7.9]. From the variation of the Raman frequency with excitation photon energy the exciton effective mass $m_{\mathrm{e}} + m_{\mathrm{h}}$ was determined.

To illustrate these results we show first in Fig. 7.38 the measured dependence of the frequencies of multiphonon Raman modes in Cu_2O on incident photon frequency, reproduced from [7.9]. Except for the three peaks labeled X, Y and Z, most of the other Raman peaks have been identified as multi-

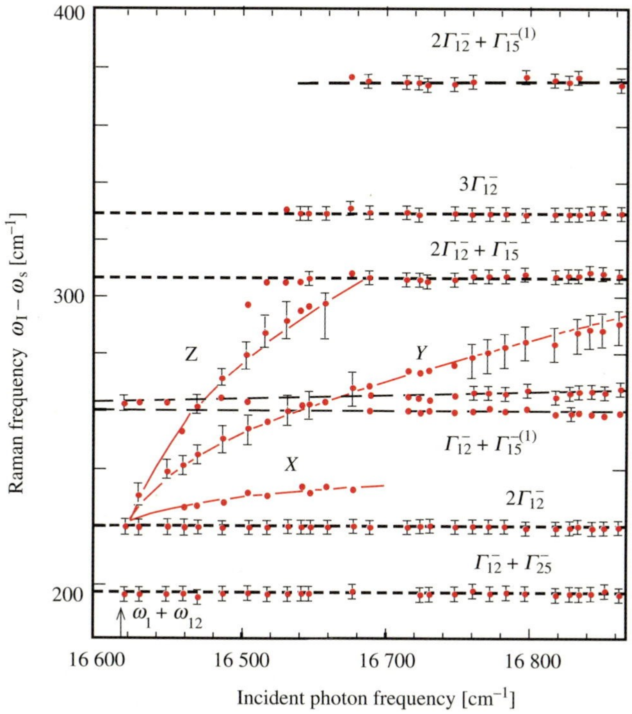

Fig. 7.38. Dependence of the frequencies of Raman modes in Cu_2O on incident photon frequency. The dispersive modes are labeled X, Y, and Z and have been identified, respectively, with $2\Gamma_{12}^-$ + TA, $2\Gamma_{12}^-$ + LA and $2\Gamma_{12}^-$ + 2LA multiphonon scattering. The *solid curves* drawn through these dispersive modes have been calculated assuming a 1s exciton with mass equal to 3.0 times the free electron mass. (From [7.9])

optical-phonon modes based on the known zone-center optical phonon frequencies [7.117, 118]. Unlike the multi-optical-phonon peaks, X, Y and Z exhibit considerable dispersion when the excitation photon frequency is changed. *Yu* and *Shen* [7.9] attributed X and Y to three-phonon scattering by $2\Gamma_{12}^-$ + TA and $2\Gamma_{12}^-$ + LA modes. The corresponding resonant Raman processes are shown schematically in Fig. 7.39. Because of the similarity between Figs. 7.37 and 7.39 we shall use the cascade model to write down their scattering probability as

$$P_{\mathrm{ph}}^{(m)} \propto \alpha_1 \left(\frac{\tau_{\mathrm{rel}}}{\Gamma_1(\boldsymbol{K}_1)} \right) \left(\frac{\tau_{\mathrm{rad}}}{\Gamma_1(\boldsymbol{K}_2)} \right) . \tag{7.72}$$

Here α_1 is the probability of the incident photon exciting a 1s exciton with wavevector $\boldsymbol{K}_1$ with the assistance of a dispersionless Γ_{12}^- optical phonon (with wavevector $-\boldsymbol{K}_1$ because of wavevector conservation), τ_{rel} the rate of relaxation of the 1s exciton into a lower energy state with wavevector $\boldsymbol{K}_2$ and τ_{rad} the rate of radiative recombination of the 1s exciton with emission of a second Γ_{12}^- optical phonon (with wavevector $-\boldsymbol{K}_2$). The energy and wavevector of the photoexcited 1s exciton are determined by energy conservation:

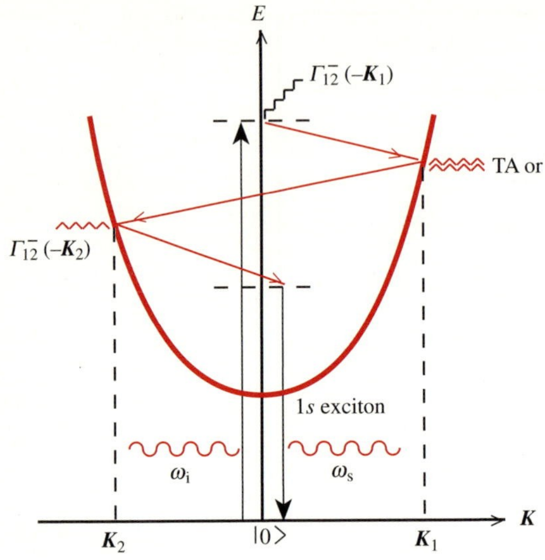

Fig. 7.39. Schematic diagram of the three-phonon scattering in Cu_2O involving emission of an acoustic phonon with linear dispersion and two dispersionless optical phonons of Γ_{12}^{-} symmetry. This diagram, and also (7.72), imply that the intermediate states that are reached after emission of the first and the second acoustic phonon are *real*, i.e., that the corresponding processes which involve virtual (i.e., energy nonconserving) states are negligible

$$E_{1s}(K_1) = E_{1s}(0) + \frac{\hbar^2 K_1^2}{2M} = \hbar(\omega_i - \omega_0), \tag{7.73}$$

where M is the mass of the exciton, and $\hbar\omega_i$ and $\hbar\omega_0$ are the energies of the incident photon and the Γ_{12}^{-} phonon, respectively. Notice that while energy conservation is in general unnecessary for transitions to virtual intermediate states during Raman scattering, this case is different because of the assumption of real excitation of the 1s exciton. The damping constant (Γ_1) of the 1s exciton is so small that the scattering probability is greatly enhanced whenever this exciton is at resonance. Similarly, not only wavevector, but also energy is conserved in the next step when the exciton relaxes via emission of an LA phonon:

$$\boldsymbol{q}_{\mathrm{LA}} = (\boldsymbol{K}_1 - \boldsymbol{K}_2) \tag{7.74}$$

(where $\boldsymbol{q}_{\mathrm{LA}}$ is the LA phonon wavevector) and

$$\omega_{\mathrm{LA}} = \hbar(K_1^2 - K_2^2)/(2M) \tag{7.75}$$

(ω_{LA} being the LA phonon frequency).

Assuming a linear dispersion for the LA phonon,

$$\omega_{\mathrm{LA}}(\boldsymbol{q}) = v_{\mathrm{LA}}|\boldsymbol{q}|, \tag{7.76}$$

with phonon velocity v_{LA}, (7.74 and 75) can be combined into the equation

$$v_{\mathrm{LA}}|\boldsymbol{K}_1 - \boldsymbol{K}_2| = \hbar(\boldsymbol{K}_1^2 - \boldsymbol{K}_2^2)/(2M). \tag{7.77}$$

For a given $\boldsymbol{K}_1$, the solutions to (7.77) range from $\boldsymbol{K}_2 = \boldsymbol{K}_1$ (i. e., $\boldsymbol{q}_{\mathrm{LA}} = 0$) to $\boldsymbol{K}_2$ being diagonally opposite $\boldsymbol{K}_1$ but with magnitude $K_2 = K_1 - (2Mv_{\mathrm{LA}}/\hbar)$. In the latter case q_{LA} reaches the maximum value $q_{\mathrm{LA}}(\max)$:

$$q_{\mathrm{LA}}(\max) = 2K_1 - (2Mv_{\mathrm{LA}}/\hbar). \tag{7.78}$$

Using the fact that the exciton–LA interaction squared matrix element $|\langle 1s|\mathcal{H}_{X,LA}|1s\rangle|^2$ is proportional to q_{LA}, *Yu* and *Shen* [7.9] showed that the relaxation rate τ_{rel} is proportional to q_{LA}^2 (see also Problem 7.10). On the basis of this result they were able to explain quantitatively the experimental lineshape of the resonantly enhanced $2\Gamma_{12}^- + LA$ peak and also obtain the theoretical Raman frequency versus excitation photon frequency curves shown in Fig. 7.38. The theoretical curve for the peak Y is essentially a plot of $2\omega_0 + v_{LA}q_{LA}(\max)$ with $q_{LA}(\max)$ calculated from (7.78). The values of K_1 used in this equation are determined by the incident photon frequency via (7.73). Since the acoustic phonon velocities can be calculated from the elastic constants, the only unknown quantity is the exciton mass M. *Yu* and *Shen* [7.9] obtained M equal to three times the free electron mass by fitting the $2\Gamma_{12}^-$+LA mode. Once M was known they were able to predict the dependence of the Raman frequencies on ω_i for the $2\Gamma_{12}^-$ + TA and $2\Gamma_{12}^-$ + 2LA modes (peaks X and Z in Fig. 7.38) with no adjustable parameters. The excellent agreement between the theoretical curves and the experimental points strongly supports their interpretation, in particular the assumption of real intermediate states, and shows that the exciton effective mass can be directly determined by resonant Raman scattering. In the next section we shall demonstrate that the exciton-polariton dispersion can also be determined via resonant scattering by acoustic phonons.

c) Resonant Brillouin Scattering: Exciton-Polaritons

In the previous section we have considered resonant Raman scattering where excitonic effects are important. In high quality crystals where the exciton damping is smaller than or comparable to the exciton-radiation interaction strength, we expect exciton-polariton effects to be important (Sects. 6.3.2 and 7.1.4). The role of the exciton-polariton (to be abbreviated as polariton in the rest of this section) in an understanding of resonant Raman scattering was pointed out by several researchers around 1969–1970 [7.119–121]. Here we are interested mainly in the determination of polariton dispersion curves by resonant Brillouin scattering. Techniques such as emission [7.122] and transmission through ultrathin plates [7.123] have also been shown to be capable of measuring polariton dispersion (Fig. 6.22). However, so far the most direct method for achieving this is resonant Brillouin scattering (RBS).

The idea of studying polaritons by RBS, first proposed by *Brenig* et al. [7.124], can be understood with the help of Fig. 7.40. Suppose a photon is incident on a medium from the left as shown in Fig. 6.1. At the surface, polaritons propagating to the right are excited. The number of polariton branches excited depends on the frequency of the incident photon. Only the lower branch polariton (Fig. 6.22) will be excited if the incident photon frequency is less than the longitudinal exciton frequency ω_L. Otherwise two polaritons, one from each branch, will be simultaneously excited. Within the polariton picture, Brillouin scattering is simply scattering of polaritons by acoustic phonons. In a backscattering geometry, the polaritons propagating to the right are scattered

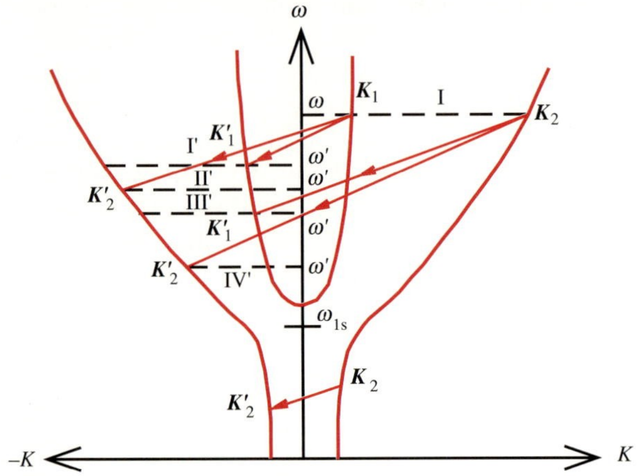

Fig. 7.40. Schematic diagram of the dispersion curves of a polariton and the Brillouin backscattering processes for polaritons with different frequencies. For polaritons with frequency less than the exciton frequency ω_{1s} only one scattering process is possible. However, for polaritons with high enough frequency ($> \omega_{1s}$) to excite both the *upper* and lower branch, four scattering processes are possible. The final polariton states in these four processes are indicated by I′, II′, III′ and IV′. In principle the *arrows* indicating the scattering processes to states I′ and II′ and to III′ and IV′ should overlap. For the sake of clarity they have been separated

by acoustic phonons into states traveling to the left. As shown in Fig. 7.40, two such polaritons going to the right can be scattered into as many as four propagating to the left (labeled I′, II′, III′ and IV′) by phonons belonging to one particular acoustic branch. The arrows in Fig. 7.40 represent the scattering processes and have been drawn so as to conserve both energy and wavevector. The slopes of the arrows are determined by the acoustic phonon velocity. In this respect these arrows are very similar to the one in Fig. 7.39 representing the scattering of the 1s exciton in Cu_2O with emission of an LA phonon. The main difference between the exciton and polariton picture lies in the existence of two polariton branches with the same energy. As a result, for each acoustic phonon branch, polariton scattering can produce as many as four Brillouin peaks, compared to only one for the simpler exciton picture. As in the case of the $2\Gamma_{12}^{-}$ + LA mode in Cu_2O, the Brillouin frequencies in scattering of polaritons by acoustic phonons vary with the excitation frequency. Polariton dispersion curves have been deduced from these variations.

The theoretical predictions of *Brenig* et al. [7.124] were first verified experimentally by *Ulbrich* and *Weisbuch* [7.125]. Using a cw tunable dye laser they excited polaritons in GaAs at low temperature and observed enhanced Brillouin peaks with a double monochromator. The number and frequencies of these peaks (both Stokes and anti-Stokes) varied with the incident laser frequency in good agreement with the predictions of *Brenig* et al. Their results are reproduced in Fig. 7.41. The theoretical curves in this figure were calculated with the polariton dispersion obtained by solving:

$$\frac{c^2k^2}{4\pi\varepsilon_0\varepsilon_\infty\omega^2} = 1 + \frac{\omega_L^2 - \omega_T(0)^2}{\omega_T(0)^2 + \omega_T(0)(\hbar k^2/M) - \omega^2} \tag{7.79}$$

where ε_∞ is the background dielectric constant (related to electronic transitions, other than the exciton being considered), M the exciton mass, and $\omega_T(0)$ the transverse exciton frequency (or ω_T in Fig. 6.22). We notice that this expression is very similar to that for phonon-polaritons discussed in Sect. 6.4. In fact one can obtain this result by replacing the transverse optical phonon frequency in (6.110b) by the corresponding k-dependent exciton frequency: $\omega_T(k) = \omega_T(0) + (\hbar k^2/M)$. The theoretical polariton curves in Fig. 7.41a were obtained by fitting the experimental points in Fig. 7.41b with these adjustable parameters: $\varepsilon_\infty = 12.55$; $M = 0.6$ times the free electron mass, $\hbar\omega_T(0) = 1.5150$ eV, and $\hbar\omega_L - \hbar\omega_T(0) = 0.08$ meV. Although a fairly large number of parameters are used to fit the experimental results, these parameters can all be determined quite accurately because of the large amount of information contained in the experimental data. As an example, the broken curve in Fig. 7.41b was obtained by changing M to 0.3 times the free electron mass while keeping the other parameters unchanged. This shows that the $2 \to 2'$ curve in Fig. 7.41b is quite sensitive to the value of M.

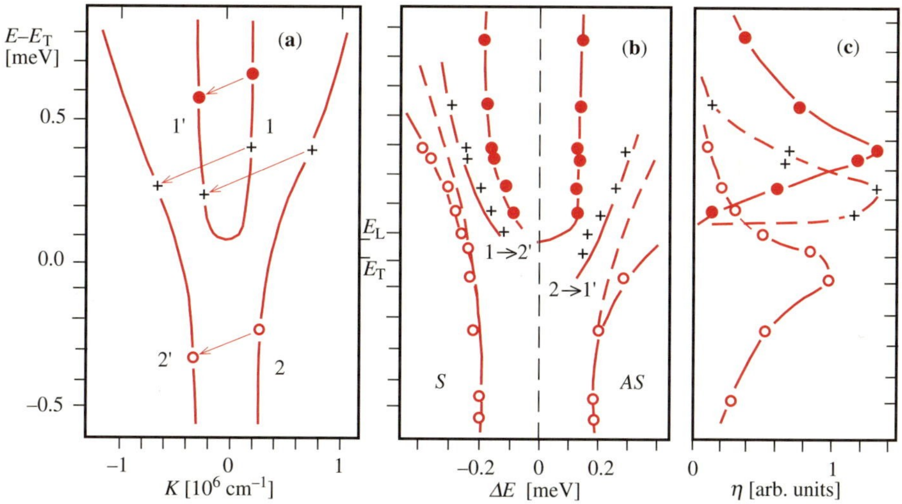

Fig. 7.41. (**a**) Polariton dispersion curve of GaAs calculated with the polariton parameters given in the text. The *arrows* indicate the four Stokes processes responsible for the Brillouin peaks in (**b**). (**b**) Experimental (*points*) and theoretical (*solid curves*) Brillouin peak shifts as a function of the incident laser energy. The theoretical curves are calculated with the polariton dispersion in (**a**). The LA phonon velocity is assumed to be 4.8×10^5 cm/s. As an indication of how sensitive the results are to M, the *broken curves* have been calculated with $M = 0.3$ times the free electron mass. (**c**) Dependence of the scattering efficiencies of the Brillouin peaks on incident photon energy. (From [7.125])

Since the pioneering work of *Ulbrich* and *Weisbuch* [7.125], RBS has been studied in almost all semiconductors that show well-defined exciton-polaritons. It is beyond the scope of this book to review all the subsequent work in this area. Interested readers are referred to the excellent review articles [7.93, 126, 127].

PROBLEMS

7.1 *Einstein's A and B Coefficients*

Assume a collection of atoms with two nondegenerate levels n and m which are in thermal equilibrium with radiation from a blackbody radiator. The energies of these levels are, respectively, E_n and E_m ($E_n > E_m$). Let N_n and N_m be the population of electrons in the levels n and m, respectively. At thermal equilibrium $N_m/N_n = \exp[-(E_n - E_m)/(k_B T)]$. Use the principle of detailed balance to derive (7.1). The frequency of the radiation is ν with the photon energy $h\nu = E_n - E_m$. The velocity of light in the medium is c/n_r, where n_r is the refractive index. Let $u(\nu)\cdot\Delta\nu$ represent the average energy per unit volume of blackbody radiation with frequency between ν and $\nu + \Delta\nu$. $u(\nu)$ is given by the well-known Planck radiation law (derive this law from the Bose–Einstein distribution function):

$$u(\nu) = \frac{8\pi h\nu^3 n_r^3}{c^3\{\exp[h\nu/(k_B T)] - 1\}}, \tag{7.80}$$

where k_B is the Boltzmann constant and T the temperature. The amount of energy absorbed per unit time from the radiation field by the atoms is given by $B_{mn}u(\nu) \cdot N_m$. One may call this "stimulated absorption". At the time Einstein proposed the A and B coefficients, emission was assumed to be always spontaneous (the rate is denoted by A_{nm}). Einstein realized that if emission does not depend on $u(\nu)$ while absorption depends on $u(\nu)$ it would be impossible to achieve thermal equilibrium between the absorber and the thermal radiation. As a result he proposed the concept of stimulated emission (rate denoted by B_{nm}). One important consequence of this new idea is the laser. The total amount of energy radiated per unit time by the atoms is now given by $N_n \cdot [B_{nm}u(\nu) + A_{nm}]$.

7.2 *Band-to-Band Emission Lineshape*

From (7.12) calculate the peak frequency and full width at half maximum (FWHM) of the band-to-band PL spectrum in an intrinsic semiconductor. Calculate and sketch the shape of the luminescence spectrum for a heavily doped (degenerate) n- or p-type semiconductor. Show that (7.7) does not apply in this case and discuss why.

7.3 *Donor–Acceptor Pair Distributions*

a) To produce a type I DAP spectrum in a zinc-blende lattice both the donors and the acceptors are located on the same face-centered cubic sublat-

Table 7.3. Lattice vector and degeneracy in a type I DAP spectrum with the donor at the origin and the acceptor in the first ten shells

Shell number	Lattice vector in units of $a_0/2$	Degeneracy
1	$\langle 1,1,0\rangle$	12
2	$\langle 2,0,0\rangle$	6
3	$\langle 2,1,1\rangle;\langle -2,-1,-1\rangle$	12
4	$\langle 2,2,0\rangle$	12
5	$\langle 3,1,0\rangle$	24
6	$\langle 2,2,2\rangle;\langle -2,-2,-2\rangle$	4
7	$\langle 3,2,1\rangle;\langle -3,-2,-1\rangle$	24
8	$\langle 4,0,0\rangle$	6
9	$\langle 4,1,1\rangle;\langle -4,-1,-1\rangle$	12
9	$\langle 3,3,0\rangle$	12
10	$\langle 4,2,0\rangle$	24

tice. Suppose that a donor is located at the origin. Show that the lattice vectors and degeneracies for the acceptor located at the first ten shells are those given in Table 7.3.

b) Repeat the calculation for a type II distribution.

7.4 *Raman Selection Rules for the Γ_4 Optical Phonon in Zinc-Blende-Type Crystals*

a) Starting with the Raman tensors for the optical phonons in (7.39) derive the selection rules in Table 7.2. Follow the example given in the text. From the directions of $\boldsymbol{k}_i$ and $\boldsymbol{k}_s$ determine the phonon wavevector $\boldsymbol{q}$. Calculate the polarization directions of the TO and LO components. Use (7.39) to calculate the Raman tensor components corresponding to these phonon polaritons. Use (7.38) to arrive at the scattered intensities.

b) Derive the selection rules for the following scattering geometries: $x(z,y)y$; $x(y,z)y$; $x(z,x)y$; $x(y,x)y$.

c) Derive the selection rules for backscattering from a (311) face of (i) Si and (ii) GaAs (by both TO and LO phonons).

7.5 *Raman Tensors of Wurtzite Crystals*

Derive the forms of the irreducible components of the Raman tensor in a wurtzite-type crystal such as CdS (refer to Problem 3.7 for its structure and symmetry) with the point group C_{6v}. The answer can be found in [7.58].

7.6 *Raman Scattering Mediated by Polaritons*

a) From Fig. 7.26b, which shows the scattering geometry used by *Henry* and *Hopfield* [7.73] to determine the polariton dispersion in GaP, calculate the minimum value of polariton wave vector ($q_{\min}$) that can be measured for a given scattering angle Θ.

b) Show that the wavevector of the scattered photon is given by

$$k_s = k_L \cos\Theta - (q^2 - q_{\min}^2)^{1/2}. \tag{7.81}$$

c) Sketch the dependence of the Raman frequency ($\omega_L - \omega_s$) as a function of q for $\Theta = 0$, 2, and 4. Use either some fictitious values for k_L and the refractive index of the medium or values appropriate for GaP and the He-Ne laser used by *Henry* and *Hopfield* [7.73].

7.7 *Selection Rules for Brillouin Scattering*
The Brillouin tensor is obtained from the derivatives of χ with respect to strain, which we designate here for a cubic crystal by the three independent constants $P_{11}(\partial\chi_{xx}/\partial e_{xx})$, $P_{12}(\partial\chi_{xx}/\partial e_{yy})$, and $P_{44}(\partial\chi_{xy}/\partial e_{xy})$. Calculate the selection rules for forward and backward scattering by LA and TA phonons on a (100) face of a cubic semiconductor.

7.8 *Practice with Feynman Diagrams*
a) Translate the Feynman diagrams in Fig. 7.28 into the corresponding terms in (7.50).

b) Draw Feynman diagrams for processes that contribute to phonon-assisted optical transitions in an indirect-bandgap semiconductor. Translate these diagrams into an expression for the optical transition probability and compare with the results in (6.61). Fill in the missing terms in (6.61) to obtain the complete expression for the transition probability R_{ind}.

c) Draw Feynman diagrams for all possible two-phonon Raman scattering processes. Be sure to introduce a new vertex for the electron–two-phonon interaction Hamiltonian. In this interaction one electron or one hole can emit two phonons simultaneously.

7.9 *Symmetry of the Raman Tensor*
Starting with (7.50), show that the Raman tensor satisfies the condition (7.47). (*Hint*: see [7.57].)

7.10 *Raman Scattering of Odd-Parity Phonons in Cu_2O*
In Sect. 7.2.2 the selection rules for Raman scattering mediated by electric dipole transitions was discussed. These results are modified when one of the optical transitions involves an electric quadrupole transition. To derive the symmetry of the corresponding "allowed" phonons in Cu_2O one can start with the expression (7.67) and note that the matrix element theorem requires that the symmetry of the state excited by the electric dipole transition has the symmetry of an electric dipole (Γ_{15}^-). Similarly, the state involving the quadrupole transition must have the symmetry of the quadrupole (Γ_{25}^+). Use the matrix element theorem to determine the symmetries of the Cu_2O phonons whose $\mathcal{H}_{\mathrm{e-ion}}$ matrix elements coupling those states are nonzero.

7.11 *Raman Scattering by Acoustic Phonons in Cu_2O*
a) Calculate the final state wavevectors $\boldsymbol{K}_2$ of the $1s$ exciton (with initial wavevector $\boldsymbol{K}_1$) in Cu_2O after the emission of an LA phonon by solving (7.77). Show that the magnitude of the LA phonon wavevector q_{LA} varies between 0 and $2K_1 - (2Mv_{\mathrm{LA}}/\hbar)$.

b) Use the Golden Rule to calculate the exciton relaxation rate τ_{rel} due to emission of an LA phonon with wavevector q_{LA}. Show that the exciton–LA-phonon interaction matrix element is proportional to q_{LA} while the density of final states introduces another term proportional to q_{LA}. As a result τ_{rel} is proportional to q_{LA}^2.

7.12 *Phonons Under Uniaxial Stress*
Discuss the pattern of splitting of the optical phonons of Si and GaAs at the Γ point under the action of a uniaxial stress along either [001] or [111]. Calculate the form of the corresponding Raman tensors and the selection rules for backscattering and also for forward scattering on a (100) face (for [001] strain) and on a $(1\bar{1}0)$ face (for [111] strain). For help see [7.128].

7.13 *Second Order Raman Scattering*
Consider the second-order Raman spectra of silicon. They can be separated into components of Γ_1, Γ_{12}, and $\Gamma_{25'}$, symmetry. Describe a procedure to determine experimentally these three components by performing measurements with several polarization configurations.

7.14 *Deformation Potential and Fröhlich Interaction Scattering*
Calculate the scattering probability (in arbitrary units) for backscattering by LO phonons on a (100) surface of GaAs for incident and scattered polarizations parallel (both) either to [011] or to $[0\bar{1}1]$ using either the ordinary Raman tensor or the (diagonal) Raman tensor for scattering via Fröhlich interaction.

7.15 *"Reduced" Raman cross section and Phonon Density-of-States*
Explain qualitatively why the "reduced" Raman cross section of overtone scattering in Ge shown in Fig. 7.23 is *smaller* than the phonon density-of-states at low frequencies. (*Hint:* the electron–phonon interaction matrix element for the low frequency acoustic phonon modes is proportional to the phonon wave vector).

7.16 *Brillouin Scattering and Acoustic Phonon Velocity*
Brillouin scattering can be utilized to determine the acoustic phonon velocity v_{ac} by using (7.53) provided the refractive index n_{i} of the medium at the incident photon frequency is known. Show that it is possible to choose an experimental scattering geometry for which v_{ac} can be determined from Brillouin scattering without knowing n_{i}. (Hint: When a photon enters one medium from another, the component of its wave vector *parallel* to the interface has to be

conserved. If the scattering geometry is such that the component of the photon wave vector *perpendicular* to the interface is not changed during the scattering then the phonon wave vector will be determined entirely by the change in the component of the photon wave vector parallel to the interfaces of the sample. These parallel components of the incident and scattered photons inside the sample can be determined without knowing the refractive index of the sample.)

SUMMARY

In this chapter we have studied light emission processes in semiconductors. In *photoluminescence*, external radiation excites electron–hole pairs in the sample. These relax to lower energy states by giving up their excess energy to phonons. As a result, the emission produced by the relaxed electron–hole pairs is characteristic of the bandgap of the semiconductor or of gap states associated with defects. Therefore, luminescence is a very useful technique for studying excitons, bound excitons, donors, acceptors and even deep centers (such as isoelectronic traps). Some of the radiation passing through a medium is always scattered by fluctuations in the medium. Such light scattering can also be understood in terms of spontaneous emission from polarizations induced in the medium by the incident radiation. When the induced polarization is modulated by phonons (both optical and acoustic) the incident light is inelastically scattered. These emission processes, known as *Raman and Brillouin scattering*, are very powerful tools for determining the frequency and symmetry of vibrational modes in condensed media. Their excitation spectroscopies (known as *resonant Raman* or *resonant Brillouin scattering*), in which one measures the scattering cross section as a function of the incident photon energy, are also extremely useful. We have shown that they can be used to determine electronic excitation energies, electron–phonon interaction and dispersion of excitons. Since real electron–hole pairs are excited in resonant Raman and Brillouin scattering as well as in photoluminescence, the distinction between the two starts to blur, leading to the suggestion that resonant light scattering processes, especially multiphonon ones, can be regarded as a form on nonthermalized luminescence or *hot luminescence*.

8. Photoelectron Spectroscopy

CONTENTS

The reader will have noticed, especially in Chaps. 6 and 7, that a great deal of the information thus far presented has been obtained by *spectroscopic* techniques. By this is meant experiments in which the number of elementary excitations in a given infinitesimal energy interval (density of excitations) is measured. Among the excitations we have discussed are phonons, which have low energies, in the range of zero to 0.1 eV. Excitations of electrons from occupied valence to the empty conduction bands, and the corresponding excitons, have energies in the 0.1–10 eV range, an energy range which includes visible photons (1.8–3.5 eV).

We may ask whether excitations at higher energies play any role in semiconductor physics. Indeed they do. As an example we mention the collective plasma oscillations of all the valence electrons which occur at the angular frequency (known as the **plasma frequency**, see Problem 6.3)

$$\omega_p^2 \simeq \frac{4\pi N_v e^2}{4\pi\varepsilon_0 m}, \tag{8.1}$$

where N_v is the density of valence electrons and m the *free electron* mass.

The plasma oscillations with frequency defined by (8.1) lead to quantized excitations, known as **plasmons**, with a quantum of energy $E_p = \hbar\omega_p$.[1] For a typical tetrahedrally coordinated semiconductor with $N_v = 4$/atom we find from (8.1) a plasmon energy for the valence electrons in the 15–16 eV range. This range of photon energy is known as the **vacuum ultraviolet** (VUV) region because air absorbs UV photons with energies higher than 6 eV. Thus, optical experiments in this photon energy range require spectrometers with their components in vacuum. The constituent atoms in the semiconductors also possess *core electrons*, which have a much higher *binding energy* than their valence counterparts (e. g., 100 eV for the 2*p* electrons of Si, 30 eV for the 3*d* electrons

[1] Note that these *valence electron* plasmons differ from the low frequency free carrier plasmons defined in Sect. 6.5.1

of Ge; for a table of core-electron binding energies see [8.1,2]). These core levels have properties very similar to those in the isolated atoms and therefore contain relatively little (but nonnegligible) information about the properties of semiconductors. Their spectroscopic strength and small shifts from one material to another containing the same atoms, however, can be used for chemical analysis and materials characterization [8.3,4].

The excitations just mentioned are usually investigated with spectroscopic techniques involving photons and/or electrons. A beam of one of these types of particles, usually with a well-defined energy (i.e., monochromatic), impinges on the sample. The energy of these particles is typically in the range 10–1500 eV. As a result of the interaction of one of these particles with the sample an electron or a photon whose energy is less than that of the incident particle is emitted. The difference in energy between the incoming and outgoing particles represents the energy of one or more *elementary* excitations created by the incident particle in the solid.

Figure 8.1 presents a schematic diagram of these types of spectroscopies. The **photon** $\rightarrow$ **0** case represents the annihilation of a photon by creating an excitation, such as the absorption of light by the solid discussed in Chap. 6 for photons in the infrared (IR), visible, and near UV range. In the vacuum UV ($\hbar\omega > 6$ eV) conventional gas discharge sources are not as bright as the light emitted by electrons when bent in the magnetic field of an electron synchrotron or storage ring [8.5, 6], which is known as **synchrotron radiation**. This radiation, available nowadays at a few dedicated centers throughout the world, extends as a continuum from the far IR to the X-ray region, the upper photon energy cutoff (up to several tens of keV) depending on the energy of the emitting electrons and the bending magnetic fields. It is strongly collimated and can be extracted with either linear or circular polarization. Figure 8.2

photon $\rightarrow$ 0
absorption, reflection
photon $\rightarrow$ photon
Raman scattering
Compton scattering
X-ray fluorescence
(XES)
photon $\rightarrow$ electron
photoelectron spectroscopy
(XPS, UPS)
photon $\rightarrow$ electron + electron
Auger spectroscopy
electron $\rightarrow$ electron
characteristic energy loss
(EELS)
electron $\rightarrow$ electron + electron
Auger electron spectroscopy
electron $\rightarrow$ photon
inverse photoemission
(BIS)
appearance potential spectroscopy
(APS)

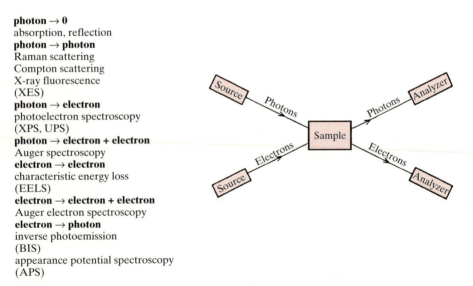

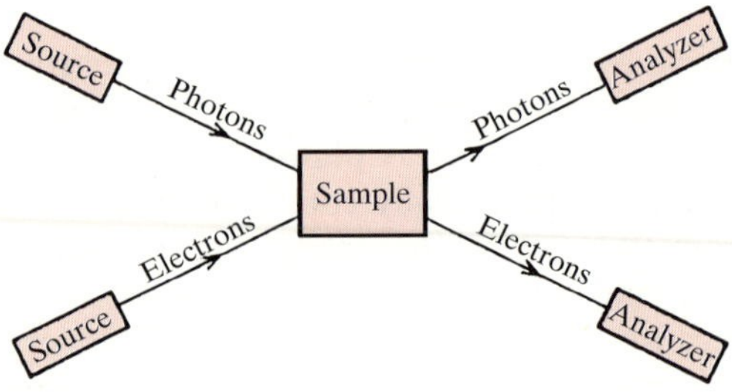

Fig. 8.1. Schematic diagram of spectroscopic methods involving photons and electrons

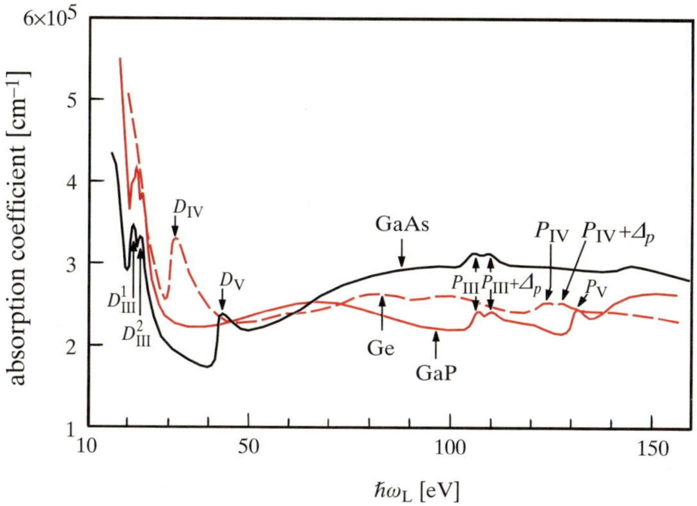

Fig. 8.2. Absorption spectra of GaP, GaAs, and Ge obtained with synchrotron radiation. The *arrows* labelled D and P indicate transitions from either d- or p-like core levels. The subscripts III, IV, V represent elements of group III, IV, V, respectively, and Δ_p stands for spin–orbit splitting [8.7]

shows the absorption spectra of Ge, GaAs, and GaP in the 20–160 eV region [8.7] as an example of the use of synchrotron radiation in photon absorption spectroscopy of semiconductors. At around 20 eV these spectra display the tail of the valence-to-conduction band transitions and transitions involving the 3d core levels of Ga(D_{III}). The transitions originating at the 3d core levels of Ge and As are seen at $\sim$ 32 eV (D_{IV}) and 43 eV (D_{V}), respectively. At higher energies, peaks related to the 3p core levels of Ge, Ga, and P are observed (denoted by P_{IV}, P_{III}, and P_{V}, respectively). These core levels are often investigated by *photoelectron spectroscopies*, as discussed below.

Monochromatized synchrotron radiation is also often used as a source in spectroscopies in which a photon is absorbed and an electron is emitted, a process known as **photoemission**. The peaks in the energy spectra of the emitted electrons exhibit shifts with respect to the energy of the incident photon corresponding to the *binding energies* of core electrons in the respective atoms. For small loss energies, details of the occupied valence electron states are also obtained.

A complementary family of spectroscopies is **bremsstrahlung isochromat spectroscopy** (BIS), also called **inverse photoemission** (Fig. 8.1). This involves bombarding the sample with monochromatic electrons, which are injected with negligible energy loss into corresponding energy states in the empty conduction bands. They then make a transition to empty states of lower energy, emitting a photon in order to conserve energy. The spectrum of the emitted photons thus yields detailed information on the *empty* conduction states, complementary to the information on occupied valence states obtained in photoemission.

Information on elementary excitations involving valence and core levels can also be obtained by bombarding a sample with electrons and examining the energy spectrum of the electrons that are emitted in this process. Some of the incident electrons manage to emerge after having lost energy to a single elementary excitation. The energy distribution of the transmitted or reflected electrons thus contains peaks which correspond to excitations in the solid. This technique is called **electron energy loss spectroscopy** (EELS) [8.8] and is particularly suitable for investigating the valence plasmons mentioned at the beginning of this section. In general, plasmons cannot be excited in optical absorption since they are *longitudinal excitations* (i.e., the electrons vibrate along the direction of propagation) while photons are *transverse* (the electric field is perpendicular to the direction of propagation, see Problem 6.3). Note that the EELS spectra are proportional to the imaginary part of ε^{-1} (the loss function), where ε is the dielectric function, which determines the optical behavior [8.9] as shown already in Chap. 6. Figure 8.3 shows the EELS spectrum of GaP obtained with 56 keV electrons. The strongest feature is the peak at $\Delta E \simeq 16$ eV, which corresponds to one-plasmon excitations. Excitations of two plasmons are observed at $\Delta E \simeq 32$ eV. This spectrum also displays peaks resulting from excitation of electrons to the conduction bands from the valence bands and from the Ga 3*d* core levels.

Among the other spectroscopic techniques sketched in Fig. 8.1 we mention explicitly **Auger electron spectroscopy** (AES) and **X-ray fluorescence** (or **emission) spectroscopy** (XES). In AES a hole is produced in a core level when an electron is ejected by an incoming electron or photon (process ① in Fig. 8.4). This hole is then filled by an electron from a higher occupied level (process ②), the liberated energy being used up by the emission of another electron (process ③). The Auger spectrum is basically the energy distribution of the electrons emitted in process 3. It contains information on the binding energy

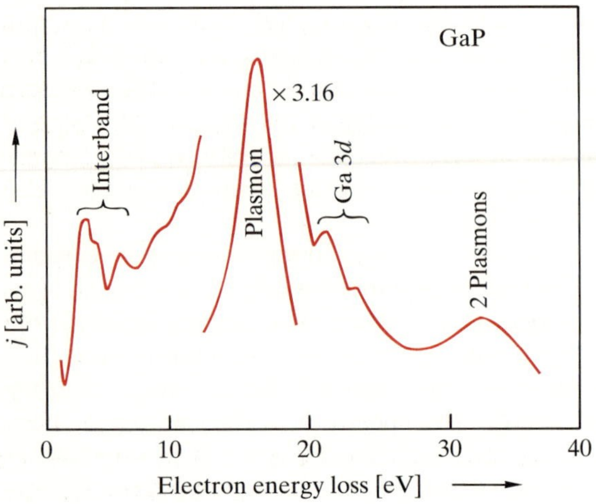

Fig. 8.3. EELS spectrum of GaP showing structures due to interband transitions, transitions involving core levels, and those corresponding to the creation of one and two plasmons [8.9]

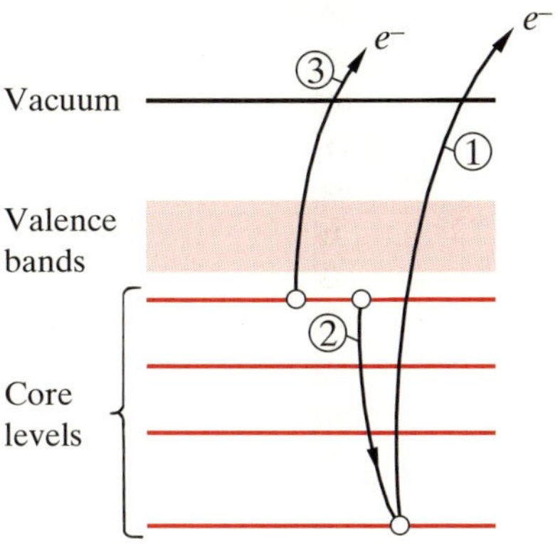

Fig. 8.4. Schematic diagram of the processes involved in Auger electron spectroscopy. The electrons in step (*1*) are expelled by either an impinging electron or a photon. The Auger spectral information is obtained by analyzing the energy of the electrons emitted via process (*3*). Step (*3*) may involve either a valence or a core electron (the latter has been assumed in the figure)

of the core electrons, somewhat modified by the complex many-body interactions between electrons occurring during the process [8.10].

In XES, a core or valence hole is produced in a process similar to ① in Fig. 8.4 (excitation by either photons or electrons). The hole is then filled by an electron with higher energy, the energy difference being released by emission of a photon whose energy is typically in the X-ray range. That photon contains information on binding energies of electrons in the valence band or core levels. Both AES and XES are standard characterization methods for semiconductors, in particular thin films and other microstructures.

A parameter of paramount importance in the spectroscopies under discussion is the **sampling depth**. In spectroscopies involving the emission of photons, the sampling depth is given by the inverse of the absorption coefficient of those photons, usually on the order of or larger than $\sim$ 200 Å (Fig. 8.2). Hence, the sampling depth encompasses several hundred monolayers and the spectra should be representative of the bulk solid provided a few precautions are taken to keep the surface clean. If the emitted particles are electrons, the sampling depth becomes the **escape depth** of those electrons, which depends on their energy. The escape depth of such electrons plotted against energy falls on a nearly *universal curve* for most solids, as illustrated in Fig. 8.5, and can be as small as 5 Å for energies in the 50–100 eV range. For such small escape depths the information obtained is characteristic of the surface rather than the bulk. Ultraclean, well-characterized surfaces [8.11, 12] are required if meaningful information is to be obtained.

8.1 Photoemission

Photoemission is an old trade. In 1887 Hertz observed that a spark between two electrodes occurs more easily if the negative electrode is illuminated by UV radiation [8.13a]. A few years later J.J. Thompson [8.13b] demonstrated

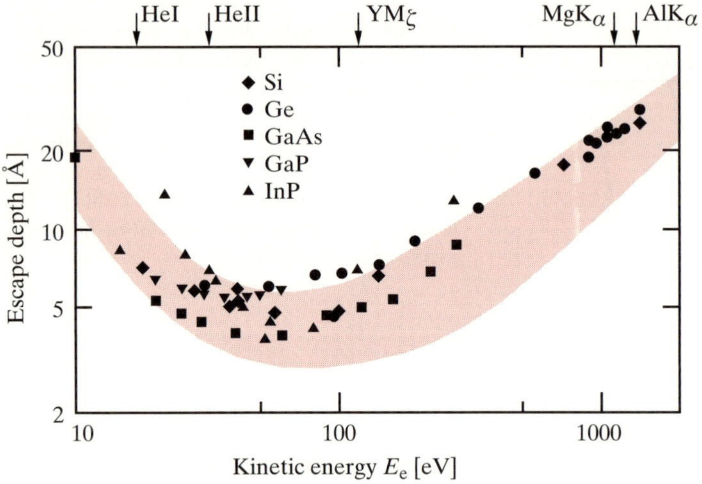

Fig. 8.5. The universal dependence of electron escape depth on energy in various solids, drawn as a shaded band that encompasses most of the existing experimental data. The energies of several laboratory photon sources are shown for reference [8.10]. Specific values for some of the most common semiconductors are also shown [8.11]

that the effect was due to emission of electrons by the electrode while under illumination. The correct interpretation of this effect was given by Einstein [8.14] (who was awarded the Nobel Prize in 1921 especially for his photoemission work). He postulated that light was composed of discrete quanta of energy E_L, this energy being proportional to ω_L, the frequency of the light: $E_L = \hbar\omega_L$. This energy, minus a binding energy I required to escape the solid, is taken up by the photoemitted electron, which thus escapes with a *maximum* energy E_e,

$$E_e = \hbar\omega_L - I, \quad (8.2)$$

where I is called the **photothreshold** (or **ionization**) **energy**. The accuracy of (8.2) has been demonstrated by many researchers by surrounding the photoemitter (a conductor) by a screen to which a negative (retarding) potential V_r is applied. For $|V_r| = E_e$ the photoinduced current should vanish. A plot of the potential at which the photocurrent vanishes versus ω_L should yield a straight line if (8.2) is valid. An early proof of this assertion is shown in Fig. 8.6.

The photothreshold energy of semiconductors with clean surfaces lies typically in the range of 5–7 eV. It can be lowered considerably by depositing a monolayer of cesium on the clean surface. In the case of GaAs such procedure leads to values of I as low as 1.4 eV, thus making this semiconductor useful as the photocathode in photomultiplier tubes operating in the near IR and visible region [8.16].

In recent years very sensitive devices have been built which enable one to measure the photocurrent for ω_L near the photothreshold ($\hbar\omega_L \simeq I$) [8.17].

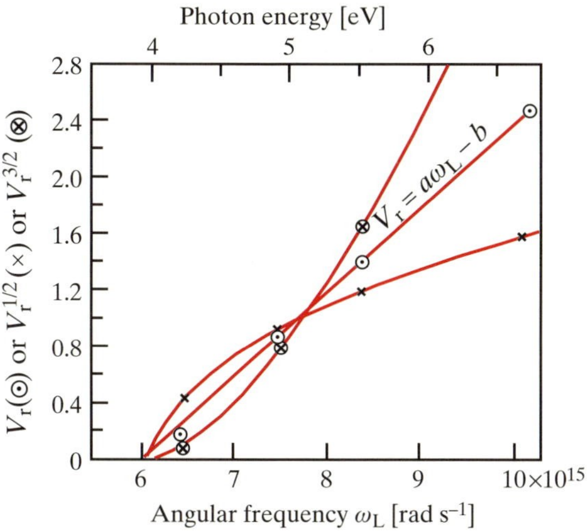

Fig. 8.6. Experimental test of (8.2) given by *Hughes* and *DuBridge* [8.15]. When the experimental data are plotted as the theshold retarding potential (in arbitrary units) V_r against the photon energy $\hbar\omega_L$, a curve much closer to a straight line is obtained, as required by (8.2), than when the data are plotted as $V_r^{1/2}$ or $V_r^{3/2}$ against $\hbar\omega_L$

For ω_L above that threshold the measured photocurrent $j(\omega_L)$ can usually be represented by a threshold function ([8.18]; see also Fig. 1.4 in [8.1])

$$j(\omega_L) = A(\hbar\omega_L - I)^{\alpha} \quad (8.3)$$

where α has values of the order of 1, depending on the type of transitions involved (direct or indirect) and on surface conditions (clean, specular, rough). Small photocurrents observed for $\hbar\omega_L < I$ are attributed to defects in the case of crystalline materials. In amorphous semiconductors they correspond to residual states that exist within the gap. Photothreshold spectroscopy is probably the most precise and direct way to investigate such states: spectrometers have been built with a dynamic range of 10^8 in the measurement of $j(\omega_L)$. Such large dynamical ranges enable one to measure densities of gap states as low as 10^{15} $(\mathrm{eV\,cm^3})^{-1}$ [8.19]. Typical **photoyield** spectra near threshold are shown in Fig. 8.7 for a cleaved Si surface.

It has been mentioned above that a monolayer of an alkali metal, for example Cs, can produce enormous changes in I. The same applies to other surface contaminants. Hence it is necessary to maintain the sample in **ultrahigh vacuum** (**UHV**) in all photoemission experiments. Assuming that a gas surrounds the sample at a pressure p and that any gas molecule impinging on the sample sticks to it (*sticking coefficient* equal to 1) it is easy to show that a *coverage* of one *monolayer* is obtained after 1 s for a pressure of 10^{-6} torr (see Problem 8.1). This unit of exposure, 10^{-6} torr, is called a **Langmuir** (L), after I. Langmuir, the winner of the 1932 Nobel Prize for chemistry. The importance of a very good vacuum is illustrated in Fig. 8.7: the yield curve extrapolates to a threshold $I = 4.87$ eV for a Si(111) surface within 1 min after *cleavage* at a pressure of 3.6×10^{-10} torr (0.02 L exposure). An hour of exposure ($\approx$1 L) to this pressure lowers I by 0.2 eV.

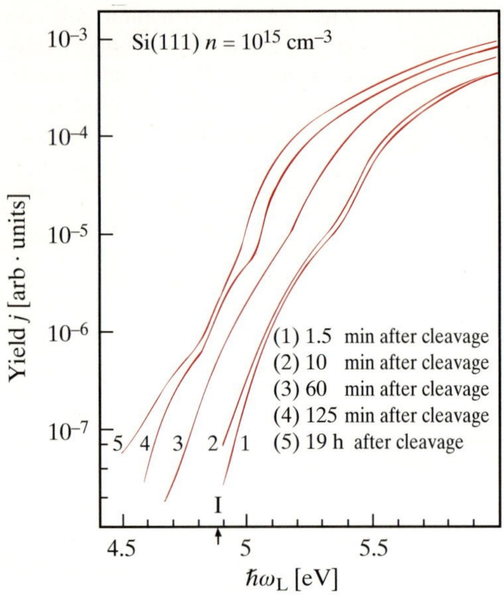

Fig. 8.7. Photocurrent (yield) near threshold measured for a Si(111) surface 1.5 min after cleavage (i. e., practically clean under the vacuum of 4×10^{-10} torr used[2]) and at several later times. Note the effects of surface contamination. (From [8.17])

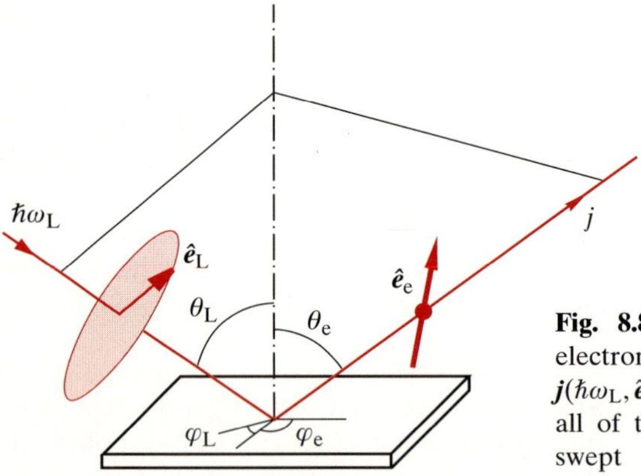

Fig. 8.8. Diagram of a typical photoelectron spectroscopy measurement of $\boldsymbol{j}(\hbar\omega_{\rm L}, \hat{\boldsymbol{e}}_{\rm L}, \theta_{\rm L}, \varphi_{\rm L};\ E_{\rm e}, \hat{\boldsymbol{e}}_{\rm e}, \theta_{\rm e}, \varphi_{\rm e})$ in which all of the variables can, in principle, be swept

The photothreshold spectrometer just mentioned is one of many types of photoemission spectrometers available for specific functions. The variety of possible instruments becomes clear when one considers that the photoelectron current is a function of a large number of variables (Fig. 8.8),

$$j = j(\hbar\omega_{\rm L}, \hat{\boldsymbol{e}}_{\rm L}, \theta_{\rm L}, \varphi_{\rm L}; E_{\rm e}, \hat{\boldsymbol{e}}_{\rm e}, \theta_{\rm e}, \varphi_{\rm e}), \tag{8.4}$$

[2] The commonly used units of pressure are the pascal (1 Pa = 1 N/m^2) and the bar (1 bar = 10^5 Pa). However, for vacuum work the preferred unit is the torr (1 torr = 1 mm of Hg = 133 Pa).

where $\hat{e}_L, \theta_L, \varphi_L$ represent the polarization, the angle of incidence, and the polar angle of the photons incident on the sample while $\hat{e}_e, \theta_e$, and φ_e are the corresponding parameters ($\hat{e}_e$ is the spin polarization) for the emitted electrons and E_e is the energy of these electrons. The number of variables in (8.4) and the information thus contained in j when measured as a function of all of them is a challenge to the imagination of both theorists and equipment manufacturers. While the theorist – with the help of modern computers – can, in principle, calculate j as a function of all the variables in (8.4), practical spectrometers can usually select only a reduced set of independent variables from the eight in (8.4). Equipment allowing measurements of electron spin polarization is very rare and will not be discussed here [8.20][3]. Measurements with polarized photons are most easily performed with synchrotron radiation [8.5], which is usually linearly polarized but also has useful circularly polarized components. Often the angles θ_L and φ_L are kept fixed while the angles θ_e and φ_e are varied, resulting in **angle-resolved photoelectron spectroscopy** (**ARPES**). Some spectrometers collect a wide range of θ_e's and φ_e's while keeping a fixed average value of these angles; this technique is referred to as **angle-integrated** [8.22].

Figure 8.9 shows the essential parts of a spectrometer that analyzes the energy E_e of the emitted electrons. Other than the UHV system (vacuum usually better than 10^{-10} torr), its basic parts are the light source, the E_e-analyzer and the detector. We now give a few details about these components.

Sources: The ideal light source is an **electron storage ring** (for synchrotron radiation) (see [8.5,6]) followed by a *grating monochromator*. The spectral resolution of the latter is typically between 0.1 and 0.01 eV: nothing is gained by making it higher than that of the energy analyzer. Since such rings are avail-

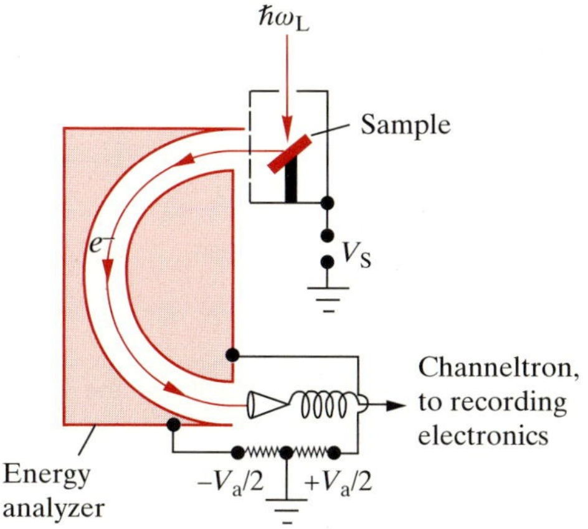

Fig. 8.9. Diagram of a photoelectron spectrometer using a hemispherical analyzer, a sample in a *Faraday cage* with a retarding/accelerating potential V_S, which is swept to obtain the spectra, and a channeltron as detector

[3] Note that GaAs, and other semiconductors, can be used as sources of spin polarized photoelectrons. See [8.21], where a beautiful example of spin-resolved photoelectron spectroscopy in GaAs is presented.

able at only a few places in the world, alternative laboratory sources, of the gas discharge or X-ray type, are used. The former employ discrete lines of a discharge in a rare gas, most commonly the HeI line at 21.22 eV and also HeII at 40.82 eV (spectroscopy with such UV photons is called **UV photoemission spectroscopy** or **UPS**). The width of these lines is on the order of 1 meV and therefore, when using them, the overall resolution is determined by that of the analyzer.[4]

X-ray tubes deliver photons with larger $\hbar\omega_L$'s (**X-ray photoemission spectroscopy** or **XPS**). Again, characteristic emission lines, this time of the metallic anodes, are used, most commonly the K_α line of aluminum at 1486.6 eV. However, these lines are much broader ($\sim$ 1 eV) than those of the gas discharge lamps and therefore their width determines the rather poor overall resolution. The Al K_α line has a number of satellites and an asymmetric shape (Fig. 8.10), which often leads to spurious signals. The use of an X-ray monochromator can reduce the width to $\sim$ 0.1 eV and thus solve these problems (Fig. 8.10). This is done, however, at the expense of considerable loss in intensity and deterioration of the signal-to-noise ratio. A list of the most common gas discharge and X-ray lines is given in Table 1.7 of [8.1].

What aspects influence the choice of either the UV gas discharge lamp or the X-ray source? We have already mentioned the resolution. Another aspect is the fact that electrons photoemitted by UV photons have energies near the minimum of Fig. 8.5 and therefore escape depths around 5 Å. Such electrons sample a region very close to the surface. This may be an advantage if surface effects are to be investigated but may distort spectra related to bulk processes. Aluminum K_α photons produce electrons with energies around 1.4 keV and escape depths of about 40 Å. The spectra are then representative of the bulk. Such photon energies are needed to investigate core levels, such as the $2p$ levels of Si (binding energy $\sim$100 eV) or Ge ($\sim$1250 eV). Angle-resolved photoemission is often performed with gas discharge lamps although, ideally, one should use synchrotron radiation to achieve a better signal-to-noise-tratio.

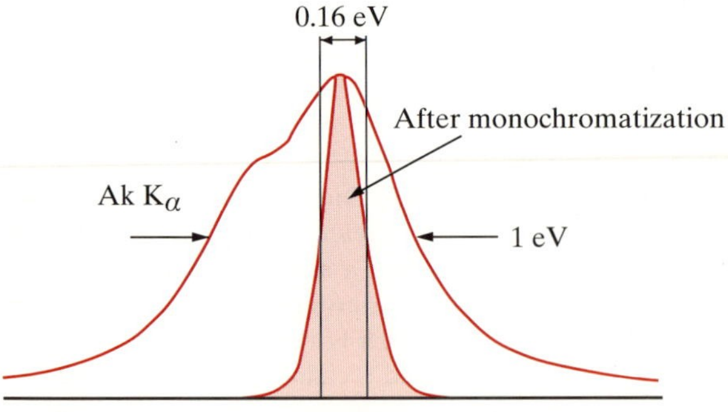

Fig. 8.10. Profile of the AlK_α line used in XPS as produced by the X-ray tube (width 1 eV) and after monochromatization (width 0.16 eV) [8.24]

[4] For the purpose of investigating the gaps of high T_c superconductors ($\sim$50 meV) by photoelectron spectroscopy, very high resolution ($\leq$10 meV) monochromators have been developed [8.23].

Electron Analyzers. A wide variety of analyzers are used in photoelectron spectroscopy, most of them based on electrostatic forces. A hemispherical analyzer is sketched in Fig. 8.9. *Cylindrical mirror analyzers* and *retarding grid systems* are also used. The hemispherical analyzer of Fig. 8.9 admits electrons through a slit and bends them by means of an electrostatic potential V_a applied between the two hemispherical metallic walls. The electrons of a given energy E_e, determined by V_a, are focused on the exit slit and, after exiting, fall on the detector. A typical relative resolution of such a system is $\Delta E_e/E_e = 10^{-2}$, independent of the electron energy. In order to keep the absolute resolution high (i. e., ΔE_e small) and constant it is customary to keep E_e at a constant, low value. This is achieved by accelerating or decelerating the electrons by means of a potential V_S applied between a *Faraday cage* surrounding the sample and the entrance slit. Photoelectron spectra are obtained by measuring the current in the detector (or the number of counts if individual electrons are counted) versus V_S.

Detector. The sketch in Fig. 8.9 is meant to indicate that the detector is a *channeltron*, which consists of a glass funnel followed by a thin tube whose interior has been coated with *a secondary emitter*; the latter produces several electrons for each of the incident ones. A channeltron is basically an *electron multiplier*. Note that the system of Fig. 8.9 is a single-channel system: only one electron energy is recorded at a given time; most of the signal (electrons) impinging on the inside of the exit slit is thus lost. Even worse, only one set of angles θ_e, φ_e is collected at a time. Multichannel systems, using a multichannel plate followed by a phosphor and a TV camera, allow the simultaneous collection of the whole spectrum, with considerable improvement (factor of $\sim$100) in signal-to-noise ratio. We have also mentioned, and shown in Fig. 8.7, how sensitive photoelectron spectra are to surface conditions. Multichannel systems decrease the measurement time and thus help to keep the surface clean while the data are being gathered (and also to keep the beam-time charges within the budget if synchrotron radiation is being used!).

Surface Preparation. The best surface preparation technique is probably cleavage, provided the material cleaves well and large enough samples are available. Ge and Si cleave relatively well along (111) surfaces while zinc-blende-type materials cleave, even better, along (110). If cleaving is not possible, surfaces can be cleaned by scraping in vacuum with a diamond file. They can also be cleaned by ion bombardment followed by thermal annealing. Surfaces of growth can be measured for samples prepared by epitaxial deposition techniques.

Origin of Energies. We have often discussed the photoelectron energy E_e without mentioning the origin that is chosen to measure this energy. One could take as the origin the energy of an electron at rest at "infinity" (i. e., away from the spectrometer, in a region without electric fields).

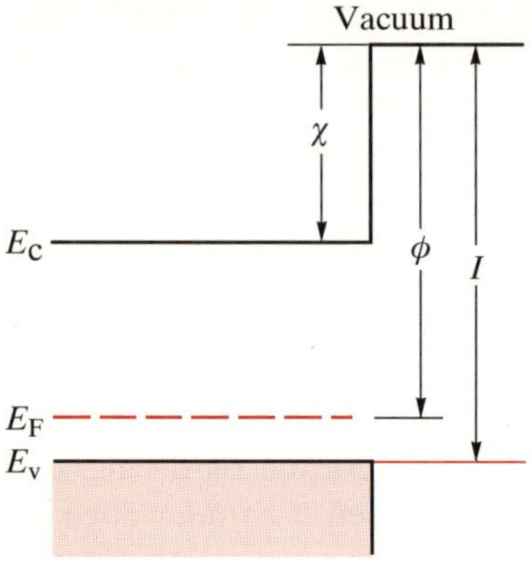

Fig. 8.11. Band diagram of a semiconductor near the surface, showing the definitions of *electron affinity* χ, *work function* ϕ, and *photothreshold energy I*

Photoelectron spectrometers, however, offer a more easily accessible, natural origin of energies if conducting samples (such as semiconductors) are measured. The Fermi levels of two conductors in contact must line up with each other, hence the Fermi level of the sample and that of the metal of the sample holder become the same. By replacing the sample by a metal (or even simply measuring photoemission from the metallic sample holder) one obtains a spectrum of photoelectrons versus V_S with a sharp step at the Fermi level of the metal, which must be the same as that of the sample to be measured. Hence the natural origin of energies in photoelectron spectroscopy is the common **Fermi level**. The energies can also be referenced to the **vacuum level** (a point outside the sample but close to it) provided the position of the Fermi level with respect to the vacuum level is known. For $\hbar\omega_L$ equal to the photothreshold energy I the photoelectrons exit with zero energy with respect to the vacuum level. If the photoelectrons arise from the bulk valence band (i. e., if no occupied surface states exist, see Sect. 8.3), I represents the energy of the vacuum level with respect to the top (i. e., the highest energy state) of the valence band E_V. The vacuum level can also be referenced to the Fermi level. Its energy is then called the **work function** and will be represented here by ϕ. The **electron affinity** χ is defined as the energy of the vacuum level referenced to the bottom of the conduction band E_c (Fig. 8.11).

Figure 8.12 shows a schematic photoemission spectrum of a semiconductor versus the potential V_S. The potential V_{S0} at which photoemission disappears corresponds to the energy of the vacuum level since no electrons can be emitted having an energy in vacuum below that of the vacuum level (this cutoff gives rise to the peak above V_{S0} in Fig. 8.12). The top of the valence band corresponds to V_{SV} while the Fermi level corresponds to V_{SF}. As shown in Fig. 8.12, V_{S0} and V_{SV} can be identified with reasonable accuracy in the pho-

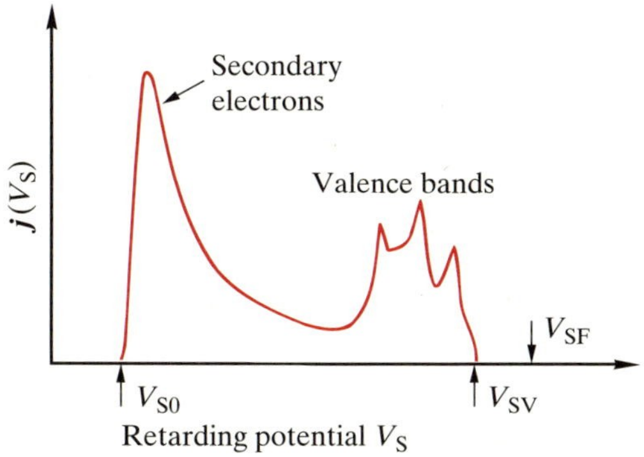

Fig. 8.12. Schematic diagram of a typical angle-integrated photoelectron spectrum $j(V_S)$ in a semiconductor versus the potential V_S of Fig. 8.9. Note the points which define the vacuum level (V_{S0}) and the top of the valence band (V_{SV}). The Fermi level (V_{SF}), as determined from the spectrum of a metal, is also indicated

toelectron spectra. V_{SF} is determined with a typical accuracy of 0.01 eV by replacing the sample by a metal as discussed above. Then I, ϕ, and χ are obtained with the equations

$$\begin{aligned} I &= \hbar\omega_L - e(V_{SV} - V_{S0}), \\ \phi &= \hbar\omega_L - e(V_{SF} - V_{S0}), \\ \chi &= I - E_0, \end{aligned} \tag{8.5}$$

where E_0 is the energy gap of the material and $\hbar\omega_L$ the impinging photon energy.

The most general description of the photoemission process consists of the excitation of an occupied electronic state inside the solid, by absorption of the incident photon, into an empty state outside the solid. This is the so-called **one-step model**. The excited electron must have its velocity pointing away from the solid so that it can be collected by the detector. This process is constrained by conservation laws. One of them is obviously energy conservation: the energy of the initial electron plus that of the photon must be equal to the energy of the emitted electron. As in most processes in crystals, there is also wavevector conservation, but only for the two components parallel to the emitting surface ($\boldsymbol{k}_\parallel$). Parallel to the plane, and assuming a clean, *unreconstructed* surface, the translational symmetry is preserved. Reconstruction may lower the translational symmetry but usually (not always!) a two-dimensional translational lattice is left. An example of surface reconstruction is shown in Fig. 8.24. Perpendicular to the surface, however, there is no translational symmetry and the corresponding component of $\boldsymbol{k}$ ($k_\perp$) need not be conserved.

The one-step photoemission process, simple as it sounds, poses considerable computational problems when quantitative evaluation is attempted. (The interested reader should consult [8.25].) One therefore often resorts to an approximation, the so-called **three-step model** [8.26]. This model assumes excitation of an electron, by the photon, from an occupied valence state to an empty conduction state (step 1) followed by *ballistic transport* (i. e., without scattering) to the surface (step 2) and transmission across the surface (step 3). Step 1 is thus closely related to the optical absorption processes described in Sect. 6.2. The conservation laws mentioned above also apply to the three-step model, although conservation of all *three* components of $\boldsymbol{k}$ is usually assumed for the first and second step.

8.1.1 Angle-Integrated Photoelectron Spectra of the Valence Bands

For angle-integrated photoelectron spectroscopy the analyzer must accept electrons over a wide range of θ_e's and φ_e's, so that, the conservation of $\boldsymbol{k}_{\|}$ does not restrict the valence states being probed. It is easy to see (Problem 8.3) that in the case of XPS (e. g., excitation with Al K_α radiation at 1487 eV), even for the small ranges of θ_e and φ_e ($\sim$5) used in angle-resolved UPS, $\boldsymbol{k}_{\|}$ conservation does not play any role, since $\boldsymbol{k}_e$ is smeared out over the whole Brillouin zone by the angular resolution. Angle-integrated photoemission is also obtained automatically when measuring polycrystalline samples. In the interpretation of angle-integrated photoemission spectra, the three-step model is usually assumed. If a small energy range of initial electrons is considered, such as that corresponding to the width of the valence bands of semiconductors ($\sim$10 eV), one can often assume that the transport to and the transmission through the surface are independent of the initial energy. In this case, because one collects electrons with energies between E_e and $E_e + \Delta E_e$ after the analyzer (where ΔE_e is the overall spectral width of the instrument, i. e., photon linewidth plus spectrometer width), the photoelectron current j must be proportional to the density of initial (i. e., valence) states $N_V(E)$ multiplied by a transition probability $P(E)$:

$$j(E_e - \hbar\omega_L) = N_V(E_e - \hbar\omega_L)P(E_e), \tag{8.6}$$

where E_e is the photoelectron energy measured with respect to the Fermi level or the top of the valence band or the vacuum level. The energy in the argument of N_V and P must, of course, be referenced to the same origin. Experience and explicit calculations (Fig. 8.13) show that $P(E)$ is not strongly dependent on E. (This is not true when d-electrons are involved in the valence bands, such as for CuCl. See [8.27].)

We show in Fig. 8.13 the photoemission spectra of germanium measured with monochromatized Al K_α radiation (XPS) and with 25 eV monochromatized synchrotron radiation (UPS). The electron energy scale has been chosen to represent initial (valence) state energies with the zero at the top of the valence band. Also included in this figure are the results of two band structure calculations of the XPS spectrum, one (dashed line) is simply based on the density of valence states, i. e., assumes that P is independent of E_V, while the

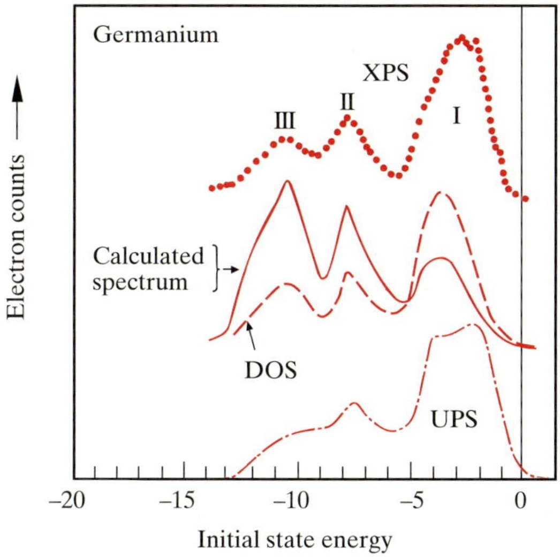

Fig. 8.13. XPS and UPS spectra of the valence bands of germanium. The *dashed line* represents the calculated density of states while the *solid line* gives the density of states multiplied by the transition probability of step 1 in the three-step model for XPS [Ref. 8.2, p. 56]

other (solid line) includes the calculated dependence of P on E_V [8.28]. Comparison of the two calculated curves gives an idea of possible effects of the energy dependence of $P(E_V)$. Note, however, that the curve which includes $P(E_V)$ actually agrees less well with the experimental XPS spectrum than that obtained under the assumption $P(E_V)$ = constant.

We compare next the spectra of Fig. 8.13 with the valence band structure of germanium. For the sake of reference we use Fig. 2.24, obtained for silicon, which is very similar to the corresponding germanium results (see also Figs. 2.25 and 8.19 for the band structure of germanium). Three sets of bands can be identified in Fig. 2.24, leading to three well-separated peaks in the *density of states* (DOS). The uppermost peak is related to the two uppermost valence bands, of nearly pure 4p atomic composition (Sect. 2.7). A second lower peak follows, related to a strongly hybridized 4p–4s band. The lowest peak (band) is 4s-like, with some p admixture away from Γ. These peaks correspond rather well to the structures labeled I, II, and III in Fig. 8.13. Note that peak I has twice the weight of either peak II or peak III. This is related to the double degeneracy of the two uppermost valence bands along the [111] and [100] directions (and near degeneracy along most of the Brillouin zone). In the tight-binding scheme (Sect. 2.7.2) this corresponds to the double degeneracy of the p orbitals *perpendicular* to those k directions. Note that for $\boldsymbol{k}$ along [111] these orbitals do not mix with the s-orbitals of the nearest neighbor along $\boldsymbol{k}$. A p-orbital pointing along the $\boldsymbol{k}$ direction strongly interacts with the s-orbital in the nearest neighbor atom, thus leading to the two lower bands and DOS peaks (Fig. 2.23).

As further examples we show in Fig. 8.14 angle-integrated XPS (Al K_α) and UPS (HeI) spectra of four III–V semiconductors.

Note that the energy scale here represents "binding energy", i. e., the initial energy of Fig. 8.13 with the opposite sign. These two types of energy scales are used at random in the literature.

These also have three peaks, similar to those of Fig. 8.13 but with one important difference: the splitting between peaks II and III is larger (~5 eV for GaAs, 3 eV for Ge). This is related to the splitting of the doubly degenerate valence band at the X_1 point of the germanium structure (Fig. 2.24) by the antisymmetric or ionic potential of the zinc-blende structure (X_1–X_3 splitting in Fig. 2.28). Note that the II–III splittings of Fig. 8.14 are about 5 eV, very close to the pseudopotential theory values of Fig. 2.28. It has been suggested that this splitting, after subtracting that of the corresponding group IV material, can be taken as a measure of the ionicity of the compound [8.29].

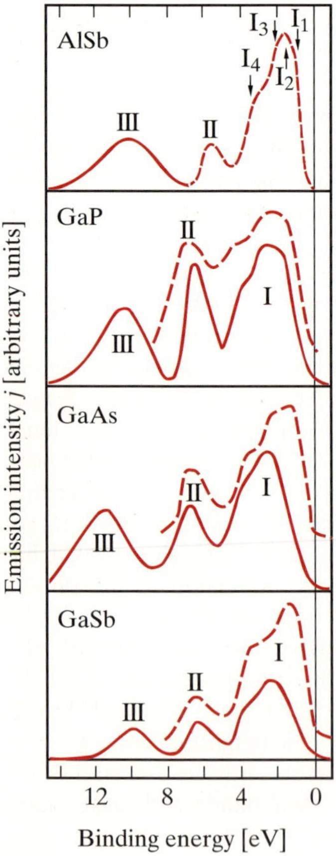

Fig. 8.14. XPS spectra of GaP, GaAs, GaSb, and AlSb (*solid lines*) and UPS angle-integrated spectra ($\hbar\omega_L$ = 21.2 eV, *dashed lines*). The notation I_i denotes structures corresponding to critical points [Ref. 8.2, p. 57]

8.1.2 Angle-Resolved Photoelectron Spectra of the Valence Bands

In order to perform *angle-resolved photoelectron spectroscopy* (ARPES) it is necessary to measure $j(\hbar\omega_L, E_L, \theta_e, \varphi_e)$. As usual in spectroscopy, j cannot be measured for exact values of θ_e and φ_e but the detector accepts a range of these parameters represented by $\theta_e \pm \Delta\theta_e/2$ and $\varphi_e \pm \Delta\varphi_e/2$. The angular resolutions $\Delta\theta_e$ and $\Delta\varphi_e$ are typically a few degrees.

Similarly to what was discussed in Sect. 8.1.1 for the dependence of j on E_L, two types of schemes are possible: single channel and multichannel. The single-channel detectors gather at a given time only the electrons within the range $\theta_e \pm \Delta\theta_e/2$ and $\varphi_e \pm \Delta\varphi_e/2$, for a fixed pair of angles θ_e, φ_e which is varied after each complete $j(E_e)$ run. Hence, most of the signal generated at a given time is lost. Nevertheless, most available data have been obtained with single-channel systems.

The standard single-channel detector consists of a miniature electrostatic analyzer with a channeltron detector attached to it (Fig. 8.9), mounted on a rotatable table so that θ_e can be changed. The azimuthal angle φ_e is varied by rotating the sample about an axis perpendicular to the photoemitting surface [8.30]. ARPES spectrometers based on movable hemispherical analyzers are available commercially.

Multichannel systems measuring simultaneously θ_e and φ_e deliver a very large set of data in a short time. The data collection time is thus reduced to a minimum, shifting emphasis to the nontrivial task of the analysis of the large number of data being simultaneously gathered. This feature can be important when using synchrotron radiation (beam time is often scarce and/or expensive) and when surface contamination has to be kept to a minimum.

A diagram of an angle-resolving multichannel system based on retarding grids, a channel plate, and a television camera or another similar two-dimensional detector is shown in Fig. 8.15. The *angular resolution* is determined by the distance between pixels in the channel plate plus crosstalk between pixels and aberrations in the electron optics. Toroidal energy analyzers, used in conjunction with synchrotron radiation, have been employed for simultaneous collection of all θ_e's [8.31].

The usefulness of ARPES is based on conservation of $\boldsymbol{k}_\parallel$, the wave vector component parallel to the emitting surface. Like $\boldsymbol{k}$ conservation in a three-dimensional crystal, $\boldsymbol{k}_\parallel$ conservation must be understood modulo $\boldsymbol{G}_\parallel$, where $\boldsymbol{G}_\parallel$ is a vector of the reciprocal lattice of the two-dimensional space group that leaves the surface invariant.[5] $\boldsymbol{G}_\parallel$ need not be considered if we confine the emitted $\boldsymbol{k}_\parallel$ to the first Brillouin zone of the surface. In principle, the component of $\boldsymbol{k}$ perpendicular to the surface ($k_\perp$) is not conserved in the general one-step model. In this case we have two equations expressing the conservation of the two components of $\boldsymbol{k}_\parallel$ and one expressing conservation of energy. The experimental data are the energy E_e of the emitted electrons and the

[5] Here we assume that the surface space group is a subgroup of that of the bulk. Difficulties may arise in defining such a group if the surface is *reconstructed*.

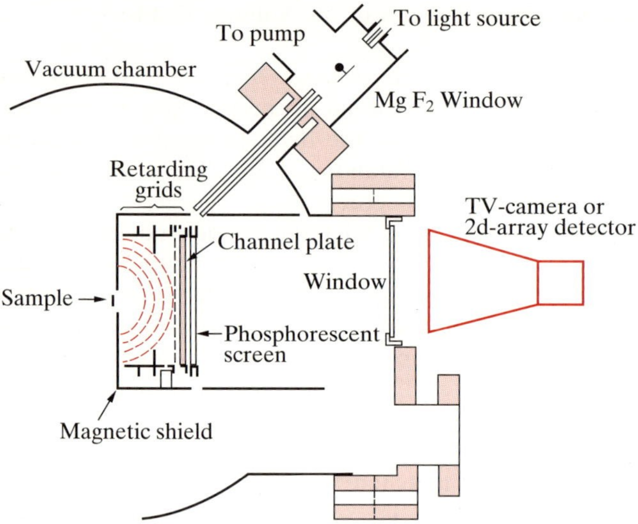

Fig. 8.15. Angle-resolved multichannel (φ_e, θ_e) photoemission system based on hemispherical *low energy electron diffraction* (LEED) electron optics [8.31]

three components of their wave vector, which we express as $\boldsymbol{K}$ (lowercase letters are reserved for $\boldsymbol{k}$ vectors inside the solid). The three components of $\boldsymbol{K}$ can be obtained from θ_e, φ_e, and E_e under the reasonable assumption of a free-electron-like dispersion relation $[E_e = \hbar^2 K^2/(2m)]$ for the electrons in vacuum. With these three conservation equations it is possible to determine three unknowns, namely E_e, θ_e, and φ_e, i. e., the photoemission spectrum, if the band structure of the solid is known. Since $k_\perp$ does not appear in the conservation equations, no information about it is obtained in this way and the bulk valence band structure $E_V(\boldsymbol{k})$ cannot be determined from the ARPES spectrum of *one given* surface.

We should mention at this point the role of the surface reciprocal lattice vectors $\boldsymbol{G}_\parallel$. For a given ω_L and E_e the photoemitted electrons form a cone, which can be expressed as $\theta_e = F_0(\hbar\omega_L, E_e, \varphi_e)$, where we assume that the function F_0 corresponds to $\boldsymbol{G}_\parallel = 0$. Since θ_e and φ_e are obtained from the direction of $\boldsymbol{K}_\parallel$, a different relationship between θ_e and φ_e is found for each nonvanishing vector $\boldsymbol{G}_\parallel$:

$$\theta_e = F_{\boldsymbol{G}_\parallel}(\hbar\omega_L, E_e, \varphi_e). \tag{8.7}$$

We thus obtain a manifold of photoemission cones, which are known as **Mahan cones** [8.32]. Equation (8.7) is based only on conservation laws, which yield no information about the strength of the photoemission current $j(\hbar\omega_L, E_e, \theta_e, \varphi_e)$. An evaluation of the relative strengths of $j_{\boldsymbol{G}_\parallel}$ for the various **Mahan cones** is a rather involved theoretical problem that will not be treated here. Fortunately, as might be expected, the first few cones (corresponding to $\boldsymbol{G}_\parallel = 0$ and the lowest values of $|\boldsymbol{G}_\parallel|$) dominate. Moreover, large values of $|\boldsymbol{G}_\parallel|$ (i. e., large θ_e) cannot be reached owing to mechanical and photon energy limitations.

In spite of the essential impossibility of determining the bulk band structure from the ARPES spectra, progress can be made in some cases by making appropriate approximations. Such is the case of *layered structures*, i. e., three-dimensional solids composed of layers well separated from each other but held together through *van der Waals interaction* (Chap. 1). The interaction between electrons in neighboring layers is very weak so that the layers in a stack can be treated as independent, i. e., as two-dimensional crystals. The electron energy is independent of $k_\perp$ and the resulting two-dimensional bands can be directly obtained from the angle-resolved spectra.

In truly three-dimensional crystals, additional information, allowing the determination of $E_\upsilon(\boldsymbol{k})$, can be found by measuring the photoemission from two or more different surfaces. The component of $\boldsymbol{k}$ perpendicular to one surface has a component parallel to the other surface. Additional conservation equations are obtained that lead to the full band structure $E_\upsilon(\boldsymbol{k})$ provided one can identify peaks corresponding to the same initial state in the photoelectron spectra of the two surfaces. This technique, however, is cumbersome and has been used rarely for semiconductors. A more fruitful approach arises from the three-step model, in which the *three-dimensional* $\boldsymbol{k}$ is conserved in the first step (a valence to conduction band excitation). The additional conservation equation for $k_\perp$ enables one to determine $k_\perp$ provided that the final state is free-electron-like. This technique has been widely and rather successfully used to determine the $\boldsymbol{k}$-dependent band structure of semiconductors. We present below a few examples of band structure determinations with ARPES (Sect. 8.3.2).

Two-Dimensional Crystals. Two-dimensional semiconductors (Chap. 1) are ideally suited for band structure determinations by ARPES. Among those that have been investigated we mention the transition metal dichalcogenides (e. g., $TaSe_2$); the In and Ga chalcogenides (InS, GaSe, InSe); GeS, GeSe, SnS, SnSe, and the semimetal graphite. We should also point out that the surfaces of semiconductors have electron states confined to the first one or two monolayers (*surface states*). They correspond to two-dimensional energy bands (*surface bands*) which can also be measured by ARPES.

The procedure for the determination of the band structure of a two-dimensional semiconductor is as follows. The spectrum of $j(E_e)$ is measured for a fixed set of all other parameters in (8.4), in particular $\hbar\omega_L$, θ_e, and φ_e. Peaks in this $j(E_e)$ spectrum are assumed to correspond to emission from a valence band state of well-defined $k_\parallel$ and E_υ. From the peak (final state) energy E_e referred to the *vacuum level* we find $K_\parallel$ with the relation

$$E_e(\hbar\omega_L, \theta_e, \varphi_e) = \frac{\hbar^2(K_\perp^2 + K_\parallel^2)}{2m} = \frac{\hbar^2 K_\parallel^2}{2m}(1 + \cot^2\theta_e). \tag{8.8}$$

This equation determines the magnitude of $\boldsymbol{K}_\parallel$. Its direction is given by the azimuthal angle φ_e. The vector $\boldsymbol{K}_\parallel$ is equal to $\boldsymbol{k}_\parallel$ after subtraction of the reciprocal lattice vector $\boldsymbol{G}_\parallel$ required to bring $\boldsymbol{k}_\parallel$ into the reduced two-dimensional

Brillouin zone. We have thus determined $\boldsymbol{k}_{\parallel}$ for the initial (valence) state. The full valence band structure

$$E_{\mathrm{v}}(\boldsymbol{k}_{\parallel}) = \hbar\omega_{\mathrm{L}} - E_{\mathrm{e}}(\hbar\omega_{\mathrm{L}}, \theta_{\mathrm{e}}, \varphi_{\mathrm{e}})$$

can be obtained by varying θ_{e} and φ_{e} with fixed $\hbar\omega_{\mathrm{L}}$. Hence, the procedure is particularly appropriate for the *discrete* lines of a gas discharge lamp. Note that, because of the existence of more than one *Mahan cone*, i.e., the possibility of obtaining the same $\boldsymbol{k}_{\parallel}$ with more than one set of θ_{e}'s and φ_{e}'s (by subtracting the required $\boldsymbol{G}_{\parallel}$'s), the measurement even possesses redundancy: the same valence bands should be obtained for different $\boldsymbol{G}_{\parallel}$'s, a fact which offers a stringent test of the accuracy of the procedure. In order to illustrate this, Fig. 8.16 shows the peaks in $j(E_{\mathrm{e}}, \theta_{\mathrm{e}})$ observed for the two-dimensional semimetal graphite. Figure 8.17 displays the band structure of graphite obtained by this procedure, compared with the results of band structure calculations [8.34]. The σ_1 bands result from the s orbitals of carbon, σ_2 and σ_3 are generated by in-plane bonding p-orbitals while π_1 bands are generated by out-of-plane bonding p-orbitals. The material is a semimetal because the bonding and antibonding π bands (only the former are shown in Fig. 8.17) are degenerate by symmetry at the P point (Problem 8.7). For data specific to semiconductors consult [8.22, 35]. This technique has been successfully applied to high T_{c} superconductors, which are also two-dimensional crystals [8.36].

Three-Dimensional Semiconductors. The most commonly used method of determining valence band structures of three-dimensional semiconductors with ARPES is based on the assumption of conservation of the three components of $\boldsymbol{k}$, which follows from the three-step model. A large number of diamond-

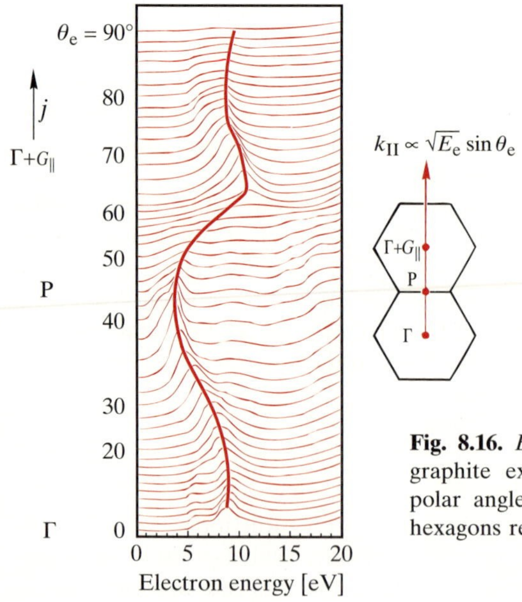

Fig. 8.16. *Energy distribution curves* (EDCs) of graphite excited with HeII radiation with the polar angle θ_{e} as a parameter [8.33]. The two hexagons represent adjacent 2*d* Brillouin zones

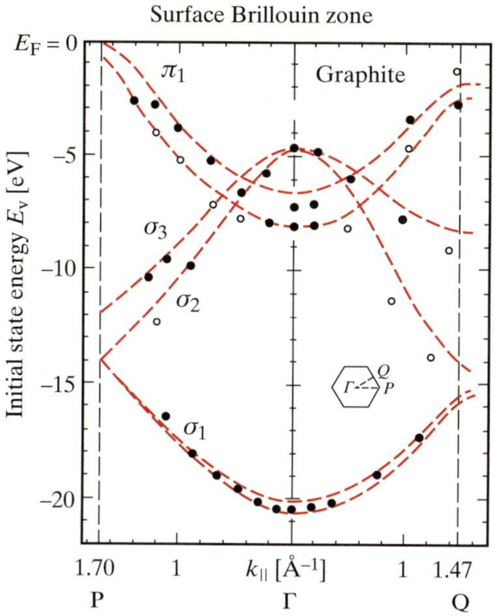

Fig. 8.17. $E_v(\boldsymbol{k}_{\|})$ dispersion of two-dimensional graphite bands measured with angle-resolved UPS (*points*), compared with the theoretical predictions (*dashed lines*) [8.34]. The inset represents the two-dimensional Brillouin zone

and zinc-blende-type semiconductors have been measured this way [8.22, 37]. The additional assumption is made that the conduction bands (i.e., the final states) are free-electron-like at the high energies resulting in the $\hbar\omega_L$-induced transitions, and thus can be expressed by the parabola $E_c = \hbar^2 k^2/(2m)$. Since the parabola represents simply a fit to the higher conduction bands, its bottom does not correspond to the bottom of the real conduction bands. Around this bottom the bands deviate strongly from free-electron behavior, as discussed in Chap. 2. Hence the bottom of the free electron parabola defined above is simply a fit parameter without any clear physical significance.

This type of ARPES is usually performed for $\theta_e = 0$ (normal emission). Hence $\boldsymbol{k}_{\|}$ of the initial states is either zero or one of the two-dimensional reciprocal lattice vectors $\boldsymbol{G}_{\|}$. The energy E_e is converted into an energy inside the solid, referenced to the bottom of the fitted internal free-electron parabola by subtracting a constant E_0. We reference E_e to the top of the valence band (i.e., to the point where the emission spectra start), so that E_0 represents the energy of the bottom of the free-electron parabola for electrons in vacuum, with respect to the top of the valence band (Fig. 8.18). The value of $k_\perp$ that corresponds to the energy of a given peak in E_e, i.e., to $E_v = \hbar\omega_L - E_e$, is obtained from

$$k_\perp = \left(\frac{2m}{\hbar^2}\right)^{1/2} (E_e - E_0)^{1/2} - G_\perp, \tag{8.9}$$

where $G_\perp$ is the component perpendicular to the surface of a reciprocal lattice vector required to bring $k_\perp$ to the reduced Brillouin zone. Equation (8.9) and

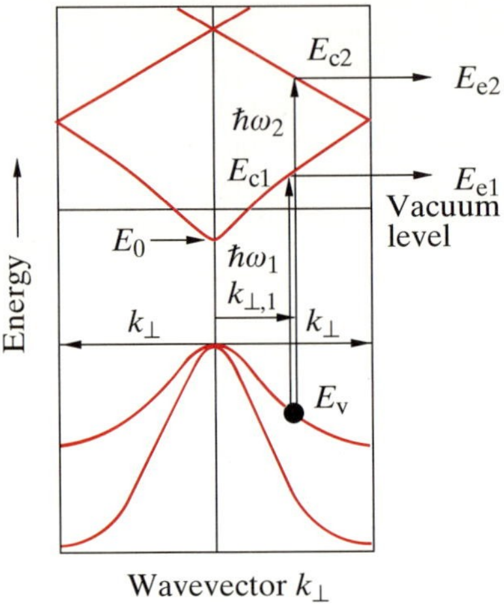

Fig. 8.18. Diagram of the method used to determine $k_\perp$ under the assumption of free-electron-like conduction bands. Two different photons ($\hbar\omega_1$ and $\hbar\omega_2$), leading to transitions with the same E_υ but different E_{c1} and E_{c2} (extended zone $k_\perp$'s differing by a reciprocal lattice vector), are considered. Note that the reduced zone $k_\perp$ increases with increasing $\hbar\omega_1$ while it decreases with increasing $\hbar\omega_2$. This fact can be used to select the correct value of the adjustable parameter E_0

$$E_\upsilon = \hbar\omega_L - E_e \tag{8.10}$$

determine the valence band structure $E_\upsilon(k_\perp)$ in the direction of the Brillouin zone perpendicular to the surface, provided E_0 is known. This E_0 can be used as an adjustable parameter so as to fit a calculated band structure when it is available. Values of E_0 in the 5–9 eV range are obtained for diamond- and zinc-blende-type materials although, as already mentioned, not much physical meaning is to be attached to this parameter, which results from having made the free-electron-like final state approximation. An analysis of normal-emission ARPES data using the corresponding calculated conduction band structure instead of the free-electron-like final states is given in [8.38] for several III–V semiconductors.

The determination of $E_\upsilon(k_\perp)$ using (8.9) and (8.10) requires the availability of several photon energies $\hbar\omega_L$. Each frequency will typically lead to a few peaks in E_e (usually three or four) and thus to a corresponding number of band structure points. A reasonably dense mesh is obtained by using all available rare gas discharge lines (about 20 lines in the 10–50 eV range) provided the light is passed through a monochromator. Better results are, of course, obtained with continuously tunable synchrotron radiation.

In the procedure just discussed one has to make use of available band structures to determine E_0. If a wide $\hbar\omega_L$ range of continuously tunable radiation is available, one can obtain the whole $E_\upsilon(k_\perp)$ from two conduction band branches differing by a $G_\perp$ (E_{c1} and E_{c2} in Fig. 8.18). We see from Fig. 8.18 that if E_0 is decreased $k_\perp$ increases for the E_{c1} state (i. e., the $\hbar\omega_1$ transition) while it increases for the E_{c2} state (i. e., $\hbar\omega_2$ transition). By trial and error it is possible to determine a value of E_0 which leads, from the measured spectrum, to the same $k_\perp$ for a given initial E_υ, independently of the E_c branch used. This procedure, first followed by *Middlemann* et al. [8.39], leads to determination of $E_\upsilon(k_\perp)$ without any prior knowledge of E_0 and the band structure.

The method just sketched has been impressively implemented by *Chen* et al. [8.40] for Ge, using a cylindrical ([110]-axis) crystal, so as to be able to measure several directions of $k_\perp(\Delta, \Lambda, \Sigma)$ by simply rotating the cylinder about its axis. The results are shown in Fig. 8.19. They agree extremely well with the pseudopotential band structure calculations of *Chelikowsky* and *Cohen* [8.41].

As discussed above, many valence band structures of semiconductors have been determined by the normal emission, $\boldsymbol{k}$-conservation method based on the three-step model. This scheme is referred to as the **direct-transitions scheme** because of the assumption of $\boldsymbol{k}$-conservation in the photon absorption step (step 1), (see Sect. 6.2). This assumption, however, cannot be exact: the presence of the surface implies at least partial lifting of $k_\perp$ conservation. Hence, we now consider the opposite case, in which $k_\perp$ is not conserved at all, while staying within the three-step model. The new situation is called the **indirect-transitions model** by analogy to the $\boldsymbol{k}$-nonconserving indirect transitions of Sect. 6.2. For initial states with a given $\boldsymbol{k}_\parallel$, which will be conserved in the photoemission process, $k_\perp$ need not be conserved. To a first approximation we thus assume that we detect all valence electrons which have a given $\boldsymbol{k}_\parallel$ (determined by E_e, φ_e, and θ_e) and all possible values of $k_\perp$. Since $E_\upsilon = \hbar\omega_\mathrm{L} - E_\mathrm{e}$, for constant resolutions $\Delta\omega_\mathrm{L}$ and ΔE_e we collect all electrons with the energy $E_\upsilon \pm \Delta E_\upsilon/2$, where ΔE_υ is assumed to be constant. The collected electrons, all with fixed $\boldsymbol{k}_\parallel$[6] and variable $k_\perp$, can be represented by a one-dimensional band structure $E_{\upsilon,\boldsymbol{k}_\parallel}(k_\perp)$. The number of electrons collected corresponding to

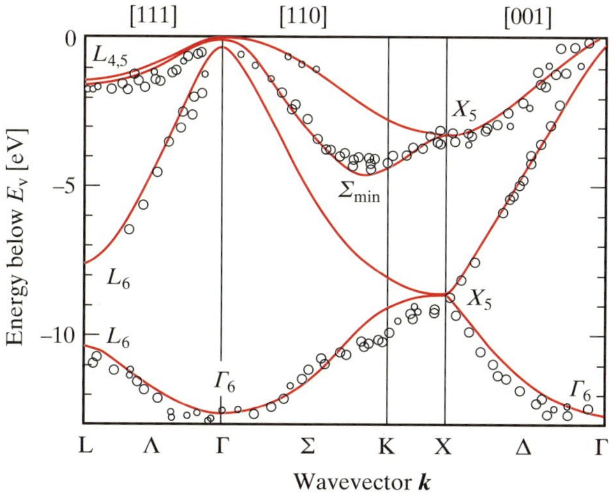

Fig. 8.19. Valence band structure of Ge determined by angle-resolved UPS, compared with the results of a theoretical prediction (*solid lines*) [8.41]. Large (*small*) circles denote large (*small*) peaks [8.40]

[6] Note that, when sweeping E_e, strictly speaking we must vary θ_e in order to keep $k_\parallel \propto E_\mathrm{e}^{1/2} \sin\theta_\mathrm{e}$ constant. This is easy in a computer-controlled system.

initial states within the interval ΔE_{υ} around E_{υ} will thus be proportional to the *one-dimensional density of states*, $N_{\upsilon,\boldsymbol{k}_{\parallel}}(k_{\perp})$:

$$j(E_{\mathrm{e}},\varphi_{\mathrm{e}},\theta_{\mathrm{e}}) \propto N_{\upsilon,\boldsymbol{k}_{\parallel}}(k_{\perp}) = \frac{1}{\pi}\left(\frac{\mathrm{d}E_{\upsilon,\boldsymbol{k}_{\parallel}}(k_{\perp})}{\mathrm{d}k_{\perp}}\right)^{-1} . \tag{8.11}$$

$N_{\upsilon,\boldsymbol{k}_{\parallel}}$ can be easily computed if the band structure is known. It will show singularities wherever the derivative of (8.11) vanishes, i. e., wherever the slope of $E_{\upsilon,\boldsymbol{k}_{\parallel}}$ is zero. This happens at the *critical points* of the one-dimensional band structure, e. g., for $k_{\parallel} = k_{\perp} = 0$, and also, sometimes, when $k_{\perp}$ crosses the edges of the Brillouin zone. Fortuitous one-dimensional critical points can also occur inside the reduced zone. Under these assumptions the measured spectrum will show peaks at the values of E_{e} calculated using (8.10) from the E_{υ}'s of such one-dimensional critical points.

We have considered two extreme cases: full conservation of $k_{\perp}$ and no conservation at all. The truth will, of course, lie somewhere in the middle: $k_{\perp}$ will be conserved to within a "resolution" $\Delta k_{\perp}$, i. e., the final state *in step one* will be within the range $k_{\perp} \pm \Delta k_{\perp}/2$. Let us now estimate $\Delta k_{\perp}$. Photoemitted electrons with the energy E_{e} have a decay length δ inside the crystal given in Fig. 8.5. Hence these electrons can be represented by the wave functions

$$\varphi(\boldsymbol{r}) \sim \mathrm{e}^{\mathrm{i}(\boldsymbol{k}_{\parallel}\cdot\boldsymbol{r}_{\parallel})}\mathrm{e}^{\mathrm{i}(k_{\perp}-\mathrm{i}k'_{\perp})r_{\perp}} = \mathrm{e}^{\mathrm{i}(\boldsymbol{k}\cdot\boldsymbol{r})}\mathrm{e}^{k'_{\perp}r_{\perp}}, \tag{8.12}$$

where $\boldsymbol{r}_{\parallel}$ and $r_{\perp}$ are components of r parallel and perpendicular to the sample surface, $r_{\perp}$ being positive when pointing towards vacuum. Equation (8.12) shows that the finite escape depth δ can be represented by an imaginary part of $k_{\perp}$, $k'_{\perp}$, with $2k'_{\perp}$ equal to δ^{-1}. It is easy to prove that such an imaginary part of $k_{\perp}$ blurs the requirement of $k_{\perp}$ conservation as represented by the delta function $\delta(k^{\mathrm{f}}_{\perp} - k^{\mathrm{i}}_{\perp})$ into the Lorentzian probability distribution

$$L(k^{\mathrm{i}}_{\perp},k^{\mathrm{f}}_{\perp}) \sim \frac{k'_{\perp}}{(k^{\mathrm{i}}_{\perp} - k^{\mathrm{f}}_{\perp})^2 + k'_{\perp}k'_{\perp}}, \tag{8.13}$$

which has the *full-width at half-maximum* (FWHM) $\Delta k_{\perp} = 2k'_{\perp} = \delta^{-1}$ (note that $2k_{\perp} \approx k^{\mathrm{i}}_{\perp} + k^{\mathrm{f}}_{\perp}$). For $E_{\mathrm{e}} \simeq 40$ eV, $\delta = 5$ Å and $\Delta k_{\perp} \simeq 0.2$ Å^{-1}, which represents one-fifth of the Γ–X distance of the Brillouin zone of germanium. A one-dimensional critical point falling within this range around one of the $k_{\perp}$'s of (8.9) is thus expected to dominate the energy distribution curve. Whether this is the case can be easily tested by changing $\hbar\omega_{\mathrm{L}}$. For direct $k_{\perp}$-conserving transitions the E_{υ} calculated with (8.10) from the E_{e} of a given spectral peak will vary with $\hbar\omega_{\mathrm{L}}$. For indirect transitions the spectra reproduce singularities in the one-dimensional density of states which occur at fixed values of E_{υ}, hence (8.10) will give values of E_{υ} independent of $\hbar\omega_{\mathrm{L}}$ (E_{e} must change so as to exactly cancel the change in $\hbar\omega_{\mathrm{L}}$).

The ARPES spectra of real semiconductors contain, of course, a mixture of $k_{\perp}$-conserving and one-dimensional density-of-state peaks. In Ge and zincblende semiconductors mainly the former are seen, while the latter explain most (but not all) of the peaks observed in the lead chalcogenides [8.42].

It is not completely clear why opposite models should apply to these two families of semiconductors. However, one may guess that the indirect transition model will apply better the larger the number of one-dimensional critical points within the Brillouin zone. Higher symmetry leads to a larger number of such critical points (Problem 8.9). The lead chalcogenides have the O_h point group while the point group of zinc-blende is T_d, which has fewer symmetry operations than O_h. Also, for zinc-blende usually a [110] cleavage face is measured while for the lead chalcogenides a [100] face, the cleavage face of this structure, is investigated. The latter has more one-dimensional critical points than the former because of the higher symmetry. Finally, the lead chalcogenides have accidental critical points along the Δ and Σ directions (see Fig. 1 in [8.42]) which increases the probability of indirect transitions. In order to illustrate these points we show in Fig. 8.20 the $\boldsymbol{k}$-space region (involving two adjacent bulk Brillouin zones) that is accessible to photoemission from a (100) surface of PbS for $\boldsymbol{k}_{\|}$ along (001). Figure 8.20c shows the corresponding $\boldsymbol{k}_{\|}$ dependence of the measured one-dimensional critical points, while Fig. 8.20b gives the results of band structure calculations, which are in quite satisfactory agreement with the experimental data of (c).

8.1.3 Core Levels

The first core levels of Ge ($3d$) in order of increasing *binding energy* (BE) are $\sim$30 eV below the top of the valence band (see Table in [8.1,2]). Silicon has no d core levels; its first core levels ($2p$) have BE of $\sim$100 eV. In the III–V compounds, the uppermost d core levels split into those of the cation (BE $\sim$18 eV) and those of the anion (BE $\sim$30 to 40 eV). In the II–VI compounds that splitting increases (cation $\sim$10 eV; anion $\sim$40 to 55 eV). Note that for the II–VI compounds the $3d$ and $4d$ levels touch the s-like bottom of the valence bands; the d–s mixing is, however, negligible because of the small extent of the d wave functions. In the I–VII compounds (e. g., CuCl) the $3d$ levels of the copper overlap the upper (p-like) parts of the conduction bands, resulting in a considerable mixing which is responsible for a number of interesting effects, such as the sign reversal of the spin–orbit splitting at Γ in CuCl [8.27]. Note that Al, P and S, like Si, have no d core levels.

The p-like core levels split under the action of spin–orbit interaction into atomic-like $j = \frac{3}{2}$ and $j = \frac{1}{2}$ components (splitting $\Delta_p \simeq 0.6$ eV for $2p$ levels in Si, 4 eV for $3p$ levels in Ge; note that these splittings increase rapidly with atomic number). The d core levels of the *atoms* have an orbital degeneracy of 5 ($l = 2$). They thus split into $j = \frac{5}{2}$ and $j = \frac{3}{2}$ under the action of spin–orbit interaction. When the atom is placed in a cubic (or tetrahedral) environment, however, the fivefold *orbital* degeneracy has to split since the O_h or T_d point group only allows at most threefold degeneracy. In the case of zinc-blende the d orbital levels split into Γ_3 (twofold) and Γ_4 (threefold, see Problem 8.10). This splitting, however, is at most a few tenths of an eV in the case of core levels, i. e., smaller than the spin–orbit splitting ($\sim$1 eV). The

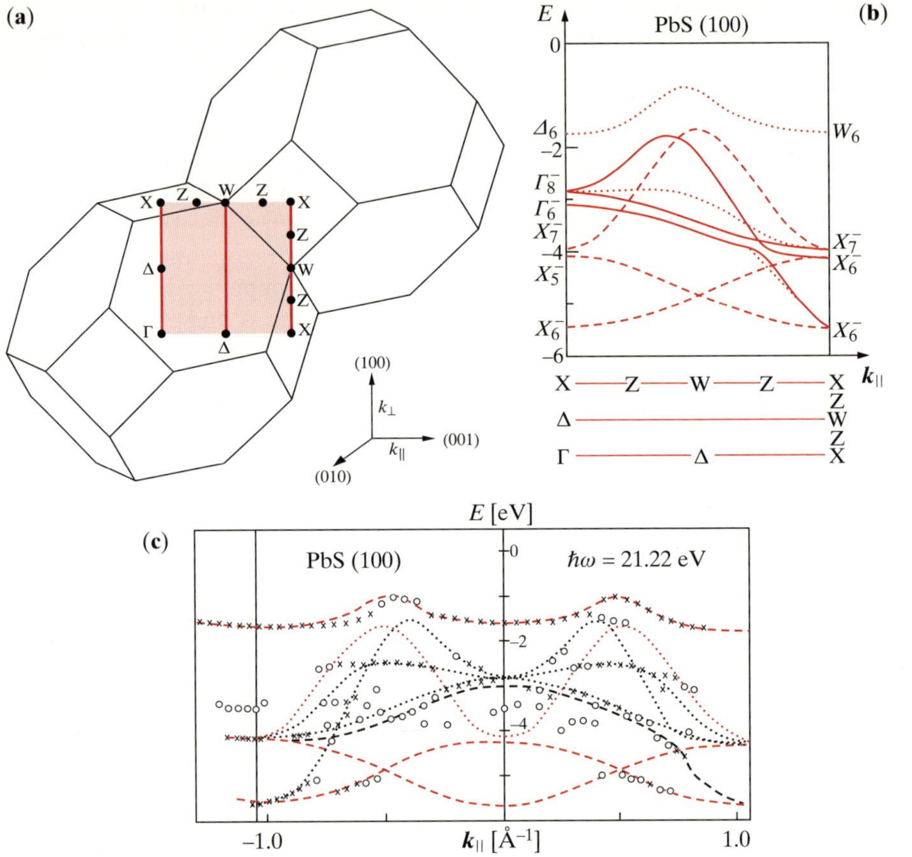

Fig. 8.20. (**a**) Extended zone scheme of the *fcc* structure appropriate to the photoemission studies of PbS described in (**c**) where critical points obtained along the Γ–X, Δ–W and X–W–X *vertical red lines* are represented. (**b**) Calculated dependence of the energies of critical points along the *three vertical red lines* of (**a**). (**c**) The right frame represents experimental data for the critical points measured by angular resolved photoemission (HeI lamp, $\hbar\omega = 21.2$ eV) with $\boldsymbol{k}_\parallel$ along [001]; curves have been drawn througth the data as similar as possible to the calculations in (**b**). The left frame describes data for $\boldsymbol{k}_\parallel$ along [011] [8.42]

crystal-field splitting must therefore be applied to the $j = \frac{5}{2}$ (sixfold) and the $j = \frac{3}{2}$ (fourfold) levels after splitting them through spin–orbit interaction. The $j = \frac{3}{2}$ level does not split under the action of the crystal field. The $j = \frac{5}{2}$ level, however, splits into Γ_8 and Γ_7 (Problem 8.10), this crystal-field splitting being, of course, rather small.

Note that the outermost cation d levels of germanium and zinc-blende-type semiconductors can all be investigated with conventional gas discharge lamps (except for Sb, the anions require synchrotron radiation or X-rays

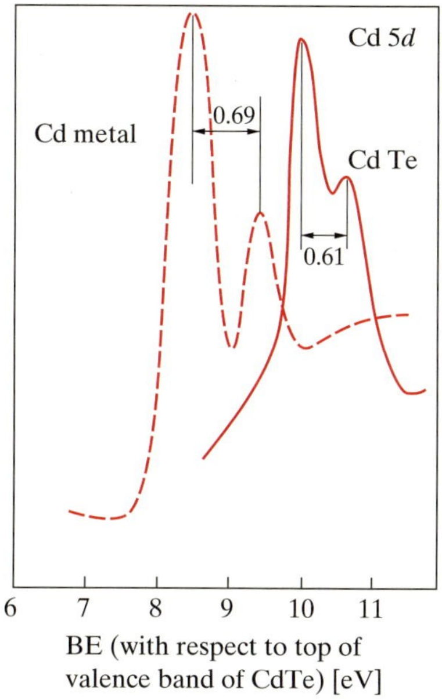

Fig. 8.21. 4*d* core level spectra of metallic Cd (*dashed line*) and CdTe (*solid line*) obtained by means of UPS (HeI). Both BE's are referenced to the top of the valence band of CdTe. This figure reveals a core shift of 1.44 eV (the Cd levels are deeper in CdTe than in metallic Cd) [8.43]

sources). Figure 8.21 shows angle-integrated photoelectron spectra of the 4*d* levels of Cd in CdTe and in the metallic element, with the binding energies referenced to the top of the valence band of CdTe. Two peaks ($j = \frac{5}{2}$ and $\frac{3}{2}$) are observed with the expected $\frac{3}{2}$ intensity ratios. Note that the spin–orbit splitting seems to be larger in the case of the element than for the compound, for which it is actually close to that of the isolated atom. This has been interpreted as a result of Cd 4*d* – Cd 4*d* overlap among the nearest neighbors in metallic Cd (i. e., band formation). Such overlap is not expected for CdTe, where the nearest neighbor of Cd is Te. A recent experimental and theoretical study of band formation by the Cd 4*d* states of CdSe can be found in [8.44].

The most interesting feature of Fig. 8.21 is the shift of the cation core levels of CdTe towards higher BE with respect to the element. Likewise, the levels of Te (anion) are shifted towards lower BE [8.43]. (These so-called **core shifts** are usually the same for all the core levels of one given atom.) A simple qualitative explanation of this rather general fact is based on the ionicity of the compound: the cation loses electrons to the anion. The removal of electrons from the neighborhood of the cation lowers the *energy* of repulsion seen by the core electrons of the cation, hence the binding energy increases. The opposite can be said for the core levels of the anions. This simple argument, however, encounters some difficulties when examined more quantitatively. If we remove a charge q (negative for cations, positive for anions) from an atom we expect a change in the BE of its core levels equal to $-q/\bar{r}$, where $\bar{r}$ represents an average distance from the valence electrons to the core. For a diatomic crystal the removed electrons are placed on the other sublattice site, giving rise to a **Madelung energy** contribution to the BE. Thus we find the total core BE shift with respect to the neutral element to be

$$\Delta E = q\left(\frac{\alpha_M}{R} - \frac{1}{\bar{r}}\right), \tag{8.14}$$

where α_M is the Madelung constant ($\alpha_M = 1.64$ for the zinc-blende structure) and R the nearest neighbor distance. Taking for the case of zinc-blende $\bar{r} \simeq R/2$, we find from (8.14)

$$\Delta E = -5.2q/R \tag{8.15}$$

with q expressed in electron charges, R in angstroms, and ΔE in eV. For a typical value of $q \simeq 1$ (as expected for II–VI compounds, see [8.45]) and R = 2.8 Å we find from (8.15) the core shift $\Delta E = +1.8$ eV for the core levels of Cd in CdTe, in rather good agreement with the data of Fig. 8.21 ($\Delta E = +1.44$ eV). This agreement is, however, somewhat deceptive: because of the near cancellation between the two terms in (8.14), a small change in the rather arbitrarily chosen value of $\bar{r}$ results in a large variation of ΔE, possibly even a sign reversal.

In order to quantify more precisely the value of $\bar{r}$, it has been suggested that

$$\bar{r} = r_m/A(\Gamma) \tag{8.16}$$

be used [Ref. 8.2, p. 128], where r_m is the interatomic distance in the cation in metallic elemental form and $A(\Gamma)$ a geometrical factor obtained by assuming that the relevant valence electrons are evenly distributed in a spherical shell of outer and inner radii $r_m/2$ and $\Gamma r_m/2$ ($\Gamma \leq 1$). The derivation of $A(\Gamma)$ is left for Problem 8.11. Here we simply state that for $\Gamma = 0.5$ we obtain $A(\Gamma) = 2.6$. This Γ gives values of the cation core shifts of zinc-blende-type semiconductors (2.2 eV for CdTe) in semiquantitative agreement with the experiments. We should mention that the charge transferred away from the cation can be related to an ionicity f_i through [Ref. 8.2, p. 129]

$$q = N - 8\left(1 - \frac{1}{\varepsilon_0}\right)\left(\frac{1 - f_i}{2}\right) \tag{8.17}$$

where $N = 3, 2, 1$ for the III–V, II–VI, and I–VII compounds, respectively, and ε_0 is the low frequency (static) dielectric constant.

We have often mentioned the sensitivity of UPS to surface conditions. Around the minimum in Fig. 8.5 the escape depth under *normal takeoff* is ~5 Å, hence for a typical semiconductor the first two monolayers should contribute about half to the total emission. For those monolayers α_M should be smaller than for the bulk since several terms are missing in the slowly convergent Madelung sum. Hence, according to (8.14), the BE of the cations will be larger at the surface, and that of the anions smaller. Consequently, it is possible to decompose the observed core spectra into two components, one corresponding to the surface, the other to the core. This is illustrated in Fig. 8.22 by plotting the spectra of the $3d$ levels of As (a spin–orbit doublet) in

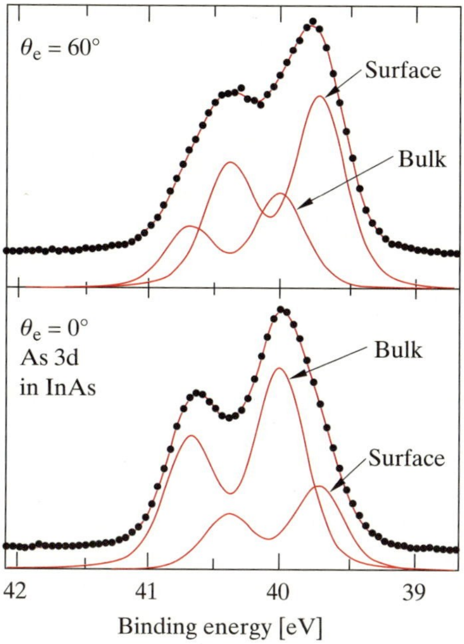

Fig. 8.22. Spectra of the 3d core levels of As measured on an InAs(110) surface with synchrotron radiation (105 eV excitation) using two different azimuthal angles, $\theta_e = 0$ and 60. Note the decomposition of the $j = (5/2, 3/2)$ doublet into a surface and a bulk component: the former is stronger for the larger θ_e [8.46]

InAs, as measured with synchrotron radiation ($\hbar\omega_L = 105$ eV). Note that these spectra can be decomposed into two doublets of comparable strengths. In order to ascertain which of the two is due to the surface, one varies the takeoff angle θ_e. The escape depth should be that of Fig. 8.5 multiplied by $\cos\theta_e$, hence for $\theta_e = 60$ the spectra should contain a larger surface contribution than that for $\theta_e = 0$. By inspection of Fig. 8.22 we identify the surface doublet as the lower BE doublet, in agreement with the above predictions based on a reduction of α_M at the surface.

We mention next the effects of surface oxidation on core level spectra. Oxygen is very strongly electronegative and therefore oxidation leads to large values of $q(\sim 2)$ when estimated with (8.14) or simply assuming that oxygen is always in the O^{2-} configuration. The signature of surface oxidation is the appearance of cation satellites at larger binding energy. If the oxidation proceeds very far, a shift of the whole core level peaks, by about 1 eV, is found. For a detailed, more quantitative discussion of this topic see [8.47].

We close this section by mentioning that measurements of core level energies with respect to the top of the valence band can be very useful in the determination of *valence band offsets* in semiconductor *heterojunctions* (Chap. 9). For details see [8.48].

8.2 Inverse Photoemission

In the previous sections we have shown how photoelectron spectroscopy can be used to determine the band structure of occupied states. The dispersion of the empty conduction bands in $\boldsymbol{k}$-space can also be determined by using the inverse procedure, i. e., **inverse photoemission**. Several versions of this technique exist. In the simplest one, electrons with a well-defined energy E_e impinge on the sample and place themselves at the corresponding conduction band energy. They then scatter inelastically to a point of lower energy in the conduction band. The energy difference between the initial and the final state of an electron is taken up by the emission of a photon. A narrow-band-pass photon detector, with a fixed central energy of $\sim$9.2 eV, is then used while the incident electron energy E_e is swept to obtain a spectrum. For this reason, the spectra are known as **bremsstrahlung isochromat spectra** (BIS).[7] In a more elaborate version, the emitted photons are dispersed by a grating and detected by a multichannel detector (e. g., a television camera or similar). The latter version is used in an angle-resolved manner by varying the angles (θ_e, φ_e) of the incident electron beam. In the spectrometer used by Ortega and Himpsel [8.28] an energy resolution of 0.3 eV is obtained in the 8–20 eV photon energy range, with an angular resolution corresponding to $\Delta k_{\|} \sim 0.1$ Å.

The basic principles of inverse photoemission measurements are the same as those discussed in Sect. 8.1.2, but reversed. One can also in this case observe either direct or indirect transitions, the latter related to singularities in the one-dimensional density of *conduction* states. The direct transitions are treated by fitting a free-electron parabola to the *initial* state, with the bottom at the adjustable energy E_0. This free-electron parabola, when folded to the reduced zone, determines the possible values of $k_\perp$ for a given, conserved $\boldsymbol{k}_{\|}$. Also here, $\boldsymbol{k}_{\|}$ is often taken to be zero (normally incident electrons). Because of the folding of the extended to the reduced zone, the value of $k_\perp$ is not uniquely determined: the various possible values correspond to various reciprocal lattice vectors $\boldsymbol{G}_{\|}$, i. e., to **inverse Mahan Cones**. Normally the primary cone ($\boldsymbol{G}_{\|} = 0$) is preferentially excited, a fact that is used to simplify the analysis of the spectra. Once $k_\perp$ is determined, the conduction band structure $E_c(\boldsymbol{k}_{\|}, k_\perp)$ is found,

$$E_c = E_e - \hbar\omega_L, \tag{8.18}$$

where $\hbar\omega_L$ is the energy of a given feature in the spectrum of emitted photons. Care has to be taken to properly reference E_e and E_c to the same origin of energies, usually the Fermi level or the top of the valence band.

An example of the power of this method is given in Fig. 8.23, which shows the complete (conduction and valence) band structure of Ge obtained by in

[7] Bremsstrahlung means in German "braking radiation". This term refers to the electromagnetic radiation emitted when electrons are stopped (braked) by a solid.

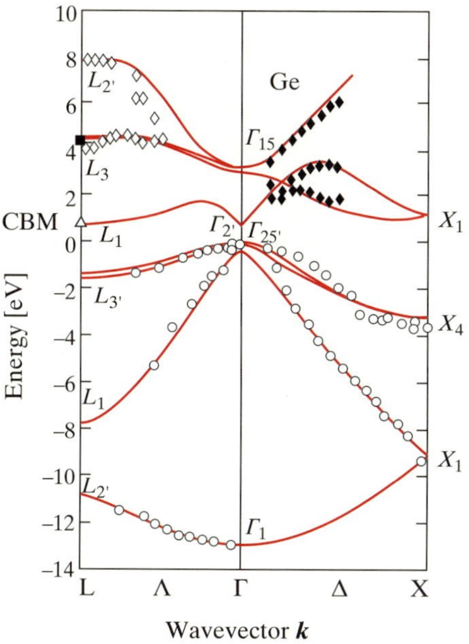

Fig. 8.23. Band structure of Ge obtained with UPS (valence band) and inverse photoemission (conduction band) compared with theoretical results. CBM indicates the conduction band minimum. For the irreducible representations, the single group notation is used [8.49]

verse photoemission plus angle-resolved photoemission, compared with the results of *ab initio* band structure calculations which take the *gap problem* mentioned in Sect. 2.5 properly into account. The agreement between experimental and theoretical results is rather convincing.

8.3 Surface Effects

We have already shown in Fig. 8.22 that electronic states near the surface, if different from those of the bulk, can significantly affect the photoemission spectra. Therefore, direct and inverse photoelectron spectroscopy in their various forms are ideal techniques for investigating and characterizing surfaces: recall for instance the extreme sensitivity of UPS to surface contamination (Fig. 8.7). In this book we do not discuss surfaces in depth. The interested reader will find such discussion in [8.11]. We do, however, present in this section a few concepts and facts related to surfaces that are necessary in order to understand direct and inverse photoelectron spectroscopy and other subjects covered in this book for which surfaces are relevant.

8.3.1 Surface States and Surface Reconstruction

We first consider a clean semiconductor surface. The basic effect of cleaving and removing half of the crystal is to break bonds, which are left as singly occupied atomic orbitals known as **dangling bonds**. If one focuses on a monolayer of such bonds they should form two-dimensional bands $E_s(\boldsymbol{k}_{\|})$ because of the translational symmetry along the surface. If these bands overlap in energy with the continuum of either empty or occupied bulk states they are broadened by interaction with those bulk states. They may, however, still be identifiable as **surface resonances** using two-dimensional photoelectron techniques, provided their broadening is not too large.

Those surface bands may lie, at least in part, in the forbidden energy gap. In this case one talks about **surface states**. Dangling bonds occupied by only one electron should give rise to partly occupied, i.e., metallic, surface bands. In the one-dimensional case half-filled (metallic) bands usually undergo the so-called **Peierls transition**: they split into two, one fully occupied the other empty, by doubling the size of the primitive cell (**reconstruction**). In doing so, the energy is lowered since the occupied bands go down in energy while the empty ones go up: the metal becomes an insulator.

The Peierls transition occurs, strictly speaking, only in one-dimensional metals. The higher the dimensionality, the less likely it is to find a similar phenomenon. In two-dimensional systems (such as surfaces), however, it is sometimes also possible to lower the energy through reconstruction (**surface reconstruction**). This reconstruction, which can be rather involved, is specific to the system under consideration and can be observed directly, in real space, by **scanning tunneling microscopy** (STM).[8] As an illustration, Fig. 8.24 shows the c(8×2) reconstruction as directly observed for a (111) Ge surface by STM. (The notation 8×2 refers to the fact that this unit cell contains 8×2 unit cells of the unreconstructed surface; c means that it is centered. Note that it has orthorhombic symmetry.) Also shown is a schematic diagram of this reconstruction, which takes place by a principle somewhat different from the Peierls distortion mentioned above. The cleaved (111) surface has a large number of *dangling bonds*, one per atom at the first monolayer. Many of these dangling bonds can be saturated by capping groups of three atoms of the first monolayer with *adatoms* (thick rings in Fig. 8.24b), thus producing the reconstructed structure of Fig. 8.24a,b. In this manner six of the eight atoms are capped with two adatoms in the primitive cell of Fig. 8.24b, thus eliminating six out of eight dangling bonds but also introducing two new ones associated with the adatoms, which are assumed to retain a tetrahedral bonding configuration. Hence by the reconstruction with adatoms we eliminate half of the dangling bonds, lowering the energy by the bond energy of one half of the surface atoms ($\sim$4 eV/atom).

8.3.2 Surface Energy Bands

The surface energy bands $E_s(\boldsymbol{k}_{\|})$ can be measured by means of photoemission (occupied states) and inverse photoemission (empty states), as discussed in Sects. 8.1 and 8.2. Inverse photoemission, however, has thus far not been as fruitful for surface studies as its direct counterpart.

[8] Invented by G. Binnig and H. Rohrer, Nobel Laureates for Physics 1987.

(a)

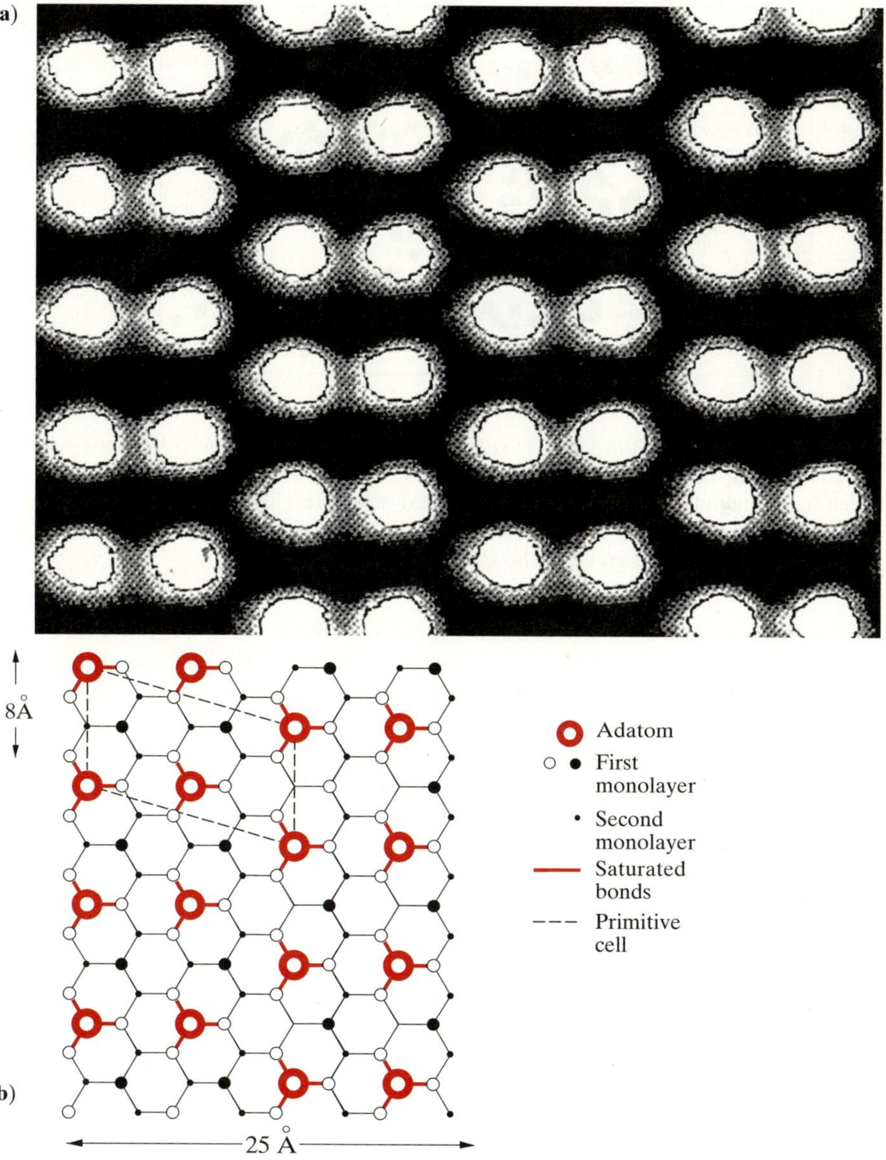

(b)

Fig. 8.24. (**a**) Scanning tunneling microscope image of a Ge(111)-c(2 × 8) surface. (**b**) Diagram of the reconstruction implicit in (**a**) drawn to the same scale. Note that the *red* circles in (**b**) and the *bright spots* in (**a**) correspond to adatoms [8.50]

As an example, Fig. 8.25 shows direct photoemission results obtained by several investigators for the c(2 × 8)(111) surface of Ge plotted in the Brillouin zone of the unreconstructed surface. We note in these data that all occupied surface bands lie below the top of the valence bands and thus represent

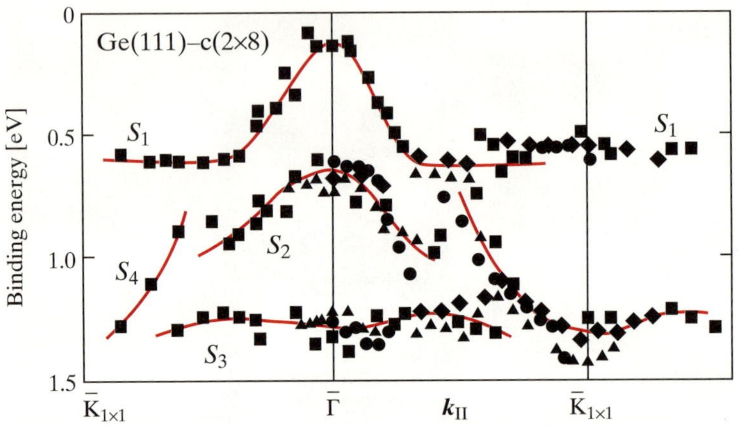

Fig. 8.25. Dispersion of the Ge(111)-c(2 × 8) surface bands measured by angle-resolved UPS. Note that the results are plotted in the BZ of the unreconstructed (1 × 1) unit cell, which is an extended zone of the reconstructed structure. (The different symbols represent results from different groups) [Ref. 8.11, p. 186]. The red lines are drawn through the measured points as a guide to the eye

surface resonances. It is worthwhile noting (see next section) that the widths of these bands (~0.5 eV) are narrower than those of the corresponding bulk bands (~2 eV). Because of Fermi level pinning (see next section) there must be empty surface states located ~0.1 eV above the top of the valence band. A few points in the surface band structure have been calculated in [8.51]. They agree qualitatively with the more complete data of Fig. 8.25.

The c(2 × 8) structure just discussed is obtained after cleaving *and* annealing the Ge(111) surface. Immediately after cleaving, the metastable 2 × 1 surface is obtained. For this reconstruction, which we do not discuss here in detail, band structure calculations, ARPES, and inverse photoemission measurements are available. These results are compared in Fig. 8.26. We see that a downshift in the energy of the calculated bands by ~0.5 eV suffices to bring them in reasonable agreement with the experiments. A surface gap between occupied and empty states of about 0.5 eV appears. This gap has been observed in Si by means of unconventional optical absorption techniques [8.52].

8.3.3 Fermi Level Pinning and Space Charge Layers

We have seen Sect. 8.3.2 that surface energy bands may or may not exist within the gap. If the gap is free of surface states (this is the case for a clean, cleaved GaAs surface) the position of the bulk bands with respect to the Fermi energy is constant from inside the crystal all the way to the surface, as shown in Fig. 8.27a. If surface bands exist in the gap they are rather narrow (Figs. 8.25 and 8.26) and consequently their *surface density of states* must be very high. This is depicted in Fig. 8.27b for a p-type and Fig. 8.27c for

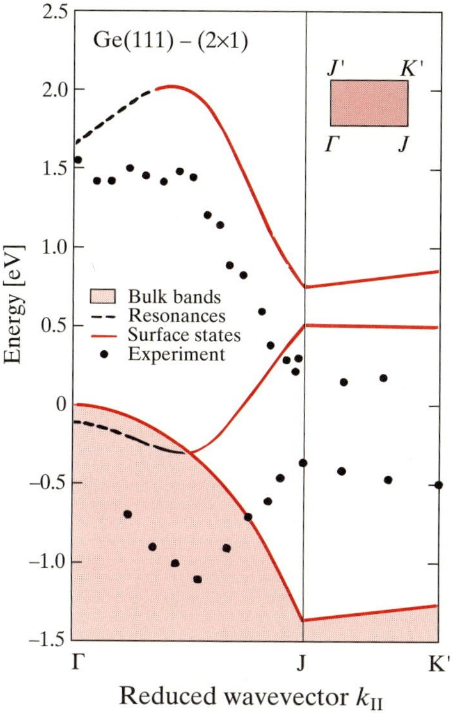

Fig. 8.26. Measured (*dots*) and calculated (*solid lines*) dispersion of the occupied and empty *dangling bond* surface bands of a Ge(111)-(2 × 1) surface. The rectangular inset represents 1/4 of the (2 × 1) surface Brillouin zone and defines the notation used for the $\boldsymbol{k}_{\parallel}$ points at the edge of this zone [Refs. 8.11, Fig. 10.6; 12, Fig. 6.36]

an n-type semiconductor. Because of the high density of surface states (10^{15}–10^{16}/eV cm^2) the surface states *pin* the Fermi level in the manner shown in these figures. An n-type doping producing $\sim 10^{18}$ electrons/cm^3 leads to a carrier density at the surface of $\sim 10^{10}$ electrons/cm^2. Hence the Fermi level can only penetrate into the surface bands by an amount

$$\frac{10^{10}\ \text{electrons/cm}^2}{10^{15}\ \text{eV/cm}^2} = 10^{-5}\ \text{eV}.$$

This means that the Fermi level barely penetrates into the band of empty surface states. This is the origin of the phenomenon known as Fermi level *pinning* illustrated in Fig. 8.27c. Likewise, for a bulk p-type material (Fig. 8.27b) the Fermi level is pinned against the top of the occupied surface states (or the top of the bulk valence bands if these states are resonances inside the bulk valence band). Because of the small surface gap ($\sim$0.2 eV) one often speaks, in both the n- and p-type cases, of a Fermi level pinned at a unique pinning energy. A list of pinning energies measured for germanium- and zinc-blende-type semiconductors can be found in [Ref. 8.11, Table 1.1].

The phenomenon of Fermi level pinning leads to the **band bendings** displayed in Figs. 8.27b–d (note that in Fig. 8.27a, with no pinning, there is no

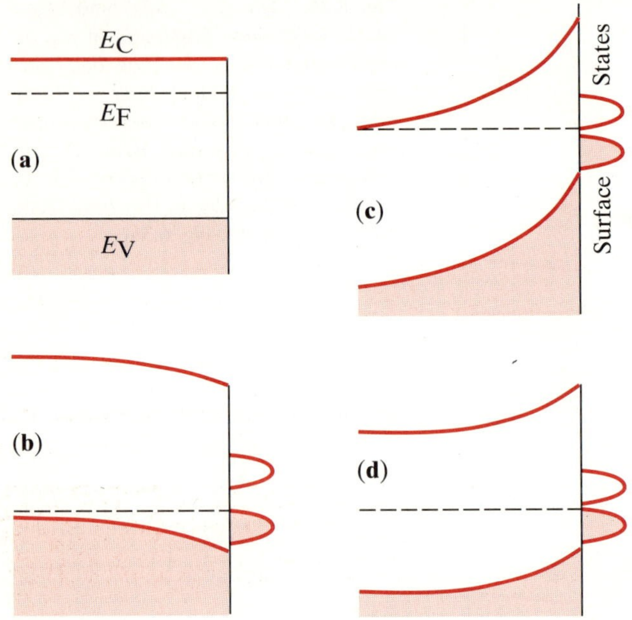

Fig. 8.27a–d. Examples of surface depletion and enrichment layers. (**a**) No Fermi level (E_F) pinning: no enrichment/depletion layer. (**b**) p-Type semiconductor with E_F pinning: hole depletion layer. (**c**) n-Type semiconductor with E_F pinning: electron depletion layer. (**d**) Near-intrinsic semiconductor with E_F pinning close to the valence band: hole enrichment layer

band bending). These surface band bendings result from the fact that in equilibrium the Fermi level E_F must remain constant from bulk to surface: the band edges must then vary so as to be compatible with the E_F pinning at the surface and the bulk E_F well away from it. A **surface depletion** layer results in the cases of Fig. 8.27b and c. Figure 8.27d represents a surface enrichment layer: the surface has, at finite temperature, more carriers (holes) than the bulk. The band edge profiles of Figs. 8.27b–d represent electrostatic potentials, which can be calculated by solving the equations of semiconductor statistics together with Poisson's equation for the relationship between potential and charge. This calculation can be greatly simplified by assuming a space charge layer with a constant charge density ϱ extending all the way from the surface to a point at a distance d from the surface and being zero beyond d. Poisson's equation can then be written as

$$\frac{d^2\phi}{dz^2} = \frac{-4\pi\varrho}{4\pi\varepsilon_0\varepsilon_0}, \tag{8.19}$$

where ϕ is the electrostatic potential and ε_0 the static dielectric constant. Equation (8.19) can be rewritten as

$$d\left(\frac{d\phi(z)}{dz}\right)^2 = -\frac{8\pi\varrho}{4\pi\varepsilon_0\varepsilon_0}d\phi \tag{8.20}$$

and integrated to yield

$$\left(\frac{d\phi(z)}{dz}\right)^2 = -\frac{8\pi}{4\pi\varepsilon_0\varepsilon_0}\varrho\phi(z). \tag{8.21}$$

The integration constant in (8.21) has been determined by setting $\phi(d) = 0$ and the electric field $\mathscr{E}(d) = -(d\phi/dz)_{z=d} = 0$. A second integration leads to

$$\phi = \frac{2\pi\varrho}{4\pi\varepsilon_0\varepsilon_0}(z-d)^2, \tag{8.22}$$

from which we obtain by setting $\phi(0) = \phi_0$ the thickness of the space-charge layer as a function of the charge density:

$$d = \left(\frac{4\pi\varepsilon_0\varepsilon_0\phi_0}{2\pi\varrho}\right)^{1/2}, \tag{8.23}$$

where ϕ_0 represents the total band bending.

Equations (8.22) and (8.23) enable us to plot the potential profile provided we make a reasonable assumption for the average charge density ϱ. In moderately doped semiconductors with pinning away from the band edges, it is conventional to assume that ϱ in the space charge layer is equal to the bulk carrier density. For the typical values $\varrho \simeq 10^{16}$ electrons/cm^3, $\varepsilon_0 = 10$, and $\phi_0 = 1$ eV, we obtain with (8.23) $d = 1300$ Å. For the typical highest carrier concentrations of 10^{20} cm^{-3} found in semiconductor we calculate with the same parameters $d = 13$ Å.

The reader may wonder whether the variation of $\phi(z)$, and thus of the bulk band energies $E_{c,v}(z)$, near the surface will smear out the photoelectron spectra of both valence bands and core levels. For not too high doping levels (up to $\sim 10^{18}$ cm^{-3} when using UPS, 10^{17} cm^{-3} for XPS) no such smearing occurs since the escape depth of the electrons is much smaller than d. We thus measure in these cases the bands and core levels referenced to the *Fermi level at the surface*. At higher dopings than these, the escape depth becomes close to or even larger than d. The measured bands and core levels should shift with respect to E_F and tend, at the highest, possible doping levels, towards the bulk values. These facts offer the possibility of determining the profiles $\phi(z)$ sketched in Fig. 8.27 by measuring the spectra of sharp core levels for several $\hbar\omega_L$'s corresponding to a range of escape energies. For this purpose, tunable synchrotron radiation is a must.

This method of measuring $\phi(z)$ is illustrated in Figs. 8.28 and 8.29. Figure 8.28 displays $2p$ core levels of Si (BE $\approx$ 100 eV), measured with respect to the Fermi level, for three differently doped samples (intrinsic, 8×10^{18} electrons/cm^3, 1.6×10^{20} holes/cm^3) with a range of $\hbar\omega_L$'s. While the peaks of the intrinsic sample do not shift with $\hbar\omega_L$, those of n-type sample shift clearly towards lower binding energies when we increase $\hbar\omega_L$ and thus approach the minimum of Fig. 8.5. The opposite is seen for the heavily doped p-type sample. The binding energies obtained from the data of Fig. 8.28 are plotted in Fig. 8.29 versus $\hbar\omega_L$ and also versus the corresponding escape depth ($\sim z$). A clear replica of the dependence of ϕ on z depicted in Fig. 8.27b, c, is obtained. In Fig. 8.29 we see a confirmation of the existence of a gap, of about 0.3 eV, between the occupied and the empty surface states of a cleaved Si(111) surface.

We conclude this chapter by mentioning the obvious fact that the surface Fermi level pinning depends strongly on the type of surface and surface reconstruction, quality of cleave, and surface contamination. A GaAs freshly cleaved (110) surface, for instance, shows no pinning. Upon exposure to oxygen, a small fraction of a monolayer coverage suffices to produce pinning with $E_F \approx 0.5$ eV above the top of the valence band at the surface.

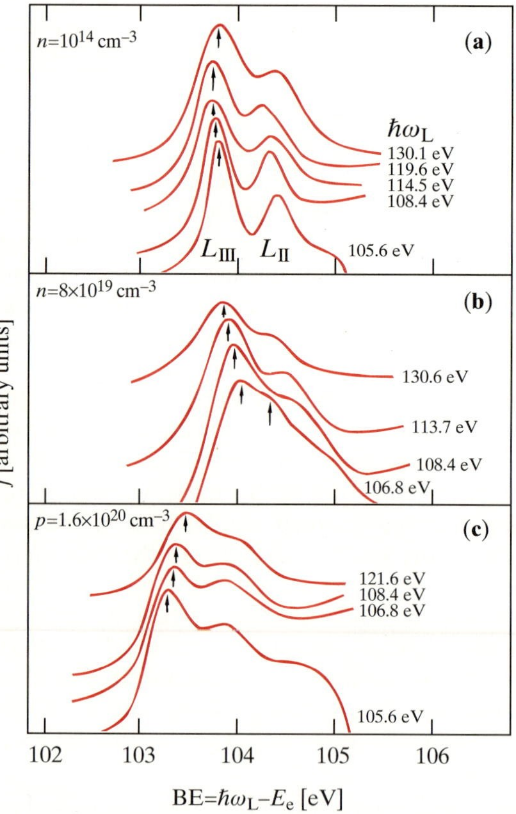

Fig. 8.28a–c. Angle-resolved UPS spectra of the core levels of near-intrinsic and heavily doped Si obtained for several $\hbar\omega_L$'s with synchrotron radiation. The data illustrate the shift of the bulk bands with respect to the Fermi level due to surface pinning [8.53]

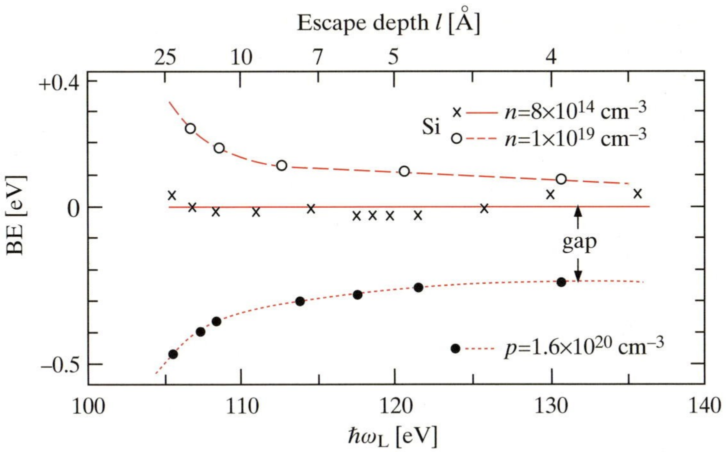

Fig. 8.29. The peak positions from Fig. 8.28 plotted versus the energy of the exciting photons $\hbar\omega_L$ and the corresponding escape depth (see Fig. 8.5). The data reproduce the behavior of the potential in the space-charge layer and reveal the existence of a surface energy gap of about 0.3 eV [8.53]

PROBLEMS

8.1 *Surface Exposure to Gasses: Langmuir*
Calculate the number of air molecules (O_2, N_2) that impinge per second on a solid surface of unit area at 300 K and a pressure of 10^{-6} torr. This unit of exposure is referred to as one Langmuir (L). Assuming a sticking coefficient of one for the atoms in the first surface monolayer (zero for all others), calculate the coverage (in terms of numbers of monolayers) of silicon (111), (100), and (110) clean surfaces after 1 L exposure to air.

8.2 *Spherical Electron Analyzer*
Calculate the band pass energy E_e (in eV) and the resolution ΔE_e (in eV/mm) of a hemispherical analyzer (Fig. 8.9) with sphere radii $R_a + \Delta R_a$ and $R_a - \Delta R_a$.

8.3 *Angle Integrated Photoemission*
Show that angle-integrated photoemission from valence bands is obtained for Al K_α radiation even when using a detector with angular resolution $(\theta_e/\Delta\theta_e) \simeq 5$ (take $\theta_e = 45°$).

8.4 *Fourier Transform of a δ-Function*
Show that

$$\int_{-\infty}^{+\infty} e^{ikr} dr = 2\mathrm{Re}\left\{\int_0^{\infty} e^{ikr} dr\right\} = 2\pi\delta(k)$$

where $\delta(k)$ is the Dirac delta function. Show also that

$$\Re\left\{\int_0^\infty e^{ikr}e^{-\alpha r}dr\right\} = \frac{\alpha}{k^2+\alpha^2}.$$

8.5 *Surface Space Groups and Surface Brillouin Zones*
Discuss the two-dimensional space groups of unreconstructed (100) and (111) surfaces of the diamond and zinc-blende structures. Draw the corresponding two-dimensional Brillouin zones. Give the irreducible representations of the groups of the $\boldsymbol{k}$ vector at the center of the two-dimensional Brillouin zone.

8.6 *Angular Resolved Photoelectron Spectroscopy (ARPES) of GaAs*
Consider ARPES from the (100) surface of GaAs for $\hbar\omega_L = 25$ eV. Using the band structure of Fig. 2.14 plot the dependence of θ_e on the energy E_e of the photoemitted electrons for $\boldsymbol{k}_\parallel$ along [010] and along [011]. Assume that the free-electron parabola has its bottom 9 eV above the top of the valence band.

8.7 *LCAO Band Structure of Graphite*
Draw the primitive cell and the Brillouin zone and write down the Hamiltonian matrix of the electrons for a two-dimensional graphite layer in the LCAO basis (Sect. 2.7) which includes one $2s$ and three $2p$ orbitals for each atom and only nearest-neighbor interactions. If you have a PC with a matrix diagonalization subroutine, calculate the corresponding energy bands using the tight binding parameters given in [8.45] (for additional help see Sect. 2.7 of this book).

8.8 *UPS Spectra of Germanium*
Using the band structure of germanium given in Fig. 8.19 and $E_0 = 8.8$ eV, calculate the dependence of the E_e of photoemission peaks on $\hbar\omega_L$ obtained for $k_\perp$ along [100] and [111] in the 20–40 eV range of $\hbar\omega_L$.

8.9 *One-Dimensional Critical Points in Photoemission*
Investigate which regions of the square and the hexagonal faces of the fcc Brillouin zone have symmetry-induced one-dimensional critical points [zero slope of $E(k_\perp)$, where $k_\perp$ is the component of $\boldsymbol{k}$ perpendicular to the face under consideration] in the case of (a) the zinc-blende structure, (b) the rocksalt structure, and (c) the germanium structure.

8.10 *Spin-Orbit Splittings of d-States in Cubic or Tetragonal Fields*
Investigate the splitting of orbital d levels under the action of a field of either cubic or tetragonal symmetry. Investigate also the effect of such fields on the spin–orbit-split $j = \frac{5}{2}$ and $j = \frac{3}{2}$ components of such levels.
Hint: The characters for the point group of all rotations in three-dimensional space are given by

$$\chi_j(\alpha) = \frac{\sin\left(j + \frac{1}{2}\right)\alpha}{\sin \frac{1}{2}\alpha}, \tag{8.24}$$

where α is the rotation angle (see Problem 4.2).

8.11 *Potential at the Center of a Uniformly Charged Shell*
Derive the potential seen at the center of a uniformly charged shell of charge 1 and inner and outer radii equal to Γr_{m} and r_{m} ($\Gamma < 1$). The origin of potential is taken to be at infinity.

8.12 *Potential at a Space-Charge Surface Layer*
Consider an air-exposed surface of a degenerate n-type GaAs sample (10^{18} electrons/cm^3) at 300 K. The Fermi level should be pinned at 1 eV below the bottom of the conduction band. Write down the equations which determine the profile of the potential $\phi(z)$ at the surface depletion layer without making the constant charge density approximation used to derive (8.22) for $T = 0$ [hint: express ϱ as a function of ϕ in (8.20) using the relation for the carrier density versus Fermi energy]. Solve the equations (numerically if need be) and plot $\phi(z)$. How should the equations be modified for temperatures high enough that the condition $kT < E_{\mathrm{F}}$ is no longer valid?

8.13 *Atomic Scattering from Surfaces*
Neutral atoms and ions with kilovolts of energy are strongly scattered by atoms so they have penetration depths of typically less than a monolayer into a crystal. As a result, they are very useful in determining the structure of surfaces via elastic scattering and surface phonon dispersions via inelastic scattering.

a) Calculate the de Broglie wavelength of He atoms with a kinetic energy of 10 keV and discuss why their scattering by surface atoms can be treated classically.

b) Calculate the *minimum* kinetic energy of a mono-energetic beam of He atoms required to study the surface phonon of an un-reconstructed [111] Ge crystal at the Brillouin zone edge.

c) Suppose electrons are going to be used to determine the surface dispersion of phonons via high resolution electron energy loss spectroscopy (HREELS). Discuss what range of kinetic energies one should choose for the incident electrons.

Hints: see the following references:
H. Lüth: *Surfaces and Interfaces of Solid Materials*, 3rd Edition (Springer, Berlin, Heidelberg, 1998) Chapters 4 & 5.
J.P. Toennies: The study of the forces between atoms of single crystal surfaces from experimental phonon dispersion curves, in Solvay Conference on Surface Science, edited by F.W. de Wette (Springer, Berlin, Heidelberg, 1988) p. 248.

SUMMARY

We have briefly discussed a wide range of spectroscopic techniques that involve the use of electrons and/or photons. These techniques yield very detailed information about occupied and empty electron energy bands and also core levels of semiconductors. The angle-resolved versions of photoemission and inverse photoemission have produced convincing pictures of the $E(\boldsymbol{k})$ dependence of bulk electronic states. They also have yielded information on surface states. We presented spectra of excitations of core levels and discussed the information that can be obtained from them. We also introduced the concepts of surface reconstruction, electronic surface states, and surface energy bands, and presented a few phenomena related to them, such as Fermi level pinning. This led to a brief discussion of the technologically important concepts of charge depletion and enrichment layers at semiconductor surfaces.

9. Effect of Quantum Confinement on Electrons and Phonons in Semiconductors

CONTENTS

In Chap. 5 we studied the Gunn effect as an example of negative differential resistance (NDR). This effect is observed in semiconductors, such as GaAs, whose conduction band structure satisfies a special condition, namely, the existence of higher conduction minima separated from the band edge by about 0.2–0.4 eV. As a way of achieving this condition in any semiconductor, *Esaki* and *Tsu* proposed in 1970 [9.1] the fabrication of an artificial periodic structure consisting of alternate layers of two dissimilar semiconductors with layer thicknesses of the order of nanometers. They called this synthetic structure a **superlattice**. They suggested that the artificial periodicity would fold the Brillouin zone into smaller Brillouin zones or "mini-zones" and therefore create higher conduction band minima with the requisite energies for Gunn oscillations.

With the development of sophisticated growth techniques such as molecular beam epitaxy (MBE) and metal–organic chemical vapor deposition (MOCVD) discussed in Sect. 1.2, it is now possible to fabricate the superlattices (to be abbreviated as SLs) envisioned by *Esaki* and *Tsu* [9.1]. In fact, many other kinds of nanometer scale semiconductor structures (often abbreviated as **nanostructures**) have since been grown besides the SLs. A SL is only one example of a planar or two-dimensional nanostructure. Another example is the quantum well (often shortened to QW). These terms were introduced in Sects. 1.2 and 7.1.5 but have not yet been discussed in detail. The purpose of this chapter is to study the electronic and vibrational properties of these two-dimensional nanostructures. Structures with even lower dimension than two have also been fabricated successfully and studied. For example, one-dimensional nanostructures are referred to as **quantum wires**. In the same spirit, nanometer-size crystallites are known as **quantum dots**. There are so many different kinds of nanostructures and ways to fabricate them that it is impossible to review them all in this introductory book. In some nanostructures strain may be introduced as a result of lattice mismatch between a substrate and its overlayer, giving rise to a so-called **strained-layer superlattice**. In this chapter we shall consider only the best-studied nanostructures. Our purpose is to introduce readers to this fast growing field. One reason why nanostructures are of great interest is that their electronic and vibrational properties are modified as a result of their lower dimensions and symmetries. Thus nanostructures provide an excellent opportunity for applying the knowledge gained in the previous chapters to understand these new developments in the field of semiconductors physics.

Due to limitations of space we shall consider in this chapter only the effects of spatial confinement on the electronic and vibrational properties of nanostructures and some related changes in their optical and transport properties. Our main emphasis will be on QWs, since at present they can be fabricated with much higher degrees of precision and perfection than all other structures. We shall start by defining the concept of **quantum confinement** and discuss its effect on the electrons and phonons in a crystal. This will be followed by a discussion of the interaction between confined electrons and phonons. Finally we shall conclude with a study of a device (known as a *resonant tunneling device*) based on confined electrons and the **quantum Hall effect** (QHE) in a two-dimensional electron gas. The latter phenomenon was discovered by *Klaus von Klitzing* and coworkers in 1980 and its significance marked by the award of the 1985 Nobel Prize in physics to von Klitzing for this discovery. Together with the **fractional quantum Hall effect** it is probably the most important development in semiconductor physics within the last two decades.

9.1 Quantum Confinement and Density of States

In this book we have so far studied the properties of electrons, phonons and excitons in either an infinite crystal or one with a periodic boundary condition (the cases of surface and interface states in Chap. 8 being the only exceptions). In the absence of defects, these particles or excitations are described in terms of Bloch waves, which can propagate freely throughout the crystal. Suppose the crystal is finite and there are now two infinite barriers, separated by a distance L, which can reflect the Bloch waves along the z direction. These waves are then said to be *spatially confined*. A classical example of waves confined in one dimension by two impenetrable barriers is a vibrating string held fixed at two ends. It is well-known that the normal vibrational modes of this string are standing waves whose wavelength λ takes on the discrete values given by

$$\lambda_n = 2L/n, \quad n = 1,2,3\ldots \,. \tag{9.1}$$

Another classical example is a Fabry-Perot interferometer (which has been mentioned already in Sect. 7.2.6 in connection with Brillouin scattering). As a result of multiple reflections at the two end mirrors forming the cavity, electromagnetic waves show maxima and minima in transmission through the interferometer at discrete wavelengths. If the space inside the cavity is filled with air, the condition for constructive interference is given by (9.1). At a transmission minimum the wave can be considered as "confined" inside the interferometer.

For a free particle with effective mass m^* confined in a crystal by impenetrable barriers (i. e., infinite potential energy) in the z direction, the allowed wavevectors k_z of the Bloch waves are given by

$$k_{zn} = 2\pi/\lambda_n = n\pi/L, \quad n = 1,2,3\ldots \,, \tag{9.2}$$

and its ground state energy is increased by the amount ΔE relative to the unconfined case:

$$\Delta E = \frac{\hbar^2 k_{z1}^2}{2m^*} = \left(\frac{\hbar^2}{2m^*}\right)\left(\frac{\pi^2}{L^2}\right). \tag{9.3}$$

This increase in energy is referred to as the **confinement energy** of the particle. It is a consequence of the uncertainty principle in quantum mechanics. When the particle is confined within a distance L in space (along the z direction in this case) the uncertainty in the z component of its momentum increases by an amount of the order of $\hbar/L$. The corresponding increase in the particle's kinetic energy is then given by (9.3). Hence this effect is known also as **quantum confinement**. In addition to increasing the minimum energy of the particle, confinement also causes its excited state energies to become quantized. We shall show later that for an infinite one-dimensional "square well" potential the excited state energies are given by $n^2\Delta E$, where $n = 1,2,3\ldots$ as in (9.2).

It is important to make a distinction between confinement by barriers and localization via scattering with imperfections. Free carriers in semiconductors are scattered by phonons and defects within an average scattering time $\langle\tau\rangle$ introduced in Sect. 5.2. We can define their **mean free path** $\langle l \rangle$ as the product of their average velocity and $\langle\tau\rangle$. Such scattering can also decrease the uncertainty in a particle's position and hence increase its momentum uncertainty. This results in an uncertainty in its energy of an amount given by (9.3) with $L^2 \simeq \langle l^2\rangle$. This effect is typically associated with defects or disorder in solids and is not the same as the quantum confinement effects of interest in this chapter. One way to distinguish between these two cases is to examine the wavevector k_z along the confinement direction. The wavevector of a particle confined in a quantum well, without scattering, is discrete as it corresponds to a standing wave, and is given by (9.2). Scattering at defects dephases a wave so that its amplitude decays exponentially within the mean free path $\langle l \rangle$. The Fourier transform of such a damped wave involves $\boldsymbol{k}_z$, which is not discrete but has a Lorentzian distribution with a width equal to $1/\langle l \rangle$. *Tiong* et al. [9.2a] have proposed a model to estimate $\langle l \rangle$ for phonons from the frequency shift and broadening of optical phonons localized by defects introduced by ion implantation.

Most excitations have a finite lifetime. Optical phonons, for instance, decay via interactions with other phonons (through anharmonicity) [3.3] or defects. As a result, their energies have an imaginary part represented by the damping constant Γ (see for example Sect. 6.4). The effect of Γ is to introduce a width to the energy levels. Therefore, in order to see confinement effects it is necessary for the confinement energy to be at least Γ. Equivalently, this translates, via (9.3), into a maximum value in L for observing confinement effects. In other words, when L is too large the excitation will decay before even reaching the barrier. Since the confinement energy is inversely proportional to m^*, it is more difficult to observe quantum confinement effects in heavier particles. Typically the sample has to be cooled to low temperature (so as to decrease Γ) in order to observe a small confinement energy.

The confinement behavior of excitons is different from both electrons and phonons since they consist of an electron plus a hole separated from each other by a Bohr radius a_0 (Sect. 6.3.1). When L is much larger than a_0, the exciton can move between the barriers like a free particle with total mass M (equal to the sum of the electron and hole masses [9.2b]). The maximum value of L for confinement is determined by the exciton mean free path. When L is smaller than a_0, the exciton properties are modified by the confinement of its constituent electron and hole. For example, the exciton binding energy will be increased since the electron and hole are forced to be closer to each other. In the limit of a two-dimensional exciton, the binding energy is increased by a factor of four relative to the three-dimensional case [(6.95) in Sect. 6.3.3]. Sometimes the confinement potential (if assumed to be infinite) has a larger effect on the two constituent particles than their Coulomb interaction. In such cases it is more convenient to regard the heavier of the two particles as being trapped inside the potential well (since its wavefunction will be more confined in the center of the well) while the other particle is attracted to it via the Coulomb interaction. Similarly, we expect donor and acceptor binding energies to be enhanced when such impurities are confined to a distance smaller than their Bohr radii.

In addition to changing the energies of excitations, confinement also modifies their density of states (DOS). We have already considered the effect of dimensionality on DOS in the vicinity of critical points in Chap. 6 (Table 6.1). In general, reducing the dimensionality "enhances" the singularity in the DOS at a critical point. For instance, on reducing the dimension from three in bulk samples to two in a QW, the electronic DOS at the bandgap E_g changes from a threshold depending on photon energy $\hbar\omega$ as $(\hbar\omega - E_g)^{1/2}$ to a step function. Since the transition probabilities calculated using the Fermi Golden Rule involve the density of final states, confinement can have an important impact on the dynamics of scattering processes in semiconductor devices. For example, it has been demonstrated that laser diodes fabricated from QWs have higher efficiency and smaller threshold current than corresponding bulk laser diodes [9.3]. It has been predicted that quantum dot lasers (zero-dimensional) should have even smaller threshold currents. In addition, their lasing frequencies will be much less sensitive to temperature change. In this book we shall not consider these effects of confinement on devices. Interested readers should refer to other books specializing on this topic [e.g. 9.4, 5].

9.2 Quantum Confinement of Electrons and Holes

As an illustration of how electrons are confined in semiconductors and how to calculate their properties, we shall consider the case of a single QW. Its structure is a "sandwich" consisting of a thin layer (thickness L) of a semiconductor material (denoted by A) between two layers of another semiconductor B (of equal thicknesses L'). The direction perpendicular to these layers will be referred to as the z axis. There are more complex structures, consisting of several repeating units of the form B/A/B/A/B/A/B/A... (where $L' \gg L$), which are known as **multiple quantum wells** or MQWs. Superlattices and MQWs are similar in construction except that the well separations in a MQW are large enough to prevent electrons from **tunneling** from one well to another. The barrier width L' in a SL is thin enough for electrons to tunnel through so that the electrons "see" the alternating layers as a *periodic potential* in addition to the crystal potential.

We shall assume that the bandgap of the well A (E_{gA}) is smaller than that of the barriers B ($E_{gB} > E_{gA}$) in a single QW. Owing to this bandgap difference, the conduction and valence band edges of A and B do not align with each other. The difference between their band edges is known as the **band offset** and has already been introduced in Sect. 5.3. This band offset produces the potential responsible for confining the carriers in one layer only. Thus the control and understanding of this band offset is crucial in the fabrication of quantum confinement devices. While our understanding of what determines the band offset of two dissimilar semiconductors is still not perfect, great progress has been made in the fabrication techniques to control the shape of the bandgap discontinuity. For example, in the well-studied GaAs (=A) / GaAlAs (=B) system the interfaces between A and B have been shown by high resolution transmission electron microscopy to be as narrow as one monolayer (Fig. 9.1). Extensive comparisons between experimental results and theoretical calculations have also shown that the band edge discontinuities can be rather abrupt, making a simple **square well** a good approximation for the confinement potential in most QWs. As a result we shall not discuss further the various theories proposed to explain the band offset, instead we shall assume that its value is known from experiment. *Dingle* et al. [9.6] have defined a factor Q equal to the ratio between the conduction band offset (ΔE_c) and the

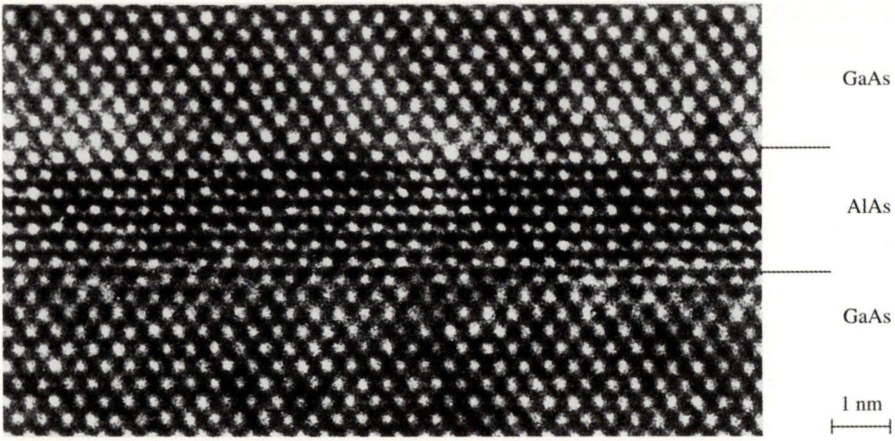

Fig. 9.1. High resolution transmission electron micrograph (TEM) showing a GaAs/AlAs superlattice for a [110] incident beam. (Courtesy of K. Ploog, Paul Drude Institute, Berlin.) In spite of the almost perfect interfaces, try to identify possible Al atoms in Ga sites and vice versa

bandgap difference (ΔE_g) as a way to characterize the band offset. For example, in the technologically important GaAs/GaAlAs and InGaAs/InP QWs the values of Q have been determined to be 0.6 [9.7] (this was pointed out already in Sect. 5.3) and 0.3 [9.8], respectively.

9.2.1 Semiconductor Materials for Quantum Wells and Superlattices

Although a square confinement potential is not the only kind existing in nanostructures, it is nevertheless the most common one. The achievement of a sharp interface imposes very stringent requirements on the growth conditions, such as purity of the source materials, substrate temperature and many others too numerous to list here. However, ultimately the quality of the interface between two dissimilar materials A and B, known as a heterojunction (Sect. 5.3), is determined by their chemical and physical properties. Of these perhaps the most important is the difference in their lattice constants. When these are nearly the same, it is easy for all the atoms of A to be aligned perfectly with those of B. This lattice alignment is known as **pseudomorphic growth** (see Chap. 1) and is highly desirable for achieving high-quality heterojunctions. There are only a few such lattice-matched systems. Figure 9.2 plots the low temperature energy bandgaps of a number of semiconductors with the diamond and zinc-blende structures versus their lattice constants. The shaded vertical regions show the groups of semiconductors with similar lattice constants. Materials within the same shaded region but having different bandgaps can, in principle at least, be combined to form heterojunctions with a particular band offset. This choice of band offsets can be widened by

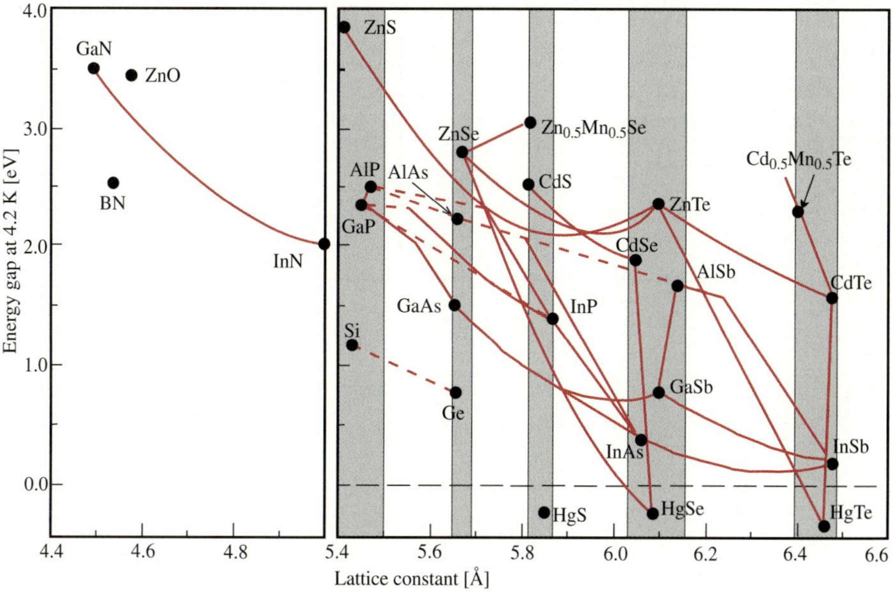

Fig. 9.2. A plot of the low temperature energy bandgaps of a number of semiconductors with the diamond and zinc-blende structure versus their lattice constants. The *shaded regions* highlight several families of semiconductors with similar lattice constants. Semiconductors joined by *solid lines* form stable alloys. [Chen A.B., Sher A.: *Semiconductor Alloys* (Plenum, New York 1995) Plate 1.] Note that the negative gap of HgSe is controversial [2.28]. Broken lines indicate that the bandgap is indirect

growing binary (such as SiGe), ternary (such as AlGaAs) and quarternary (such as GaInAsP) alloys. The solid lines in Fig. 9.2 joining together some of the semiconductors indicate that these materials form stable alloys over the entire alloy range (such as InGaAs, GaAlAs and InGaP). Using Fig. 9.2 as a guide it is possible to "tailor" a heterojunction to have a desired band offset or a QW to have a given shape of confinement potential. This is the basis of what *Capasso* [9.9] has called **bandgap engineering**.

One example of such band gap engineering which has resulted in a device far superior to the conventional device without the use of heterojunctions is the Double Heterojunction (DH) laser. The semiconductor laser diode, as first invented by three groups independently, was based on a junction formed between p-type and n-type GaAs (such junctions formed between the same material with different dopings are known as **homojunctions**). The efficiency of such lasers was not very high and consequently they were difficult to operate in the continuous wave (CW) mode and had also short lifetimes. A major improvement in the design of the laser diode was the use of two heterojunctions in the active region. It resulted in highly efficient CW lasers with long lifetimes. These DH lasers have become part of home and office appliances like compact disc players, laser printers and CD-ROM players in computers.

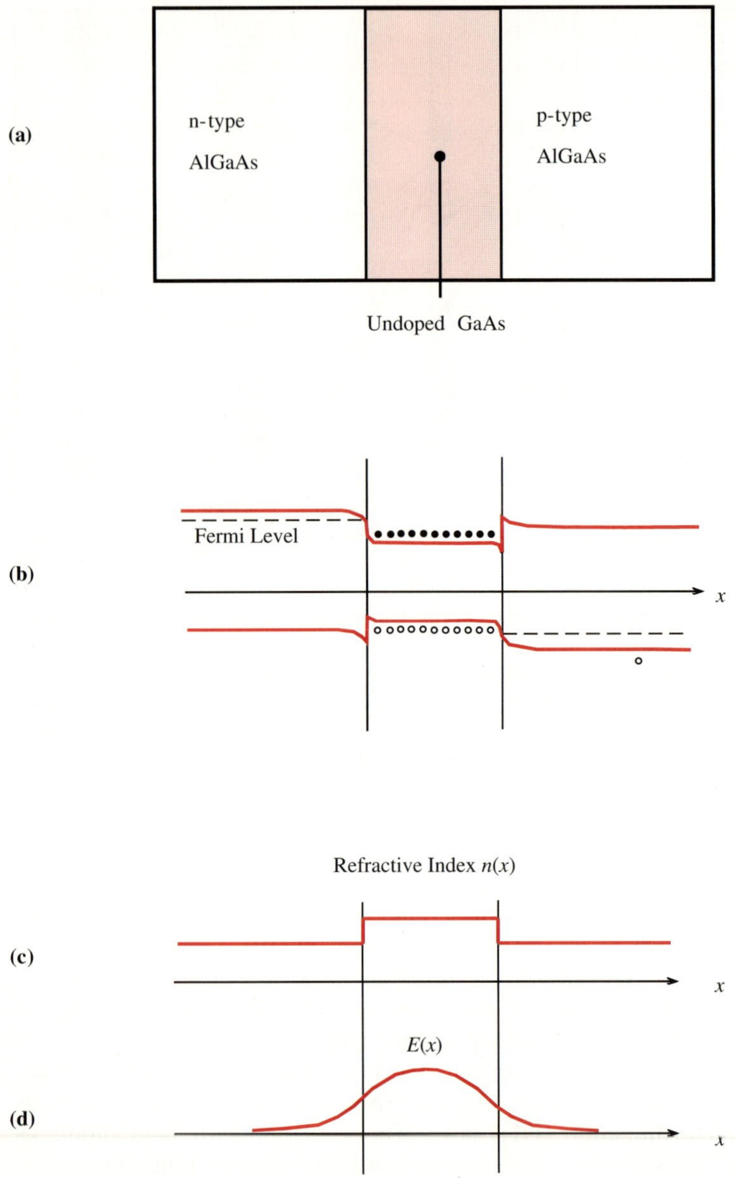

Fig. 9.3. (**a**) Structure of a double heterojunction (DH) laser. (**b**) Dependence of the band edge energies along the growth direction x. (**c**) Dependence of the refractive index along x. (**d**) x-dependence of the electric field confined mainly to the GaAs layer

The idea of using heterojunctions, such as the one formed between GaAs and AlGaAs, as efficient injectors of carriers was proposed by H. Kroemer in the 1960's [9.10]. The idea of using two such heterojunctions to form a DH laser was championed by Zh. I. Alferov and his colleagues at St. Petersburg, Russia [9.11, 12]. For their contributions to the DH laser, Kroemer and Alferov shared the 2000 Nobel Prize in Physics with J.S. Kilby (who is credited as being the co-inventor of the integrated circuits). The basic idea behind the DH laser is shown in Fig. 9.3. Fig. 9.3(a) shows the schematic structure of the laser. Fig. 9.3(b) shows schematically the energy band diagram in the vicinity of the two heterojunctions. Note that the junction is under forward bias (the potential difference across the diode is equal to the difference in energy between the Fermi levels on the two sides of the junction) to inject electrons and holes from AlGaAs into the GaAs layer. This figure shows how the electrons and holes are confined inside the GaAs region by the larger band gap AlGaAs layers. Such confinement has the effect of preventing carriers from diffusing away from the junction area and hence increase their chance of recombining radiatively. The advantage of the DH idea is not limited to the confinment of carriers in the active region. Fig. 9.3(c) shows that a discontinuity also exists in the refractive index at the heterojunctions. The refractive index of GaAs is *larger* than that of AlGaAs. As a result, photons impinging at the junction with angles larger than the critical angle suffer total internal reflection. Thus the AlGaAs/GaAs/AlGaAs DH forms also a waveguide which channels photons within the GaAs layer where they help to stimulate radiative recombination of injected electron-hole pairs. The enhancement in the electric field of the electromagnetic wave within the DH is shown schematically in Fig. 9.3(d). The combined effects of the DH on the electronic and optical properties of the diode contribute to its much lower threshold current density for laser operation.

Having the same lattice constants is not a necessary condition for pseudomorphic growth of one semiconductor (epitaxial layer or **epilayer** in short) on another (substrate). It is possible to force the epilayer to have the same lattice constant parallel to the substrate even though it may be different in the bulk. The result is a strained but otherwise perfect epilayer. However, there is a limit to the thickness of a strained layer one can grow while maintaining a perfect lattice. As the strain energy increases with the thickness, beyond some thickness known as the **critical layer thickness**, the epilayer can lower its total energy by relieving the strain via the creation of **misfit dislocations**. In a simple-minded approach, one can assume that the dislocation energy is proportional to the number of atoms in the dislocation. Since dislocations tend to start at the interface and then propagate upwards, their energy is proportional to the layer thickness. For very thin layers, the strain energy may be smaller than the dislocation energy and the epilayer grows pseudomorphically while maintaining a perfect lattice without dislocations. As the strain energy increases with the volume of the strained epilayer, for a thicker film it will cost less energy to generate misfit dislocations than to strain the entire film. Hence, films thicker than the critical layer thickness are dislocated but not strained. The critical layer thickness for two dissimilar materials obviously depends on their lattice constant difference. By keeping the superlattice layers thinner than the critical layer thickness it is possible, in principle, to grow a strained-layer superlattice (SLS) from any two semiconductors without regard to their lattice constants. One example is the Si/Ge SLS, where the lattice mismatch is about 4%. We shall not consider SLSs further in this book. Readers are referred to [9.13] and [6.88] for more details of the properties of SLSs.

Finally, if there are interface states within the gap of one of the two materials forming the QW they can pin the Fermi level and give rise to band bending (Sect. 8.3.3). While band bending is sometimes desirable or unavoidable (say, as a consequence of doping), in general it makes the confinement potential more complicated to treat theoretically and therefore we shall assume that it is absent unless otherwise noted.

9.2.2 Classification of Multiple Quantum Wells and Superlattices

Since the idea of growing SLs and QWs was introduced in order to make electronic devices, it is now customary to classify these structures according to the confinement energy schemes of their electrons and holes. Given two semiconductors A and B (with $E_{gB} > E_{gA}$) forming the MQW B/A/B/A/B/A/B/A..., one can obtain three scenarios as shown schematically in Fig. 9.4. These confinement schemes are usually labeled **type I** and **type II**. In type I MQWs or SLs the electrons and holes are both confined (the energy of the confined particles is represented by the red broken lines) within the same layers A (forming the well). Type I MQWs are formed by GaAs/GaAlAs provided the GaAs layer thickness is larger than 2nm or the Al mole fraction less than 0.3.

A type IIA MQW can be called "*spatially indirect bandgap*" semiconductor: While the electrons are confined in one layer, the holes are confined in a different layer. Type IIA MQWs or SLs are formed by GaAs (A) and AlAs (B) when the GaAs layer thickness is smaller than 2 nm. The small well width causes the electron energy in GaAs to lie above the weakly confined (because of the large effective mass) conduction band minimum in AlAs (which occurs at the X point of the Brillouin zone). As a result, the electrons are confined in the AlAs layers while the holes are still confined in the GaAs layers.

A type IIB MQW behaves like zero-gap or small-gap semiconductors since there is either only a very small energy gap between the electrons in layer A and the holes in layer B or none at all. The supperlattice InAs/GaSb is an example of Type IIB behavior.

One may also find occasionally in the literature a superlattice formed by a semiconductor and a zero-gap semiconductor. This is referred to as a **type III superlattice** [9.14]. Finally MQWs and SLs can also be formed by two identical

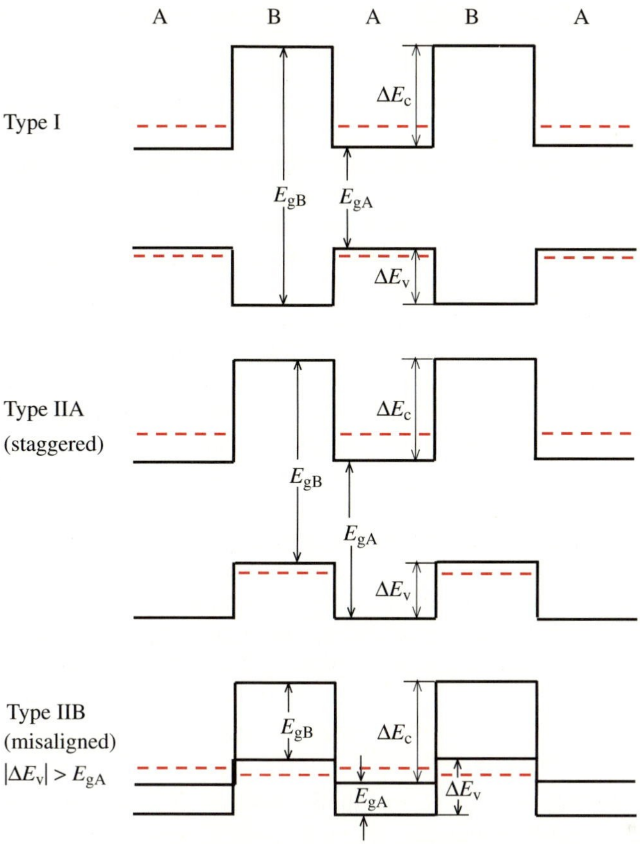

Fig. 9.4. Schematic diagrams of three arrangements of the confinement of electrons and holes in MQWs and superlattices formed by two semiconductors A and B with bandgaps E_{gA} and E_{gB}, respectively. In type I samples both the electrons and holes are confined in the same layer A. The energies of the confined particles are represented by *red lines*. In type IIA systems the electrons and holes are confined in different layers. Type IIB samples are a special case of type IIA behavior. They are either small gap semiconductors or semimetals

semiconductors but doped differently. For example, a SL formed by doping the layer A n-type and the layer B p-type is known as a **doping superlattice** (or **nipi structure**, where i stands for an intrinsic layer between the n- and p-type layers) [9.15]. Isotope superlattices have also been investigated [9.16].

9.2.3 Confinement of Energy Levels of Electrons and Holes

To understand the electronic levels of a QW we shall start by assuming a simple square well potential defined by

$$V(z) = \begin{cases} 0 & \text{for } (-L/2) < z < (L/2), \\ V_0 & \text{for } z > (L/2) \text{ or } z < (-L/2), \end{cases} \tag{9.4}$$

where L and V_0 are, respectively, the width and height of the well. Notice that the height of the well is equal on both sides and, consequently, the well potential (9.4) has reflection symmetry with respect to its center. To take advantage of this symmetry we have chosen the origin at the center of the well. As a result, the eigenstates of electrons in the well have definite parity under the operation of reflection with respect to the $z = 0$ plane. Consequently there are also selection rules (in addition to those imposed by the crystal symmetry) governing the interaction of electrons and phonons in QWs with external perturbations. Care should be exercised when using these results based on the reflection symmetry. In the most common case of a GaAs QW surrounded by AlAs barriers grown on a (001)-oriented GaAs substrate, strictly speaking, the noncentrosymmetric GaAs crystal has no mirror planes perpendicular to the growth axis. Nevertheless, a parity-like symmetry exists with respect to the two-fold rotations about the [100] and [010] axes (Sect. 2.3.2). The assumption of a mirror plane through the center of a symmetric QW remains, however, valid within the effective mass approximation commonly used in the literature (which neglects the details of the microscopic electron band structure). Asymmetric QWs, in which parity is not a good quantum number, can be fabricated by choosing two different barriers on the two sides of the QW, but they will not be studied here.

a) Confinement of Electrons in Quantum Wells

In order to start with the simplest case, we shall consider an electron in a Type I QW constructed of two semiconductors with similar parabolic conduction bands. Furthermore, we assume that the growth direction (z axis) is parallel to one of the principal axes of the effective mass tensor in both materials. As we discussed in detail in Sect. 4.2.1, the energy and wavefunction of this electron can be calculated with the effective mass approximation provided the potential $V(z)$ satisfies the conditions enumerated in Sect. 4.2.1. This means, in principle, that the well width has to be much larger than one monolayer (for calculations of energy levels in ultrathin QWs see [9.17, 18]). With this approximation the electron wavefunction can be written in a form similar to (4.12). The Wannier functions and envelope functions inside the well will be denoted by $a_{n\mathrm{A}}$ and $C_{n\mathrm{A}}$, respectively, while the corresponding functions in the barrier will be denoted by $a_{n\mathrm{B}}$ and $C_{n\mathrm{B}}$. This approximation is often referred to as the **envelope function approximation** in the QW literature. The electron wavefunction in the QW can be expressed as

$$\Psi(\boldsymbol{r}) = \begin{cases} N^{-1/2} \sum C_{n\mathrm{A}}(\boldsymbol{R}_i) a_{n\mathrm{A}}(\boldsymbol{r} - \boldsymbol{R}_i) & \text{for } (-L/2) \leq z \leq (L/2), \\ N^{-1/2} \sum C_{n\mathrm{B}}(\boldsymbol{R}_i) a_{n\mathrm{B}}(\boldsymbol{r} - \boldsymbol{R}_i) & \text{for } z > (L/2) \text{ or } z < (-L/2). \end{cases} \tag{9.5}$$

Similarly, the equations of motion for the envelope functions inside and outside the well are different. For isotropic effective masses (to be denoted by

m^*_{A} and m^*_{B} for the well and barriers, respectively) these equations are

$$\left[-\left(\frac{\hbar^2}{2m^*_{\mathrm{A}}}\right)\frac{\partial^2}{\partial \boldsymbol{R}^2}\right]C_{\mathrm{A}}(\boldsymbol{R}) \approx [E - E_{\mathrm{cA}}]C_{\mathrm{A}}(\boldsymbol{R}) \quad \text{for} \quad (-L/2) \le z \le (L/2) \tag{9.6a}$$

and

$$\left[-\left(\frac{\hbar^2}{2m^*_{\mathrm{B}}}\right)\frac{\partial^2}{\partial \boldsymbol{R}^2} + V_0\right]C_{\mathrm{B}}(\boldsymbol{R}) \approx [E - E_{\mathrm{cA}}]C_{\mathrm{B}}(\boldsymbol{R}) \quad \text{for} \quad z > (L/2) \quad \text{or} \quad z < (-L/2), \tag{9.6b}$$

where E_{cA} is the conduction band edge in the well (without loss of generality it will be taken to be zero).

Since the confinement potential depends on z only, (9.6a and b) are separable into two equations, one depending on z and the other depending on x and y only. The wavefunctions $C_{(\mathrm{A\ or\ B})}(x,y,z)$ can then be expressed as the products of the solutions of these two equations:

$$C_{(\mathrm{A\ or\ B})}(x,y,z) = \phi_{(\mathrm{A\ or\ B})}(x,y)\psi_{(\mathrm{A\ or\ B})}(z). \tag{9.7a}$$

The equation for $\phi_{(\mathrm{A\ or\ B})}(x,y)$ is that of a free particle, and hence its solutions are plane waves of the form

$$\phi_{(\mathrm{A\ or\ B})}(x,y) \propto \exp[\pm \mathrm{i}(k_x x + k_y y)], \tag{9.7b}$$

where k_x and k_y are components of the Bloch wavevectors parallel to the well. Since translational invariance in the xy plane is preserved, all theorems concerning $\boldsymbol{k}$-conservation derived for bulk crystals apply to k_x and k_y but not to the z component.

Notice that the mass changes from m^*_{A} inside the well to m^*_{B} outside it. However, the boundary condition that the wavefunctions must be continuous across the QW interface requires that *both k_x and k_y be the same inside and outside the well.* The equations for $\psi_{(\mathrm{A\ or\ B})}(z)$ are

$$-\left(\frac{\hbar^2}{2m^*_{\mathrm{A}}}\right)\left(\frac{d^2}{dz^2} + k_x^2 + k_y^2\right)\psi_{\mathrm{A}}(z) \approx E\psi_{\mathrm{A}}(z) \quad \text{for} \quad (-L/2) \le z \le (L/2) \tag{9.8a}$$

and

$$\left[-\left(\frac{\hbar^2}{2m^*_{\mathrm{B}}}\right)\left(\frac{d^2}{dz^2} + k_x^2 + k_y^2\right) + V_0\right]\psi_{\mathrm{B}}(z) \approx E\psi_{\mathrm{B}}(z) \quad \text{for} \quad z > (L/2) \quad \text{or} \quad z < (-L/2). \tag{9.8b}$$

Except for the fact that the masses m^*_{A} and m^*_{B} may be different, (9.8) at $k_x = k_y = 0$ is identical to the quantum mechanics textbook case of a particle confined in a one-dimensional square well. The solutions for the case $m^*_{\mathrm{A}} = m^*_{\mathrm{B}}$ can be found, for example, in [Ref. 9.19, p. 34]. Here we shall try to summarize the results only.

In general there are two kinds of solutions to (9.8). When $\{E - [\hbar^2/(2m_B^*)](k_x^2 + k_y^2)\} > V_0$, the solutions are plane waves and their energy spectrum is continuous. In this case the particle has sufficient kinetic energy to overcome the barrier and is therefore not confined inside the well. These continuum solutions, while important, are not of interest to us at this point.

For $\{E - [\hbar^2/(2m_B^*)](k_x^2 + k_y^2)\} < V_0$ the solutions to (9.8b) are exponential functions of the form

$$\psi_B(z) = \alpha_1 e^{\tau z} + \alpha_2 e^{-\tau z}, \tag{9.9a}$$

where τ is a positive real number given by

$$\left[E - \left(\frac{\hbar^2}{2m_B^*}\right)(k_x^2 + k_y^2 - \tau^2)\right] = V_0. \tag{9.10}$$

In order that $\psi_B(z)$ be finite at $z = \pm\infty$,

$$\psi_B(z) = \begin{cases} \alpha_1 e^{\tau z} & \text{for } z < (-L/2), \\ \alpha_2 e^{-\tau z} & \text{for } z > (L/2). \end{cases} \tag{9.9b}$$

The wavefunction $\phi_B(x, y)\psi_B(z)$ describes a wave traveling parallel to the well but exponentially decaying into the barriers from the interfaces. Such waves are known as **evanescent waves**. The z component of its wavevector is equal to $\pm i\tau$ and therefore is imaginary. Within the well, the solutions to (9.8a) can be expressed as linear combinations of symmetrized (with respect to reflection onto the $z = 0$ plane) wavefunctions, such as cosine (symmetric) and sine (antisymmetric) functions:

$$\psi_A(z) = \beta_1 \cos(k_z z) \quad \text{or} \quad \beta_2 \sin(k_z z) \quad \text{for} \quad (-L/2) < z < (L/2). \tag{9.9c}$$

The allowed values of E are discrete in this case. These bound state solutions (i. e., the four coefficients α_1, α_2, β_1 and β_2) are determined by the usual requirement that the wavefunctions and their first derivatives be continuous across the two QW interfaces. In general there are no analytic expressions for the eigenvalues except when V_0 is infinite. In this special case the particle is completely confined inside the well (and hence m_B^* becomes irrelevant) and the values of k_z are given by the classical expression for standing waves:

$$k_z = n\pi/L \quad \text{where} \quad n = 1, 2, 3, \ldots \,. \tag{9.11}$$

The corresponding energies are

$$E_n(k_x, k_y) = \left(\frac{\hbar^2}{2m_A^*}\right)\left[\left(\frac{n\pi}{L}\right)^2 + k_x^2 + k_y^2\right] \quad n = 1, 2, 3, \ldots \,. \tag{9.12a}$$

At $k_x = k_y = 0$, the energy levels are equal to

$$E_n = \frac{\hbar^2}{2m_A^*}\left(\frac{n\pi}{L}\right)^2. \tag{9.12b}$$

Figure 9.5 shows the electron energy levels in the infinite barrier QW. To distinguish them from the electron energy bands of the corresponding bulk crys-

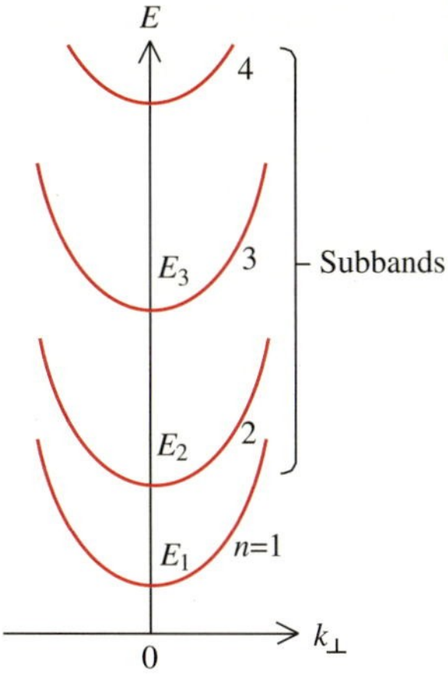

Fig. 9.5. Schematic diagram of the energy subbands in an infinite barrier quantum well versus in-plane wavevector $k_\perp$

tal A, these energy bands of electrons confined in two dimensions are known as **subbands**.

When V_0 is finite, the subband energies cannot be expressed analytically. They can be determined either graphically or numerically with the help of a computer. The former procedure for the special case of $m_A^* = m_B^*$ can be found in quantum mechanics textbooks. In the case $m_A^* \neq m_B^*$ the continuity condition on the derivatives $\partial\psi_A(z)/\partial z$ and $\partial\psi_B(z)/\partial z$ must be modified to the so-called **Bastard boundary condition** [9.20a]

$$\left(\frac{1}{m_A^*}\right)\left(\frac{\partial\psi_A}{\partial z}\right) = \left(\frac{1}{m_B^*}\right)\left(\frac{\partial\psi_B}{\partial z}\right) \quad \text{at} \quad z = \pm(L/2). \tag{9.13}$$

This condition ensures that the particle flux is continuous across the interface between A and B.

Figure 9.6 shows the calculated bound state energies (for $k_x = k_y = 0$) of an electron in a $Ga_{0.47}In_{0.53}As$–$Al_{0.48}In_{0.52}As$ QW (grown on a lattice-matched InP substrate) as a function of well width L. The well depth V_0 is equal to 0.5 eV. Notice that the number of bound states lying within the well decreases as L becomes smaller. For $L < 30$ Å only one bound state exists. It should be noted that the "unbound" or continuum solutions are modified by the confinement potential and differ from those in the bulk crystals. Discussions of the case when the effective mass is parabolic but not isotropic can be found in [9.21].

b) Confinement of Holes in Quantum Wells

As we showed in Sect. 4.2.4, the equation of motion of holes in diamond- and zinc-blende-type semiconductors under the influence of a Coulomb potential is rather complicated even within the effective mass approximation. In the case of shallow acceptors discussed in that section, the solutions are simplified by the spherical symmetry of the potential. Since the QW confinement potential

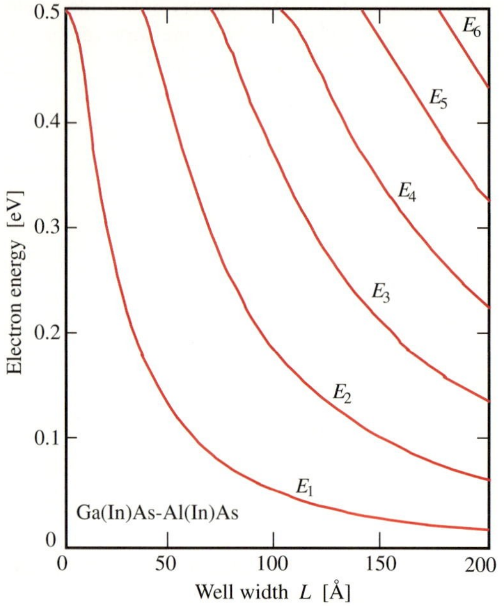

Fig. 9.6. Calculated electron confinement energies in a $Ga_{0.47}In_{0.53}As/Al_{0.48}In_{0.52}As$ QW grown on a lattice-matched InP substrate as a function of well width L. From [9.20b]. Binding energies higher than 0.5 eV are not represented since they correspond to non-confined states which overlap with the continuum of the barrier. Note that for $L < 30$Å only one bound state appears

$V(z)$ is not spherical we expect the problem to be more difficult. Indeed numerical calculations are necessary to obtain meaningful results. An example of hole subbands obtained by such calculations is shown by the solid curves in Fig. 9.7 for a $GaAs/Ga_{0.7}Al_{0.3}As$ QW with well widths of 100 and 150 Å [9.22]. The labeling of the bands suggests that they arise from either the heavy hole (HH) or light hole (LH) bands of the bulk. We shall see that, strictly speaking, this is not the case since the bulk valence bands are heavily mixed by $V(z)$. These subbands can be qualitatively understood in the following way.

Let us assume that the spin–orbit splitting of the hole bands is so large that only the $J = 3/2$ heavy ($J_z = \pm 3/2$) and light ($J_z = \pm 1/2$) hole bands need be considered. Furthermore, the axis of quantization for J is chosen to be along the growth direction. Their dispersions in the bulk crystal can be calculated from the Luttinger Hamiltonian $\mathcal{H}_{\mathrm{L}}$ in (2.70), to which the confinement potential $V(z)$ must be added. The resultant Hamiltonian contains, in general, terms of the form $J_x J_y(\partial/\partial x)(\partial/\partial y)$ etc. from the expansion of $(\boldsymbol{J} \cdot \nabla)^2$. The Schrödinger equation corresponding to this Hamiltonian is not separable. Since these "off-diagonal" terms arise from the warping of the valence bands and are not important for understanding qualitatively the effect of confinement on the valence band dispersion, we shall make the *ad hoc* assumption that they are negligible. In most semiconductors these terms are small (exceptions being Si, SiC, and diamond; see Table 4.3) and can be introduced later as weak perturbations. Within this approximation the hole Hamiltonian in a QW can be written as:

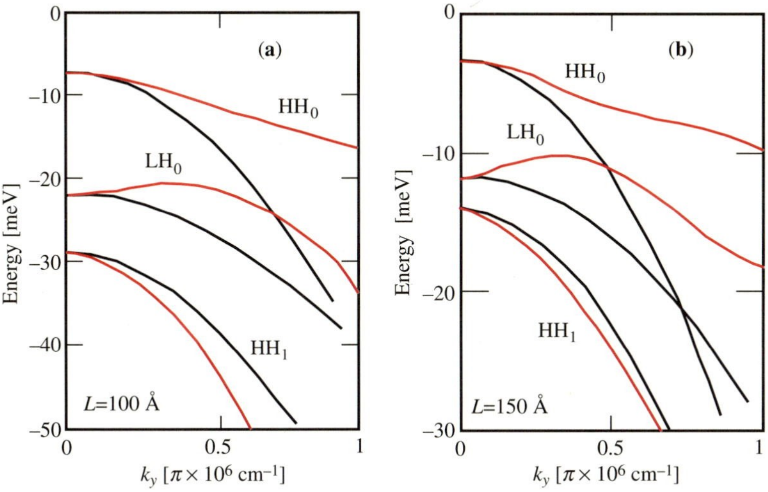

Fig. 9.7. Hole subband energies in a $GaAs/Ga_{0.7}Al_{0.3}As$ QW calculated by *Bastard* and *Brum* [9.22] within the envelope-function approximation for well widths of (**a**) 100 and (**b**) 150 Å, respectively. The labels HH and LH denote the subbands arising from the heavy and light hole valence bands in the bulk, respectively (*red curves*). The *black curves* represent the subband energies calculated when the mixing between the heavy and light hole bands is neglected

$$\mathcal{H}_{\mathrm{QW}} = \left(\frac{\hbar^2}{2m}\right)\left[\left(\gamma_1 + \frac{5\gamma_2}{2}\right)\nabla^2 - 2\gamma_2(\nabla_x^2 J_x^2 + \mathrm{c.p.})\right] + V(z) \tag{9.14}$$

The corresponding Schrödinger equation is now separable into two equations as in (9.6). The Hamiltonian for motion in the z direction becomes

$$\left(\frac{\hbar^2}{2m}\right)\left[\left(\gamma_1 + \frac{5\gamma_2}{2}\right) - 2\gamma_2 J_z^2\right]\left(\frac{\partial}{\partial z}\right)^2 + V(z) \tag{9.15a}$$

This equation suggests that the $J_z = 3/2$ state (we shall avoid using the labels "heavy" and "light" because they are no longer meaningful as we shall see later) behaves, for the purpose of calculating the confinement properties with the Hamiltonian in (9.15a), as if its effective mass $m_{\mathrm{h}z}$ were equal to [cf. (2.67)]

$$(m_{\mathrm{h}z})^{-1} = (\gamma_1 - 2\gamma_2)/m, \tag{9.16a}$$

while the $J_z = 1/2$ state acts as if it had the lighter mass $m_{\mathrm{l}z}$

$$(m_{\mathrm{l}z})^{-1} = (\gamma_1 + 2\gamma_2)/m. \tag{9.16b}$$

Since the confinement energy is inversely proportional to the effective mass, see (9.12), it is larger for the $J_z = 1/2$ state than for the heavy $J_z = 3/2$ state. This situation is shown schematically in Fig. 9.8a.

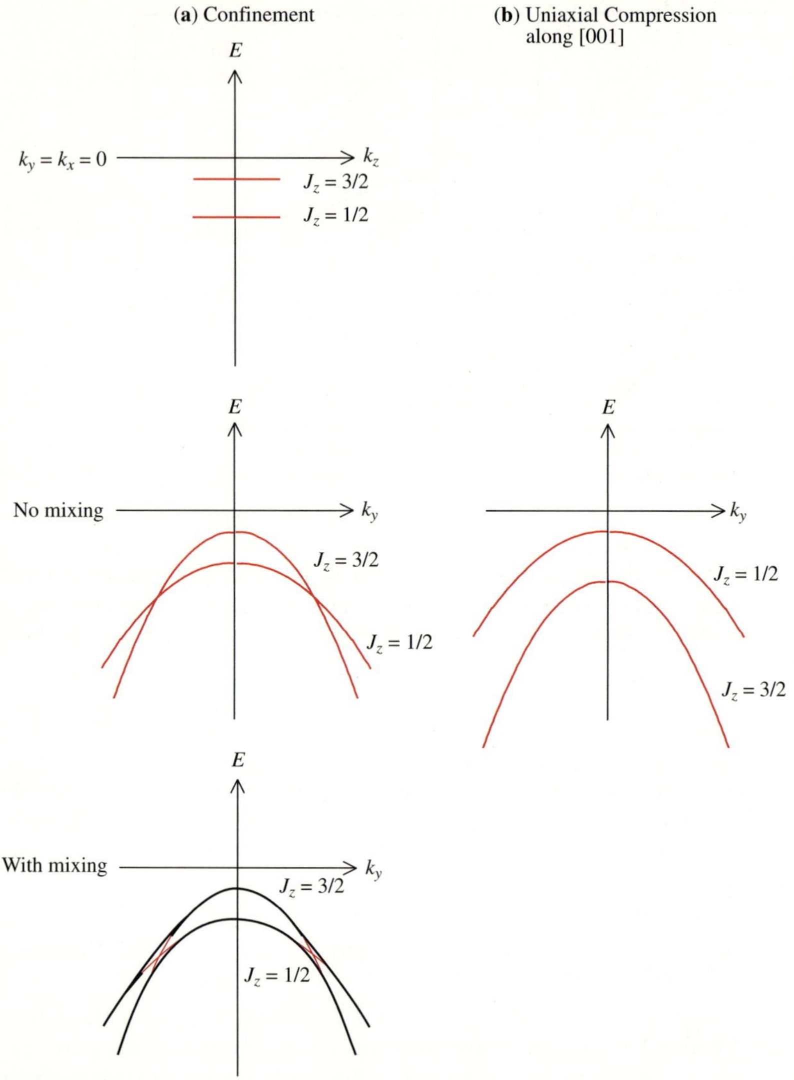

Fig. 9.8. Schematic dispersion of the $J_z = 3/2$ and $J_z = 1/2$ valence bands in the presence of (**a**) a "square well" confinement potential in the z direction both when the "off-diagonal" terms are neglected (without mixing) and included (with mixing), and (**b**) a compressive uniaxial stress along the z direction is added

The Hamiltonian for the motion perpendicular to the confinement potential (say along the y axis) is given by

$$\left(\frac{\hbar^2}{2m}\right)\left[\left(\gamma_1 + \frac{5}{2}\gamma_2\right) - 2\gamma_2 J_y^2\right]\left(\frac{\partial}{\partial y}\right)^2. \tag{9.15b}$$

The expectation values $\langle 3/2, 3/2|J_y^2|3/2, 3/2\rangle$ and $\langle 3/2, 1/2|J_y^2|3/2, 1/2\rangle$ are equal to 3/4 and 7/4, respectively. When these values are used to calculate the expectation values of the Hamiltonian in (9.15b), we find that the $J_z = 3/2$ state behaves as if it had the mass

$$(m_{\mathrm{hy}})^{-1} = (\gamma_1 + \gamma_2)/m \tag{9.17a}$$

in the y direction while the $J_z = 1/2$ state has the mass

$$(m_{\mathrm{ly}})^{-1} = (\gamma_1 - \gamma_2)/m. \tag{9.17b}$$

Table 4.3 shows that $\gamma_1 > \gamma_2$ and both are positive in most zinc-blende-type semiconductors. Thus we arrive at the somewhat paradoxical result that the $J_z = 3/2$ state (heavy hole in the bulk along the direction of quantization and also in calculating the confinement properties) has a *smaller* mass in the x and y directions than the $J_z = 1/2$ state (or light hole in the bulk). This phenomenon is known as **mass reversal**. It was first encountered in cyclotron resonance experiments in p-type Si under compressive [001] uniaxial stress [9.23]. In that case the uniaxial stress splits the degeneracy of the heavy and light hole bands at the Brillouin zone center (see Problem 3.8c) and the bands become parabolic. The sign of the deformation potential b is such that the $J_z = 1/2$ state (with the larger mass now!) has a higher energy than the $J_z = 3/2$ state [3.30]. As a result the two bands do not cross, as shown in Fig. 9.8b. However, in a QW the hole bands cross each other as shown in Fig. 9.8a and also in Fig. 9.7 as the black curves. When the "off-diagonal" terms are included, the two bands are mixed together and level-anticrossing (see also Sect. 6.4) results whenever they might intersect each other. This is shown schematically in Fig. 9.8a. While compressive uniaxial stress along high-symmetry directions of diamond- and zinc-blende-type semiconductors simplifies the $J = 3/2$ valence bands in the $\boldsymbol{k}$ direction perpendicular to the stress, by lifting their degeneracy and removing their warping, the confinement potential of a QW complicates the valence band structures by inducing mixing and level-anticrossing. In spite of the ambivalent meaning of heavy and light masses, the terms "heavy hole" and "light hole" are often used in the literature to designate the subbands in QWs arising from the $J = 3/2$ valence bands.

c) Electrons and Holes in Superlattices

In principle, the energy of electrons and holes in superlattices can be calculated in the same way as for QWs provided one uses the appropriate boundary conditions. The periodicity of the superlattice (with period equal to the sum of the widths of a well and a barrier and represented by d) imposes the same boundary conditions on the electron and phonon wavefunction as the periodic potential of a one-dimensional crystal. As a result, it should be possible to express the envelope wave function $\psi(z)$ for an electron in the conduction band in the form of Bloch waves propagating along the direction of growth of the SL. For the simple case of $k_x = k_y = 0$ in a GaAs/$\mathrm{Al}_{1-x}\mathrm{Ga}_x\mathrm{As}$ SL (where for $0.15 < x < 0.30$ one can approximately assume that the electron has isotropic effective masses $m_{\mathrm{A}}^* = m_{\mathrm{B}}^*$) the wave equation of the electron in the z direction reduces to that of a one-dimensional periodic square-well potential, better known as the **Kronig–Penney model**. The solutions of this problem are discussed in many quantum mechanics textbooks (see, e. g. [9.24]). We shall not go into the details of the solutions of the wave equation for the envelope function. It suffices to point out that for a SL consisting of wells and barriers with widths a and b, respectively, and barrier height V_0, the electron energy E is related to the Bloch wavevector $k_z = k$ (dispersion) via transcendental equations of the form

$$\cos(kd) = \cos(k_1 a)\cos(k_2 b) - \frac{k_1^2 + k_2^2}{2k_1 k_2}\sin(k_1 a)\sin(k_2 b) \tag{9.18a}$$

for $E > V_0$

and

$$\cos(kd) = \cos(k_1 a)\cosh(\kappa b) - \frac{k_1^2 - \kappa^2}{2k_1 \kappa}\sin(k_1 a)\sinh(\kappa b) \tag{9.18b}$$

for $E < V_0$,

where k_1, k_2 and κ are defined by

$$E = \hbar^2 k_1^2/(2m_{\mathrm{A}}^*), \tag{9.19a}$$

$$E - V_0 = \hbar^2 k_2^2/(2m_{\mathrm{A}}^*) \quad \text{for } E > V_0, \tag{9.19b}$$

$$V_0 - E = \hbar^2 \kappa^2/(2m_{\mathrm{A}}^*) \quad \text{for } E < V_0. \tag{9.19c}$$

The transcendental equations (9.18a and b) are usually solved numerically. As a comparison of the energies in a QW and in a SL we show in Fig. 9.9 the evolution of electron energies in a one-dimensional MQW (shown schematically in the inset of Fig. 9.9) with a potential barrier height of 0.4 eV as a function of the barrier width (chosen to be equal to the well width). As the barrier width decreases the neighboring quantum wells interact with each other and the MQW becomes a SL: its discrete energy levels broaden into bands known as **minibands**. It should be remembered that the envelope-function approximation breaks down when the layers are only a few monolayers thick. To calculate the electron energies in these so-called **short-period superlattices**, one has to apply standard band structure calculation techniques to an "artificial" three-dimensional crystal generated with a "supercell" consisting of one period of the superlattice. The electron energies in short-period SLs have been calculated this way using either pseudopotential or tight-binding methods as described in Chap. 2. Figure 9.10 shows the hole miniband energies in a GaAs/$Ga_{0.7}Al_{0.3}As$ SL calculated by *Chang* and *Schulman* [9.26] using a LCAO model (see Sect. 2.7), although the GaAs and GaAlAs layers (68 and 71 monolayers thick, respectively) are more than a few monolayers thick. Note that one monolayer of GaAs has a thickness of 2.827 Å, very close to that of GaAlAs. Figure 9.10 clearly displays several lh–hh anticrossings of the type discussed above.

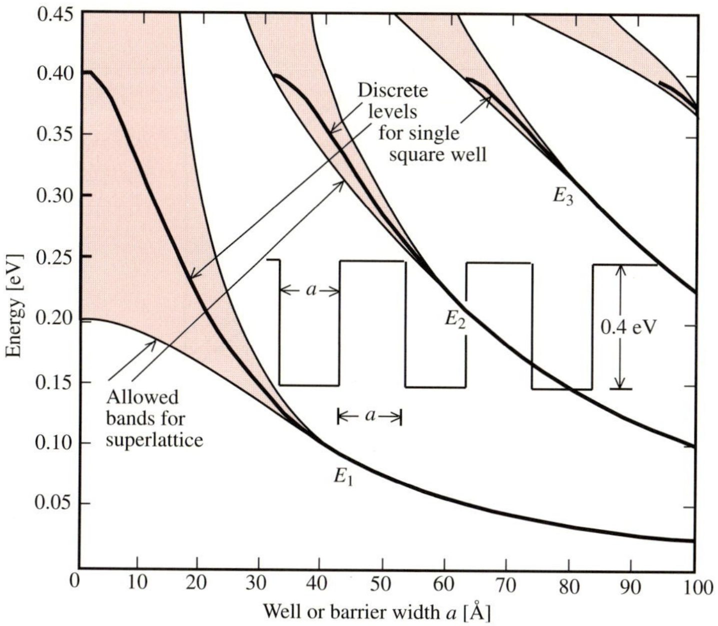

Fig. 9.9. Evolution of energy levels in a QW into minibands in a SL as a function of the barriers widths (assumed to be equal to the well width) calculated for a particle in a Kronig–Penney potential (shown in the *inset*). The height of the square potentials is 0.4 eV. Notice that there are gaps between these minibands even when they are *above* the potential barriers. From [9.25]

9.2.4 Some Experimental Results

At the time of writing of this book the theoretically predicted energies of subbands and minibands in QWs and SLs had been extensively verified experimentally. It is impossible to review all the results in this chapter. Instead, we shall present some typical results for type I QWs based on the well-studied GaAs/AlGaAs system. The excellent agreement between theory and experiment provides another stringent proof of the high degree of perfection achieved in growing these nanostructures. Many of these results have been obtained by transport (tunneling experiments to be described later in this chapter) and optical techniques (such as absorption, both interband and intersubband, photoluminescence, luminescence excitation and Raman spectroscopies, which have been described in Chap. 6 and 7).

The first convincing evidence of confinement of the electrons and holes in QWs was provided by optical absorption measurements. In Fig. 7.16a we have already shown such a spectrum for a 20 nm GaAs QW as reported by *Dingle* [9.27]. While absorption spectra are representative of the joint density of states of electrons and holes (often modified by excitonic effects which are enhanced in two dimensions as mentioned in Sect. 9.1), the spectrum in

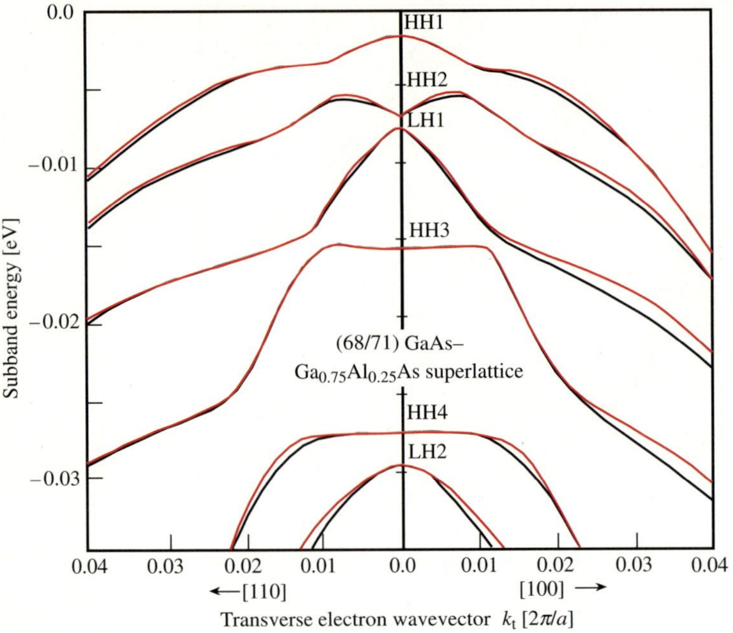

Fig. 9.10. Hole miniband energies as a function of wavevector transverse to the growth axis in a GaAs (68 monolayers)/$Ga_{0.7}Al_{0.3}As$ (71 monolayers) SL calculated by *Chang* and *Schulman* [9.26] using an LCAO model

Fig. 7.16a nevertheless shows two important features of quantum confinement. The first is a confinement-induced blue shift of the exciton peaks with respect to those in bulk GaAs. This may not be obvious from a comparison of Fig. 6.25 with Fig. 7.16a because of the rather large width of the QW in Fig. 7.16a. It becomes clearer in the narrower QW shown in Fig. 7.16b. The second feature is the appearance of a series of peaks (corresponding to the quantized levels with quantum numbers $n = 1, 2$, etc.) whose separation changes with the well width as predicted by (9.12b). These exciton peaks (labeled $n = 1$ and 2) exhibit a doublet structure due to the splitting of the valence bands heavy and light hole components caused by the confinement potential. Figure 9.11 shows a more detailed comparison of the predicted and observed optical transition energies between the subbands of the heavy and light holes and those of the electron in a 316 Å wide GaAs/AlGaAs QW measured by *Dingle* [9.27]. The interpretation of the experimental spectrum is simplified by the electric dipole selection rules. As pointed out by *Dingle* et al. [9.6], for infinite well depth the electron and hole subband quantum numbers have to be the same in an electric dipole transition because of the orthogonality of the envelope functions. This rule is relaxed in wells of finite height where the Hamiltonians for electrons and holes become different and their respective wavefunctions need not be orthogonal to each other. In symmetric wells, however, parity conservation

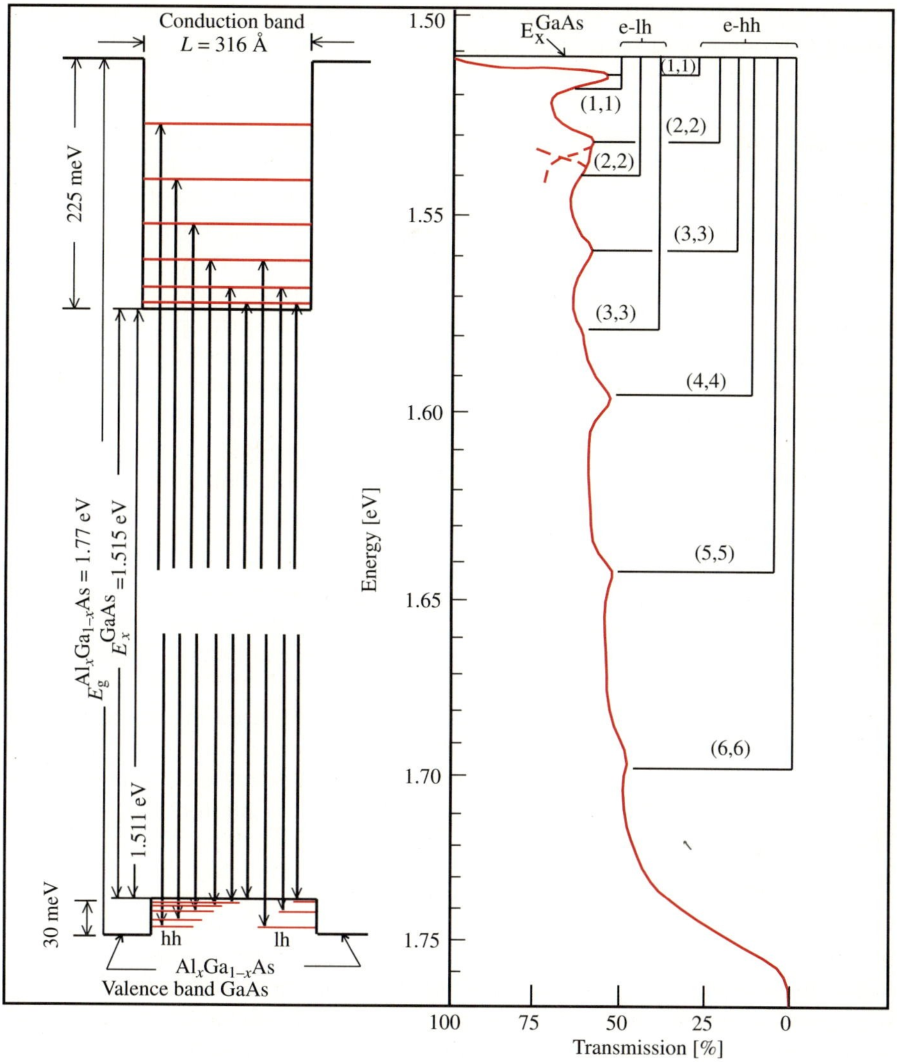

Fig. 9.11. Transmission spectrum of 316 Å GaAs/$Al_{0.2}Ga_{0.8}As$ QW measured as a function of photon energy at low temperature (*right panel*). The peaks labeled (n,n) have been identified with optical transitions from the nth heavy hole (hh) and light hole (1h) subbands to the nth conduction subband as shown by *arrows* in the band diagram in the *left panel*. From [9.27]. Note that the value of the band offset assumed by Dingle for these calculations has since been revised [9.28]

(Sect. 9.2.3) still implies allowed transitions only between subbands with the same parity. However, transitions between subbands with the same quantum number tend to be stronger even for finite wells (Fig. 9.11). Similar excitonic peaks associated with the quantized electron and hole levels in QWs have also been observed by photoluminescence and the corresponding excitation spectroscopy. Examples of such PLE spectra in two QWs of widths equal to 5 and 10 nm have already been shown in Fig. 7.16b and discussed in Sect. 7.1.5.

The results just mentioned exemplify the separate quantization of electrons and holes. Evidence for confinement of excitons, of the type mentioned in Sect. 9.1, has also been reported. See [9.2b] and references therein.

In both absorption and PLE measurements the intersubband energy separations are measured rather than the energy of an individual subband. Furthermore, no information on subband dispersion is obtained. *Kash* et al. [9.29] have utilized a kind of "hot luminescence excitation spectroscopy" to determine directly the hole subband dispersion in QWs. The principle behind their experiment was first demonstrated in bulk GaAs by *Fasol* et al. [9.30]. We shall consider here only the case of a QW, although similar results are obtained for bulk materials. In Fig. 9.12a an incident tunable laser ($\hbar\omega_{\text{laser}}$) excites an electron from the heavy hole subband into the lowest energy electron subband. Energy and wavevector conservation requires that

$$\hbar\omega_{\text{laser}} = E_{1\text{e}}(\boldsymbol{k}_{\|}) + E_{1\text{h}}(-\boldsymbol{k}_{\|}) + E_{\text{g}}, \tag{9.20a}$$

where $E_{1\text{e}}(\boldsymbol{k}_{\|})$ and $E_{1\text{h}}(\boldsymbol{k}_{\|})$ are the kinetic energies of motion of the electron and hole, respectively, in the plane of the QW, $\boldsymbol{k}_{\|}$ is their wavevector parallel to the QW (equal for electrons in the valence and conduction bands in vertical transitions) and E_{g} is the energy gap separating the electron and hole subbands. The GaAs layers in the GaAs/AlGaAs QW used in the experiment of *Kash* et al. [9.29] have been doped with Be (concentration of the order of 10^{18} acceptors/cm^3). These acceptors capture radiatively electrons from the subband before they relax. The energy of the emitted photon ($\hbar\omega_{\text{PL}}$) in this "hot luminescence" (Sect. 7.1) process is given by

$$\hbar\omega_{\text{PL}} = E_{1\text{e}}(\boldsymbol{k}_{\|}) + E_{\text{g}} - E_{\text{a}}, \tag{9.20b}$$

where E_{a} is the binding energy of the Be acceptor level in GaAs. In principle, the energy $E_{1\text{e}}(\boldsymbol{k}_{\|})$ of the "hot" electron can be deduced from the hot luminescence spectra. From this energy the wavevector $\boldsymbol{k}_{\|}$ of the electron (and hence that of the hole) can be calculated using (9.12), assuming that the electron effective mass m^* in GaAs is known. Using $E_{1\text{e}}(\boldsymbol{k}_{\|})$, the corresponding hole energy $E_{1\text{h}}(-\boldsymbol{k}_{\|})$ can be calculated from (9.20a) as a function of $\boldsymbol{k}_{\|}$. In this way both the heavy and light hole dispersions can be determined from the variation of $\hbar\omega_{\text{PL}}$ with $\hbar\omega_{\text{laser}}$.

In practice, this procedure is complicated by the fact that the heavy and light hole dispersions are not isotropic. The incident laser photon will excite electrons with a distribution of $\boldsymbol{k}_{\|}$ subject to the energy conservation condition. In bulk semiconductors the electron wavevector $\boldsymbol{k}$ in (9.20b) will be averaged over all directions so that only a "spherically averaged" valence band dispersion is determined this way [9.30]. To overcome this difficulty in QWs, *Kash* et al. [9.29] measured the *difference* $I_{\|} - I_{\perp}$ between the intensities for emission polarized parallel ($I_{\|}$) and perpendicular ($I_{\perp}$) to the incident laser.

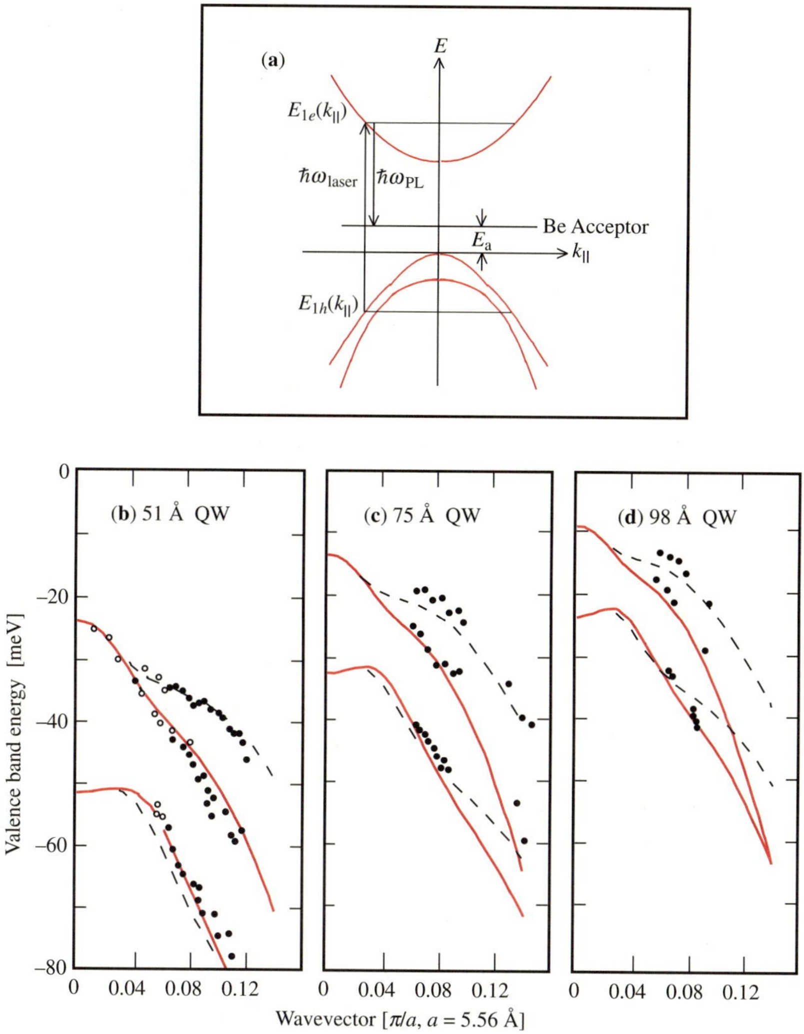

Fig. 9.12. (**a**) Schematic subband structure in a GaAs/AlGaAs QW showing hot luminescence processes which occur as a result of excitations from a hole subband to the corresponding electron subband followed by radiative capture of the electron into a Be acceptor level. (**b**–**d**) The hole subband dispersions deduced by Kash et al. [9.29] in three GaAs/$Al_{0.315}Ga_{0.685}As$ QWs of different width using the schematic band structure shown in (**a**). The *solid* and *dashed curves* represent, respectively, the theoretical dispersions along the [10] and [11] QW directions. The filled circles are the experimentally determined data points. The open circles in (**b**) were measured in an asymmetrically coupled double-QW sample consisting of a 51 Å GaAs well which was separated by a thin barrier from an adjacent undoped 153 Å QW. The latter QW acts as a sink for the thermalized electrons and holes and hence suppresses the thermalized luminescence background from the 51 Å well with the higher energy band gap

By considering the optical transition probabilities for light polarized along the [100] and [110] directions (in the plane of the QW), they showed that $I_{\|} - I_{\perp}$ is dominated by transitions originating from electrons with $\boldsymbol{k}_{\|}$ parallel to the laser polarization. This result allowed them to determine the anisotropic hole dispersion in three GaAs/$Al_{0.315}Ga_{0.685}As$ QWs with different well thicknesses. Their results for $\boldsymbol{k}_{\|}$ parallel to the [10] and [11] QW directions are shown in Figs. 9.12b–d as the filled circles. The corresponding theoretical dispersion curves are shown as the solid and dashed curves in the same figures. For small values of $\boldsymbol{k}_{\|}$ the weak hot luminescence is masked by the strong thermalized luminescence. To eliminate this background *Kash* et al. [9.29] grew also a 51 Å well with an adjacent undoped 153 Å thick QW. This thicker well has subbands with lower electron energy than those in the doped one. If these two QWs are separated by a thin (26 Å) barrier then carriers can tunnel (to be discussed in more detail in Sect. 9.5) from the Be-doped well into the undoped well. This provides an efficient nonradiative decay path for photoexcited carriers in the doped well and suppresses its thermalized emission (Sect. 7.1.1). The hot luminescence in the doped well is not affected since these emission processes occur on a much shorter time scale. The data represented by open circles in Fig. 9.12 were measured in this asymmetrically coupled QW. We see that the experimental points reproduce well the "mass reversal" predicted by theory as a result of the anticrossing between the heavy and light hole bands.

In doped QWs (where some of the electron or hole subbands are already populated) it is possible to observe the subband structure via infrared (intersubband) absorption and Raman spectroscopies. The principle of the former technique is analogous to the infrared absorption of electrons bound to donors and acceptors discussed in Sect. 6.5.2. These absorption processes can be exploited in the fabrication of infrared detectors in a way similar to those corresponding to carriers bound to impurities in bulk materials. In the case of a QW there is the additional advantage that the spectral range of the detector response can be easily tuned by changing the well width. The subband separation in QWs can be determined directly via electronic Raman scattering, a topic which is omitted from this book because of its specialized nature. Readers interested in this topic are referred to the review article by *Pinczuk* and *Abstreiter* [9.31].

9.3 Phonons in Superlattices

9.3.1 Phonons in Superlattices: Folded Acoustic and Confined Optic Modes

In Sect. 9.2 we discussed the effect of confinement on electrons in a quantum well. This effect appears when the allowed electronic energies in the well correspond to forbidden energies in the barrier (i. e. gap states). Like electrons, phonons can also be represented as Bloch waves with energy (i. e. frequency) dispersion relations which give the values of the allowed energies versus Bloch wavevector. The question thus arises as to whether confinement effects can also appear for phonons in QWs, MQWs and superlattices.

In order to answer this question we reproduce in Fig. 9.13 the phonon dispersion relations of GaAs and AlAs, the most common constituents of such structures. We note that the optical branches are rather similar to electronic bands. The energy, for instance, varies quadratically with $\boldsymbol{k}$-vector around critical points, including the Γ point (i. e. the center of the Brillouin zone, abbreviated as BZ). The acoustic branches, however, have rather different features: they extend down to zero frequency for any material at $\boldsymbol{k} = 0$, a result of the fact that an acoustic phonon with $\boldsymbol{k} = 0$ (i. e. infinite wavelength) corresponds to a uniform translation of the crystal. For such a translation no restoring forces appear since the distances between atoms do not change. Another feature peculiar to acoustic phonons is that their dispersion relations around Γ are linear (not quadratic) in $\boldsymbol{k}$.

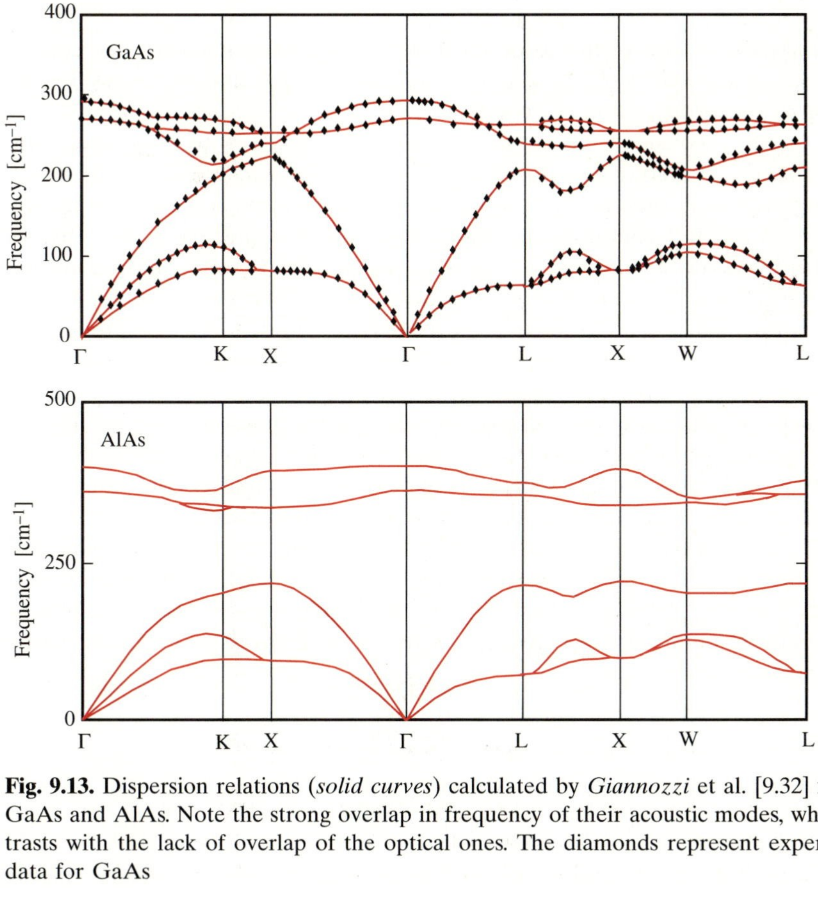

Fig. 9.13. Dispersion relations (*solid curves*) calculated by *Giannozzi* et al. [9.32] for bulk GaAs and AlAs. Note the strong overlap in frequency of their acoustic modes, which contrasts with the lack of overlap of the optical ones. The diamonds represent experimental data for GaAs

These peculiarities in the dispersion relations of acoustic phonons usually prevent their confinement: for any system of two given materials there are propagating acoustic modes in both of them within the range of frequencies from zero to the maximum frequency of the elastically softer of the two of them (GaAs in the case of Fig. 9.13). Note that for GaAs/AlAs systems the maximum acoustic frequency of both constituents is nearly the same, so that there is hardly any frequency range in which propagating acoustic modes exist only for one of the two components. In the limit in which elasticity theory holds, the acoustic phonons in a superlattice will correspond to elastic waves propagating with a dispersion relation $\omega = \overline{v}k$ where $\overline{v}$ is an average of the sound velocity of the two media. For $\boldsymbol{k}$ along the axis of growth, and a superlattice consisting of a repetition of a thickness d_A of medium A followed by d_B of medium B (the period $d = d_A + d_B$), it is easy to see what this average is. The time of propagation along d is $t = d_A/v_A + d_B/v_B$, hence the average speed of sound is

$$\overline{v} = \frac{d}{t} = d\left(\frac{d_A}{v_A} + \frac{d_B}{v_B}\right)^{-1}. \tag{9.21}$$

The details of the propagation of acoustic phonons in layered structures have been discussed by *Rytov* [9.33], who pointed out that the work is also relevant to the propagation of *seismic waves*.

Let us now consider the optic modes in Fig. 9.13. Note that they form narrow bands, centered around 280 cm^{-1} in GaAs and 380 cm^{-1} in AlAs. For many of the optical mode frequencies of GaAs there are no propagating modes in AlAs and confinement effects must result. Contrary to the case of electrons, however, in the case of Fig. 9.13 we can have modes that propagate in any one of the two constituents and not in the other: there are GaAs-like modes, for which the AlAs layers act as barriers, and AlAs-like modes, for which GaAs acts as the barrier. Such confined optical modes have been profusely studied in recent years [9.34, 35].

In order to illustrate these effects in the simplest possible case let us consider a superlattice with a period composed of two layers of element A of atomic mass m_A and two of element B of mass m_B (A and B could be Si and Ge, respectively). The unit cell of such a periodic structure is shown in Fig. 9.14. We shall treat only modes that propagate along the axis of growth and assume that we only have restoring forces between neighboring planes with the same restoring force constant f between Ge–Ge, Ge–Si, and Si–Si planes. The equations of motion for phonons with wavevector $\boldsymbol{k}$ propagating along the superlattice axis are

$$\begin{aligned}
-m_A\omega^2 v &= f[(x - v) + (u\mathrm{e}^{-\mathrm{i}dk} - v)],\\
-m_B\omega^2 x &= f[(y - x) + (v - x)],\\
-m_B\omega^2 y &= f[(u - y) + (x - y)],\\
-m_A\omega^2 u &= f[(v\mathrm{e}^{\mathrm{i}dk} - u) + (y - u)].
\end{aligned} \tag{9.22}$$

In order for this set of homogeneous linear equations (in the displacements

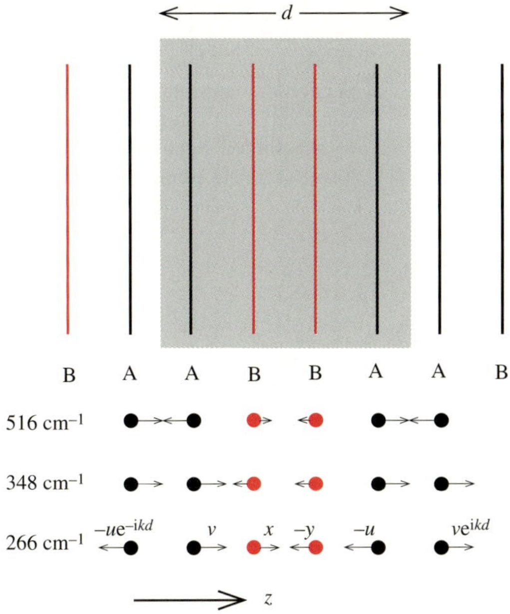

Fig. 9.14. *Top*: Unit cell of an A_2B_2 superlattice (it could be, for instance, Si_2Ge_2). *Bottom*: The unit cell of the displacement pattern for the three modes at the center of the Brillouin zone, as calculated for the Si_2Ge_2 system

v, x, y, u) to have nonvanishing solutions its determinant must vanish. This leads to a secular equation of fourth degree in ω^2 for a general k. This equation can be broken up in two quadratic equations which can be easily solved algebraically for $k = 0$ (this can also be done at the edge of the so-called mini-BZ, $k = \pi/d$, see Problem 9.3). The four allowed frequencies at $k = 0$ include $\omega^2 = 0$ (the acoustic modes of the superlattice) and

$$\omega^2 = f\left(\frac{1}{m_A} + \frac{1}{m_B}\right), \tag{9.23}$$

$$\omega^2 = f\frac{3(m_A + m_B) \pm \sqrt{9(m_A - m_B)^2 + 4m_A m_B}}{2m_A m_B}. \tag{9.24}$$

The two separate quadratic eigenvalue equations are most easily obtained by considering that the phonon displacements (eigenvectors) are either odd or even with respect to the center of the supercell in Fig. 9.14. Since odd and even eigenvectors do not mix, one can separate (9.22) into two uncoupled sets of equations, leading to (9.23) for the odd parity eigenvector (plus $\omega^2 = 0$) and to (9.24) for the even parity one (see Problem 9.3). We have evaluated (9.23) and (9.24) for $m_B = 2.6m_A$ as appropriate to a Ge_2Si_2 superlattice and choosing $f/m_A = (520\,\text{cm}^{-1})^2$, so as to approximately reproduce the phonon frequency at $\boldsymbol{k} = 0$ for the bulk materials. One finds from (9.23) $\omega = 348\,\text{cm}^{-1}$ and from (9.24) $\omega = 516\,\text{cm}^{-1}$ and $\omega = 266\,\text{cm}^{-1}$. The uppermost mode is very close to the $\boldsymbol{k} = 0$ mode of bulk Si ($520\,\text{cm}^{-1}$) while the two lower ones are close to that of bulk Ge ($300\,\text{cm}^{-1}$). The dispersion relations obtained for this superlattice are plotted in Fig. 9.15 for an arbitrary value of k in the first

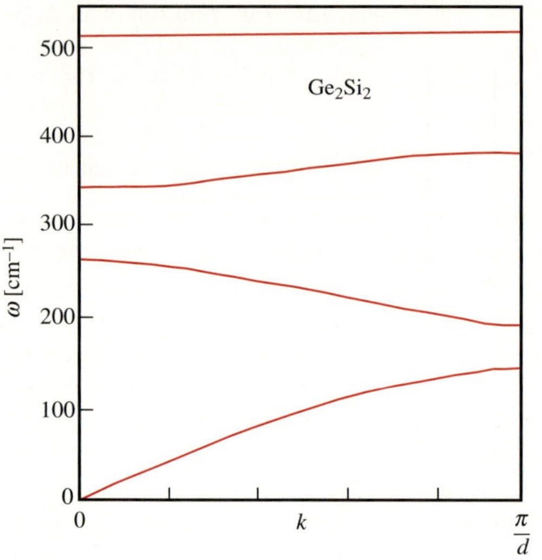

Fig. 9.15. Dispersion relations calculated for the Si_2Ge_2 superlattice of Fig. 9.14. Note the nearly dispersionless nature of the two upper (confined) branches in contrast to the folded behavior of the two lower (acoustic) ones

BZ of the superlattice (the mini-BZ). Note that the $\boldsymbol{k}$-vector at the edge of this BZ is half that for the corresponding X point of Si ($2\pi/d$, where $d \equiv a_0$). Thus, there are twice as many branches in the dispersion relations of Fig. 9.15 as in those of either Si (Fig. 3.1) or Ge. One expresses this fact by saying that the dispersion relations (or the BZ) have been **folded**. For $m_A = m_B$, instead of Fig. 9.15 one would obtain the Γ–X dispersion relation of the bulk longitudinal phonons folded through the middle of the Γ–X line. The existence of a superlattice with $m_A \neq m_B$ introduces splittings of the folded bands at k equal to 0 and π/d. These splittings are similar to those introduced in the free-electron bands (Fig. 2.8) by the presence of a periodic potential: the mass modulation along the growth direction in the superlattice is the counterpart of the periodic potential.

It follows from the discussion above that the lowest two branches of Fig. 9.15 can be described as the folded LA dispersion relation of the two bulk constituents, averaged according to (9.21) and with a gap at $k = \pi/d$ due to the mass modulation. The upper two branches, which could be labeled as **folded optic phonons**, cannot be described as average optic bands of the two constituents: the upper one is nearly flat and its frequency corresponds to that of bulk Si at Γ while the lower one is rather close to that of bulk Ge at Γ. Hence no averaging seems to take place for optic phonons, a phenomenon encountered rather generally in superlattices, the more so the larger the thicknesses of the individual layers. This illustrates rather vividly the conjecture made above that the optic modes exist in one of the components but not in the other. They are thus referred to as **confined modes**. Such confinement is clearly evidenced by the displacement pattern shown in Fig. 9.14 for the 516 cm^{-1} mode: the Ge atoms barely move. The confinement is less complete for the Ge-like modes, which become more confined when the Si layers, which act as barriers, are thicker. For detailed calculations of phonon spectra of more complex Ge-Si superlattices see [9.36, 37].

9.3.2 Folded Acoustic Modes: Macroscopic Treatment

The dispersion relations of folded acoustic modes, including the splittings at the center and the edge of the mini-BZ, can be easily calculated in the so-called elastic limit, which corresponds to the frequency range in which the dispersion relations of the constituents can be assumed to be linear. We have already shown that for waves propagating along the growth direction (which we shall assume to be [001] unless otherwise specified) the velocity of the long wavelength folded modes is given by (9.21). For the purpose of calculating the full dispersion relations we consider in both constituent media elastic waves with the same frequency ω, of the type given in (3.17) for the atomic displacement and (3.19) for the strain tensor. We note that the components of the wavevector q perpendicular to the growth axis (q_x, q_y) must be the same in both media, as in the case of electron wavefunctions discussed in Sect. 9.2.3. The components of q along $z(q_z)$, however, must change from one medium into the other so as to obtain the same ω for different speeds of sound. However, the superlattice as a whole has translational symmetry along the z direction with translational vectors of length nd $(n = \pm 1, \pm 2, \ldots)$. We can therefore express the displacement (δR) of atoms in each layer in the form of a Bloch wave. For each period containing two layers A and B at $n = 0$

$$\delta R_0(x,y,z) = \begin{cases} \delta R_{\mathrm{A}} \exp[\mathrm{i}(q_x x + q_y y)] \exp[\mathrm{i}(q_{z\mathrm{A}} z - \omega t)], & 0 < z < d_{\mathrm{A}}, \\ \delta R_{\mathrm{B}} \exp[\mathrm{i}(q_x x + q_y y)] \exp[\mathrm{i}(q_{z\mathrm{B}} z - \omega t)], & d_{\mathrm{A}} < z < d_{\mathrm{B}}, \end{cases} \tag{9.25a}$$

and, correspondingly, for a period placed at a finite value of n

$$\delta R_n(x,y,z) = \delta R_0(x,y,z - nd)\mathrm{e}^{\mathrm{i}(nkd)}, \quad nd < z < (n+1)d, \tag{9.25b}$$

where $d_{\mathrm{A,B}}$ are the thicknesses of the constituent layers, $d = d_{\mathrm{A}} + d_{\mathrm{B}}, v_{\mathrm{A,B}}$ the respective sound velocities, $q_{z\mathrm{A}} = \omega / v_{\mathrm{A}}$ and $q_{z\mathrm{B}} = \omega / v_{\mathrm{B}}$. We shall now sketch how to obtain the dispersion relation $\omega(q_x, q_y, k)$ for the folded acoustic modes.

We take in each medium two waves with the same frequency ω and wavevectors $\pm(q_x, q_y, q_{z\mathrm{A}})$, $\pm(q_x, q_y, q_{z\mathrm{B}})$ in media A and B, respectively. We thus have as unknowns the four amplitudes of these waves. For a general propagation direction we must consider all three possible polarizations for the acoustic waves, i. e., a total a 12 amplitude coefficients, which have to be determined by appropriate boundary conditions. At an interface the displacements must be continuous (three conditions) and also the strain components perpendicular to the interface, which are related to the gradient of the displacements (three additional conditions). We must apply these boundary conditions at one AB and also at one BA interface, hence a total of 12 linear homogeneous equations in the twelve amplitude coefficients results. It is easy to see that the boundary conditions at all other equivalent AB and BA interfaces are automatically satisfied because of the choice of Bloch waves along z. The resulting 12×12 secular equation leads to the dispersion relation for $\omega(q_x, q_y, k)$.

The 12×12 secular equation is usually rather complicated and must be solved numerically in the general case. In particular, in cases of high symmetry, such as for either $q_x = q_y = 0$ or for $q_y = k = 0$ in the [001] superlattices treated here, the longitudinal and transverse modes do not mix and the set of 12 equations breaks up into three uncoupled sets of 4 equations, leading to 4×4 secular equations which, for $q_x = q_y = 0$, can be expressed in the implicit form [9.33, 34]

$$\begin{aligned}\cos(kd) &= \cos\left(\frac{\omega d_{\mathrm{A}}}{v_{\mathrm{A}}}\right)\cos\left(\frac{\omega d_{\mathrm{B}}}{v_{\mathrm{B}}}\right) \\ &\quad -\frac{1}{2}\left(\frac{\varrho_{\mathrm{B}}v_{\mathrm{B}}}{\varrho_{\mathrm{A}}v_{\mathrm{A}}}+\frac{\varrho_{\mathrm{A}}v_{\mathrm{A}}}{\varrho_{\mathrm{B}}v_{\mathrm{B}}}\right)\sin\left(\frac{\omega d_{\mathrm{A}}}{v_{\mathrm{A}}}\right)\sin\left(\frac{\omega d_{\mathrm{B}}}{v_{\mathrm{B}}}\right),\end{aligned} \tag{9.26}$$

where $\varrho_{\mathrm{A,B}}$ are the mass densities of the layers A and B, respectively.

Notice the similarity of (9.26) to (9.18a), the secular equation which determines the energy bands of electrons in a Kronig–Penney potential within the effective mass approximation. The similarity between (9.26) and (9.18a) is typical of problems involving wave propagation in stratified media. These equations allow us to obtain the Bloch vector k along the propagation direction as a function of the frequency ω.

Equation (9.26) can be rewritten in the slightly modified form

$$\cos(kd) = \cos\left[\omega\left(\frac{d_{\mathrm{A}}}{v_{\mathrm{A}}}+\frac{d_{\mathrm{B}}}{v_{\mathrm{B}}}\right)\right]-\frac{\varepsilon^2}{2}\sin\left(\frac{\omega d_{\mathrm{A}}}{v_{\mathrm{A}}}\right)\sin\left(\frac{\omega d_{\mathrm{B}}}{v_{\mathrm{B}}}\right), \tag{9.27}$$

where

$$\varepsilon = \frac{\varrho_{\mathrm{B}}v_{\mathrm{B}}-\varrho_{\mathrm{A}}v_{\mathrm{A}}}{(\varrho_{\mathrm{B}}v_{\mathrm{B}}\varrho_{\mathrm{A}}v_{\mathrm{A}})^{1/2}}, \tag{9.28}$$

which clearly shows that whenever the **acoustic impedances** $\varrho_{\mathrm{B}}v_{\mathrm{B}}$ and $\varrho_{\mathrm{A}}v_{\mathrm{A}}$ are equal, i. e. for $\varepsilon = 0$, the dispersion relation is simply that of a medium with the average velocity given in (9.21).

9.3.3 Confined Optical Modes: Macroscopic Treatment

The confinement of optic modes is mathematically expressed by imposing the boundary condition requiring that the vibrational amplitudes vanish in the immediate vicinity of the layer boundaries A–B and B–A. Under these conditions we find A-like and B-like confined optical modes whose frequencies can be read off from the corresponding bulk dispersion relations as a function of the discrete effective k-vectors:

$$k_m = \frac{\pi}{d_{\mathrm{A,B}}}m, \quad m = 1,2,3,\ldots\,. \tag{9.29}$$

This equation can be applied to either LO or TO modes. The corresponding atomic displacements alternate in sign between one atomic layer and the next, as required for optical modes, with their magnitudes determined by the envelope functions

$$
\begin{aligned}
u_m(z) &= \cos k_m z, \quad m = 1,3,5,\dots\,, \\
u_m(z) &= \sin k_m z, \quad 2,4,6,\dots\,.
\end{aligned}
\tag{9.30}
$$

where z indicates the position of an atomic plane along the superlattice axis and $z = 0$ is taken at the mid-point of a layer.

While (9.29) usually gives a good approximation to the effective wavevectors, in the special case of short superlattices (only a few atomic monolayers) one must consider whether the best possible approximation is obtained by making the wavefunction zero *exactly* at the boundary of each material layer. A glance at Fig. 9.16 shows that a more reasonable choice would be to make the envelope function zero at the first atoms of the "other kind" (i. e. the A atoms for the B-like vibrations and vice versa). *Molinàs-Mata* and *Cardona* have made a detailed study of this point for Ge_nSi_m superlattices [9.36]. The case of GaAs is illustrated in Fig. 9.16. The vibrational amplitudes follow the patterns of (9.30) but the effective k_m's must be made slightly smaller than those of (9.29): the amplitudes do not vanish at the nominal AlAs layer boundary but at the first Ga layer outside that boundary. This is easy to understand: The boundary is formed by As-layers, which should still vibrate at

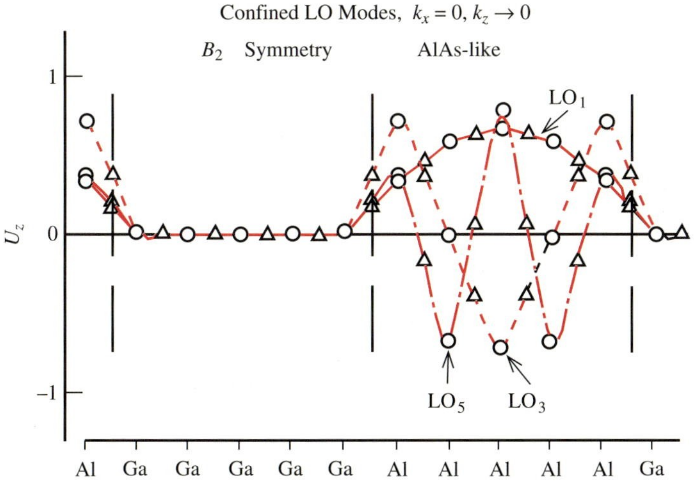

Fig. 9.16. Displacement pattern for AlAs-like confined LO modes in a $(GaAs)_5/(AlAs)_5$ superlattice. The triangles represent the magnitude of the displacement of the As atoms while the circles represent that for Ga and Al. From [9.38]

AlAs frequencies, while Ga, with a much heavier mass than Al, will not. This fact can be accounted for by replacing the layer thickness $d_{A,B}$ in (9.29) by $d_{A,B} + a_0/4$, where a_0 is the bulk lattice constant, which corresponds to four atomic layers for diamond and zinc-blende-type materials [9.39].

9.3.4 Electrostatic Effects in Polar Crystals: Interface Modes

The existence of modes confined around the surface of a polar crystal (surface plasmons, surface optical phonons) was mentioned in Problem 6.11. Here we discuss analogous phenomena which arise at interfaces between two semiconductors (heterojunctions) of which at least one is polar. We treat first the case of a single planar heterojunction, then a double heterojunction (i. e., a QW) and finally the case of a periodic MQW or a superlattice. Note what while in the case of electrons a clear distinction can be made between MQWs (with negligible dispersion for wavevectors along the growth direction) and SLs, this distinction cannot be made with respect to phonons: the folded acoustic branches always show dispersion, even for large periods, while the optical ones do not, even for relatively small periods. We thus refer only to the electronic behavior when classifying periodic structures as either MQWs or superlattices.

In our treatment of phonons for such two-dimensional structures we shall neglect the so-called **retardation effects** (see Problem 6.11b for the effect of retardation on surface modes). This implies that the wavelengths of the excitations under consideration are assumed to be small compared with the wavelengths of photons having the same frequency and thus the wave nature of the electric fields which accompany the ionic motion can be neglected. These fields are therefore assumed to be of electrostatic origin, i. e., they obey only the Gauss equation (6.106) and the corresponding Maxwell equation (6.112c) in which the time derivative is set equal to zero in order to neglect electromagnetic wave propagation effects:

$$\nabla \times \boldsymbol{E} = 0. \tag{9.31}$$

Equation (9.31) implies that the field $\boldsymbol{E}$ can be derived from a scalar potential $\phi(\boldsymbol{r})$, which, on account of (6.106), must fulfill Laplace's equation

$$\nabla^2 \phi(\boldsymbol{r}) = 0 \tag{9.32}$$

and the standard boundary conditions of continuity of $D_\perp = \varepsilon E_\perp$ for the components of the electric displacement $D_\perp$ normal to the interface and the continuity of the parallel component of the field $E_\parallel$. We treat next a single interface between two semiconductors A and B with isotropic dielectric functions $\varepsilon_A(\omega)$ and $\varepsilon_B(\omega)$:

$$E_{\parallel A} = E_{\parallel B}, \tag{9.33a}$$

$$\varepsilon_A E_{\perp A} = \varepsilon_B E_{\perp B}. \tag{9.33b}$$

We neglect for the time being the **mechanical boundary conditions** on the vibrational amplitude $\boldsymbol{u}$ and the corresponding stress (the justification will be given at end of this section), solve (9.32) in both media and apply the electrostatic boundary conditions (9.33) keeping in mind that $\boldsymbol{E} = -\nabla\phi$. We assume that the interface is the plane $z = 0$, with $z < 0$ representing medium A, and consider a solution of (9.32) of the form

$$\phi_{\mathrm{A}} = A\mathrm{e}^{\mathrm{i}q_x x}\mathrm{e}^{+q_x z}, \quad \text{for } z \leq 0 \tag{9.34a}$$

$$\phi_{\mathrm{B}} = B\mathrm{e}^{\mathrm{i}q_x x}\mathrm{e}^{-q_x z}, \quad \text{for } z \geq 0, \tag{9.34b}$$

where we have chosen the x axis to be along the direction of the in-plane $\boldsymbol{q}$ and use $q_z^2 = q_x^2$ as follows from (9.32). Note that (9.34) reflects the existence of translational symmetry (i. e. the fact that the ϕ's must be Bloch functions) along x and y and the absence along z. In (9.34) ϕ_{A} and ϕ_{B} are *concentrated* near the interface since they decay exponentially away from $z = 0$. Modes described by such functions are known as **interface modes**.

Boundary condition (9.33a) applied to (9.34) leads to $A = B$, while (9.33b) leads to the "secular equation" for the frequency ω_{IF} of the interface modes

$$\varepsilon_{\mathrm{A}}(\omega_{\mathrm{IF}}) = -\varepsilon_{\mathrm{B}}(\omega_{\mathrm{IF}}). \tag{9.35}$$

This equation has solutions only if there are frequencies for which the two dielectric functions ε_{A} and ε_{B} have *opposite signs*.

As discussed in Sect. 6.5.1, ε can be negative in heavily doped semiconductors for $0 < \omega < \omega_{\mathrm{p}}$. In this case (9.35) leads to **interface plasmons**. For a discussion of plasmons in QWs and MQWs see [9.31]. The dielectric constant also becomes negative in polar semiconductors for $\omega_{\mathrm{TO}} < \omega < \omega_{\mathrm{LO}}$, see (6.110). In this case (9.35) can have solutions provided ω is within this range ($\varepsilon_{\mathrm{A}} < 0$) in material A while for material B $\varepsilon_{\mathrm{B}} > 0$. Hence in the case of interfaces between a polar (e. g. GaAs) and a nonpolar (e. g. Ge, Si) semiconductor we expect one interface mode at a frequency $\omega_{\mathrm{TO}} < \omega_{\mathrm{IF}} < \omega_{\mathrm{LO}}$ (Fig. 9.17) while for an interface between, for example GaAs and AlAs we expect two, one GaAs-like and the other AlAs-like.

We consider next the case of a QW of material B with material A as a barrier (Fig. 9.18). Because of the reflection symmetry about the bisecting plane (labeled R in Fig. 9.18), the solution of (9.32) must be either odd or even with respect to that reflection. The parities are indicated in Fig. 9.18: the plus (minus) sign corresponds to the even (odd) function. Because of this choice of $\phi(\boldsymbol{r})$, the boundary conditions only need to be applied at one of the interfaces, the other being then automatically satisfied. We thus find for the AB interface at the left of Fig. 9.18:

$$\begin{aligned} A &= B[1 \pm \mathrm{e}^{-q_x d}], \\ A\varepsilon_{\mathrm{A}}(\omega) &= -B\varepsilon_{\mathrm{B}}(\omega)[\mp\mathrm{e}^{-q_x d}] \end{aligned} \tag{9.36}$$

(where A and B are defined in Fig. 9.18). The corresponding secular equation leads to the two branches

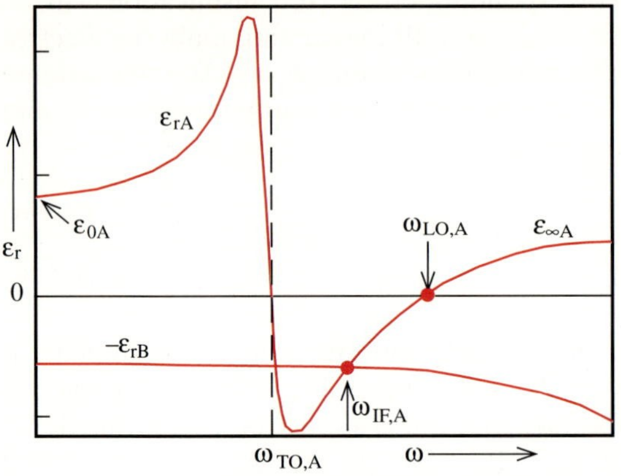

Fig. 9.17. Schematic diagram showing how the solution ω_{IF} of (9.35) is obtained for an interface between a polar semiconductor A (e. g. GaAs) and a nonpolar one B (e. g. silicon) from the plots of $\varepsilon_{\mathrm{rA}}$ and $-\varepsilon_{\mathrm{rB}}$ as a function of ω

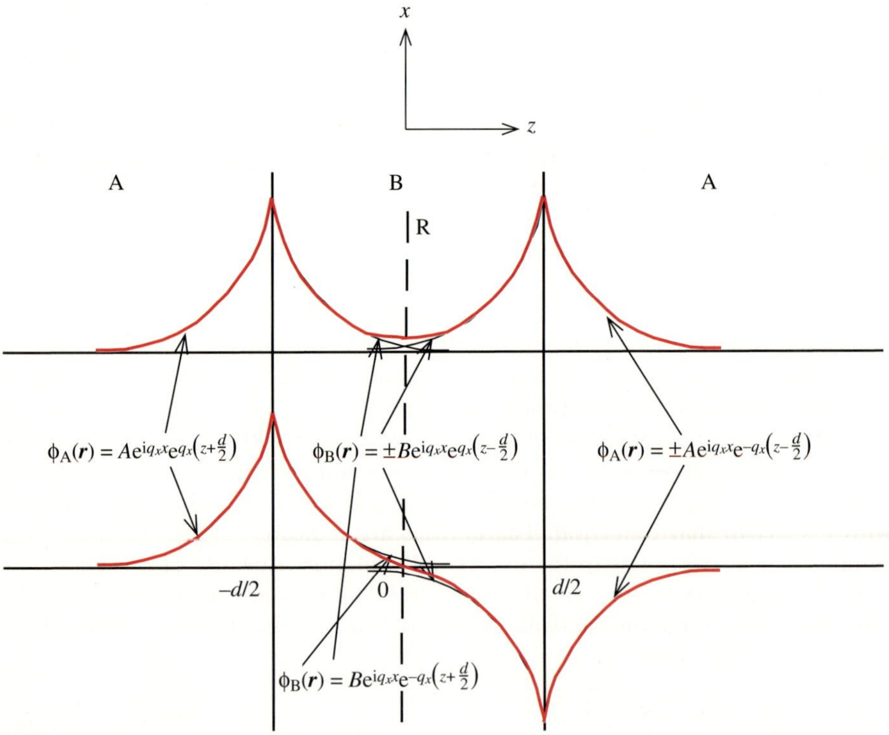

Fig. 9.18. Sketch of a single quantum well B with barriers of material A and the wavefunctions (*solid curve*) needed to obtain the even (+ sign) and odd (− sign) interface mode potentials

$$\varepsilon_A(\omega) = -\varepsilon_B(\omega) \begin{cases} \tanh \dfrac{q_x d}{2} & (9.37a) \\ \coth \dfrac{q_x d}{2} & (9.37b) \end{cases}$$

These modes were predicted by *Fuchs* and *Kliewer* [9.40]. Note that (9.37) reduces to (9.35) for $q_x \to \infty$ [in this case the two interfaces do not interact with each other because of the strong surface confinement of $\phi(z)$]. For $q_x \to 0$ we recover as solutions of (9.37) $\omega = \omega_{\mathrm{TO,B}}$ and $\omega = \omega_{\mathrm{LO,B}}$. Dispersion relations similar to those resulting from (9.37), but in this case for two different barriers: air and AlAs surrounding a GaAs QW, are plotted in Fig. 9.19 together with experimental results obtained using Raman spectroscopy by *Nakayama* et al. [9.41]. For a discussion of the corresponding $\phi(x, z)$ see Problem 9.6.

We discuss next the "electrostatic" modes of a periodic MQW (or superlattice) made out of layers of thickness d_A of material A and d_B of material B. As in the treatment of folded acoustic phonons, see (9.25), we impose the Bloch condition of the potential $\phi(x, z)$ given for one layer B and its A neighbors in Fig. 9.18. In this manner we only have to deal with two boundary conditions (for $E_\parallel$ and $D_\perp$) at an AB and two at a BA interface. They suffice to determine the two coefficients related to the $\exp(\pm q_x z)$ waves of materials A plus two similar coefficients for material B. One obtains the typical secular equation (see (3.41) of [9.34] and references therein):

$$\cos(kd) = \cosh(q_x d_A)\cosh(q_x d_B) + D(\omega)\sinh(q_x d_A)\sinh(q_x d_B) \qquad (9.38)$$

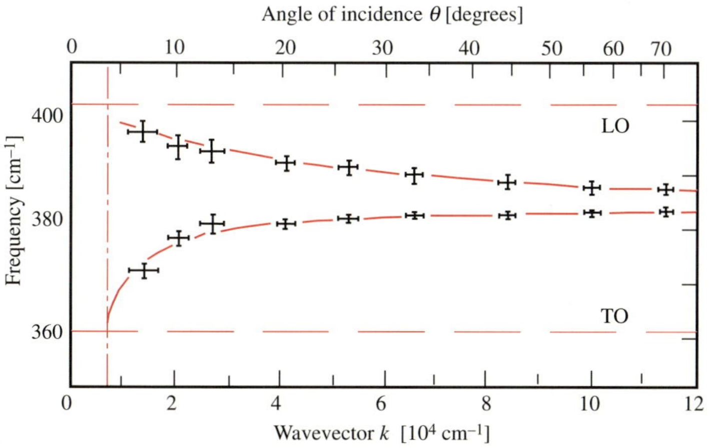

Fig. 9.19. AlAs-like interface modes of an air/GaAs (60 nm)/AlAs (500 nm) two-heterojunction system. The points were obtained in Raman measurements in which the laser was incident at an angle Θ to the normal while the scattered light was collected in the normal direction. The *solid lines* were obtained with an expression similar to (9.37) but taking into account the asymmetry of the system (air/GaAs/AlAs). From [9.41]

with

$$D(\omega) = \frac{1}{2}\left(\frac{\varepsilon_A(\omega)}{\varepsilon_B(\omega)} + \frac{\varepsilon_B(\omega)}{\varepsilon_A(\omega)}\right).$$

Equation (9.38) enables us to calculate the magnitude k of the superlattice Bloch wavevector (along the direction of growth) as a function of ω and q_x. We recall that (9.38) has resulted only from the electrostatic boundary conditions and does not guarantee the fulfillment of the appropriate mechanical boundary conditions. We shall come back to this point below. We should also point out the close similarity between (9.38), (9.26) and (9.18a). They have a canonical structure common to all problems involving wave propagation in stratified media.

It is instructive to consider several limiting cases of (9.38), first of all the limit of in-plane propagation in which $k = 0$ but $q_x \neq 0$. It is easy to see that (9.38) reduces to the two branches

$$-\frac{\varepsilon_A(\omega)}{\varepsilon_B(\omega)} = \begin{cases} \tanh\left(\frac{q_x d_A}{2}\right) \coth\left(\frac{q_x d_B}{2}\right), & (9.39a) \\ \tanh\left(\frac{q_x d_B}{2}\right) \coth\left(\frac{q_x d_A}{2}\right). & (9.39b) \end{cases}$$

For an evaluation and discussion of these equations see Problem 9.7.

We consider next the case in which the magnitude of the *total* wavevector $(q_x^2 + k^2)^{1/2}$ tends to zero. We define an angle Θ between that wavevector and the superlattice axis: $\tan\Theta = q_x/k$. In this case the secular equation (9.38) reduces to

$$\langle\varepsilon(\omega)\rangle \langle\varepsilon^{-1}(\omega)\rangle = -\tan^2\Theta, \tag{9.40}$$

where $\langle\ldots\rangle$ represents the average of the functions over one period of the MQW, weighted with the corresponding thicknesses d_A and d_B. Replacing $\varepsilon_A(\omega)$ and $\varepsilon_B(\omega)$ from (6.110) into (9.40), we find a quadratic equation for $\omega^2(\Theta)$ leading to two branches of the dispersion relation. The solutions are represented schematically in Fig. 9.20 for the three cases $d_B > d_A$, $d_B = d_A$ and $d_B < d_A$. This figure reveals that even for $(q_x, k) \to 0$ the frequency is dispersive as a function of the angle Θ, a fact which is related to the singular nature of the Coulomb interaction encountered already in Chap. 3.

Let us consider the case of in-plane propagation, i. e., $\Theta = \pi/2$ in (9.40). This equation then breaks up into [9.42]

$$\langle\varepsilon(\omega)\rangle = \frac{d_A\varepsilon_A(\omega) + d_B\varepsilon_B(\omega)}{d} = 0, \tag{9.41a}$$

$$\langle\varepsilon(\omega)^{-1}\rangle = \frac{1}{d}\left(\frac{d_A}{\varepsilon_A} + \frac{d_B}{\varepsilon_B}\right) = 0. \tag{9.41b}$$

For $d_A = d_B$ both equations (9.41a, b) reduce to (9.35) and the two resonance

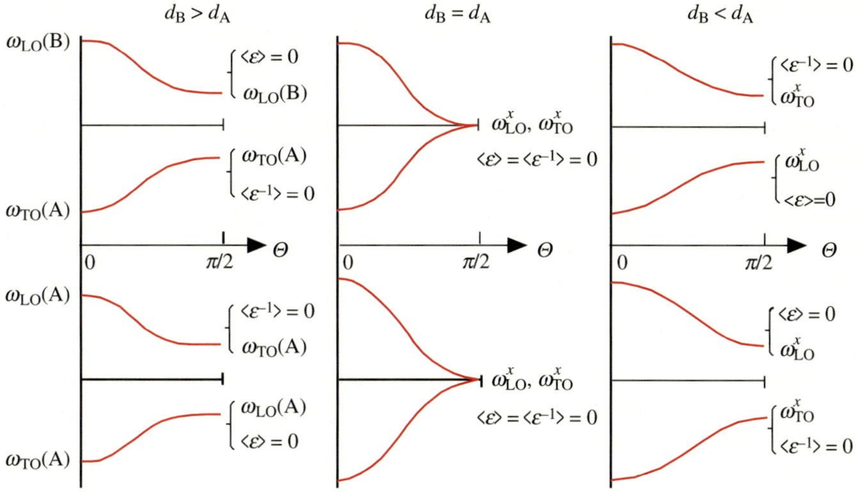

Fig. 9.20. Schematic diagram of the angular dispersion of "interface" electrostatic modes of superlattices as obtained with (9.40) for $d_B > d_A$, $d_B = d_A$ and $d_B < d_A$. Here ω^x_{LO} and ω^x_{TO} represent in-plane propagating modes

frequencies, in the case of vanishing wavevector with $\Theta = \pi/2$, become equivalent to that of the interface phonons for a single interface, as shown in Fig. 9.20. The degenerate frequencies split for $d_A \neq d_B$, following the pattern given in Fig. 9.20.

It is interesting to note that (9.41a, b), valid for $\Theta = \pi/2$, can be obtained directly by applying the $E_\parallel$ and $D_\perp$ boundary conditions to the case $k_x \to 0$ (which implies infinite wavelengths along x). In this case, and using the $\phi(x, z)$ given in Fig. 9.18 for layer B, we can see that E_x is uniform for the even solution while for the odd one E_z is uniform. In the former case we find from the boundary conditions for $E_\parallel$

$$E_{\parallel A} = E_{\parallel B},$$

which, on averaging over A and B (a period), leads to

$$\langle D_\parallel \rangle = \langle \varepsilon(\omega) \rangle E_\parallel. \tag{9.42a}$$

Hence the MQW can be regarded as a crystal with an effective dielectric function $\langle \varepsilon(\omega) \rangle$. We find the "longitudinal" modes of the **effective medium** at the frequency defined by (9.41a). In the "odd" case $E_\perp$ is uniform and the continuity of $D_\perp$ leads to

$$\langle E_\perp \rangle = \left\langle \frac{1}{\varepsilon(\omega)} \right\rangle E_\perp. \tag{9.42b}$$

The corresponding frequency is indeed given by (9.41b). This leads us to consider the MQW as a crystal with the bulk symmetry of the constituents lowered by the layering: For GaAs/AlAs systems grown along (001) the T_d cubic point group of the bulk is lowered to the tetragonal D_{2d}. The transverse effective charge e^* of (6.119) is thus expected to be anisotropic, different along [001] and [100]. This is evidenced by the different LO–TO splittings [see (6.108), where $Q^2 = e^{*2}/4\pi\varepsilon_0\varepsilon_\infty$], which are always smaller for $\Theta = \pi/2$ than for $\Theta = 0$. Figure 9.21 gives a simple qualitative explanation for this fact. In (a) the depolarizing field effect, which leads to an LO–TO splitting equal to that of the bulk, is illustrated for propagation along z. In (b) we depict the case of LO modes propagating along x. We see that while depolarizing fields are generated at layer A for A-like modes, they are not at B layers. Consequently, the depolarizing field and the concomitant LO–TO splitting are decreased with respect to the bulk, in agreement with the angular dispersion of Fig. 9.20.

The picture of the angular dispersion of the LO–TO splittings given above, similar to that found in conventional noncubic polar crystals, enables us to reach a few more plausible conclusions. The "interface" effects should only appear for IR-active modes. In terms of (9.30) these modes correspond to odd values of m: for even modes the displacements, and therefore the induced dipole moments (i. e., the average effective charges) average to zero in one layer. The largest effective charges ($\simeq$ the bulk ones) are obtained for $m = 1$. For odd $m \geq 3$ the effective charge is reduced by a factor of $1/m$ and the corresponding LO–TO splitting by a factor of $1/m^2$. Hence the "interface mode" effects are expected to be very small, often negligible, for $m \geq 3$.

We finally come back to the justification of having neglected the mechanical boundary conditions and the possible effects of this assumption. For $\Theta = \pi/2$ and $q_x \to 0$ the even mode of Fig. 9.18 leads to $\phi(z)$ and u_x constant

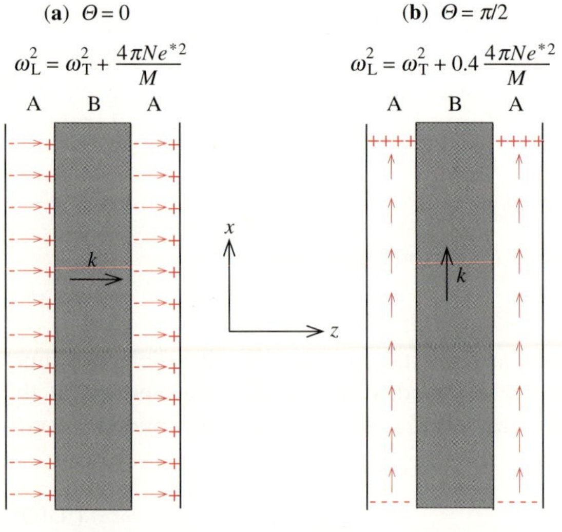

Fig. 9.21. Schematic diagram of electrostatic effects on phonons propagating (**a**) along and (**b**) perpendicular to the MQW axis for A-like modes. The effects are much smaller in case (**b**) because no polarization charges appear in material B in the frequency range of the A-like modes

throughout the B-layer (LO-like mode). The odd mode leads to a constant u_z (TO-like). Since the atomic displacement $\boldsymbol{u}$ must be zero in the A-layer for B-like modes, we are confronted with a flagrant violation of the required continuity of $\boldsymbol{u}$ at the interfaces. Actually, microscopic calculations (and also macroscopic ones using the techniques of *Trallero-Giner* et al. [9.43]) show that in the interior of a layer $\boldsymbol{u}$ is nearly constant for the $m = 1$ interface modes, while it becomes rounded off towards zero when approaching the layer boundary (Fig. 9.22b). The physical origin of this change can be understood in the following way. Let us consider the bulk TO dispersion relation and expand it around $\boldsymbol{q} = 0$ to terms of order q^2:

$$\omega_{\mathrm{TO}} = \omega_{\mathrm{TO},0} - C(q_x^2 + q_y^2 + q_z^2). \tag{9.43}$$

Taking an imaginary $q_z = \pm\mathrm{i}|q_z|$ allows us to obtain values of ω_{TO} equal to ω_{IF} for phonon envelope functions which decay rapidly (within a couple of monolayers) exponentially around the interfaces. Their amplitudes can be chosen to bring the nearly flat envelope function of ω_{IF} to zero, thus fulfilling one of the mechanical boundary conditions.

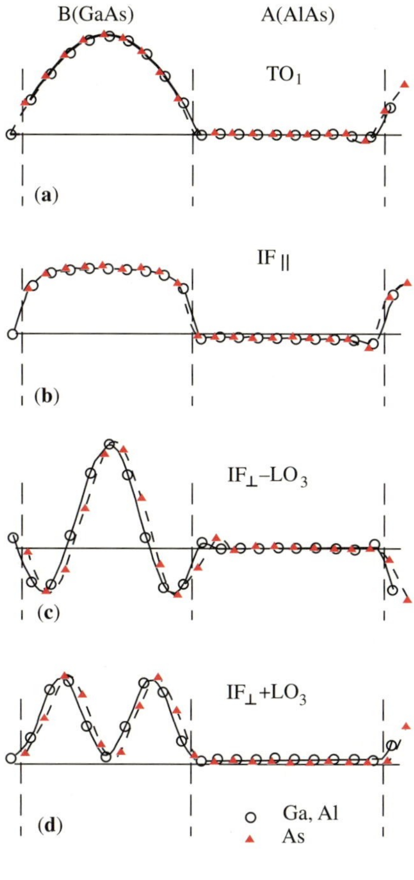

Fig. 9.22. (**a**) Envelope function for the TO_1 confined modes (i. e., $\Theta = 0$) of a $(\mathrm{GaAs})_9(\mathrm{AlAs})_{10}$ superlattice grown along [111]. (**b**) The equivalent modes for $\Theta = \pi/2$ showing the typical flat-top pattern of these so-called interface modes. [Note that, in spite of their name, these modes are not confined to the neighborhood of the interfaces since q_x is infinitesimally small. However, the "interface" label may still be justified since they come closer to the interfaces than the confined mode shown in (**a**).] (**c, d**) The envelope function for the $\Theta = \pi/2$ LO-like, IF and $m = 3$ modes, showing that they can be decomposed into a mixture of a flat-top mode of the (**b**) type and an $m = 3$ confined mode. From [9.44]

For details about the stress boundary conditions and the corresponding analytic envelope function see [9.45]. It is obvious that the dispersion relation of interface modes obtained from (9.38) is going to be changed somewhat when the mechanical boundary conditions are imposed and the exponentially decaying TO modes (plus oscillating LO modes of large q_z which may also occur at the frequency ω_{IF}) are included. From the discussion above it is easy to discern the qualitative nature of the changes (for an illustration see Fig. 9.23). when Θ is varied continuously from 0 to $\pi/2$, the long wavelength $m = 1$ modes of (9.30) become the main interface modes and the envelope function evolves from the sinusoidal one of Fig. 9.16 to the flat top one typical of IF modes, shown in Fig. 9.22b. This evolution is due to the long-range electrostatic fields. Hence the IF modes will mix, through the boundary conditions, with all IR-active confined modes, i. e., those with $m = 3, 5, 7, \ldots$. As a result

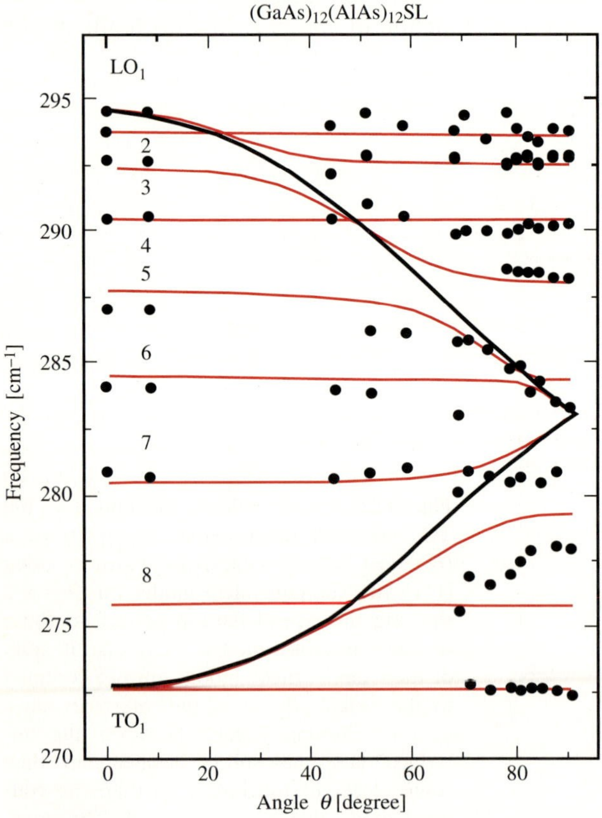

Fig. 9.23. The *thick black curves* represent the IF modes obtained with (9.38), i. e., neglecting the mechanical boundary conditions (BCs). The *red curves* were obtained with due consideration of mechanical BCs. The experimental points were obtained with Raman spectroscopy. Note the weak anticrossings with even confined modes due to having assumed a small nonvanishing component of in-plane wavevector. From [9.46]

of the mixing, anticrossings will appear in the dispersion relations, as shown in Fig. 9.23 for a $(GaAs)_{12}(AlAs)_{12}$ MQW. The black dots in this figure indicate experimental data measured by Raman spectroscopy [9.46]. The black curve was calculated with (9.38), i. e., without mechanical boundary conditions, the red curves display full calculations including confined modes and mechanical boundary conditions. They display rather strikingly the anticrossings mentioned above (LO_1–LO_3, LO_3–LO_5, TO_1–LO_7) and represent the experimental points rather well.

9.4 Raman Spectra of Phonons in Semiconductor Superlattices

Raman spectroscopy has proven to be by far the most powerful technique for investigating vibrational excitations in superlattices. These types of samples do not usually have enough volume to allow inelastic neutron scattering experiments. Infrared spectroscopy has also been used by a few researchers but it is seriously hampered by the fact that at the main Reststrahlen band the reflectivity is close to one, and weaker structures (such as those corresponding to $m = 3, 5, 7, \ldots$) cannot be easily seen. To a first approximation the reflection spectra can be understood as those of a uniaxial crystal with the average dielectric constants which follow from (9.42), see also (9.41) [9.47, 48];

$$\begin{aligned} \langle \varepsilon_z \rangle^{-1} &= \frac{1}{d}\left(\frac{d_A}{\varepsilon_A} + \frac{d_B}{\varepsilon_B}\right), \\ \langle \varepsilon_x \rangle &= \frac{1}{d}(d_A \varepsilon_A + d_B \varepsilon_B). \end{aligned} \tag{9.44}$$

Reflectance derivative spectroscopy, see (6.61), however, has helped to resolve weak structures in the IR spectra produced by the $m = 3, 5, \ldots$ confined modes [9.49]. Because of the relatively small amount of information obtained from the IR measurements, as compared with overwhelming Raman results, we shall not discuss the former any further. Let us, however, note that IR measurements may be useful in cases in which the TO phonons are forbidden by symmetry, such as that of backscattering on [001] superlattices.

9.4.1 Raman Scattering by Folded Acoustic Phonons

We mentioned in Sects. 9.1 and 9.2 that the acoustic phonons of a MQW or superlattice can be obtained from those of the bulk materials by performing the following operations:

- averaging the acoustic dispersion relations for each $\boldsymbol{q}$,
- folding them, as many times as needed, to the mini-BZ,
- opening the gaps induced at the zone center and zone edge by the mismatch of acoustic impedances.

In the case of Ge_2Si_2 (Fig. 9.15) only one gap appears at the edge of the BZ. Note that this edge corresponds, in general, to a longitudinal $\boldsymbol{k}$ vector of magnitude π/d, while in the bulk crystal for a [001] direction the corresponding wavevector is $2\pi/a_0$, twice as large as that in Fig. 9.15.

As implied in (7.36), in the bulk case one can usually assume that the wavevector of Raman-active phonons is very small, i. e., only zone-center phonons can be Raman active. The justification given was the usual one within the dipole approximation: the wavelength of the light is much larger than the relevant characteristic lengths of the material (i. e., the exciton radius and/or the lattice constant). The $\boldsymbol{q} \approx 0$ restriction remains valid in short period superlattices but must be modified for d sufficiently large. We define the reduced magnitude of the scattering wavevector $\tilde{q}$ for backscattering as

$$\tilde{q} = \frac{4\pi n/\lambda_L}{\pi/d} = \frac{4nd}{\lambda_L}. \tag{9.45}$$

Note that for typical values of the refractive index $n \simeq 3.5$ and a laser wavelength $\lambda_L = 500$ nm we have $\tilde{q} = 1$ for $d = 36$ nm, which corresponds to 130 monolayers of GaAs. In this case the scattering will be produced by phonons *at the edge* of the mini-BZ. By varying either d or λ_L it is thus possible to cover the whole range of reduced wavevector $\tilde{q}$ and even to reach the edge of the mini-BZ. This fact is illustrated in Fig. 9.24 for an [001] supperlattice with the complicated period $(AlAs)_{99}[(GaAs)_{17}(AlAs)_4]_5$, measured in the range $730 \leq \lambda_L \leq 830$ nm, which exhibits the lowest and the first folded LA (FLA)

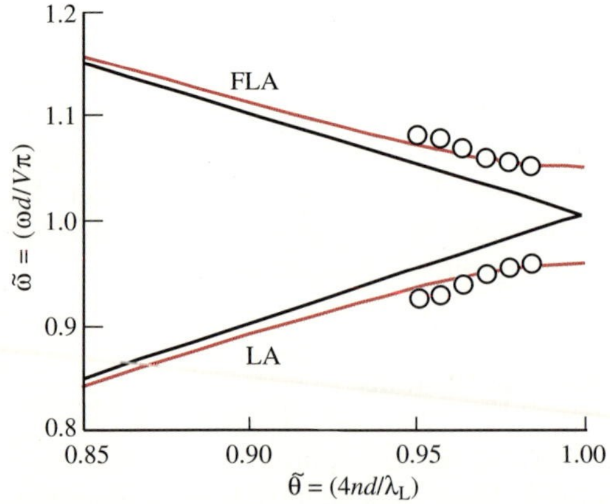

Fig. 9.24. Phonon dispersion relations of a superlattice with the period $AlAs_{99}[(GaAs)_{17}$-$(AlAs)_4]_5$ obtained by Raman spectroscopy. The rather complicated period was chosen so as to increase the splitting at the edge of the mini-BZ. The reduced energy is plotted versus reduced wavevector. The *red line* represents the dispersion calculated with Rytov's equation, (9.26). The *black line* gives the dispersion obtained with an average sound velocity, and hence exhibits no gap. From [9.50]

phonon branches and clearly demonstrates the existence of a gap at the edge of the mini-BZ [9.50]. Note the excellent agreement with the data points of the fit with the theoretical curve in which only the elastic constant C_{11} of AlAs was adjusted. (It was not well known experimentally when this work was performed. Hence the spectra provide a way to determine elastic constants.)

We discuss next the mechanisms responsible for the scattering efficiency of folded acoustic phonons. In principle the Raman tensor is related to (7.36), where instead of derivatives with respect to phonon amplitude Q we should consider derivatives with respect to strain e_{ij}. Note that for long wavelength acoustic phonons $e_{ij} \propto \mathrm{i} q_i Q_j$. The derivative of χ with respect to Q_j is linear in q_i on account of translational invariance; the derivatives of χ with respect to e_{ij} are thus $\boldsymbol{q}$-independent in the elastic limit.

The derivatives of χ with respect to the strain are the components of a fourth-rank tensor, the so-called **elasto-optic** coefficients. As in the case of $\delta\varepsilon^{(3)}_{ijkl}$ (see Problem 6.14), in a cubic crystal there are three independent elasto-optic coefficients. We must point out, however, that the "Raman tensor" of (7.36) for folded acoustic modes is not simply a derivative with respect to a uniform strain but rather involves an integral over the period d, since the strain is not uniform within one period. This is clearly illustrated for the 266 cm^{-1} acoustic folded mode of Fig. 9.14: the strain between Ge and Si planes is nearly zero while that between Ge and Ge has the opposite sign to that between Si and Si. The corresponding Raman tensor is thus obtained by averaging the photoelastic response over a superlattice period. We write, in short-hand notation, the corresponding Raman tensor $\boldsymbol{R}_{\mathrm{FA}}$ as [9.51, 52]

$$\boldsymbol{R}_{\mathrm{FA}} \propto \int_0^d \frac{\partial \chi}{\partial \boldsymbol{e}} \cdot \boldsymbol{e}(z) dz. \tag{9.46}$$

Let us consider acoustic phonons at the folded gap for $\tilde{q} = 0$. The corresponding strains are either odd or even with respect to the bisecting plane of one of the layers. The same fact holds true for $\tilde{q} = 1$, but in this case the phonons are odd (even) with respect to the A bisector or even (odd) with respect to the B bisector plane (see Problems 9.4 and 9.8). For the odd phonons (9.46) vanishes while for the even ones it does not. Hence, in the case of phonons with $\tilde{q} = 0$, only one of the two components of a folded doublet should appear in the Raman spectrum. It is easy to see that for $\tilde{q} \neq 0$ both components appear (Problem 9.8).

We close this discussion by mentioning that a nonvanishing $\boldsymbol{R}_{\mathrm{FA}}$ may result, in (9.46), from either a difference in elasto-optic coefficients of the two constituents or from deviations from the simple plane-wave behavior of $\boldsymbol{e}(z)$ expected for homogeneous media. The latter will be large only if large differences in the density ϱ or in the elastic constants exist. This is seldom the case when the constituent materials grow well epitaxially one on top of the other (the largest differences may appear for ϱ in the Ge–Si systems). Because of interband resonances in one material but not in the other, however, the elasto-optic coefficients may differ by orders of magnitude. Hence they usually determine the $\boldsymbol{R}_{\mathrm{FA}}$ (which will, in general, have complex components!). The observed Raman efficiencies can thus be used to determine elasto-optic coefficients [9.53, 54]. Differences in the dielectric functions of the two media can also contribute to the scattering by folded acoustic modes [9.52]. This effect, however, is also expected to be small.

We discuss next resonance effects in the scattering by folded acoustic modes. Figure 9.25 displays scattering spectra for a $(\mathrm{GaAs})_{16}(\mathrm{AlAs})_{15}$ MQW. Each curve is labeled by an energy $\Delta\hbar\omega = \hbar(\omega_{\mathrm{L}} - \omega_{\mathrm{e1,h1}})$ which represents the detuning between the laser frequency and the transition frequency between the first confined heavy hole and electron states ($\hbar\omega_{\mathrm{e1,h1}} = 1.7$ eV at 10 K). Surprisingly, we do not see a resonance enhancement of the folded LA doublet (the two gray-shaded peaks of Fig. 9.25). Instead we see the emergence of a broad resonant background (shaded red in Fig. 9.25). This phenomenon has been attributed to **inhomogeneous broadening** of the $\omega_{\mathrm{e1,h1}}$ frequency by random fluctuations of the layer thicknesses [9.55].

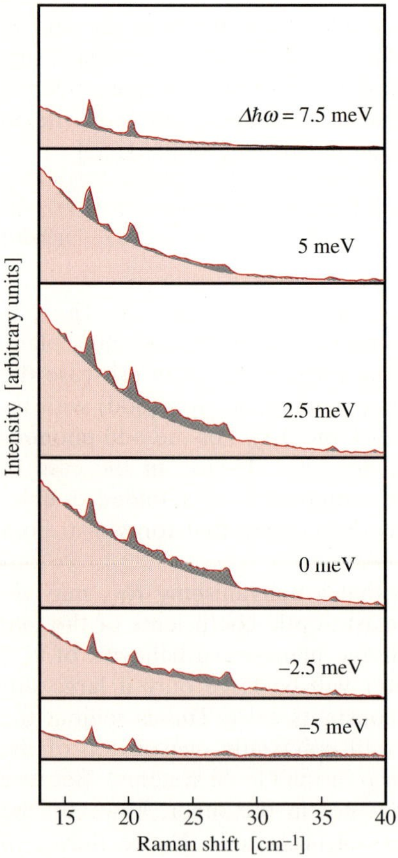

Fig. 9.25. Effect of resonance on the LA phonons of a $(\mathrm{GaAs})_{16}(\mathrm{AlAs})_{16}$ MQW around the lowest 1h → 1e GaAs-like confined transition. The *red-shaded background* is due to LA scattering without wavevector conservation (because of the inhomogeneous broadening). The *gray-shaded sharp* peaks correspond to folded doublets. Note the strong resonance of the background as opposed to that of the folded doublets. The measurements were performed in the presence of a magnetic field in order to enhance the resonant behavior. From [9.55]

This is a good time to make a digression and introduce the concepts of **homogeneous** and **inhomogeneous** energy (or frequency) **broadening**. By homogeneous broadening we understand that which appears as an imaginary energy (or frequency) in the denominator of a Lorentzian resonance such as the γ in (6.117), Γ_q in (7.54) and Γ_a in (7.60). It is the result of a finite lifetime $\tau = \gamma^{-1}$ of the corresponding excitation, usually attributed to decay into other excitations. An inhomogeneous broadening, in contrast, results from a collection of oscillators such as that in (6.117) but with a distribution in frequencies around a central one ω_g. In the simplest case of a random distribution, a Gaussian law

$$p(\omega) \propto \exp\left(-\left|\frac{\omega-\omega_g}{g}\right|^2\right) d\omega_g \tag{9.47}$$

is likely to be followed by $p(\omega)$, the number of oscillators of frequency between ω and $\omega + d\omega_g$. In (9.47) g represents the inhomogenous width. The oscillators with probability $p(\omega)$ may also be affected by a homogeneous width γ, often smaller than g. MQWs are expected to show considerable inhomogenous broadening of their electronic transition frequencies (such as $\omega_{\mathrm{e1,h1}}$) on account of the random fluctuations of the layer thicknesses which can occur along the direction of growth or also within the plane of individual layers. A careful examination of the two interfaces in Fig. 9.1 shows that while they are very flat, atoms of both kinds, Al and Ga, can be found on both sides of a given interface. It is therefore reasonable to assume that there is a layer thickness fluctuation of the order of one monolayer of GaAs, i. e., $a_0/2$ or $\delta = \pm a_0/4$. Equation (9.3) enables us to estimate the corresponding inhomogeneous broadening g (under the assumption of an infinite potential barrier):

$$g \simeq 2\frac{\delta}{d}\Delta E. \tag{9.48}$$

For $\Delta E \simeq 180$ meV, $d = 44$ Å, and $\delta = \pm 1.4$ Å we find with (9.48) that $g \simeq 10$ meV.

We return to Fig. 9.25. As we shall see below, the homogeneous width γ of the 1h $\rightarrow$ 1e transitions is $\simeq 0.8$ meV, much smaller than the inhomogeneous one g. Under these conditions, and for a laser frequency in the range

$$\omega_{\mathrm{e1,h1}} - g < \omega_{\mathrm{L}} < \omega_{\mathrm{e1,h1}} + g, \tag{9.49}$$

(7.50a) will be resonant for a subset of GaAs wells *distributed at random*. To a first approximation, only these random layers should be included in the sum of (7.50a). As a result of the "randomicity" of the terms in the sum, there is *effectively* no translational invariance and no reason why the wavevector should be conserved in the scattering process. Hence, acoustic phonons of an arbitrary $\boldsymbol{q}$ can participate in the scattering, and the scattering continuum shown by the red-shaded portions of Fig. 9.25 is obtained. Outside the range of (9.49) the gap fluctuations g are relatively unimportant and $\boldsymbol{q}$-conservation is recovered.

Figure 9.26 displays the inverse ratio of the discrete folded doublet intensities I_{F} to that of the background I_{B} at the same Raman shift. The red curve

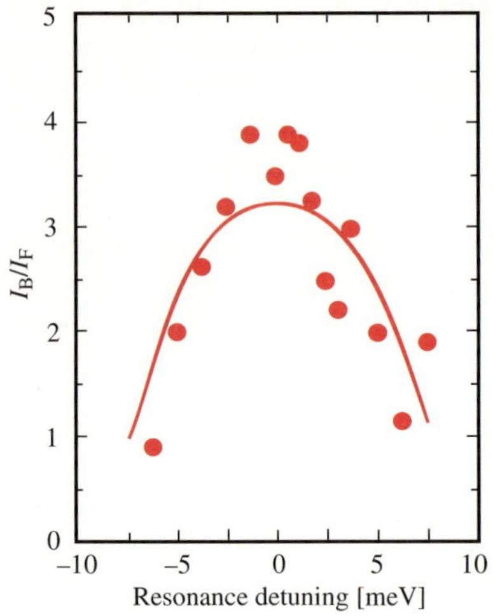

Fig. 9.26. Ratio of the background intensity I_B to that of the folded LA doublets of Fig. 9.25 as a function of detuning from the 1h → 1e transition. The *curve* represents a fit to the experimental points from which an inhomogeneous broadening $g = 2.50$ meV and a homogeneous one $\gamma = 0.76$ meV are found. From [9.55]

represents a fit to the experimental points performed by *Ruf* et al. [9.55]. One can show (Problem 9.9b) that the width of this fitting curve ($\approx$12 meV) is approximately 4g, i. e., $g \simeq 3$ meV. From the detailed fit *Ruf* et al. [9.55] obtained $g = 2.5$ meV. By dividing g by the maximum value of $I_B/I_F \simeq 3.3$ one obtains the homogeneous width of the transition $\gamma = 0.76$ meV.

9.4.2 Raman Scattering by Confined Optical Phonons

In order to discuss the Raman scattering by confined optical modes, we display in Fig. 9.27 the displacement patterns and electrostatic potentials of LO modes in a [001] polar superlattice for $m = 1, 2, 3, 4$ and $k_{x,y} = 0$. Note the out-of-phase character of ϕ and u_z. Note also that u_z vanishes at the layer interfaces as required by the mechanical boundary conditions.

As shown in (7.5a), phonon Raman scattering takes place through electron–phonon interaction. Two kinds of interactions are possible: one, induced by the atomic displacement, does not require the material to be polar. For MQWs it is derived from the analogous effect in the corresponding bulk material. In the Ge/Si or GaAs/AlAs systems, it is determined by tensors of the form (7.39) and for backscattering on a [001] face allows only LO scattering. If the laser and scattered polarizations $\hat{\boldsymbol{e}}_L$ and $\hat{\boldsymbol{e}}_S$ are parallel to the crystal axes (either $\hat{\boldsymbol{x}}$ or $\hat{\boldsymbol{y}}$), this LO scattering requires crossed polarizations. For even m the oscillations in u_z cancel exactly in the calculation of the deformation potential matrix element of the electron–phonon interaction and the scattering remains forbidden. For $m = 1$, however, no cancellation occurs while for $m = 3, 5, \ldots$ the cancellation is only partial: the scattering intensities for the $m = 1, 3, 5, \ldots$ confined modes should be approximately proportional to $1, 3^{-2}, 5^{-2}, \ldots$.

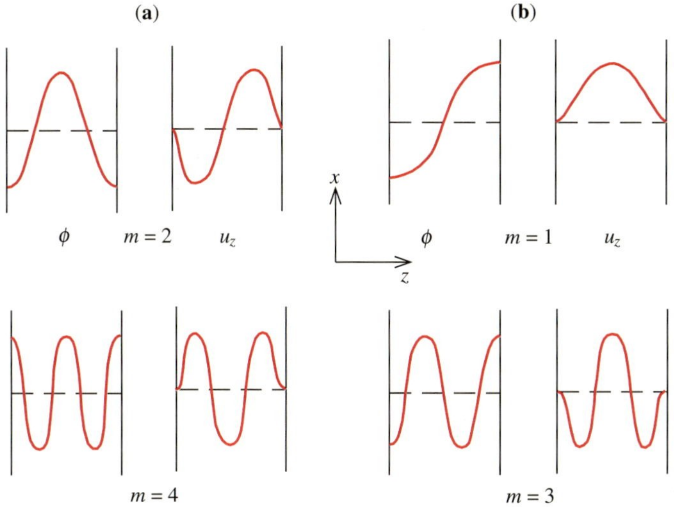

Fig. 9.27a, b. Schematic diagrams of the dependence on z of the displacement u_z and the electrostatic potential ϕ for confined LO_m ($m = 1$ to 4) modes with $k_{x,y} = 0$ in a [001] polar superlattice

The coupling via matrix elements of the electrostatic potential ϕ (Fröhlich interaction) has been discussed in Chap. 7 in connection with forbidden scattering by LO phonons for parallel $\hat{e}_L$ and $\hat{e}_S$ see (7.70b). In that case, which applies to bulk materials, the electron–phonon interaction resulted from the long-range nature of the Coulomb interaction. The corresponding q^{-1} singularity was lifted by the combined effects of the LO phonon on the conduction and valence bands, which are equal and have to be subtracted from each other. A nonsingular residual effect is left, which is proportional to the difference of the electron and hole inverse masses and to the transverse charge of the phonons e^*, see (7.70), (3.39) and (6.119). Since superlattices can also be regarded as crystals, this type of “forbidden” scattering should also appear for the IR-active modes, i. e., for the LO_m phonons with m odd. The effect should decrease with increasing m ($\propto m^{-2}$) and should vanish for backscattering along the axis of growth in the large period case since the energy bands along z become completely flat (the inverse masses along z vanish). For backscattering with in-plane propagation (i. e., light incident on the edge of the MQW) a similar effect should also exist: the Raman tensor should be proportional to the difference of the in-plane electron and hole inverse masses which does not vanish even in the large period case. Very few experiments have been performed in this configuration (for examples see [9.56, 57]).

So far we have not found a mechanism that allows scattering by the LO modes of even m. It is easy to see, however, that these modes belong to the completely symmetric representation A_1 of the D_{2d} point group of a GaAs/AlAs superlattice. The corresponding Raman tensor has the form (Problem 9.10)

$$R_{\mathrm{LO}}(A_1) = \begin{pmatrix} a & 0 & 0 \\ 0 & a & 0 \\ 0 & 0 & b \end{pmatrix}. \tag{9.50}$$

The tensor (9.50) vanishes in the case of the optical phonon in bulk GaAs. It can be seen that its origin in the MQWs is the electron–phonon coupling induced by the matrix elements of $\phi(z)$ (Fig 9.17): the diagonal matrix element of $\phi(z)$ with respect to confined electronic wavefunctions $\varphi_n(z)$

$$M_{\mathrm{ep}} = \int_0^d \varphi_n^*(z)\phi_m(z)\varphi_n(z)dz, \tag{9.51}$$

vanishes for m odd but not for m even. However, it decreases rapidly with increasing m. The matrix element in (9.51) has been derived from the Fröhlich electron–phonon interaction although the corresponding potential in QW is not long range: it is nonzero only inside of a monolayer of one type of material and vanishes outside as a result of the oscillations in $\phi(z)$. Note that the Raman tensor (7.50a) will contain two cancelling terms including matrix elements of the type (9.51): one for the conduction band and one for the valence band. The cancellation will be exact for infinite electron barriers. However, since these barriers are not infinite a net coupling will usually be obtained.

From the preceding discussion we conclude that for $\hat{\boldsymbol{e}}_{\mathrm{L}} \parallel \hat{\boldsymbol{x}}$, $\hat{\boldsymbol{e}}_{\mathrm{S}} \parallel \hat{\boldsymbol{y}}$ (crossed polarizations), the $m = 1,3,5,\ldots$ modes should be observed while for parallel $\hat{\boldsymbol{e}}_{\mathrm{L}} \parallel \hat{\boldsymbol{e}}_{\mathrm{S}}$ polarizations $m = 2,4,\ldots$ should appear (Problem 9.12). The former case is illustrated in Fig. 9.28 for a $(\mathrm{GaAs})_{16}(\mathrm{AlAs})_{16}$ superlattice. Notice the peaks which correspond to the $m = 1,3,\ldots,11$ phonons. Their frequencies have been plotted in the inset versus effective wavevector k_m, see (9.29), and compared with the dispersion relation of bulk GaAs; the agreement is excellent. The circles in the inset represent peaks obtained for parallel polarization. Those with the two highest frequencies correspond to LO_2 and LO_4 confined modes. The rest is believed to be related to anticrossings of interface modes with confined modes [9.58]. This will be discussed in the next section.

9.4.3 Raman Scattering by Interface Modes

The observation of interface modes by Raman spectroscopy requires an in-plane component of the scattering wavevector, so as to operate with an angle $\Theta > 0$ (Fig. 9.20). While this has been performed recently in a controlled manner with light incident on the edge of the superlattice (see experimental points in Fig. 9.23) the first observations of interface modes [9.42, 59] were performed in a backscattering configuration along the z direction, the required in-plane wavevector being supplied by the roughness of individual interfaces (which is responsible for the violation of the in-plane translational invariance). Figure 9.29 shows the observations of *Sood* et al. [9.59] for the AlAs-like modes of GaAs/AlAs MQWs. The broad band centered between the bulk LO and TO frequencies is characteristic of interface modes in symmetric (i. e.,

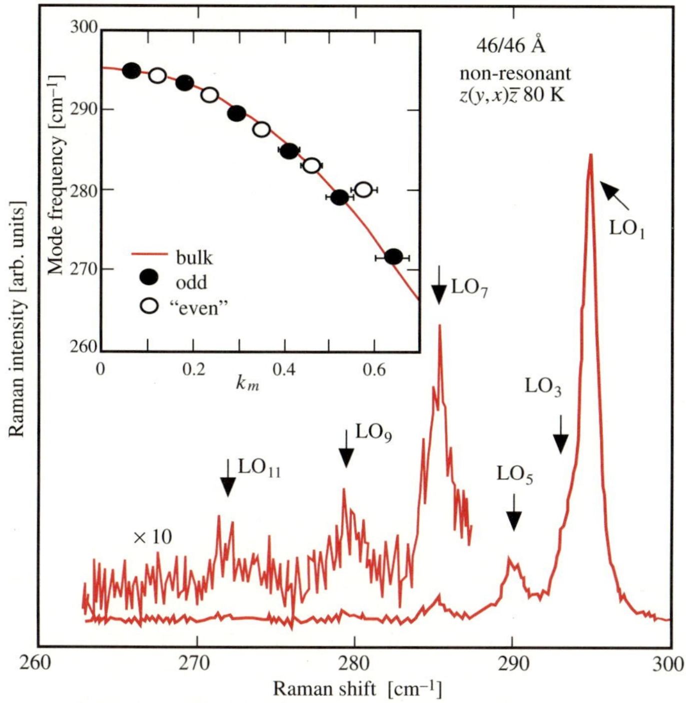

Fig. 9.28. Raman spectrum of a $(GaAs)_{16}(AlAs)_{16}$ superlattice obtained for cross polarization under nonresonant conditions. The *solid points* in the *inset* represent the frequencies of these modes versus equivalent bulk wavevector k_m. The *solid line* represents the LO dispersion of bulk GaAs. The circles represent points obtained for parallel polarizations, possibly related to even m modes. From [9.58]

$d_A = d_B$) structures: the peak occurs at the solution of (9.35). In the asymmetric case one must use (9.41) in order to obtain the long-wavelength solutions for $\Theta = \pi/2$. Contrary to the symmetric ($d_A = d_B$) case, two different modes appear for $d_A \neq d_B$. They are responsible for the two structures seen in the spectra shown in the lower part of Fig. 9.29. Either the upper or the lower of the two structures dominates, depending on whether $d_B > d_A$ or $d_B < d_A$. The dominating structure corresponds to that labeled ω_{LO}^x in Fig. 9.20.

As we have seen, the spectra of Fig. 9.29 can be qualitatively explained on the basis of simple electrostatic interface modes of the type predicted by (9.38) without interaction with the m = odd confined modes. The reason why this interaction can be neglected is that the dispersion relations of optical modes (both LO and TO) in AlAs are very flat (Fig. 9.13) and therefore anticrossings are only important in the close neighborhood of the bulk $\Gamma = 0$ TO and LO phonons (note that some structures appear in this region in Fig. 9.29). In the case of GaAs, however, the bulk dispersion relations are broader. [The difference in the dispersion relations of GaAs and AlAs (Fig. 9.13) is due to the

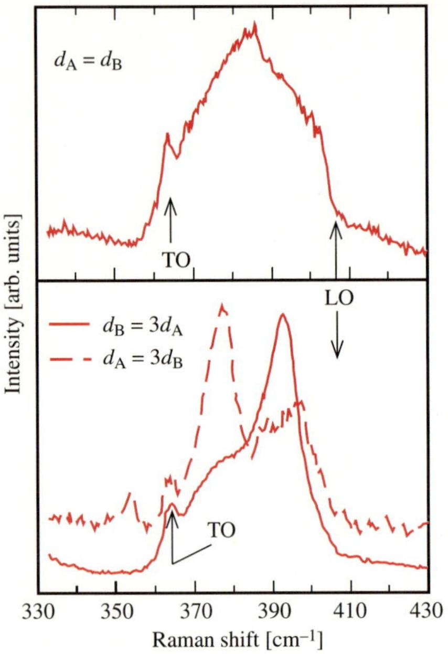

Fig. 9.29. Raman spectra in the frequency range of the optical phonons of bulk AlAs obtained at resonance in the $\hat{\boldsymbol{e}}_L \parallel \hat{\boldsymbol{e}}_S$ configuration for three GaAs/AlAs (A/B) with different layer thickness ratios [9.59]. $T = 10$ K

fact that the masses of Ga and As are very close while those of Al and As are rather different. This raises the optical phonons and lowers the acoustic ones for AlAs at the edge of the BZ.] The broader dispersion relations cause a number of anticrossings to occur throughout the whole region between the bulk TO and LO frequencies, as shown in Fig. 9.23 for the GaAs-like optical modes of a $(GaAs)_{12}(AlAs)_{12}$ MQW. As a result of the gaps related to the anticrossings, a number of structures appear between the bulk LO and TO peaks; they are displayed in Fig. 9.30. The experimental spectra in this figure were obtained in backscattering configuration for $\hat{\boldsymbol{e}}_L \parallel \hat{\boldsymbol{e}}_S \parallel [100]$. They are thus, in principle, expected to show the confined modes for even values of m with an intensity which should decrease strongly with increasing m. The LO_2 and LO_4 modes appear clearly in the experimental spectra of Fig. 9.30 and their intensities have the expected behavior. The other peaks (labeled IF in the figure) have also been assigned in the past to LO_6, LO_8, LO_{10}, ... They have been plotted in the inset of Fig. 9.28 under this assumption: except for the lowest one, they fit the bulk dispersion relation rather well. Their intensities, however, do not decay with increasing m in the expected manner. This has led *Shields* et al. [9.58] to conjecture that they are not related to even-m confined modes but to anticrossing IF-odd-m confined modes of the type of Fig. 9.23. The regions in this figure where the dispersion relations are flat act as "van Hove singularities", giving rise to very large densities of states. This is exhibited clearly by the calculated curves of Fig. 9.30; the peaks labeled IF can be shown to correspond to flat regions of the anticrossing dispersion curves [9.58].

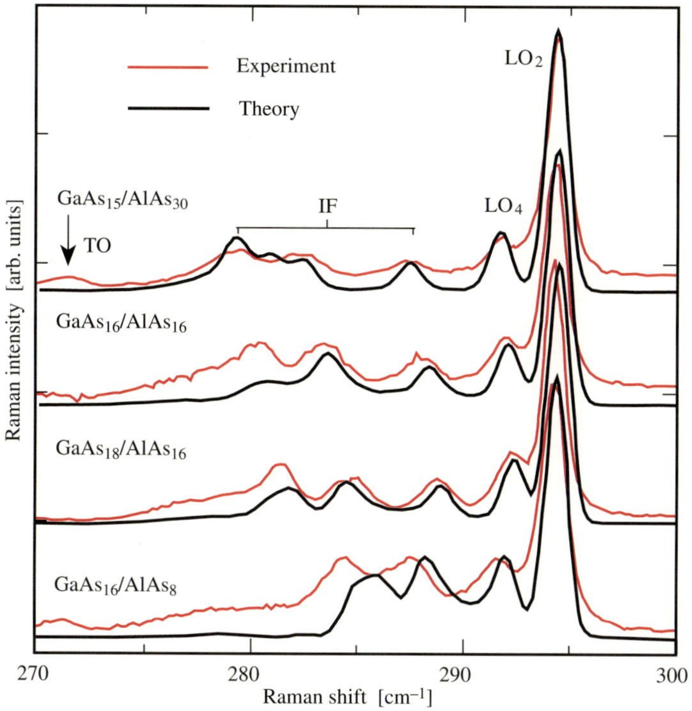

Fig. 9.30. Comparison of calculated (*black lines*) and measured (*red lines*) Raman spectra due to GaAs-like optical modes of GaAs/AlAs MQWs with different layer dimensions, for outgoing resonance with the e1−lh1 transitions at 10 K, in parallel polarization geometry. The spectra are normalized to have the same height and shifted vertically for clarity. From [9.58]

9.4.4 Macroscopic Models of Electron–LO Phonon (Fröhlich) Interaction in Multiple Quantum Wells

In Sect. 9.3.3 we discussed how optical phonons in MQWs and SLs formed by two materials with nonoverlapping optical phonon energies, such as GaAs and AlAs, are confined in their respective layers. In addition, interface modes are formed whose displacement decays exponentially (although in some cases very slowly) away from interfaces. In Chap. 5 we studied the important role played by the Fröhlich electron–LO phonon interaction in determining the mobility of electrons in semiconductors such as GaAs. It is natural now to ask how the Fröhlich interaction is modified in MQWs and in SLs. To answer this question, we shall consider only the case of MQWs and SLs constructed from GaAs and AlAs.

As shown in Sect. 3.3.5, the Hamiltonian for the Fröhlich interaction can be expressed essentially in terms of a scalar electrostatic potential multiplied by the electronic charge. In principle, this potential can be calculated using microscopic lattice dynamical models with suitable boundary conditions. In practice these calculations are very time consuming and require high speed supercomputers. In addition, the results are specific to a particular well width and barrier thickness. Thus it is desirable to find simple, albeit approximate, expressions of the scalar potential that can be applied to samples with different well widths. Several such "macroscopic" models have been proposed. These models are called macroscopic because at their starting point they often assume the sample to be a continuum. Their main difference lies in their treatment of the boundary conditions imposed on the optical phonons at interfaces of the MQW or SL. Depending on the boundary conditions they adopt, some of these macroscopic models have been labeled as "mechanical models" or "dielectric continuum models".

Roughly speaking the "mechanical models" require that the atomic displacements of the confined LO phonon vanish at the interface, even if this condition may lead to violations of Maxwell's equations at that interface. An example of the displacement pattern and electrostatic potential of confined LO phonons that satisfy such mechanical boundary conditions is shown in Fig. 9.27. Notice that the electrostatic potential ϕ does not vanish at the interface. For a phonon confined in medium A with nonzero q_x, we can write the scalar potential as

$$\begin{aligned}\phi(x,z) &= \phi_0 \exp(\mathrm{i}q_x x)\cos(k_m z) \quad \text{when } m \text{ is even, or} \\ &= \phi_0 \exp(\mathrm{i}q_x x)\sin(k_m z) \quad \text{when } m \text{ is odd.}\end{aligned} \tag{9.52}$$

The component of the electric field parallel to the interface (E_x) is given by

$$E_x = -d\phi/dx = (-\mathrm{i}q_x)\phi(x,z) \tag{9.53}$$

and does not vanish at the interface (since ϕ is not zero there) as required by the continuity of the tangential components of the electric field across the boundary of two dielectric media (note that E_x is zero in medium B if the phonon is confined to A). Since such models do not take into account Maxwell's equations, they do not include the details of interface modes without additional *ad hoc* assumptions.

On the other hand, the "dielectric continuum models" use Maxwell's equations as their starting point and obtain the interface modes as part of the solutions of the Laplace equation (9.32). As pointed out in Sect. 9.3.4, while such models violate the mechanical boundary conditions (requiring the atomic displacements of confined phonons to vanish at the interface), for the interface modes they provide rather good approximations to the microscopic results because the atomic displacements in reality are nonzero until very close to the interfaces.

A macroscopic model which tries to mimic the results of microscopic calculations has been proposed by *Huang* and *Zhu* [9.60]. They performed a model microscopic lattice dynamical calculation to determine the atomic displacement and electrostatic potential. They noticed that the dielectric continuum model provided a rather good approximation to the microscopic result, except for the violation of the mechanical boundary conditions. Especially the interface modes were very well reproduced by that model. To take care of both the mechanical and Maxwell's boundary conditions it is necessary for both ϕ and its derivative $d\phi/dz$ to vanish at the interfaces. *Huang* and *Zhu* [9.60] noted that this can be achieved by subtracting an appropriate constant from those ϕ with even symmetry (with respect to reflection onto a plane passing through the center of a layer) or an appropriate term linear in z from those that have odd symmetry. In this manner they arrive at the following approximate expressions for ϕ:

$$\phi_m(z) = \cos\left(\frac{m\pi z}{(n+1)a}\right) - (-1)^{m/2}, \quad m \text{ even}, \tag{9.54a}$$

$$\phi_m(z) = \sin\left(\frac{\mu_m \pi z}{(n+1)a}\right) + \left(\frac{C_m z}{(n+1)a}\right), \quad m \text{ odd}. \tag{9.54b}$$

In these expressions a represents the distance between atomic planes in either GaAs or AlAs and na the layer thickness d_A or d_B in (9.29). The constants μ_m and C_m are determined by the conditions that both ϕ_m and $d\phi_m/dz$ vanish at the interface taken to be at $z = \pm(n+1)a/2$. These boundary conditions are satisfied when μ_m and C_m are solutions of the two equations

$$\tan(\mu_m\pi/2) = \mu_m\pi/2 \tag{9.55a}$$

and

$$\sin(\mu_m\pi/2) = -C_m/2. \tag{9.55b}$$

Some solutions of these two transcendental equation are

$\mu_3 = 2.86064$; $\mu_5 = 4.918$; $\mu_7 = 6.95$ and

$C_3 = 1.9523$; $C_5 = -1.983$; $C_7 = 1.992$.

The electron–LO phonon interaction obtained with the potentials in (9.54) will be referred to as the **Huang–Zhu model**.

In Fig. 9.31 the *potentials* associated with the lowest order confined LO phonons and interface modes in a GaAs/AlAs superlattice obtained with the three macroscopic models discussed above are compared with those obtained by an *ab initio* microscopic model [9.61]. We see that the Huang–Zhu model does approximate the microscopic model best, with the dielectric continuum model a close second. *Rudin* and *Reinecke* [9.62] have computed the total scattering rate of electrons in GaAs/AlAs quantum wells by the electrostatic potential of LO phonons as a function of well width using these three macroscopic models. Their results are shown in Fig. 9.32. While there are differences between the scattering rates computed from these models, the difference between the dielectric continuum model and the Huang-Zhu model disappears for small well widths. For very small well widths the scattering rate is dominated by the interface modes and these modes are nearly identical in these two models. Using a rather involved optical technique, *Tsen* et al. [9.63] have studied experimentally the relative strength of electron LO–phonon

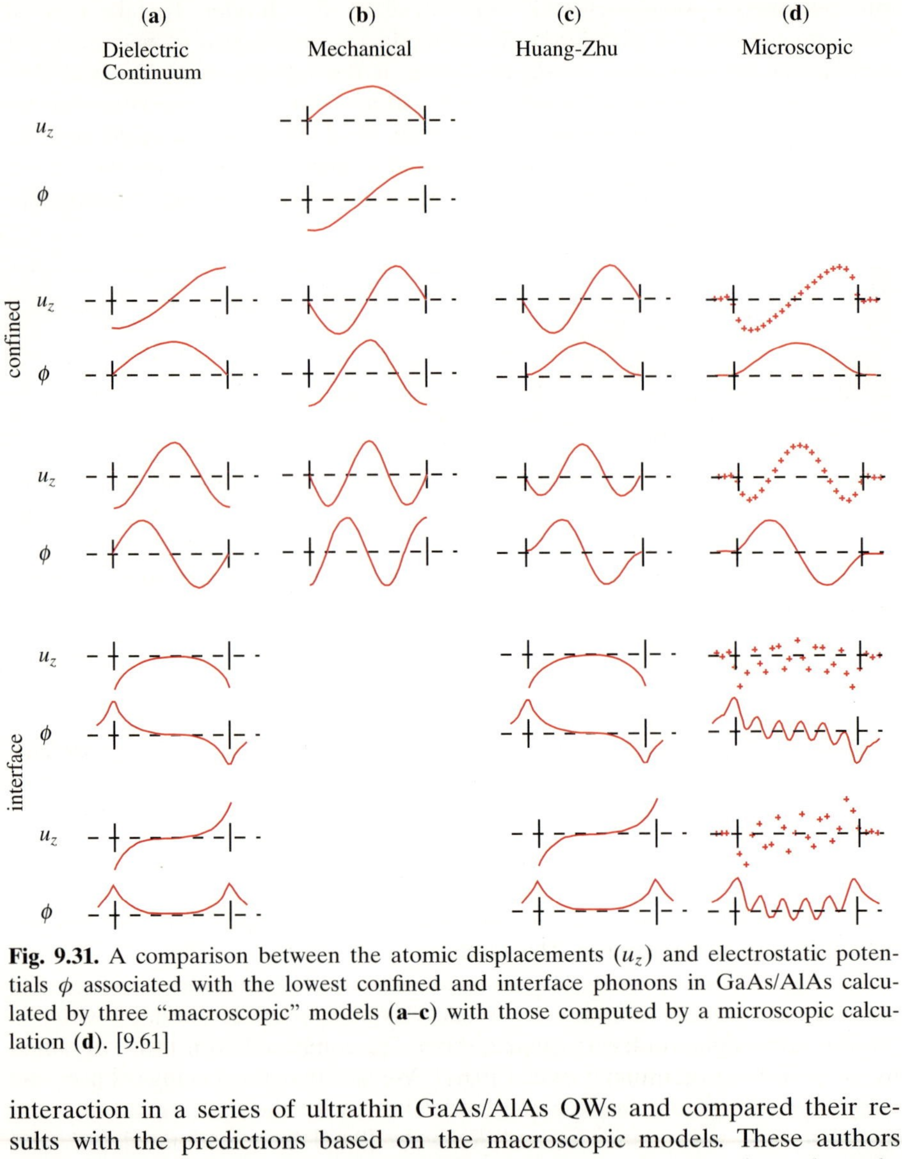

Fig. 9.31. A comparison between the atomic displacements (u_z) and electrostatic potentials ϕ associated with the lowest confined and interface phonons in GaAs/AlAs calculated by three "macroscopic" models (**a–c**) with those computed by a microscopic calculation (**d**). [9.61]

interaction in a series of ultrathin GaAs/AlAs QWs and compared their results with the predictions based on the macroscopic models. These authors used a beam of picosecond laser pulses to excite electrons into the subband states with 200 meV of excess energy above the subband minimum. These energetic electrons relax via electron–LO phonon interaction and emit a nonequilibrium population of LO phonons with lifetimes of the order of 10 ps. Using a second, weaker, and delayed picosecond laser pulse train, tuned to resonate with the lowest energy subband, these authors probed the phonon population of various confined and interface LO modes via Raman scattering. They found that their experimental results agreed best with the predictions of the Huang–Zhu model.

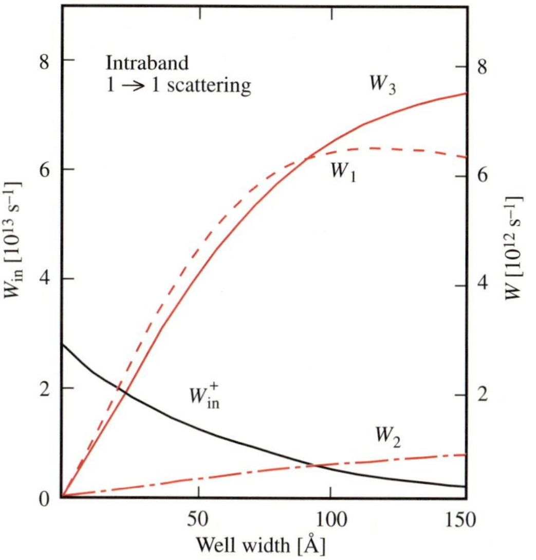

Fig. 9.32. The total intrasubband (1 → 1 subbands) scattering rate of an electron by confined and interface LO phonons as a function of well width in a GaAs/AlAs quantum well calculated with the three "macroscopic" models shown in Fig. 9.31 and discussed in the text. W_{in}^+ (*black curve*) represents scattering by interface modes. W_1, W_2, and W_3 (*red curves*) represent contributions from confined modes calculated with the three different models discussed in the text: W_3 corresponds to the Huang-Zhu model, W_1 to electrostatic boundary conditions and W_2 to the use of mechanical boundary conditions only. From [9.62]. For additional information see [9.61] and Problem 9.13

9.5 Electrical Transport: Resonant Tunneling

In the final two sections we shall discuss some electrical transport measurements involving quantum wells and heterojunctions. Since this is a very active and prolific field we have to limit ourselves to only two topics: resonant tunneling and the quantum Hall effect. They are chosen because they both specifically make use of the properties of quantum confined electrons. In this section we shall discuss the phenomenon of resonant tunneling while leaving the quantum Hall effect to the last section.

Tunneling of a particle through a barrier is one of the most studied phenomena in quantum mechanics. For a particle tunneling in one dimension the theory can be found in almost all quantum mechanics textbooks. Tunneling plays an important role in many semiconductor devices. In particular, the tunnel diode or Esaki diode discovered by *Esaki* in 1958 [9.64] involves tunneling through a forward-biased heavily doped (degenerate) junction in germanium.

One important characteristic of the Esaki diode is that it exhibits negative differential resistance (NDR), making possible its application as a high frequency (microwave) oscillator (Sect. 5.4.2). The properties of the original Esaki diode were determined (and hence also limited) mainly by the band structure of the bulk semiconductor. In 1973 *Tsu* and *Esaki* [9.65] suggested that NDR can also be achieved in a superlattice. However, it took more than ten years before high quality QW samples exhibiting NRD could be fabricated [9.66]. Even in that case the sample involved a QW rather than a superlattice. NDR in a GaAs/AlAs superlattice was reported several years later [9.67]. Since this pioneering work NRD has been observed in many structures involving QWs and superlattices. To illustrate the physics we shall consider only the simplest structure consisting of a QW sandwiched between two barriers, often referred to as a **double-barrier QW structure**. Readers interested in transport in superlattices should consult the review by *Palmier* [9.68].

9.5.1 Resonant Tunneling Through a Double-Barrier Quantum Well

Figure 9.33a shows schematically the band diagram of a typical double-barrier QW structure in the growth direction (z axis). In this case the QW formed by a layer of lightly doped (carrier concentration N_{D2}) GaAs is surrounded by two undoped GaAlAs barriers. It is assumed that the barrier height and well width (W_2) are such that only one subband (E_1) is formed in the well. This three-layer structure is further "sandwiched" by two more heavily n-doped GaAs layers (E_F represents the Fermi energy) to provide electrical contact. The device shown in Fig. 9.33a is an n-type device, although p-type tunneling devices can be constructed similarly. Figures 9.33b and c show the band scheme when a bias voltage is applied to the device. Under the applied field electrons can tunnel from the GaAs layer on the left (the **emitter**) to that on the right (the **collector**). Qualitatively we expect that the tunnel current will be small initially and will increase with the applied voltage. This is shown schematically in Fig. 9.33d near the origin. As the bias voltage reaches the value $2E_1/e$, E_F in the emitter is resonant with the subband E_1 in the well (this statement is true only when the two barriers have equal width). This is shown in Fig. 9.33b. Under this condition electrons which tunnel into the well can, in principle, be captured in the well to be released again so as to tunnel through the second barrier. We may expect the tunnel current to increase strongly at this voltage. This phenomenon is known as **resonant tunneling**. Once the voltage exceeds $2E_1/e$ (Fig 9.33c) the tunnel current will decrease drastically, giving rise to a region with a negative differential resistance as shown in Fig. 9.33d.

The tunnel current in the above double-barrier QW structure can be calculated as a function of bias voltage by using the following approximate treatment. For simplicity we shall assume that the emitter, well, and collector are fabricated from the same material (such as GaAs in Fig. 9.33) and electrons in both behave like free carriers with isotropic effective mass m^*_A. An electron with wavevector $\boldsymbol{k}$ in the emitter will tunnel through the barrier into the collector without scattering, i. e., there will be neither loss in energy nor change in in-plane wavevector (note that the wavevector along the growth direction z is not conserved since the device is not translationally invariant along this direction). We shall neglect the Coulomb interaction between electrons, which produces an "image potential" on the tunneling electron. We shall also ignore band-bending induced by the bias voltage at the interface of the emitter and collector with the barriers (this has been assumed in drawing Figs. 9.33b and c). These simplifying assumptions allow us to reduce the three-dimensional problem to a one-dimensional one.

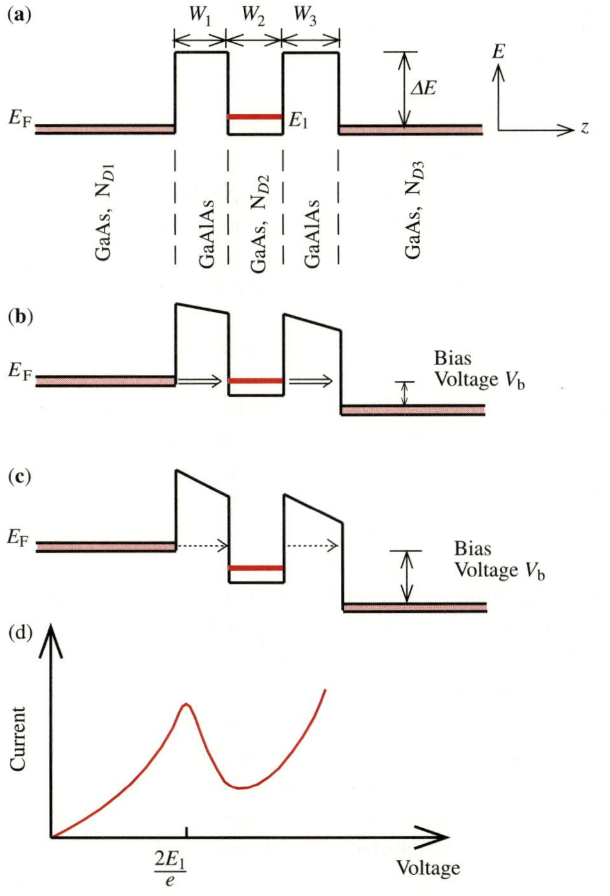

Fig. 9.33a–d. Spatial variation of the electron energy in a GaAs/GaAlAs/GaAs/GaAlAs/-GaAs double-barrier QW resonant tunneling device for three bias conditions: (**a**) zero bias; (**b**) bias voltage $V_b = 2E_1/e$ and (**c**) $V_b > 2E_1/e$, E_1 being the energy of the electron subband inside the GaAs QW. (**d**) Schematic I–V characteristic of the device depicted in (**a**) showing a region of NDR for bias voltage just above $2E_1/e$. [9.66]

Let the growth direction be denoted as the z direction, like in Fig. 9.33. Since the potential $V(z)$ (which is a function of the bias voltage V_b) seen by the tunneling electron in Fig. 9.33 depends on z only, the Schrödinger equation becomes separable into two equations by expressing the wavefunction as the product of two functions like in (9.7a). The solutions in the x and y directions are plane waves as in (9.7b) and will not be considered further. The eigenvalues for these solutions are given by $E_{x,y} = [\hbar^2/(2m_A^*)](k_x^2 + k_y^2)$. The Schrödinger equation for motion in the z direction is

$$\left[-\left(\frac{\hbar^2}{2m_A^*}\right)\left(\frac{d^2}{dz^2}\right) + V(z)\right]\psi_A(z) = E_z\psi_A(z) \tag{9.56a}$$

for z outside the barriers and

$$\left[-\left(\frac{\hbar^2}{2m_B^*}\right)\left(\frac{d^2}{dz^2}\right) + V(z)\right]\psi_B(z) = E_z\psi_B(z) \tag{9.56b}$$

for z inside the barriers, m_B^* being the electron effective mass inside the two barriers layers (assumed to be isotropic). The total energy E of the tunneling electrons is given by $E = E_{x,y} + E_z$. Except for the fact that the electron mass is different inside and outside the barriers, (9.56) corresponds to the familiar one-dimensional tunneling problem whose solutions can be found in many quantum mechanics textbooks. Instead of repeating these calculations here, we shall simply outline the procedure below. The details can be found in [9.5, p. 524] for example.

We are interested only in the case where the electron energy E_z is smaller than the height of both barriers even in the presence of a positive bias voltage. Under such circumstances the electron wavefunction can be written as the sum of incident and reflected plane waves in the emitter and well regions. Within the barriers the wavefunctions have purely imaginary wavevectors, i. e., they are exponential functions. In the collector region the wavefunction consists of a plane wave traveling to the right only, since it is assumed to extend to infinity on the right and therefore there is no reflected wave. At the interface one applies the usual continuity conditions to the wavefunction and its first derivative with respect to z, i. e. (9.13). From these conditions the coefficients of the incident and reflected wavefunctions in one region are related to those of the adjacent regions via a 2×2 matrix, known as a **transfer matrix** [9.69].

As an example we shall assume that the potential can be divided into $n+1$ regions defined by $z = [-\infty, z_1], [z_1, z_2], \ldots, [z_n, \infty]$ such that the potential V_i inside region i is constant. The emitter and the collector correspond, respectively, to the regions 1 and $n + 1$. In real situations where the potential is not constant within a certain region, one divides this region into many smaller regions until the potential can be approximated by constants inside each small region. Let A_i and B_i represent the amplitudes of the waves traveling to the right and to the left, respectively, in region i. We shall denote the generalized wavevector in the region i by k_i:

$$\frac{\hbar^2 k_i^2}{2m_i^*} = E_z - V_i, \tag{9.57}$$

where m_i^* is the electron mass in region i. From (9.57) it is clear that k_i is imaginary or real depending on whether region i is a barrier ($V_i > E_z$) or not. When k_i is imaginary the wave is evanescent as discussed in Sect. 9.2.3. The coefficients (A_1, B_1) and (A_{n+1}, B_{n+1}) in the emitter and collector regions, respectively, are related by

$$\begin{pmatrix} A_1 \\ B_1 \end{pmatrix} = M_1 M_2 \dots M_n \begin{pmatrix} A_{n+1} \\ B_{n+1} \end{pmatrix}, \tag{9.58}$$

where the elements $M_p(\alpha, \beta)$ ($\alpha, \beta = 1$ or 2) of the *transfer matrices* are given by

$$M_p(1,1) = \left(\frac{1}{2} + \frac{k_{p+1} m_p^*}{2k_p m_{p+1}^*} \right) \exp[\mathrm{i}(k_{p+1} - k_p) z_p], \tag{9.59a}$$

$$M_p(1,2) = \left(\frac{1}{2} - \frac{k_{p+1} m_p^*}{2k_p m_{p+1}^*} \right) \exp[-\mathrm{i}(k_{p+1} + k_p) z_p], \tag{9.59b}$$

$$M_p(2,1) = \left(\frac{1}{2} - \frac{k_{p+1} m_p^*}{2k_p m_{p+1}^*} \right) \exp[\mathrm{i}(k_{p+1} + k_p) z_p], \tag{9.59c}$$

$$M_p(2,2) = \left(\frac{1}{2} + \frac{k_{p+1} m_p^*}{2k_p m_{p+1}^*} \right) \exp[-\mathrm{i}(k_{p+1} - k_p) z_p]. \tag{9.59d}$$

From these results the transmission coefficient of the potential for an electron with energy E_z can be calculated:

$$T(E_z) = \left| \frac{A_{n+1}}{A_1} \right|^2 \tag{9.60}$$

Figure 9.34 shows the dependence of $T(E_z)$ on E_z for an electron tunneling through a double barrier structure with a barrier height of 1.2 V at zero bias and at an applied bias of 0.1 V. Note that the potential is no longer constant inside the barriers when a bias voltage is applied (Figs. 9.33b and c). As pointed out above, the transfer matrix method can still be applied by approximating the slowly varying potential with several constant potential steps. Notice that, in the zero bias case, the transmission coefficient reaches unity at the E_z values of 0.25 and 0.83 eV. At these electron energies resonant tunneling occurs. At a nonzero bias voltage the transmission coefficient is no longer unity even under resonant tunneling conditions.

9.5.2 I–V Characteristics of Resonant Tunneling Devices

Experimentally one does not measure directly the tunneling probability represented by $T(E_z)$. Instead, the dependence of the tunneling current on the bias voltage (the so-called I–V characteristics of a **resonant tunneling device**) is usually obtained. However, if $T(E_z)$ is known as a function of E_z, the

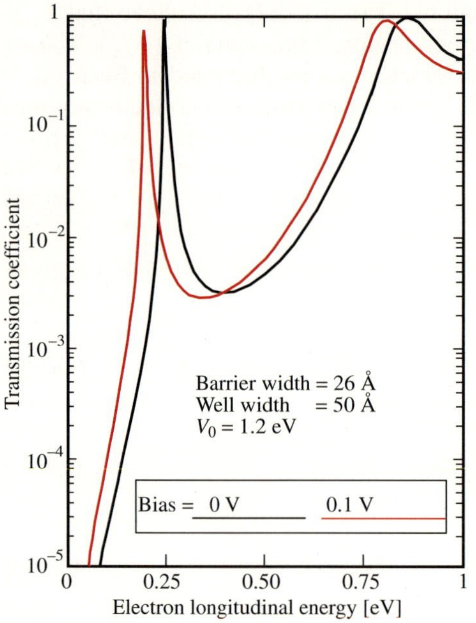

Fig. 9.34. Calculated energy dependence of the transmission coefficient of an electron through a double barrier structure at zero bias (*black curve*) and at an applied bias voltage of 0.1 eV (*red curve*). The width of the barrier and the well are 26 and 50 Å, respectively. The height of the barriers relative to the bottom of the well is 1.2 eV. [9.5]

total tunneling current I can be calculated by summing the tunneling probability over the electron distribution in the emitter using the following expression:

$$I = \frac{e}{4\pi^3\hbar} \int_0^\infty dk_x dk_y \int_0^\infty dk_z T(E_z)[f(E) - f(E')] \left(\frac{\partial E}{\partial k_z}\right), \tag{9.61}$$

where e is the absolute value of the charge of the electron, $f(E)$ the electron occupancy (given by the Fermi–Dirac distribution function under equilibrium conditions) in the emitter region, and $f(E')$ the corresponding occupancy in the collector region. The term $[f(E) - f(E')]$ ensures that the electron is tunneling from an occupied state to an empty state. Under the assumption of no scattering, the electron energy E' in the collector region is related to that in the emitter by

$$E' = E + eV_\mathrm{b}. \tag{9.62}$$

In the above discussions we have assumed the bias voltage V_b to be positive. For negative bias the role of the emitter and collector is reversed.

Figure 9.35 shows the I–V characteristics (a) and the conductance (b), dI/dV, of the resonant tunneling diode fabricated by *Sollner* et al. [9.66] and depicted in Fig. 9.33a. The parameters of this double-barrier structure are $W_1 = W_2 = W_3 = 50$ Å, $N_{\mathrm{D}1} = N_{\mathrm{D}3} = 10^{18}\,\mathrm{cm}^{-3}$ and the bulk equivalent electron concentration inside the well $N_{\mathrm{D}2} = 10^{17}\,\mathrm{cm}^{-3}$. The $Ga_{0.75}Al_{0.25}As$ barriers are supposed to be undoped and presumed to be semi-insulating due

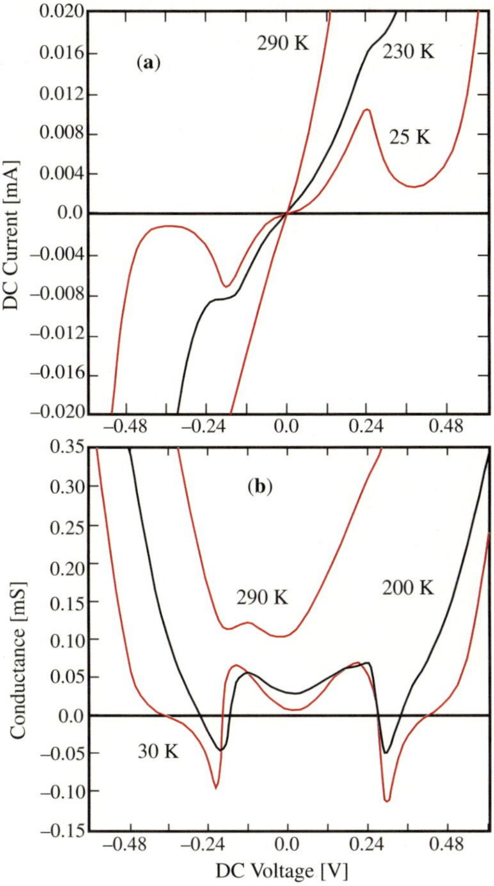

Fig. 9.35. (**a**) Current versus voltage and (**b**) conductance (dI/dV) versus voltage curves at three temperatures obtained by *Sollner* et al. [9.66] from the double barrier QW device shown in Fig. 9.33. The parameters of this double-barrier structure are $W_1 = W_2 = W_3 =$ 50 Å, $N_{D1} = N_{D3} = 10^{18}$ cm^{-3} and the bulk equivalent electron concentration inside the well $N_{D2} =$ 10^{17} cm^{-3}. The $Ga_{0.75}Al_{0.25}As$ barriers are assumed to be undoped

to compensation of background shallow donors by other defects located close to the middle of the gap. Only the 25 K curve clearly exhibits a distinct region of NDR. At room temperature a hint of a NDR region is shown by the conductance curve under reverse bias. The I–V characteristic is not completely symmetric with respect to zero bias, although it should be if there were no band bending. While the above theory gives a qualitative explanation of the experimental results in Fig. 9.35, obtaining a good quantitative agreement is much more difficult.

One experimental parameter relevant to device applications is the so-called **peak-to-valley current ratio**. This is defined as the ratio of the current at the resonant tunneling peak energy to that at the minimum (or valley) before the current starts to increase again with voltage. This ratio is about 6 for a negative bias and 4 for a positive voltage at 25 K in the device illustrated in Fig. 9.35. The magnitude of this ratio is determined by scattering of the tunneling electrons within the well by phonons, interface roughness and other defects. The importance of scattering by phonons is shown by the rapid degradation of the peak-to-valley ratio in Fig. 9.35 as the temperature is increased. Scattering by interface roughness invalidates the one-dimensional approximation discussed above. Its effect on resonant tunneling devices fabricated from GaAs/GaAlAs has recently been modeled numerically [9.70]. Much larger peak-to valley ratios have been achieved in resonant tunneling devices based on other materials. For example, Fig. 9.36 shows a device constructed out of $In_{0.53}Ga_{0.47}As$ (emitter and collector), AlAs (barriers) and InAs (well). Its peak-to-valley ratio is equal to 30 at room temperature and as large as 63 at 77 K (Fig. 9.37).

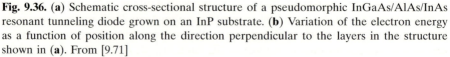

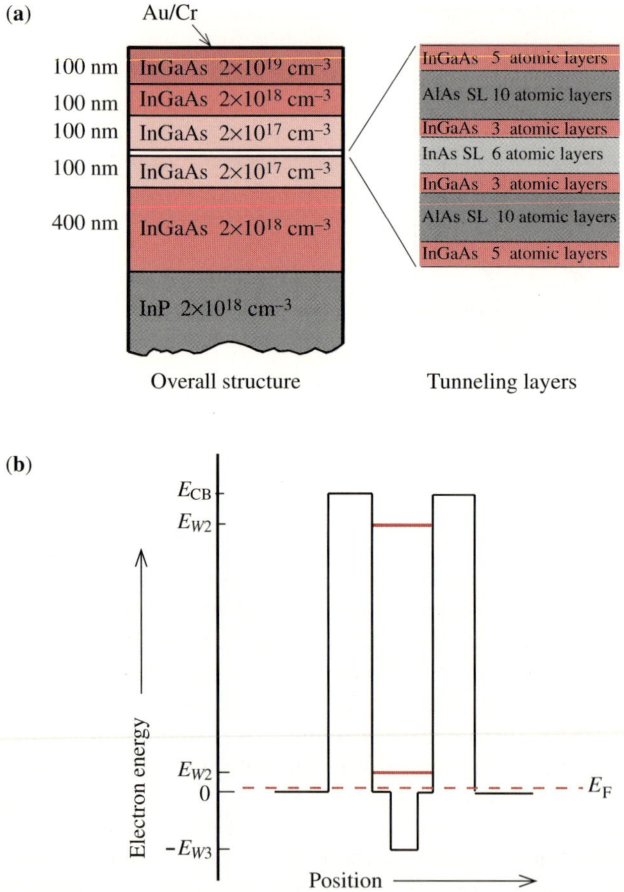

Fig. 9.36. (**a**) Schematic cross-sectional structure of a pseudomorphic InGaAs/AlAs/InAs resonant tunneling diode grown on an InP substrate. (**b**) Variation of the electron energy as a function of position along the direction perpendicular to the layers in the structure shown in (**a**). From [9.71]

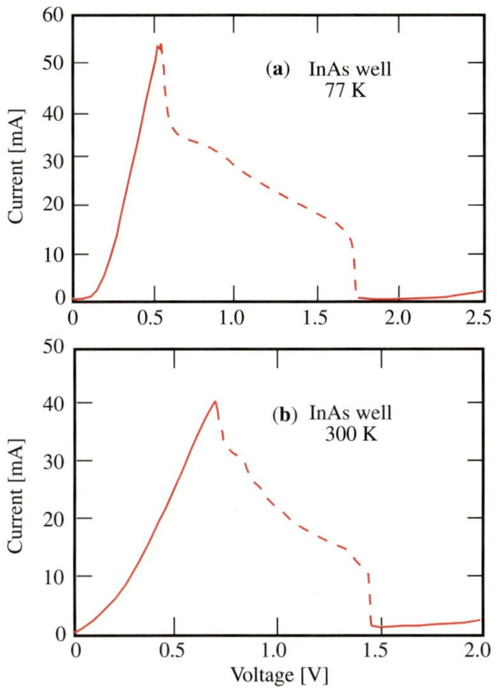

Fig. 9.37. The I–V characteristics of the 30 × 30 (μm)^2 pseudomorphic InGaAs/AlAs/InAs resonant tunneling diode shown in Fig. 9.36, measured at (**a**) 77 K and (**b**) 300 K. Notice the very large peak-to-valley current ratio in this device found even at room temperature. From [9.71]

9.6 Quantum Hall Effects in Two-Dimensional Electron Gases

Since the discovery of the *quantum Hall effect* (QHE) in a two-dimensional electron gas in a Si metal-oxide-semiconductor (MOS) structure by *von Klitzing* et al. in 1980 [9.72], the study of the properties of lower dimensional electrons in heterostructures under high magnetic fields has "mushroomed" into one of the most exciting areas of semiconductor physics. The QHE was soon joined by the discovery of what is now known as the **fractional QHE** by *Tsui* et al. in 1982 [9.73] (as a result the original QHE is referred to as the normal QHE or **integral QHE**). There are still many unanswered questions regarding the fractional QHE at the time of writing this book and the field continues to evolve at a fast pace. Hence we shall limit our discussions to the more established and better known integral QHE. In the rest of this section QHE will be understood to mean the integral QHE unless otherwise stated.

In Sect. 5.4 we discussed the transport properties of a three-dimensional free electron gas in the presence of an applied electric and magnetic field using a classical approach. In Sect. 5.5.2 we described the Hall effect and defined the Hall coefficients R_H in (5.81). We shall now consider this Hall effect again but under quite different conditions. Firstly, the electron gas is free to move only in a plane (two dimensions) perpendicular to the applied magnetic field

B. Secondly, the magnetic field and the sample quality satisfy the strong field condition which requires that $\omega_c\tau \gg 1$ (Sect. 5.5 for definitions of ω_c and τ). Under these conditions, the classical approximations of Sect. 5.5 are no longer valid. In this section we shall begin by describing quantum mechanically the properties of a three-dimensional electron gas in a magnetic field using the theory developed by *Landau* [9.74]. This will be followed by a description of the experiment of *von Klitzing* et al. [9.72]. A theoretical interpretation of these remarkable results is presented in a short note written by *Dung-Hai Lee* (Sect. 9.6.4).

9.6.1 Landau Theory of Diamagnetism in a Three-Dimensional Free Electron Gas

Let us assume that a three-dimensional free electron gas, with an isotropic effective mass m^*, is moving in a time-independent and uniform magnetic field $\boldsymbol{B}$ applied along the z direction. The magnetic field affects both the orbital motion and the spin dynamics of the electrons. For simplicity we shall neglect the interaction between the electron spin and the magnetic field since it is not relevant for the QHE (although this is not true for the fractional QHE). The Hamiltonian for an electron moving under the combined influence of external electric and magnetic fields has already been given in (6.25). For a uniform dc magnetic field (6.25) simplifies into

$$\mathcal{H} = \frac{1}{2m^*}\left(\boldsymbol{p} + \frac{e\boldsymbol{A}}{c}\right)^2, \tag{9.63}$$

where e is the magnitude of the electronic charge, $\boldsymbol{p}$ is the electron momentum operator and $\boldsymbol{A}$ is the vector potential associated with the magnetic field (for SI units in this and following equations delete c). The solution of this Hamiltonian can be found in textbooks on either quantum mechanics (e. g. [9.75]) or solid state physics (e. g. [9.77]). Here we shall just summarize the results.

Landau simplified (9.63) by introducing the **Landau gauge** in which:

$$\boldsymbol{A} = (0, Bx, 0) \tag{9.64}$$

[in (6.23) we have defined another gauge; namely, the Coulomb gauge]. One can readily show that (9.64) satisfies the requirement that $\boldsymbol{B} = \mathrm{curl}\boldsymbol{A}$. Using (9.64) the Schrödinger equation corresponding to the Hamiltonian $\mathcal{H}$ becomes

$$\left[\frac{\partial^2}{\partial x^2} + \left(\frac{\partial}{\partial y} + \frac{\mathrm{i}eB}{\hbar c}x\right)^2 + \frac{\partial^2}{\partial z^2} + \frac{2m^*E}{\hbar^2}\right]\Psi(x,y,z) = 0. \tag{9.65}$$

This equation is separable into two equations, one for the motion along the z direction and the other for that in the xy plane. The motion in the z direction is that of a free particle with energies and wavefunctions given, respectively, by

$$E_z = \frac{\hbar^2 k_z^2}{2m^*} \tag{9.66a}$$

and

$$\psi(z) = \exp(\pm \mathrm{i} k_z z). \tag{9.66b}$$

This result is consistent with the classical result that the electron motion parallel to the magnetic field is unchanged and remains free-electron-like. The equation for motion in the xy plane (i. e., in the plane perpendicular to the magnetic field) can be solved by writing the wavefunction in the form

$$\phi(x, y) = u(x)\exp(\mathrm{i}k_y y) \tag{9.67}$$

Substituting (9.66b and 67) into (9.65), the wave equation for $u(x)$ can be expressed as

$$\left(-\frac{\hbar^2}{2m^*}\right)\frac{\partial^2 u(x)}{\partial x^2} + \left(\frac{m^*}{2}\right)\left(\frac{eB}{m^* c}x - \frac{\hbar k_y}{m^*}\right)^2 u(x) = E' u(x), \tag{9.68}$$

where $E' = E - E_z$. Equation (9.68) resembles the Schrödinger equation for a one-dimensional simple harmonic oscillator with resonant frequency ω_c (i. e., the cyclotron frequency) and equilibrium position

$$x_0 = \frac{\hbar k_y}{m^* \omega_\mathrm{c}}. \tag{9.69}$$

The eigenvalues E' of (9.68) are given by the well-known expression for simple harmonic oscillators

$$E' = (n - \tfrac{1}{2})\hbar\omega_\mathrm{c} \quad \text{with } n = 1, 2, \ldots \,. \tag{9.70}$$

For reasons which will become clear later we have chosen the quantum number n to start from 1, rather than from zero as is usually done in the case of simple harmonic oscillators. These quantized energy levels are known as **Landau levels**. In the three-dimensional case the electron energy $E = E' + E_z$ forms bands (the energies are independent of k_y, i. e., one dimensional in reciprocal space) as shown in Fig. 9.38a. The corresponding density of states is shown in the same figure.

Since the electron bands in the absence of magnetic fields are three dimensional in reciprocal space and the total number of degrees of freedom of an electron is not affected by the magnetic field, each Landau level must be strongly degenerate. The degeneracy ξ (which is equal for all Landau levels) multiplied by the number of Landau levels must be equal to the number of

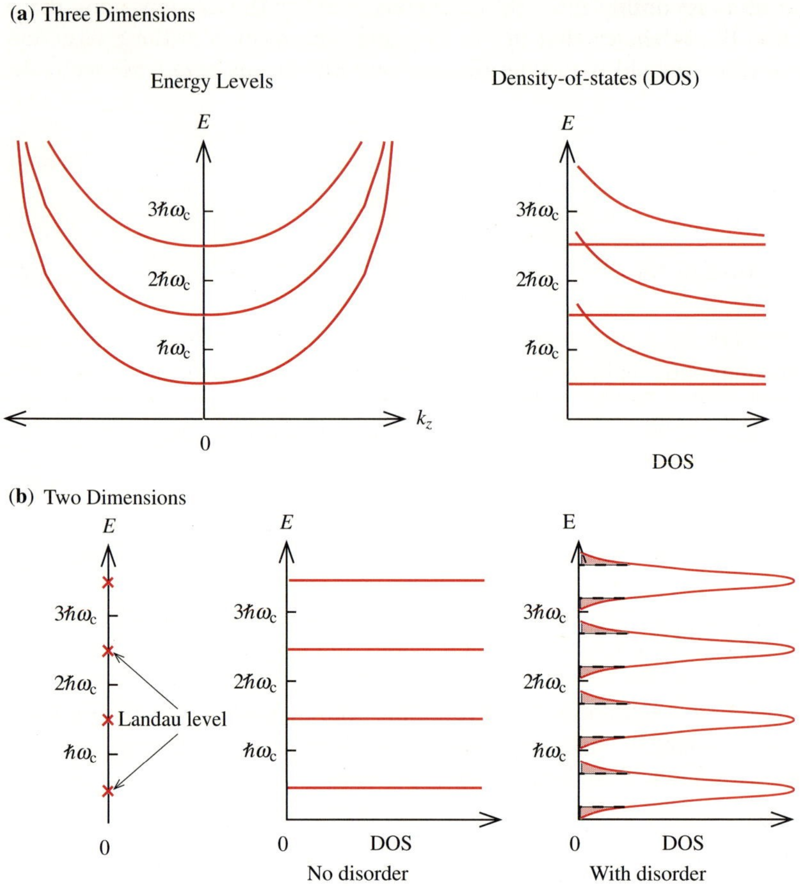

Fig. 9.38. (**a**) Schematic energy bands (or Landau levels) and the corresponding density of states (DOS) of a free electron in an external magnetic field in three dimensions. (**b**) The corresponding energy levels and DOS for a free electron confined to move in a plane perpendicular to the magnetic field. The DOS of the 2DEG is shown for cases involving no disorder and also when disorder is present. In the latter case the shaded regions denote the localized states while the mobility edges (*dashed black lines*) separate them from the delocalized ones

degrees of freedom in the absence of the magnetic field. It can be shown that ζ is equal to

$$\zeta = \frac{1}{2\pi\hbar} m^* \omega_c L_x L_y = \frac{1}{2\pi} \frac{L_x L_y}{l_B^2}, \tag{9.71}$$

where L_x and L_y are the dimensions of the sample in the x and y directions, respectively, and $l_B = [\hbar c/(eB)]^{1/2}$ ($c = 1$ for SI units) is the so-called *Landau magnetic length*. (The reader should show that this length equals the radius of the classical orbit which corresponds to the $n = 1$ Landau level).

The results just obtained can also be expressed in the following, somewhat different way [9.76]: the cyclotron orbits of electrons in the presence of a magnetic field are quantized in k-space into orbits that contain integral multiples of the area

$$\mathcal{A} = 2\pi eB/(c\hbar). \tag{9.72}$$

In summary, the net quantum mechanical effect of a uniform DC magnetic field on the motion of an electron in three dimensions can be described as the quantization of the cyclotron orbits into "simple-harmonic-oscillator-like" Landau levels. Each cyclotron orbit encloses in k-space integral multiples of the area defined by (9.72) in the plane perpendicular to the magnetic field. The electron moves like a free particle only along the direction parallel to the magnetic field. As a result, its density of states shown in Fig. 9.38a resembles that of a one-dimensional particle (Fig. 6.9) for each Landau level. The degeneracy of a Landau level n is independent of n and given by (9.71).

In the presence of a weak applied electric field in addition to the strong magnetic field, we expect the fast cyclotron motion of the electron to be unchanged in the lowest order of approximation. The main effect of the electric field is to induce a drift motion in the equilibrium position x_0 or "**guiding center**" of the cyclotron orbits. The drift motion of this guiding center referred to as the "guiding center orbit", under the combined effect of the electric and magnetic fields, can be treated by the classical approach given in Sect. 5.5. With this simplified approach we can again obtain the same magneto-conductivity tensor and Hall coefficient as in Sect. 5.5.

9.6.2 Magneto-Conductivity of a Two-Dimensional Electron Gas: Filling Factor

If the electrons are now constrained (i. e., confined) to move only in a plane perpendicular to the magnetic field, their "free-electron-like" motion parallel to the magnetic field will be suppressed. Their allowed energies become discrete and are given by E' (9.70). The corresponding density of states would be a collection of delta functions separated by energy gaps equal to $\hbar\omega_c$ as shown in Fig. 9.38b. The existence of these energy gaps between Landau levels means that when the Fermi level lies between two Landau levels this two-dimensional electron gas (2DEG) behaves like a semiconductor with a bandgap equal to $\hbar\omega_c$. At temperatures $T \ll (\hbar\omega_c/k_B)$ this "semiconductor" is insulating, i. e., the diagonal elements of the magneto-conductivity tensor $\sigma_{xx} = \sigma_{yy}$ tend to zero. It is interesting to note that when the diagonal elements of the magneto-conductivity matrix vanish, the diagonal elements of its inverse matrix (known as the **magneto-resistivity tensor** ϱ) also vanish (Problem 9.14). However, when the diagonal elements of the resistivity tensor ϱ_{xx} and ϱ_{yy} of a 2DEG are equal to zero this does not mean that the sample has become *superconducting*. If the off-diagonal elements ϱ_{xy} and ϱ_{yx} are nonzero the sample is still dissipative.

Let us now assume that $T = 0$ and the concentration N_e of the 2DEG is such that the Fermi level lies just above the Landau level with quantum number n. Since the degeneracy of each Landau level is ζ (remember that we have neglected the spin of the electron), the total number of electrons in these n Landau levels is

$$N_e = n\zeta. \tag{9.73}$$

Substituting in the expression for ζ from (9.71) we obtain

$$N_e = nBL_xL_y(e/hc). \tag{9.74}$$

The quantity hc/e ($c = 1$ for SI units) is the smallest unit of magnetic flux which can be enclosed by an electron orbit, known as the **flux quantum** (usually represented as Φ_0), while BL_xL_y is the total magnetic flux Φ passing through the 2DEG of area BL_xL_y. Thus (9.74) can be rewritten as

$$N_e = n(\Phi/\Phi_0). \tag{9.75}$$

Let $N_\Phi = \Phi/\Phi_0$ denote the total number of flux quanta enclosed by the 2DEG. It is customary to define $\nu = N_e/N_\Phi$ as the **filling factor** of the 2DEG. According to (9.75), ν equals the n of the maximum occupied Landau level when the Fermi level lies between n and $n + 1$. The reciprocal $(\nu)^{-1}$, is the average number of flux quanta enclosed by each electron in the sample. Experiments in which $\nu \geq 1$ are said to be in the integral QHE regime; otherwise ($\nu < 1$) they are in the **fractional QHE** regime. When the filling factor of a 2DEG equals an integer n, its Fermi level lies between the nth and $(n + 1)$th Landau levels (if we had labeled the lowest Landau level as having the quantum number $n = 0$ rather than $n = 1$, the filling factor would differ from n by 1) and the diagonal elements of its magneto-resistivity tensor vanish. This simple-minded picture (which assumes no disorder in the sample) suggests that the diagonal elements of the magneto-resistivity tensor of a 2DEG should vanish periodically as a function of magnetic field, a fact which has been experimentally observed by *von Klitzing* et al. [9.72]. However, their results also contain many interesting surprises, as will be described in the next section. These experimental results cannot be explained without including the effects of disorder on the electron transport in a 2DEG.

In the derivation of (9.70) we have neglected the electron spin which, in the case of free electrons in a magnetic field, induces a splitting of the Landau levels by $\Delta E = \pm\frac{1}{2}g\mu_B B$, where μ_B is the **Bohr magneton** and $g \simeq 2$, the electron g-factor. In a semiconductor ΔE is determined by an **effective *g*-factor** g^* which can differ considerably from $g \simeq 2$ ($g^* \simeq 0.4$ for electrons in GaAs). See Problem 9.16.

9.6.3 The Experiment of von Klitzing, Pepper and Dorda

The geometry of the sample used by *von Klitzing* et al. [9.72] is shown in Fig. 9.39a. It consisted of a Si **metal–oxide–semiconductor field effect transistor** (or **MOSFET**). The cross section of the sample is shown schematically in

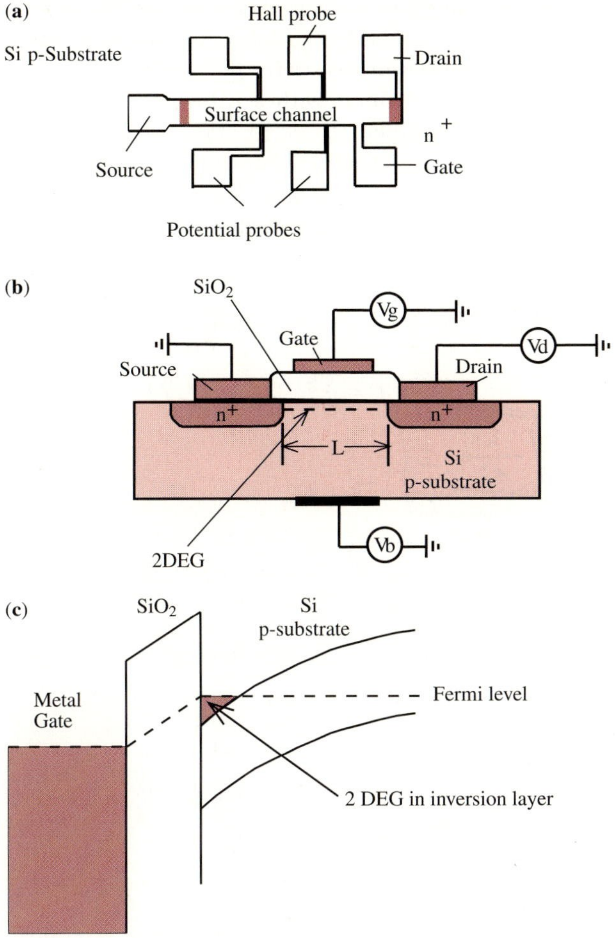

Fig. 9.39. (**a**) Top view of the MOSFET Hall "bar" used in the experiment of *von Klitzing* et al. [9.72] (**b**) Cross-sectional view of the sample in (**a**) along the surface channel showing the two-dimensional electron gas (2DEG) under the gate. (**c**) The spatial variation in electron energy across the MOSFET when the gate voltage is biased such that an inversion layer is formed at the Si substrate and the oxide interface

Fig. 9.39b. The two-dimensional electron gas which is the "heart" of the experiment is confined in an inversion layer (Chap. 8) at the interface between the silicon dioxide (SiO_2) and the p-type Si substrate. The band bending at this interface, when the substrate is biased to produce an inversion layer, is shown in Fig. 9.39c. The substrate, SiO_2 layer and top metal electrode (known as the gate) form a parallel plate capacitor. The total amount of charge on these electrodes is proportional to the gate voltage V_g. As a result, the areal charge density N_e can be varied continuously by changing V_g. In the experiment a constant DC magnetic field was applied perpendicular to the sample along the z direction. A constant current was maintained in the sample in the

x direction via an applied potential while the voltage drops across the sample in the x and y directions (denoted by U_{xx} and U_{xy}, respectively) were measured. U_{xx} is proportional to the **longitudinal resistivity** ϱ_{xx} while U_{xy} is proportional to the **transverse resistivity** ϱ_{xy} (albeit with a different constant of proportionality) and hence to the Hall coefficient, see (5.81).

The raw data obtained by *von Klitzing* et al. are reproduced in Fig. 9.40. Note that U_{xx}, U_{xy} and N_e are denoted by U_x, U_H and N respectively in this figure. We notice that U_{xx} (U_x in the figure) vanishes regularly for certain val-

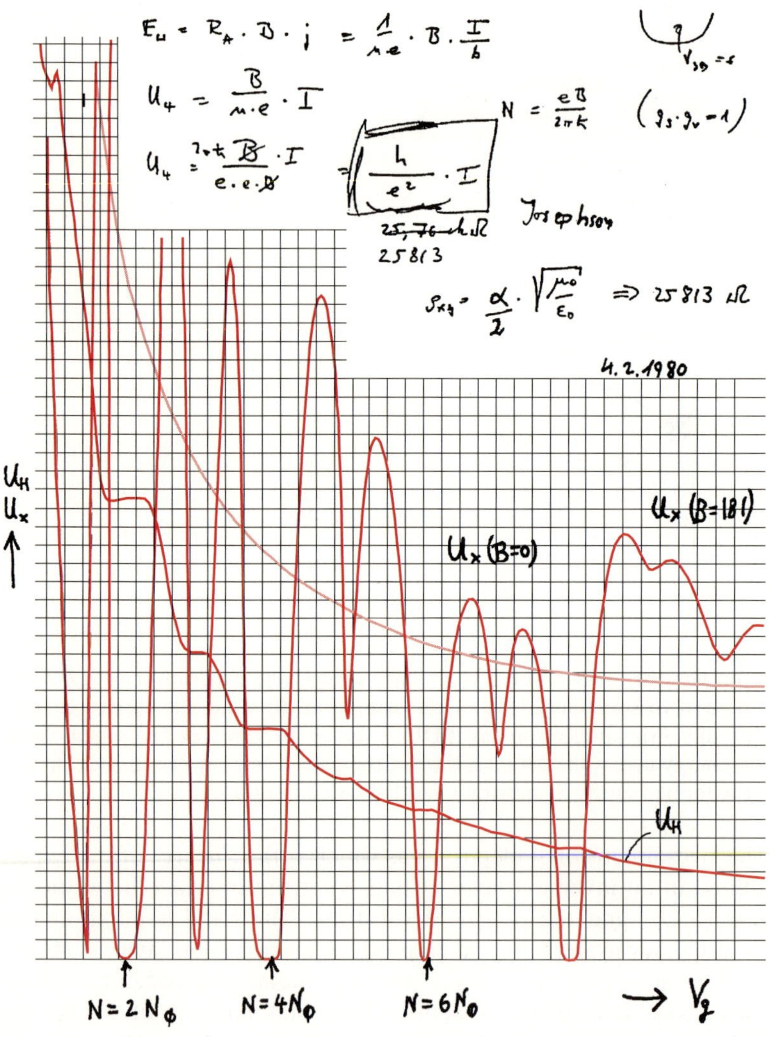

Fig. 9.40. Original curves showing QHE plateaus and notes which led to the discovery of the quantum Hall effect. The Hall voltage U_H and the voltage U_x in the current direction of a silicon MOSFET are measured as a function of the gate voltage V_g at a fixed magnetic field. The gate voltages for integer filling factors 2, 4, and 6 are marked by *arrows*

ues of the gate voltage, which can be interpreted as the result of the filling factor being equal to integral values 2, 4 and 6, in agreement with the theory in Sect. 9.6.2. The unexpected result is that U_{xy} (U_{H} in Fig. 9.40) exhibits plateaus whenever U_{xx} vanishes. Furthermore, the values of the transverse resistance (also known as the **Hall resistance**) at these plateaus equals 25.813 kΩ divided by consecutive integers. As indicated by the handwritten notes of *von Klitzing* in Fig. 9.40 this "quantum of resistance" is defined in terms of the fundamental constants h/e^2. Since 1980 the value of h/e^2 has been measured to an accuracy of 2 parts in 10^7 and is now used as a standard for resistance. The explanation of these plateaus in the Hall resistance (or **Hall plateaus**) is not obvious from the magneto-transport properties of a 2DEG discussed in the previous section. In the following section we have invited *Dung-Hai Lee* to present a simplified but physical explanation of this intriguing phenomenon.

9.6.4 Explanation of the Hall Plateaus in the Integral Quantum Hall Effect

The explanation of the integral QHE that we are going to present is based on a model of a noninteracting 2DEG. Our current understanding is that while electron–electron interaction is crucial to the fractional QHE, the integral QHE will survive even if this interaction is absent. What is indispensable for explaining the integral QHE is the presence of **disorder**. In real samples some defects and imperfections will always be present. As a result of the disorder-induced potential, the delta functions in the DOS of the 2DEG shown in Fig. 9.38b will be broadened into peaks with nonzero widths. Nonzero temperature and finite sample size also affect this broadening. The tails of these peaks correspond to tail states known to exist within the bandgap of a bulk disordered semiconductor. Electrons near the main peaks are mobile or **delocalized** and can be considered analogous to the free carriers in the bands of three-dimensional semiconducting crystals. Electrons in the tails of the peaks are immobile or **localized** and therefore do not contribute to the conductivity. The boundary between the localized and delocalized states is known as the **mobility edge**. The mobility edges and the localized states (shaded regions) associated with Landau levels of electrons in two dimensions are shown schematically in Fig. 9.38b. Mobility edges are important only for states induced within energy gaps by disorder. Since there are no energy gaps between Landau levels in a three-dimensional electron gas, disorder is not as important in three-dimensional magneto-transport as in two dimensions. Note that the concept of a mobility edge had been used earlier for amorphous (i. e., extremely disordered) semiconductors [9.77, 78].

To understand the effect of disorder on the motion of electrons in the presence of a strong magnetic field we shall assume, for simplicity, that the potentials induced by the defects V_{dis} vary slowly in space, i. e.,

$$|\nabla V_{\mathrm{dis}}| l_{\mathrm{B}} \ll \hbar\omega_{\mathrm{c}}, \qquad (9.76)$$

where l_{B} is the *magnetic length* defined below (9.71). Within this approximation, the electronic motion can be understood classically as consisting of two parts: a very fast cyclotron motion and a slow drift of the guiding center (Sect. 9.6.1) along the equipotential contours defined by the total potential V (the sum of any applied electric potential and V_{dis}). The direction of this drift is determined by that of the Lorentz force: $\nabla V \times \boldsymbol{B}$.

A simple way to convert this classical picture to a quantum mechanical one is via the Bohr–Sommerfeld quantization condition [9.76] which requires that the areas enclosed by the cyclotron orbits in $\boldsymbol{k}$-space be in multiple units of $\mathcal{A}$, see (9.72), while in real space they enclose an integral number of flux quanta. These conditions are not affected by the presence of disorder within the approximation of (9.76). For simplicity we shall assume that the electron concentration (and hence E_{F}) is fixed while the magnetic field strength is varied. Although this is different from the experimental conditions of *von Klitzing* et al. (Fig. 9.40), it has been shown that identical Hall plateaus are observed in this experimental configuration. This is not surprising considering the fact that the Hall field, defined in (5.80a), depends on the magnetic field and the inverse of the carrier concentration in a similar manner.

We shall further assume that the fluctuations in V_{dis} are less than $(1/2)\hbar\omega_{\mathrm{c}}$. Note that because of broadening in the DOS introduced by disorder the Fermi energy E_{F} may lie *anywhere between two Landau levels*, and not just midway between them as in the case of no disorder. The guiding centers are assigned the quantum number (n_0) of the Landau level *nearest* to E_{F}. Except for the special case where E_{F} lies *exactly midway* between two Landau levels, n_0 is determined unambiguously. At $T = 0$, potential contours with $E < E_{\mathrm{F}}$ contain guiding center orbits which are occupied by electrons while those with $E > E_{\mathrm{F}}$ contain empty orbits. Thus, near a minimum in the potential V all equipotential contours with energy value smaller than E_{F} will be occupied. In the literature, a region filled with occupied guiding center orbits is known as a **quantum Hall droplet**. Electrons that reside on the outermost orbits are known as **edge-electrons**. Since these electrons are located at the Fermi level they are the only ones significant for charge transport. The effect of increasing B is to increase $\hbar\omega_{\mathrm{c}}$ and decrease l_{B} hence shrinking the size of the quantum Hall droplets. In Fig. 9.41a we depict the edge electrons (indicated by arrows) and the corresponding Hall droplets (shaded region) in the case that the magnetic field is strong enough for the Fermi level to lie in the localized region below the first Landau level, as shown in the DOS diagram in Fig. 9.41a. When the field is decreased sufficiently for E_{F} to lie at energies immediately above the delocalized regime, as shown in the DOS plot in Fig. 9.41c, the size of the Hall droplets has grown to the point where they fill most of the sample, leaving only small pockets of the sample (**hole droplets**) containing no occupied orbits.

The schematic pictures of the edge electron orbits in Fig. 9.41 allow us to understand the appearance of the Hall plateaus. For large B, the quantum Hall droplets in Fig. 9.41a are small and well separated from each other in the valleys of the disorder potentials. Hence the orbits of the edge-electrons do not

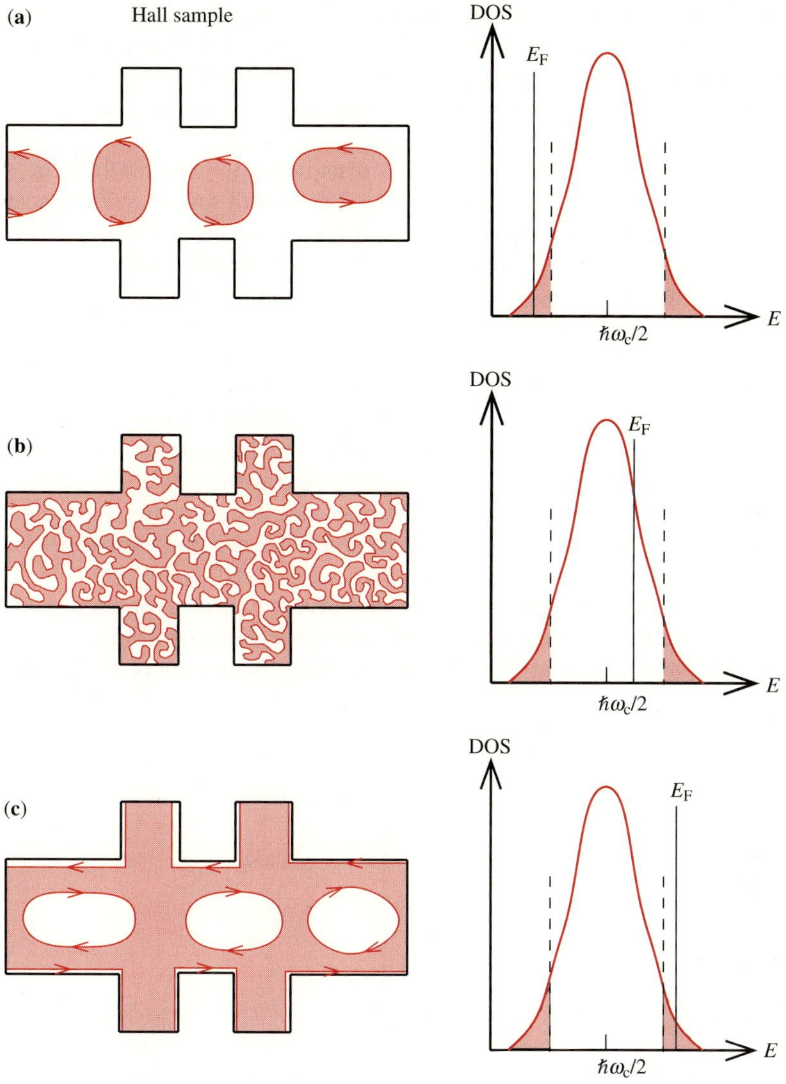

Fig. 9.41a–c. Schematic picture of the quantum Hall droplet (*shaded areas*), orbit of the edge electrons (*curve with arrows*) and density of states (DOS) for three positions of the Fermi energy E_F relative to the mobility edge. (**a**) E_F below the mobility edge in the localized region (*shaded region* in the DOS plots); (**b**) E_F in the delocalized regime and (**c**) E_F in the localized region again but now above the delocalized states. The DOS are representative of a finite size sample at $T > 0$

overlap, these electrons are localized and $\sigma_{xx} = \sigma_{xy} = 0$. As B is decreased the quantum Hall droplets expand and eventually some of the droplets will overlap. However, σ_{xx} and σ_{xy} remain zero as long as the number of droplets that overlap is too small for a conducting path to form between the electrodes (the latter situation is called *percolation*). Finally, when B is decreased sufficiently that a large number of droplets have merged, the electrons can percolate from one electrode to the other. At the same time, the outermost guiding center orbit migrates to the physical boundary of the sample. This is shown in Fig. 9.41c. When this happens the longitudinal resistivity ϱ_{xx} (and similarly σ_{xx}) becomes zero while σ_{xy} takes the value e^2/h. As a matter of fact, $\sigma_{xy}/(e^2/h)$ acts as a counter of the number of edge channels in this percolation process. Thus the transverse conductivity σ_{xy} switches from the $\nu = 0$ plateau where $\sigma_{xy} = 0$ to the $\nu = 1$ plateau where $\sigma_{xy} = e^2/h$ as shown in Fig. 9.42. The regions labeled a and c in this figure correspond to the situations shown in Figs. 9.41a and c, respectively. The corresponding behaviour in σ_{xy} (or U_{H}) as a function of electron density N_{e} (or V_{g} in the experiment of *von Klitzing* et al.) is shown in Fig. 9.42b. The regions labeled a and c in this figure correspond to those in Fig. 9.42a. Repeating the above process for the $\nu = z$ Landau level explains the Hall plateaus associated with higher integral values of ν.

The transition region between two Hall plateaus is quite interesting. At the edge of the $\nu = 0$ plateau, when σ_{xx} is about to become nonzero, the largest electron droplet size diverges. On the other hand, near the $\nu = 1$ plateau, when σ_{xx} is about to become zero, the largest "hole" droplet size diverges. We have tried to show schematically the edge electron orbits in Fig. 9.41b. For very large samples at $T = 0$ this transition is extremely sharp and

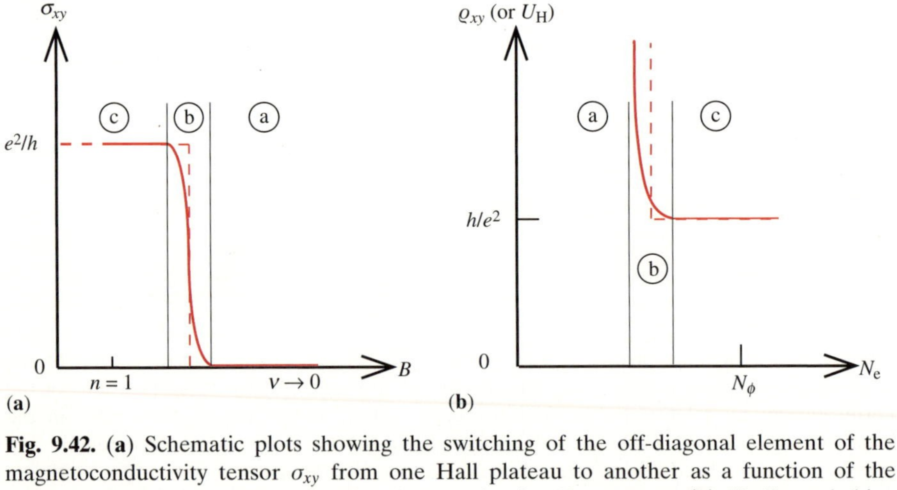

Fig. 9.42. (**a**) Schematic plots showing the switching of the off-diagonal element of the magnetoconductivity tensor σ_{xy} from one Hall plateau to another as a function of the magnetic field (B) while keeping the electron density N_{e} constant. (**b**) Similar switching of the Hall resistance ϱ_{xy} as a function of the electron density (N_{e}) (or the gate voltage V_{g} as in experiments using MOSFETs) under constant magnetic field. The regions labeled a, b and c in these figures are supposed to correspond to the situations depicted in Fig. 9.41a, b and c, respectively. The *broken curves* correspond to $T = 0$ and an infinitely large two-dimensional sample while the *solid curves* are valid for the more realistic situation involving a finite sample size and a temperature $T > 0$

should be treated as a **phase transition**. The boundary between the Hall electron and hole droplets should be **fractal** and cannot be represented by the simple curves shown in Fig. 9.41b. The corresponding transitions between the Hall plateaus should also be infinitely sharp, as shown by the broken curves in Figs. 9.42a and b. However, due to finite sample size and nonzero temperature effects these transitions are broadened (region labeled b in Figs. 9.42a and b) into the solid curves in these figures. The preceding percolation picture de-emphasized the importance of quantum tunneling in the process of transition between plateaus. It turns out that ignoring these tunneling processes does not jeopardize our understanding of the existence of the Hall plateaus, although it does influence our ability to predict the precise nature of the transition. Moreover, according to our present understanding, the details of these transition regions may also be affected by electron–electron interaction. A discussion of this topic is beyond the simple, heuristic approach adopted here.

The above picture constitutes our basic paradigm for explaining the Hall plateaus in a two-dimensional electron gas. We emphasize that this picture ignores electron–electron interaction. Since this interaction is responsible for the plateaus in the fractional quantum Hall regime, our model cannot account for the plateaus observed for *fractional* filling factors or the transitions between them. At the time of writing of this book, there is still no detailed theory of the critical nature of the transitions between the fractional Hall plateaus. One way to understand these fracitonal Hall plateaus is presumably to imagine a new kind of fractional quantum Hall droplets which also possess edge states. The fractional QHE would then be attributed to the emergence of these edge states at the boundary of the sample as in the case of the integral QHE [9.79, 80][1].

9.7 Concluding Remarks

The discussion of the QHE is an appropriate point to conclude this book. Its discovery illustrates how new, unexpected and exciting physics can arise in the field of semiconductors. Whenever new semiconducting materials become available, whether these new materials are purer or belong to a new family or have an artificial structure, they have often led to the discovery of new phenomena and also novel applications. The fabrication of synthetic layered structures, such as quantum wells and superlattices discussed in this chapter, is an excellent example. While these structures were originally proposed for device applications, they turned out to have impact far beyond anyone's imagination. They have become the driving force for many new developments outside semiconductor physics and devices, in materials science, surface physics, molecular physics, and chemistry. Considering the fact that many new methods of growing and fabricating semiconductors are being explored and developed at major

[1] For the discovery [9.81] and the development of the theory [9.82] of the fractional QHE Störmer, Tsui and Laughlin were awarded the Physics Nobel Prize in 1998.

laboratories around the world, it is safe to predict that semiconductor physics has not yet reached saturation but is still growing at a healthy rate. Hopefully this book will serve as a link between the past, centered on the bulk properties of semiconductors, and the future in which various nanostructures, such as quantum wires and quantum dots, are bound to play an increasingly important role.

PROBLEMS

9.1 *Particle in a One-Dimensional Square Well Potential*

With the help of a personal computer and a conputational program such as *Mathcad*[2] or *Mathematica*[3], calculate:

a) The energies of the bound states of a particle with isotropic mass m^* equal to 0.045 times the free electron mass (to emulate the effective mass of a conduction band electron in a $Ga_{0.47}In_{0.53}As$ alloy) confined in a one-dimensional square well potential of height 0.5 V. Assume that the mass of the particle in the barrier is the same as in the well. Perform the calculation for well widths equal to 50, 100 and 150 Å and compare your results with those shown in Fig. 9.7 based on a more rigorous calculation for $GaAs/Ga_{0.7}Al_{0.3}As$ QW by Bastard and Brum [9.20].

b) The probabilities of finding the particle inside the well for these bound states.

c) Repeat the calculation for a 50 Å well but now assuming that the particle mass in the barrier is equal to only one–half of its value inside the well. Can you predict whether the binding energies of the bound states are now larger or smaller than before based on (9.13) alone?

9.2 *Particle in a One-dimensional Periodic Square Well (or Kronig–Penney) Potential*

With the help of a personal computer and a computational program such as *Mathcard* or *Mathematica*, calculate the eigenvalues of a one-dimensional periodic square well potential of height 0.4 eV as a function of the wavevector k for a few discrete values of k between 0 and π/a, where $a = 30$ Å is the width of both the well and the barrier. Plot the band structure of the minibands and compare the energies of the band extrema and the widths you obtain with those given in Fig. 9.9.

9.3 *Transverse Phonons in Tetrahedral Structures*

Discuss why (9.22) is valid for the longitudinal but not for the transverse phonons of a Si_2Ge_2 superlattice grown along [100]. What modifications are

---

[2] Mathcad is a trademark of MathSoft, Inc.

[3] Mathematica is a trademark of Wolfram Research, Inc.

needed in order to obtain the corresponding equations for the transverse modes?

Hint: examine a stick-and-ball model of the GaAs structure.

9.4 *Longitudinal Phonons in Superlattices*
Derive (9.23 and 24) by substituting into (9.22) displacement patterns either odd or even with respect to the B-B midpoint of Fig. 9.14.

Show that for $k = \pi/d$ the displacement patterns are also either odd or even with respect to the B-B midpoint but, contrary to the $k = 0$ case, a pattern odd (even) with respect to the B-B midpoint is even (odd) with respect to the A-A midpoint. Derive the following expression for the four eigenfrequencies at $k = \pi/d$:

$$\omega^2 = f\frac{(3m_\mathrm{B} + m_\mathrm{A}) \pm \sqrt{(3m_\mathrm{B} + m_\mathrm{A})^2 - 8m_\mathrm{A}m_\mathrm{B}}}{2m_\mathrm{A}m_\mathrm{B}}$$

and a similar expression obtained by permuting m_A and m_B.

9.5 *Splittings of Acoustic Modes in Superlattices*
Using (9.27) find an expression for the splitting of the folded acoustic phonon bands at the center and edge of the mini-BZ as a function of ε.

9.6 *Electrostatic Potential in Interface Modes*
Plot the potential $\phi(x)$ versus x for the interface modes which correspond to the QW described in the caption of Fig. 9.19 for several values of k_x. Discuss the dependence on k_x. Show that the corresponding field $\boldsymbol{E}$ is neither longitudinal nor transverse with respect to the (in-plane) propagation direction.

9.7 *Angular Dispersion of Interface Modes*
a) Plot the dispersion relations $\omega(q_x)$ obtained from (9.39a) for GaAs/AlAs MQWs using as the abscissa the reduced variable $\tilde{q} = dq_x/\pi$ instead of q_x and $d_\mathrm{A}/d_\mathrm{B}$ as a parameter. Discuss the dependence of these dispersion relations on $d_\mathrm{A}/d_\mathrm{B}$.

b) Repeat part (a) but for Ge/GaAs MQWs. [Remember that Ge has no infrared active phonons and hence its $\varepsilon(\omega)$ is not dispersive in the infrared.]

9.8 *Selection Rules for Scattering by Folded Acoustic Modes*
Show that while at $\tilde{q} = 0$ only one component of a folded acoustic doublet should be observed in Raman scattering, for $\tilde{q} \neq 0$ both components should appear. What happens for $\tilde{q} = 1$? (Fig. 9.24).

9.9 *Effects of Thickness Fluctuations on Resonant Light Scattering*
In Sect. 9.4.1 we showed that the scattering by acoustic phonons in an inhomogeneously broadened MQW displays mainly a broad, luminescence-like background (Fig. 9.25).

a) Discuss why a similar phenomenon does not appear in the case of scattering by optic phonons.

b) Derive an expression for the fitting curve in Fig. 9.26 as a function of the inhomogeneous gap fluctuations. Show that the width of this curve (≈ 12 meV) is approximately four times the inhomogeneous width g.

9.10 *Point Group and Selection Rules for [001] Superlattices*
Find the irreducible representations to which LO and TO phonons belong in the case of GaAs/AlAs superlattices grown along [001] for both even and odd m's. The point group is in this case D_{2d} and its character table is

	E	$2S_4^z$	C_2^z	$2C_2$	$2\sigma_d$
A_1	1	1	1	1	1
A_2	1	1	1	−1	−1
B_1	1	−1	1	1	−1
B_2	1	−1	1	−1	1
E	2	0	−2	0	0

Show that the LO modes for even m have the Raman tensor given in (9.50).

Hint: See [9.83].

9.11 *Point Groups of AB/CD Zincblende-type Superlattices*
Find the point group to which a GaAs/AlSb superlattice grown along [001] belongs. Note that there are three possible inequivalent structures of this type depending on whether: the interfaces between GaAs and AlSb layers are (a) GaSb layers, (b) AlAs layers, (c) alternately AlAs and GaSb layers.

9.12 *Selection Rules for Backscattering by LO-Phonons*
Derive the selection rules for backscattering by LO phonons for $\hat{\boldsymbol{e}}_{\mathrm{L}}$ and $\hat{\boldsymbol{e}}_{\mathrm{S}}$ polarized (parallel and crossed) along [110] and [$1\bar{1}0$] in a GaAs/AlAs MQW grown along [001].

9.13 *Dependence of Electron–Phonon Interaction of Quantum Well Width*
a) Explain qualitatively the following features found in the theoretical curves of Fig. 9.32, which show the dependence of the strength of the interaction between electrons in quantum wells with various interface and confined optical phonon modes:

1) There is very little difference between the curves labeled W_1 and W_3 which were calculated, respectively, using the "dielectric continuum model" and the "Huang-Zhu model". On the other hand, the curve labeled W_2, which has been calculated with the "mechanical model", corresponds to much smaller scattering rates than both the W_1 and W_3 curves.

2) The theoretical scattering rates in all the three curves labeled W_1–W_3 for confined optical phonon modes *decrease* as the quantum well width is decreased. On the other hand, the scattering rate for the interface phonon mode (curve labeled W_{in}^{+}) increases as the quantum well width is decreased.

b) Based on the deformation potential type of electron–phonon interaction and its dependence on phonon wave vector in bulk semiconductors, explain qualitatively how you expect the deformation potential electron–phonon interaction to depend on quantum well width for

1) confined transverse optical phonons and
2) folded longitudinal acoustic phonons.

9.14 *Tunneling Through a One-Dimensional Barrier*

Use the transfer matrix formalism in (9.58, 59) and a computer to calculate the transmission coefficient of an electron tunneling through the one-dimensional barrier in Fig. 9.34 as a function of its energy. Do this first for the case of zero bias and then for the case of a 0.1 V bias. In the latter case you have to approximate the biased potential by a series of square potentials. Start by choosing a relatively small number of square wells (say four) and find out how accurately you can reproduce the result in Fig. 9.34. Then double the number of wells and see how closely your results have converged to those in Fig. 9.34.

9.15 *Magnetoconductivity Tensor and Magnetoresistivity Tensor of a 2DEG*

Show that the magnetoconductivity tensor of a two-dimensional electron gas (2DEG) in the *xy*-plane with a magnetic field perpendicular to that plane is given by

$$\sigma = \begin{pmatrix} 0 & \sigma_{xy} \\ \sigma_{xy}^{*} & 0 \end{pmatrix}$$

when the filling factors of the Landau levels are integers. Show that the corresponding magnetoresistivity tensor is given by

$$\varrho = \begin{pmatrix} 0 & 1/\sigma_{xy}^{*} \\ 1/\sigma_{xy} & 0 \end{pmatrix}.$$

9.16 *Effective g-Factor of Conduction Band Electrons in Zinc-Blende-Type Semiconductors*

In studying the quantum Hall effect in Sect. 9.6 we have completely neglected the electron spin. In principle, an applied magnetic field will act on the magnetic moment of the electron. In the case of electrons in atoms their magnetic moment has two origins: one due to the orbital motion of the electron and the other due to its spin. This interaction between the applied magnetic field and the magnetic moment of the electron results in splittings of the electron energy levels, an effect known as the Zeeman effect. For electrons in semiconductors like GaAs the magnetic-field-induced Zeeman splitting of the electron energy level can be greatly reduced as result of spin–orbit coupling. This

problem asks you to show how a conduction electron will appear to have an **effective g-factor** whose magnitude can be larger or smaller than that of a free electron ($g = 2$). Before attempting this problem the reader should consult the Appendix: *Appendix 9.1 Effective g-Factor (g^*) of Electrons in Semiconductors* to be found at the Web Site:
http://Pauline.Berkeley.edu/textbook/Appendices.

Assume that a zinc-blende-type semiconductor can be modeled by three bands around its band gap at $k = 0$. One of them is a spin-degenerate conduction band of Γ_6 symmetry. The other two are valence bands of symmetries: Γ_8 (corresponding to the $J = 3/2$ bands) and Γ_7 (corresponding to the $J = 1/2$ split-off band). The band gap energy (i.e., the separation between the conduction band and the Γ_8 valence band) will be denoted by E_0 while the spin–orbit splitting between the two valence bands is Δ_0 (see Table 6.2).

a) Using the effective mass approximation, show that the conduction electron behaves under an applied magnetic field as if its g-factor had the effective value:

$$g^* = 2 - \frac{4P^2\Delta_0}{3mE_0(E_0 + \Delta_0)}$$

The matrix element P^2 is defined in (2.42).

b) Use (2.43) to express g^* in terms of the conduction effective mass m_c^* instead of P^2. Look up the values of E_0, Δ_0 and m_c^* for GaAs, GaN, InSb and ZnSe in Table 2.2. From these values calculate the theoretical values of g^* and compare them with the experimental values:

Semiconductor	Experimental Value of g^*
GaAs	0.44
GaN	1.98
InSb	-51
ZnSe	1.15

For more recent work on the g^*-factor in semiconductors, see: M. Willatzen, M. Cardona, N.E. Christensen: Spin–orbit coupling parameters and electron g factor of II-VI zinc-blende materials. Phys. Rev. B**51**, 17992–17994 (1995).

SUMMARY

In this chapter we studied the effect of quantum confinement on electrons and phonons in semiconductors in synthetic layered structures, known as quantum wells and superlattices, that are usually fabricated with the technique of molecular beam epitaxy. Due to limited space, we have considered mainly the most studied systems composed of lattice-matched GaAs, AlAs and their alloys. However, this system is versatile enough to demonstrate much of the physics involved, such as formation of electronic subbands and minibands, the confinement of optical phonons, folding of acoustic phonons and the introduction of interface modes. We also illustrated the effect of confinement on the transport properties of carriers in these materials by studying the phenomena of resonant tunneling and the integral quantum Hall effect. The fractional quantum Hall effect has become one of the most exciting areas of current research.

Appendix: Pioneers of Semiconductor Physics Remember...

Semiconductor physics has a long and distinguished history. The early developments culminated in the invention of the transistor by Bardeen, Shockley, and Brattain in 1948. More recent work led to the discovery of the laser diode by three groups independently in 1962. Many prominent physicists have contributed to this fertile and exciting field. In the following short contributions some of the pioneers have recaptured the historic moments that have helped to shape semiconductor physics as we know it today. They are (in alphabetical order):

Elias Burstein
Emeritus Mary Amanda Wood Professor of Physics,
University of Pennsylvania, Philadelphia, PA, USA.
Editor-in-chief of Solid State Communications 1969–1992;
John Price Wetherill Medal, Franklin Institute 1979;
Frank Isakson Prize, American Physical Society, 1986.

Marvin Cohen
Professor of Physics, University of California, Berkeley, CA, USA.
Oliver Buckley Prize, American Physical Society, 1979;
Julius Edgar Lilienfeld Prize, American Physical Society, 1994.

Leo Esaki
President, Tsukuba University, Tsukuba, Japan.
Nobel Prize in Physics, 1973.

Eugene Haller
Professor of Materials Science and Mineral Engineering,
University of California, Berkeley, CA, USA.
Alexander von Humboldt Senior Scientist Award, 1986.
Max Planck Research Award, 1994.

Conyers Herring
Professor of Applied Physics, Stanford University, Stanford, CA, USA.
Oliver Buckley Prize, American Physical Society, 1959;
Wolf Prize in Physics, 1985.

Charles Kittel
Emeritus Professor of Physics, University of California, Berkeley, CA, USA.
Oliver Buckley Prize, American Physical Society, 1957;
Oersted Medal, American Association of Physics Teachers, 1978.

Neville Smith
Scientific Program Head, Advanced Light Source,
Lawrence Berkeley Laboratory, Berkeley, CA, USA.
C.J. Davisson and L.H. Germer Prize, American Physical Society, 1991.

Jan Tauc
Emeritus Professor of Physics and Engineering, Brown University,
Providence, RI, USA.
Alexander von Humboldt Senior Scientist Award, 1981;
Frank Isakson Prize, American Physical Society, 1982.

Klaus von Klitzing
Director, Max-Planck-Institut für Festkörperforschung, Stuttgart, Germany.
Nobel Prize in Physics, 1985.

Ultra-Pure Germanium: From Applied to Basic Research or an Old Semiconductor Offering New Opportunities

Eugene E. Haller
University of California, Berkeley, USA

Imagine arriving one morning at the laboratory and somebody comes to ask you if single crystals of germanium with a doping impurity concentration in the 10^{10}–10^{11} cm^3 range can be grown! You quickly compare this concentration with the number of Ge atoms per cm^{-3}, which is close to 4×10^{22}. Well, you pause and wonder how anybody can ask if a 99.999999999% pure substance can be made. The purest chemicals available are typically 6 or 7 nines pure. Robert N. Hall of the General Electric Company proposed in 1968 [1] that such crystals could be grown and that they would be most useful in fabricating very large volume (up to 400 cm^3) p-i-n junctions working as gamma-ray detectors [2].

When I arrived at Berkeley as a postdoc I joined the group of F.S. (Fred) Goulding, who headed one of the leading groups of semiconductor detector and electronics experts at the Lawrence Berkeley Laboratory (LBL), then called the Radiation Laboratory. There I met W.L. (Bill) Hansen, who had started the race towards the ultra-pure Ge single-crystal goal believed to be attainable by Hall. Bill was extremely knowledgeable in chemistry, physics, and general laboratory techniques. In addition, he was the fastest-working experimentalist I had ever encountered. Somewhat overwhelmed, I started to work with Bill and Fred on these Ge crystals. When Bill tried out various Czochralski crystal growth configurations [3], he rigorously pursued ultra-purity by using the simplest crystal growth design, the purest synthetic silica (SiO_2) container for the Ge melt, and hydrogen gas purified in a Pd diffusion system. I, on the other hand, tried to build up an arsenal of characterization techniques which would allow us to find out within hours the purity and crystalline perfection we had achieved. The IEEE meetings on nuclear science, which were held every fall, provided the forum where we "crossed swords" with Hall [4–7]. It was a close race. Hall had the advantage of enormous experience, which started way back when Ge was first purified and single crystals were grown for transistors. We had the advantage of blissful ignorance but also excellent and helpful colleagues. Furthermore, nobody could match Bill's agility in trying out new purification and crystal growth methods. One major development for us was learning, through Hall, about a super-sensitive photoconductivity technique which was capable of identifying extremely small numbers of impurities in Ge single crystals. The technique had been discovered by Russian scientists at the Institute of Radio-engineering and Electronics in Moscow [8, 6.85]; see Figs. 6.39 and 6.40. They found that a two-step ionization process of shallow hydrogenic donors or acceptors in a very cold crystal would lead to photoconductivity peaks which were very sharp and unique for each dopant species. Paul Richards, of the Physics Department at the University of California at Berkeley, had a home-built Fourier-transform far-infrared spectrometer and the necessary liquid helium temperature dewar. By the end of the first day of experimenting we had a spectrum of a p-type high-purity Ge crystal with only 10^{10} cm^{-3} net amount of acceptors and we knew also that phosphorus and aluminum were the major residual impurities.

In parallel with a number of novel and interesting physics studies we fabricated gamma-ray detectors at LBL. We broke records in the resolution of the gamma-ray photopeaks with our ultra-pure crystals [2]. Soon the commercial detector manufacturers became interested and started their own ultra-pure Ge crystal-pulling programs. In a few years several companies in the US and in Europe succeeded in developing large-diameter ($\simeq$ 8 cm) single crystals with incredibly good yield, excellent purity ($< 2 \times 10^{10}$ cm^{-3}) and very small concentrations (10^8 cm^{-3}) of deep-level defects which would detrimentally affect the charge collection in large-size coaxial p-i-n diodes. In order to achieve the best spectral resolution, electrons and holes had to have mean-free-paths of up to several meters. Most semiconductor physicists simply shook their heads and could not comprehend these numbers.

How pure is ultra-pure Ge? The person who cares only about electrically active impurities would say that crystals with a few 10^{10} cm^{-3} of impurities are routinely grown. But are there other inactive impurities? Yes, of course there are. Hydrogen, oxygen, silicon and carbon are usually present at concentrations of up to 10^{14} cm^{-3}, depending on the crystal growth conditions. These impurities do not interfere with Ge's operation as radiation detectors provided certain rules are followed: no heating to temperatures above 350°C and no rapid temperature changes. Can we reduce the concentration of these four electrically inactive impurities? Yes, we can, but we pay a price. Eliminating hydrogen by growing in vacuum leads to the introduction of impurities which can no longer be "flushed" out of the crystal puller. Furthermore, hydrogen will passivate the very small concentrations of deep-level defects and impurities which are always present. Free oxygen and silicon are generated by the reduction of the ultra-pure silica crucible by the liquid Ge. We do not know of any substance which can replace silica with, perhaps, the exception of graphite. Numerous attempts to grow ultra-pure Ge in graphite crucibles have failed so far because the resultant crystals contain too many Al acceptors.

Most recently, the interest in Ge has sharply increased because isotopically pure Ge can be obtained from Russia. Isotopically pure Ge bulk crystals [9–12] and isotope superlattices [13] have been grown. New phonon physics and electronic transport studies are currently being pursued by several groups with these isotopically controlled crystals and multilayers.

Have we arrived at the ultimately ideal material: isotopically and chemically pure and crystallographically perfect Ge single crystals? Perhaps the answer is no, but I certainly do not know of another parameter that can be controlled.

References

1 R.N. Hall: in *Proc. of the 12th Int. Conf. on Physics of Semiconductors*, ed. by M.H. Pilkuhn (Teubner, Stuttgart 1974), p. 363
2 E.E. Haller, F.S. Goulding: *Handbook on Semiconductors*, Vol. 4, ed. by C. Hilsum (Elsevier, New York 1993), Chap. 11, p. 937–963
3 W.L. Hansen, E.E. Haller: Mater. Res. Soc. Proc. **16**, 1 (1983)
4 R.N. Hall, T.J. Soltys: IEEE Trans. Nucl. Sci. **NS-18**, 160 (1971)
5 E.E. Haller, W.L. Hansen, F.S. Goulding: IEEE Trans. Nucl. Sci. **NS-20**, 481 (1973)
6 E.E. Haller, W.L. Hansen, G.S. Hubbard, F.S. Goulding: IEEE Trans. Nucl. Sci. **NS-23**, 81 (1976)
7 E.E. Haller, W.L. Hansen, F.S. Goulding: Adv. Phys. **30**, 93 (1981)
8 E.E. Haller: Physics **146B**, 201 (1987)
9 E.E. Haller: Semicond. Sci. Technol. **5**, 319 (1990)
10 E.E. Haller: Solid State Phenom. **32–33**, 11 (1993)
11 G. Davies, J. Hartung, V. Ozhogin, K. Itoh, W.L. Hansen, E.E. Haller: Semicond. Sci. Technol. **8**, 127 (1993)
12 H.D. Fuchs, P. Etchegoin, M. Cardona, K. Itoh, E.E. Haller: Phys. Rev. Lett. **70**, 1715 (1993)
13 J. Spitzer, T. Ruf, M. Cardona, W. Dondl, R. Schorer, G. Abstreiter, E.E. Haller: Phys. Rev. Lett. **72**, 1565 (1994)

Two Pseudopotential Methods: Empirical and Ab Initio

Marvin L. Cohen
University of California, Berkeley, USA

It took a relatively long time to develop methods capable of determining the detailed electronic structure of solids. In contrast, for gases, unraveling the mysteries of atomic energy levels went hand in hand with the development of quantum theory. Atomic optical spectra yielded sharp lines that could be interpreted in terms of excitations of electrons from occupied to empty states. These studies provided important tests of the theory. However, compared to atomic spectra, solid-state spectra are broad, since the interactions between the atoms spread the allowed occupied and empty energy levels into energy bands. This made interpretation of spectra in terms of electronic transitions very difficult. Trustable precise electronic energy band structures were needed to interpret solid-state spectra, but these were difficult to obtain.

In principle, the Schrödinger equation can describe the behavior of electrons in solids; but without approximations, solutions for the electronic energy levels and wavefunctions are extremely difficult to calculate. Despite considerable effort, the situation around 1960 was still unsatisfactory. Creative models of solids had been introduced to explain many physical phenomena such as electronic heat capacities and superconductivity with spectacular success. However, calculations capable of yielding band structures and other properties for specific materials were not available.

An important intermediate step was the introduction of the empirical pseudopotential model (EPM). Pseudopotentials had been around since 1934, when Fermi introduced the concept to examine the energy levels of alkali atoms. Since he was interested in highly excited atoms, he ignored the oscillations of the valence electron wavefunctions in the regions near the nucleus. By assuming a smooth wavefunction responding to a weak potential or pseudopotential, Fermi could easily solve for the outer electron energy levels.

Since most solid-state effects, such as bonding, are principally influenced by the changes in the outermost electrons, this picture is appropriate. For the EPM it is assumed that the solid is composed of a periodic array of positive cores. Each core has a nucleus and core electrons. Each of the outer valence electrons moves in the electrostatic potential or pseudopotential produced by the cores and by the other valence electrons. In this one-electron model, each electron is assumed to respond to this average periodic crystalline pseudopotential. The periodicity allows Fourier decomposition of the potential and the EPM fits data to obtain Fourier coefficients. Usually only three coefficients per atom are needed.

The EPM stimulated interactions between theorists and experimentalists and the result was one of the most active collaborations in physics. Not only were optical and photoemission spectra of solids deciphered, the activities resulted in new experimental techniques and a much deeper understanding of the behavior of electrons in solids. The meeting ground between experiment and theory is usually response functions such as dielectric functions or reflectivity. In the early phases of this work the actual energy band structures, which are plots of energy versus wavevector, were the domain of theorists. However, the introduction of angular resolved photoemission spectroscopy (ARPES) gave energy bands directly and provided further tests of the EPM.

The EPM band structures obtained in the 1960s and 1970s are still used today. In addition, the EPM produced the first plots of electronic charge density for crystals. These plots displayed covalent and ionic bonds and hence gave considerable structural information. Optical constants, densities of states, and many other crystal properties were obtained with great precision using EPM-derived energy levels and wavefunctions.

Despite the success of the EPM, there was still considerable motivation to move to a first-principles or *ab initio* model. The approach chosen was similar to Fermi's. Instead of an EPM potential, the interaction of the valence electron with the core was described using an *ab initio* pseudopotential constructed from a knowledge of atomic wavefunctions. The valence electron–electron interactions were modeled using a density functional theory which, with approximations, allows the development of an electron–electron potential using the electronic charge density. However, the latter approach is appropriate only for calculating ground-state properties. Excited states such as those needed to interpret atomic spectra require adjustments to this theory. These adjustments are complex and require significant computer time compared to the EPM, but they are successful in reproducing the experimental data and the approach is completely *ab initio*.

One of the most important applications of the *ab initio* pseudopotential model was the determination of structural properties. It became possible to explain pressure-induced solid–solid structural transitions and even to predict new structural phases of solids at high pressure using only atomic numbers and atomic masses. Bulk moduli, electron–phonon coupling constants, phonon spectra, and a host of solid-state properties were calculated. The results allowed microscopic explanations of properties and predictions. An example was the successful prediction that semiconducting silicon would become a superconducting hexagonal metal at high pressure.

The two types of pseudopotential approaches, empirical and *ab initio*, have played a central role in our conceptual picture of many materials. Often the resulting model is referred to as the "standard model" of solids. Unlike the standard model of particle physics, which is sometimes called a theory of everything, the standard model of solids is most appropriate for those solids with reasonably itinerant electrons. Despite this restriction, the model is extremely useful and a triumph of quantum theory.

The Early Stages of Band-Structures Physics and Its Struggles for a Place in the Sun

Conyers Herring
Stanford University, Stanford, USA

It is universally recognized today that among the components necessary for a theoretical understanding of the properties of semiconductors, their specific electronic band structures have an extremely fundamental place. Textbooks on semiconductors typically have, among their earliest chapters, one on band structure, which contains diagrams of energy versus wavevector for important semiconductors, usually obtained from first-principles numerical calculations. But obviously these calculations would not be so conspicuously featured if they did not agree with a great body of experimental information. What the present-day student may not realize is that, despite the spurt of activity in the early post-transistor years – roughly 1948–1953 – the workers of this period had almost no knowledge of band structures, and had to muddle through as best they could without it. The evolution of this aspect of semiconductor physics provides a thought-provoking perspective on how science moves toward truth by erratic diffusional steps, rather than with military precision.

The possible range of band structures had, of course, long been known in principle. The standard generalities about Bloch waves and their energy spectra had been known for a couple of decades; symmetry-induced degeneracies had been classified; early band-structure calculations, though not quantitatively reliable, had suggested that degenerate and multi-valley band edges might often occur. The trouble lay elsewhere. When so many possibilities for exciting work were opening up, people tended to avoid projects that would be tedious and time-consuming. Band-structure theorists, equipped only with mechanical calculators, often opted to use incomplete boundary conditions or limited basis sets. Experimentalists, despite rapid improvements in purity and perfection of materials, continued to focus mostly on properties whose interpretation did not depend critically on anisotropies and other special features of the energy bands. Much of the blame for this neglect must be cast on the theorists, not only for their failure to agree on calculated band structures, but also because, for too long, they shied away from the tedium of making detailed calculations of properties such as magnetoresistance for various kinds of nonsimple band structures.

My own experience provides a typical example. In December 1953 I delivered an invited paper at an APS meeting with the title "Correlation of Electronic Band Structures with Properties of Silicon and Germanium". In it I tried to reason as logically as possible from the existing experimental and theoretical literature, to draw plausible conclusions about the possible band-edge symmetries for these elements. While I got a few things right, it was distressing to learn over the next year or so that most of my inferences were wrong. How did I go astray?

My first step, safe enough, was to classify the possible types of band-edge points: those at wavevector $\boldsymbol{k} = 0$, and those at $\boldsymbol{k} \neq 0$ (multi-valley); for each of these the states could be degenerate (two or more states of the same energy and $\boldsymbol{k}$) or nondegenerate. In surveying the experimental and theoretical evidence bearing on the choices among these numerous alternatives, I began by trying to limit the possible choices to those that could occur for band structures qualitatively similar to that newly calculated by Herman [1] for diamond, which seemed more reliable than any others that had been made for any material with this crystal structure. Using the "$\boldsymbol{k} \cdot \boldsymbol{p}$ method" for qualitative estimations of the energy-band curvatures on moving away from $\boldsymbol{k} = 0$, this meant that I neglected perturbations of the p-like $\boldsymbol{k} = 0$ states $\Gamma_{25'}$, Γ_{15} by the anti-bonding s-like level $\Gamma_{2'}$, which is quite high in diamond but, contrary to my assumption, much lower in silicon and germanium. This neglect turned out to make me omit the possibility of conduction-band edges on the [111] axes in $\boldsymbol{k}$-space for n-germanium, and to retain the possibility of valence-band edges on the [100] axes for p-silicon.

From this flawed start I tried to narrow the possibilities further by appealing to experimental evidence, and especially to magnetoresistance. The near-vanishing of longitudinal magnetoresistance in [100]-type directions was obviously consistent with multi-valley band-edge regions centered on the [100]-type axes in $\boldsymbol{k}$-space, and this proved to be the correct identification for n-type silicon. But, lacking explicit calculations, I assumed that the energy surfaces of a degenerate hole band at $\boldsymbol{k} = 0$ would be so strongly warped as to preclude the near-zero [100] longitudinal magnetoresistance observed for p-silicon. So my predictions were all wrong here. Finally, I had the tedious task of calculating the complete anisotropy of magnetoresistance for multi-valley models, which a few months later were shown to give strong evidence for [111]-type valleys for n-germanium.

What all this illustrates is that to achieve an acceptable understanding of band structures, each of three types of information sources had to reach a certain minimum level of sophistication. Band calculations from first principles had to be made with accuracy and self-consistency in an adequately large function space. Experimental measurements of properties sensitive to band structure had to be made under well-controlled conditions. And theoretical predictions of these properties for different band structure models had to be available. There were gaps in all three of these sources up to the end of 1953; it is thus not surprising that Shockley, in writing what was intended as a basic text for the coming semiconductor age [2], stated, in spite of his awareness of the diversity of possible band structures, that the theoretical reasoning in the book would all be based on the simple model with an isotropic effective mass. Remarkably, in a year or so starting in 1954, each of the three sources filled itself in sufficiently so that they could pull together (e. g., better theoretical bands [3], cyclotron resonance [4], magnetoresistance theory [5]) and band-structure physics became a solid and accepted component of basic knowledge.

References

1 F. Herman: Phys. Rev. **88**, 1210 (1952)

2 W. Shockley: *Electrons and Holes in Semiconductors* (Van Nostrand, New York 1950), esp. p. 174

3 For an early review of progress 1953–1955, see, for example, F. Herman: Proc. IRE **43**, 1703 (1955)

4 See the following contribution by C. Kittel

5 B. Abeles, S. Meiboom: Phys. Rev. **95**, 31 (1954); M. Shibuya, Phys. Rev. **95**, 1385 (1954)

Cyclotron Resonance and Structure of Conduction and Valence Band Edges in Silicon and Germanium

Charles Kittel
University of California, Berkeley, USA

A prime objective of the Berkeley solid-state physics group (consisting of Arthur Kip and myself) from 1951 to 1953 was to observe and understand cyclotron resonance in semiconductors. The practical problems were to gain reliable access to liquid helium, and to obtain an adequate magnet and sufficiently pure crystals of Ge and Si. The liquid helium was obtained from the Shell Laboratories and later from the Giauque laboratory on campus. The magnet was part of a very early cyclotron (from what one may call the Ernest O. Lawrence collection), and the dc current for the magnet came from recycled US Navy submarine batteries. The semiconductor crystals were supplied by the Sylvania and Westinghouse Research Laboratories, and later by the Bell Telephone Laboratories. I think the microwave gear came from war surplus at MIT Radiation Laboratory. Evidently, very little of the equipment was purchased.

The original experiments were on Ge [1], both n-type and p-type. There were too few carriers from thermal ionization at 4 K to give detectable signals, but the carriers that were present were accelerated by the microwave electric field in the cavity up to energies sufficient to produce an avalanche of carriers by impact ionization. This was true cyclotron resonance! A good question is, why not work at liquid hydrogen temperature, where the thermal ionization would be adequate? Hydrogen was then, and perhaps is still now, considered to be too hazardous (explosive) to handle in a building occupied by students.

A better question is, why not work at liquid nitrogen temperature, where there are lots of carriers and the carrier mobilities are known to be much higher than at the lower temperatures? Cyclotron resonance at liquid nitrogen temperature had been tried at several other laboratories without success. The reason for the failures is that the plasma frequencies, being mixed with the cyclotron frequencies to produce a magnetoplasma frequency, are too high at the higher carrier concentrations – you are not measuring a cyclotron resonance but instead a magnetoplasma resonance [2]. Indeed, one can follow the plasma displacement of the original cyclotron lines when the cavity is allowed to warm up. In radio wave propagation in the ionosphere this effect is called magneto-ionic reflection, a subject I had learnt from the lectures of E.V. Appleton at Cambridge.

A better way to produce carriers at 4 K was suggested by the MIT group. They irradiated the crystal with weak light sufficient to excite both electrons and holes. With this method both electrons and holes could be excited in the same crystal. Alternatively, one can excite a known carrier type by infrared irradiation of n- or p-type material. By modulating the optical excitation the detection of the absorption signal was made highly sensitive [3]. In addition, if there is any doubt about the sign of the carriers, circularly polarized microwaves can be (and were) used to distinguish the sense of rotation of the carriers in the magnetic field.

The most surprising result of the original experiments was the observation of two effective masses (m^*) for the Ge holes: $m^*/m_0 = 0.04$ and 0.3, both approximately isotropic. Frank Herman and Joseph Callaway had calculated that the top of the valence band in Ge occurs at the center of thr Brillouin zone and is threefold degenerate (sixfold with spin), corresponding to p bonding orbitals on the Ge atoms. This would have given rise to three hole masses. We suggested [4,5] that the spin–orbit (s.o.) interaction splits the p orbitals into fourfold degenerate (related to $p_{3/2}$ orbitals) and twofold degenerate (related to $p_{1/2}$ orbitals) bands at the zone-center. We found that the most general form of the energy of the upper valence bands in the diamond structure to second order in wavevector $\boldsymbol{k}$ is (2.62)

$$E(\boldsymbol{k}) = Ak^2 \pm [B^2k^4 + C^2(k_x^2k_y^2 + k_y^2k_z^2 + k_z^2k_x^2)]^{1/2}.$$

This was perhaps the first application of the spin–orbit interaction in semiconductors.

The "s.o. split-off" or lower band in Ge is 0.30 eV below the top of the valence band edge. This s.o. splitting and the lower band itself are explored best by optical absorption. The analysis by Kahn [6] of the available experiments was an important confirmation of our model developed from cyclotron resonance.

One of the early applications of the results of cyclotron resonance experiments in Si and Ge was to the theory of the ionization energies of the shallow donor and acceptor states in these materials. The approximate ionization energies are 0.04 eV for electrons and 0.05 eV for holes in Si, and 0.01 eV for both electrons and holes in Ge. The near equality of the ionization energies for both electrons and holes was astonishing, at the time, because their band edge structures were known to be completely different (thanks to cyclotron resonance). The problem was discussed in the summer of 1954 with visitors to Berkeley, notably Freeman Dyson and Joaquin Luttinger. The near equality turns out to be merely a matter of coincidence after the electron and hole ionization energies are calculated separately.

The donor ionization energy was calculated first at Berkeley [7]. We used the hamiltonian for an ellipsoidal energy surface at any of the degenerate band edges for electrons and the dielectric constant of the bulk crystal. The calculated energies are in good agreement with experiment, at least for donors with atomic numbers close to that of the host crystal. For heavier donors, central cell corrections must be made. The acceptor problem is more difficult because of the fourfold degeneracy of the valence band edges at the zone center, and is reviewed by Walter Kohn [8], with satisfying results.

References

1 G. Dresselhaus, A.F. Kip, C. Kittel: Phys. Rev. **92**, 827 (1953)
2 G. Dresselhaus, A.F. Kip, C. Kittel: Phys. Rev. **100**, 618 (1955)
3 G. Dresselhaus, A.F. Kip, C. Kittel: Phys. Rev. **98**, 368 (1955)
4 G. Dresselhaus, A.F. Kip, C. Kittel: Phys. Rev. **95**, 568 (1954)
5 R.J. Elliot: Phys. Rev. **96**, 266 (1954)
6 A.H. Kahn: Phys. Rev. **97**, 1647 (1955)
7 C. Kittel, A.H. Mitchell: Phys. Rev. **96**, 1488 (1954)
8 W. Kohn, in *Advances in Solid State Physics*, Vol. 5, ed. by F. Seitz, D. Turnbull (Academic, New York 1957), p. 257–320

Optical Properties of Amorphous Semiconductors and Solar Cells

Jan Tauc
Brown University, Providence, USA

In the early 1960s the foundations for an understanding of the optical properties of crystalline semiconductors were established. They were based on the existence of long-range order and ***k***-vector conservation, which led to sharp structures in the spectra associated with the Van Hove singularities. My group, working in the Institute of Solid State Physics of the Czechoslovak Academy of Sciences in Prague, was making contributions to this ongoing effort which flourished throughout the 1960s. While on leave at Harvard in 1961–1962, I started thinking about what the optical properties should be like when long-range order is absent, and I began working on this problem after my return to Prague.

There is a huge group of materials, called glasses, that lack long-range order; they are produced by quenching the melt, which of course does not have long-range order. In these materials the liquid has the same short-range order as the solid phase. This is not the case for semiconductors with tetrahedral bonding. The efficiency of this bond in lowering energy depends on the geometrical rigidity of the structure; once it is loosened at high temperature, the energetically favorable phase is a metallic one (some kind of close packing). So even if it were possible to quench, say, liquid Ge (which it is not), the short-range order of this "glass" would be completely different from crystalline Ge, and therefore a comparison of the optical properties would be meaningless. There are, however, ways to prepare amorphous Ge (a-Ge) (and other tetrahedral semiconductors) with the same short-range order as crystalline Ge (c-Ge) as a thin film, for example by condensing evaporated Ge on a cold substrate.

In Prague, we first worked on the optical properties of some liquids, which clearly demonstrated the fact that the main optical properties depend on the short-range order. A breakthrough came when we learned that Radu Grigorovici had prepared thin films of a-Ge at the Institute of Physics of the Romanian Academy of Sciences. A close, fruitful and friendly collaboration soon developed, and for some years Radu regularly visited Prague. We spent long hours and days discussing amorphous semiconductors with tetrahedral bonding. At that time, we did not know of anyone else who would be interested in amorphous semiconductors of this kind (there was an important group in Leningrad which had been studying amorphous chalcogenide semiconductors since the 1950s, but these are real glasses and very different from tetrahedral semiconductors). Radu was interested in the preparation, structure and electronic transport, while we in Prague worked on the optical properties.

From the reflection spectra, using Kramers-Kronig analysis, we determined the optical constants of a-Ge in the spectral range up to 12 eV and confirmed the expectation that there should be no sharp structures [1]. Instead of the three prominent peaks in the fundamental absorption band observed in c-Ge, there is just one band, which has the remarkable feature of having much larger absorption in the low energy region (a "red shift"). From the transmission spectra we determined the dependence of the absorption coefficient α on photon energy in the absorption edge region. The data gave a straight line when $\sqrt{\omega\alpha}$ was plotted as a function of photon energy $\hbar\omega[\sqrt{\omega\alpha} = \text{const.} \times (\hbar\omega - E_g)]$. This plot defines an energy E_g, which it is natural to call the optical gap. Of course, it was the most obvious plot to try: if the k-vector is not conserved, if the density of electron states close to the valence and conduction band extrema is proportional to the square root of energy as in the crystal, and if the matrix element is a constant then $\alpha \propto (\hbar\omega - E_g)^2/\omega$, as is the case for phonon-assisted indirect transitions in crystalline semiconductors. In fact, in amorphous semiconductors there was no rigorous theoretical justification for this law at that time (and there is no generally accepted one today), so it must be considered as empirical. It is, however, most amazing that this plot works in many amorphous semiconductors. In the literature, this kind of edge is sometimes referred to as a "Tauc edge" and used as a definition of the "optical" gap, which is usually somewhat different from the gap determined from electrical conductivity measurements ("electrical gap").

The "red shift" mentioned above is observed also in a-Si and is the basis for the usefulness of this material for solar cell. Although Radu and I, during our walks in Prague (which was run down at that time but still beautiful), considered various possible applications of these materials, the truth is that they are useless as electronic materials because they are full of defects which act as traps, preventing n- and p-type doping. A prominent defect is a Si atom with only three neighbors, i. e., with an unpaired electron (a "dangling bond"). Our walks ended in 1968 after the tragic political events which put an end to what has since become known as the "Prague Spring".

In the 1970s the oil crisis hit the world, and thinking about renewable energy sources became popular. Among these, solar cells appeared very attractive. Cells made of c-Si are very good but too expensive for large scale deployment. The reason is that c-Si is an indirect-gap semiconductor and the absorption coefficient is small in the spectral region of the solar flux. To absorb it, the cell must be relatively thick ($\sim$ 100 μm), which requires a large amount of a rather expensive material, in addition to the expensive technology (crystal growing, wafer cutting, polishing, etc.). Because of the red shift, a-Si absorbs solar light much more efficiently: the cells can be made much thinner, and thin film technology is much cheaper.

A discovery dramatically improved the electronic properties of a-Si. It started with the work of Chittick and coworkers at Standard Telecommunications Laboratories in England in the late 1960s. A standard procedure for the crystal growth of a silicon layer on a Si substrate is the decomposition of SiH_4 gas by the high temperature of the substrate. Instead, Chittick et al. [2]

studied the decomposition of SiH_4 by radio frequency glow discharge, which enabled them to deposit silicon on a cool noncrystalline substrate. They produced thin amorphous Si films whose electronic properties were radically improved through a reduced defect state density in the gap. They showed that the resistivity of these films could be lowered by two orders of magnitude by adding PH_3 gas to SiH_4 – the first demonstration of doping of an amorphous semiconductor. Their company did not let them continue the work. What is quite amazing is that Chittick told many of us about this work in 1969 and no one grasped the enormous significance of his result except Spear and Le Comber at the University of Dundee. They reported in 1975 [3] n- and p-doping and in 1976 production of p-n junctions. It was first believed that the good properties were due to an exceptionally gentle deposition technique, but the work of W. Paul and others showed that they were due to the presence of hydrogen in the films. Hydrogen in a-Si:H reduces the defect state density by compensating the dangling bonds.

The gap of a-Si:H (about 1.8 eV) is larger than that of a-Si (1.4 eV) but the spectrum is also red-shifted with respect to c-Si, and therefore the films can be quite thin (1 μm) and still absorb a substantial part of the solar spectrum. One would think that with all these clues in hand someone would go ahead and design an a-Si:H solar cell. It did happen, but not in this way. Carlson and Wronski [4] discovered such cells independently at RCA in thin-film solar cells made of polycrystalline Si. They observed that when the substrate was cold enough the cells had a better efficiency and found that these better cells were amorphous rather than polycrystalline; only then did they realize the connection of their discovery to the current research on a-Si:H. These cells are today produced for small-scale applications and still remain a primary candidate for large-scale photovoltaic energy conversion plants which may be needed someday.

References

1 J. Tauc, A. Abraham, L. Pajasova, R. Grigorovici,A. Vancu: *Non-Crystalline Solids* (North-Holland, Amsterdam 1965), p. 606
2 R.C. Chittick, J.H. Alexander, H.F. Sterlin: J. Electrochem. Soc. **116**, 77 (1969)
3 W.E. Spear, P.G. Le Comber: Solid State Commun. **17**, 1193 (1975)
4 D.E. Carlson, C.R. Wronski: Appl. Phys. Lett. **28**, 671 (1976)

Optical Spectroscopy of Shallow Impurity Centers

Elias Burstein
University of Pennsylvania, Philadelphia, USA

In the fall of 1948, Frank Isakson, head of the Physics Section of the Office of Naval Research, was a frequent visitor at the Naval Research Laboratory, where I was a member of the Crystal Branch. During one of our frequent discussions of projects of mutual interest, he informed me about the Navy's interest in developing an infrared (IR) photoconductor with a response beyond 7 μm, the long wavelength limit of PbS films, an intrinsic photoconductor developed in Germany during World War II. The properties of the III–V semiconductors were still unknown at that time. In the summer of 1949 I had the good fortune of being able to attend the annual Modern Physics Symposium at the University of Michigan, one of a series of symposia that started in 1928. The lecturers that summer were Luis Alvarez (High Energy Physics), Richard Feynman (Path Integral Method), Frederick Seitz (Solid State Physics) and Gordon B.B. Sutherland (Infrared Spectroscopy of Solids).

In his lectures on semiconductors, Seitz discussed the nature of the impurity levels in Si and Ge and summarized the thermal ionization energies of group III acceptors and group V donors that had been obtained by Pearson and Bardeen at Bell Telephone Laboratories [1] from data on the temperature dependence of the carrier densities derived from resistivity and Hall measurements. He also discussed their conclusions that the ionization energies of the group III acceptors (0.048 eV) and group V donors (0.045 eV) were in reasonable agreement with a simple effective-mass hydrogen model. It was at that point in the lecture that the idea came to me to make use of the photoionization of un-ionized hydrogenic impurity centers in Si and Ge as the basis for IR detectors.

Shortly after returning to Washington, DC, I went to see John Bardeen, who provided me with several Si samples. Together with John J. Oberly, James W. Davisson and Bertha Henvis, I started measurements of the low temperature IR absorption spectra of the Si samples. I wanted to study the absorption spectra associated with photoionization of un-ionized impurity centers before making an effort to observe the photoconductive response. Our first measurements, using a Perkin-Elmer model 12C spectrometer with interchangeable NaCl, KBr, KSR-5(TlBr+I) prisms and mirror optics, were carried out at 77 K, since a simple calculation based on the thermal ionization energy of impurities indicated that over 90% of the impurity centers would remain un-ionized at this temperature.

The observed spectrum for a boron-doped Si sample was quite striking, since it exhibited peaks corresponding to transitions from the ground state to excited states of the acceptor centers, as well as the onset of a photoionization continuum [2,3] (Fig. 1). Moreover, the positions of the excitation peaks corresponded closely to the 1*s*–2*p*, 1*s*–3*p* and 1*s*–4*p* transitions of a hydrogenlike center and yielded an ionization energy of 0.046 eV, in good agreement with the thermal ionization data [4]. However, the oscillator strengths of the absorption peaks are markedly different from those for a hydrogenic center. In particular, the oscillator strength of the 1*s*–2*p* peaks is an order of magnitude smaller than that for a hydrogenic center. The widths of the excitation peaks decrease on cooling to liquid helium temperature, but there is no appreciable shift in the peak positions, indicating that Franck–Condon effects are small. Our data showed no obvious evidence of transitions from the ground state of

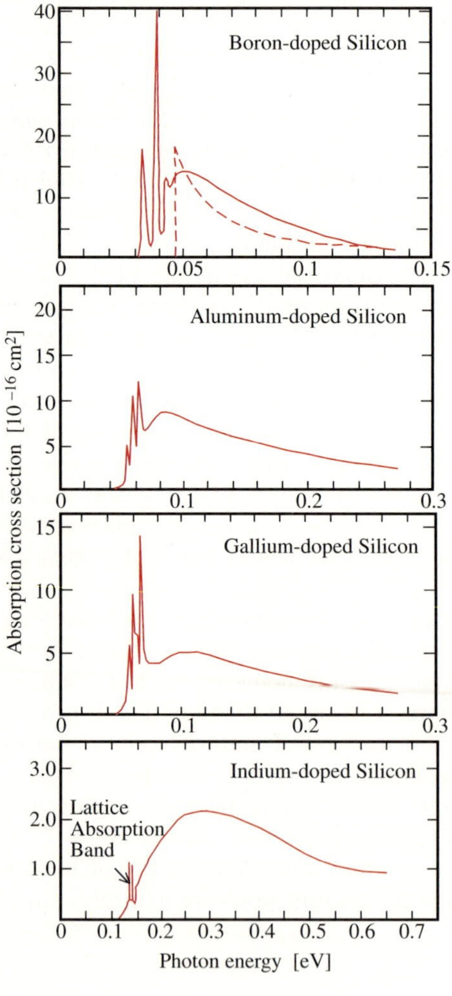

Fig. 1. Photoexcitation and photoionization absorption spectra of group III acceptors in Si at liquid helium temperature [6]. For boron-doped Si, the *dashed line* is the theoretical photoionization absorption spectrum of the corresponding hydrogenic model

the split-off valence band, which is not unexpected since the spin–orbit interaction is small in Si and the optical spectra are broadened appreciably.

Efforts to detect photoconductivity in n- and p-type Si at 77 K were unsuccessful, due to the presence of large numbers of thermally excited carriers. However, a photoconductive response was observed at liquid helium temperature. The spectral response of relatively pure n-Si is shown in Fig. 2 [5]. The dips in the photoconductive response between 8 and 24 μm correspond to lattice vibration absorption peaks. The data yielded a donor optical ionization energy of 0.04 eV. Photoconductivity studies were later carried out at liquid helium temperature on Ge doped with group III and V impurities. The photoconductive response was found to extend out to 38 μm, the limit of measurement at that time [6].

Fairly complete optical studies were carried out for the group III acceptors (B, Al, Ga and In) and for the group V donors [7,8]. Absorption spectra for the group III centers are shown in Fig. 1. The variations in the ionization energy (Fig. 3) are accompanied by changes in the character of the excitation and photoionization absorption spectra. The positions of the excitation bands for Al, Ga and In centers, unlike B, do not correspond to a hydrogenic model, their oscillator strengths also differ appreciably from those of a hydrogenic model (Fig. 3). These deviations, which become more pronounced on going from B to In, are due to central cell corrections. The states with *s* character have their energies and wavefunctions rather strongly modified, since their wavefunctions are relatively large at the impurity atoms. The states with *p* characters, whose wavefunctions are small at the center of the impurity atom, are affected to a lesser degree. The agreement between the experimental ionization energy for B and the predictions of the hydrogenic model is probably due to a cancellation of different effects.

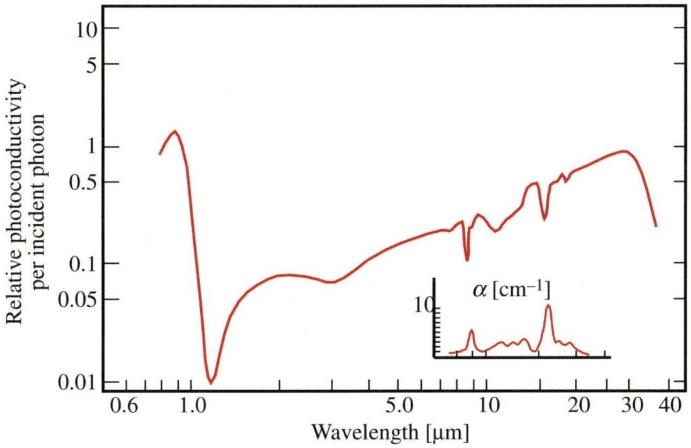

Fig. 2. The relative photoconductive response per incident photon of a relatively pure n-Si sample [4]. The dips in the photoconductive response between 8 and 24 μm correspond to the peaks in the optical absorption due to lattice vibrations

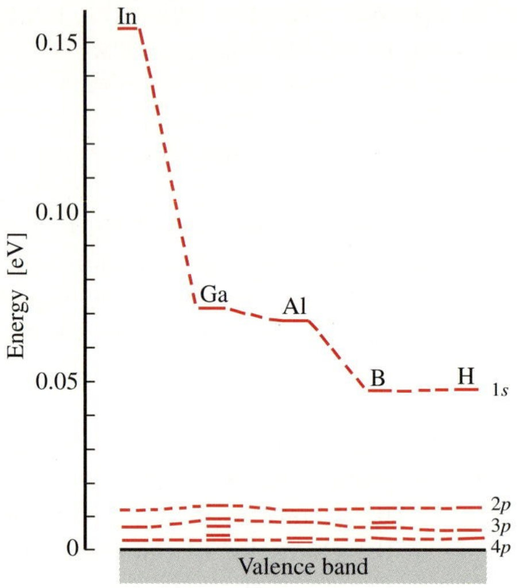

Fig. 3. Suggested term scheme for group III acceptors in Si showing the levels whose energies are derived from the low temperature absorption spectra [6]

The optical ionization energies for donors in Si were found to be 10% larger than the thermal ionization energies, due in part to the presence of low-lying excited states that were not taken into account in the calculation of the activation energies. The positions of the ground state relative to the conduction band are appreciably different for the three donors P, As and Sb, again due to central cell effects. However, their excited p-states are observed at about the same positions relative to the conduction band (see Fig. 3 and [7]). Moreover, they are in good agreement with the results of the effective mass formulation of the donor p levels by Kohn and Luttinger [9], which takes into account the fact that the conduction band of Si has six nondegenerate minima along the [100] and equivalent directions.

The data obtained in these early investigations were limited by the relatively low quality of the Si Samples, the poor resolution of the spectrometers and by the electronics. There has been major progress in the IR spectroscopy of shallow impurity levels in semiconductors since then, made possible by significant improvements in crystal quality, spectrometers and detectors, by the development of photothermal conductance spectroscopy, which has much higher sensitivity than IR detectors, and by the availability of tunable IR lasers [10].

References

1 J. Bardeen, G.L. Pearson: Phys. Rev. **75**, 865 (1949)
2 E. Burstein, J.J. Oberly, J.W. Davisson, B.W. Henvis: Phys. Rev. **82**, 764 (1951)
3 E. Burstein, E.E. Bell, J.W. Davisson, M. Lax: J. Phys. Chem. **57**, 849 (1953)

4 F.J. Morin, J.P. Maita, R.G. Schulman, N.B. Hannay: Phys. Rev. **97**, 833 (1954)
5 E. Burstein, J.J. Oberly, J.W. Davisson, Phys. Rev. **89**, 331 (1953)
6 E. Burstein, J.W. Davisson, E.E. Bell, W.J. Turner, H.G. Lipson: Phys. Rev. **93**, 65 (1954)
7 E. Burstein, G. Picus, B.W. Henvis, R.F. Wallis: J. Phys. Chem. Solids **1**, 65 (1956)
8 G. Picus, E. Burstein, B.W. Henvis: J. Phys. Chem. Solids **1**, 75 (1956)
9 W. Kohn, J.M. Luttinger: Phys. Rev. **97**, 883 (1954); ibid. **98**, 915 (1955)
10 See the review by A.K. Ramdas, S. Rodriguez: Rep. Prog. Phys. **44**, 1297 (1981)

On the Prehistory of Angular Resolved Photoemission

Neville V. Smith
Lawrence Berkeley Laboratory, Berkeley, USA

Band mapping using angle-resolved photoemission started in the early 1970s. Interest in the angular dependence of the photoelectric effect, however, goes back much further. Figure 1 shows an apparatus used in the 1920s by Herbert Ives and coworkers [1] at the Bell Telephone Laboratories. These workers were definitely *not* concerned with band structures. Wave mechanics was a newfangled concept, and solid-state physics had yet to be invented. They were concerned rather with optimizing the efficiency of photocathodes for use in television and eventually videotelephony.

The sample (C) sits at the center of a spherical collector (B). Application of retarding potentials to the collector permits measurement of the photoelectron energy spectra. A finger (F) moving around a slot in the collector permits measurements as a function of angle of emission. We recognize here a resemblance to modern experimental methods. More striking is the resemblance to the apparatus used by Clinton Davisson and Lester Germer in establishing the wave nature of the electron [2]. This is not surprising. These scientists were all working at the same time in the same building in Manhattan.

It is diverting to speculate on the interactions between Ives and Davisson. It seems likely, on the artistic evidence, that they were using the services of the same glass blower! But what did they talk about? Would they have been pleased to know that their separate lines of research would converge half a century later into the indispensable technique of band mapping?

Evan Kane proposed in a prescient paper published in 1964 that bands could in principle be mapped using the angular dependence of photoemission spectra [3]. A decade elapsed, however, before bands were actually mapped [4]. Mort Traum and I approached this problem in the early 1970s but with some hesitance. There were persuasive proponents of the view that photoelectrons would be so thoroughly scattered before emerging from the sample that all memory of angular information would be lost. We were so intimidated by this that we built only a minimal apparatus, essentially the same as that of Ives but with a channel electron multiplier in place of the finger F. To circumvent the indeterminacy of $k_\perp$, we looked at two-dimensional materials, the layer compounds TaS_2 and $TaSe_2$. Frank DiSalvo was manufacturing single crystals of these compounds in his laboratory a few doors down the corridor. Len Mattheiss was calculating their band structures just a few further doors down the corridor, and we found beautiful agreement with his predictions [5].

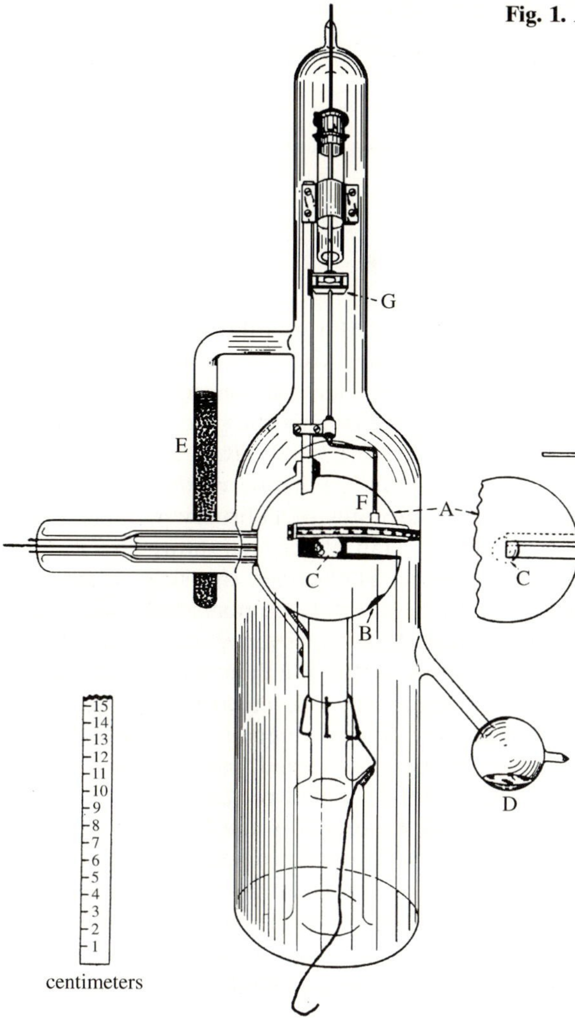

Fig. 1. Apparatus used by Ives et al. [1]

With these shortcuts and fine collaborators we were able to perform the first demonstration of band mapping [4]. In hindsight, it is embarrassing to contemplate our hesitance and timidity. It is all now so obvious and commonplace.

References

1 H.E. Ives, A.R. Olpin, A.L. Johnsrud: Phys. Rev. **32**, 57 (1928)
2 C.J. Davisson, L.H. Germer: Phys. Rev. **30**, 705 (1927)
3 E.O. Kane: Phys. Rev. Lett. **12**, 97 (1964)
4 N.V. Smith, M.M. Traum, F.J. DiSalvo: Solid State Commun. **15**, 211 (1974)
5 L.F. Mattheiss: Phys. Rev. B **8**, 3719 (1973)

The Discovery and Very Basics of the Quantum Hall Effect

Klaus von Klitzing
Max-Planck-Institut für Festkörperforschung, Stuttgart, Germany

The discovery of the quantum Hall effect (QHE) was the result of basic research on silicon field effect transistors – the most important device in microelectronics. Unlike in other conductors, the electron concentration in these devices can be varied in a wide range just by changing the gate voltage. Therefore this system is ideal for an investigation of the Hall effect at different carrier densities by analyzing the Hall voltage as a function of the gate voltage. The experimental curves together with the notes of February 4, 1980, which characterize the birthday of the quantum Hall effect, are shown in Fig. 9.39. As expected qualitatively from the classical Hall effect, the Hall voltage U_{H} varies (at a fixed magnetic field $B = 18$ T) inversely proportional to the number N of free electrons (or gate voltage V_{g}). However, plateaus are visible if the ratio of the number N of electrons to the number N_{Φ} of flux quanta within the area of the device is an integer. For one electron per flux quantum (this corresponds to a fully occupied lowest Landau level with the filling factor 1) the Hall voltage divided by the current has the fundamental value $R_{\mathrm{K}} = h/e^2 = (25812.807 \pm 0.005)\ \Omega$. This Hall plateau is barely visible in the upper left corner of Fig. 9.39 and distorted by the large device resistance due to localization phenomena at this relatively small electron density. The plateaus at 2 or 4 times larger electron concentration are much better resolved. Today, electronic systems with higher quality are available so that measurements at much smaller electron densities with filling factors smaller than one are possible. This is the region where the fractional quantum Hall effect is observed [9.70].

A special situation seems to be present if two flux quanta are available for one electron (filling factor 1/2): Quasiparticles (composite fermions) are formed which behave like electrons moving in an effective magnetic field $B^* = 0$. The Shubnikov–de Haas oscillations of these composite fermions are equivalent to the structures of the fractional quantum Hall effect.

Already the first publication on the QHE [1] with the original title "Realization of a Resistance Standard Based on Fundamental Constants" indicated that an application similar to the Josephson effect may be possible. Today, it is known that different materials (silicon field effect transistors, GaAs/AlGaAs heterostructures) show the same value for the quantized Hall resistance within the experimental uncertainty of 3.5×10^{-10}, and since 1990 all calibrations of resistances are based on the quantum Hall effect with a fixed value $R_{\mathrm{K-1990}} = 25812.807\ \Omega$ for the von Klitzing constant R_{K}.

Different approaches can be used to deduce a quantized value for the Hall resistance. The calculation shown in Fig. 9.39, which led to the discovery of the QHE, is simply based on the classical expression for the Hall effect. A quantized Hall resistance h/e^2 is obtained for a carrier density corresponding to the filling factor one. It is surprising that this simple calculation leads to the correct result. Laughlin was the first to try to deduce the result of the QHE in a more general way from gauge invariance principles [2]. However, his device geometry is rather removed from the real Hall effect devices with metallic contacts for the injection of the current and for the measurement of the electrochemical potential.

The Landauer–Büttiker formalism, which discusses the resistance on the basis of transmission and reflection coefficients, is much more suitable for analyzing the quantum Hall effect [3]. This formalism was very successful in explaining the quantized resistance of ballistic point contacts [4] and, in a similar way, the quantized Hall resistance is the result of an ideal one-dimensional electronic transport. In a classical picture this corresponds to jumping orbits of electrons at the boundary of the device. In the future, the textbook explanation of the QHE will probably be based on this one-dimensional edge channel transport (see Fig. 9.40).

References

1 K. v. Klitzing, G. Dorda, M. Pepper: Phys. Rev. Lett. **45**, 494 (1980)

2 R.B. Laughlin: Phys. Rev. B **23**, 5632 (1981)

3 M. Büttiker: Phys. Rev. Lett. **57**, 1761 (1986)

4 B.J. von Wees, H. van Houten, S.W.J. Beenakker, J.G. Williamson, L.P. Kouwenhoven, D. van der Marel, C.T. Foxon: Phys. Rev. Lett. **60**, 848 (1988);
D.A. Wharam, T.J. Thornton, R. Newbury, M. Pepper, H. Ahmed, J.E.F. Frost, D.G. Hasko, D.C. Peacock, D.A. Ritchie, G.A.C. Jones: J. Phys. C **21**, L 209 (1988)

The Birth of the Semiconductor Superlattice

Leo Esaki
University of Tsukuba, Tsukuba, Japan

In 1969, research on artificially structured materials was initiated when Tsu and I [1,2] proposed an engineered semiconductor superlattice with a one-dimensional periodic potential. In anticipation of advances in controlled epitaxy of ultrathin layers, two types of superlattices were envisioned: doping and compositional, as shown in Fig. 1.

Before arriving at the superlattice concept, we had been examining the feasibility of structural formation of potential barriers and wells that were thin enough to exhibit resonant tunneling [3]. A resonant tunnel diode [4,5] appeared to have more spectacular characteristics than the Esaki tunnel diode [6], the first quantum electron device consisting of only a single tunnel barrier. It was thought that advanced technologies with semiconductors might be

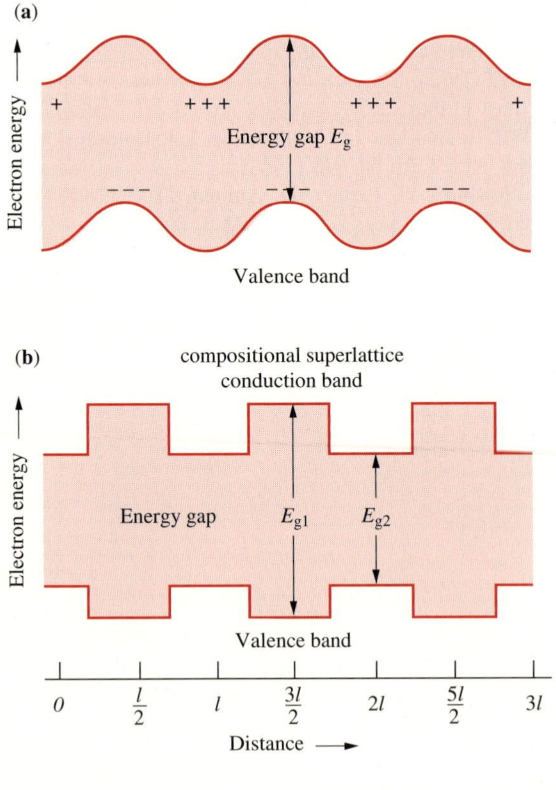

Fig. 1a,b. Spatial variations of the conduction and valence band edges in two types of superlattices: (**a**) doping, (**b**) compositional

ready for demonstration of the de Broglie electron waves. Resonant tunneling (see Sect. 9.5) can be compared to the transmission of an electromagnetic wave through a Fabry–Perot resonator. The equivalent of a Fabry–Perot resonant cavity is formed by the semiconductor potential well sandwiched between the two potential barriers.

The idea of the superlattice occurred to us as a natural extension of double-, triple- and multiple-barrier structures: the superlattice consists of a series of potential wells coupled by resonant tunneling. An important parameter for the observation of quantum effects in the structure is the phase-coherence length, which approximates the electron mean free path. This depends on the bulk quality as well as the interface quality of crystals, and also on the temperatures and values of the effective mass. As schematically illustrated in Fig. 2, if characteristic dimensions such as superlattice periods or well widths are reduced to less than the phase-coherent length, the entire electron system will enter a mesoscopic quantum regime of low dimensionality, on a scale between the macroscopic and the microscopic. Our proposal was to explore quantum effects in the mesoscopic regime.

The introduction of the one-dimensional superlattice potential perturbs the band structure of the host materials, yielding a series of narrow subbands and forbidden gaps which arise from the subdivision of the Brillouin zone into a series of minizones. Thus, the superlattice was expected to exhibit unprecedented electronic properties. At the inception of the superlattice idea, it was recognized that the long, tailormade lattice period provided a unique opportunity to exploit electric-field-induced effects. The electron dynamics in the super-

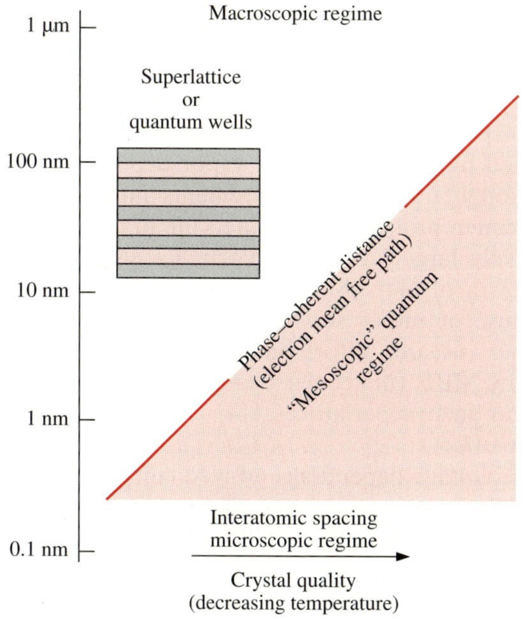

Fig. 2. Schematic illustration of a "mesoscopic" quantum regime (*shaded*) with a superlattice of quantum wells (*inset*)

lattice direction was analyzed for conduction electrons in a narrow subband of a highly perturbed energy–wavevector relationship. The result led to the prediction of a negative differential resistance at a modestly high electric field, which could be a precursor of Bloch oscillations. The superlattice allows us to enter the regime of electric-field-induced quantization: the formation of Stark ladders [7,8], for example, can be proved in a (one-dimensional) superlattice [9], whereas in natural (three-dimensional) crystals the existence and nature of these localized states in a high electric field have been controversial [10,11].

This was, perhaps, the first proposal which advocated using advanced thin-film growth techniques to engineer a new semiconductor material designed by applying the principles of quantum theory. The proposal was made to the US Army Research Office (ARO), a funding agency, in 1969, daringly stating, with little confidence in a successful outcome at the time, "the study of superlattices and observations of quantum mechanical effects on a new physical scale may provide a valuable area of investigation in the field of semiconductors".

Although this proposal was favorably received by ARO, the original version of the paper [1] was rejected for publication by *Physical Review* on the referee's unimaginative assertion that it was "too speculative" and involved "no new physics". The shortened version published in *IBM Journal of Research and Development* [2] was selected as a Citation Classic by the Institute for Scientific Information (ISI) in July 1987. Our 1969 proposal was cited as one of the most innovative ideas at the ARO 40th Anniversary Symposium in Durham, North Carolina, 1991.

At any rate, with the proposal we launched a program to make a "Gedanken-experiment" a reality. In some circles, the proposal was criticized as close to impossible. One of the objections was that a man-made structure with compositional variations on the order of several nanometers could not be thermodynamically stable because of interdiffusion effects. Fortunately, however, it turned out that interdiffusion was negligible at the temperatures involved.

In 1970, Chang, Tsu and I [12] studied a $GaAs–GaAs_{0.5}P_{0.5}$ superlattice with a period of 20 nm synthesized by CVD (chemical vapor deposition) by Blakeslee and Aliotta [13]. Although transport measurements failed to reveal any predicted effect, the specimen probably constituted the first strained-layer superlattice having a relatively large lattice mismatch. Early efforts in our group to obtain epitaxial growth of $Ge_{1-x}Si_x$ and $Cd_{1-x}Hg_xTe$ superlattices were soon abandoned because of rather serious technical problems at that time. Instead, we focused our research effort on compositional $GaAs–Ga_{1-x}Al_xAs$ superlattices grown by MBE (molecular beam epitaxy). In 1972, we found a negative resistance in such superlattices [14], which was interpreted in terms of the superlattice effect.

Following the derivation of the voltage dependence of resonant tunnel currents [5], Chang, Tsu and I observed current–voltage characteristics with a negative resistance [15]. Subsequently, Chang and I measured quantum transport properties in a superlattice with a narrow bandwidth, which exhibited an oscillatory behavior [16]. Tsu et al. performed photocurrent measurements on

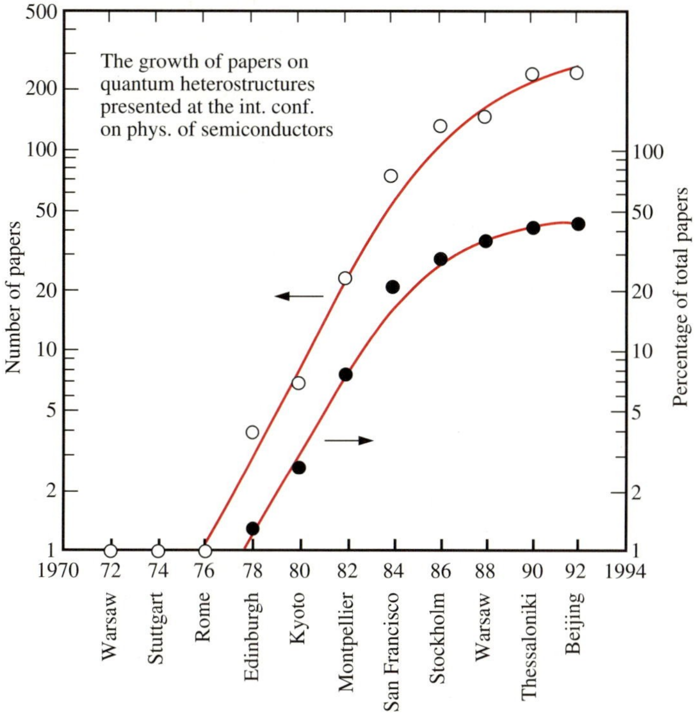

Fig. 3. Growth in relevant papers at the biennial International Conference on the Physics of Semiconductors

superlattices subjected to an electric field perpendicular to the plane layers with the use of a semitransparent Schottky contact, which revealed their miniband configurations [17].

Heteroepitaxy is of great interest for the growth of compositional superlattices. Innovations and improvements in epitaxial techniques such as MBE and MOCVD (metal-organic chemical vapor deposition) have made it possible to prepare high-quality heterostructures with predesigned potential profiles and impurity distributions having dimensional control close to interatomic spacing. This great precision has cleared access to the mesoscopic quantum regime [18,19].

Since a one-dimensional potential can be introduced along with the growth direction, famous examples in the history of one-dimensional mathematical physics, including the above-mentioned resonant tunneling [3], Kronig–Penney bands [20], Tamm surface states [21], Zener band-to-band tunneling [22], and Stark ladders including Bloch oscillations [7–9], all of which had remained textbook exercises, could, for the first time, be practiced in a laboratory. Thus, do-it-yourself quantum mechanics is now possible, since its principles dictate the details of semiconductor structures [23].

Our original proposal [1] and pioneering experiments have triggered a wide spectrum of experimental and theoretical investigations on superlattices and quantum wells over the last two decades. A variety of engineered structures now exhibit extraordinary transport and optical properties which do not exist in any natural crystal. This new degree of freedom offered in semiconductor research through advanced materials engineering has inspired many ingenious experiments, resulting in observations of not only predicted effects but also totally unknown phenomena. As a measure of the growth of the field, Fig. 3 shows the number of papers related to the subject and the percentage of the total presented at the biennial International Conference on the Physics of Semiconductors. Following 1972, when the first paper [14] was presented, the field went through a short period of incubation before experiencing a phenomenal expansion in the 1980s. It appears that nearly half of all semiconductor physicists in the world are working in this area. Activity at this new frontier of semiconductor physics has in turn given immeasurable stimulus to device physics, provoking new ideas for applications. Thus, a new class of transport and opto-electronic devices has emerged.

References

1 L. Esaki, R. Tsu: IBM Res. Note RC-2418 (1969)
2 L. Esaki, R. Tsu: IBM J. Res. Devel. **14**, 61 (1970)
3 D. Bohm: *Quantum Theory* (Prentice Hall, Englewood Cliffs, NJ 1951), p. 283
4 L.V. Iogansen, Zh. Eksp. Teor. Fiz. **45**, 207 (1963) [Sov. Phys. – JETP **18**, 146 (1964)]
5 R. Tsu, L. Esaki: Appl. Phys. Lett. **22**, 562 (1973)
6 L. Esaki: Phys. Rev. **109**, 603 (1958)
7 H.M. James: Phys. Rev. **76**, 1611 (1949)
8 G.H. Wannier: *Elements of Solid State Theory* (Cambridge University Press, Cambridge 1959), p. 190; Phys. Rev. **117**, 432 (1960)
9 W. Shockley: Phys. Rev. Lett. **28**, 349 (1972)
10 J. Zak: Phys. Rev. Lett. **20**, 1477 (1968); Phys. Rev. B **43**, 4519 (1991)
11 A. Rabinovitch, J. Zak: Phys. Rev. B **4**, 2358 (1971)
12 L. Esaki, L.L. Chang, R. Tsu: Proc. 12th Int. Conf. Low Temp. Phys., Kyoto, Japan 1970, p. 551
13 A.E. Blakeslee, C.F. Aliotta: IBM J. Res. Devel. **14**, 686 (1970)
14 L. Esaki, L.L. Chang, W.E. Howard, V.L. Rideout: Proc. 11th Int. Conf. Phys. Semiconductors, Warsaw, Poland 1972, p. 431
15 L.L. Chang, L. Esaki, R. Tsu: Appl. Phys. Lett. **24**, 593 (1974)
16 L. Esaki, L.L. Chang: Phys. Rev. Lett. **33**, 495 (1974)
17 R. Tsu, L.L. Chang, G.A. Sai-Halasz, L. Esaki: Phys. Rev. Lett. **34**, 1509 (1975)
18 L. Esaki: IEEE J. Quantum Electron. **QE-22**, 1611 (1986)
19 L. Esaki: in *Highlights in Condensed Matter Physics and Future Prospects*, ed. by L. Esaki (Plenum, New York 1991), p. 55
20 R. de L. Kronig, W.G. Penney: Proc. R. Soc. London A **130**, 499 (1931)
21 I. Tamm: Phys. Z. Sowjetunion **1**, 733 (1932)
22 C. Zener: Proc. R. Soc. London **145**, 523 (1934)
23 L. Esaki, Phys. Scr. **T42**, 102 (1992)

References[1]

Chapter 1

1.1 J. I. Pankove, T. D. Moustakas (eds.): *Gallium Nitride I*, Semiconductors and Semimetals, Vol. 50 (Academic Press, San Diego 1998)
S. Nakamura, G. Fasol: *The Blue Laser Diode. GaN Based Light Emitters and Lasers* (Springer, Berlin, Heidelberg 1997)

1.2 L. Gao, Y. Y. Xue, F. Chen, Q. Xiong, R. L. Meng, D. Ramirez, C. W. Chu, J. H. Eggert, H. K. Mao: Superconductuvity up to 164 in $HgBa_2Ca_{m-1}Cu_mO_{2m+2-\delta}$ ($m = 1,2,3$), under pressure. Phys. Rev. B**50**, 4260–4263 (1994)

1.3 R. Saito, G. Dresselhaus and M. S. Dresselhaus: *Physical Properties of Carbon Nanotubes* (Imperial College Press, London 1998)

1.4 J. Shinan, Z. Vardeny and Z. Kapati (ed.): *Optical and Electronic Properties of Fullerenes and Fullerene-Band Materials* (Marcel Dekker, New York, 1999)

1.5 C. N. R. Rao and B. Raveau (ed.): *Colossal Magnetoresistance, Charge Ordering and Related Properties of Manganese Oxides* (World Scientific, Singapore, 1998)

1.6 T. Ruf, R. W. Henn, M. Asen-Palmer, E. Gmelin, M. Cardona, H.-J. Pohl, G. G. Devyatyel, P.G. Sennikov: Thermal conductivity of isotopically enriched silicon. Solid State Commun. **115**, 243 (2000)

1.7 J. Czochralski: A new method for measuring the crystallization velocity of metals (in German). Z. Phys. Chem. **92**, 219–221 (1918)

1.8 O. Madelung, M. Schulz, H. Weiss (eds.): *Landolt-Börnstein*, Series III, Vol. 17c (Semiconductors) (Springer, Berlin, Heidelberg 1984) p. 29. This series contains comprehensive references on the growth techniques and properties of individual semiconductors up to 1984

1.9 W. C. Dash: Growth of silicon crystals free of dislocations. J. Appl. Phys. **30**, 459 (1959)

1.10 C. T. Foxon, B. A. Joyce: Growth of thin films and heterostructures of III–V compounds by molecular beam epitaxy, in *Growth and Characterization of Semiconductors* ed. by R. A. Stradling, P. C. Klipstein (Hilger, Bristol 1990) p. 35

1.11 S. Nakamura and G. Fasol: The Blue Laser Diode: GaN Based Light Emitters and Lasers (Springer-Verlag, Berlin, 1997) p. 36–37

1.12 I. N. Stranski and L. Krastanow: Sitzungsberichte d. Akad. d. Wissenschaften in Wien, Abt. 11B, Band **146**, 797 (1937)

1.13 D. W. Pashley: The Basics of Epitaxy, in *Growth and Characterization of Semiconductors* ed. by R. A. Stradling, P. C. Klipstein (Hilger, Bristol 1990) p. 1

1.14 F. C. Frank and J. H. van der Merwe: One-dimensional dislocations I. Static theory, Proc. Royal Society A**198**, 205–216 (1949); One-dimensional disloca-

[1] In case several references are listed under one number, it is tacitly assumed that the first reference is a, the second one b, etc. The reference numbers highlighted in red correspond to proceedings of International Conferences on the Physics of Semiconductors (*ICPS*). See Preface to the Second Edition.

tions II. Misfitting monolayers and oriented overgrowth, Proc. Royal Society A**198**, 216–225 (1949); One-dimensional dislocations III. Influence of the second harmonic term in the potential representation on the properties of the model, Proc. Royal Society A**200**, 125–134 (1949)

1.15 See, for example, D. Bimberg, M. Grundmann, and N. N. Lebentsov: Growth, Spectroscopy, and Laser Application of Self-ordered III-V Quantum Dots. Bulletin of the Materials Research Society, **23**, 31 (1998)

1.16 J. Bohm, A. Lüdge, W. Schröder: Crystal Growth by Floating Zone Melting, in *Handbook of Crystal Growth, Vol. 2* ed. by D. T. H. Hurle (North-Holland, Amsterdam, 1994) p. 213–258

General Reading

Bernard, J.C., M. Sugawara (ed.): *Self-Assembled in GaAs/GaAs Quantum Dots*. Volume 60 of Semiconductors and Semimetals (Academic Press, New York, 1999)

Bimberg, D., M. Grundmann and N. Ledentsov: *Quantum Dot Heterostructures* (John Wiley & Son, New York, 1999)

Chernov A. A.: *Modern Crystallography III – Crystal Growth*, Springer Ser. Solid-State Sci., Vol. 36 (Springer, Berlin, Heidelberg 1984)

Gilman J. J. (ed.): *The Art and Science of Growing Crystals* (Wiley, New York 1963)

Hermann M. A., H. Sitter: *Molecular Beam Epitaxy*, 2nd edn., Springer Ser. Mater. Sci., Vol. 7 (Springer, Berlin, Heidelberg 1996)

Kittel C.: *Introduction to Solid State Physics*, 7th edn. (Wiley, New York 1995)

Laudies R. A.: *The Growth of Single Crystals* (Prentice-Hall, New York 1970)

Matthews J. W. (ed.): *Epitaxial Growth*, Pts. a & b (Academic, New York 1975)

Panish M. B., H. Temkin: *Gas Source Molecular Beam Epitaxy*, Springer Ser. Mater. Sci., Vol. 26 (Springer, Berlin, Heidelberg 1993)

Williams, J. O.: Metal Organic Chemical Vapour Deposition (MOCVD) for the preparation of semiconductor materials and devices, in *Growth and Characterization of Semiconductors*, ed. by R. A. Stradling, P. C. Klipstein (Hilger, Bristol 1990) p. 17

Chapter 2

2.1 *Quantum Theory of Real Materials* (eds. Chelikowsky, J.R., Louie, S.G.) (Kluwer, Dordrecht, 1996)

2.2 C. Kittel: *Introduction to Solid State Physics*, 7th edn. (Wiley, New York 1995) p. 37

2.3 L. M. Falicov: *Group Theory and its Physical Applications* (Univ. Chicago Press, Chicago 1966)

2.4 G. F. Koster: Space groups and their representations, in *Solid State Physics* **5**, 173–256 (Academic, New York 1957)

2.5 G. Lucovsky, A comparison of the long wavelength optical phonons in trigonal Se and Te, Phys. Stat. Sol. (b) **49**, 633 (1972)

2.6 D. M. Greenaway, G. Harbeke: *Optical Properties and Band Structure of Semiconductors* (Pergamon, New York 1968) p. 44

2.7 H. Jones: *The Theory of Brillouin Zones and Electronic States in Crystals*, 2nd edn. (North-Holland, Amsterdam 1975)

2.8 M. L. Cohen, J. Chelikowsky: *Electronic Structure and Optical Properties of Semiconductors*, 2nd edn., Springer Ser. Solid-State Sci., Vol. 75 (Springer, Berlin, Heidelberg 1989)

2.9 J.R. Chelikowsky, D.J. Chadi, M.L. Cohen: Calculated valence band densities of states and photoemission spectra of diamond and zinc-blende semiconductors. Phys. Rev. B **8**, 2786–2794 (1973)

2.10 C. Varea de Alvarez, J.P. Walter, R.W. Boyd, M.L. Cohen: Calculated band structures, optical constants and electronic charge densities for InAs and InSb. J. Chem. Phys. Solids **34**, 337–345 (1973)

2.11 P. Hohenberg, W. Kohn: Inhomogeneous electron gas. Phys. Rev. B **863**, 136 (1964)

2.12 W. Kohn, L. Sham: Self-consistent equations including exchange and correlation effects. Phys. Rev. A **113**, 140 (1965)

2.13 M.S. Hybertsen, S.G. Louie: Electron correlation in semiconductors and insulators. Phys. Rev. B **34**, 5390–5413 (1986)

2.14 N. Trouillier, J.L. Martins: Efficient pseudopotentials for plane wave calculations. Phys. Rev. B **43**, 1993–2006 (1991)

2.15 E.O. Kane: Band structure of indium antimonide. J. Phys. Chem. Solids **1**, 249–261 (1957)

2.16 M. Cardona, F.H. Pollak: Energy-band structure of germanium and silicon. Phys. Rev. **142**, 530–543 (1966); see also Vol. 41B

2.17 M. Cardona: Band parameters of semiconductors with zincblende, wurtzite, and germanium structure. J. Phys. Chem. Solids **24**, 1543–1555 (1963); erratum: ibid. **26**, 1351E (1965)

2.18 O. Madelung, M. Schulz, H. Weiss (eds.): *Landolt-Börnstein*, Series III, Vol. 17a–h (Semiconductors) (Springer, Berlin, Heidelberg 1987)

2.19 E.O. Kane: The $\mathbf{k} \cdot \mathbf{p}$ method. *Semiconductors and Semimetals* **1**, 75–100 (Academic, New York 1966)

2.20 M. Cardona, N.E. Christensen, G. Fasol: Relativistic band structure and spin-orbit splitting of zincblende-type semiconductors. Phys. Rev. B **38**, 1806–1827 (1988)

2.21 G. Dresselhaus, A.F. Kip, C. Kittel: Cyclotron resonance of electrons and holes in silicon and germanium crystals. Phys. Rev. **98**, 368–384 (1955)

2.22 M. Willatzen, M. Cardona, N.E. Christensen: LMTO and $\boldsymbol{k}\cdot\boldsymbol{p}$ calculation of effective masses and band structure of semiconducting diamond. Phys. Rev. B**50**, 18054 (1994)

2.23 J.M. Luttinger: Quantum theory of cyclotron resonance in semiconductors: General theory. Phys. Rev. **102**, 1030–1041 (1956)

2.24 W.A. Harrison: *Electronic Structure and the Properties of Solids: The Physics of the Chemical Bond* (Dover, New York 1989)

2.25 D.J. Chadi, M.L. Cohen: Tight-binding calculations of the valence bands of diamond and zincblende crystals. Phys. Stat. Solidi B **68**, 405–419 (1975)

2.26 W.A. Harrison: The physics of solid state chemistry, in *Festkörperprobleme* **17**, 135–155 (Vieweg, Braunschweig, FRG 1977)

2.27 F. Herman: Recent progress in energy band theory, in *Proc. Int'l Conf. on Physics of Semiconductors* (Dunod, Paris 1964) pp. 3–22

2.28 T. Dietl, W. Dobrowolski, J. Kosut, B.J. Kowalski, W. Szuskiewicz, Z. Wilamoski, A.M. Witowski: HgSe: Metal or Semiconductor? Phys. Rev. Lett. **81**, 1535 (1998);
D. Eich, D. Hübner, R. Fink, E. Umbach, K. Ortner, C.R. Becker, G. Landwehr, A. Flezsar: Electronic structure of HgSe investigated by direct and inverse photoemission. Phys. Rev. B**61**, 12666–12669 (2000)

2.29 T.N. Morgan: Symmetry of electron states in GaP. Phys. Rev. Lett. **21**, 819–823 (1968)

2.30 R.M. Wentzcovitch, M. Cardona, M.L. Cohen, N.E. Christensen: X_1 and X_3 states of electrons and phonons in zincblende-type semiconductors. Solid State Commun. **67**, 927–930 (1988)
2.31 S.H. Wei, A. Zunger: Band gaps and spin-orbit splitting of ordered and disordered $Al_xGa_{1-x}As$ and $GaAs_xSb_{1-x}$ alloys. Phys. Rev. B **39**, 3279–3304 (1989)

General Reading

Group Theory and Applications

Burns G.: *Introduction to Group Theory and Applications* (Academic, New York 1977)
Evarestov R.A., V.P. Smirnov: *Site Symmetry in Crystals*, Springer Ser. Solid-State Sci., Vol. 108 (Springer, Berlin, Heidelberg 1993)
Falicov L.M.: *Group Theory and Its Physical Applications* (Univ. Chicago Press, Chicago 1966)
Heine V.: *Group Theory in Quantum Mechanics* (Pergamon, New York 1960)
Inui T., Y. Tanabe, Y. Onodera: *Group Theory and Its Applications in Physics*, 2nd edn. Springer Ser. Solid-State Sci., Vol. 78 (Springer, Berlin, Heidelberg 1996)
Jones H.: *Groups, Representations, and Physics* (Hilger, Bristol 1990)
Koster G.F.: Space groups and their representations. *Solid State Physics* **5**, 173–256 (Academic, New York 1957)
Lax M.: *Symmetry Principles in Solid State and Molecular Physics* (Wiley, New York 1974)
Ludwig W., C. Falter: *Symmetries in Physics*, 2nd edn., Springer Ser. Solid-State Sci., Vol. 64 (Springer, Berlin, Heidelberg 1996)
Tinkham M.: *Group Theory and Quantum Mechanics* (McGraw-Hill, New York 1964)
Vainshtein B.K.: *Fundamentals of Crystals*, 2nd edn., Modern Crystallography, Vol. 1 (Springer, Berlin, Heidelberg 1994)

Electronic Band Structures

Cohen M.L., Chelikowsky, J.: *Electronic Structure and Optical Properties of Semiconductors*, 2nd edn., Springer Ser. Solid-State Sci., Vol. 75 (Springer, Berlin, Heidelberg 1989)
Greenaway D.L., Harbeke, G.: *Optical Properties and Band Structure of Semiconductors* (Pergamon, New York 1968)
Harrison W.A.: *Electronic Structure and the Properties of Solids: The Physics of the Chemical Bond* (Dover, New York 1989)
Jones H.: *The Theory of Brillouin Zones and Electronic States in Crystals* (North-Holland, Amsterdam 1962)
Phillips J.C.: *Covalent Bonding in Crystals, Molecules, and Polymers* (Univ. Chicago Press, Chicago 1969)

Chapter 3

3.1 D.N. Talwar, M. Vandevyver, K. Kunc, M. Zigone: Lattice dynamics of zinc chalcogenides under compression: Phonon dispersion, mode Grüneisen, and thermal expansion. Phys. Rev. B **24**, 741–753 (1981)

A. Debernardi, M. Cardona: Isotope effects on the lattice constant by perturbation theory: an *ab initio* calculation, Phys. Rev. B **54**, 11305–11310 (1996)
3.2 H. Goldstein: *Classical Mechanics* (Addison-Wesley, Reading 1950) p. 329
3.3 A. Debernardi: Phonon Linewidth in III–V semiconductors from density functional perturbation theory. Phys. Rev. B **57**, 12847–12858 (1998)
3.4 G. Nilsson, G. Nelin: Study of the homology between silicon and germanium by thermal-neutron spectroscopy. Phys. Rev. B **6**, 3777–3786 (1972)
3.5 W. Weber: Adiabatic bond charge model for the phonons in diamond, Si, Ge, and α-Sn. Phys. Rev. B **15**, 4789–4803 (1977)
3.6 D. Strauch, B. Dorner: Phonon dispersion in GaAs. J. Phys.: Condens. Matter **2**, 1457–1474 (1990)
3.7 T. Ruf, J. Serrano, M. Cardona, P. Pavone, M. Pabst, M. Krisch, M. D'Astuto, T. Suski, I. Grzegory and M. Leszczynski: Phonon Dispersion curves in wurtzite-structure GaN determined by inelastic x-ray scattering. Phys. Rev. Lett. **86**, 906–909 (2001)
3.8 R. Zallen, R.M. Martin, V. Natoli: Infrared Activity in Elemental Crystals. Phys. Rev. B**49**, 7032 (1994)
3.9 M. Born, K. Huang: *Dynamical Theory of Crystal Lattices* (Oxford Univ. Press, Oxford 1988, reprint of the original 1954 edition)
3.10 M. Born: The space lattice theory of diamond. Ann. Physik **44**, 605–642 (1914) in German
3.11 Y.C. Hsieh: The vibrational spectrum and the specific heat of germanium and silicon. J. Chem. Phys. **22**, 306–311 (1954)
3.12 F. Herman: Lattice vibrational spectrum of germanium. J. Phys. Chem. Solids **8**, 405–418 (1959)
3.13 W. Cochran: Theory of the lattice vibrations of germanium. Proc. R. Soc. (London) Ser.A **253**, 260–276 (1959)
3.14 G. Dolling, R.A. Cowley: The thermodynamics and optical properties of germanium, silicon, diamond, and gallium arsenide, Proc. Phys. Soc. **88**, 463–494 (1966)
3.15 J.C. Phillips: Covalent bonds in crystals. I. Elements of a structural theory. Phys. Rev. **166**, 832–838 (1968); II. Partially ionic bonding. ibid. **168**, 905–911 (1968)
3.16 M.J.P. Musgrave, J.A. Pople: A general valence force field for diamond. Proc. R. Soc. (London) Ser.A **268**, 474–484 (1962)
3.17 M.A. Nusimovici, J.L. Birman: Lattice dynamics of Wurtzite: CdS. Phys. Rev. **156**, 925–938 (1967)
3.18 A. Debernardi, N.M. Pyka, A. Göbel, T. Ruf, R. Lauck, S. Kramp, M. Cardona: Lattice Dynamics of Wurtzite CdS, Solid State Commun. **103**, 297–301 (1997)
3.19 J.M. Rowe, R.M. Nicklow, D.L. Price, K. Zanio: Lattice dynamics of cadmium telluride. Phys. Rev. B **10**, 671–675 (1974)
3.20 F. Widulle, S. Kramp, N.M. Pyka, A. Göbel, T. Ruf, A. Debernardi, R. Lauck, M. Cardona: The phonon dispersion of wurtzite CdSe. Physica B**263–264**, 448–451 (1998)
3.21 G. Lang, K. Karch, M. Schmitt, P. Pavone, A.P. Mayer, R.K. Wehner, D. Strauch: Anharmonic lineshift and linewidth of the Raman mode in Ge and Si. Phys. Rev. B**59**, 6182 (1999);
S. Shobhana, D. Vanderbilt: Anharmonic self-energies of phonons in silicon. Phys. Rev. B**43**, 4541 (1991)
3.22 P.N. Keating: Effect of invariance requirements on the elastic strain energy of crystals with application to the diamond structure. Phys. Rev. **145**, 637–645 (1966)

3.23 R.M. Martin: Elastic properties of ZnS structure semiconductors. Phys. Rev. B **1**, 4005–4011 (1970)
R.M. Martin: Dielectric screening model for lattice vibrations of diamond-structure crystals. Phys. Rev. **186**, 871 (1969)
3.24 J. Noolandi: Theory of crystal distortions in $A^{II}B^{IV}C_2{}^V$ and $A^IB^{III}C_2{}^{VI}$ chalcopyrite semiconductors. Phys. Rev. B **10**, 2490–2494 (1974)
3.25 S. Göttlicher, E. Wolfel: X-ray determination of the electron distribution in crystals (in German). Z. Elektrochem. **63**, 891–901 (1959)
3.26 L.W. Yang, P. Coppens: On the experimental electron distribution in silicon. Solid State Commun. **15**, 1555–1559 (1974)
3.27 J. Chelikowsky, M.L. Cohen: Nonlocal pseudopotential calculations for the electronic structure of eleven diamond and zincblende semiconductors. Phys. Rev. B **14**, 556-582 (1976)
3.28 P. Pavone, K. Karch, O. Schütt, W. Windl, D. Strauch, P. Gianozzi, S. Baroni: *Ab initio* lattice dynamics of diamond. Phys. Rev. B **48**, 3156–3163 (1993)
M. Schwoerer-Bohning, A.T. Macrauder, D.A. Arms: Phonon Dispersion in Diamond measured by inelastic X-ray scattering. Phys. Rev. Lett. **80**, 5572–5575 (1998)
3.29 G.P. Srivastava: *The Physics of Phonons* (Hilger, Bristol 1990)
3.30 A. Blacha, H. Presting, M. Cardona: Deformation potentials of k=0 states of tetrahedral semiconductors. Phys. Stat. Solidi b **126**, 11–36 (1984)
3.31 D.D. Nolte, W. Walukiewicz, E.E. Haller: Critical criterion for axial modes of defects in as-grown n-type GaAs. Phys. Rev. B **36**, 9374–9377 (1987)
3.32 M. Cardona, N.E. Christensen: Acoustic deformation potentials and heterostructure band offsets in semiconductors. Phys. Rev. B **35**, 6182–6194 (1987)
3.33 E.O. Kane: Strain effects on optical critical-point structure in diamond-type crystals. Phys. Rev. **178**, 1368–1398 (1969)
3.34 G.E. Pikus, G.L. Bir: Effect of deformation on the hole energy spectrum of germanium and silicon. Sov. Phys. – Solid State **1**, 1502–1517 (1960)
3.35 G.E. Pikus, G.L. Bir: *Symmetry and Strain Induced Effects in Semiconductors* (Wiley, New York 1974)
3.36 E.L. Ivchenko and G.E. Pikus: Superlattices and other Heterostructures, (Springer, Heidelberg, 1997), p. 71
3.37 C. Herring, E. Vogt: Transport and deformation-potential theory for many-valley semiconductors with anisotropic scattering. Phys. Rev. **101**, 944–961 (1956)
3.38 H. Brooks: Theory of the electrical properties of germanium and silicon. *Advances in Electronics and Electron Physics* **7**, 85–182 (Academic, New York 1955)
3.39 J.F. Nye: *Physical Properties of Crystals* (Oxford Univ. Press, Oxford 1969)
3.40 G.D. Mahan, J.J. Hopfield: Piezoelectric polaron effects in CdS. Phys. Rev. Lett. **12**, 241–243 (1964)
3.41 K. Hübner: Piezoelectricity in zincblende- and wurtzite-type crystals. Phys. Stat. Solidi B **57**, 627–634 (1973)
3.42 W.A. Harrison: *Electronic Structure and the Properties of Solids: The Physics of the Chemical Bond* (Dover, New York 1989) p. 224
3.43 O. Madelung, M. Schulz, H. Weiss (eds.): *Landolt-Börnstein*, Series III, Vol. 22 (Semiconductors), Subvolume a. Intrinsic Properties of Group IV Elements, III-V, II-VI and I-VII Compounds (Springer, Berlin, Heidelberg 1987)
3.44 S. Adachi: GaAs, AlAs, and $Al_xGa_{1-x}As$: Materials parameters for use in research and device applications. J. Appl. Phys. **58**, R1–29 (1985)

3.45 L. Kleinmann: Deformation potentials in Si: I. Uniaxial strain. Phys. Rev. **128**, 2614–2621 (1962)
3.46 E. Anastassakis, M. Cardona: Internal strains and Raman-active optical phonons. Phys. Stat. Solidi B **104**, 589–600 (1981)
3.47 W. Plötz, P. Vogl: Theory of optical-phonon deformation potentials in tetrahedral semiconductors. Phys. Rev. B **24**, 2025–2037 (1981)
3.48 M. Cardona, M. Grimsditch, D. Olego: Theoretical and experimental determinations of Raman scattering cross sections in simple solids, in *Light Scattering in Solids*, ed. by J. L. Birman, H. Z. Cummins, K. K. Rebane (Plenum, New York 1979) pp. 249–256
3.49 S. Zollner, S. Gopalan, M. Cardona: Intervalley deformation potentials and scattering rates in zincblende semiconductors. Appl. Phys. Lett. **54**, 614–616 (1989)
3.50 C. Carabatos, B. Prevot: Rigid ion model lattice dynamics of cuprite (Cu_2O). Phys. Status Solid B **44**, 701–712 (1971)
3.51 P. Molinàs-Mata, M. Cardona: Planar force-constant models and internal strain parameter of Ge and Si. Phys. Rev. B **43**, 9799–9809 (1991)
3.52 P. Molinàs-Mata, A. J. Shields, M. Cardona: Phonons and internal stresses in IV-VI and III-V semiconductors: The planar bond-charge model. Phys. Rev. B **47**, 1866–1875 (1993)
3.53 C. H. Xu, C. Z. Wang, C. T. Chan, K. M. Ho: Theory of the thermal expansion of Si and Diamond. Phys. Rev. B**43**, 5024–5027 (1991)
3.54 A. Debernardi and M. Cardona: Isotopic effects on the lattice constant in compound semiconductors by perturbation theory: an ab initio calculation. Phys. Rev. B**54**, 11305–11310 (1996)

General Reading

Lattice Dynamics

Bilz, H., W. Kress: *Phonon Dispersion Relations in Insulators*, Springer Ser. Solid-State Sci., Vol. 10 (Springer, Berlin, Heidelberg 1979). This is touted as a "phonon atlas" by its authors. It presents a collection of phonon dispersion curves and densities of states for more than a hundred insulators, including all the well-known semiconductors.

Born, M., K. Huang: *Dynamical Theory of Crystal Lattices* (Oxford Univ. Press, Oxford 1988), reprint of the original 1954 edition

Horton G. K., Maradudin, A. A. (eds.): *Dynamical Properties of Solids*, Vols. 1–5 (North-Holland, Amsterdam 1974)

Sinha S. K.: Phonons in semiconductors. CRC Critical Reviews in Solid State Sciences **3**, 273–334 (1973)

Srivastava G. P.: *The Physics of Phonons* (Hilger, Bristol 1990)

Properties Related to Phonons

Harrison W. A.: *Electronic Structure and the Properties of Solids: The Physical of the Chemical Bond* (Dover, New York, 1989)

Kittel C.: *Introduction to Solid State Physics*, 7th edn. (Wiley, New York 1995) Chap. 4

Madelung O.: *Introduction to Solid-State Theory*, Springer Ser. Solid-State Sci., Vol. 2 (Springer, Berlin, Heidelberg 1978)

Madelung O., Schulz, M., Weiss, H. (eds.): *Landolt-Börnstein*, Series III, Vol. 22 (Semiconductors), Subvolume a) Intrinsic Properties of Group IV Elements, III-V, II-VI and I-VII Compounds (Springer, Berlin, Heidelberg 1987)

Nye J.F.: *Physical Properties of Crystals* (Oxford Univ. Press, Oxford 1969)
Pikus G.E., Bir, G.L.: *Symmetry and Strain Induced Effects in Semiconductors* (Wiley, New York 1974)

Chapter 4

4.1 G. Wannier: *Elements of Solid State Theory* (Cambridge Univ. Press, Cambridge 1959), for discussions of Wannier functions
4.2 W. Kohn: Shallow impurity states in silicon and germanium. *Solid State Physics* **5**, 257–320 (Academic, New York 1957)
4.3 J.M. Ziman: *Principles of the Theory of Solids*, 2nd edn (Cambridge Univ. Press, Cambridge 1972), for discussion of the effective-mass approximation, see Chap. 6, pp. 147–176
4.4 R.K. Watts: *Point Defects in Crystals* (Wiley-Interscience, New York 1977)
4.5 N. Chand, T. Henderson, J. Klem, W.T. Masselink, R. Fischer, Y.-C. Chang, H. Morkoç: Comprehensive analysis of Si-doped $Al_xGa_{1-x}As$ (x=0 to 1): Theory and experiment. Phys. Rev. B **30**, 4481–4492 (1984)
4.6 M. Mizuta, M. Tachikawa, H. Kukimoto, S. Minomura: Direct evidence for the DX center being a substitutional donor in AlGaAs alloy system. Jpn. J. Appl. Phys. **24**, L143–146 (1985)
4.7 D.J. Chadi, K.J. Chang: Energetics of DX-center formation in GaAs and $Al_xGa_{1-x}As$ alloys. Phys. Rev. **39**, 10063–10074 (1989)
4.8 J. Dabrowski, M. Scheffler: Defect metastability in III-V compounds. Mater. Sci. Forum **83–87**, 735–750 (1992)
4.9 W. Kohn, J.M. Luttinger: Theory of donor levels in silicon. Phys. Rev. **97**, 1721 (1955); Theory of donor states in silicon. ibid. **98**, 915–922 (1955)
4.10 R.A. Faulkner: Higher donor excited states for prolate-spheroid conduction bands: A re-evaluation of silicon and germanium. Phys. Rev. **184**, 713–721 (1969)
4.11 S. Pantelides, C.T. Sah: Theory of localized states in semiconductors. I. New results using an old method. Phys. Rev. B **10**, 621–637 (1974) and II. The pseudo impurity theory applications to shallow and deep donors in silicon. ibid. 638–658 (1974)
4.12 N. Lipari, A. Baldereschi: Interpretation of Acceptor Spectra in Semiconductors. Solid State Commun. **25**, 665–668 (1978)
4.13 W. Kohn. D. Schechter: Theory of acceptor levels in germanium. Phys. Rev. **99**, 1903–1904 (1955)
4.14 A. Baldereschi, N.O. Lipari: Spherical model of shallow acceptor states in semiconductors. Phys. Rev. B **8**, 2697–2709 (1973)
4.15 A. Baldereschi, N.O. Lipari: Cubic contributions to the spherical model of shallow acceptor states. Phys. Rev. B **9**, 1525–1539 (1974)
4.16 N.O. Lipari, A. Baldereschi: Interpretation of acceptor spectra in semiconductors, Solid State Commun. **25**, 665, 668 (1978)
4.17 M. Willatzen, M. Cardona, N.E. Christensen: Spin-orbit coupling parameters and g-factors of II–VI zincblende materials. Phys. Rev. B **51**, 17992–17994 (1995)
4.18 M.A. Hasse, J. Qiu, J.M. DePuydt, H. Cheng: Blue-green laser diode. Appl. Phys. Lett. **59**, 1272–1274 (1991)
4.19 H. Jeon, J. Ding, W. Patterson, A.V. Nurmikko, W. Xie, D.C. Grillo, M. Kobayashi, R.L. Gunshor: Blue-green injection laser diodes in (Zn,Cd)Se/ZnSe quantum wells. Appl. Phys. Lett. **59**, 3619–3621 (1991)

4.20 J. Serrano, M. Cardona, T. Ruf: Spin-orbit splitting of acceptor-related states, Solid State Commun. **113**, 411–414 (2000)

4.21 J. Serrano, A. Wysnedek, T. Ruf, M. Cardona: Spin-orbit splitting of acceptors in Si and C, Physica B**273–274**, 640–644 (1999)

4.22 S. Pantelides: The electronic structure of impurity and defect states in semiconductors. Rev. Mod. Phys. **50**, 797–858 (1978)

4.23 P.M. Mooney: Deep donor levels (DX centers) in III–V semiconductors. J. Appl. Phys. **67**, R1–26 (1990)

4.24 N.F. Mott: *Metal-Insulator Transition* (Taylor and Francis, London 1990) p. 76

4.25 D.J. Chadi: Doping in ZnSe, ZnTe, MgSe, and MgTe wide-band-gap semiconductors. Phys. Rev. Lett. **72**, 534–537 (1994)

4.26 E.N. Economou: *Green's Functions in Quantum Physics*, 2nd edn., Springer Ser. Solid-State Sci., Vol.7 (Springer, Berlin, Heidelberg 1983) pp. 97–125

4.27 M. Lannoo, J. Bourgoin: *Point Defects in Semiconductors I, Theoretical Aspects*, Springer Ser. Solid-State Sci., Vol.22 (Springer, Berlin, Heidelberg (1981) pp. 68–152

4.28 P.A.M. Dirac: *The Principles of Quantum Mechanics* (Oxford Univ. Press, Oxford 1967) pp. 58–61

4.29 H.P. Hjalmarson, P. Vogl, D.J. Wolford, J.D. Dow: Theory of substitutional deep traps in covalent semiconductors. Phys. Rev. Lett. **44**, 810–813 (1980)

4.30 J.C. Phillips: *Covalent Bonding in Crystals, Molecules, and Polymers* (Univ. Chicago Press, Chicago 1969) p. 232, Table E. 1

4.31 R.A. Faulkner: Toward a theory of isoelectronic impurities in semiconductors. Phys. Rev. **175**, 991–1009 (1968)

4.32 D.G. Thomas, J.J. Hopfield: Isoelectronic traps due to nitrogen in GaP. Phys. Rev. **150**, 680–703 (1966)

4.33 D.J. Wolford, J.A. Bradley, K. Fry, J. Thompson: The nitrogen isoelectronic trap in GaAs, in *Physics of Semiconductors 1984*, ed. by J.D. Chadi, W.A. Harrison (Springer, New York 1984) pp. 627–630

4.34 E. Cohen, M.D. Sturge: Excited states of excitons bound to nitrogen pairs in GaP. Phys. Rev. B **15**, 1039–1051 (1977)

4.35 W.Y. Hsu, J.D. Dow, D.J. Wolford, B.G. Streetman: Nitrogen isoelectronic trap in $GaAs_{1-x}P_x$. II. Model calculation of the electronic states N_Γ and N_X at low temperature. Phys. Rev. B **16**, 1597–1615 (1977)

4.36 S. Pantelides (ed.): *Deep Centers in Semiconductors, A State of the Art Approach* (Gordon and Breach, New York 1986) Chaps. 1 and 7

4.37 A. Messiah: *Quantum Mechanics* (North-Holland, Amsterdam 1961), pp. 1054–1060

General Reading

Economou E.N.: *Green's Functions in Quantum Physics*, 2nd edn., Springer Ser. Solid-State Sci., Vol.7 (Springer, Berlin, Heidelberg 1983)

Lannoo M., J. Bourgoin: *Point Defects in Semiconductors I, Theoretical Aspects*, Springer Ser. Solid-State Sci., Vol.22 (Springer, Berlin, Heidelberg 1981)

Pantelides S.: The Electronic Structure of Impurity and Defect States in Semiconductors. Rev. Mod., Phys. **50**, 797–858 (1978)

Pantelides S. (ed.): *Deep Centers in Semiconductors, A State of the Art Approach* (Gordon and Breach, New York 1986)

Schubert E. F.: *Doping in III–V Semiconductors* (Cambridge Univ. Press, Cambridge 1993)

Wannier G.: *Elements of Solid State Theory* (Cambridge Univ. Press, Cambridge 1959), for discussions of Wannier functions

Watts R. K.: *Point Defects in Crystals* (Wiley-Interscience, New York 1977)

Ziman J. M.: *Principles of the Theory of Solids*, 2nd edn. (Cambridge Univ. Press, Cambridge 1972), for discussions of the effective-mass approximation

Chapter 5

5.1 J. M. Ziman: *Principles of Theory of Solids*, 2nd edn. (Cambridge Univ. Press, Cambridge 1972) pp. 129–178

5.2 B. K. Ridley: *Quantum Processes in Semiconductors*, 2nd edn. (Clarendon, Oxford 1988)

5.3 H. S. Robertson: *Statistical Thermophysics* (Prentice Hall, Englewood Cliffs, NJ 1993) pp. 445–449

5.4 C. Jacoboni, P. Lugli: *The Monte Carlo Method for Semiconductor Device Simulation* (Springer, Wien 1989) pp. 104–160

5.5 D. K. Ferry: *Semiconductors* (Macmillan, New York 1991)

5.6 S. S. Devlin: Transport properties, in *Physics and Chemistry of II–VI Compounds*, ed. by M. Aven, J. S. Prener (North-Holland, Amsterdam 1967)

5.7 C. Kittel: *Introduction to Solid State Physics*, 7th edn. (Wiley, New York 1995)

5.8 E. M. Conwell, M. O. Vassel: High-field distribution function in GaAs. IEEE Trans. ED-**13**, 22–27 (1966)

5.9 C. L. Collins, P. Y. Yu: Nonequilibrium phonon spectroscopy: A new technique for studying intervalley scattering in semiconductors. Phys. Rev. B **27**, 2602–2604 (1983)

5.10 D. L. Rode: Low field electron transport. *Semiconductors and Semimetals* **10**, 1–89 (Academic, New York 1982)

5.11 D. Long: Scattering of conduction electrons by lattice vibrations in silicon. Phys. Rev. **120**, 2024–2032 (1960)

5.12 J. L. Birman, M. Lax, R. Loudon: Intervalley-scattering selection rules in III–V semiconductors. Phys. Rev. **145**, 620–622 (1966)

5.13 D. K. Ferry: First-order optical and intervalley scattering in semiconductors. Phys. Rev. B **14**, 1605–1609 (1976)

5.14 H. Brooks: Scattering by ionized impurities in semiconductors. Phys. Rev. **83**, 879 (1951)

5.15 E. M. Conwell, V. Weisskopf: Theory of impurity scattering in semiconductors. Phys. Rev. **77**, 388–390 (1950)

5.16 R. L. Liboff: *Quantum Mechanics* (Addison-Wesley, Reading, MA 1980) p. 625

5.17 S. M. Sze: *Semiconductor Devices* (Wiley, New York 1985) p. 33

5.18 G. E. Stillman, C. M. Wolfe, J. O. Dimmock: Hall coefficient factor for polar mode scattering in n-type GaAs. J. Phys. Chem. Solids **31**, 1199–1204 (1970)

5.19 K. Fletcher, P. N. Butcher: An exact solution of the linearized Boltzmann equation with applications to the Hall mobility and Hall factor of n-GaAs. J. Phys. C **5**, 212–224 (1972)

5.20 H. L. Störmer, R. Dingle, A. C. Gossard, W. Wiegmann, R. A. Logan: Electronic properties of modulation-doped GaAs-$Al_xGa_{1-x}As$ Superlattices, in *Physics of Semiconductors 1978*, ed. by B. L. H. Wilson (Inst. Phys., Bristol 1979) pp. 557–560

5.21 W. Walukiewicz, H.E. Ruda, J. Lagowski, H.C. Gatos: Electron mobility in modulation-doped heterostructures. Phys. Rev. B **30**, 4571–4582 (1984)

5.22 S. Wang: *Fundamentals of Semiconductor Theory and Device Physics* (Prentice Hall, Englewood Cliffs, NJ 1989)

5.23 E.M. Conwell: *High Field Transport in Semiconductors*. Solid State Physics, Suppl. 9 (Academic, New York 1967)

5.24 E.J. Yoffa: Dynamics of dense laser-induced plasmas. Phys. Rev. B **21**, 2415–2425 (1980)

5.25 W.H. Knox, C. Hirlimann, D.A.B. Miller, J. Shah, D.S. Chemla, C.V. Shank: Femtosecond excitation of nonthermal carrier populations in GaAs Quantum Wells. Phys. Rev. Lett. **56**, 1191–1193 (1986)

5.26 K. Seeger: *Semiconductor Physics*, 5th edn., Springer Ser. Solid-State Sci., Vol. 40 (Springer, Berlin, Heidelberg 1991)

5.27 B. Carnez, A. Cappy, A. Kaszynski, E. Constant, G. Salmer: Modeling of a submicrometer gate field-effect transistor including effects of nonstationary electron dynamics. J. Appl. Phys. **51**, 784–790 (1980)

5.28 J. Singh: *Physics of Semiconductors and Their Heterostructures* (McGraw-Hill, New York 1993) pp. 524–531

5.29 J.S. Blakemore: Semiconducting and other major properties of gallium arsenide. J. Appl. Phys. **53**, R123–181 (1982)

5.30 J. Shah, B. Deveaud, T.C. Damen, W.T. Tsang, A.C. Gossard, P. Lugli: Determination of intervalley scattering rates in GaAs by subpicosecond luminescence spectroscopy. Phys. Rev. Lett. **59**, 2222–2225 (1987)

5.31 D.S. Kim, P.Y. Yu: Hot-electron relaxation and hot phonons in GaAs studied by subpicosecond Raman scattering. Phys. Rev. B **43**, 4158–4169 (1991)

5.32 P.J. Vinson, C. Pickering, A.R. Adams, W. Fawcett, G.D. Pitt: The band structure of GaAs from transferred electron effects at high pressure, in: *Physics of Semiconductors 1976*, ed. by. F.G. Fumi (Tipografia Marves, Rome 1976) pp. 1243–1246

5.33 J.B. Gunn: Microwave oscillations of current in III–V semiconductors. Solid State Commun. **1**, 88–91 (1963)

5.34 J.B. Gunn: Microwave oscillations of current in III–V semiconductors. IBM J. Res. Dev. **8**, 141–159 (1964)

5.35 R. Dalven: *Introduction to Applied Solid State Physics*, 2nd edn. (Plenum, New York 1990) pp. 158–165

5.36 K. Seeger: *Semiconductor Physics*, 5th edn., Springer Ser. Solid-State Sci., Vol. 40 (Springer, Berlin, Heidelberg 1991) pp. 217–272

5.37 C. Herring, E. Vogt: Transport and deformation potential theory for many-valley semiconductors with anisotropic scattering. Phys. Rev. **101**, 944–961 (1956); erratum **105**, 1933 (1956)

5.38 E.H. Hall: On a new action of the magnet on electric current. Am. J. Math. **2**, 287–292 (1879)

5.39 L. van der Pauw: A method of measuring specific resistivity and Hall effect of discs of arbitrary shape. Philips Res. Rep. **13**, 1–9 (1958)

General Reading

Transport Properties

Dalven R.: *Introduction to Applied Solid State Physics* 2nd edn. (Plenum, New York 1990)

Ferry D. K.: *Semiconductors* (Macmillan, New York 1991)
Rode D. L.: Low field electron transport. *Semiconductors and Semimetals* **10**, 1–89 (Academic, New York 1982)
Kittel C.: *Introduction to Solid State Physics* 7th edn. (Wiley, New York 1995)
Nag B. R.: *Electron Transport in Compound Semiconductors*, Springer Ser. Solid-State Sci., Vol. 11 (Springer, Berlin, Heidelberg 1980)
Ridley B. K.: *Quantum Processes in Semiconductors*, 2nd edn. (Clarendon, Oxford 1988)
Seeger K.: *Semiconductor Physics*, 5th edn., Springer Ser. Solid-State Sci., Vol. 40 (Springer, Berlin, Heidelberg 1991);
J. Shah: *Ultrafast Spectroscopy of Semiconductors and Semiconductor Nanostructures*, 2nd edn., Springer Ser. Solid-State Sci., Vol. 115 (Springer, Heidelberg, 1999)
Wiley J. D.: Mobility of holes in III–V compounds. *Semiconductors and Semimetals* **10**, 91–174 (Academic, New York 1982)
Ziman J. M.: *Principles of the Theory of Solids*, 2nd edn. (Cambridge Univ. Press, Cambridge 1972)

Hot Carriers

Conwell E. M.: *High Field Transport in Semiconductors*, Solid State Physics, Suppl. 9 (Academic, New York 1967)
Conwell E. M.: In *Handbook of Semiconductors* (North-Holland, Amsterdam 1982) Vol. 1, pp. 513–561
Jacoboni C., P. Lugli: *The Monte Carlo Method for Semiconductor Device Simulation* (Springer, New York 1989)

Devices

Singh J.: *Physics of Semiconductors and Their Heterostructures* (McGraw-Hill, New York 1993)
Sze S. M.: *Semiconductor Devices* (Wiley, New York 1985)
Wang, S.: *Fundamentals of Semiconductor Theory and Device Physics* (Prentice Hall Englewood Cliffs, NJ 1989)

Chapter 6

6.1 L. Brillouin: Scattering of light and X-rays by a transparent homogeneous body: Influence of the thermal agitation (in French). Ann. Pysique **17**, 88–122 (1922)
L. I. Mandelstam: On light scattering by an inhomogeneous medium (in Russian). Zh. Russko Fiz. Khim. Obshch. (J. Russian Physico-Chemical Soc.) **58**, 381 (1926)
6.2 Y. R. Shen: *The Principles of Nonlinear Optics* (Wiley, New York 1989)
D. L. Mills: *Nonlinear Optics* (Springer, Berlin, Heidelberg 1991)
6.3 H. Haug, S. W. Koch: *Quantum Theory of Optical and Electronic Properties of Semiconductors* (World Scientific, Singapore 1990)
6.4 F. Henneberger, S. Schmidt-Rink, E. O. Göbel (eds.): *Optics of Semiconductor Nanostructures* (Akademie, Berlin 1993);
J. Shah: Ultrafast Spectroscopy of Semiconductors and Semiconductor Nanostructures, (Springer, Heidelberg, 1999)
6.5 J.D. Joannopoulos, R.D. Meade, J.N. Winn: *Photonic Crystals* (Princeton University Press, 1995)

6.6 L. D. Landau, I. M. Lifshitz: *Statistical Physics*, 3rd edn. (Addison-Wesley, Reading, MA 1980)
6.7 P. Y. Yu, M. Cardona: Spatial dispersion in the dielectric constant of GaAs. Solid State Commun. **9**, 1421–1424 (1971)
6.8 P. Etchegoin, M. Cardona: Stress induced optical activity in zincblende-type semiconductors. Solid State Commun. **82**, 655–661 (1992)
6.9 V. M. Agranovich, V. Ginzburg: *Crystal Optics with Spatial Dispersion*, Springer Ser. Solid-State Sci., Vol. 42 (Springer, Berlin, Heidelberg 1984)
6.10 J. D. Jackson: *Classical Electrodynamics*, 2nd edn. (Wiley, New York 1975)
6.11 M. Cardona: *Modulation Spectroscopy*, Solid State Physics, Suppl.11 (Academic, New York 1969) pp. 55–65
6.12 D. E. Aspnes, A. A. Studna: Dielectric functions and optical parameters of Si, Ge, GaP, GaAs, GaSb, InP, InAs, and InSb from 1.5 to 6.0 eV. Phys. Rev. B **27**, 985–1009 (1983)
6.13 P. S. Hauge: Recent developments in instrumentation in ellipsometry. Surf. Sci. **96**, 108–140 (1980)
6.14 H. R. Philipp, H. Ehrenreich: Ultraviolet optical properties – optical properties of III–V compounds. *Semiconductors and Semimetals* **3**, 93–124 (Academic, New York 1967)
6.15 C. Kunz: *Synchrotron Radiation, Techniques and Applications*, Topics Curr. Phys., Vol. 10 (Springer, Berlin, Heidelberg 1979)
6.16 E. E. Koch: *Handbook of Synchrotron Radiation* (North-Holland, Amsterdam 1983)
6.17 W. Heitler: *The Quantum Theory of Radiation*, 3rd edn. (Oxford Univ. Press, Oxford 1954) pp. 56–64
6.18 M. Cardona, F. H. Pollak: Energy-band structure of germanium and silicon: The $\mathbf{k} \cdot \mathbf{p}$ method. Phys. Rev. **142**, 530–543 (1966)
6.19 M. Cardona, N. E. Christensen, G. Fasol: Relativistic band structure and spinorbit splitting of zincblende-type semiconductors. Phys. Rev. B **38**, 1806–1827 (1988)
6.20 L. Van Hove: The occurence of singularities in the elastic frequency distribution of a crystal. Phys. Rev. **89**, 1189–1193 (1953)
6.21 M. L. Cohen, J. R. Chelikowsky: *Electronic Structure and Optical Properties of Semiconductors*, 2nd edn., Springer Ser. Solid-State Sci., Vol. 75 (Springer, Berlin, Heidelberg 1989)
6.22 L. X. Benedict, E. L. Shirley, R. B. Bohn: Theory of optical absorption in diamond, Si, Ge, and GaAs. Phys. Rev. B **57**, R9385–9387 (1998)
6.23 S. Albrecht, L. Reining, R. Del Sole, G. Onida: *Ab initio* calculation of excitonic effects in the optical spectra of semiconductors. Phys. Rev. Lett. **80**, 4510–4153 (1998)
Note that the peak calculated to appear in Si at 3.8 eV is not observed experimentally and must be due to a shortcoming in the calculation;
6.24 M. Cardona, L.F. Lastras-Martínez, D.E. Aspues: Comment on "Ab initio calculations of excitonic effects in the optical spectra of semiconductors", Phys. Rev. Lett. **83**, 3970 (1999)
6.25 C. W. Higginbotham: Band structure and optical properties of semiconductors, PhD Thesis, Brown University, Providence, RI (1970)
6.26 M. Cardona, NE. Christensen: Spin-Orbit Splittings in AlN, GaN, and InN, Solid State Commun. **116**, 421–425 (2000)
6.27 G. Ramírez-Flores, H. Navarro-Contreras, A. Lastras-Martínez, R.C. Powell, J.E. Greene: Temperature dependence of the optical band gap of the metastable zincblende structure β-GaN, Phys. Rev. B**50**, 8433–8438 (1994)

6.28 S. Logothefidis, J. Petalas, M. Cardona, T.D. Moustakas: Optical properties and temperature dependence of the interband transitions of cubic and hexagonal GaN, Phys. Rev. B**50**, 18017, 18029 (1994)
6.29 O. Madelung, M. Schulz, H. Weiss (eds.): *Landolt-Börnstein*, Ser.III, Vol. 22 (Semiconductors), Subvol. a: Intrinsic properties of group-IV elements, III–V, II–VI and I–VII compounds (Springer, Berlin, Heidelberg 1987)
6.30 R. B. Schoolar, J. R. Dixon: Optical constant of lead sulfide in the fundamental absorption edge region. Phys. Rev. **137**, A667–670 (1965)
6.31 G. W. Gobeli, H. Y. Fan: Semiconductor Research, Second Quartertly Rept., Purdue University, Lafayette, IN (1956), as reproduced in [6.25]
6.32 E. J. Johnson: Absorption near the fundamental edge, optical properties of III–V compounds. *Semiconductor and Semimetals* **3**, 153–258 (Academic, New York 1967)
6.33 G. G. MacFarlane, V. Roberts: Infrared absorption of germanium near the lattice edge. Phys. Rev. **97**, 1714–1716 (1955); Infrared absorption of silicon near the lattice edge. ibid. **98**, 1865–1866 (1955)
6.34 G. G. MacFarlane, T. P. McLean, J. E. Quarrington, V. Roberts: Fine structure in the absorption-edge spectrum of Ge. Phys. Rev. **108**, 1377–1383 (1957)
6.35 M. Gershenzon, D. G. Thomas, R. E. Dietz: Radiative transitions near the band edge of GaP, in *Proc. Int'l Conf. on the Physics of Semiconductors*, ed. by A. C. Stickland (IOP, London 1962) p. 752–759
6.36 S. Nikitine: Excitons, in *Optical Properties of Solids*, ed. by S. Nudelman, S. S. Mitra (Plenum, New York 1969) pp. 197–237
6.37 G. H. Wannier: The structure and electronic excitation levels in insulating crystals. Phys. Rev. **52**, 191–197 (1937)
6.38 N. F. Mott: Conduction in polar crystals: II. The conduction band and ultra-violet absorption of alkali-halide crystals. Trans. Faraday Soc. **34**, 500–506 (1938)
6.39 R. S. Knox: *Theory of Excitons*, Solid State Physics, Suppl.5, ed. by F. Seitz, D. Turnbull (Academic, New York 1963)
6.40 D. L. Dexter, R. S. Knox: *Excitons* (Wiley, New York 1965)
6.41 K. Cho (ed.): *Excitons*, Topics Curr. Phys., Vol. 14 (Springer, Berlin, Heidelberg 1974)
6.42 J. J. Hopfield: Theory of contribution of excitons to the complex dielectric constant of crystals. Phys. Rev. **112**, 1555–1567 (1958)
6.43 Y. Toyozawa: On the dynamical behavior of an exciton. Suppl. Prog. Theor. Phys. **12**, 111–140 (1959)
6.44 R.J. Elliott: Intensity of optical absorption by excitons. Phys. Rev. **108**, 1384–1389 (1957)
6.45 M. Steube, K. Reimann, D. Fröhlich and S.J. Clarke: Free excitons with $n = 2$ in bulk GaN. Appl. Phys. Lett. **71**, 948–949 (1997)
B. Monemar: Optical Properties of GaN, in *Semiconductors and Semimetals*, Vol. 50 (Academic, N.Y., 1998) p. 305–363
S. J. Clarke: Free Excitons with $n = 2$ in bulk GaN, Appl. Phys. Lett. **71**, 948–949 (1991)
6.46 A. Baldereschi, N. O. Lipari: Energy levels of direct excitons in semiconductors with degenerate bands. Phys. Rev. B **3**, 439–450 (1971)
6.47 A. Baldereschi, N. O. Lipari: Spherical model of shallow acceptor states in semiconductors. Phys. Rev. B **8**, 2697–2709 (1973)
6.48 S. Adachi: *Physical Properties of III–V Semiconductor Compounds* (Wiley, New York 1992)

6.49 B. Segall, D.T.F. Marple: Intrinsic exciton absorption, in *Physics and Chemistry of II–VI Compounds*, ed. by M. Aven, J.S. Prener (North-Holland, Amsterdam 1967) pp. 319–378

6.50 M. Altarelli, N.O. Lipari: Exciton dispersion in semiconductors with degenerate bands. Phys. Rev. B **15**, 4898–4906 (1977)

6.51 P.Y. Yu: Resonant Brillouin scattering of exciton polaritons. Comments Solid State Phys. **9**, 37–48 (1979)

6.52 J.J. Hopfield: Aspects of polaritons. J. Phys. Soc. Jpn., Suppl. **21**, 77–88 (1966)

6.53 L.D. Landau, I.M. Lifshitz: *Quantum Mechanics, Nonrelativistic Theory*, 3rd edn. (Addison-Wesley, Reading, MA 1965) pp. 116–125 and 600–610

6.54 M.D. Sturge: Optical Absorption of gallium arsenide between 0.6 and 2.75 eV. Phys. Rev. **127**, 768–773 (1962)

6.55 P.W. Baumeister: Optical absorption of cuprous oxide. Phys. Rev. **121**, 359–362 (1961)

6.56 Y. Petroff, P.Y. Yu, Y.R. Shen: Study of photoluminescence in Cu_2O. Phys. Rev. B **12**, 2488–2495 (1975)

6.57 Y. Petroff, M. Balkanski: Coulomb effects at saddle-type critical points in CdTe, ZnTe, ZnSe, and HgTe. Phys. Rev. B **3**, 3299–3301 (1971)

6.58 J.P. Walter, M.L. Cohen, Y. Petroff, M. Balkanski: Calculated and measured reflectivity of ZnTe and ZnSe. Phys. Rev. B **1**, 2661–2667 (1970)

6.59 E.O. Kane: Coulomb effects at saddle-type critical points. Phys. Rev. **180**, 852–858 (1969)

6.60 S. Antoci, G.F. Nardelli: Evidence for a high sensitivity of Λ_2-Λ_1 resonant absorption structures on energy-band behavior for CdTe and ZnTe. Phys. Rev. B **6**, 1311–1314 (1972)

6.61 S. Flügge, H. Marschall: *Rechenmethoden der Quantentheorie* (Springer, Berlin, Heidelberg 1952) p. 80

6.62 J.C. Phillips: Excitons, in *The Optical Properties of Solids*, ed. by J. Tauc (Academic, New York 1966) pp. 155–184

6.63 U. Fano: Effects of configuration interaction on intensities and phase shifts. Phys. Rev. **124**, 1866–1878 (1961)

6.64 W. Hanke, L.J. Sham: Many-particle effects in the optical spectrum of a semiconductor. Phys. Rev. B **21**, 4656–4673 (1980)

6.65 N. Meskini, H.J. Mattausch, W. Hanke: Many-body effects in the absorption spectrum of a heteropolar crystal. Solid State Commun. **48**, 807–809 (1983)

6.66 B. Velický, J. Sak: Excitonic effects in the interband absorption of semiconductors. Phys. Status Solidi **16**, 147–157 (1966)

6.67 F.A. Jenkins, H.E. White: *Fundamentals of Optics*, 3rd edn. (McGraw-Hill, New York 1957) p. 472

6.68 M. Born, K. Huang: *Dynamical Theory of Crystal Lattices* (Oxford Univ. Press, Oxford 1988, reprint of the original 1954 edition) pp. 228–238

6.69 M. Hass: Lattice reflections, in *Optical Properties of III–V Compounds*, Semiconductors and Semimetals, Vol. 3 (Academic, New York 1967) pp. 3–16

6.70 A. Debernardi: Phonon linewidth in III-V semiconductors from density functional perturbation theory. Phys. Rev B**57**, 12847–12858 (1998)

6.71 A. Debernardi: Anharmonic effects in the phonons of III-V semiconductors: first principles calculations. Solid State Commun. **113**, 1–10 (2000)

6.72 H. Siegle, L. Eckey, A. Hoffmann, C. Thomsen, B.K. Meyer, D. Schikora, M. Hankeln, K. Lischka: Quantitative determination of hexagonal minority phase in GaN using Raman spectroscopy, Solid State Commun. **96**, 943–949 (1995)

6.73 H. D. Fuchs, C. H. Grein, R. I. Devlin, J. Kuhl, M. Cardona: Anharmonic decay time, isotopic scattering time, and inhomogeneous line broadening of optical phonons in ^{70}Ge, ^{76}Ge and natural Ge crystals. Phys. Rev. B **44**, 8633–8642 (1991)
6.74 R. J. Collins, H. Y. Fan: Infrared lattice absorption bands in germanium, silicon and diamond. Phys. Rev. **93**, 674–678 (1954)
6.75 F. A. Johnson: Lattice absorption bands in silicon. Proc. Phys. Soc. (London) **73**, 265–272 (1959)
6.76 M. Lax, E. Burstein: Infrared lattice absorption in ionic and homopolar crystals. Phys. Rev. **97**, 39–52 (1955)
6.77 J. L. Birman: Space group selection rules: Diamond and zinc blende. Phys. Rev. **127**, 1093–1106 (1962)
6.78 W. G. Spitzer: Multiphonon lattice absorption, in *Optical Properties of III–V Compounds*. Semiconductors and Semimetals **3**, 17–69 (Academic, New York 1967)
6.79 H. Bilz, R. Geik, K. F. Renk: Lattice vibrations and infrared absorption of germanium, silicon, and diamond, in *Proc. Int'l Conf. on Lattice Dynamics*, ed. by R. F. Wallis (Pergamon, Oxford 1963) p. 355–368
6.80 A. Debernardi, M. Bernasconi, M. Cardona, M. Parrinello: Infrared absorption in α-Si from *ab initio* molecular dynamics. Appl. Phys. Lett. **71**, 2692–2694 (1997)
6.81 J. C. Phillips: *Covalent Bonding in Crystals, Molecules, and Polymers* (Univ. Chicago Press, Chicago 1969)
6.82 W. A. Harrison: *Electronic Structure and the Properties of Solids: The Physics of the Chemical Bond* (Dover, New York 1989)
6.83 S. K. Sinha: Phonons in semiconductors. CRC Crit. Rev. Solid State Sci. **3**, 273–334 (1973)
6.84 S. de Gironcoli, S. Baroni, R. Resta: Piezoelectric properties of III–V semiconductors from first-principles linear-response theory. Phys. Rev. Lett. **62**, 2853–2856 (1989)
6.85 J. A. Sanjurjo, E. López-Cruz, P. Vogl, M. Cardona: Dependence on volume of the phonon frequencies and their IR effective charges of several III–V semicondutors. Phys. Rev. B **28**, 4579–4584 (1983)
6.86 P. Vogl: In *Physics of Nonlinear Transport in Semiconductors*, ed. by D. K. Ferry, J. R. Barker, C. Jacoboni (Plenum, New York 1980) pp. 75–116
6.87 R. D. King-Smith, D. Vanderbilt: Theory of polarization of crystalline solids. Phys. Rev. B **47**, 1651–1654 (1993)
6.88 E. Anastassakis, M. Cardona: Phonons, strains, and pressure in semiconductors, in *High Pressure in Semiconductor Physics II* ed. by T. Suski and W. Paul (Academic, San Diego 1998) p. 196
6.89 H. Ibach, H. Lüth: *Solid-State Physics*, 2nd edn. (Springer, Berlin, Heidelberg 1995) Chap. 9
6.90 C. Kittel: *Introduction to Solid State Physics*, 7th edn. (Wiley, New York 1995)
6.91 H. Y. Fan: Effects of free carries on the optical properties, in *Optical Properties on III–V Compounds. Semiconductors and Semimetals* **3**, 405–419 (Academic, New York 1967)
6.92 J. R. Dixon: Optical absorption mechanism in InAs, in *Proc. Int'l on the Physics of Semiconductors* (Czech. Acad. Sci., Prague; and Academic, New York 1961) pp. 366–368
6.93 R. Braunstein, E. O. Kane: The valence band structure of the III–V compounds. J. Phys. Chem. Solids **23**, 1423–1429 (1962)
6.94 W. G. Spitzer, H. Y. Fan: Determination of optical constants and carrier effective mass of semicondutors. Phys. Rev. **106**, 882–890 (1957)

6.95 G. Herzberg: *Atomic Spectra and Atomic Structure* (Dover, New York 1944)
6.96 C. Jagannath, Z.W. Grabowski, A.K. Ramdas: Linewidth of the electronic excitation spectra of donors in silicon. Phys. Rev. B **23**, 2023–2098 (1981)
6.97 R.L. Aggarwal, A.K. Ramdas: Optical determination of the symmetry of the ground states of group-V donors in silicon. Phys. Rev. **140**, A1246–1253 (1965)
6.98 T.M. Lifshits, F. Ya. Nad': Photoconductivity in germanium doped with group-V impurities at photon energies below the impurity ionization energy. Sov. Phys. – Dokl. **10**, 532–533 (1965)
6.99 K. Seeger: *Semiconductor Physics*, 5th edn., Springer Ser. Solid-State Sci., Vol. 40 (Springer, Berlin, Heidelberg 1991) pp. 6-70
6.100 E.E. Haller, W.L. Hansen: High resolution Fourier transform spectroscopy of shallow acceptors in ultra-pure germanium. Solid State Commun. **15**, 687–692 (1974)
6.101 A.K. Ramdas, S. Rodriguez: Spectroscopy of the solid state analogues of the hydrogen atom: Donors and acceptors in semiconductors. Rep. Prog. Phys. **44**, 1297–1387 (1981)
6.102 O.J. Glembocki, B.V. Shanabrook, N. Bottka, W.T. Beard, J. Comas: Photoreflectance characterization of interband transitions in GaAs/AlAs multiple quantum wells and modulation doped heterojunctions. Appl. Phys. Lett. **46**, 970–972 (1985)
6.103 A. Frova, P. Handler: Electric field shift of the absorption edge in the space charge region of a Ge p-n junction, in *Physics of Semiconductors*, ed. by M. Hulin (Dunod, Paris 1964) p. 157–164
6.104 B.O. Seraphin. R.B. Hess: Franz-Keldysh effect above the fundamental edge in germanium. Phys. Rev. Lett. **14**, 138–140 (1965)
6.105 D.E. Aspnes: Modulation spectroscopy/electric field effects on the dielectric function of semiconductors, in *Handbook of Semiconductors*, ed. by M. Balkanski (North-Holland, Amsterdam 1980) Vol. 2, pp. 109–154
6.106 F.H. Pollak, H. Shen: Modulation spectroscopy of semiconductors: Bulk/thin films, microstructures, surfaces/interfaces and devices. Mater. Sci. Eng. R **10**, 275–374 (1993)
6.107 D.D. Sell. E.E. Stokowski: Modulated piezoreflectance and reflectance studies of GaAs, in *Proc. 10th Int'l Conf. on the Physics of Semiconductors*, ed. by S.P. Keller, C. Hensel, F. Stern (Nat'l Bureau of Standards, Springfield, VA 1970) pp. 417–422
6.108 R.R.L. Zucca, Y.R. Shen: Wavelength-modulation spectra of some semiconductors. Phys. Rev. B **1**, 2668–2676 (1970)
6.109 B.O. Seraphin, N. Bottka: Band structure analysis from electroreflectance studies. Phys. Rev. **145**, 628–636 (1966)
6.110 S. Gopalan, P. Lautenschlager, M. Cardona: Temperature dependence of the shifts and broadenings of the critical points in GaAs. Phys. Rev. B **35**, 5577–5584 (1987)
6.111 P.B. Allen, M. Cardona: Temperature dependence of the direct gap of Si and Ge. Phys. Rev. B **27**, 4760–4769 (1983)
6.112 E. Matatagui, A. Thomson, M. Cardona: Thermoreflectance in semiconductors. Phys. Rev. **176**, 954–960 (1968)
6.113 A. Gavini, M. Cardona: Modulated piezoreflectance in semiconductors. Phys. Rev. B **1**, 672–682 (1970)
6.114 M. Abramowitz, I.A. Stegun: *Handbook of Mathematical Functions*, NBS Math. Ser. No. 44 (US GPO, Washington, DC 1970) Sect. 10
6.115 K. Thamarlingham: Optical absorption in the presence of uniform electric field. Phys. Rev. **130**, 2204–2206 (1963)

6.116 D. E. Aspnes: Electric field effects on the dielectric constant of solids. Phys. Rev. **153**, 972–982 (1967)
6.117 W. Franz: Influence of an electric field on an optical absorption edge (in German). Z. Naturforsch. **13a**, 484–489 (1958)
6.118 L. V. Keldysh: Effect of a strong electric field on the optical properties of insulating crystals. Sov. Phys. – JETP **34**, 788–790 (1958)
6.119 J. P. Estrera, W. M. Duncan, R. Glosser: Complex Airy analysis of photoreflectance spectra for III–V semiconductors. Phys. Rev. B **49**, 7281–7294 (1994)
6.120 D. E. Aspnes, A. A. Studna: Schottky barrier electroreflectance: Application to GaAs. Phys. Rev. B **7**, 4605–4625 (1973)
6.121 D. E. Aspnes. Schottky barrier electroreflectance of Ge: Non-degenerate and optically degenerate critical points. Phys. Rev. B **12**, 2297–2310 (1975)
6.122 M. Kuball: Effects of hydrogen exposure, doping, and electric fields on the properties of GaAs surfaces. Dissertation, University of Stuttgart (1995)
6.123 E. Gharhamani, D. J. Moss, J. E. Sipe: Linear and nonlinear optical properties of $(GaAs)_m/(AlAs)_n$ superlattices. Phys. Rev. B **43**, 9269–9272 (1991)
6.124 D. F. Blossey: Wannier excitons in an electric field: II. Electroabsorption in direct-band-gap solids. Phys. Rev. **3**, 1382–1391 (1971)
6.125 J. S. Kline, F. H. Pollak, M. Cardona: Electroreflectance in the Ge-Si alloys. Helv. Phys. Acta **41**, 968–976 (1968)
6.126 M. Cardona, K. L. Shaklee, F. H. Pollak: Electroreflectance at a semiconductor-electrolyte interface. Phys. Rev. **154**, 696–720 (1967)
6.127 J. G. Gay: Screening of excitons in semiconductors. Phys. Rev. B **4**, 2567–2575 (1971)
6.128 C. Parks, A. K. Ramdas, S. Rodriguez, K. M. Itoh, E. E. Haller: Electronic band structure of isotopically pure Ge: Modulated transmission and reflectivity study. Phys. Rev. B **49**, 14244–14250 (1994)
6.129 S. Zollner, M. Cardona, S. Gopalan: Isotope and temperature shifts of direct and indirect band gaps in diamond-type semiconductors. Phys. Rev. B **45**, 3376–3385 (1992)
6.130 X. Yin, X. Guo, F. H. Pollak, Y. Chan, P. A. Mantau, P. D. Kirchner, G. D. Petit, J. M. Woodal: Photoreflectance study of the surface Fermi level at a (001) n- and p-type GaAs surface. J. Vac. Sci. Technol. A **10**, 131–136 (1992)
6.131 I. Kamiya, D. E. Aspnes, L. T. Florez, J. P. Harbison: Reflectance difference spectroscopy on (001) GaAs surfaces in ultrahigh vacuum. Phys. Rev. B **46**, 15894–15904 (1992)
6.132 W. Richter: Optical in-situ control during MOVPE and MBE growth. Philos. Trans. R. Soc. (London) A **344**, 453–467 (1993)
6.133 M. Cardona, F. H. Pollak, K. L. Shaklee: Electroreflectance in semiconductors. J. Phys. Soc. Jpn. Suppl. **21**, 89–94 (1966)
6.134 D. E. Aspnes. A. A. Studna: Anisotropies in the above-band-gap optical spectra of cubic semiconductors. Phys. Rev. Lett. **54**, 1956–1959 (1985)
6.135 W. L. Mochan, R. G. Barrera: Local field on the surface conductivity of adsorbed overlayers. Phys. Rev. Lett. **56**, 2221–2224 (1986)
6.136 Y. Chang. D. E. Aspnes: Theory of the dielectric function anisotropy of (001) GaAs (2×1) surface. Phys. Rev. B **41**, 12002–12012 (1990)
6.137 D. E. Aspnes: Observation and analysis of epitaxial growth with reflectance-difference spectroscopy. J. Electron. Mater. B **30**, 109–119 (1995)
6.138 L. X. Benedict, T. Wethkamp, K. Wilmers, C. Cobet, N. Esser, E L. Shirley, W. Richter and M. Cardona: Dielectric function of wurtzite GaN and AlN thin films. Sol. Stat. Commun. **102**, 129 (1999)

6.139 C. Cobet: private communications
6.140 D.R. Penn: Wavenumber-dependent dielectric function of semiconductors. Phys. Rev. **128**, 2093–2097 (1962)
6.141 M. Cardona: Infrared dielectric constants and ultraviolet optical properties of solids with diamond, zincblende, wurtzite and rocksalt structures. J. Appl. Phys. **36**, 2181–2186 (1965)
6.142 P.Y. Yu, M. Cardona: Temperature coefficient of the refractive index of zincblende and diamond-type semiconductors. Phys. Rev. B **2**, 3193–3197 (1970)
6.143 M. Cardona: Fresnel reflection and surface plasmons. Am. J. Phys. **39**, 1277 (1971)
6.144 M. Cardona: Electronic optical properties of solids, in *Solid State Physics, Nuclear Physics and Particle Physics*, ed. by I. Saavedra (Benjamin, New York 1968) pp. 737–816
6.145 P. Wickbold, E. Anastassakis, R. Sauer, M. Cardona: Raman phonon piezospectroscopy in GaAs: Infrared measurements. Phys. Rev. B **35**, 1362–1368 (1987)
6.146 J.F. Nye: *Physical Properties of Crystals* (Oxford Univ. Press, Oxford 1969)
L.A. Shuvalov (ed.): *Modern Crystallography IV*, Springer Ser. Solid-State Sci., Vol. 37 (Springer, Berlin, Heidelberg 1988)
6.147 P. Lautenschlager, P.B. Allen, M. Cardona: Temperature dependence of band gaps in Si and Ge. Phys. Rev. B **31**, 2163–2171 (1985)
6.148 Y.R. Shen: *The Principles of Nonlinear Optics* (Wiley, New York 1984) pp. 86–93
6.149 J.A. Van Vechten, M. Cardona, D.E. Aspnes, R.M. Martin: Theory of the 3rd-order susceptibility, in *Proc. 10th Int'l Conf. on the Physics of Semiconductors*, ed. by S.P. Keller, C. Hensel, F. Stern (Nat'l Bureau of Standards, Springfield, VA 1970) pp. 82–86

General Reading

Azzam R.M.A., N.M. Bashara: *Ellipsometry and Polarized Light* (North-Holland, Amsterdam 1977)
Bassani F.: *Electronic States and Optical Transitions in Solids* (Pergamon, London 1975)
Burstein E.: *Atomic Structure and Properties of Solids* (Academic, New York 1972)
Greenaway D., G. Harbeke: *Optical Properties and Band Structure of Semiconductors* (Pergamon, London 1968)
Ill'inskii Yu.A., L.V. Keldysh: *Electromagnetic Response of Material Media* (Plenum, New York 1994)
Kalt H.: *Optical Properties of III–V Semiconductors*, Springer Ser. Solid-State Sci., Vol. 120 (Springer, Berlin, Heidelberg 1996)
Landau L., I.M. Lifshitz: *The Classical Theory of Fields* (Addison-Wesley, Reading, MA 1958)
Pollak F.H., H. Shen: Modulation Spectroscopy of Semiconductors, Bulk, Thin Film, Microstructures, Surfaces/Interfaces and Devices. Mater. Sci. Eng. R **10**, 275–374 (1993)
Palik E.A.: *Handbook of Optical Constants* (Academic, Orlando, FL 1985)
Pankove J.: *Optical Processes in Semiconductors* (Dover, New York 1971)
Tauc J.: *The Optical Properties of Solids* (Academic, New York 1966)
Willardson R.K., A.C. Beer (eds.): *Optical Properties of III–V Compounds*, Semiconductors and Semimetals, Vol. 3 (Academic, New York 1967)

Chapter 7

7.1 W. Heitler: *The Quantum Theory of Radiation* (Oxford Univ. Press, Oxford 1954)
7.2 A. Einstein: Emission and absorption of radiation in the quantum theory (in German). Verh. Dtsch. Phys. Ges. **18**, 318–323 (1916)
7.3 A. Einstein: The quantum theory of radiation (in German). Phys. Z. **18**, 121–128 (1917)
7.4 W. van Roosbroeck, W. Schockley: Photon-radiative recombination of electrons and holes in germanium. Phys. Rev. **94**, 1558–1560 (1954)
J. Pankove: *Optical Processes in Semiconductors* (Dover, New York 1971) p. 110
7.5 E. Burstein: Anomalous optical absorption limit in InSb. Phys. Rev. **93**, 632–633 (1954)
7.6 T.S. Moss: The interpretation of the properties of indium antimonide. Proc. Phys. Soc. (London) B **67**, 775–782 (1954)
7.7 S. Wang: *Fundamentals of Semiconductor Theory and Device Physics* (Prentice Hall, Englewood Cliffs, NJ 1989) pp. 760–766
7.8 N. Caswell, J.S. Weiner, P.Y. Yu: A study of non-thermalized luminescence spectra: The case of Cu_2O. Solid State Commun. **40**, 843–846 (1981)
7.9 P.Y. Yu, Y.R. Shen: Resonance Raman studies in cuprous oxide. II. The yellow and green excitonic series. Phys. Rev. B **17**, 4017–4030 (1978)
7.10 L.T. Canham: Silicon quantum wire array fabrication by electrochemical and chemical dissolution of wafers. Appl. Phys. Lett. **57**, 1046–1048 (1990)
7.11 L.T. Canham, W.Y. Leong, T.I. Cox, M.I. Beale, K.J. Nash, P. Calcott, D. Brumhead, L.L. Taylor, K.J. Marsh: Efficient visible photoluminescence and electroluminescence from highly porous silicon, in *The Physics of Semiconductors*, ed. by P. Jiang, H.-Z. Zheng (World Scientific, Singapore 1992) Vol. 2, pp. 1423–1430
7.12 K.H. Jung, S. Shih, D.L. Kwong: Developments in luminescent porous Si. J. Electrochem. Soc. **140**, 3046–3064 (1993)
7.13 J. Zeman, M. Zigone, G.L.J.A. Rikken, G. Martinez: Hydrostatic pressure effects on the porous silicon luminescence. J. Phys. Chem. Solids **56**, 655–661 (1995)
For reviews on the luminescence of porous silicon see: L. Brus: Luminescence of silicon materials: Chains, sheets, nanocrystals, microcrystals, and porous silicon. J. Phys. Chem. **98**, 3575–3581 (1994)
7.14 C. Kittel: *Introduction to Solid State Physics*, 7th edn. (Wiley, New York 1995)
7.15 J. Pankove. *Optical Processes in Semiconductors* (Dover, New York 1971)
7.16 P.Y. Yu, B. Welber: High pressure photoluminescence and resonance Raman studies of GaAs. Solid State Commun. **25**, 209–211 (1978)
7.17 H.Y. Fan: Temperature dependence of the energy gap in semiconductors. Phys. Rev. **82**, 900–905 (1951)
7.18 J.C. Inkson: The effect of electron interaction on the band gap of extrinsic semiconductors. J. Phys. C **9**, 1177–1183 (1976)
7.19 W.P. Dumke: Optical transitions involving impurities in semiconductors. Phys. Rev. **132**, 1998–2002 (1963)
7.20 F.E. Williams: Theory of the energy levels of donor-acceptor pairs. J. Phys. Chem. Solids **12**, 265–275 (1960)
7.21 P.J. Dean: Inter-impurity recombinations in semiconductors. *Progress in Solid State Chemistry* **8**, 1–216 (Pergamon, Oxford 1973)

7.22 D. G. Thomas, M. Gershenzon, F. A. Trumbore: Pair spectra and "edge" emission in gallium phosphide. Phys. Rev. **133**, A269–279 (1964)

7.23 P. J. Dean, E. G. Schönherr, R. B. Zetterstrom: Pair spectra involving the shallow acceptor Mg in GaP. J. Appl. Phys. **41**, 3474–3479 (1970)

7.24 D. G. Thomas, J. J. Hopfield, W. N. Augustyniak: Kinetics of radiative recombination of randomly distributed donors and acceptors. Phys. Rev. **140**, A202–220 (1965)

7.25 D. D. Sell, S. E. Stokowski, R. Dingle, J. V. DiLorenzo: Polariton reflectance and photoluminescence in high-purity GaAs. Phys. Rev. B **7**, 4568–4586 (1973)

7.26 F. Askary, P. Y. Yu: Polariton luminescence and additional boundary conditions: Comparison between theory and experiment. Solid State Commun. **47**, 241–244 (1983)

7.27 Y. Toyozawa: On the dynamical behavior of an exciton. Prog. Theor. Phys. Suppl. **12**, 111–140 (1959)

7.28 J. J. Hopfield: Theory of the contribution of excitons to the complex dielectric constant of crystals. Phys. Rev. **112**, 1555–1567 (1958)

7.29 J. D. Jackson: *Classical Electrodynamics*, 2nd edn. (Wiley, New York 1975) pp. 17–22; p. 396

7.30 S. I. Pekar: The theory of electromagnetic waves in a crystal in which excitons are produced. Sov. Phys. – JETP **6**, 785–796 (1958); ibid. **7**, 813–822 (1958)

7.31 C. S. Ting, M. J. Frankel, J. L. Birman: Electrodynamics of bounded spatially dispersive media: The additional boundary conditions. Solid State Commun. **17**, 1285–1289 (1975)

7.32 K. Cho (ed.): *Excitons*, Topics Curr. Phys. Vol. 14 (Springer, Berlin, Heidelberg 1979)

7.33 A. Stahl, I. Balslev: *Electrodynamics of the Semiconductor Band Edge*, Springer Tracts Mod. Phys., Vol. 110 (Springer, Berlin, Heidelberg 1987)

7.34 J. Tignon, T. Hasche, D. Chaula, H. C. Schneider, F. Jahnke, S. W. Koch: Unified Picture of Polariton Propagation in Bulk GaAs Semiconductors. Phys. Rev. Lett. **84**, 3382 (2000)

7.35 T. Steiner, M. L. Thewalt, E. S. Koteles, J. P. Salerno: Effect of neutral donor scattering on the time-dependent exciton-polariton photoluminescence line shape in GaAs. Phys. Rev. B **34**, 1006–1013 (1986)

7.36 D. G. Thomas, J. J. Hopfield: Optical properties of bound exciton complexes in cadmium sulfide. Phys. Rev. **128**, 2135–2148 (1962)

7.37 R. McWeeny: *Coulson's Valence*, 3rd edn. (Oxford Univ. Press, Oxford 1979) p. 90

7.38 J. J. Hopfield: The quantum chemistry of bound exciton complexes, in *Proc. 7th Int'l Conf. on the Physics of Semiconductors*, ed. by M. Hulin (Dunod, Paris 1964) pp. 725–735

7.39 L. Pauling, E. B. Wilson: *Introduction to Quantum Mechanics* (McGraw-Hill, New York 1935) p. 225

7.40 D. D. Thornton, A. Honig: Shallow-donor negative ions and spin-polarized electron transport in silicon. Phys. Rev. Lett. **30**, 909–912 (1973)

7.41 S. Huant, S. P. Najda, B. Etienne: Two-dimensional D^- centers. Phys. Rev. Lett. **65**, 1486–1489 (1980)

7.42 M. L. Lambert: Mobile and Immobile Effective-Mass-Particle Complexes in Nonmetallic Solids. Phys. Rev. Lett. **1**, 450–453 (1958)

7.43 G. Finkelstein, H. Shtrikmann and I. Bar-Joseph: Optical spectroscopy of a two-dimensional electron gas near the metal-insulator transition. Phys. Rev. Lett. **74**, 976 (1993)

7.44 R.T. Cox, V. Huard, K. Kheng, S. Lovisa, R.B. Miller, K. Saminadayar, A. Arnoult, J. Cibert, S. Tatarenko and M. Potemski: Exciton Trions in II-VI Heterostructures. Acta Physica Polonica A, **94**, 99 (1998)

7.45 M. Stuke (ed.): *Dye Lasers: 25 Years*, Topics Appl. Phys., Vol. 70 (Springer, Berlin, Heidelberg 1992)

7.46 W. Gellermann: Color center lasers. J. Phys. Chem. Solids **52**, 249–279 (1991)

7.47 Y.A. Carts: Titanium sapphire's star rises. Laser Focus World **25**, 73–88 (September 1989)

7.48 A. Sanchez, R.E. Fahey, A.J. Strauss, R.L. Aggarwal: Room temperature CW operation of the Ti:Al_2O_3 laser, in *Tunable Solid State Lasers II*, ed. by A.B. Budgor, L. Esterowitz, G. DeShazer, Springer, Ser. Opt. Sci., Vol. 52 (Springer, Berlin, Heidelberg 1986) p. 202

7.49 R. Dingle: Optical properties of semiconductors superlattices, in *Proc. 13th Int'l Conf. on the Physics of Semiconductors*, ed. by F. Fumi (Tipografia Marves, Rome 1976) pp. 965–974

7.50 H.W. Liu, R. Ferreira, G. Bastard, C. Delalande, J.F. Palmier, B. Etienne: Optical evidence of assisted tunneling in a biased double quantum well structure. Appl. Phys. Lett. **54**, 2082–2084 (1989)

7.51 P.S. Kop'ev, N.N. Ledentsov, B. Ya. Meltser, I.N. Uraltsev, A.L. Efros, D.R. Yakovlev: Exciton luminescence probe of interface quality in multi-quantum well structures, in *The Physics of Semiconductors*, ed. by O. Engström (World Scientific, Singapore 1986) pp. 219–222

7.52 H. Tews, H. Venghaus, P.J. Dean: Excited states of shallow acceptors in ZnSe. Phys. Rev. B **19**, 5178–5184 (1979)

7.53 L. Brillouin: Scattering of light and X-rays by a transparent homogeneous body: Influence of the thermal agitation (in French). Ann. Physique **17**, 88–122 (1922)

7.54 L.I. Mandelstam: On light scattering by an inhomogeneous medium (in Russian). Zh. Russko Fiz. Khim. Obshch. (J. Russian Physico-Chemical Soc.) **58**, 381 (1926)

7.55 C.V. Raman: A new class of spectra due to secondary radiation. Ind. J. Phys. **2**, 387–400 (1928)
C.V. Raman: A change of wavelength in light scattering. Nature **121**, 619–620 (1928)

7.56 A.K. Ramdas, S. Rodriguez: Raman Scattering in Diluted Semiconductor in *Light Scattering in Solids VI*, ed. by M. Cardona and G. Güntherodt (Springer, Heidelberg 1991) pp. 137–206; see also Light scattering in Solids VII, ed. by M. Cardona and G. Güntherodt (Springer, Heidelberg, 1999)

7.57 R. Loudon: Theory of the first-order Raman effect in crystals. Proc. Roy. Soc. (London) A **275**, 218–232 (1963)

7.58 R. Loudon: Raman effect in crystals. Adv. Phys. **13**, 423–482 (1964); erratum ibid. **14**, 621 (1965)

7.59 E. Burstein, A. Pinczuk: Light scattering by collective excitation in dielectrics and semiconductors, in *The Physics of Opto-electronic Materials*, ed. by W.A. Albers Jr (Plenum, New York 1971) pp. 33–79

7.60 Z.P. Su, T. Ruf, K.R. Wald and P.Y. Yu: Resonant Generation of Nonequilibrium Phonons in GaAs/AlAs Quantum Wells via Intersubband Scattering. *Proc. of 22nd Int. Conf. on The Physics of Semicond.*, et. by D.J. Lockwood, (World Scientific, Singapore, 1995) p. 963–966

7.61 Y.R. Shen: *The Principles on Nonlinear Optics* (Wiley, New York 1984) pp. 141–186
7.62 M.R. Zatzick: How to make every photon count. Electro-Opt. Syst. Design **20–23**, 26–27 (1972)
7.63 R.K. Chang, M.B. Long: Optical multichannel detection, in *Light Scattering in Solids II*, ed. by M. Cardona, G. Güntherodt, Topics Appl. Phys., Vol. 49 (Springer, Berlin, Heidelberg 1983) Chap. 3, pp. 179–205
7.64 J.C. Tsang: Multichannel detection and Raman sprectroscopy of surface layers and interfaces, in *Light Scattering in Solids V*, ed. by M. Cardona, G. Güntherodt, Topics Appl. Phys., Vol. 66 (Springer, Berlin, Heidelberg 1989) Chap. 6, pp. 233–284
7.65 C. Firmani, E. Ruiz, C.W. Carlson, M. Lampton, F. Paresce: High-resolution imaging with a two-dimensional resistive anode photon counter. Rev. Sci. Instrum. **53**, 570–574 (1982)
7.66 C. Firmani, L. Gutierrez, E. Ruiz, G.F. Bisiacchi, L. Salas, F.E. Paresce, C.W. Carlson, M. Lampton: High dispersion spectrophotometry with the mepsicron photon counter. Astron. Astrophys. **134**, 251–257 (1984)
7.67 A. Mooradian, G.B. Wright: First order Raman effect in III–V compounds. Solid State Commun. **4**, 431–434 (1966)
7.68 P.A. Temple, C.E. Hathaway: Multiphonon Raman spectrum of silicon. Phys. Rev. B **7**, 3685–3697 (1973)
7.69 J.L. Birmann: Space group selection rules: Diamond and zinc blende. Phys. Rev. **127**, 1093–1106 (1965)
7.70 L. Kleinman: Polarization effects in two-phonon Raman scattering in cubic crystals. Solid State Commun. **3**, 47–53 (1965)
7.71 B.A. Weinstein, M. Cardona: Second-order Raman spectrum of germanium. Phys. Rev. B **7**, 2545–2551 (1973)
7.72 G. Nelin, G. Nilsson: Phonon density of states in germanium at 80 K measured by neutron spectrometry. Phys. Rev. B **5**, 3151–3160 (1972)
7.73 C.H. Henry, J.J. Hopfield: Raman scattering by polaritons. Phys. Rev. Lett. **15**, 964–966 (1965)
7.74 L.B. Humphreys, A.A. Maradudin: Ionic Raman effect. III. First and second-order ionic Raman effect. Phys. Rev. B **6**, 3886–3897 (1972)
7.75 V.B. Berestetskii, E.M. Lifshitz, L.P. Pitaevskii: *Quantum Electrodynamics* (Pergamon, Oxford 1980)
7.76 W.E. Parry: *The Many-Body Problem* (Clarendon, Oxford 1973)
7.77 V.I. Belitsky, A. Cantarero, M. Cardona, I.G. Lang, S.T. Pavlov: Spatial correlation of electrons and hole in multiphonon resonant Raman scattering in a high magnetic field. Phys. Rev. B **46** 15767–15788 (1992)
7.78 R.M. Martin, L.M. Falicov: Resonance Raman scattering, in *Light Scattering in Solids I*, ed. by M. Cardona, 2nd edn., Topics Appl. Phys., Vol. 8 (Springer, Berlin, Heidelberg 1983) Chap. 3, pp. 70–145
7.79 A. Pine: Brillouin scattering in semiconductors, in *Light Scattering in Solids I*, ed. by M. Cardona, 2nd edn., Topics Appl. Phys., Vol. 8 (Springer, Berlin, Heidelberg 1983) Chap. 6, pp. 253–273
7.80 A. Dervisch, R. Loudon: Theory of Brillouin scattering lineshape in an opaque material. J. Phys. C **9**, L669–673 (1976)
7.81 A.S. Pine, G. Dresselhaus: Lineshape asymmetries in light scattering from opaque materials, in *Light Scattering in Solids*, ed. by M. Balkanski, R.C.C. Leite, S.P.S. Porto (Flammarion, Paris 1976) p. 138

7.82 J. R. Sandercock: Brillouin scattering measurements on silicon and germanium. Phys. Rev. Lett. **8**, 237–240 (1972)
7.83 J. R. Sandercock: Light scattering from surface acoustic phonons in metals and semiconductors. Solid State Commun. **26**, 547–551 (1978)
7.84 J. R. Sandercock: The design and use of a stabilized multipassed interferometer of high contrast ratio, in *2nd Int'l Conf. on Light Scattering in Solids*, ed. by M. Balkanski (Flammarion, Paris 1971) pp. 9–12
7.85 J. M. Ralston, R. L. Wadsack, R. K. Chang: Resonant cancellation of Raman scattering from CdS and Si. Phys. Rev. Lett. **25**, 814–818 (1970)
7.86 J. Menéndez, M. Cardona: Interference between allowed and forbidden Raman scattering by longitudinal-optical phonons in GaAs. Phys. Rev. Lett. **51**, 1297–1299 (1983)
7.87 R. Zeyher, H. Bilz, M. Cardona: On the ω^4 law in microscopic theories of inelastic light scattering. Solid State Commun. **19**, 57 (1976)
R. M. Martin: Resonance Raman scattering near critical points. Phys. Rev. B **10**, 2620 (1974)
7.88 B. A. Weinstein, M. Cardona: Resonant first- and second-order Raman scattering in GaP. Phys. Rev. B **8**, 2795–2809 (1973)
7.89 A. Cantarero, C. Trallero-Giner, M. Cardona: Excitons in one-phonon resonant Raman scattering. Solid State Commun. **69**, 1183–1186 (1989)
7.90 A. K. Ganguly, J. L. Birman: Theory of lattice Raman scattering in insulators. Phys. Rev. **162**, 806–816 (1967)
7.91 B. Bendow, J. L. Birman: Polariton theory of Raman scattering in insulating crystals. Phys. Rev. B **4**, 1678–1686 (1971)
7.92 P. Y. Yu: Study of excitons and exciton-phonon interactions by resonant Raman and Brillouin spectroscopies, in *Excitons*, ed. by K. Cho, Topics Curr. Phys., Vol. 14 (Springer, Berlin, Heidelberg) 1979) p. 211
7.93 C. Weisbuch, R. G. Ulbricht: Resonant light scattering mediated by excitonic polaritons in semiconductors, in *Light Scattering in Solids III*, ed. by M. Cardona, G. Güntherodt, Topics Appl. Phys., Vol. 51 (Springer, Berlin, Heidelberg 1982) Chap. 7
7.94 J. Reydellet, P. Y. Yu, J. M. Besson, M. Balkanski: Cascade resonant Raman processes in GaSe, in *Physics of Semiconductors 1978*, ed. by B. L. H. Wilson (Inst. of Phys., Bristol, UK 1979) pp. 1271–1274
7.95 E. I. Rashba, G. Gurgenishvili: Edge absorption theory in semiconductors. Sov. Phys. – Solid State **4**, 759–760 (1962)
7.96 A. S. Barker Jr., A. J. Sievers: Optical studies of the vibrational properties of disordered solids. Rev. Mod. Phys. **47**, S1–178 (1975)
P. Y. Yu, M. H. Pilkhun, F. Evangelisti: Resonant Raman scattering of Cl in CdS. Solid State Commun. **25**, 371–373 (1978)
7.97 M. A. Washington, A. Z. Genack, H. Z. Cummins, R. H. Bruce, A. Compaan, R. A. Forman: Spectroscopy of excited yellow exciton states in Cu_2O by forbidden resonant Raman scattering. Phys. Rev. B **15**, 2145–2153 (1977)
7.98 J. S. Weiner, N. Caswell, P. Y. Yu, A. Mysyrowicz: Ortho- to para-exciton conversion in Cu_2O: A subnanosecond time-resolved photoluminescence study. Solid State Commun. **46**, 105–108 (1983)
7.99 M. Cardona, M. Grimsditch, D. Olego: Theoretical and experimental determination of Raman scattering cross sections in simple solids, in *Light Scattering in Solids*, ed. by J. L. Birman, H. Z. Cummins, K. K. Rebane (Plenum, New York 1979) pp. 249–256

7.100 A. Cantarero, C. Trallero-Giner, M. Cardona: Excitons in one-phonon resonant Raman scattering: Deformation potential interaction. Phys. Rev. B **39**, 8388–8397 (1989)

7.101 A. Cantarero, C. Trallero-Giner, M. Cardona: Excitons in one-phonon resonant Raman scattering: Fröhlich and interference effects. Phys. Rev. B **40**, 12290–12295 (1989)

7.102 M. Cardona: Resonance phenomena, in *Light Scattering in Solids II*, ed. by M. Cardona, G. Güntherodt, Topics Appl. Phys., Vol. 50 (Springer, Berlin, Heidelberg 1982) Chap. 2, pp. 19–176

7.103 R. M. Martin: Theory of the one-phonon resonance Raman effect. Phys. Rev. B **4**, 3676–3684 (1971)

7.104 R. M. Martin, T. C. Damen: Breakdown of selection rules in resonance Raman scattering. Phys. Rev. Lett. **26**, 86–89 (1971)

7.105 P. J. Colwell, M. V. Klein: Wave vector dependence and numerical value of the scattering efficiency for the resonant Raman effect in CdS. Solid State Commun. **8**, 2095–2100 (1970)

7.106 S. Permogorov, A. Reznitzky: Wave vector conservation in the first order resonant Raman scattering. Solid State Commun. **18**, 781–784 (1976)

7.107 E. F. Gross, S. A. Permogorov, V. V. Travnikov, A. V. Sel'kin: Inelastic scattering of polaritons in CdS crystals, in *Light Scattering in Solids*, ed. by M. Balkanski (Flammarion, Paris 1971) pp. 238–243

7.108 R. C. C. Leite, J. C. Scott, T. C. Damen: Multiple-phonon resonant Raman scattering in CdS. Phys. Rev. Lett. **22**, 780–782 (1969)

7.109 M. V. Klein, S. P. S. Porto: Multiple-phonon-resonance Raman effect in CdS. Phys. Rev. Lett. **22**, 782–784 (1969)

7.110 R. M. Martin, C. M. Varma: Cascade theory of inelastic scattering of light. Phys. Rev. Lett. **26**, 1241–1244 (1971)

7.111 R. Zeyher: Theory of multiphonon Raman spectra above the energy gap in semiconductor. Solid State Commun. **16**, 49–52 (1975)

7.112 J. Menéndez, M. Cardona: Interference effects: A key to unterstanding forbidden Raman scattering by LO phonons in GaAs. Phys. Rev. B **31**, 3696–3706 (1985)

7.113 M. V. Klein: The equivalence of resonance Raman scattering in solids with absorption followed by luminescence. Phys. Rev. B **8**, 919–921 (1973)

7.114 Y. R. Shen: Distinction between resonance Raman scattering and hot luminescence. Phys. Rev. B **9**, 622–626 (1974)

7.115 A. García-Cristóbal, A. Cantarero, C. Trallero-Giner, M. Cardona: Excitonic model for second-order resonance Raman scattering. Phys. Rev. B **49**, 13430–13445 (1994)

7.116 D. Olego, M. Cardona: Raman scattering by two LO-phonons near Γ in GaAs. Solid State Commun. **39**, 1071–1075 (1981)

7.117 K. Huang: The long wave modes of the Cu_2O lattice. Z. Physik **171**, 213–225 (1963)

7.118 C. Carabatos, B. Prevot: Rigid ion model lattice dynamics of cuprite (Cu_2O). Phys. Status Solidi (b) **44**, 701–712 (1971)

7.119 E. Burstein, D. L. Mills, A. Pinczuk, S. Ushioda: Exciton-enhanced Raman scattering by optical phonons. Phys. Rev. Lett. **22**, 348–352 (1969)

7.120 J. J. Hopfield: Resonant scattering of polaritons as composite particles. Phys. Rev. **182**, 945–952 (1969)

7.121 B. Bendow, J. L. Birman: Polariton theory of resonant Raman scattering in insulating crystals. Phys. Rev. B **1**, 1678–1686 (1970)

7.122 D.C. Reynolds, R.N. Euwema, T.C. Collins: Evidence for spatial dispersion in emission from CdS platelets, in *Proc. Int'l Conf. on the Physics of Semiconductors*, ed. by S.M. Ryvkin (Nauka, Moscow 1968) pp. 210–216

7.123 V.A. Kiselev, B.S. Razbirin, I.N. Uraltsev: Additional waves and Fabry-Perot interference of photoexcitation (polaritons) in thin II–VI crystals. Phys. Status Solidi B **72**, 161–172 (1975)

7.124 W. Brenig, R. Zeyher, J.L. Birman: Spatial dispersion effects in resonant polariton scattering. II. Resonant Brillouin scattering. Phys. Rev. B **6**, 4617–4622 (1972)

7.125 R.G. Ulbrich, C. Weisbuch: Resonant Brillouin scattering of excitonic polaritons in gallium arsenide. Phys. Rev. Lett. **38**, 865–868 (1977)

7.126 R.G. Ulbrich: Resonant Brillouin scattering in semiconductors. *Festkörperprobleme* (Adv. Solid State Phys.) **18**, 217–240 (Vieweg, Braunschweig 1978)

7.127 E.S. Koteles: Investigation of exciton-polariton dispersion using laser techniques, in *Excitons*, ed. by E.I. Rashba, M.D. Sturge (North-Holland, Amsterdam 1982) pp. 85–140

7.128 P. Wickboldt, E. Anastassakis, R. Sauer, M. Cardona: Raman phonon piezospectroscopy in GaAs: Infrared measurements. Phys. Rev. B **35**, 1362–1368 (1987)

General Reading

Agrawal, G.P.: Semiconductor Lasers: post, present, and future. (AIP Press, Woodburry, New York, 1995)

Balkanski M. (ed.): *Light Scattering in Solids* (Flammarion, Paris 1971)

Bebb H.B., E.W. Williams: Photoluminescence: I. Theory, II. Gallium arsenide. *Semiconductors and Semimetals* **8**, 182–329 (Academic, New York 1972)

Cardona M., G. Güntherodt (eds.): *Light Scattering in Solids I–VI*, Topics Appl. Phys., Vols. 8, 50, 51, 54, 66, 68 (Springer, Berlin, Heidelberg 1975–91)

Chapter 8

8.1 M. Cardona. L. Ley (eds.): *Photoemission in Solids I*, Topics App. Phys., Vol.26 (Springer, Berlin, Heidelberg 1978)

8.2 L. Ley, M. Cardona (eds.): *Photoemission in Solids II*, Topics Appl. Phys., Vol.27 (Springer, Berlin, Heidelberg 1979)

8.3 K. Siegbahn, C. Nordling, A. Fahlman, R. Nordberg, K. Hamrin, J. Hedman, G. Johansson, T. Bergman, S.-E. Karlsson, I. Lindgren, B. Lindberg: *ESCA – Atomic, Molecular and Solid State Structure* (Almqvist and Wiksels, Uppsala 1967)

8.4 L.E. Davis, N.C. MacDonald, P.W. Palmberg, G.E. Riach, R.E. Weber: *Handbook of Auger Electron Spectroscopy* (Physical Electronics Industries, Eden Prairie, MN 1979)

8.5 C. Kunz (eds.): *Synchrotron Radiation Techniques and Applications*, Topics Curr. Phys., Vol.10 (Springer, Berlin, Heidelberg 1979)

8.6 E.E. Koch: *Handbook of Synchrotron Radiation* (North-Holland, Amsterdam 1983)

8.7 M. Cardona, W. Gudat, B. Sonntag, P.Y. Yu: Optical absorption of semiconductors from 12 to 200 eV, in: *Proc. 10th Int'l Conf. on the Physics of Semiconductors*, Cambridge MA (National Bureau of Standards, Washington, DC 1971) pp. 209–212

8.8 J. Daniels, C. v. Festenberg, H. Raether, K. Zeppenfeld: Optical constants of solids by electron spectroscopy. *Springer Tracts Mod. Phys.* **54**, 77–135 (Springer, Berlin, Heidelberg 1970)

8.9 C. v. Festenberg: Z. Phys. **227**, 453 (1969)

8.10 D.A. Shirley: In *Photoemission in Solids I*, ed. by M. Cardona. L. Ley, Topics Appl. Phys., Vol. 26 (Springer, Berlin, Heidelberg 1978) Chap. 4

8.11 W. Mönch: *Semiconductor Surfaces and Interfaces*, 2nd edn., Springer Ser. Surf. Sci., Vol. 26 (Springer, Berlin, Heidelberg 1995)

8.12 H. Lüth: *Surfaces and Interfaces of Solids*, 3rd edn. (Springer, Berlin, Heidelberg 1995)

8.13 H. Hertz: The effect of ultraviolet light on an electrical discharge. Ann. Phys. (Leipzig) **31**, 983–1000 (1887) in German
J.J. Thompson: Cathode rays. Phil. Mag. **44**, 293 (1897)

8.14 A. Einstein: The production and transformation of light: A heuristic point of view. Ann. Phys. (Leipzig) **17**, 132 (1905) in German

8.15 A.L. Hughes, L.A. DuBridge: *Photoelectric Phenomena* (McGraw-Hill, New York 1932) p. 16, Fig. 2–7

8.16 R.I. Bell: *Negative Electron Affinity Devices* (Clarendon, Oxford 1973)

8.17 C. Sebenne, D. Bolmont, G. Guichar, M. Balkanski: Surface states from photoemission threshold measurements on a clean, cleaved Si (111) surface. Phys. Rev. B **12**, 3280–3285 (1983)

8.18 E.O. Kane: Theory of photoelectric emission from semiconductors. Phys. Rev. **127**, 131–141 (1962)

8.19 K. Winer, I. Hirabayashi, L. Ley: Distribution of occupied near-surface bandgap states. Phys. Rev. B **38**, 7680–7693 (1988)

8.20 J. Kirschner: *Polarized Electrons at Surfaces*, Springer Tracts Mod. Phys., Vol. 106 (Springer, Berlin, Heidelberg 1985)
J. Kessler: *Polarized Electrons*, 2nd edn., Springer Ser. Atoms Plasmas, Vol. 1 (Springer, Berlin, Heidelberg 1985)

8.21 H. Riechert, S.F. Alvarado, A.N. Titkov, V.I. Safarev: Precession of the spin polarization of photoexcited electrons in the band bending region of GaAs (110). Phys. Rev. Lett. **52**, 2297–2300 (1984)

8.22 F.J. Himpsel: Measurements of the photoemission in the study of solids. Adv. Phys. **32**, 1–51 (1983)

8.23 T. Valla, A.V. Fedorov, P.D. Johnson, B.O. Wells, S.L. Hulbert, Q. Li, G.D. Gu, N. Koshizuka: Evidence for Quantum critical behavior in optimally doped $Bi_2Sr_2CaCu_2O_{8+\delta}$. Science **285**, 2110–2113 (1999). Here a commercial hemispherical analyzer with a resolution of 10 meV was used

8.24 H. Feldner-Feldegg, U. Gelius, B. Wannberg, A.G. Nilsson, E. Basilier, K. Siegbahn: New developments in ESCA instrumentation. J. Electron Sprectrosc. **5**, 643–689 (1974)

8.25 W.L. Schaich: Theory of photoemission: Independent praticle model, in *Photoemission in Solids I*, ed. by M. Cardona, L. Ley, Topics Appl. Phys., Vol. 26 (Springer, Berlin, Heidelberg 1978) Chap. 2

8.26 C.N. Berglund, W.E. Spicer: Photoemission studies of Cu and Ag: Theory. Phys. Rev. A **136**, 1030–1044 (1964)

8.27 A. Goldmann, J. Tejeda, N.J. Shevchik, M. Cardona: Density of valence states of CuCl, CuBr, CuI, and AgI. Phys. Rev. B **10**, 4388–4402 (1974)

8.28 W.V. Nemoshkalenko, V.G. Aleshin, Yu.N. Kucherenko: Theoretical study of X-ray photoelectron spectrum of germanium valence electrons. Solid State Commun. **20**, 1155–1157 (1976)

8.29 S.P. Kowalczyk, L. Ley, F.R. McFeeley, D.A. Shirley: An ionicity scale based on X-ray photoemission valence band spectra of $A^N B^{8-N}$ and $A^N B^{10-N}$ type crystals. J. Chem. Phys. **61**, 2850–2856 (1974)

8.30 N.V. Smith: Angular dependent photoemission. In *Photoemission in Solids I*, ed. by. M. Cardona, L. Ley, Topics Appl. Phys., Vol. 26 (Springer, Berlin, Heidelberg 1978) Chap. 6

8.31 R. Leckey, J.D. Riley, A. Stampfl: Angle resolved photoemission using a toroidal energy analyser. J. Electron Spectrosc. **52**, 855–866 (1990)

8.32 G.D. Mahan: Theory of Photoemission in Simple Metals. Phys. Rev. B**2**, 4334–4350 (1970)

8.33 P.M. Williams: Direct evalution of electronic band structures of layered solids using angle resolved photoemission. Nuovo Cimento B **38**, 216–225 (1977)

8.34 I.T. McGovern, W. Eberhardt, E.W. Plummer, J.E. Fischer: The band structures of graphite and graphite intercalation compounds as determined by angle resolved photoemission using synchrotron radiation. Physica B **99**, 415–419 (1980)

8.35 L.F. Ley: Angular resolved photoemission and the band structure of solids. J. Electron Spectrosc. **15**, 329–347 (1979)

8.36 M.C. Schabel, C.H. Park, A. Matsuura, Z.X. Shen, D.A. Bonn, R. Liang, W.N. Hardy: Angle Resolved Photoemission on untwinned $YBa_2Cu_3O_{6.95}$. Phys. Rev. B **57**, 6107–6115 (1998)

8.37 F.J. Himpsel: Electronic structure of solids: photoemission spectra and related data. *Landolt-Börnstein*, Vol. 23 (Springer, Berlin, Heidelberg 1989)

8.38 G. Williams, F. Cerrina, G.J. Lapeyre, J.R. Anderson, R.J. Smith, J. Hermanson: Experimental study of the band structure of GaP, GaAs, GaSb, InP, InAs, InSb. Phys. Rev. B **34**, 5548–5557 (1986)

8.39 H.U. Middleman, L. Sorba, U. Kinkel, K. Horn: Valence-band-structure determination of InSb by angle resolved photoemission. Phys. Rev. B **34**, 957–962 (1986)

8.40 H.X. Chen, W. Ranke, E. Schröder-Bergen: Complete band structure of germanium determined by photoemission. Phys. Rev. B **42**, 7429–7433 (1990)

8.41 J.R. Chelikowsky and M.L. Cohen: Nonlocal pseudopotential calculations for the electronic structure of eleven diamond and zinc-blende semiconductors. Phys. Rev. B **14**, 556–582 (1976)

8.42 T. Grandke, L. Ley, M. Cardona: Angular resolved UV photoemission and electronic band structure of the lead chalcogenides. Phys. Rev. B. **18**, 3847–3871 (1978)

8.43 N.J. Shevchik, J. Tejeda, M. Cardona, D.W. Langer: Photoemission and density of valence states of II–VI compounds. Phys. Stat. Solidi B **59**, 87–100 (1973)

8.44 K.O. Magnusson, G. Neuhold, K. Horn, D.A. Evans: Electronic band structure of cubic CdSe by angle resolved photoemission: Cd 4*d* and valence states. Phys. Rev. B **57**, 8945–8950 (1998)

8.45 W.A. Harrison: *Electronic Structure and the Properties of Solids: The Physics of the Chemical Bond* (Dover, New York 1989)

8.46 J.N. Anderson, U.O. Karlsson: Surface core level shifts of InAs (110). Phys. Rev. B **41**, 3844–3846 (1990)

8.47 T. C. Chiang: Core level photoemission of surfaces, interfaces, and overlayers. Crit. Rev. Solid State Mater. Sci. **14**, 275–317 (1988)
8.48 S. P. Kolwalczyk, E. A. Kraut, J. R. Waldrop, R. W. Grant: Measurement of ZnSe/GaAs and ZnSe/Ge heterojunctions band discontinuities by X-ray photoelectron spectroscopy. J. Vac. Sci. Technol. **21**, 482–485 (1982)
8.49 J. E. Ortega, F. J. Himpsel: Inverse-photoemission study of Ge (110), Si (100), and GaAs (110): Bulk bands and surface states. Phys. Rev. B **47**, 2130–2137 (1993)
8.50 P. Molinàs i Mata: Atomic structure and electronic properties of pure and Ga-covered Ge(111) surfaces as seen with STM (in German). Dr. Sci. Dissertation, University of Stuttgart (1993)
8.51 N. Takeuchi, A. Selloni, E. Tosatti: Do we know the true structure of Ge(111)-c(2×8)? Phys. Rev. Lett. **69**, 648–651 (1992)
8.52 M. A. Olmstead, N. Amer: Direct measurement of the polarization dependence of Si (111) 2×1 surface-state absorption by use of photothermal displacement spectroscopy. Phys. Rev. Lett. **52**, 1148 (1984)
8.53 W. Eberhardt, G. Kalkoffen, C. Kunz, D. Aspnes, M. Cardona: Surface band bending, EDC and Yield spectroscopy from 2p core levels in heavily doped silicon. Phys. Stat. Solidi B **88**, 135–143 (1978)

General Reading

Electron Spectrocopies

Cardona M., L. Ley: *Photoemission in Solids I*, Topics Appl. Phys., Vol. 26 (Springer, Berlin, Heidelberg 1978)
Daniels J., C. v. Festenberg, H. Raether, K. Zeppenfeld: Optical constants of solids by electron spectroscopy. *Springer Tracts Mod. Phys.* **54**, 77–135 (Springer, Berlin, Heidelberg 1970)
Davis E., N. C. MacDonald, P. W. Palmberg, G. E. Riach, R. E. Weber: *Handbook of Auger Electron Spectroscopy* (Physical Electronics Industries, Eden Prairie, MN 1976)
Himpsel F. J.: Electronic structure of solids: Photoemission spectra and related data, in *Landolt-Börnstein*, Vol. 23 (Springer, Berlin, Heidelberg 1989)
Hüfner S.: *Photoelectron Spectroscopy*, 2nd edn., Springer Ser. Solid-State Sci., Vol. 82 (Springer, Berlin, Heidelberg 1996)
Ley L., M. Cardona: *Photoemission in Solids II*, Springer Ser. Solid-State Sci., Vol. 35 (Springer, Berlin, Heidelberg 1979)
Siegbahn K., C. Nordling, A. Fahlman, R. Nordberg, K. Hamrin, J. Hedman, G. Johansson, T. Bergman, S.-E. Karlsson, I. Lindgren, B. Lindberg: *ESCA-Atomic, Molecular and Solid State Structure* (Almqvist and Wiksels, Uppsala 1967)
Thompson M., M. D. Barker, A. Christio, J. F. Tysan: *Auger Electron Spectroscopy* (Wiley, New York 1985)

Synchrotron Radiation

Eberhardt W. (ed.): *Applications of Synchrotron Radiation*, Springer Ser. Surf. Sci., Vol. 35 (Springer, Berlin, Heidelberg 1995)
Koch E. E.: *Handbook of Synchrotron Radiation* (North-Holland, Amsterdam 1983)
Kunz C.: *Synchrotron Radiation*, Topics Curr. Phys., Vol. 10 (Springer, Berlin, Heidelberg 1979)

Electronic and Surface Structure

Bai C.-L.: *Scanning Tunneling Microscopy and Related Techniques*, Springer Ser. Surf. Sci., Vol. 32 (Springer, Berlin, Heidelberg 1995)

Bell R. I.: *Negative Electron Affinity Devices* (Clarendon, Oxford 1973)

Güntherodt H.-J., R. Wiesendanger (eds.): *Scanning Tunneling Microscopy I*, 2nd edn., Springer Ser. Surf. Sci., Vol. 20 (Springer, Berlin, Heidelberg 1994)

Harrison W. A.: *Electronic Structure and the Properties of Solids: The Physics of the Chemical Bond* (Dover, New York 1989)

Lannoo M., P. Friedel: *Atomic and Electronic Structure of Surfaces*, Springer Ser. Surf. Sci., Vol. 16 (Springer, Berlin, Heidelberg 1991)

Lüth H.: *Surfaces and Interfaces of Solids* 3rd edn. (Springer, Berlin, Heidelberg 1995)

Mönch W.: *Semiconductor Surfaces and Interfaces*, 3rd ed., Springer Ser. Surf. Sci., Vol. 26 (Springer, Berlin, Heidelberg 2001)

Wiesendanger R., H.-J. Güntherodt (eds.): *Scanning Tunneling Microscopy II and III*, 2nd edn., Springer Ser. Surf. Sci., Vols. 28 and 29 (Springer, Berlin, Heidelberg 1995 and 1996)

Chapter 9

9.1 L. Esaki, R. Tsu: Superlattice and negative differential conductivity in semiconductors. IBM J. Res. Devel. **14**, 61–65 (1970)

9.2 K.K. Tiong, P.M. Amirtharaj, F.H. Pollak, D.E. Aspnes: Effects of As^+ ion implantation on the Raman spectra of GaAs: "Spatial correlation" interpretation. Appl. Phys. Lett. **44**, 122–124 (1984)
N. Tomassini, A. d'Andrea, R. del Sole, H. Tuffigo-Ulmer, R.T. Cox: Center of mass quantization of excitons in $CdTe/Cd_{1-x}Zn_xTe$ quantum wells. Phys. Rev. B. **51**, 5005–5012 (1995)

9.3 M. Asada, Y. Migamoto, Y. Suematsu: Gain and the threshold of three dimensional quantum-box lasers. IEEE J. QE–**22**, 1915–1921 (1986)

9.4 C. Weisbuch, B. Vinter: *Quantum Semiconductor Structures, Fundamentals and Applications* (Academic, San Diego, CA 1991)

9.5 J. Singh: *Physics of Semiconductors and their Heterostructures* (McGraw-Hill, New York 1993) p. 524

9.6 R. Dingle, W. Wiegmann, C.H. Henry: Quantum states of confined carriers in very thin $Al_xGa_{1-x}As$–GaAs–$Al_xGa_{1-x}As$ heterostructures. Phys. Rev. Lett. **33**, 827–830 (1974)

9.7 M.H. Meynadier, C. Delalande, G. Bastard, M. Voss. F. Alexandre, J.L. Lievin: Size quantization and band-offset determination in GaAs–GaAlAs separate confinement heterostructure. Phys. Rev. B **31**, 5539–5542 (1985)

9.8 M.S. Skolnick, P.R. Tapster, S.J. Bass, A.D. Pitt, N. Apsley, S.P. Aldredy: Investigation of InGaAs–InP quantum wells by optical spectroscopy. Semicond. Sci. Technol. **1**, 29–40 (1986)

9.9 F. Capasso: Band-gap engineering: From physics and materials to new semiconductor devices. Science **235**, 172–176 (1987)

9.10 H. Kroemer: Theory of a wide gap emitter for transistors. Proc. IRE **45**, 1535 (1957); A proposed class of heterojunction injection lasers. Proc. IEEE **51**, 1782 (1963); a general review on heterojunctions: Problems in the theory of heterojunction discontinuities Crit. Rev. SSC **5**, 555 (1975).

9.11 Zh. I. Alferov and D. Z. Gabruzov: Recombination radiation spectrum of GaAs with current excitation via p-n heterojunctions of GaP-GaAs. [English translation] Sov. Phys. Solid State **7**, 1919–1921 (1966).
9.12 Zh. Alferov, V. M. Andreev, V. I. Korol'kov, E. L. Portnoi, D. N. Tret'yakov: Coherent Radiation of Epitaxial Heterojunction Structures in the AlAs-GaAs system. [English translation] Sov.Phys. Semiconductors **2**, 1289 (1969).
9.13 T. P. Pearsall (ed.): *Strain-Layer Superlattices: Physics*, Semiconductors and Semimetals, Vol. 32 (Academic, New York, NY 1990)
9.14 L. Esaki: Semiconductor superlattices and quantum wells, in *Proc. 17th Int'l Conf. on The Physics of Semiconductors*, ed. by J. Chadi, W. A. Harrison (Springer, New York, Berlin 1984) pp. 473–483
9.15 K. Ploog, G. H. Döhler: Compositional and doping superlattices in III–V semiconductors. Adv. Phys. **32**, 285–359 (1983)
9.16 J. Spitzer, T. Ruf, W. Dondl, R. Schorer, G. Abstreiter, E. E. Haller: Optical Phonons in Isotopic ^{70}Ge^{74}Ge Superlattices. Phys. Rev. Lett. **72**, 1565–1568 (1994)
9.17 J. Ihm: Effect of the layer thickness on the electronic character in GaAs–AlAs superlattices. Appl. Phys. Lett. **50**, 1068–1070 (1987)
9.18 S. Gopalan, N. E. Christensen, M. Cardona: Band edge states in short period $(GaAs)_m(AlAs)_m$ superlattices. Phys. Rev. B **39**, 5165–5174 (1989)
9.19 L. I. Schiff: *Quantum Mechanics* (McGraw-Hill, New York 1955)
9.20 G. Bastard: Superlattice band structure in the envelope-function approximation. Phys. Rev. **24**, 5693–5697 (1981)
G. Bastard, J. A. Brum, R. Ferreira: Electronic states in semiconductor heterostructures. *Solid State Physics* **4**, 229–415 (Academic, San Diego, CA 1991)
9.21 E. L. Ivchenko, G. Pikus: *Superlattices and Other Heterostructures: Symmetry and Optical Phenomena*, Springer Ser. Solid-State Sci., Vol. 110 (Springer, Berlin, Heidelberg 1995)
9.22 G. Bastard, J. A. Brum: Electronic states in semiconductor heterostructures. IEEE J. QE–**22**, 1625–1644 (1986)
9.23 J. C. Hensel, G. Feher: Cyclotron resonance experiment in uniaxially stressed silicon: Valence band inverse mass parameters and deformation potentials. Phys. Rev. **129**, 1041–1062 (1963)
9.24 L. Liboff: *Introductory Quantum Mechanics* (Addison-Wesley, Reading, MA 1980) pp. 267–279
9.25 L. Esaki: A perspective in superlattice development, in *Symp. on Recent Topics in Semiconductor Physics* (Prof. Y. Uemura's Festschrift) (World Scientific, Singapore 1983) pp. 1–71
9.26 Y. C. Chang, J. N. Schulman: Interband optical transitions in GaAs–$Ga_{1-x}Al_xAs$ and InAs–GaSb superlattices. Phys. Rev. B **31**, 2069–2079 (1985)
9.27 R. Dingle: Optical properties of semiconductor superlattices, in *Proc. 13th Int'l Conf. on the Physics of Semiconductors*, ed. by F. G. Fumi (Tipografia Marves, Rome 1976) pp. 965–974
9.28 M. Cardona, N. E. Christensen: Heterostructure band offsets in semiconductors. Phys. Rev. B **35**, 6182–6194 (1987)
9.29 J. A. Kash, M. Zachau, M. A. Tischler: Anisotropic valence bands in quantum wells: Quantitative comparison of theory and experiment. Phys. Rev. Lett. **69**, 2260–2263 (1992)
9.30 G. Fasol, W. Hackenberg, H. P. Hughes, K. Ploog, E. Bauser, H. Kano: Continous-wave spectroscopy of femtosecond carrier scattering in GaAs. Phys. Rev. B **41**, 1461–1478 (1990)

9.31 A. Pinczuk, G. Abstreiter: Spectroscopy of free carrier excitations in semiconductor quantum wells, in *Light Scattering in Solids V*, ed. by M. Cardona, G. Güntherodt, Topics Appl. Phys., Vol. 66 (Springer, Berlin, Heidelberg 1989) pp. 153–211

9.32 P. Giannozzi, S.R. Gironcoli, P. Pavone, S. Baroni: *Ab initio* calculation of phonon dispersions in semiconductors. Phys. Rev. B **43**, 7231–7242 (1991)

9.33 S.M. Rytov: Acoustic properties of a thinly laminated medium. Sov. Phys. – Acoust. **2**, 68–80 (1956)

9.34 B. Jusserand, M. Cardona: Raman spectroscopy of vibrations in superlattices, in *Light Scattering in Solids V*, ed. by M. Cardona, G. Güntherodt, Topics Appl. Phys., Vol. 66 (Springer, Berlin, Heidelberg 1989) Chap. 3

9.35 J. Menéndez: Phonons in GaAs–$Al_xGa_{1-x}As$ superlattices. J. Luminesc. **44**, 285–314 (1989)

9.36 P. Molinàs-Mata, M. Cardona: Planar force-constant models and internal strain parameter of Ge and Si. Phys. Rev. B **43**, 9959–9961 (1991)

9.37 R.A. Ghanbari, J.D. White, G. Fasol, C.J. Gibbings, C.G. Tuppen: Phonon frequencies for Si–Ge strained-layer superlattices calculated in a three-dimensional model. Phys. Rev. B **42**, 7033–7041 (1990)

9.38 W. Richter, D. Strauch: Lattice dynamics of GaAs/AlAs superlattices. Solid-State Commun. **64**, 867–872 (1987)

9.39 E. Molinari, A. Fasolino, K. Kunc: Superlattice effects on confined phonons. Phys. Rev. Lett. **56**, 1751 (1986)

9.40 R. Fuchs, K.L. Kliewer: Optical modes of vibration in an ionic crystal slab. Phys. Rev. A **140**, 2076–2088 (1965)

9.41 M. Nakayama, M. Ishida, N. Sano: Raman scattering by interface-phonon polaritons in GaAs–AlAs heterostructures. Phys. Rev. B **38**, 6348–6351 (1988)

9.42 R. Merlin, C. Colvard, M.V. Klein, H. Morkoc, A.Y. Cho, A.C. Gossard: Raman scattering in superlattices: Anisotropy of polar phonons. Appl. Phys. Lett. **36**, 43–45 (1980)

9.43 C. Trallero-Giner, F. García-Moliner, V.R. Velasco, M. Cardona: Analysis of the phenomenological models for long-wavelength polar optical modes in semiconductor layered systems. Phys. Rev. B **45**, 11944–11948 (1992)

9.44 Z.V. Popović, M. Cardona, E. Richter, D. Strauch, L. Tapfer, K. Ploog: Phonons in GaAs/AlAs superlattices grown along the (111) direction. Phys. Rev. B **41**, 5904–5913 (1990)

9.45 M.P. Chamberlain, M. Cardona, B.K. Ridley: Optical modes in GaAs/AlAs superlattices. Phys. Rev. B **48**, 14356–14364 (1993)

9.46 M. Zunke, R. Schorer, G. Abstreiter, W. Klein, G. Weimann, M.P. Chamberlain: Angular dispersion of confined optical phonons in GaAs/AlAs superlattices studied by micro-Raman spectroscopy. Solid State Commun. **93**, 847–851 (1995)

9.47 G. Scarmarcio, L. Tapfer, W. König, A. Fischer, K. Ploog, E. Molinari, S. Baroni, P. Giannozzi, S. de Gironcoli: Infrared reflectivity by transverse-optical phonons in $(GaAs)_m/(AlAs)_n$ ultrathin-layer superlattices. Phys. Rev. B **43**, 14754–14757 (1991)

9.48 E. Jahne, A. Röseler, K. Ploog: Infrared reflectance and ellipsometric studies of GaAs/AlAs superlattices. Superlattices Microstruct. **9**, 219–222 (1991)

9.49 Yu. Pusep, A. Milekhin, A. Poropov: FTIR spectroscopy of $(GaAs)_n/(AlAs)_m$ superlattices. Superlattices Microstruct. **13**, 115–123 (1992)

9.50 M. Krieger, H. Sigg: Zone-edge gap in the folded acoustic phonon dispersion of an AlAs–GaAs semiconductor superlattice, in *The Physics of Semiconductors*, ed. by J. Lockwood (World Scientific, Singapore 1995) pp. 959–962
9.51 C. Colvard, T.A. Gant, M.V. Klein, R. Merlin, R. Fischer, H. Morkoc, A.C. Gossard: Folded acoustic phonons and quantized optic phonons in (GaAl)As superlattices. Phys. Rev. B **31**, 2080–2091 (1985)
9.52 J. He, B. Djafari Rouhani, J. Sapriel: Theory of light scattering by longitudinal-acoustic phonons in superlattices. Phys. Rev. B **37**, 4086–4098 (1988)
9.53 J. He, J. Sapriel, H. Brugger: Semiconductor photoelastic constants measured by light scattering in superlattices. Phys. Rev. B **39**, 5919–5923 (1989)
9.54 J. Spitzer, Z.V. Popovic, T. Ruf, M. Cardona, R. Nötzel, K. Ploog: Folded acoustic phonons in GaAs/AlAs superlattices grown on non-(100)-oriented surface. Solid State Electron. **37**, 753–756 (1994)
9.55 T. Ruf, J. Spitzer, V.F. Sapega, V.I. Belitsky, M. Cardona, K. Ploog: Interface roughness and homogeneous linewidths in quantum wells and superlattices studies by resonant acoustic-phonon Raman scattering. Phys. Rev. B **50**, 1792–1806 (1994)
9.56 J.E. Zucker, A. Pinczuk, D.S. Chemla, A.C. Gossard, W. Wiegmann: Optical vibrational modes and electron-phonon interaction in GaAs quantum wells. Phys. Rev. Lett. **53**, 1280–1283 (1984)
9.57 A. Fainstein, P. Etchegoin, M.P. Chamberlain, M. Cardona, K. Tötemeyer, K. Eberl: Selection rules and dispersion of GaAs/AlAs multiple quantum well optical phonons studied by Raman scattering in right angle, forward and backscattering in-plane geometries. Phys. Rev. B **51**, 14448–14458 (1995)
9.58 A.J. Shields, M.P. Chamberlain, M. Cardona, K. Eberl: Raman scattering due to interface optical phonons in GaAs/AlAs multiple quantum wells. Phys. Rev. B **51**, 17728–17739 (1995)
9.59 A.K. Sood, J. Menéndez, M. Cardona, K. Ploog: Interface vibrational modes in GaAs–AlAs superlattices. Phys. Rev. Lett. **54**, 2115–2118 (1985)
9.60 K. Huang, B. Zhu: Dielectric continuum models and Fröhlich interaction in superlattices. Phys. Rev. B **38**, 13377–13386 (1988)
9.61 H. Rücker, E. Molinari, P. Lugli: Microscopic calculation of the electron-phonon interaction in quantum wells. Phys. Rev. B **45**, 6747–6756 (1992)
9.62 S. Rudin, T. Reinecke: Electron-LO-phonon scattering rates in semiconductor quantum wells. Phys. Rev. B **41**, 7713–7717 (1991)
9.63 K.T. Tsen, K.R. Wald, T. Ruf, P.Y. Yu, H. Morkoc: Electron-optical phonon interactions in ultrathin multiple quantum wells. Phys. Rev. Lett. **67**, 2557–2560 (1991)
9.64 L. Esaki: New phenomenon in narrow germanium p-n junctions. Phys. Rev. **109**, 603–604 (1958)
9.65 R. Tsu, L. Esaki: Tunneling in a finite superlattice. Appl. Phys. Lett. **22**, 562–564 (1973); erratum **43**, 9288 (1991)
9.66 T.C.L.G. Sollner, W.D. Goodhue, P.E. Tannenwald, C.D. Parker, D.D. Peck: Resonant tunneling through quantum well at frequencies up to 2.5 THz. Appl. Phys. Lett. **43**, 588–590 (1983)
9.67 A. Silbille, J.F. Palmier, H. Wang, F. Mollot: Observation of Esaki-Tsu negative differential velocity in GaAs/AlAs superlattices. Phys. Rev. Lett. **64**, 52–55 (1990)

9.68 J. P. Palmier: Miniband transport and resonant tunneling in superlattices, in *Resonant Tunneling in Semiconductors*, ed. by. L. L. Chang (Plenum, New York 1991) pp. 361–375
E. E. Mendez, W. I. Wang, B. Ricco, L. Esaki: Resonant tunneling of holes in GaAs–GaAlAs superlattices. Appl. Phys. Lett. **47**, 415–417 (1985). For this and related work Mendez was awarded the Prince of Asturias Prize, 1998
9.69 E. Merzbacher: *Quantum Mechanics* (Wiley, New York 1961) pp. 94–97
9.70 D. Z.-Y. Ting, S. K. Kirby, T. C. McGill: Interface roughness effects in resonant tunneling structures. Appl. Phys. Lett. **64**, 2004–2006 (1994)
9.71 T. P. E. Broekaert, W. Lee, C. G. Fonstad: Pseudomorphic $In_{0.53}Ga_{0.47}As$/AlAs/InAs resonant tunneling diodes with peak-to-valley current ratios of 30 at room temperature. Appl. Phys. Lett. **53**, 1545–1547 (1988)
9.72 K. von Klitzing, G. Dorda, M. Pepper: New method for high-accuracy determination of the fine-structure constant based on quantum Hall resistance. Phys. Rev. Lett. **45**, 494–497 (1980)
9.73 D. C. Tsui, H. L. Störmer, A. C. Gossard: Two dimensional magnetotransport in the extreme quantum limit. Phys. Rev. Lett. **45**, 1559–1562 (1982)
9.74 L. D. Landau: Diamagnestism of metals (in German). Z. Physik. **64**, 629–637 (1930)
9.75 L. D. Landau, E. N. Lifshitz: *Quantum Mechanics*, 3rd edn. (Pergamon, Oxford 1977)
9.76 J. M. Ziman: *Principles of the Theory of Solids*, 2nd edn. (Cambridge Univ. Press, Cambridge 1972) pp. 313–326
9.77 M. Brodsky (ed.): *Amorphous Semiconductors*, 2nd edn., Topics Appl. Phys, Vol. 36 (Springer, Berlin, Heidelberg 1985)
9.78 J. Tauc: *Amorphous and Liquid Semiconductors* (Plenum, New York 1974)
9.79 H. Störmer: Fractional Quantum Hall effect today, Solid State Commun. **107**, 617–622 (1998)
9.80 S. M. Girvin: Exotic quantum order in two-dimensional systems, Solid State Commun. **107**, 623–628 (1998)
9.81 D. C. Tsui, H. Störmer, A. C. Gossard: Two-dimensional magnetotransport in the extreme quantum limit. Phys. Rev. Lett. **48**, 1559–1562 (1982).
9.82 R. B. Laughlin: Anomalous QHE: An incompressible fluid with fractionally charged excitations. Phys. Rev. Lett. **50**, 1395–1398 (1983)
9.83 R. Loudon: Raman effect in crystals. Adv. Phys. **13**, 423–482 (1964); erratum *ibid.* **14**, 621 (1965)

General Reading

Growth of Quantum Wells and Superlattices

Hermann M. A., H. Sitter: *Molecular Beam Epitaxy*, 2nd. edn., Springer Ser. Mater. Sci., Vol. 7 (Springer, Berlin, Heidelberg 1996)
Milnes A. G., D. L. Feucht: *Heterojunctions and Metal-Semiconductor Junctions* (Academic, NY, 1972)
Panish M. B., H. Temkin: *Gas Source Molecular Beam Epitaxy*, Springer Ser. Mater. Sci., Vol. 26 (Springer, Berlin, Heidelberg 1993)
Ploog K.: Formation of semiconductor interfaces during molecular beam epitaxy, in *Semiconductor Interfaces: Formation and Properties*, ed. by G. LeLay, J. Denien, N. Boccara (Springer, Berlin, Heidelberg 1987)

Stradling R. A., P. C. Klipstein (eds.): *Growth and Characterization of Semiconductors* (Hilger, Bristol, UK 1990)
Wang S.: *Fundamentals of Semiconductor Theory and Device Physics*, (Prentice Hall, Englewood Cliffs, New Jersey, 1989)

Properties of Quantum Wells and Superlattices

Ando T., Y. Arakawa, K. Furuya, S. Komiyama, H. Nakashima (eds.): *Mesoscopic Physics and Electronics*, NanoScience and Technology (Springer, Berlin, Heidelberg 1998)
Dekker A. J.: *Solid State Physics* (Prentice Hall, Englewood Cliffs, NJ 1957) p. 244
Ehrenreich H., D. Turnball (eds.): *Semiconductor Heterostructures and Nanostructures* Adv. Res. Appl., Vol. 44 (Academic, San Diego, CA 1991)
Ivchenko E. L., G. Pikus: *Superlattices and Other Heterostructures: Symmetry and Other Optical Phenomena*, Springer Ser. Solid-State Sci., Vol. 110 (Springer, Berlin, Heidelberg 1995)
Jaros M.: *Superlattices of Semiconductor Microstructures* (Clarendon, Oxford 1985)
Jusserand B., M. Cardona: Raman spectroscopy of vibrations in superlattices, in *Light Scattering in Solids V*, ed. by M. Cardona, G. Güntherodt, Topics Appl. Phys., Vol. 66 (Springer, Berlin, Heidelberg 1989) pp. 49–152
Kroemer H.: Theory of heterojunction: A critical review, in *Molecular Beam Epitaxy and Heterostructures*, ed. by L. L. Chang, K. Ploog (Nijhoff, Dordrecht 1985) pp. 331–379
Menéndez J.: Phonons in GaAs–$Al_xGa_{1-x}As$ superlattices. J. Luminesc. **44**, 285–314 (1989)
Pearsall, T. P. (ed.): *Strain-Layer Superlattices: Physics*, Semiconductors and Semimetals, Vol. 32 (Academic, Orlando, FL 1990)
Pinczuk A., G. Abstreiter: Spectroscopy of free carriers excitations in semiconductor quantum wells, in *Light Scattering in Solids V*, ed. by M. Cardona, G. Güntherod, Topics Appl. Phys., Vol. 66 (Springer, Berlin, Heidelberg 1989) Chap. 4, pp. 153–211
Sapriel J., B. Djafari Rouani: Vibrations in superlattices. Surf. Sci. Rept. **10**, 189–275 (1989)
Singh J.: *Physics of Semiconductors and their Heterostructures* (McGraw-Hill, New York 1993)
Wang S.: *Fundamentals of Semiconductor Theory and Device Physics* (Prentice Hall, Englewood Cliffs, NJ 1989)
Weisbuch C., B. Vinter: *Quantum Semiconductor Structures, Fundamentals and Applications* (Academic, San Diego, CA 1991)

Quantum Hall Effect

Chakraborty T., P. Pietiläinen: *The Quantum Hall Effects – Fractional and Integral*, 2nd edn., Springer Ser. Solid-State Sci., Vol. 85 (Springer, Berlin, Heidelberg 1995)
Jain, J. K.: The composite fermion: a quantum particle and its quantum fluids, Physics today, April 2000, pp. 39–45
Janssen M., O. Viehweger, U. Fastenrath, J. Hajdu: *Introduction to the Theory of the Integer Quantum Hall Effect* (VCH, Weinheim 1994)
McDonald A. H. (ed.): *Quantum Hall Effect: A Perspective* (Kluwer, Boston 1989)
Prange R. E., S. M. Girvin (eds.): *The Quantum Hall Effect* (Springer, New York, Berlin 1990)

Subject Index

C

E

F

H

J

K

L

N

O

P

Q

R

T

U

V

W

Table of Fundamental Physical Constants

Quantity	Symbol	Value	CGS	SI
Velocity of light in vacuo	c	2.997925	10^{10} cm/s	10^{8} m/s
Electronic charge	e	1.60219	–	10^{-19} C
		4.803250	10^{-10} esu	–
Electron rest mass	m_0	9.10956	10^{-28} g	10^{-31} kg
Proton rest mass	M_p	1.67261	10^{-24} g	10^{-27} kg
Neutron rest mass	M_n	1.674920	10^{-24} g	10^{-27} kg
Planck's constant	h	6.62620	10^{-27} erg s	10^{-34} J s
	$\hbar = h/(2\pi)$	1.05459	10^{-27} erg s	10^{-34} J s
Boltzmann constant	k_B	1.38062	10^{-16} erg/K	10^{-23} J/K
Bohr radius $[\hbar^2/(m_0 e^2)]$	a_B	5.29177	10^{-9} cm	10^{-11} m
Rydberg constant $[m_0 e^4/(2\hbar^2)]$	Ry	2.17991	10^{-11} erg	10^{-18} J
Classical radius of electron $[e^2/(mc^2)]$	r_e	2.81794	10^{-13} cm	10^{-15} m
Reciprocal of the fine structure constant $(\hbar c/e^2)$	$1/\alpha$	137.036	–	–
Magnetic flux quantum (hc/e)	Φ_0	4.135708	10^{-7} gauss cm^2	10^{-15} Wb
Photon Energies				
1 electron volt	eV	1.60219	10^{-12} erg	10^{-19} J
eV/h		2.41797	10^{14} Hz	10^{14} Hz
eV/(hc) (wavenumber)		8.06546	10^{3} cm^{-1}	10^{5} m^{-1}
eV/(k_B)		1.16048	10^{4} K	10^{4} K
Permittivity of vacuum	ε_0	8.854		10^{-12} m

From B.N. Taylor, W.H. Parker, D.N. Langenberg: Rev. Mod Phys. **41**, 375 (1969)